MONOGRAPHS ON THE
PHYSICS AND CHEMISTRY OF
MATERIALS

MONOGRAPHS ON THE PHYSICS AND CHEMISTRY OF MATERIALS

Neutron diffraction (3rd edn) G. E. Bacon

Strong solids (3rd edn) A. Kelly and N. H. Macmillan

Optical spectroscopy of inorganic solids B. Henderson and G. F. Imbusch

Chemistry of the metal–gas interface M. N. Roberts and C. S. McKee

Quantum theory of collective phenomena G. L. Sewell

Experimental high-resolution electron microscopy (2nd edn) J. C. H. Spence

Theory of defects in solids A. M. Stoneham

Experimental techniques in low-temperature physics Guy K. White

Optical Spectroscopy of Inorganic Solids

B. HENDERSON and G. F. IMBUSCH

University of Strathclyde *University College, Galway*

CLARENDON PRESS · OXFORD

This book has been printed digitally and produced in a standard specification in order to ensure its continuing availability

OXFORD
UNIVERSITY PRESS

Great Clarendon Street, Oxford OX2 6DP

Oxford University Press is a department of the University of Oxford.
It furthers the University's objective of excellence in research, scholarship,
and education by publishing worldwide in

Oxford New York

Auckland Cape Town Dar es Salaam Hong Kong Karachi
Kuala Lumpur Madrid Melbourne Mexico City Nairobi
New Delhi Shanghai Taipei Toronto

With offices in

Argentina Austria Brazil Chile Czech Republic France Greece
Guatemala Hungary Italy Japan South Korea Poland Portugal
Singapore Switzerland Thailand Turkey Ukraine Vietnam

Oxford is a registered trade mark of Oxford University Press
in the UK and in certain other countries

Published in the United States
by Oxford University Press Inc., New York

Reprinted 2010

ISBN 978-0-19-929862-4

To
Sheila and Mary

PREFACE

The development of optical spectrometers in the first half of the nineteenth century allowed scientists to examine the abundant optical spectra emitted by matter, from the Sun's dark Fraunhofer lines, to the colours of flames, and later, to the rich sharp-line spectra of atomic gas discharges. With the invention of the diffraction grating the emission spectra of molecular discharges were found to be composed of complex series of sharp lines, and additional fine structure was observed in the emission spectra of the elements. Spectroscopists of the time recorded these myriad spectra in considerable detail although there were no adequate theoretical explanations for them. Bohr's early and very simple quantum model of the hydrogen atom, which gave such good agreement with measurement, gave clear indications that the optical spectra of the atomic gases were a fundamental manifestation of their quantum nature and atomic spectroscopy became the experimental proving-ground for developments in, and understanding of, quantum mechanics. Optical spectroscopy of condensed matter presented a more daunting challenge, requiring in addition an understanding of the structure of solids, perfect and imperfect, of the behaviour of electrons in solids, and of the nature of elementary excitations in solids. One of the attractions of the study of this spectroscopy is the insight it affords us into fundamental solid state processes. Being also an area of major technological interest ensures a good measure of research support. Thus it attracts creative research workers and continues to be a dynamic field of scientific enquiry.

We have written this book with the postgraduate student and beginning experimental research worker in mind in the hope that it will serve as an introduction to the *art* of solid state optical spectroscopy. To an extent it reflects our experiences in teaching this type of material to undergraduate and postgraduate students in Dublin, Galway, and Glasgow these past two decades. It attempts to present the theoretical basis of the subject and then to illustrate the various topics with relevant experimental data. In the theoretical discussions we have consciously chosen simple models, and elementary, if sometimes inelegant, mathematical methods, cognizant of the need of laboratory research workers to relate theoretical models and their predictions to experimental measurements.

Chapter 1 is a brief overview of the field. Chapters 2 to 5 form the theoretical base for the topics developed in later chapters. In Chapter 2 we begin to set down the quantum basis for the description of atoms, ions, and defect centres in solids. In the latter cases only the simplest models are described in order to emphasize the diagnostic aspects of defect structure analysis. Considerations of symmetry are an important aspect of a spectroscopist's skills; such considerations are formalized in the theory of groups. Chapter 3 presents the elements of point group theory, with particular emphasis on the symmetry properties of electronic centres in solids. Electromagnetic radiation and its interaction with electronic centres form the topics of Chapter 4. Lattice vibrations can strongly affect the spectroscopic properties of the optically-active centres in solids. The various spectroscopic manifestations of the electron–lattice interaction are described in Chapter 5, with particular use being made of the single configurational coordinate model. The remaining chapters are concerned with experimental aspects of solid state spectroscopy. Chapter 6 discusses the physical basis of the commonly-used experimental techniques. Chapters 7, 8, and 9 are concerned with the applications of these techniques to the spectroscopic properties of colour centres, dopant rare earth ions, and dopant transition metal ions in solids, respectively. When the concentration of optically-active centres is sufficiently high, excitation can be transferred non-radiatively between adjacent centres. In addition, more complex aggregate centres are formed with distinct spectroscopic properties. These are the topics of Chapter 10. The optical processes operating in solid state lasers based on these optically-active materials are described in Chapter 11. Chapter 12 describes the method of optical detection of magnetic resonance (ODMR) and discusses in particular the application of this technique to the elucidation of the optical properties of insulating and semiconducting materials. The topics we have chosen constitute but a subset of the field of solid state optical spectroscopy. They reflect our own interests and experiences. We have concentrated on the area of inorganic insulating solids; in the field of semiconductor luminescence we have limited our discussion to studies involving the application of ODMR techniques.

In a book of this nature we are only able to reference a limited number of original publications, and we are conscious that this does not do justice to many of the workers who have contributed to the present understanding of the subject. Similarly, we selected from the great wealth of available spectroscopic information only a limited number of examples to illustrate the topics under discussion. We wish to express our appreciation to the authors who have allowed us to make use of measurements and figures from their publications. The permission of the editors of these publications is also appreciated. We especially wish to express our gratitude to Professors Sugano, Tanabe, and Kamimura for allowing us to use numerous tables from their book *Multiplets of Transition Metal Ions in Crystals* (Academic Press, New York and London 1970).

Several of our friends have read and made detailed comments on various chapters; to Max Glasbeek, Yves Merle d'Aubigné, Sara McMurry, Kevin O'Donnell, Marshall Stoneham, and Mitsuo Yamaga we are greatly indebted.

For their labour in preparing many typescripts we are also indebted to Mai Kyne, Nora Kelso, Lee Guckian, and especially Joanna Calder; and we acknowledge with thanks the support and patience of the editors of Oxford University Press.

Strathclyde and Galway B. H.
1988 G. F. I.

CONTENTS

1

Spectroscopy and electronic structure of inorganic solids

1.1 Historical comments

THE optical properties of solids have been of scientific and technological interest for over three centuries, and made use of for decorative purposes from the times of the earliest civilizations. Early cave artists produced pictures of their times and lifestyles, in which spectacular colour variations were achieved by intimate mixtures of different, naturally-occurring, inorganic pigments. Good examples of inorganic pigments are mercuric sulphide (a brilliant scarlet usually called vermilion), the beautiful green chromic oxide, and the bright blue cobaltous oxide. Pottery was coloured by burning such pigments into the surface; glasses were tinted by the addition of coloured metal oxides to the melt. Naturally our unsophisticated forebears could not explain the colours they saw, nor the absence of such colours in the dark. Today most people realize that colour has to do with what happens when light falls on matter, as is readily appreciated when the colours of objects in a well-darkened room are revealed under the illumination of a beam of white light. A little more thought (or persuasion) convinces one that the perceived coloration is due to *colour subtraction*; i.e. the colour observed is the *complementary colour* to that absorbed from white light by the solid. Our understanding of the colour in our everyday environment derives from Newton's observation that sunlight contains all the colours of the rainbow with wavelengths ranging over the visible spectrum (approximately 400–700 nm).

Newton's impact on the study of optics was profound. Both prior to and subsequent to his era, continuing up to the present day, scientists have studied the interaction of light with condensed matter for the information it gives concerning the nature of light. Indeed this remained a dominant interest of many physical scientists until the beginning of the twentieth century. Measurements were made of the ability of matter to reflect, refract, and polarize light. Newton measured the variation of the index of refraction, n, with wavelength, λ, which leads to dispersion. Although his *corpuscular theory* provided an explanation of the laws of reflection and refraction it failed to account for the phenomena of polarization, interference, and diffraction. Young was one of the principal proponents of the *wave theory*, which was used both to explain the coloured fringes of Newton's rings and the interference pattern of the celebrated double-slit experiment. However, detailed interpretation of these phenomena awaited Fresnel's developments in wave theory, in particular the use of transverse waves required to explain double and conical refraction (see Preston 1895). The nature of the transverse waves was not understood, however, and they were thought to be the vibrations of an elastic aether. Evidently there were problems with both corpuscular and wave theories. Thus it was almost from the very beginning that human ingenuity failed to produce in a single model an explanation of both the particle-like and wave-like properties of radiation. Since the direction of Fresnel's transverse oscillations is the direction of the electric field in Maxwell's electromagnetic theory (1865, 1873), the elementary charged constituents of matter are set into oscillation transverse to the direction of propagation of the light beam. Maxwell's theory paved the way for the classical resonance models of Lorentz (1909) and Drude (1902), which accounted for anomalous dispersion and for the spectral variation of the refractive index. The quantitative interdependence of the refractive index and the dielectric constant may also be derived from Maxwell's theory. Coincident with developments from this, measurements were made of the optical spectra of different atoms, and of the Zeeman and Stark effects. We now perceive the gradually changing emphasis of optical research towards the use of radiation to probe the electronic structure of matter, which in due course provided the impetus and the testing ground for the development of the quantum theory of atomic structure.

During the present century there have been many significant developments in optical spectroscopy, fuelled perhaps in equal measure by theoretical and instrumental innovation. Einstein (1905) had introduced the photon, the quantum of radiant energy—a particle of zero mass travelling with the speed of light—in a short section of a (now) celebrated paper explaining the photoelectric effect. In so doing he re-awakened the controversy of particle–wave duality, which had lain dormant since Fresnel's times. It is a controversy which thrives to this day although in a somewhat different formalism (Pike and Sarkar 1986). Following on from Bohr's theory of the hydrogen atom, physicists began the serious quest for a quantum theory to describe the more complex atoms. The seminal idea of de Broglie that matter has wave-like properties provided the impetus for Schrödinger's wave mechanics, laying the foundation of the modern quantum theory of atomic processes. One of the first applications was to the calculation of the energy levels of atoms and the strength of spectroscopic transitions. This led to a quantitative description of the mechanisms of optical absorption and emission. Extension of spectroscopy into the realms of molecular and solid state physics more or less coincided with the first application of quantum ideas in these areas (e.g. Born and Oppenheimer 1927; Heitler and London 1927; Pauling 1939, 1967; Slater 1930; Van Vleck 1932). Of course the coloration of minerals and of gemstones was of practical interest long before this. Indeed, absorption bands in diamond, ruby, and blue fluorspar had been known since the eighteenth century, although the atomic mechanisms responsible remained a mystery until relatively recently. Before these mechanisms and other atomic processes in solids could be described quantum mechanically, the symmetry properties of the solid system had to be taken into account. In 1929 Bethe published a classic paper on crystal field theory, treating the electrostatic field of the host lattice as a static perturbation on the spectral terms of free ions, and then classifying the new levels as representations of the symmetry group of the crystal field. Extensions of Bethe's theory provide the theoretical basis for most transition metal and rare-earth ion spectroscopy (Sugano *et al.* 1970; Judd 1963). In its present form, solid state spectroscopy is an intimate mélange of quantum mechanics, group theory, spectroscopic measurement, and experimental innovation. Before we embark on detailed exposition of these subjects there are a number of basic concepts which it may be helpful to discuss, and these are the topics which concern us, briefly, in the remainder of this chapter. However, the emphasis of this book is on spectroscopy, albeit as applied to problems in solid state physics. With one exception, i.e. lattice vibrations, only a rather cursory account is given of the more general aspects of solid state physics.

1.2 The classical optical constants

For the most part this text is concerned with spectroscopy in the wavelength range 200–3000 nm. Such wavelengths are much greater than the spacings between atoms in solids. In consequence, radiation *propagating* in solids may be treated classically while the solid is viewed as a continuous medium. The response of the solid is then described in terms of macroscopic fields, which are averages of the rapidly varying atomic fields over volumes large on the atomic scale but small on the scale of optical wavelengths. Furthermore, if the solid behaves isotropically the dielectric constant (κ), relative magnetic permeability (μ_r) and electrical conductivity (σ) are related linearly to the induced polarization ($\boldsymbol{P}$), magnetization ($\boldsymbol{M}$) and current density, $\boldsymbol{J}$, by

$$\begin{aligned} \boldsymbol{P} &= \varepsilon_0(\kappa-1)\boldsymbol{E}, \\ \boldsymbol{M} &= (\mu_r-1)\boldsymbol{H}, \\ \boldsymbol{J} &= \sigma\boldsymbol{E}. \end{aligned} \tag{1.1}$$

The responses may also be related to atomic properties. For example, the macroscopic polarization may be related to the atomic polarization (α_{at}) by

$$\boldsymbol{P} = N\alpha_{at}\langle \boldsymbol{E}_{loc}\rangle \tag{1.2}$$

where the local (microscopic) electric field E_{loc} differs from the total macroscopic field E. Such linear relationships, applicable at low radiation intensities, must be modified at the high intensities available from high-powered lasers.

Solids may contain both *free* and *bound* charges which will oscillate at the frequency of an applied radiation field. Both types of charge contribute to the polarization of the material. However, since metals are not the concern of this text we consider only the effects of the applied field on the bound charge. In this case Maxwell's fourth equation for non-magnetic materials reads

$$\nabla \times \boldsymbol{H} = \varepsilon_0 \kappa \frac{\partial \boldsymbol{E}}{\partial t} \tag{1.3}$$

where κ is the dielectric constant, which may be determined from some assumed model of the solid. If the assumed time dependence of the radiation field has the form $\exp(i\omega t)$ then the dielectric constant is complex and we write

$$\kappa = \kappa_1 - i\kappa_2 \tag{1.4}$$

where κ_1 and κ_2 are real quantities. We may then use Maxwell's equations to relate κ to more directly measured quantities such as the real refractive index, n, reflectivity, R, and the absorption coefficient, α. For example, the refractive index, given by $\sqrt{\kappa}$, is now a complex number, $n - ik$, in which both n and k are real, k being the extinction coefficient. From this we obtain

$$n^2 - k^2 = \kappa_1, \tag{1.5}$$

$$2nk = \kappa_2. \tag{1.6}$$

The electric field, $\boldsymbol{E}(\boldsymbol{r}, t)$ of a plane wave propagating through a solid along the z-direction with frequency ω is

$$\boldsymbol{E}(\boldsymbol{r}, t) = \boldsymbol{E}_0 \exp(-\omega kz/c) \exp i\omega(t - nz/c). \tag{1.7}$$

The term in $\exp(-\omega kz/c)$ represents the exponential damping of the wave as it penetrates the solid, in which the extinction coefficient, k, is related to the absorption coefficient α by $\alpha = 2(\omega/c)k$. The boundary conditions at the surface of the solid determine the reflectivity, R, which is given by

$$R = \frac{(1-n)^2 + k^2}{(1+n)^2 + k^2} \tag{1.8}$$

at normal incidence. The so-called *optical constants* n, k, and R are defined classically in terms of the electrical properties. In consequence, there is no

Table 1.1
Spectral variation of optical properties of solids

Solid → Wavelength range	Metal (e.g. Al)	Semiconductor (e.g. Si)	Insulator (e.g. SiO_2)
Ultraviolet $\lambda \simeq 50$–300 nm	$R \simeq 0$–92%	$R \simeq 40$–60% Large α	Onset of fundamental electronic absorption
Visible, near infrared 300–3000 nm	$R \simeq 92$–95%	Partially reflecting and absorbing	Transparent
$\lambda > 10\,000$ nm	$R \simeq 100\%$	Transparent	Reststrahlen region, high absorption and reflection

contribution from the atomic structure of the solid. And yet, as Table 1.1 shows, the spectral variations of these constants depend markedly on the predominant type of bonding forces in the solids. As we will see shortly, measurement of the frequency dependences of the optical constants reflects the quantum structure of the solid.

1.2.1 Resonance models for the optical constants

Atomistic models of the optical properties of solids require some knowledge of the interatomic bonding and of whether or not electrons and ions are free or bound to a fixed site in the solid. In metals the conduction electrons are 'free' to migrate through the material, whereas in ionic solids the valence electrons are bound to the electronegative ions. Semiconductors have both free and bound charges. In consequence, the most general model of the interaction of radiation with solids requires the solution of a differential equation for a forced damped oscillator:

$$\ddot{x} + \Gamma\dot{x} + \omega_0^2 x = (q/m)E_0 \exp(i\omega t) \tag{1.9}$$

where q and m are the charge and mass of the particles in question, and $\gamma = m\Gamma$ is the damping constant. The restoring force constant $m\omega_0^2$ is zero for a metal, as the electrons are free to move through the solid, and non-zero for dielectric materials, since both outer electrons and ions are bound to their positions in the solid by Coulombic forces. On solving eqn (1.9) for the time-varying electric dipole moment per unit field, i.e. the polarizability, we obtain

$$\alpha_{\text{at}} = \frac{qx(t)}{E} = \frac{q^2}{m}\left(\frac{1}{\omega_0^2 - \omega^2 + i\Gamma\omega}\right) \tag{1.10}$$

where we take $|q| = e$. The inverse proportionality on m in eqn (1.10) implies that electronic polarizabilities are important throughout the optical spectrum, whereas ionic polarizabilities are important only in regions close to their resonances. Electronic polarizabilities determine the index of refraction in the transparent region. The real and imaginary parts of the refractive index, n and k are complicated functions of the radiation frequency (ω), resonance frequency (ω_0) and damping constant, Γ; simple formulae for n^2 and k^2 are obtained only in the limit that $(\kappa-1) \ll 1$. Assuming there to be N weakly-coupled oscillators per unit volume of solid, each with a set of resonances (labelled by index, i) at frequencies ω_i with damping constants Γ_i and if there are f_i electrons per oscillator which can participate fully in the ith resonance, then we can derive the following expressions for n and k, in the case that $f_i \ll 1$ (Wooten 1972):

$$n^2 = 1 + \frac{Nq^2}{\varepsilon_0 m} \sum_i \frac{f_i(\omega_i^2 - \omega^2)}{(\omega_i^2 - \omega^2)^2 + \Gamma_i^2\omega^2}, \tag{1.11}$$

$$k^2 = \frac{2Nq^2}{\varepsilon_0 m} \sum_i \frac{\Gamma_i \omega}{(\omega_i^2 - \omega^2)^2 + \Gamma_i^2\omega^2}. \tag{1.12}$$

In eqns (1.11) and (1.12) the parameter f_i is called the *oscillator* strength; in the classical model where the electron oscillates according to eqn (1.9), f_i has the

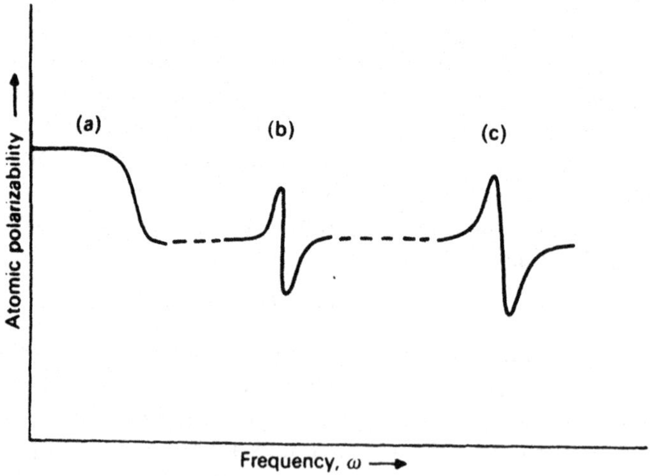

FIG. 1.1. Showing schematically resonances in the atomic polarizability of solids arising from (a) re-orientation of permanent dipoles, (b) lattice vibrations, and (c) electronic excitation.

value of unity. In real situations f_i can be very small, as is predicted in quantum mechanical models.

These equations define the spectral variation of the optical constants and, through eqn (1.8), of the reflectivity (R) also. In general, only a few major resonance phenomena are to be expected (Fig. 1.1); electronic resonances at relatively short wavelengths (visible to near ultraviolet) and ionic dipole resonances at long wavelengths (far infrared). Diatomic and more complex solids, e.g. silica, which sustain *permanent ionic dipoles*, display strong ionic resonances typically at wavelengths in the range 5–10 μm. In monatomic solids such as germanium the infrared absorption is weak, being due to *induced dipoles* produced by the action of the radiation. Figure 1.2 shows the theoretical variation of n, k, and R in the region of the electronic resonances. This model (Lorentz 1909) shows that we may identify four distinct regimes, I–IV in Fig. 1.2. In the low-frequency region I, $\omega \ll \omega_0$, where ω_0 is the electronic resonance, the solid is transparent with negligible absorption, small reflectivity and refractive index, n, increasing (normal dispersion). Region II ($\omega \simeq \omega_0$) is characterized by strong absorption, high reflectivity, and 'anomalous' dispersion, i.e. n decreasing with increasing ω. Within region III where $\omega > \omega_0$ the dispersion becomes 'normal', the absorption decreases to quite small values but the reflectivity remains relatively high. Finally, in region IV,

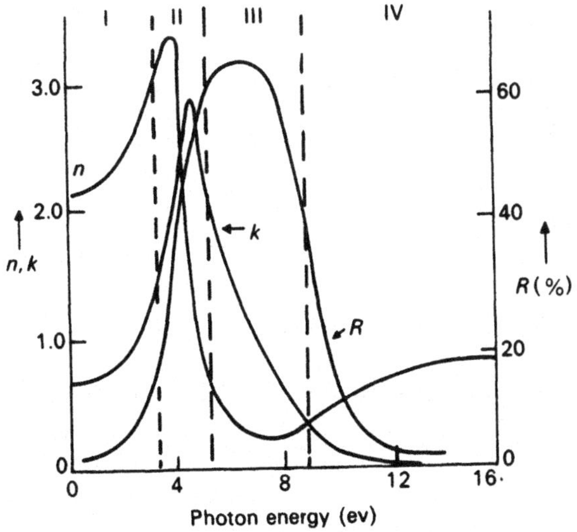

FIG. 1.2. Variation of the refractive index, n, extinction coefficient, k, and reflectivity, R, as a function of photon energy, $\hbar\omega$, calculated from eqns (1.11), (1.12) and (1.8) assuming $\hbar\omega_i = 4$ eV, $f_i = 0.1$ and $\hbar\Gamma_i = 1.0$ eV. (Adapted from Wooten 1972.)

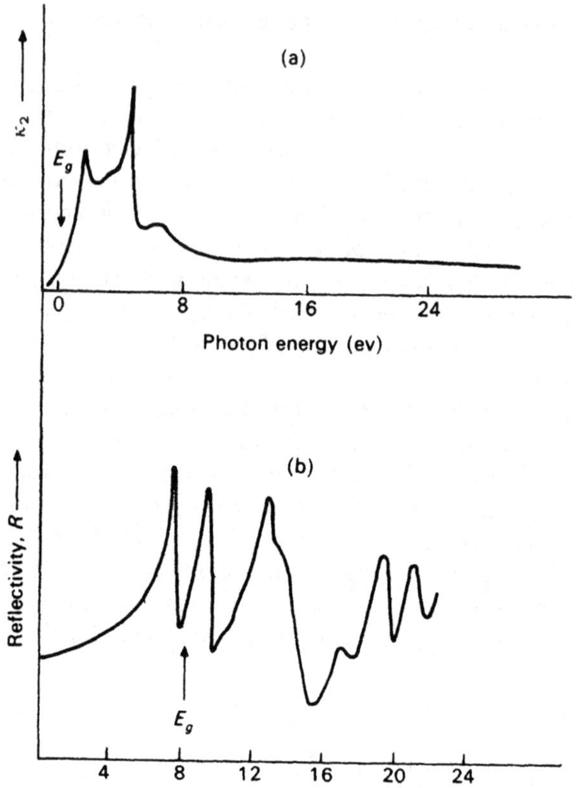

FIG. 1.3. The dependence on photon energy of κ_2 for germanium (a) and the reflectivity, R, (plotted logarithmically) of potassium chloride (b). Note that in both cases above E_g the reflectivity increases and decreases as is expected from the classical model (regions II and III in Fig. 1.2). Structure due to excitons (in potassium chloride) and band structure effects (see Section 1.5.1) is also observed. (After Phillip and Ehrenreich 1963.)

$\omega \gg \omega_0$, the binding energy of the charges is negligible relative to the photon energy, $\hbar\omega$, and they respond as though they were free. Note that the model represented by eqn (1.9) is that of an insulator. However, the behaviour of the optical constants (Fig. 1.2, regions III and IV) is rather characteristic of a metal above the plasma frequency. Indeed we can use eqn (1.9) to represent a typical metal by setting the restoring force, $m\omega_0^2$, to zero which is formally equivalent to Drude's (1902) free electron model of a metal. Figure 1.3 compares the reflectivity of potassium chloride and germanium at photon energies in the range 0–24 eV. Although superficially in agreement with classical predictions these spectra also show additional structure due to quantum mechanical effects (see Section 1.5).

1.3 Spectroscopy and quantum mechanics

Optical spectroscopy has been a principal testing ground of quantum mechanics and of models of atomic structure since the early part of this century. Bohr derived an expression for the electronic energy levels of hydrogen and hydrogen-like ions:

$$E_n = \frac{-mZ^2e^4}{2(4\pi\varepsilon_0)^2 n^2 \hbar^2} \tag{1.13}$$

which is identical with that later derived from Schrödinger's wave mechanical model. Sommerfeld's modification of the Bohr theory led to the idea of elliptical orbits, while retaining the arbitrarily introduced concept of angular momentum quantization. There was no such arbitrariness in wave mechanics, quantization of both orbital angular momentum and the electronic energy arising quite naturally out of the imposition of boundary conditions required to obtain physically realistic solutions for Schrödinger's time-independent equation. However, the Schrödinger theory could not properly account for the fine structure in atomic spectra or for the additional splittings of spectral lines where observed in a magnetic field. Two further developments of quantum mechanics were necessary for such phenomena to be interpreted. The first of these was the introduction of a relativistic wave equation for the electron by Dirac (1927). Solution of this equation led to the additional source of electronic angular momentum, of magnitude $\sqrt{\frac{1}{2}(\frac{1}{2}+1)}\hbar$, which we now refer to as *spin*. It was also determined that relativistic effects (including spin–orbit coupling) removed the l-degeneracies of the particular shells and so produced splittings in the spectral lines of atoms. These splittings have been much studied in the spectra of hydrogen-like atoms, helium, and the alkali metal atoms. Recently very high-resolution spectroscopy of the H_α-line in a low-pressure atomic hydrogen discharge tube has been used both to give a very precise determination of the Rydberg constant and to test the accuracy of quantum electrodynamics. The second important step was the development of efficient procedures for calculating and classifying the electronic energy levels of multi-electron atoms (Slater 1930; Hartree 1928). Nowadays the precision with which we calculate the atomic energy levels and, therefore, the possible positions of spectral lines is determined predominantly by the availability of time on large computer systems.

Quantum mechanics is capable of predicting not only the energies of absorbed or emitted photons in atomic transitions but also of calculating the strengths of such processes using the time-dependent Schrödinger equation. Such calculations yield the selection rules of atomic spectroscopy (Condon and Shortley 1953). The classical results given in eqns (1.11) and (1.12) are also obtained from a quantum mechanical calculation. The damping constant, Γ_i, which determines the resonance width, is related to the lifetime of the excited state through the uncertainty principle. The oscillator strength, f_i, which in the classical model is just the number of electrons per atom which can participate fully in the ith resonance, is now interpreted as a quantum mechanical entity which is related to the transition probability of the absorption process (see Chapter 4).

1.4 Atom and ions in solids

Another early application of quantum mechanics was in the calculation of the energies of electrons involved in chemical bonding. Obviously such calculations become more difficult (and more computationally tedious) with the size of the molecular unit. Calculations of molecular energy levels of the hydrogen molecule ion, hydrogen, and methane are relatively straightforward (Pauling 1967). However, quantum mechanical modelling of the vast number of interacting atoms in a solid presents a rather more daunting challenge. The simplest model assumes that a lattice of fixed positive ions provides a periodic potential in which a gas of electrons are (essentially) free to move (Bloch 1928; Kronig and Penney 1930). The periodicity of the ionic potential, with an appropriate boundary condition, requires that the only energy levels available for occupation by the electrons occur in bands of discrete, very closely spaced levels (bands); energy values between these bands are forbidden. This model leads to the familiar classification of the electronic properties of metals, semiconductors, and insulators (Gibson and Elliot 1974).

The forces which hold atoms together in solids are often classified as ionic, covalent, and metallic. There is also a variety of weaker forces, e.g. van der Waals, hydrogen bonds, etc. (Pauling 1967). In general, several bonding mechanisms are at work in any particular solid. For example, the bonding in aluminium oxide is predominantly ionic, with a small degree of covalent bonding, whereas the opposite situation is found in silica, i.e. predominantly covalent bonding with some ionic bonding. However, the unifying feature of all bonding mechanisms is that the energies of the occupied outer-shell electrons of the constituent atoms are lowered by bonding, in which the interplay of long-range attractive and short-range repulsive forces leads to there being an equilibrium separation between atoms. This interplay is expressed by a simplified expression for the potential energy, $V(R)$, of the bound system, e.g.

$$V(R)=\frac{A}{R^m}\left[1-\left(\frac{m}{n}\right)\left(\frac{R_0}{R}\right)^{n-m}\right] \tag{1.14}$$

where $n > m$, R is the instantaneous separation between the atoms and R_0 is the equilibrium separation. In an ionic solid, such as sodium chloride, we have $m = 1$, representing the fact that positive and negative charges are attracted by Coulomb interaction; the overlap repulsion ($\propto R^{-n}$) is a much shorter range potential and n lies in the range 5–10. For inert gas solids (neon, argon), $m = 6$ and $n = 12$ give accurate values of the bond energies calculated using eqn (1.14). Rather more complicated functions must be used to represent bonding in metallic solids and in strongly covalent solids. The nature of bonding forces in large measure determines the energy band structure of a solid.

Figure 1.4 shows schematically how the energy band structure of sodium develops as a large number of atoms are brought into closer and closer contact with one another. When the atoms are widely separated, (i.e. at $1/R = 0$), the electrons exist in discrete, sharply defined energy levels. Energy gaps, E_g, exist between the adjacent energy levels. As R decreases and the individual atomic charge distributions begin to interact the electronic energies are perturbed, and the energy levels shift and split into $(2l+1)N$ molecular levels, where N is the number of atoms involved in bonding and $(2l+1)$ is the orbital degeneracy of the level. Thus each atomic *ns*-level is a single orbital level, each atom *np* level splits into three molecular levels, and so on. For example, in a hypothetical Na_6 molecule (Fig. 1.4(b)) the 2p level splits into 18 molecular levels and the 3s level splits into six. The Pauli exclusion principle dictates that each level is at most occupied by two electrons with spins aligned antiparallel. However, in this Na_6 molecule only the lowest three molecular 3s levels are occupied because each sodium atom contributes only one 3s electron each. In other words, half of the 3s levels are filled and half remain empty. When N is of the order of Avogadro's number, as in solids, the splitting between adjacent levels within a band is so small that it is more useful to think of a *continuous band* of levels. The magnitude of the energy splitting for a particular level, i.e. the width of the energy band, is determined largely by the mutual electrostatic interaction between two electrons on adjacent atoms, and is of the order 5–10 eV; it is not much changed by the number of atoms bonded together. Obviously the splitting of the atomic energy levels into bands at the equilibrium value of the atomic spacing, R_0, is much larger for the outer energy levels than for the inner core electrons, as Fig. 1.4(b) shows. Thus it is that in sodium metal the 3s electrons occupy an *energy band*, the conduction band, which at most may contain $2N$ electrons. Since there are only N valence electrons provided by the sodium atoms, this band is only half-filled. This is illustrated on the left-hand side of Fig. 1.5: the box representing the 3s band of energy levels is only partially shaded, indicating that the band is half-filled. When atomic energy levels are close together in energy, the energy bands produced from them may overlap as is the case for the top of the 3s band and the bottom of the 3p band in metallic sodium. In this region the

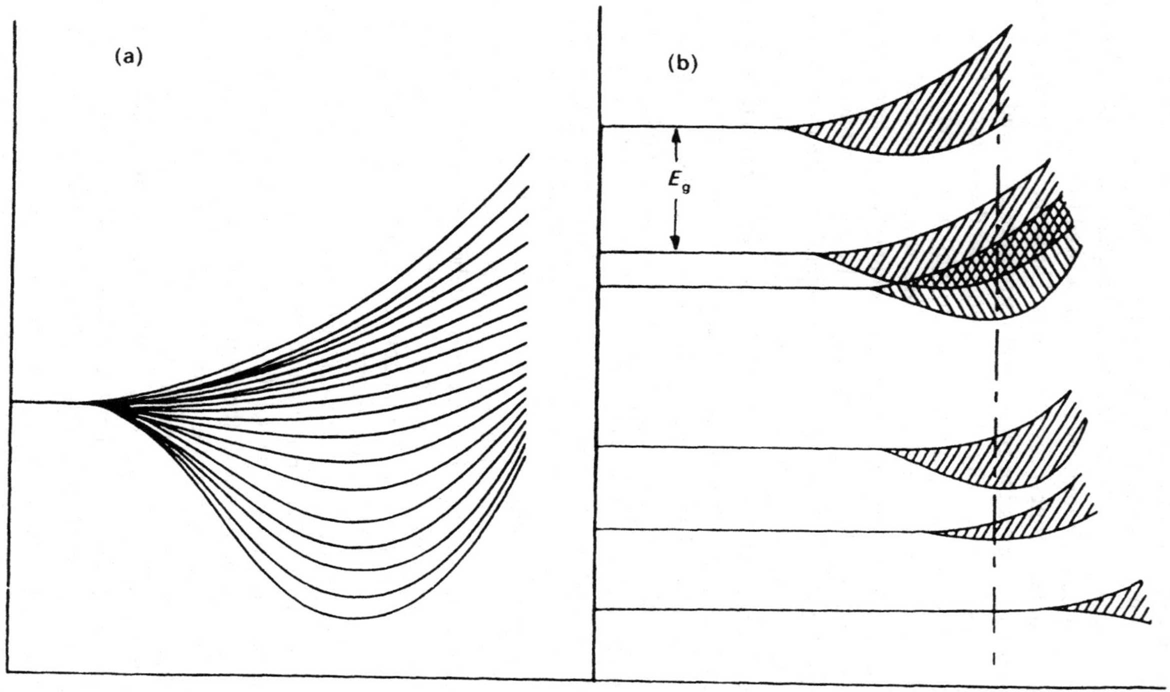

FIG. 1.4. Representing schematically the development of energy bands as atoms are brought closer together in (a) the 3p states of an hypothetical Na_6 molecule and (b) bulk sodium.

electrons occupy hybridized orbitals, the wavefunctions of which are admixtures of pure 3s and 3p wavefunction.

The picture given in Fig. 1.5 provides a qualitative account of the electrical conductivity of solids. In metals such as sodium, with a half-filled conduction band (Fig. 1.5(a)), the electrons are easily excited into higher-lying states in the band by an applied electric field. Magnesium (Fig. 1.5(b)) is also a good metallic conductor. In this case the two valence electrons per atom are sufficient to fill the 3s conduction band. However, because of the small overlap of 3s and 3p bands (Fig. 1.5b) there are a few unoccupied levels in the 3s band and a few occupied levels in the 3p band. It is these electrons which provide the conductivity in metallic magnesium. However, in a solid with a filled valence band, electronic conduction occurs only as a consequence of excitation across the bandgap. Since in an ionic crystal the bandgap, E_g, between the fully occupied valence band and an unoccupied conduction band may exceed 3 or 4 eV, such a material would be a good insulator. The case of sodium fluoride is shown in Fig. 1.5(c). This compound is almost perfectly ionic, each sodium atom transferring one 3s electron to a 3p level on one of the fluorine atoms in forming ionic bonds. As the band scheme in Fig. 1.5(c) shows, this process leaves the 3s and 3p bands localized on the Na^+ ion empty, whereas the 3s and 3p bands on the Cl^- ions are fully occupied. The bandgap, $E_g \sim 7$ eV, is the separation between the top of the Cl^- 3p band and the bottom of the Na^+ 3s band. In consequence, electrical conductivity via electron transport in an electric field does not occur in sodium chloride even close to the melting temperature. The small conductivity (*c.* 10^{-10} $(\Omega\,\text{cm})^{-1}$) is rather associated with charge transport by thermally-activated ionic motion (see e.g. Henderson 1972).

In semiconductors it is usual to regard the *ns* and *np* states as fully hybridized (Pauling 1967); such bands are fully occupied so that at low temperature these materials are insulators. However, the nearest unoccupied band may be quite close in energy, (e.g. in silicon the conduction band is only $\simeq 1$ eV above the valence band), so that electrons at the top of the valence band may be thermally excited across the bandgap, E_g, into the normally unoccupied levels in the conduction band. Such electrons are then free to sustain a current in the presence of an electric field. Obviously, the larger the bandgap the higher the temperature required to provide for the thermally-activated electrical conductivity that typifies semiconductor behaviour.

The simplest quantum mechanical description of the band structure of solids is based on a non-interacting electron approximation in which the electron moves in a potential, $V(\boldsymbol{r})$, which has the periodicity of the crystal lattice, and is caused by the fixed atomic cores plus the charge distribution of all the other outer-shell electrons. In this model the Schrödinger equation

$$[(\hbar^2/2m)\nabla^2 + E - V(\boldsymbol{r})]\psi(\boldsymbol{r}, \boldsymbol{k}) = 0 \tag{1.15}$$

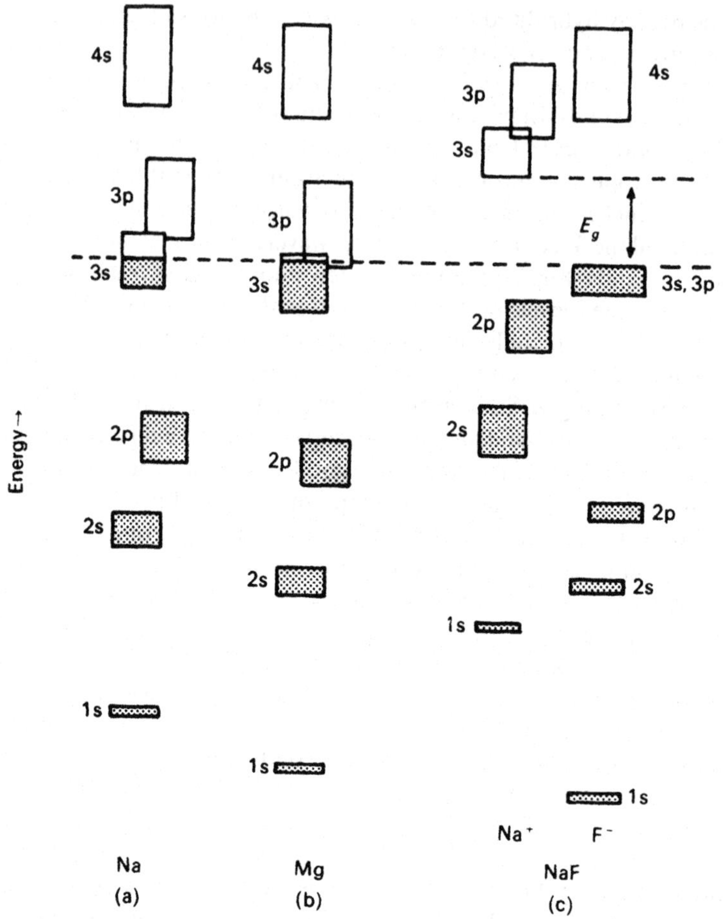

FIG. 1.5. The filling of energy bands in sodium, magnesium, and sodium fluoride (schematic). The energy scale has been adjusted so that the highest occupied level in each solid has the same energy.

has eigenfunctions, the so-called Bloch functions

$$\psi(\boldsymbol{r}, \boldsymbol{k}) = u_k(\boldsymbol{r}) \exp i\boldsymbol{k} . \boldsymbol{r} \tag{1.16}$$

where $u_k(\boldsymbol{r})$ has the periodicity of the lattice and $\boldsymbol{k}$ is a vector in reciprocal space. If the crystal sample has cubic symmetry with linear dimension L, $\boldsymbol{k}$ is restricted to the values $-\pi/L < k_i < \pi/L$, where k_i refers to any one of the components k_x, k_y, and k_z. In the simplest case $V(\boldsymbol{r})$ is assumed constant throughout the crystal; indeed it is convenient to take this constant to be zero.

Then $u_k(\mathbf{r})$ is a constant and the eigenvalues are

$$E = \frac{\hbar^2}{2m} k^2. \tag{1.17}$$

When the periodicity of the potential $V(\mathbf{r})$ is taken into account the formula for E is more complicated. Plots of E versus $\mathbf{k}$ along different directions in reciprocal space are often referred to as the band structure of the solid. In Fig. 1.6 we show such a plot for germanium; at low temperature the shaded region separates the fully occupied valence band from the empty conduction band. As we shall see, measurement of the optical constants directly probes the band structure of the solid.

The concept of an *effective mass* for the electron in an energy band is of some importance, particularly in discussion of electron transport phenomena. Then it is useful to write a simple classical equation for the motion of electrons under an external perturbation in the form external force = mass × acceleration. Since the actual force experienced by the electron consists of the external force *plus* any internal force due to periodic potential, there is not a simple relationship between the external force and acceleration. We can use

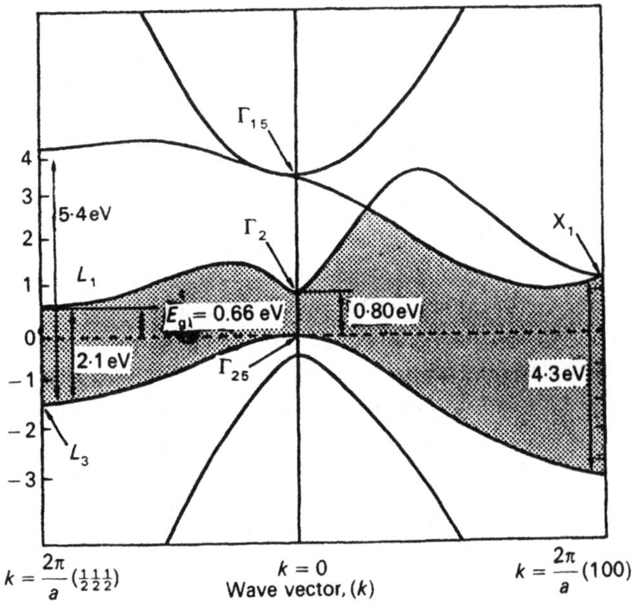

FIG. 1.6. Partial energy band diagram for germanium. The Γ's, X, and L's are points of high symmetry in the Brillouin zone. (After Kittell 1966.)

such an equation if we introduce an effective mass, m^*, to replace the free electron mass. This effective mass takes into account the action of the internal forces. Indeed m^* is actually a second-rank tensor, although for many simple applications it can be treated as a scalar. Its magnitude, however, can differ considerably from the free electron mass. Indeed, for electrons close to the top of the energy band m^* is negative (see e.g. Gibson and Elliott 1974).

We have emphasized that the energy band structure depends upon lattice periodicity, which in turn follows from definite geometrical relationships between all the atoms in the solid. In fact a crystal structure is produced by associating an atom or group of atoms with a particular three-dimensional lattice. A lattice is an abstract mathematical construction. The symmetry properties of a solid are determined by two factors, the nature of the bonding between atoms and the relative sizes of atoms involved in the solid (see e.g. Hayes and Stoneham 1985). Consider first the case of the alkali halides (MX). If we regard the atoms M and X as fully ionized, then each cation ($M^+ = Li^+$, Na^+, K^+, Rb^+) has six equidistant octahedrally arranged anions ($X^- = F^-$, Cl^-, Br^-, I^-) surrounding it. In principle, because the Coulomb interaction between anion and cation provides a non-directional attractive force between anions and cations, each cation may be surrounded by any number of anions (and vice versa). In practice, the relative sizes of the ions determine the arrangement. For example, the octahedral arrangement (Fig. 1.7(a)) is sustained only for a ratio of cation to anion radii, r_+/r_-, in the range $\sqrt{2}-1 < r_+/r_- < \sqrt{3}-1$. This rule, derivable from simple geometric arguments, applies also to the alkaline earth oxides. However, beryllium oxide, for which $r_+/r_- = 0.22$, does not have the rock-salt structure. Nor do the caesium halides ($0.732 < r_+/r_- < 1$) crystallize in this form: instead they have eightfold coordination of anions around each cation (Fig. 1.7(b)). Similar considerations apply to other simple compounds, e.g. calcium fluoride. However, there is also a need to maintain strict charge neutrality. In the calcium fluoride structure (Fig. 1.7(c)) each doubly-charged cation is at the centre of a tetrahedral arrangement of anions, there being eight such groupings in each unit cell. In this case $0.225 < r_+/r_- < 0.414$. Such ideas are readily adapted to more complex ionic crystals such as alumina, potassium magnesium fluoride, etc. (Greenwood 1968).

In contrast to the alkali halides, which are predominantly ionic, the majority of materials are of mixed bonding character. The series of compounds magnesium oxide, magnesium sulphide, magnesium selenide, and magnesium telluride all have the rock-salt structure. However, the properties become decreasingly ionic as the element from Group VI becomes less electronegative. Indeed, magnesium telluride has a high reflectivity and a temperature-dependent electrical conductivity, such as is normally to be

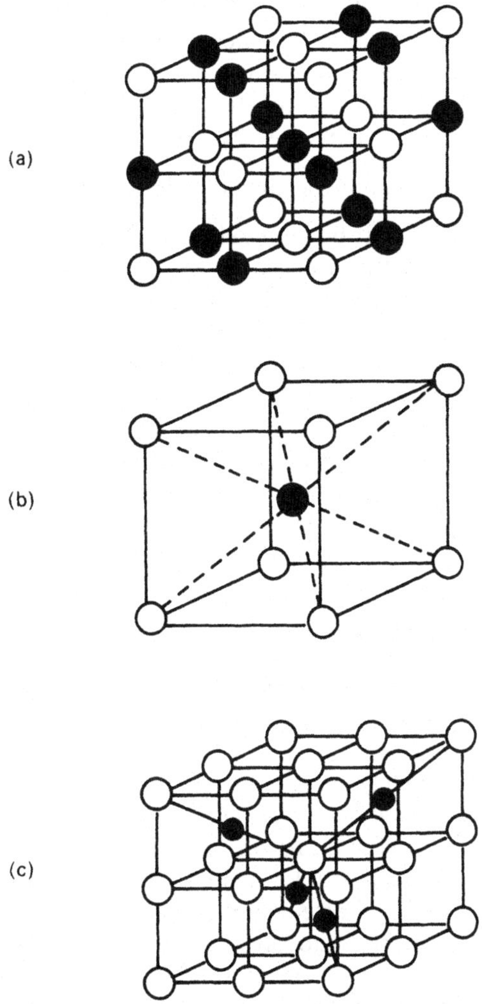

FIG. 1.7. Unit cells of (a) alkali halide, (b) caesium halide, and (c) calcium fluoride crystals.

expected from a semiconductor. The bonding in semiconductors is mainly covalent. Quantum mechanical considerations show that in covalent bonding the electron charge density is concentrated in the region between atoms, and is axially symmetric about the line joining the nuclei. This is the origin of directional bonding in semiconductors. Consider the situation in diamond where carbon atoms are bonded together in a tetrahedral arrangement. In the free carbon atom the configuration $(1s)^2(2s)^2(2p)^2$ has the lowest energy. However, in diamond the 2s and 2p states combine to form a set of four hybrid bonds per atom directed towards the corners of a regular tetrahedron. Quantum mechanically the wave function of such an (sp^3) bonding orbital is written as a linear combination of 2s and 2p atomic wavefunctions, i.e. $\alpha|2s\rangle + \sum_i \beta_i|2p_i\rangle$; there cannot be more than four such (sp^3) hybrid bonds per atom. Since each atom contributes one electron to each of the four bonds, and each bond is shared by two atoms, then on average each set of sp^3 hybrid orbitals is fully occupied. There are two electrons per bond with spins aligned antiparallel. The crystal structure of diamond (Fig. 1.8(a)) is shared with the elemental semiconductors silicon and germanium. Related tetrahedral structural arrangements in diatomic semiconductors such as gallium arsenide (Fig. 1.8(b)) and zinc sulphide (Fig. 1.8(c)) follow from the mixture of ionic and covalent bonding. For example, in gallium arsenide the partial ionic bonding requires the formation of Ga^- and As^+ ions. Since each of these ions has four outer electrons they may then enter into a diamond-like hybridized bonding structure.

1.5 Optical properties of solids

This text is concerned largely with the optical spectroscopy of inorganic insulators; semiconductors are discussed to a lesser extent and metals not at all. As discussed in Section 1.4, the highest band of occupied levels—the *valence band*—is completely occupied in an insulator and separated from the lowest band of excited states—the *conduction band*—by a significant energy gap. The distinction between insulators and semiconductors (at least electrically) is largely semantic, since at low temperatures pure semiconductors (e.g. silicon, germanium, gallium arsenide) are good insulators. As an arbitrary measure we define an insulator to have largely ionic bonding and a bandgap greater than approximately 2.5 eV. Most common semiconductors are predominantly covalent, with rather narrower bandgaps ($\leqslant 2.5$ eV). The energy band structure of a solid determines the number, strength, and spectral positions of the electronic absorption resonances implied by eqns (1.11) and (1.12). However, other resonances are also possible, since both impurities and defect centres may occur and have energy levels within the bandgap. Optical transitions may then take place either between the energy levels of the particular centre or between these levels and the valence and/or conduction band.

1.5.1 Interband optical transitions

In Section 1.4 we discussed as distinct topics the concepts of bonds between atoms and the development of energy bands. These topics are not unrelated;

(a)

(b)

(c)

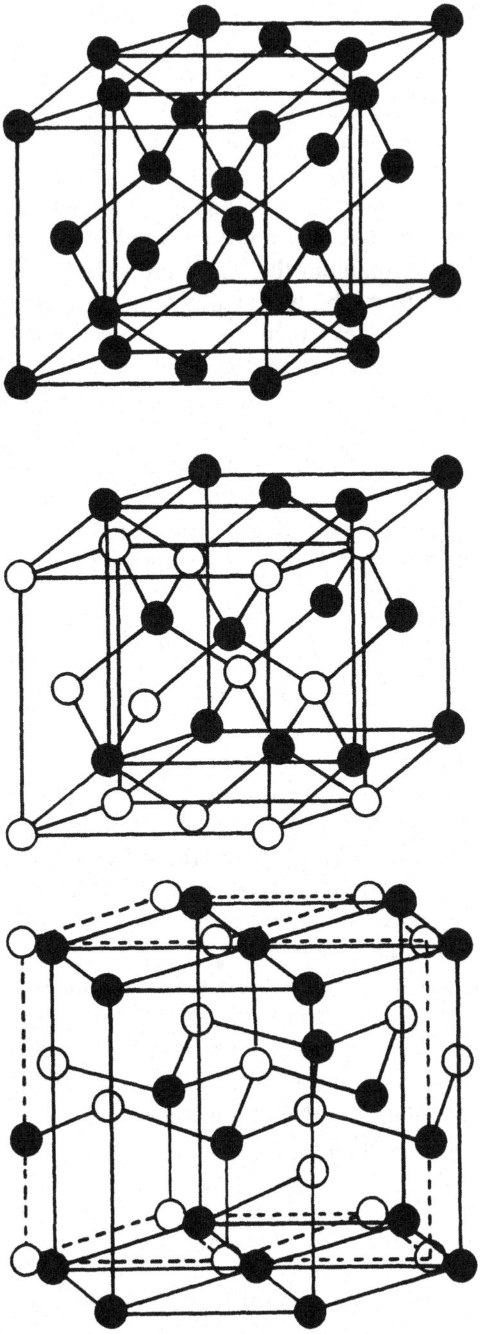

FIG. 1.8. Unit cells of (a) silicon, (b) gallium arsenide, and (c) zinc sulphide.

rather are they different aspects of the same problem—that of the cohesivity of solids. Nonetheless it is convenient to discuss some properties of materials in terms of bonding and others in terms of band theory. For the purposes of describing the optical properties of inorganic solids, band theory is particularly appropriate and from this point on will be applied almost exclusively. We illustrate this in terms of the excitation of direct and indirect interband transitions in pure semiconductors. As Fig. 1.6 shows, the energy gap between valence band and conduction is not constant for all values of the electron wavevector, $\boldsymbol{k}$. In germanium the top of the valence band at the point $\boldsymbol{k}=0$ (Γ_{25}) is separated from the lowest point in the conduction bands (L_1) at $\boldsymbol{k}=(2\pi/a)\,(\frac{1}{2}\frac{1}{2}\frac{1}{2})$ by 0.66 eV; there are four such conduction band minima situated at zone boundaries in the ⟨111⟩ directions. Germanium is an example of a semiconductor with an *indirect bandgap*. So too is silicon, although the conduction band minima occur about 85 per cent of the way between $\boldsymbol{k}=0$ and the boundary along the zone ⟨100⟩ direction. Zinc selenide and gallium arsenide are examples of *direct-gap* semiconductors. Gallium phosphide is another indirect-gap semiconductor; in the alloy system GaP_xAs_{1-x}, where $0<x<1$, there is a transition from direct gap to indirect gap at the composition $x=0.46$.

Optical transitions across the bandgap conserve total momentum. If the photon momentum is determined by its wavevector, $\boldsymbol{q}$, the selection rule for the absorption of a photon in a transition from the state $\boldsymbol{k}_v$ to $\boldsymbol{k}_c$ is

$$\hbar\boldsymbol{k}_c=\hbar\boldsymbol{k}_v+\hbar\boldsymbol{q} \qquad (1.18)$$

where $\boldsymbol{k}_c$ and $\boldsymbol{k}_v$ are the wavevectors for electrons at appropriate points in the conduction and valence bands, respectively. Since the photon wavevector $\boldsymbol{q}$ has a magnitude which is only a small fraction of the reciprocal lattice vector we may assume $\boldsymbol{q}=0$ so that we have the $\boldsymbol{k}$-selection rule $\boldsymbol{k}_c=\boldsymbol{k}_v$. Figure 1.9(a) shows a simple energy band for a direct-gap semiconductor. Absorption transitions at the bandgap energy, E_g, which mark the fundamental absorption edge, are indicated by the vertical arrow. Such *direct* interband transitions obey the $\boldsymbol{k}$-selection rule (eqn 1.18). A partial energy band diagram for an indirect-gap semiconductor is shown in Fig. 1.9(b). The fundamental absorption process at E_g, the arrow (i), is an *indirect transition* which does not satisfy the $\boldsymbol{k}$-selection rule. This radiative transition only occurs if accompanied by the absorption or emission of a phonon in order to conserve $\boldsymbol{k}$. The selection rule is now

$$\hbar\boldsymbol{k}_c=\hbar\boldsymbol{k}_v\pm\hbar\boldsymbol{k}_p, \qquad (1.19)$$

where $\boldsymbol{k}_p$ is the phonon wavevector. Since this process involves both electron–radiation and electron–phonon interactions, it has a smaller transit-

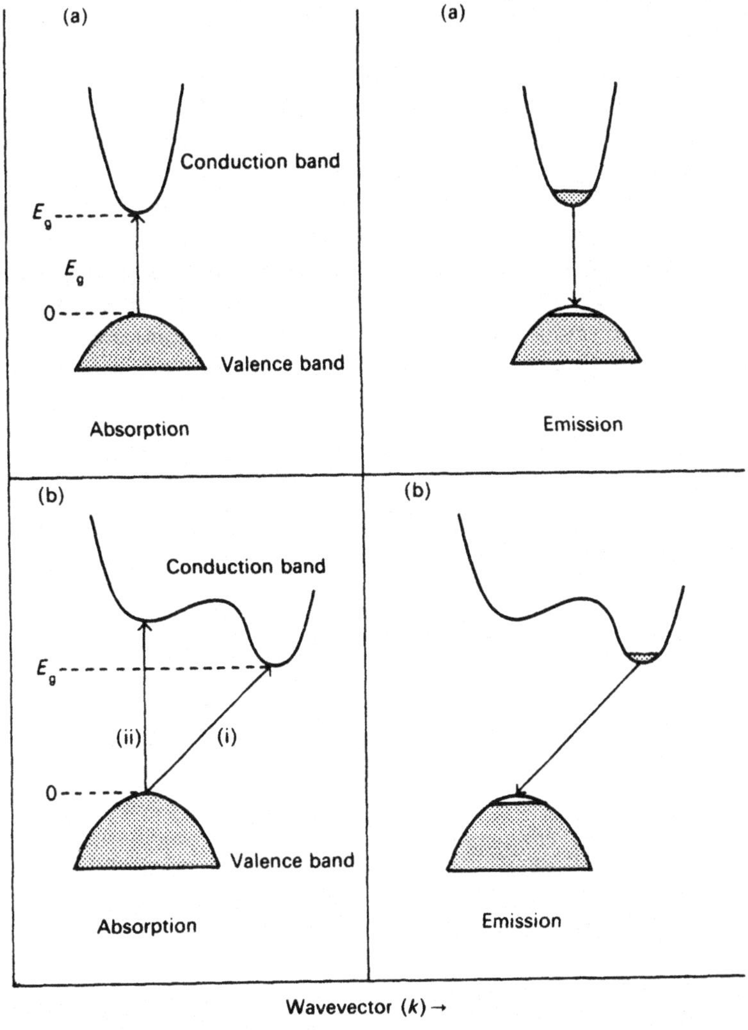

FIG. 1.9. Illustrating absorption and emission transitions in (a) direct and (b) indirect interband semiconductors.

ion probability than the direct transition. Of course, direct absorption transitions also occur (e.g. arrow (ii) in Fig. 1.9b) but at energies greater than the bandgap. Germanium is an indirect-gap semiconductor with an indirect-gap threshold between the Γ_{25} and L_1 points in the Brillouin zone (Fig. 1.6) of 0.74 eV at 4.2 K. Direct and/or indirect transitions are observed in measurements of the absorption coefficient (Section 6.1) close to the absorption edge.

Figure 1.10 shows the results of such measurements at $T = 4.2$–90 K for germanium. At low temperature, $T < 20$ K, there are no phonons to absorb, and the two components beginning at ~0.75 eV and ~0.77 eV in Fig. 1.10(a) involve excitation across the indirect bandgap accompanied by the emission of phonons at two different energies. Above 77 K phonons with these two different energies are also absorbed during the transition, resulting in absorption of radiation at somewhat lower energies. The onset of direct transitions occurs just below the photon energy of 0.89 eV, above which the absorption coefficient increases strongly. Whether a material has a direct or an indirect bandgap is of signal importance in processes of electron–hole recombination by radiative emission rather than by non-radiative multi-phonon emission. Electrons in the conduction band occupy energy states at the *bottom* of the band whereas any holes occupy states at the *top* of the

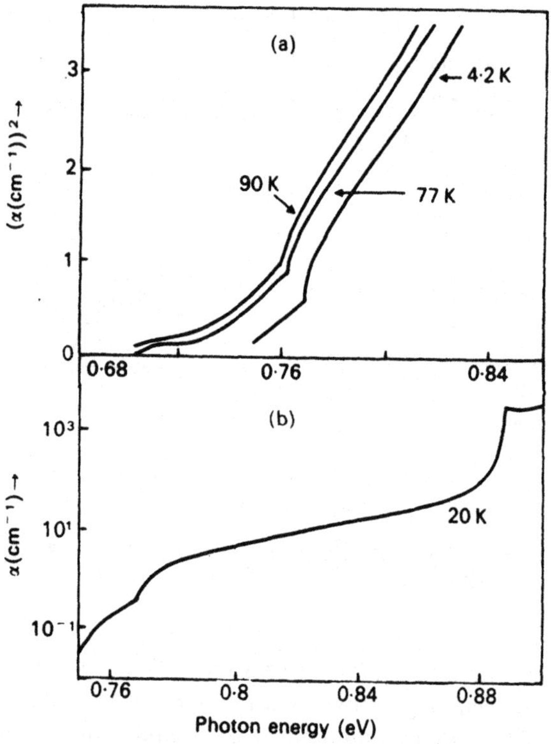

FIG. 1.10. The absorption edge spectrum of germanium; (a) portrays indirect gap transitions at $T \leqslant 90$ K (Macfarlane *et al.* 1957) and (b) shows the increasing absorption due to the onset of direct gap transitions.

valence band. Thus, in radiative emission processes electrons at the bottom of the conduction bottom transfer to empty states at the top of the valence band, releasing energy as a photon of radiation. In a direct-gap semiconductor (Fig. 1.9(a)) this is a direct radiative transition satisfying eqn (1.18) with probability likely to be greater than that of the non-radiative process. Thus the electron–hole recombination process results in luminescence with $h\nu \simeq E_g$. For indirect-gap semiconductors, however, this radiative transition is indirect (Fig. 1.9(b)) with small transition probability. It is then more likely that electron–hole recombination will occur non-radiatively. The probability of radiative recombination in an indirect-gap semiconductor may be enhanced by suitably doping the material so that recombination takes place at the impurity centre where the $\boldsymbol{k}$-selection rule may be broken. Of particular interest in the luminescence context are III–V alloy systems such as GaP_xAs_{1-x}, where $0 < x < 1$, in which the direct bandgap may be varied from 2.78 eV ($x = 1$, gallium phosphide, indirect gap)) to 1.43($x = 0$, gallium arsenide, direct gap)). Such materials are of considerable commercial interest as LEDs and semiconductor diode lasers (Berg and Dean 1976). The intrinsic luminescence spectra shifts in a predictable way that is roughly linear with x (Fig. 1.11). However, there is an abrupt change of slope of the transition at the crossover from direct-gap to indirect-gap material. In GaP_xAs_{1-x} the crossover occurs at $x = 0.46$ (Craford *et al.* 1972). This behaviour indicates that the transition energy at the luminescence peak senses the *mean* alloy composition. Contrariwise, the luminescence width varies in a manner which reflects the fact that within any particular alloy there are *fluctuations* in composition to which the transition responds. The fluctuations are a source of inhomogeneous broadening, and the luminescence bands are in consequence broader for the alloys than for the pure compounds.

The spectral variations of the optical constants give detailed information about the band structure. The spectrum for germanium (Fig. 1.3(a)) shows the effects of the broad valence band, and structure in the conduction bands above the band edge. (The data in Fig. 1.3(a) are reproduced with an expanded energy scale in Fig. 1.12(a).) The absorption rate at any photon energy above the bandgap is determined by the density of states in the valence and conduction bands separated by the energy $E \simeq \hbar\omega$. Singularities in the density of states produce edges in the absorption spectra, which may increase or decrease with an increase in frequency. Sharp peaks are observed when upward and downward edges almost coincide. This is the origin of the peaks at 1.3 eV and 4.5 eV in the spectrum of germanium (Fig. 1.3(a)); weaker structure is observed at 3.5 eV and 6.0 eV. These structures are resolved more clearly using modulation spectroscopy (Cardona 1969) in which some derivative of the optical spectrum is measured. For example, wavelength-modulation spectroscopy requires that a wavelength-modulated monochromatic

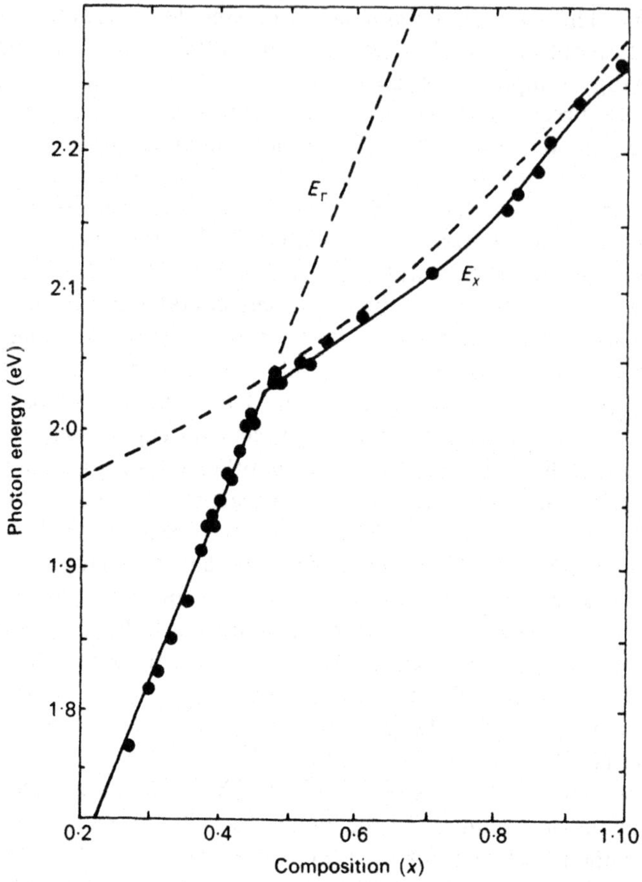

FIG. 1.11. Showing the variation of electroluminescence peak energy with composition x, at 300 K in GaP_xAs_{1-x} alloys. For $x \leqslant 0.46$, in which range these materials are direct-gap semiconductors, the peak position follows the composition linearly. Above the crossover from direct-gap to indirect-gap at $x = 0.46$ the variation becomes non-linear (adapted from Craford *et al.* 1972).

beam is slowly swept through the appropriate spectral range and the a.c. component of the reflectance measured by synchronous phase-sensitive detection; the derivative $dR/d\lambda$ or dR/dE can then be recorded directly. The resolution of such methods can be better than 10^{-3} eV. When applied to the reflectivity of germanium it becomes clear that the 1.3 eV and 6.0 eV bands are split by spin–orbit splitting of around 20 meV (Fig. 1.12(b)).

Insulators such as potassium chloride are highly transparent, with very little absorption and a small reflectivity, at low photon energies. This is shown

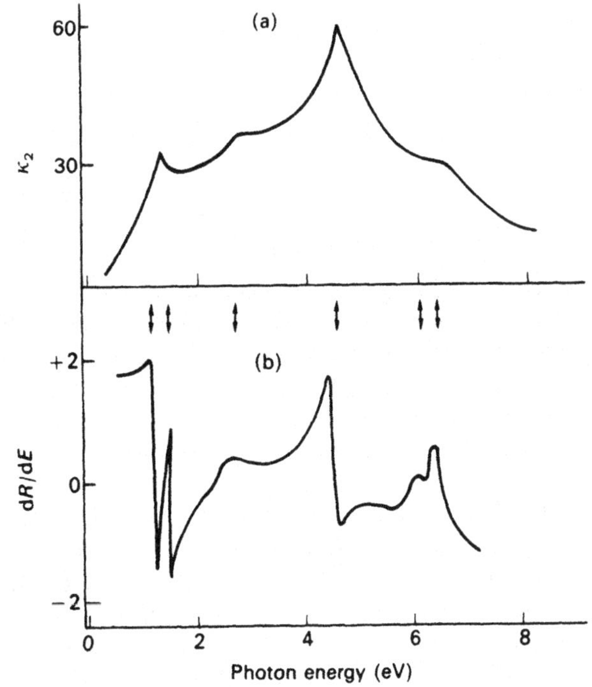

FIG. 1.12. Showing (a) the real part of the dielectric constant, κ_2, adapted from Brust *et al.* (1962) compared with (b) the modulated reflectance (dR/dE) spectrum of germanium (adapted from Zucca and Shen 1971) at photon energies of 0–6.5 eV measured at 5 K.

in Fig. 1.3(b) where the region of transparency extends to about 7.0 eV; the bandgap of potassium chloride is large. Above the region of transparency there is a number of sharp peaks related to the narrow energy bands and the formation of *Frenkel excitons.* In ionic crystals the strong Coulomb interaction between bound electron–hole pairs produces sharp energy levels, which give rise to the narrow exciton absorption bands in Fig. 1.3(b) at photon energies near to the bandgap energy. Frenkel excitons are self-trapped by the strong distortion of the lattice induced by the charged components of the exciton. Self-trapping is not effective in semiconductors because the electron–hole attraction is screened by the dielectric constant. *Wannier excitons* are free to move with a well-defined wavevector until they decay radiatively or become trapped at impurities or defects. These *free excitons* are observed in semiconductors with relatively small bandgap. Some aspects of exciton spectroscopy are discussed in Chapters 10 and 12 (see also Appendix 1A): more substantial accounts are given by Knox (1963) and in Rashba and Sturge (1982).

1.5.2 Bandgap transitions

Interband transitions are not the only spectral characteristics of interest. Indeed, there is a rich variety of spectroscopic phenomena associated with both impurities and defects which have optical absorption and emission bands at energies below the bandgap. It is the nature and characterization of transitions involving *bandgap states* originating on intrinsic lattice defects (Chapter 7), rare-earth metal ions (Chapter 8) and transition metal ions (Chapter 9) that are the predominant concern of this text. In large measure theory and experiment have developed in parallel, almost since the very beginning of quantum mechanics. Crystal field theory (Bethe 1929) and its later developments (Sugano *et al.* 1970) were used to interpret the observed spectra of different transition metal ions in ionic host crystals. Such theories determine the electronic energy levels of such systems in the *static* electric field of the crystal in terms of a few empirical parameters. The effects of ionic vibrations were treated separately. In a seminal paper Born and Oppenheimer (1927) showed that under appropriate conditions the electronic and ionic motions are separable, and from this point they went on to provide the basis for interpreting the bandshapes of electronic spectra of simple molecules. In application to solid state systems the Born–Oppenheimer approximation finds its simplest expression in terms of the configurational coordinate model, in which the electronic states are assumed to be coupled to a single vibrational mode. Such a theory is able to give a simple semi-quantitative interpretation of the various bandshapes which are to be found in the spectra of solids. The various theories used to model the optical properties of defect centres in both ionic and semiconductor crystals owe much to the use of large, efficient computers. A wide range of different spectral features—transition energies, bandshapes, and radiative lifetimes—have been reported for many different types of optical centre. It remains one of the most remarkable successes of quantum mechanics that such spectra can be interpreted, often to the finest detail and experimental nuance.

Appendix 1A: Excitons in crystalline solids

The original concepts of the exciton were promulgated by Frenkel (1931, 1936) and by Wannier (1937). These quasi-particles play important roles in the spectroscopy of solids. Under certain circumstances we can envisage the Frenkel exciton on a single atom, where the excited electron is localized by strong interaction with the hole left in the electron shell from which the electron was excited. Such Frenkel excitons are characterized by a small orbital radius for the excited electron. In strongly ionic crystals the charged nature of the electrons and holes enables them to interact with a small number

Table 1A.1
The main species of excitons (after Hayes and Stoneham (1984)).

Exciton	In essence, an electron and hole moving with a correlated motion as an electron–hole pair.
Wannier exciton	Electron and hole both move in extended orbits. Energy levels related to hydrogen atom levels by scaling using effective masses and dielectric constant. Occurs in covalent solids such as silicon.
Frenkel exciton	Electron and hole both move in compact orbits, usually essentially localized on adjacent ions. Seen in ionic solids, such as KCl, in absorption.
Self-trapped exciton	One or both carriers localized by the lattice distortion they cause. Observed in ionic solids, such as KCl, in emission.
Bound exciton	Only a useful idea when a defect merely prevents translational motion of an exciton and does not otherwise cause significant perturbation.
Core exciton	Lowest-energy electronic excitation from a core state, leaving an unoccupied core orbital (e.g., the 1s level of a heavy atom) and an electron in the conduction band whose motion is correlated with that of the core hole.
Excitonic molecule	Complex involving two holes and two electrons.
Multiple bound excitons	Complex of many holes and a similar number of electrons, apparently localized near impurities. Some controversy exists, but up to six pairs of localized carriers have been suggested.
Exciton gas	High concentration of electrons and holes in which each electron remains strongly associated with one of the holes (an insulating phase).
Electron–hole drops	High concentration of electrons and holes in which the motions are plasma-like (a metallic phase), not strongly correlated as in excitons and included here only for comparison.

of neighbouring ions, causing them to relax to new equilibrium positions. The distortion is responsible for the self-trapping mechanism that operates in ionic crystals. In other materials the excitation cannot be restricted to a single atom or group of atoms, and the quantum of energy migrates from atom to atom. Since the exciton is no longer restricted to a single atom, carrying no net charge, only energy, it may be regarded as an excited state of the whole solid.

It is in this sense that we refer to *Wannier excitons*. However, in the current literature the term 'exciton' has come to mean any one of a rather large number of excitations of the solid plus radiation system. Hayes and Stoneham (1984) detailed the various excitonic phenomena as shown in Table 1A.1.

2

Energy levels of free atoms and of optical centres in crystals

CHAPTER 1 set out to develop a familiarity with the basic concepts of spectroscopy. Here we present some formal quantum mechanics required to calculate energy levels involved in spectroscopic transitions. The atomic system is described by a simplified, yet usable, Hamiltonian operator, $\mathscr{H}$, which has eigenstates $|a\rangle$ with eigenvalues E_a; the wavefunctions describing the stationary states take the form

$$\psi_a(\boldsymbol{r}, t) = |a\rangle \exp(-\mathrm{i}E_a t/\hbar). \tag{2.1}$$

This chapter is concerned with calculating and then characterizing the eigenstates of $\mathscr{H}$, and we are assisted in this task by studying the symmetry properties of $\mathscr{H}$. We make use of the theorem that

if an operator $\mathcal{O}$ commutes with H, i.e. $[\mathscr{H}, \mathcal{O}] = 0$, it follows that the eigenstates of $\mathscr{H}$ can be chosen to be eigenstates of $\mathcal{O}$ and so can be labelled by the eigenvalues of $\mathcal{O}$.

This characterization of the wavefunctions of the electronic system according to the eigenvalues of the operator $\mathcal{O}$ allows us to make use of the following helpful theorems:

1. There are no matrix elements of $\mathscr{H}$ between eigenstates of $\mathcal{O}$ with different eigenvalues
2. There are no matrix elements of $\mathcal{O}$ between wavefunctions with different energy levels.

2.1 Electronic structure of one-electron atoms

According to the Dirac theory the Hamiltonian for an electron in an electromagnetic field characterized by potentials ϕ and A is

$$\mathscr{H} = \frac{(\boldsymbol{p} + e\boldsymbol{A})^2}{2m} - e\phi + \zeta(r)\boldsymbol{l}\,.\,\boldsymbol{s} + \frac{e}{m}\boldsymbol{s}\,.\,\boldsymbol{B} + \text{ smaller terms} \tag{2.2}$$

where the magnetic field strength is given by $\boldsymbol{B} = \nabla \times \boldsymbol{A}$, and $\boldsymbol{s}$ is the operator for electron spin. For an electron in the electrostatic field of the nucleus,

$\phi = Ze/4\pi\varepsilon_0 r$ and $A=0$, and neglecting smaller terms the Hamiltonian becomes

$$\mathscr{H} = \frac{p^2}{2m} - \frac{Ze^2}{4\pi\varepsilon_0 r} + \zeta(r)\boldsymbol{l}.\boldsymbol{s}$$
$$= \mathscr{H}_o + \mathscr{H}_{so} \tag{2.3}$$

where the spin–orbit coupling $\zeta(r)\boldsymbol{l}.\boldsymbol{s}$ is written as $\mathscr{H}_{so}$, and $\mathscr{H}_o$ is the simplified Hamiltonian for the electron in the field of a nucleus of charge Ze which is stationary at the origin. (A suitably brief introduction to the Dirac equation is given in Griffith (1961).)

2.1.1 *Orbital angular momentum*

If $\mathscr{H}_{so}$ is neglected we are left with the simplified Hamiltonian

$$\mathscr{H}_o = -\frac{\hbar^2}{2m}\nabla^2 - \frac{Ze^2}{4\pi\varepsilon_0 r}$$
$$= -\left(\frac{\hbar^2}{2m}\right)\left(\frac{\partial^2}{\partial r^2} + \left(\frac{2}{r}\right)\frac{\partial}{\partial r}\right) + \frac{\hbar^2 l^2}{2mr^2} - \frac{Ze^2}{4\pi\varepsilon_0 r}. \tag{2.4}$$

$\boldsymbol{l} = \boldsymbol{r} \times \boldsymbol{p}$ is the operator for the orbital angular momentum of the particle which in spherical polar coordinates is

$$l_x = -i\hbar\left(y\frac{\partial}{\partial z} - z\frac{\partial}{\partial y}\right)$$
$$= i\hbar\left(\sin\phi\frac{\partial}{\partial\theta} + \cos\theta\sin\phi\frac{\partial}{\partial\phi}\right)$$
$$l_y = -i\hbar\left(\cos\phi\frac{\partial}{\partial\theta} - \cos\theta\sin\phi\frac{\partial}{\partial\phi}\right)$$
$$l_z = -i\hbar\frac{\partial}{\partial\phi}$$

and

$$l^2 = -\left[\frac{1}{\sin\theta}\frac{\partial}{\partial\theta}\left(\sin\theta\frac{\partial}{\partial\theta}\right) + \frac{1}{\sin^2\theta}\frac{\partial^2}{\partial\phi^2}\right]. \tag{2.5}$$

The commutation relationships for the components of $\boldsymbol{l}$ are

$$[l_x, l_y] = i\hbar l_z, \quad [l_y, l_z] = i\hbar l_x, \quad [l_z, l_x] = i\hbar l_y$$
$$[l^2, \boldsymbol{l}] = 0. \tag{2.6}$$

Since $\boldsymbol{l}$ does not contain r, we find that $[\mathscr{H}, \boldsymbol{l}] = 0$, indicating that orbital angular momentum is conserved for the simplified one-electron atom. $\mathscr{H}_o$, $\boldsymbol{l}^2$,

l_z form a set of commuting operators; consequently energy eigenstates can be chosen which are also eigenstates of $\boldsymbol{l}^2$ and l_z. Such energy eigenstates are

$$\psi_{nlm_l}(\boldsymbol{r}) = R_{nl}(r)\, Y_l^{m_l}(\theta, \phi) \tag{2.7}$$

where the labels n, l, m_l are the quantum numbers which characterize the energy eigenstates. The $R_{nl}(r)$ function is written as the ket $|nl\rangle$, the $\psi_{nlm_l}(\boldsymbol{r})$ function as the ket $|nlm_l\rangle$, and the $Y_l^{m_l}(\theta, \phi)$ functions are the spherical harmonics being eigenstates of $\boldsymbol{l}$ and l_z, i.e.

$$\begin{aligned} \boldsymbol{l}^2 Y_l^{m_l}(\theta, \phi) &= l(l+1)\hbar^2 Y_l^{m_l}(\theta, \phi) \\ l_z Y_l^{m_l}(\theta, \phi) &= m_l \hbar Y_l^{m_l}(\theta, \phi). \end{aligned} \tag{2.8}$$

Tables of the spherical harmonics are to be found in texts on quantum mechanics (e.g. Weissbluth 1978). The energy eigenvalues

$$E_{nlm_l} = \frac{-mZ^2e^4}{2(4\pi\varepsilon_0)^2 n^2 \hbar^2} \tag{2.9}$$

depend only on the principal quantum number n, which takes positive integral values, an identical result to Bohr's (eqn 1.13). The quantum number, l, characterizing the magnitude of the orbital angular momentum also takes integral values: $l = 0, 1, 2, \ldots, (n-1)$, whereas m_l measures the z-component of the orbital angular momentum. There are $2l+1$ integral values of $m_l = l, (l-1), (l-2), \ldots -(l-1), -l$. Thus for a given value of n the several different allowed values of l and m_l contribute a large degeneracy in each level.

The simplified Hamiltonian $\mathscr{H}_0$ (eqn 2.4) is invariant under inversion of coordinates, $\boldsymbol{r} \to -\boldsymbol{r}$ (i.e. $x, y, z \to -x, -y, -z$) in Cartesian coordinates, or $r, \theta, \phi \to r, (\pi - \theta), (\pi + \phi)$ in spherical polar coordinates) as is easily seen from the Cartesian form of this equation:

$$\mathscr{H}_0 = -\frac{\hbar^2}{2m}\left(\frac{\partial^2}{\partial x^2} + \frac{\partial^2}{\partial y^2} + \frac{\partial^2}{\partial z^2}\right) - \frac{Ze^2}{4\pi\varepsilon_0(x^2+y^2+z^2)^{\frac{1}{2}}}. \tag{2.4}$$

If we write the *inversion operator* as P_i, where

$$P_i f(\boldsymbol{r}) = f(-\boldsymbol{r}) \tag{2.10}$$

for *any* function $f(\boldsymbol{r})$, then $[\mathscr{H}_0, P_i] = 0$. As a result the eigenstates of $\mathscr{H}_0$ can be chosen to be eigenstates of P_i, that is, to be *even* or *odd* under space inversion. The eigenstates defined in eqn (2.7) have this property since

$$Y_l^{m_l}(\pi - \theta, \pi + \phi) = (-1)^l Y_l^{m_l}(\theta, \phi) \tag{2.11}$$

and then we have

$$P_i |nlm_l\rangle = (-1)^l |nlm_l\rangle. \tag{2.12}$$

Wavefunctions $|nlm_l\rangle$ characterized by even values of l are said to have *even parity*, i.e. they do not change sign under inversion of coordinates, while wavefunctions characterized by odd values of l are said to have *odd parity*. The parity property of wavefunctions is of special significance in the determination of selection rules for spectroscopic transitions.

2.1.2 Electron spin

The concept of electron *spin*, an intrinsic angular momentum with an associated magnetic moment, was first introduced by Goudsmit and Uhlenbeck (1926) to account for 'anomalous' experimental results of spectra measured in a magnetic field. This intrinsic angular momentum was shown by Dirac to be a consequence of relativistic considerations. Equation (2.2) is an approximate Hamiltonian derived from Dirac's (1927) relativistic equation.

As can be seen in eqn (2.2), spin enters into the eigenvalue problem through the spin–orbit coupling term, $\mathscr{H}_{so}=\zeta(r)\boldsymbol{l}.\boldsymbol{s}$, and the Zeeman term $-(e/m)\boldsymbol{s}.\boldsymbol{B}$, where the electron magnetic moment due to spin is $-(e/m)\boldsymbol{s}=\boldsymbol{\mu}_s$. The operator $\boldsymbol{s}$ has the commutation properties of angular momentum so that we can replace $\boldsymbol{l}$ by $\boldsymbol{s}$ in eqn (2.6). $\boldsymbol{s}^2$ and $\boldsymbol{s}_z$ are a commuting pair with two common eigenstates. These eigenstates are characterized by a spin quantum number, s, analogous to l, and by a quantum number m_s, analogous to m_l, which gives the z component of spin angular momentum in units of $\hbar$. The only value of s is $\frac{1}{2}$, and the two values of m_s are $\pm\frac{1}{2}$. These two spin eigenstates may be written in ket form as $|sm_s\rangle$ or as $s(m_s)$. However, they are often given the labels $|\alpha\rangle$ and $|\beta\rangle$, where $|\alpha\rangle=|\frac{1}{2}\,\frac{1}{2}\rangle=S(\frac{1}{2}\rangle$ is the 'spin-up' state and $|\beta\rangle=|\frac{1}{2}-\frac{1}{2}\rangle=S(-\frac{1}{2}\rangle$ is the 'spin-down' state. By analogy with eqn (2.8) for orbital angular momentum we have

$$
\begin{aligned}
\boldsymbol{s}^2|\tfrac{1}{2}m_s\rangle &= \tfrac{1}{2}(\tfrac{1}{2}+1)\hbar^2|\tfrac{1}{2}m_s\rangle=\tfrac{3}{4}\hbar^2|\tfrac{1}{2}m_s\rangle \\
s_z|\tfrac{1}{2}m_s\rangle &= m_s\hbar|\tfrac{1}{2}m_s\rangle
\end{aligned}
\tag{2.13}
$$

and the two spin functions are orthonormal:

$$
\begin{aligned}
\langle\tfrac{1}{2}\tfrac{1}{2}|\tfrac{1}{2}\tfrac{1}{2}\rangle &= \langle\tfrac{1}{2}\ -\tfrac{1}{2}|\tfrac{1}{2}\ -\tfrac{1}{2}\rangle=1 \\
\langle\tfrac{1}{2}\,\tfrac{1}{2}|\tfrac{1}{2}\ -\tfrac{1}{2}\rangle &= 0
\end{aligned}
\tag{2.14}
$$

The simplest orbital-plus-spin description of the one-electron system, i.e. the product state $|nlm_l\rangle|sm_s\rangle$, can be written as the *spin–orbital* $u=|nlsm_l m_s\rangle$, the numbers inside the ket being the quantum numbers specifying the electron state. The s label is usually omitted since s always has the value $\frac{1}{2}$. If the spin–orbit coupling and Zeeman terms are ignored the eigenvalue depends only on n, which emphasizes again the high degree of degeneracy in the one-electron atom.

2.1.3 The coupled representation—total angular momentum

We now consider the energy degeneracy of the electron states. The energy depends on n only, and for a given value of n, n different values of l are allowed.

For each n and l there are $2(2l+1)$ different sets of values allowed, corresponding to the different values of the quantum numbers m_l and m_s. We can speak of a $2(2l+1)$-fold function space or vector space associated with each set of nl values (or nls values) and the $|nlm_lm_s\rangle$ states form an orthonormal basis set for this (nls) space. We can choose a different set of orthonormal basis states for this function space. Chemists tend to use a linear combination of $|nlm_l\rangle$ states which are real and can be plotted to give a pictorial interpretation of chemical bonding. Basis functions for the nls space can also be chosen which are classified according to the *total angular momentum* of the electron whose operator is

$$\boldsymbol{j} = \boldsymbol{l} + \boldsymbol{s}. \tag{2.15}$$

From the commutation properties of the $\boldsymbol{l}$ and $\boldsymbol{s}$ operators we find that $\boldsymbol{j}$ also has the commutation properties of angular momentum (eqn 2.6) where $\boldsymbol{l}$ is replaced by $\boldsymbol{j}$. We find that $\boldsymbol{l}^2$, $\boldsymbol{s}^2$, $\boldsymbol{j}^2$, and $\boldsymbol{j}_z$ form a commuting set of operators which also commute with $\mathscr{H}_0$. Hence we can choose energy eigenstates which are also eigenstates of $\boldsymbol{l}^2$, $\boldsymbol{s}^2$, $\boldsymbol{j}^2$, and $\boldsymbol{j}_z$. As before $\boldsymbol{l}^2$ and $\boldsymbol{s}^2$ are characterized by the quantum numbers l and s $(=\frac{1}{2})$ while $\boldsymbol{j}^2$ and $\boldsymbol{j}_z$ are characterized by quantum numbers j and m_j. These new energy eigenstates for the n, l, s space are expressed in ket form as $|nlsjm_j\rangle$ and are made up of linear combinations of the $|nlsm_lm_s\rangle$ states. The relationship between the two sets of basis states can be written as

$$|nlsjm_j\rangle = \sum_{m_l\, m_s} |nlsm_lm_s\rangle \langle lsm_lm_s|lsjm_j\rangle \tag{2.16}$$

in which the $\langle lsm_lm_s|lsjm_j\rangle$ are the *vector-coupling* or *Clebsch–Gordan coefficients*, tables of which are given in various quantum mechanics texts (e.g. Weissbluth 1978). Some values are listed in Appendix 3A. There are two allowed values of j, viz. $l \pm \frac{1}{2}$, and for each value of j there are $(2j+1)$ values of m_j: $j, (j-1), (j-2), \ldots -(j-1), -j$. It is a simple matter to check that there are as many different allowed sets of values of j, m_j as there are allowed sets of values of m_l, m_s, i.e. $2(2l+1)$. In this situation we say that l and s are coupled to give the total angular momentum j, and that the new total angular momentum states are eigenstates in the *coupled representation*. The coupling of two angular momentum operators to form a new angular momentum operator is treated from a more general viewpoint in Section 3.1.8. We list in Table 2.1 properties of the general angular momentum operator $\boldsymbol{J}$ and of the eigenstates of $\boldsymbol{J}^2$ and J_z. The $|nlsm_lm_s\rangle$ and $|nlsjm_j\rangle$ states are equally valid sets of energy eigenstates of the Hamiltonian $\mathscr{H}_0$.

2.1.4 Spin–orbit coupling

The largest term omitted from the Hamiltonian in eqn (2.4) is the spin–orbit coupling energy term $\mathscr{H}_{so}$. This can be pictured as the magnetic coupling

Table 2.1

Some properties of the general angular momentum operator, $\boldsymbol{J}$, and angular momentum eigenstates $|JM\rangle$. $\boldsymbol{J}$ can refer to orbital or spin angular momentum or to a sum of different angular momenta

$$[J_x, J_y]=i\hbar J_z,\ [J_y, J_z]=i\hbar J_x,\ [J_z, J_x]=i\hbar J_y$$
$$J_{\pm}=J_x \pm i J_y$$
$$J^2=J_x^2+J_y^2+J_z^2=J_+J_-+J_z^2-\hbar J_z=J_-J_++J_z^2+\hbar J_z$$
$$[J_{\pm}, J_z]=\mp\hbar J_z$$
$$J^2|JM\rangle=J(J+1)\hbar^2|JM\rangle$$
$$J_z|JM\rangle=M\hbar|JM\rangle$$
$$J_+|JM\rangle=\sqrt{(J-M)(J+M+1)}\,\hbar|J\,M+1\rangle$$
$$J_-|JM\rangle=\sqrt{(J+M)(J-M+1)}\,\hbar|J\,M-1\rangle$$

between the spinning electron (regarded as an elementary magnet) and the magnetic field due to the relative orbital motion of the nucleus and electron. The form of the function $\zeta(r)$ is given by the Dirac theory (Weissbluth 1978) as

$$\zeta(r)=\frac{1}{4\pi\varepsilon_0}\frac{Ze^2}{2m^2c^2r^3}. \tag{2.17}$$

The $\boldsymbol{l}.\boldsymbol{s}$ factor can be written as

$$\boldsymbol{l}.\boldsymbol{s}=\tfrac{1}{2}(l_+s_-+l_-s_++2l_zs_z). \tag{2.18}$$

$l_{\pm}$ and $s_{\pm}$ respectively commute with l^2 and s^2 but not with l_z or s_z, as Table 2.1 shows. Hence the $|nlsm_lm_s\rangle$ states are not eigenstates of $\boldsymbol{l}.\boldsymbol{s}$. However, $\boldsymbol{l}.\boldsymbol{s}$ commutes with $\boldsymbol{j}$. To see this note that $\boldsymbol{l}.\boldsymbol{s}$ can be written as

$$\boldsymbol{l}.\boldsymbol{s}=\tfrac{1}{2}(j^2-l^2-s^2). \tag{2.19}$$

Since $\boldsymbol{j}=\boldsymbol{l}+\boldsymbol{s}$ commutes with j^2, l^2, and s^2 it follows that $\boldsymbol{j}$ commutes with $\boldsymbol{l}.\boldsymbol{s}$, and therefore with $\mathscr{H}_{so}$. Hence $\boldsymbol{l}.\boldsymbol{s}, j^2, j_z$ are a set of commuting operators for which a common set of eigenfunctions are the coupled representation states $|nlsjm_j\rangle$. These are eigenstates of $\mathscr{H}_0$, as we saw in Section 2.1.3.

Are these eigenstates also of $\mathscr{H}_{so}=\zeta(r)\boldsymbol{l}.\boldsymbol{s}$? To answer this question we calculate the matrix elements of $\mathscr{H}_{so}$ using the basis functions of the coupled representation, i.e.

$$\left\langle n'l's'j'm_j'\right|\xi(r)\frac{j^2-l^2-s^2}{2}\left|nls\,jm_j\right\rangle$$
$$=\langle n'l'|\xi(r)|nl\rangle\hbar^2\frac{j(j+1)-l(l+1)-s(s+1)}{2}\delta_{l'l}\,\delta_{s's}\,\delta_{j'j}\,\delta_{m_j'm_j} \tag{2.20}$$

where $\delta_{l'l}$ is the Kronecker delta. Evidently $\mathscr{H}_{so}$ can mix states of the same l, s, j, m_j values but with different values of n. If the separation in energy between states belonging to different n values is much larger than the matrix elements of $\mathscr{H}_{so}$ then the mixing of states with different n values is very small and can be neglected. To that approximation the basis functions $|nlsjm_j\rangle$ are eigenstates of $\mathscr{H}_0+\mathscr{H}_{so}$. The diagonal matrix elements of $\mathscr{H}_{so}$ are obtained from eqn (2.20). For a given value of l the two values of $j=l\pm\frac{1}{2}$ have spin–orbit coupling energies of $\zeta l/2$ and $-\zeta(l+1)/2$, respectively, where

$$\zeta=\langle nl|\xi(r)|nl\rangle. \tag{2.21}$$

This splitting of the state with quantum numbers nls into two states with different energies, characterized by the two allowed j values, is shown in Fig. 2.1. That the spin–orbit coupling energy does not depend on m_j follows from the fact that $\boldsymbol{l}.\boldsymbol{s}$ commutes with $J_{\pm}$, since this means that there are no matrix elements of j_+ or j_- between eigenstates of $\mathscr{H}_{so}$ with different values of the spin–orbit coupling energy. Spin–orbit coupling causes the doublet splitting of the sodium D lines and of many other transitions in the visible spectrum of the sodium gas discharge spectrum.

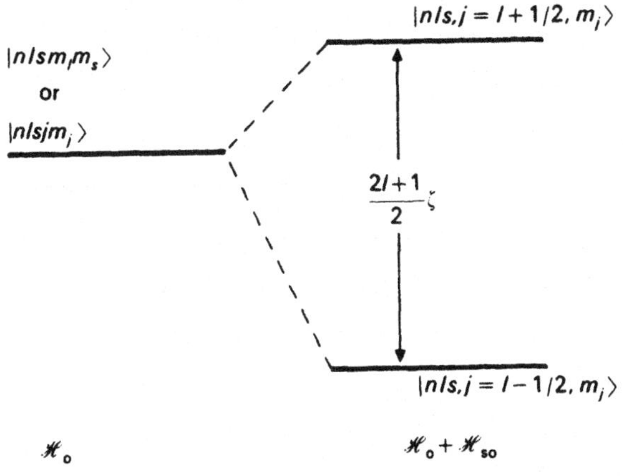

FIG. 2.1. In the absence of spin–orbit coupling the one-electron wavefunctions can be labelled by either set of quantum numbers, $nlsm_lm_s$ or $nlsjm_j$. In the presence of spin–orbit coupling only the second set are 'good' quantum numbers. This coupling separates the two different j states. Configurational mixing is assumed negligible.

2.2 Transitions between stationary states

The discussion so far has concentrated on determining the eigenstates of the Hamiltonian of the one-electron atom i.e. $\Psi_n(r,\ t)=\psi_n(r)\exp(-\mathrm{i}E_nt/\hbar)$, n representing the appropriate set of quantum numbers. These stationary states have the important property that the charge density is independent of time: $\rho(\boldsymbol{r}, t) = e|\Psi_n(\boldsymbol{r}, t)|^2 = e|\psi_n(\boldsymbol{r})|^2$. For example, consider the wavefunction (unnormalized) composed of equal amounts of 2s and 3p stationary states:

$$\begin{aligned}\Psi &= \Psi_{2s}(\boldsymbol{r}, t) + \Psi_{3p}(\boldsymbol{r}, t)\\ &= \psi_{2s}(\boldsymbol{r})\exp(-\mathrm{i}E_{2s}t/\hbar) + \psi_{3p}(\boldsymbol{r})\exp(-\mathrm{i}E_{3p}t/\hbar).\end{aligned} \tag{2.22}$$

The change density is

$$\begin{aligned}\rho(\boldsymbol{r}, t) &= (-e)|\Psi(\boldsymbol{r}, t)|^2\\ &= -e(|\psi_{2s}(\boldsymbol{r})|^2 + |\psi_{3p}(\boldsymbol{r})|^2 + \psi_{2s}^*\psi_{3p}\exp\mathrm{i}(E_{2s} - E_{3p})t/\hbar\\ &\quad + \psi_{3p}^*\psi_{2s}\exp\mathrm{i}(E_{3p} - E_{2s})t/\hbar).\end{aligned} \tag{2.23}$$

This contains an oscillatory term

$$\rho(\boldsymbol{r})\exp(-\mathrm{i}\omega t) + \rho^*(\boldsymbol{r})\exp(\mathrm{i}\omega t) \tag{2.24}$$

where $\rho(\boldsymbol{r}) = e\psi_{2s}^*\psi_{3p}$ and $\omega = (E_{3p} - E_{2s})/\hbar$. Since the states ψ_{2s} and ψ_{3p} are of opposite parity $\rho(\boldsymbol{r})$ is an odd function and has a non-zero dipole moment. Thus the atom in this non-stationary state behaves like an oscillating electric dipole, which, according to classical electromagnetic theory, can emit or absorb radiation at the oscillating frequency $\omega = (E_{3p} - E_{2s})/\hbar$. This is just the Bohr frequency condition for a transition between stationary states 2s and 3p:

$$\hbar\omega = E_{3p} - E_{2s}. \tag{2.25}$$

This simple example suggests that to have transitions between stationary states some interaction must occur to perturb the system and cause a mixing of different stationary states. And the mixing must involve states of opposite parity (the *parity selection rule*). The problem of transitions between states is addressed using time-dependent quantum mechanical perturbation theory in Chapter 4.

2.3 Multi-electron atoms

2.3.1 The central field Hamiltonian

For multi-electron atoms the Hamiltonian is a sum over all N electrons of one-electron operators (eqn 2.3) plus the inter-electron Coulomb interaction, $\mathscr{H}' = \sum_{i>j} e^2/4\pi\varepsilon_0 r_{ij}$. Hence the Hamiltonian is written as

$$\mathscr{H} = \sum_i \left(\frac{\boldsymbol{p}_i^2}{2m} - \frac{Ze^2}{4\pi\varepsilon_0 r_i} + \zeta(r_i)\boldsymbol{l}_i \cdot \boldsymbol{s}_i\right) + \sum_{i>j} \frac{e^2}{4\pi\varepsilon_0 r_{ij}}. \tag{2.26}$$

In this equation the final term poses the greatest computational difficulty. In the *central field approximation*, developed to overcome this problem, the interaction of the ith electron with the other $(N-1)$ electrons is replaced by a spherically-averaged potential energy term,

$$U_i(r_i) = \left\langle \sum_{j \neq i} \frac{e^2}{4\pi\varepsilon_0 r_{ij}} \right\rangle \tag{2.27}$$

in which the bracket signifies spherical averaging. In this approximation the Hamiltonian is now written as

$$\begin{aligned}\mathscr{H} &= \sum_i \left(\frac{\boldsymbol{p}_i^2}{2m} - \frac{Ze^2}{4\pi\varepsilon_0 r_i} + \zeta(r_i)\boldsymbol{l}_i \cdot \boldsymbol{s}_i + U_i(r_i) \right) \\ &= \sum_i \left(\frac{\boldsymbol{p}_i^2}{2m} + V_i'(r_i) + \zeta(r_i)\boldsymbol{l}_i \cdot \boldsymbol{s}_i \right) \end{aligned} \tag{2.28}$$

where $V_i'(r_i)$ is the spherically symmetric one-electron operator which represents the potential energy of the ith electron in the field of the nucleus and all other electrons. This Hamiltonian consists of a sum of one-electron terms each of which is the same as the hydrogen-like Hamiltonian in eqn (2.3) except that a more complicated radial potential energy V' is involved.

Further simplification follows from the neglect of the smaller spin–orbit coupling terms, which leaves the orbital Hamiltonian

$$\mathscr{H}_0 = \sum_i \left(\frac{\boldsymbol{p}_i^2}{2m} + V_i'(r_i) \right). \tag{2.29}$$

The angular part of each one-electron term in eqn (2.29) is identical to that in the case of hydrogen, the radial and angular parts of the wavefunction are again separable, and the wavefunction can be written as

$$|nlm_l\rangle = R_{nl}'(r_i)\, Y_l^{m_l}(\theta_i, \phi_i) \tag{2.30}$$

However, $R_{nl}'(r_i)$ is the solution of the radial equation involving the central field potential $V_i'(r_i)$. This radial function is characterized by a principal quantum number n and an angular momentum quantum number l, and the energy of the one-electron state depends on both n and l (E_{nl}). The complete spin–orbital, u, obtained by multiplication of the orbital state by the spin state $S(m_s)$, is characterized by the four quantum numbers n, l, m_l, m_s. Hence,

$$u = R_{nl}'(r)\, Y_l^{m_l}(\theta, \phi) S(m_s) = |nlm_l\rangle |sm_s\rangle = |nlm_l m_s\rangle. \tag{2.31}$$

The many-electron eigenstate of $\mathscr{H}_0$ is then written as the product state, $\Pi_{\alpha_i} u_{\alpha_i}$, where the index α_i represents all the one-electron quantum numbers of the ith state. The energy of this state, $\Sigma E_{n_i l_i}$, depends on the set of $n_i l_i$ values. This set, $(n_i l_i)$, is called the *electron configuration*. The energy does not depend on the m_l and m_s values, and as before these eigenstates have a large energy degeneracy.

The calculation of the central field potential, $V'_i(r_i)$, is exceedingly tedious. However, Hartree (1928) developed a more efficient procedure which involved searching through successive approximations for a *self-consistent field*. Starting from an assumed central field potential, $V'_i(r_i)$, one calculates $|n_i l_i m_{l_i}\rangle$, and this is repeated for all electrons. These $|n_i l_i m_{l_i}\rangle$ wavefunctions, where $i \neq k$, are used to calculate an improved $V'_k(r_k)$ for electron k. This central field is then compared with the original assumed potential; if they differ the procedure is iterated until the final results converge to a self-consistent $V'_i(r_i)$ and $|n_i l_i m_{l_i}\rangle$. These orbital wavefunctions are used in the spin–orbital $u_i = |n_i l_i m_{l_i}\rangle |s_i m_{s_i}\rangle$. The multi-electron Hartree wavefunction is then the product function

$$u_1(1)u_2(2)u_3(3)\ldots u_N(N)$$

in which the radial and spin coordinates of the electrons are represented by (1), (2), etc., and each subscript represents a set of four quantum numbers defining the spin orbital.

2.3.2 *Exchange symmetry*

The Hamiltonian $\mathscr{H}$ (eqn 2.26) is invariant under the interchange of the coordinates, spin and orbital, of any two electrons: i.e. $[\mathscr{H}, P_{ij}] = 0$, where $\boldsymbol{P}_{ij}$ is an operator which interchanges spin and orbital coordinates of electrons i and j, so that $(\boldsymbol{r}_i, s_i \leftrightarrow \boldsymbol{r}_j, s_j)$. The interchange operator, P_{ij}, may be written as the product operator

$$P_{ij} = P_{ij}^{\text{orb}} \cdot P_{ij}^{\text{spin}}, \tag{2.32}$$

P_{ij}^{orb} and P_{ij}^{spin} being operators for the interchanges $\boldsymbol{r}_i \leftrightarrow \boldsymbol{r}_j$ and $s_i \leftrightarrow s_j$, respectively. The simplified Hamiltonian $\mathscr{H}_0$, eqn (2.29), is similarly invariant under P_{ij} but it is also variant under P_{ij}^{orb}.

The eigenvalues of the interchange operator are $\gamma = \pm 1$; the eigenstates with $\gamma = +1$ and -1 being, respectively, *even* (or *symmetric*) and *odd* (or *antisymmetric*) under the interchange. Experimentally the complete wavefunctions of electrons are found to be antisymmetric under P_{ij}. However, no such antisymmetrization of the wavefunction is allowed for in the Hartree treatment. The spin orbitals may be organized into an antisymmetric N-electron wavefunction by writing them in the form of a Slater determinant (Slater 1930):

$$\psi(1, 2, 3, \ldots) = \frac{1}{\sqrt{N!}} \begin{vmatrix} u_1(1) & u_2(1) & u_3(1) & \ldots & u_N(1) \\ u_1(2) & u_2(2) & u_3(2) & \ldots & u_N(2) \\ u_1(3) & u_2(3) & u_3(3) & \ldots & u_N(3) \\ \vdots & & & & \vdots \\ u_1(N) & u_2(N) & u_3(N) & \ldots & u_N(N) \end{vmatrix} \tag{2.33}$$

Recalling that when any two rows or any two columns of a determinant are identical the determinant is zero, we recognize that when two one-electron states are identical two columns of the Slater determinant are identical so that the wavefunction is identically zero. This gives a simple statement of the

Pauli Exclusion Principle: in an atom no two electrons can occupy identical states, i.e. can have the same four quantum numbers.

The Slater determinantal wavefunctions indicate those one-electron states which are occupied by electrons, without assigning specific electrons to specific states. The Slater determinantal wavefunctions can be represented by the ket $|(n_i l_i m_{l_i} m_{s_i})\gamma = -1\rangle$, where $\gamma = -1$ signifies a properly antisymmetrized arrangement of the N electrons among the N one-electron states $n_i l_i m_{l_i} m_{s_i}$. The symmetry of the Hamiltonian under interchange (or exchange) of coordinates of any two electrons is called *exchange symmetry*. Instead of the Hartree self-consistent field approach which takes no account of the requirement of antisymmetry, one can use the Hartree–Fock approach which uses a variational technique to obtain the best Slater determinantal wavefunctions. This approach is described in detail by Tinkham (1964).

2.3.3 *Classification of wavefunctions of the central field Hamiltonian*

We have already noted that since the energy of the eigenstates of $\mathcal{H}_0$ (eqn 2.29) does not depend on the set of m_{l_i} and m_{s_i} values there is a large energy degeneracy. Consequently many different sets of energy eigenfunctions can be chosen. It is clear that $\mathcal{H}_0$ commutes with $\boldsymbol{l}_i$, and also with $\boldsymbol{L} = \Sigma_i \boldsymbol{l}_i$. In addition, since $\mathcal{H}_0$ does not involve spin $\mathcal{H}_0$ commutes both with $\boldsymbol{s}_i$ and with $\boldsymbol{S} = \Sigma_i \boldsymbol{s}_i$. In consequence $\mathcal{H}_0$ commutes with $\boldsymbol{J} = \boldsymbol{L} + \boldsymbol{S}$. Since $\mathcal{H}_0$, $\boldsymbol{l}_i^2$, $\boldsymbol{s}_i^2$, $\boldsymbol{L}^2$, $\boldsymbol{S}^2$, L_z, S_z constitute a commuting set of operators the eigenstates of $\mathcal{H}_0$ can be classified by the set of quantum numbers $(n_i l_i)\, LSM_L M_S$. Including the requirement that the eigenstate be antisymmetrized under the interchange, i.e. $\gamma = -1$, the eigenstate can be expressed as a ket $|(n_i l_i)LSM_L M_S \gamma = -1\rangle$ formed from the linear combinations of Slater determinantal wavefunctions for the specific set of $(n_i l_i)$ values. Specific examples of these eigenstates are written down in a later section. That the eigenfunctions must be antisymmetric restricts the number of possible L, S values for any given configuration. Similarly we can see that $\mathcal{H}_0$, $\boldsymbol{l}_i^2$, $\boldsymbol{s}_i^2$, $\boldsymbol{L}^2$, $\boldsymbol{S}^2$, $\boldsymbol{J}^2$, J_z, P_{ij} form a commuting set of operators, and energy eigenstates can be formed which are classified by the quantum numbers $(n_i l_i)$, L, S, J, M_J, $\gamma = -1$. These eigenstates are formed from linear combinations of the $|(n_i l_i)LSM_L M_S \gamma = -1\rangle$ functions, i.e.

$$|(n_i l_i)LSJM_J \gamma = -1\rangle = \sum_{M_L, M_S} |(n_i l_i)LSM_L M_S \gamma = -1\rangle \langle LSM_L M_S | LSJM_J\rangle \tag{2.34}$$

The $\langle LSM_L M_S | LSJM_J\rangle$ are the Clebsch–Gordan coefficients.

2.3.4 Wavefunctions and energy levels for outer electrons

Optical spèctroscopy and electron spin resonance (ESR) spectroscopy are generally concerned with the small number of electrons on the outside of the atomic centre, outside the closed electron shells. We will therefore concern ourselves only with these outer electrons. Further, in most cases of interest to us the atomic centres will be ions rather than uncharged atoms. At present we consider free ions. Subsequently such ions are considered as constituents of a solid. The free-ion Hamiltonian is

$$\mathscr{H}_{\mathrm{FI}} = \mathscr{H}_{\mathrm{o}} + \mathscr{H}' + \mathscr{H}_{\mathrm{so}} \tag{2.35}$$

$$\mathscr{H}_{\mathrm{o}} = \sum_i \left(\frac{p_i^2}{2m} + V'(r_i) \right) \tag{2.36}$$

where

$$\mathscr{H}' = \sum_{i>j} \left(\frac{e^2}{4\pi\varepsilon_0 r_{ij}} \right) \tag{2.37}$$

$$\mathscr{H}_{\mathrm{so}} = \sum_i \zeta(r_i) \boldsymbol{l}_i \cdot \boldsymbol{s}_i \tag{2.38}$$

and the summation is over the outer electrons. Consider first $\mathscr{H}_{\mathrm{o}}$. Each outer electron moves in the central field of the nucleus and of the inner closed-shell electrons $V'(r)$. The form of $\mathscr{H}_{\mathrm{o}}$ is identical with that of eqn (2.29) but now i refers only to the outer electrons, all of which experience the same central field potential $V'(r_i)$. The wavefunctions are the Slater determinants $|(n_i l_i m_{l_i} m_{s_i})\gamma = -1\rangle$.

$\mathscr{H}'$ is the energy of the Coulomb interaction between the outer electrons. The effect of $\mathscr{H}'$ is to subject the outer electrons to a non-central force, so that the individual electron orbital angular momenta, $\boldsymbol{l}_i$, are not constants of the motion, i.e. $[\mathscr{H}', \boldsymbol{l}_i] \neq 0$. However, $[\mathscr{H}', \boldsymbol{L}] = 0$. To see this we write

$$\left[\sum_{i>j} \frac{1}{|\boldsymbol{r}_i - \boldsymbol{r}_j|}, \boldsymbol{L} \right] = i\hbar \sum_{i>j} \sum_k \left(\boldsymbol{r}_k \times \boldsymbol{\nabla}_k \frac{1}{|\boldsymbol{r}_i - \boldsymbol{r}_j|} \right) \tag{2.39}$$

which is zero because

$$\boldsymbol{\nabla}_i \frac{1}{|\boldsymbol{r}_i - \boldsymbol{r}_j|} = -\boldsymbol{\nabla}_j \frac{1}{|\boldsymbol{r}_i - \boldsymbol{r}_j|}. \tag{2.40}$$

Since spin does not enter into $\mathscr{H}_{\mathrm{o}} + \mathscr{H}'$ this part of the Hamiltonian also commutes with $\boldsymbol{S}$. Hence $\mathscr{H}_{\mathrm{o}} + \mathscr{H}'$, $\boldsymbol{L}^2$, $\boldsymbol{L}_z$, $\boldsymbol{S}^2$, $\boldsymbol{S}_z$, P_{ij} form a commuting set of operators so that the eigenstates of $\mathscr{H}_{\mathrm{o}} + \mathscr{H}'$ are classified by quantum numbers L, S, M_L, M_S, $\gamma = -1$.

We have already found a set of eigenstates of $\mathscr{H}_{\mathrm{o}}$, $|(n_i l_i) LSM_L M_S \gamma = -1\rangle$ which are linear products of Slater determinants. Are these eigenstates of $\mathscr{H}_{\mathrm{o}}$ also eigenstates of $\mathscr{H}_{\mathrm{o}} + \mathscr{H}'$? To answer this question we must evaluate matrix elements of $\mathscr{H}'$ using these product states as basis functions. Since $[\mathscr{H}', \boldsymbol{L}]=0$ and $[\mathscr{H}', \boldsymbol{S}]=0$ the matrix elements of $\mathscr{H}'$ between functions of different LSM_LM_S are zero. However, non-zero matrix elements of $\mathscr{H}$ can exist between functions of the same LSM_LM_S derived from different configurations:

$$\langle (n_il_i)'LSM_LM_S|\mathscr{H}'|(n_il_i)LSM_LM_S\rangle.$$

This is *configuration mixing.* Generally the matrix elements of $\mathscr{H}'$ between functions belonging to different configurations are much smaller than the energy separation between different configurations. Consequently the effect of configuration mixing is small. Thus to the extent that configuration mixing can be neglected the eigenstates of $\mathscr{H}_0$, viz. $|((n_il_i)LSM_LM_S\gamma)=-1\rangle$, are also eigenstates of $\mathscr{H}_0+\mathscr{H}'$. And since $\mathscr{H}_0+\mathscr{H}'$ commutes with $L_\pm$ and $S_\pm$ the energy of these eigenstates does not depend on M_L, M_S.

The calculation of the diagonal matrix elements of $\mathscr{H}'$, the Coulomb interaction between the outer electrons, is described very clearly in a number of texts (see, for example, Griffith 1961). These matrix elements are expressed either in terms of the *Slater parameters*, F_0, F_2, F_4, . . . or in terms of the *Racah parameters, A, B, C*, . . . The latter are defined as linear combinations of the Slater parameters. The Coulomb interaction between the outer electrons splits the energy level of the electron configuration (n_il_i) into a number of *LS terms*. For a given *LS* level there are $(2L+1)(2S+1)$ distinct energy eigenstates. Thus the energy eigenstates are not unique. In particular, linear combinations of M_L, M_S eigenstates can be chosen which are also eigenstates of $\boldsymbol{J}^2$ and J_z. These we represent by the kets $|(n_il_i)LSJM_J\gamma=-1\rangle$ and are defined in terms of the $|(n_il_i)LSM_LM_S\gamma=-1\rangle$ states by eqn (2.34).

2.3.5 *The case of two* p *electrons*

We illustrate the above ideas by considering two outer electrons occupying the $(nl, n'l')$ configuration, and in which $l=l'=1$ (p electrons). If $n=n'$ these electrons are said to be *equivalent*. We start with simple product functions and form linear combinations which are characterized by LSM_LM_S values, and then take care of the antisymmetry requirements. The simple product orbital function is written as $|nlm_l\rangle|n'lm_l'\rangle$; linear combinations of these product functions are chosen which are eigenstates of L^2, L_z, i.e.

$$|(n1n'1)LM_L\rangle=\sum_{m_l,m_l'}|n1m_l\rangle|n'1m_l'\rangle\langle 11m_lm_l'|11LM_z\rangle \tag{2.41}$$

where $\langle 11m_l,m_l'|11LM_L\rangle$ are Clebsch–Gordan coefficients. As we show in Chapter 3 the one-electron states with $l=l'=1$ can be combined to form states of total orbital angular momentum $L=2, 1, 0$. From the table of Clebsch–Gordan coefficients (Table 3A.1) we find

$$|(nn')22\rangle=|n11\rangle|n'11\rangle \tag{2.42}$$

States of $L=2$, $M_L<2$ can be generated by operating on $|22\rangle$ with L_-. Similarly we find from the table

$$|(nn')11\rangle=\frac{1}{\sqrt{2}}[|n11\rangle|n'10\rangle-|n10\rangle|n'11\rangle] \quad (2.43)$$

and

$$|(nn')00\rangle=\frac{1}{\sqrt{3}}[|n11\rangle|n'1-1\rangle - |n10\rangle|n'10\rangle+|n1-1\rangle|n'11\rangle]. \quad (2.44)$$

Consider first the case of $n'=n$. The above orbital states can be seen to be eigenstates of P_{12}^{orb}, the D state ($L=2$) and S state ($L=0$) being symmetric under P_{12}^{orb} and the P state ($L=1$) being antisymmetric.

From the four product spin states, $\alpha(1)\alpha(2)$, $\alpha(1)\beta(2)$, $\beta(1)\alpha(2)$, $\beta(1)\beta(2)$, where (1) and (2) now refer to the spin coordinates of electrons 1 and 2, linear combinations of these spin states can be chosen which are eigenstates of $\boldsymbol{S}^2$ and S_z. These are (Table 3A.1)

$$|SM_s\rangle=|11\rangle=\alpha(1)\alpha(2)$$

$$|10\rangle=\frac{1}{\sqrt{2}}[\alpha(1)\beta(2)+\beta(1)\alpha(2)]$$

$$|1-1\rangle=\beta(1)\beta(2)$$

$$|0,0\rangle=\frac{1}{\sqrt{2}}[\alpha(1)\beta(2)-\beta(1)\alpha(2)]. \quad (2.45)$$

The $S=1$ states are symmetric and the $S=0$ state is antisymmetric under the parity operator P_{12}^{spin}. The full spin-plus-orbital states $|(np)^2LSM_LM_S\rangle$ are the product states $|(np)^2LM_L\rangle|SM_S\rangle$.

Let us now consider the requirement that the electronic wavefunction be antisymmetric under $P_{12}=P_{12}^{\mathrm{orb}}.P_{12}^{\mathrm{spin}}$. This requires that the D and S orbital states (which are symmetric) may only be combined with the spin singlet ($S=0$) state, whereas the P orbital state (which is antisymmetric) is combined with the spin triplet ($S=1$) state. In spectroscopic notation the spin orbital states are labelled as ${}^{2S+1}(L)$, where S in the superscript represents the value of the total spin, and (L) means the symbol S, P, D, . . . for $L=0, 1, 2, \ldots$ Thus the allowed LS terms in the $(\mathrm{np})^2$ configuration are ^{1}D, ^{3}P, ^{1}S. In the limit of negligible configuration mixing, these states are eigenstates of $\mathscr{H}_0+\mathscr{H}'$. The diagonal matrix elements $\langle(\mathrm{np})^2LSM_LM_S|\mathscr{H}'|(\mathrm{np})^2LSM_LM_S\rangle$ evaluated in terms of Slater parameters (Tinkham 1964; Griffith 1961) are

$$E(^1\mathrm{D})=F_0+F_2$$
$$E(^3\mathrm{P})=F_0-5F_2 \quad (2.46)$$
$$E(^1\mathrm{S})=F_0+10F_2.$$

We note that the state having the maximum spin multiplicity is lowest, as predicted by Hund's rule.

We now examine the case where $n'\neq n$, and consider in particular the D orbital state (eqn 2.42): we write this as

$$|(n1n'1)22\rangle=|n11\rangle|n'11\rangle=f(1)g(2) \quad (2.47)$$

where (1), (2) now refer to the orbital coordinates of electrons 1 and 2. For convenience we have written the *distinct* orbital states as f and g. This product state is not an eigenstate of P^{orb}. However, there is another $|22\rangle$ orbital state derived from the two same orbitals, viz. $g(1)f(2)$, and symmetric and antisymmetric $|22\rangle$ states may be formed from the f and g orbital states in the following way:

$$|(fg)22\gamma_{\mathrm{orb}}=\pm1\rangle=\frac{1}{\sqrt{2}}[f(1)g(2)\pm g(1)f(2)]. \quad (2.48)$$

These symmetric and antisymmetric orbital states must be multiplied by antisymmetric and symmetric spin states forming ^{1}D and ^{3}D states, respectively. In this way two *inequivalent* p electrons can occupy either ^{3}D or ^{1}D states. Similarly ^{3}P and ^{1}P, and ^{3}S and ^{1}S states are also allowed.

Returning to the symmetric and antisymmetric $|22\rangle$ states (eqn 2.48) the diagonal matrix elements of $\mathscr{H}'$ for the two cases are:

$$\langle(\mathrm{npn'p})22\gamma_{\mathrm{orb}}=\pm1|\mathscr{H}'|(\mathrm{npn'p})22\gamma=\pm1\rangle$$
$$=\langle f(1)g(2)|\mathscr{H}'|f(1)g(2)\rangle\pm\langle f(1)g(2)|\mathscr{H}'|g(1)f(2)\rangle$$
$$=K\pm J \quad (2.49)$$

in which K and J are known as the *direct* and *exchange* integrals, respectively. The presence of the exchange integral is a consequence of the Pauli principle requiring that the electron wavefunction be antisymmetric under P_{12}. These integrals, which for atoms are both positive, can be expressed in terms of the Slater parameters (Tinkham 1964). The higher energy $\gamma_{\mathrm{orb}}=+1$ state must be multiplied by the antisymmetric spin singlet state ($S=0$) while the $\gamma_{\mathrm{orb}}=-1$ state must be multiplied by the symmetric spin triplet state ($S=1$). Thus the ^{3}D and ^{1}D states are separated in energy by $2J$, as shown in Fig. 2.2.

We note that we can write an effective perturbation Hamiltonian to produce the same separation of the $S=0$ and $S=1$ states. This is

$$\mathscr{H}'=K-\tfrac{1}{2}J-2J s_1 . s_2. \quad (2.50)$$

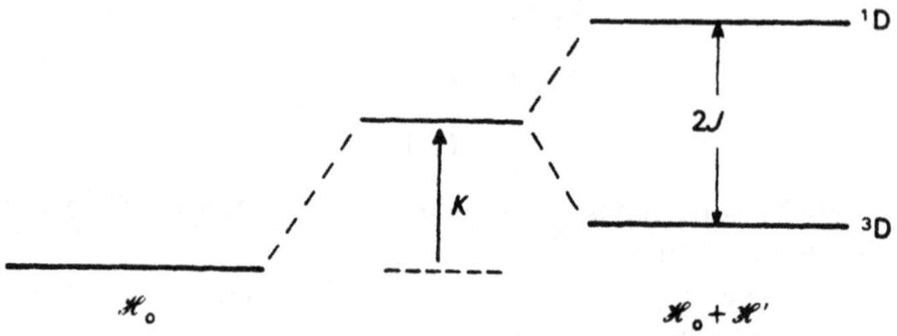

FIG. 2.2. The $L=2$ state (D state) formed from two inequivalent p-electron states is split into two by the Coulomb interaction between the electrons, $\mathscr{H}'$.

To see that this leads to the same energy levels we note that the diagonal matrix elements of $-2\mathbf{s}_1 \cdot \mathbf{s}_2$ are $-\frac{1}{2}$ and $+\frac{3}{2}$ for the spin triplet ($S=1$) and spin singlet ($S=0$) states, respectively. This form of $\mathscr{H}'$ allows us to interpret the separation in energy between spin triplet and spin singlet states as being caused by a *spin alignment operator* $-2J\,\mathbf{s}_1 \cdot \mathbf{s}_2$ which aligns the spins parallel ($S=1$) or antiparallel ($S=0$). For two electrons on the same atom J is positive and the spin triplet state is the lower. When two atoms are close enough for a Coulomb interaction to occur between the electrons on the two atoms this interaction can similarly be expressed as an apparent coupling between the spins of individual pairs of electrons of the form $-2J\,\mathbf{s}_1 \cdot \mathbf{s}_2$. In this case J may be positive or negative leading to a ferromagnetic or antiferromagnetic coupling between atoms. This is the basis for the Heisenberg theory of ferromagnetism in solids.

2.3.6 *Spin–orbit coupling for multi-electron atoms*

The operator for spin–orbit coupling is $\mathscr{H}_{so}$ (eqn 2.38). For atoms with low atomic number the effect of $\mathscr{H}'$ is greater than that of $\mathscr{H}_{so}$ and the approach is first to seek wavefunctions of $\mathscr{H}_0 + \mathscr{H}'$, as we have already done (these are the *LS* term states), and then to take $\mathscr{H}_{so}$ into account. For atoms with high atomic number the effect of spin–orbit coupling can be greater than the effect of the Coulomb interaction between the outer electrons. In that case one first considers the coupling of spin and orbital angular momentum vectors on individual electrons (to form j) and then the coupling of the j states through the Coulomb interaction between the electrons. This is the j–j coupling approach. We shall concentrate on the first approach, where $\mathscr{H}_{so}$ is smaller than $\mathscr{H}'$.

The spin–orbit coupling term is

$$\mathscr{H}_{so} = \sum_i \zeta(r_i)\mathbf{l}_i \cdot \mathbf{s}_i = \sum_i \zeta(r_i)\tfrac{1}{2}(j_i^2 - l_i^2 - s_i^2). \qquad (2.51)$$

Just as in the single-electron case this can mix states which differ in the principle quantum numbers of one-electron states; this mixing is expected to be small and is neglected. The operator $\mathcal{H}_{so}$ commutes with l_i^2 and s_i^2 but does not commute with L^2 and S^2. It commutes with j_i and thus with J. Hence there are no matrix elements of $\mathcal{H}_{so}$ between states of different JM_J but there can be matrix elements of $\mathcal{H}_{so}$ between states of the same JM_J values from different LS terms (term mixing). In the *Russell–Saunders approximation* one neglects mixing from different LS terms. To that approximation the $|(n_i l_i)LSJM_J\rangle$ states are eigenstates of $\mathcal{H}_{so}$, and so are eigenstates of $\mathcal{H}_0 + \mathcal{H}' + \mathcal{H}_{so}$. It can be shown (Tinkham 1964; Di Bartolo 1968) with the aid of the Wigner–Eckart theorem (Chapter 4) that within the same LS term the matrix elements of $\mathcal{H}_{so}$ are proportional to those of $\boldsymbol{L}.\boldsymbol{S} = \frac{1}{2}(\boldsymbol{J}^2 - \boldsymbol{L}^2 - \boldsymbol{S}^2)$. Hence the matrix elements can be written $[J(J+1) - L(L+1) - S(S+1)]\zeta(LS)/2$. Within a given LS term there are a number of $|LSJM_J\rangle$ states of different J value which, in the absence of spin–orbit coupling, have the same energy, and the separation in energy between adjacent J states is

$$E_J - E_{J-1} = \frac{\zeta}{2}[J(J+1) - J(J-1)] = \zeta J \tag{2.52}$$

indicating that the splitting is proportional to the larger J values of the two adjacent levels. This is the *Landé interval rule*. The spin–orbit coupling parameter ζ can be expressed as a series of radial integrals (Di Bartolo 1968) but it is generally regarded as a parameter whose value is obtained from the observed energy separations of the J levels. The splitting of the level of the $(\mathrm{np})^2$ electron configuration into *LS terms* under the action of $\mathcal{H}'$, and further into *J multiplets* under the action of $\mathcal{H}_{so}$ is shown schematically in Fig. 2.3. The departure of the observed splitting of an LS term from the Landé interval rule can be regarded as a measure of the failure of the Russell–Saunders coupling. We shall consider this when we discuss the energy levels of rare-earth ions in solids.

2.4 Optical centres in a static crystalline environment

There are many spectroscopically interesting electronic centres in inorganic solids, including single- and multi-vacancy centres in ionic crystals, recombination centres in semiconductors, molecular ions, and multi-electron impurities such as transition metal and rare-earth ions. It would be desirable to develop a common theoretical approach to the calculation of the energy levels of these centres, but because of differences in the nature of the centres and in the nature of the interaction of the centre with the surrounding atoms of the host solid such a common approach is not possible. In practice, significantly different methodologies are required for the different centres. Symmetry considerations, however, which are general to all centres, can greatly assist the

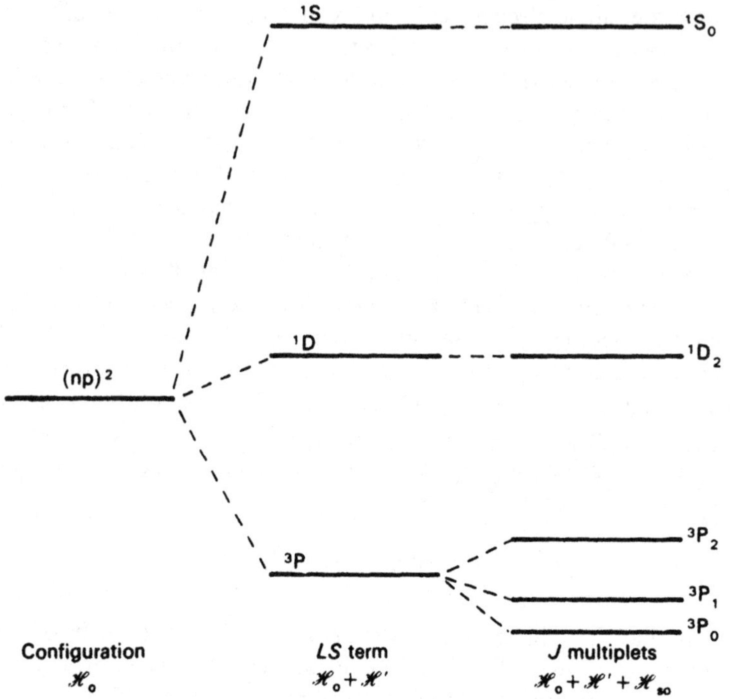

FIG. 2.3. The Coulomb interaction, $\mathscr{H}'$, splits the $(np)^2$ configuration into *LS terms*, while $\mathscr{H}_{so}$ splits the ^3P term into *J multiplets*.

analysis. Not least in the computational problem is the fact that the lattice in which the electronic centre resides is not static. Each atom or ion takes part in the cooperative vibrational motion of the solid, and the electronic levels are modulated by internal electric fields at frequencies corresponding to the lattice vibrations. It is convenient to separate the two components of this problem—the *static* crystal field due to the average lattice configuration and the *dynamic* crystal field due to the vibrating lattice—and treat them separately. The remainder of this chapter is concerned with the energies of the electronic centre in a static crystalline environment.

In the static crystal field approach each ion is assumed fixed at a mean lattice position. However, the way in which we set up the static crystal field is determined by the nature of the problem. For example, in semiconductors and vacancy centres in ionic crystals the lattice is sometimes treated as a dielectric continuum. The most notable quantitative success of this approach was due to Kohn and Luttinger (1955) in their theory of shallow impurity states in semiconductors, although the method has also found applications to F-aggregate† centres in alkali halides. Generally speaking this approach is not tenable for deeply bound states, such as F-centres† or dopant transition and rare-earth ions. For the particular case of dopant rare-earth ions the optically-active f electrons are only weakly affected by the crystal field. The effect of the crystal field can be considered as a perturbation on the $|f^n LSJM_J\rangle$ free-ion states, and it causes a splitting of the free-ion level. The symmetry of the crystal field potential is an important factor. The number and symmetry of the crystal field states are deduced in Chapter 3 by group-theoretical methods using only a knowledge of the symmetry of the crystal field potential.

In seeking a Hamiltonian operator to describe the electronic centre we take advantage of the fact that optical transitions involve only changes in the states of the outer electrons; the inner electrons occupying closed shells or subshells are normally unaffected by these transitions. We regard these inner electrons as creating the constant electrostatic *central field* (eqn 2.27) with which the optically active electrons interact. The Hamiltonian is written as

$$\begin{aligned}\mathcal{H} &= \mathcal{H}_{\mathrm{FI}} + \mathcal{H}_{\mathrm{c}} \\ &= \mathcal{H}_{\mathrm{o}} + \mathcal{H}' + \mathcal{H}_{\mathrm{so}} + \mathcal{H}_{\mathrm{c}} \end{aligned} \tag{2.53}$$

where $\mathcal{H}_{\mathrm{o}}$, $\mathcal{H}'$, and $\mathcal{H}_{\mathrm{so}}$ are defined in eqns (2.36–2.38) and $\mathcal{H}_{\mathrm{c}}$ represents the energy of interaction of the outer electrons with the electrostatic crystal field. We can write the crystal field term as

$$\mathcal{H}_{\mathrm{c}} = \sum_i \sum_l \mathcal{H}_{\mathrm{c}}(\boldsymbol{r}_i, \boldsymbol{R}_l). \tag{2.54}$$

The summations are over the electrons (i) and the neighbouring ions (l), and the latter occupy positions $\boldsymbol{R}_l$. In the first instance we are interested in the static average crystal field, so the time-average position of the lth ion is used for $\boldsymbol{R}_l$ in eqn (2.54). Methods of determining the eigenstates and eigenvalues of the Hamiltonian (eqn 2.53) depend upon the relative sizes of the various terms. We identify three different regimes.

1. *Weak crystal field*: $\mathcal{H}_{\mathrm{c}} \ll \mathcal{H}', \mathcal{H}_{\mathrm{so}}$. $\mathcal{H}_{\mathrm{c}}$ is neglected initially. The remaining terms constitute the free-ion Hamiltonian. The free-ion states are first calculated as described in Section 2.3 and then $\mathcal{H}_{\mathrm{c}}$ taken into account by perturbation theory. This weak field approach is appropriate to the trivalent rare-earth ions since for these ions the optically active 4f electrons are partially screened from the lattice ions by their outer filled $5s^2 5p^6$ subshells. Hence the weakness of the crystal field.

2. *Intermediate crystal field*: $\mathcal{H}' > \mathcal{H}_{\mathrm{c}} > \mathcal{H}_{\mathrm{so}}$. In this case $\mathcal{H}_{\mathrm{so}}$ is initially neglected. One starts with the free-ion LS term functions. Since $\mathcal{H}_{\mathrm{c}}$ is an orbital operator the free-ion L functions are used as basis functions to

† Colour centres, such as F-centres and F-aggregate centres are discussed in Chapter 7.

calculate matrix elements of $\mathscr{H}_c$ from which new crystal field orbital states are formed. These are multiplied by spin S functions, the requirement of the Pauli principle is taken into account, and the effect of $\mathscr{H}_{so}$ is then calculated.

3. *Strong crystal field*: $\mathscr{H}_c > \mathscr{H}' > \mathscr{H}_{so}$. $\mathscr{H}'$ and $\mathscr{H}_{so}$ are initially neglected. The remaining Hamiltonian is a sum of one-electron orbital terms, and the eigenstates are products of one-electron crystal field orbitals. The interaction between the electrons is next taken into account. The resulting orbital state is characterized by an orbital quantum number, Λ, which is the crystal field analogue of the free-atom L value. When the spin states are taken into account the states are characterized by Λ, S parameters. Spin–orbit coupling is then introduced as a perturbation.

2.4.1 *The crystal field*

The simplest description of the crystal field uses the point ion model, in which ligand ions causing the electrostatic crystal field are represented by point charges. This model neglects both the finite spatial extent of the ligand charge density and the wavefunction overlap of the optically active electrons with the ligands. The lth ion is represented as a charge $q_l = -Z_l e$ at the lattice point $\boldsymbol{R}_l$, which has spherical polar coordinates (a_l, θ_l, ϕ_l) and Cartesian coordinates (x_l, y_l, z_l). The electrostatic potential $\phi(r, \theta, \phi)$ due to the surrounding point charges is

$$\phi(\boldsymbol{r}) = \frac{1}{4\pi\varepsilon_0} \sum_l \frac{q_l}{|\boldsymbol{R}_l - \boldsymbol{r}|} \tag{2.55}$$

and the crystal field Hamiltonian is

$$\mathscr{H}_c = \sum_i (-e)\phi(\boldsymbol{r}_i) = \frac{1}{4\pi\varepsilon_0} \sum_i \sum_l \frac{Z_l e^2}{|\boldsymbol{R}_l - \boldsymbol{r}_i|}. \tag{2.56}$$

If we assume $r_i < a_l$ we can expand $|\boldsymbol{R}_l - \boldsymbol{r}_i|^{-1}$ in terms of spherical harmonics, and the interaction of the ith electron with the electrostatic crystal field can be written (Sugano *et al.* 1970)

$$\mathscr{H}_c(\boldsymbol{r}_i) = \frac{1}{4\pi\varepsilon_0} \sum_l \sum_{k=0}^{\infty} \sum_{t=-k}^{+k} \left(\frac{4\pi}{2k+1}\right)^{\frac{1}{2}} \frac{Z_l e^2}{a_l^{k+1}} r_i^k C_t^{(k)}(\theta_i, \phi_i)\, Y_k^t(\theta_l, \phi_l) \tag{2.57}$$

where

$$C_t^{(k)}(\theta, \phi) = \left(\frac{4\pi}{2k+1}\right)^{\frac{1}{2}} Y_k^t(\theta, \phi)$$

and $Y_k^t(\theta, \phi)$ is the spherical harmonic.

We need to evaluate matrix elements of the type

$$\langle n'l'm_l' | r^k C_t^{(k)} | nlm_l \rangle = \langle R'_{n'l'} | r^k | R'_{nl} \rangle \langle Y_{l'}^{m_l'} | C_t^{(k)} | Y_l^{m_l} \rangle. \tag{2.58}$$

The radial integral can cause configuration mixing, but, because of the large energy separation between configurations the amount of configuration mixing is small and generally can be ignored. We therefore assume that $n'l'=nl$. The radial integral can then be written $\langle r^k \rangle_{nl}$. The integral over angles, labelled $c^k(l'm_l', lm_l)$, was originally evaluated by Gaunt (1929). Tables of values are found in a number of texts, e.g. Condon and Shortley (1935), Griffith (1961). These integrals are non-zero only under the following conditions:

$$t = m_l' - m_l,$$

$$k + l + l' = \text{even integer},$$

$$|l - l'| \leqslant k \leqslant l + l'. \tag{2.59}$$

The values of $c^k(lm, l'm') = (-1)^{m-m'} c^k(l'm', lm)$ for $l = l' = 2$ given in Table 2.2 are used when we discuss crystal field effects on d electrons. Note that for d electrons ($l = 2$) the c^k values are non-zero only for $k = 0, 2, 4$.

Table 2.2

Values of $c^k(lm, l'm') = (-1)^{m-m'} c^k(l'm', lm)$ for $l = l' = 2$

m	m'	$k=0$	$k=2$ (× 1/7)	$k=4$ (× 1/21)
± 2	± 2	1	-2	1
± 2	∓ 2	0	0	$\sqrt{70}$
± 2	± 1	0	$\sqrt{6}$	$-\sqrt{5}$
± 2	∓ 1	0	0	$-\sqrt{35}$
± 2	0	0	-2	$\sqrt{15}$
± 1	± 1	1	1	-4
± 1	∓ 1	0	$-\sqrt{6}$	$-\sqrt{40}$
± 1	0	0	1	$\sqrt{30}$
0	0	1	2	6

The crystal field experienced by a centre will reflect the symmetry of the environment of the centre, and we attempt to classify each crystal field in accordance with this symmetry. Figure 2.4 shows three different arrangements of ions which give rise to electrostatic crystal fields of specific symmetry. The crystal field due to the six ions in (a) is said to have *octahedral symmetry* (indicated by the label O_h), and the electronic centre at the origin is said to occupy a site of octahedral symmetry. The six point charges of amount $-Ze$ are each a distance a from the origin along $\pm x$, $\pm y$, $\pm z$ orthogonal axes. As an example, the site of the Mg^{2+} ion in magnesium oxide is surrounded by six O^{2-} ions in an arrangement of octahedral symmetry (Fig. 2.4(a)). In this material the crystal field due to the more distant ions is of the same functional form as that of the six nearest O^{2-} ions. In (b) the crystal field is due to four

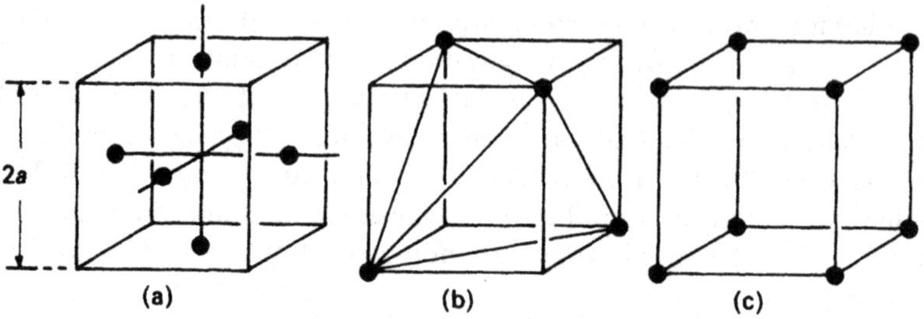

FIG. 2.4. The arrangement of ions giving rise to electrostatic crystal fields of (a) octahedral, (b) tetrahedral, and (c) cubic symmetries. The x, y, z-axes shown here are called cubic axes.

point charges of amount $-Ze$ at the alternative vertices of a regular cube, i.e. at the vertices of a regular tetrahedron. This crystal field is said to have *tetrahedral symmetry* (label T_d). In (c) the crystal field is due to eight ions with equal charge $-Ze$ at the vertices of a regular cube. This field has *cubic symmetry*. For purposes of comparison the octahedra used to describe the three arrangements of ions in Fig. 2.4 are given the same linear dimensions, $2a$.

Arrangements of six neighbouring ions are commonly found in crystals. However, there is usually a distortion from the perfect octahedron, and the crystal field is of lower symmetry. As we shall see, the crystal field may then be regarded as having two components, a strong component of octahedral symmetry and a weaker one of lower symmetry. The usual method of analysis is first to calculate the wavefunctions and energy levels in the field of octahedral symmetry and then consider the effect of the weak lower symmetry crystal field using perturbation theory. Depending upon the symmetry of the crystal field the energy levels of the electronic centre may be calculated in terms of one or more crystal field parameters. In the case of F-centres in alkali halides we can use very simple models which ignore the symmetry properties of the crystal field. Such simple calculations are discussed next, prior to a discussion of the energy levels of a single d electron in crystal fields of various symmetries.

2.5 Energy levels of F-centres and related defects

This section is concerned with electrons trapped on anion vacancy centres in ionic crystals; the simplest such centre is the F-centre, an electron trapped in a single anion vacancy.† The Hamiltonian for such a centre is then obtained from eqn (2.53) by omitting the central field and electron–electron repulsion terms, i.e.

$$\mathscr{H} = \boldsymbol{p}^2/2m + \zeta(r)\boldsymbol{l}.\boldsymbol{s} + \mathscr{H}_c(\boldsymbol{r}, \boldsymbol{R}_l). \tag{2.53a}$$

† The F-centre and its properties are discussed in detail in Chapter 7.

Most early theoretical models of the F-centre emphasized the nature of the crystal field, although almost any field emphasizing the square-well-like potential yields reasonable results for F-centre absorption energies. Less successful is the continuum theory of F-centres, in which the F-centre is treated as a hydrogen atom embedded in a dielectric continuum. Although the use of an effective mass equation for the ground state of F-centres is unjustified it has the merit of mathematical simplicity and is used when there is some empirical means of fitting the effective mass, m^*, and the dielectric constant κ. Two examples where the end results justify its use are the F_2- and F_3-centres, which consist of two and three F-centres in nearest-neighbour anion sites, respectively. In the sense that the F-centre is analogous to the hydrogen atom, then the F_2 and F_3-centres are analogous to the H_2 and H_3 molecules, respectively. Such analogies serve to define the number of electrons, the symmetry properties, and classification of electronic states of these defects. We do not intend to treat the very detailed theories of defect energy levels in all their computational glory. Such aspects of the models are to be found in Fowler (1968), Markham (1966) and Stoneham (1985).

2.5.1 F-centre models based on square-well potentials

The F-centre in alkali halide crystals consists of an electron trapped in a halide ion vacancy (Fig. 2.5(a)). In the absence of the electron the halogen ion vacancy is positively charged relative to the rest of the crystal so that the F-centre is charge-neutral. Consequently in the absence of the central field associated with the halogen atom, the trapped electron experiences only the residual Madelung interaction due to all the other positive and negative ions in the crystal. Typically the Madelung energies of alkali halide crystals are in the range 5–10 eV per ion. Although the empty volume available to the electron is not precisely defined in the crystal, as a first approximation it is assumed to be cubic with dimension $2L = 2(R_+ + R_-)$, where R refers to the univalent ionic radii of positive (+) and negative (−) ions. Since the Madelung energy of a crystal is large relative to the kinetic energy of an electron in thermal equilibrium with its surrounding, we assume it to have zero potential energy inside the vacancy and an infinite potential energy everywhere outside the cubic volume that represents the vacancy, see Fig. 2.5(b). The calculation of the F-centre energy levels reduces to finding the eigenvalues of the time-independent Schrödinger equation,

$$\left(-\frac{\hbar^2}{2m}\nabla^2 - E_{lmn}\right)\psi_{lmn} = 0 \tag{2.60}$$

under the boundary condition that the wavefunction χ_{lmn} is zero at the boundary of the potential well. Suitably normalized eigenfunctions of eqn (2.60) are

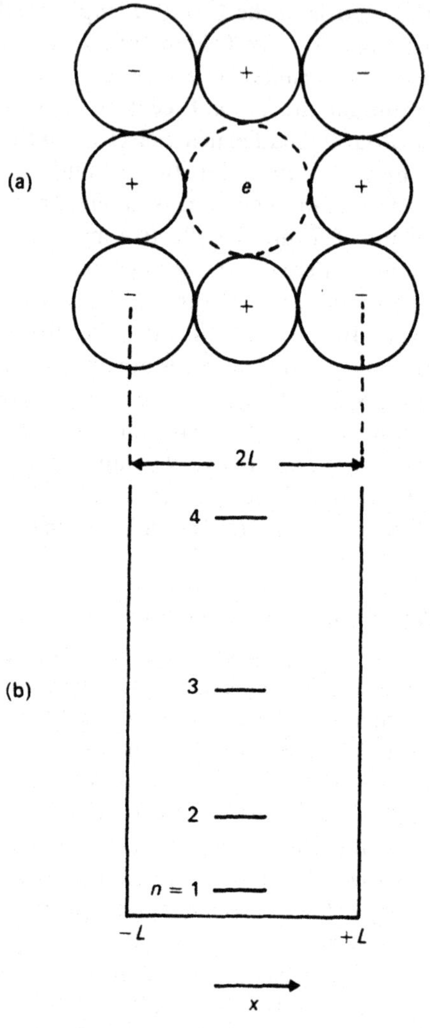

FIG. 2.5. (a) A two-dimensional sketch of the de Boer model of the F-centre on the (100) plane of an alkali halide crystal. (b) shows the one-dimensional particle-in-a-box model used to illustrate the trends of optical properties of F-centres. The potential is defined by $V(x)=0$ in the range $-L<0<+L$ and $V(x)=\infty$ for $x<-L$ and $x>+L$.

$$\chi_{lmn}=\left(\frac{1}{L^3}\right)^{\frac{1}{2}}\sin\left(\frac{l\pi x}{L}\right)\sin\left(\frac{m\pi y}{L}\right)\sin\left(\frac{n\pi z}{L}\right) \tag{2.61}$$

with corresponding energy eigenvalues,

$$E_{lmn}=\frac{\hbar^2}{8m(2L)^2}(l^2+m^2+n^2) \tag{2.62}$$

in which the quantum numbers l, m, and n may have integral values 1, 2, 3, . . . , ∞. The selection rule for allowed electric dipole transitions takes the form $\Delta l = \pm 1$, $\Delta m = \pm 1$ or $\Delta n = \pm 1$. In consequence, the first allowed electric dipole transition is from the ψ_{111} ground state to the $\psi_{211}, \psi_{121}, \psi_{112}$ excited states. Since the separation in energy of ground and first excited states is $3\hbar^2/8m(2L)^2$, this model predicts that the F-*band* occurs at a photon energy

$$E_F = \frac{1.13}{(2L)^2} \text{ eV} \tag{2.63}$$

where L is measured in nm. In this case the energy E_F corresponds to the photon energy at the peak of a rather broad band. This crude model has the merit of predicting that the F-band absorption energy scales as the inverse square of the lattice spacing, $2L$, i.e. the unit cube length, of the alkali halide crystal. It also predicts that transitions may be excited to other states occurring at higher photon energies than the F-band. Experimentally F-bands have been observed in all alkali halides; in each crystal the band is broad and structureless. Figure 2.6 shows a test of eqn (2.63) for F-centres in the alkali halides, in which the F-band peak energy is plotted logarithmically as a function of lattice spacing. The results have a best fit straight line $E_F = 0.97(2L)^{-1.772}$. The agreement between theory and experiment is per-

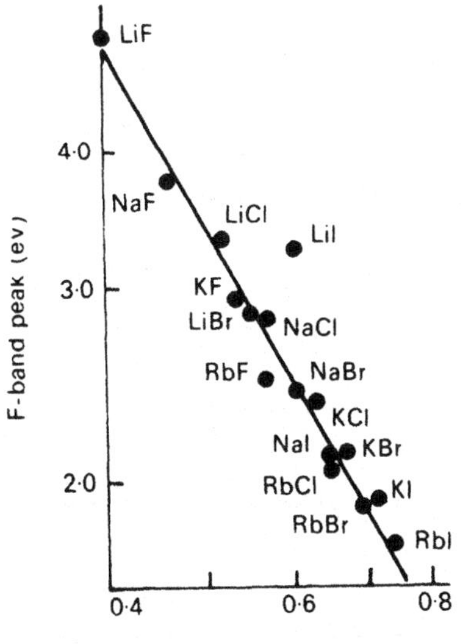

FIG. 2.6. The Mollwo-Ivey relation, eqn (2.63) and the experimental F-band energies in the alkali halides. The photon energy and lattice spacings are plotted logarithmically. (Adapted from Dawson and Pooley 1969.)

haps fortuitously good, even though the L^{-2} variation is not quite matched experimentally.

A somewhat more realistic model for the F-centre is an electron in a *finite* potential well, the potential energy of the electron in the vacancy being equal to the Madelung energy. The reduction in potential energy from infinite to finite alters the boundary conditions so that the wavefunction is finite at the vacancy boundary and falls to zero as $r \to \pm\infty$. In a one-dimensional model the potential well is defined by $V(x) = -V_0$ for $-L < x < +L$, and $V(x) = 0$ everywhere else. We are interested in bound solutions with energy between $-V_0$ and zero. If the energy is written as $-E$ then both E and V_0 are positive quantities so that E takes values in the range $0 < E < V_0$. This problem is considered in most basic quantum mechanics (e.g. Merzbacher 1970), where it is shown that the solutions of the Schrödinger equation are of either even or odd parity with respect to x. Inside the one-dimensional potential well the solutions are $\cos\beta x$ (even parity) and $\sin\beta x$ (odd parity) where $\beta = (2m(V_0 - E)/\hbar^2)^{\frac{1}{2}}$. Outside the potential well the solutions fall off exponentially as $\exp(-\alpha x)$ for $x > L$ and $\exp(\alpha x)$ for $x < -L$, where $\alpha = (2mE/\hbar^2)^{\frac{1}{2}}$. The solutions inside and outside the well must be properly matched at the well boundaries and this is only possible for discrete values of E given by the transcendental equations

$$\text{even solutions: } \tan\left(\frac{2mL^2}{\hbar^2}(V_0 - E)\right)^{\frac{1}{2}} = \left(\frac{E}{V_0 - E}\right)^{\frac{1}{2}} \tag{2.64}$$

$$\text{odd solutions: } -\tan\left(\frac{2mL^2}{\hbar^2}(V_0 - E)\right)^{\frac{1}{2}} = \left(\frac{V_0 - E}{E}\right)^{\frac{1}{2}}. \tag{2.65}$$

Graphical solutions of these separate equations, possible only for certain values of E, are easily obtained using trigonometric identities to find from eqn (2.64) the even solution

$$\cos\theta = \pm\left(\frac{V_0 - E}{V_0}\right)^{\frac{1}{2}} = \pm(1 - \varepsilon)^{\frac{1}{2}} \tag{2.66a}$$

and from eqn (2.65) the odd solution

$$\sin\theta = \pm\left(\frac{V_0 - E}{V_0}\right)^{\frac{1}{2}} = \pm(1 - \varepsilon)^{\frac{1}{2}} \tag{2.66b}$$

in which $\theta = [(2mL^2 V_0/\hbar^2)(1 - \varepsilon)]^{\frac{1}{2}}$, $E = \varepsilon V_0$ and $(1 - \varepsilon)$ measures the fractional distance of an energy level above the bottom of the potential well. Regarding θ as the variable the solutions are the points of intersection of the cosine or sine function with straight lines of positive or negative slope. In fact both even and odd solutions may be obtained from a single plot of the first quadrant of a cosine function with the line of positive slope, $+\theta/K$ where $K = 2mL^2V_0/\hbar^2$. This is illustrated in Fig. 2.7, the data used being appropriate to F-centres in lithium fluoride, for which $L = 0.201$ nm and $V_0 = 10.53$ eV. Since the points of intersection occur at specific values of θ, i.e. $\theta_k/K = (1-\varepsilon_k)^{\frac{1}{2}}$ we obtain the eigenvalues

$$E_k = V_0\left(1 - \frac{\theta_k}{K^2}\right)^2, \qquad (2.67)$$

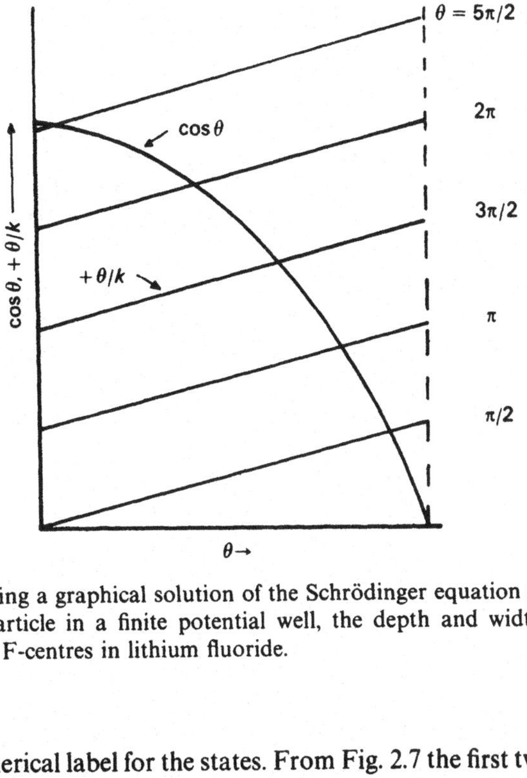

FIG. 2.7. Showing a graphical solution of the Schrödinger equation for the F-centre treated as a particle in a finite potential well, the depth and width of which are appropriate to F-centres in lithium fluoride.

k being a numerical label for the states. From Fig. 2.7 the first two eigenvalues are $E_1 = -9.77$ eV and $E_2 = -5.22$ eV. Hence the absorption peak of the F-band occurs at a photon energy $E_F = 4.55$ eV. At 4.2 K the experimental value is observed to be $E_F = 4.95$ eV so that the error is about 10 per cent. For other alkali halides, errors of order 10–20 per cent are also found; these are reasonable, bearing in mind the crudity of the model.

Note that we have given a convenient recipe for evaluation of the eigenvalues. To be certain of the parity of the eigenvalues requires a full examination of the graphical representation of eqns (2.66a) and (2.66b). For $K < \frac{1}{2}\pi$ we note that $\sin\theta$ and $\pm\theta/K$ never cross in the range 0 to $\frac{1}{2}\pi$. Hence there is no odd solution in this range. However, since $\cos\theta$ decreases monotonically with θ as $+\theta/K$ increases then an intersection will occur at some value, θ_k, which specifies the energy eigenvalue. In other words the only solution in the range $\theta = 0$ to $\frac{1}{2}\pi$ is an even solution ($\cos\theta_1$). Further examination shows that only an odd solution occurs in the angular range $\frac{1}{2}\pi \rightarrow \pi$, a further even solution between $\pi \rightarrow \frac{3}{2}\pi$, and so on. As k increases the values of $(1-\varepsilon_k)^{\frac{1}{2}}$ get ever closer to unity, so that the eigenvalues E_k get closer to the top of the potential well. This analysis implies that as $(2mL^2V_0/\hbar^2)^{\frac{1}{2}} = N\pi/2$, there is a root which alternates odd and even in the range $(1-\varepsilon_k)^{\frac{1}{2}} = 0 \rightarrow 1/N$, $1/N \rightarrow 2/N$, $2/N \rightarrow 3/N$, etc. Taking the upper bound of each range then we have a sequence

$$1-\varepsilon_k = \frac{1}{N^2}, \frac{2^2}{N^2}, \frac{3^2}{N^2} \cdots \frac{n^2}{N^2}$$

which terminates at $n = N$. This sequence of $(1-\varepsilon_k)$ values is of course just the spectrum of energy levels for a particle constrained to a box with infinitely high walls. Also, if we start with a large value of $(2mL^2(V_0)/\hbar^2)^{\frac{1}{2}}$ and decrease L then the states are essentially squeezed out of the well whereas as V is decreased the states are pushed out of the well.

The treatment of an F-centre as a particle in a finite potential well has the twin virtues of obvious simplicity and easy general applicability since the cosine function is identical for all potential wells and particles of different mass. Indeed the well depth and particle mass enter only through the slope (K^{-1}) of the straight-line graph. The wavefunctions for both square-well potentials, infinite and finite, are obviously very similar in shape. However, there are differences in detail. In the model based on the infinite square-well potential the wavefunctions fall to zero at the well boundary. However, for the F-centre modelled on the finite potential well the exponential tails of the wavefunctions outside the potential well must join smoothly to the sine functions at the boundaries. This result represents the fact that in quantum mechanics the particle (electron) may extend beyond the well boundaries. Experimental manifestations of the non-zero amplitude of ψ_k at the vacancy boundary include the hyperfine structure observed in ESR/ENDOR measurements (Henderson and Garrison 1973)† and the photocurrent observed when F-centres are excited in the F-band (Chapter 7).

There are other properties of F-centres which require theoretical examination, these include the width of the F-centre absorption and emission bands, the shift in peak position of the emission band relative to the absorption band, and the temperature dependence of the band widths and peak positions. All have a common origin in the coupling of the electronic

† Electron spin resonances (ESR) and Electron nuclear double resonance (ENDOR) are discussed in Chapters 6 and 7.

states of the centre to the vibrational motion of the crystal. Although this subject is treated in detail in Chapter 5, it is useful to consider here a simplified version of this description which accounts qualitatively for the cruder aspects of the experimental observations. We consider one particular mode of vibration of the nearest-neighbour ions of the F-centre. This so-called breathing mode involves the radial, in-phase vibrations of the six octahedrally disposed cations in the first shell of ions surrounding the F-centre. This mode may be viewed as the vibrations of the boundaries of the infinite potential well, and in view of the elasticity of simple solids we may assume that the forces which displace the ions are harmonic in the displacements. These vibrations change the total energy of the F-centre ground state by $\frac{1}{2}kx^2$, k being the spring constant and x the displacement of ions. The ground state energy is therefore

$$E_g = \frac{3\hbar^2}{8m(2L)^2} + \tfrac{1}{2}kx^2. \tag{2.68}$$

In an excited electronic state the different charge distribution causes the ions to relax away from the positions appropriate to the ground state. This relaxation is resisted by the elasticity of the crystal. The total energy of the F-centre in the first excited state is written as

$$E_e = \frac{6\hbar^2}{8m}\left[\frac{1}{2(L+x)}\right]^2 + \tfrac{1}{2}kx^2$$

Since in general $L \gg x$ we may expand $(L+x)^{-2}$ in terms of the strain, x/L, to give

$$\begin{aligned} E_e &= \frac{3\hbar^2}{16mL^2}\left[1 - \frac{2x}{L}\right] + \tfrac{1}{2}kx^2 \\ &= E_e^0 - \Delta x + \tfrac{1}{2}kx^2 \\ &= E_e^0 + \tfrac{1}{2}k\left(x - \frac{\Delta}{k}\right)^2 - \frac{\Delta^2}{2k} \end{aligned} \tag{2.69}$$

where $\Delta = 3\hbar^2/8mL^3$ and $E_e^0 = 6\hbar^2/8m(2L)^2$ is the electronic energy of the excited state in the ionic configuration appropriate to the ground state. The second term implies that the ions undergo simple harmonic oscillations about a new equilibrium position $x = \Delta/k$. Finally the electronic energy is reduced by an amount $\Delta^2/2k$. The parabolic potential wells represented by eqns (2.68) and (2.69) are shown in Fig. 2.8. The absorption transition occurs with greatest probability at $x = 0$; it is represented by the vertical arrow, AB. (We shall see in Chapter 5 that this is a Franck–Condon transition.) Once in the excited state the lattice relaxes about the defect, losing vibrational energy until point C is reached on the upper curve. De-excitation of the system is accompanied by photon emission, the vertical transition CD; further vibrational relaxation takes the lattice back to $x = 0$. Obviously if the parabolae have precisely the

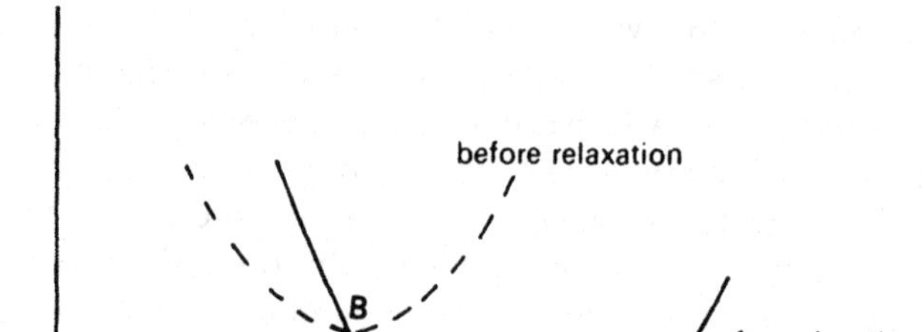

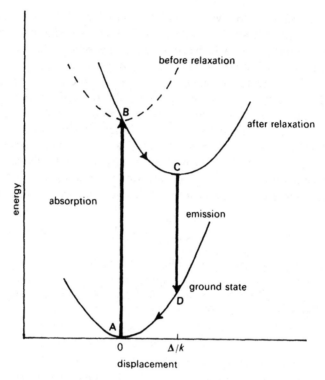

FIG. 2.8. Showing a simple model for the optical properties of defect centres in solids in which linear electron–lattice relaxation is limited by the harmonic restoring force.

same shape then the shift between the peaks in absorption and emission is given by

$$E_{abs} = E_{em} + 2\left(\frac{\Delta^2}{2k}\right). \tag{2.70}$$

This model gives sensible results for the configurational coordinate offset, Δ/k, and for the excited state energy depression $\Delta^2/2k$. In potassium chloride $E_{abs} = 2.296$ eV and $E_{em} = 1.215$ eV, hence $\Delta^2/2k \simeq 0.54$ eV. Estimating $k = 650$ eV/nm^2 from the elastic constants of the pure crystal gives $\Delta/k = 0.04$ nm. The width of the absorption band arises because the photon interacting with the centre 'catches' the lattice at some instantaneous value of the displacement, x. In consequence the energy required to excite a transition varies because of the differences in slope of the excited-state and ground-state parabolae about $x = 0$. The temperature dependence of the bandwidth follows from the greater amplitude of vibration as T increases. The infinite potential-well model also predicts that the peak energy of the emission band will follow a Mollwo–Ivey law, i.e. $E_{em} \propto (L+\Delta/k)^{-n}$, where n should be about 2. Unfortunately Δ/k cannot be directly measured. In consequence, tests of the model using the emission peak energies in terms of log–log plots of E_{em} versus L are not very satisfactory.

We have discussed this rather simple model not because of its numerical accuracy but because it gives physical insight into a wide range of optical phenomena. The model can be applied as a guide to the anticipated behaviour in what had previously been uncharted terrain. For example, one of the earliest evidences that sharp zero-phonon transitions were associated with F-centre aggregates (e.g. F_2, F_2^-, F_3 and F_3^-) in alkali halide crystal was that these lines obeyed a Mollwo–Ivey relationship in which the exponent n has a particular value for each type of centre in many different alkali halide crystals (see also Chapter 7).

2.5.2 Continuum models for defects

One of the more successful applications of the continuum theory of defects was to shallow donor states in silicon. Such pentavalent impurities as phosphorus or arsenic are easily ionized, only four electrons being necessary to satisfy the requirements of sp^3 hybrid bonding. However, the electron is still attracted to the positively charged impurity through the potential energy term $-e^2/4\pi\varepsilon_0\kappa r$, where κ is the dielectric constant of silicon. This is analogous to the hydrogen atom; the electron has bound states which have a large orbital 'radius' because of the large value of the dielectric constant. As the extra electron ranges over the semiconductor it interacts with other charges in the solid in addition to the central charge of the impurity. To take account of the real band structure of the solid requires that the free electron mass, m, be replaced by an effective mass, m^*, of an electron in the conduction band. The Hamiltonian describing this modified hydrogen atom follows from eqn (2.53a), i.e.

$$\mathcal{H} = \frac{\boldsymbol{p}^2}{2m^*} - \frac{e^2}{4\pi\varepsilon_0\kappa r} + \zeta(r)\boldsymbol{l}.\boldsymbol{s}. \tag{2.53b}$$

In consequence this electron may be represented by the product of a defect-centred 'hydrogen-like' function $\phi(\boldsymbol{r})$ modulated by a periodic Bloch function $u(\boldsymbol{r})$ for the lowest state in the conduction band, i.e. $\psi(\boldsymbol{r}) = \phi(\boldsymbol{r})u(\boldsymbol{r})$. The Bloch function, $u(\boldsymbol{r})$, takes account of the periodic charge density oscillations seen by the orbiting electron as it samples a volume of crystal containing (perhaps) several hundred atoms. In consequence, the total wavefunction $\psi(\boldsymbol{r})$ must also contain oscillations with the same lattice periodicity. If spin–orbit coupling is neglected $\phi(\boldsymbol{r})$ is a solution of the Schrödinger equation

$$\left(-\frac{\hbar^2}{2m^*}\nabla^2 - \frac{e^2}{4\pi\varepsilon_0\kappa r}\right)\phi(\boldsymbol{r}) = E\phi(\boldsymbol{r}) \tag{2.71}$$

where E is the eigenvalue measured relative to the bottom of the conduction band. The effective mass m^* is not in general isotropic: this has the effect of removing the degeneracy in the angular momentum quantum number m_l. The results of such a calculation are given in Table 2.3 for comparison with data obtained from optical spectra of silicon doped with phosphorus, arsenic, or antimony. Evidently the theory works best for p-states, which have no appreciable probability density inside the central impurity atom. However, it is not a good representation for 1s-states, in which the charge density is greatest at the impurity centre, and where the perturbing potential is certainly not described by $-e^2/4\pi\varepsilon_0\kappa r$. This is one of the problems encountered in dealing with deeply-bound states since the electron sees a potential different from the screened Coulomb term for a greater fraction of its time. Furthermore, one should not use the conduction-band effective mass for states which lie as close to the valence band as to the conduction band.

Table 2.3

A comparison of theoretical and experimental energy level splittings of shallow impurity states in silicon.

	Energy level difference (meV)			
		Optical spectroscopy		
Bound state splitting	Effective mass theory	P	As	Sb
$\|1s0\rangle - \|2p0\rangle$	18.1	34.5	42.1	31.8
$\|2p0\rangle - \|2p\pm1\rangle$	5.0	5.0	5.3	4.7
$\|2p\pm1\rangle - \|3p\pm1\rangle$	3.0	3.1	3.2	3.4

At first sight, effective mass/continuum models do not appear to hold out much promise of application to vacancy centres in ionic crystals. The ground states of such centres are located deep in the bandgap and are roughly s-like in character. The anion vacancy, which has a net charge of $+e$, will give a Coulombic field over most of the region in which the charge density is concentrated only if the electron spends most of its time outside the vacancy, since inside the vacancy the perturbing potential is neither small nor accurately Coulombic. This effect can be illustrated by calculating the electron charge density inside the anion vacancy, assuming hydrogen-like wavefunctions to represent the trapped electron. From the Schrödinger equation (2.71) the hydrogenic energy levels are given by

$$E_n = -\frac{1}{2}\frac{m^*e^4}{(4\pi\varepsilon_0\hbar)^2}\times\frac{1}{n^2\kappa^2} = -\left(\frac{e^2}{4\pi\varepsilon_0}\right)\left(\frac{m^*}{m}\right)\left(\frac{1}{2n^2\kappa^2 a_{\mathrm{H}}}\right) \qquad (2.72)$$

and the ground-state wavefunction is

$$\psi_{1s} = \pi^{-\frac{1}{2}} \left(\frac{m^*}{m\kappa a_{\mathrm{H}}} \right)^{\frac{3}{2}} \exp\left[-\left(\frac{m^* r}{m\kappa a_{\mathrm{H}}} \right) \right] \times u(\mathbf{r}) \tag{2.73}$$

where $a_{\mathrm{H}} = 0.529 \times 10^{-10}$ m is the radius of the 1s orbital of the free hydrogen atom. It is obvious that in these equations distances are scaled by κ and the energies by $1/\kappa^2$. The fractional electronic charge density outside a sphere of radius L is given by

$$\rho(L) = \int_L^\infty \psi_{1s}^2 4\pi r^2 \mathrm{d}r$$

$$= 4\left(\frac{m^*}{m\kappa a_{\mathrm{H}}} \right)^3 \int_L^\infty \exp - \left(\frac{2m^* r}{m\kappa a_{\mathrm{H}}} \right) \times r^2 \mathrm{d}r \tag{2.74}$$

assuming that $u(\mathbf{r})^2$ can be removed from the integral and replaced by its mean value of unity. Evaluating the integral gives

$$\rho(L) = \left[1 + \frac{m^* L}{m\kappa a_{\mathrm{H}}} + \frac{1}{2} \left(\frac{m^* L}{m\kappa a_{\mathrm{H}}} \right)^2 \exp - \left(\frac{m^* L}{m\kappa a_{\mathrm{H}}} \right) \right] \tag{2.75}$$

where L is the anion–cation separation along a cube edge. Since the effective mass, m^*, is not known it can be determined from eqn (2.75) using the experimentally determined Mollwo–Ivey law. Hence using eqn (2.72) the F-band energy is found to be

$$E_{\mathrm{F}} = \frac{3}{8} \left(\frac{e^2}{4\pi\varepsilon_0 \kappa^2} \right) \left(\frac{m^*}{m a_{\mathrm{H}}} \right) = 0.97 L^{-1.772}$$

and since $a_{\mathrm{H}} = 4\pi\varepsilon_0 \hbar^2/(me^2)$ we obtain

$$\frac{m^* L}{m\kappa a_{\mathrm{H}}} = 1.8\kappa L^{-0.772} \tag{2.76}$$

where both L and a_{H} are in nanometres. Hence for potassium chloride where $2L = 0.625$ nm and the high-frequency dielectric constant is 2.15, we find $\rho(L) \simeq 0.084$. In other words, some 16 per cent of the electronic charge resides outside the first cation shell. The high-frequency dielectric constant was used because the rather massive ions do not respond to the motion of the electron.

A simple test is to take the experimental values for E_{F} and κ and derive the ratio m^*/m using $E_{\mathrm{F}} = 10.2(m^*/m)/\kappa^2$. These values are shown in Table 2.4; they vary in a systematic way over the whole range of alkali halide crystals. In a really continuous medium m^*/m should be unity, whereas in a periodic crystal m^*/m should be that of an electron in the nearest band, provided that

Table 2.4

The effective mass (m^*/m_e) for F centres in alkali halides calculated using eqn. 2.72 with the experimental values of E_F and the high frequency dielectric constant κ_∞. In calculating the effective dielectric constant κ^* the effective mass m^*/m_e was assumed to be equal to unity.

	LiF	NaF	KF	NaCl	KCl	RbCl	NaBr	KBr	RbBr
E_F(eV)	5.0	3.75	2.90	2.75	2.30	2.10	2.35	2.15	1.85
κ_∞	1.92	1.74	1.85	2.25	2.13	2.19	2.62	2.33	2.33
m^*/m_e	1.51	0.93	0.81	1.13	0.86	0.83	1.32	0.96	0.83
κ^*	1.43	1.80	1.88	1.94	2.11	2.20	2.08	2.18	2.35

the wavefunction spreads over many lattice points. This seems unlikely given the calculated value of $\rho(L)$. However, measurements of $\rho(L)$ using magnetic resonance techniques (see e.g. Henderson and Garrison 1973) show that in actuality the defect electron ranges over many shells of cation and anion neighbours so that only about 70 per cent of the charge density resides within the vacancy. Although the use of the continuum model appears questionable in calculating ground-state properties it is much more reasonable in calculations related to excited states. Indeed, as we discuss in Chapter 7, the experimental studies of the F-centre support the view that excited states are spatially very diffuse and close to the bottom of the conduction band. In such excited states the continuum models are both successful and quite justifiable (Stoneham 1985).

There are many more sophisticated models of the F-centre (see e.g. Markham 1966; Fowler 1968; Stoneham 1985). It is worth noting that the *semi-continuum model* combines the best features of both the potential-well and continuum models. Within a volume of radius R centred on the F-centre the Hamiltonian is written as

$$\mathcal{H} = p^2/2m + V_0$$

where V_0 represents the sum of the Madelung energy, lattice polarization, and the electron affinity. This latter term is required to measure energies relative to the bottom of the conduction band, rather than the vacuum level. For $r > R$ an effective mass approximation is used, and the Hamiltonian is

$$\mathcal{H} = p^2/2m^* - e^2/4\pi\varepsilon_0\kappa' r \tag{2.77}$$

where m^* is the effective mass at the bottom of the conduction band and κ' is an effective dielectric constant. The form of the potential for F-centres in sodium chloride is shown in Fig. 2.9. Photon energies corresponding to the peaks of the absorption and emission bands of the F-centre in sodium chloride were computed to be 2.80 eV and 1.24 eV respectively (Fowler 1968).

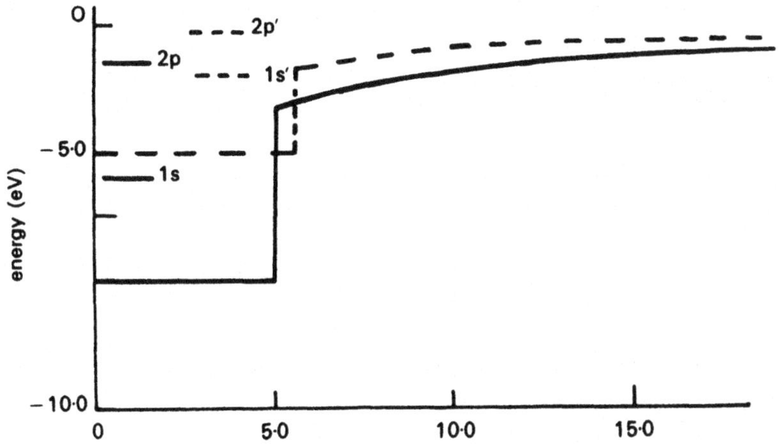

FIG. 2.9. The results of a semi-continuum calculation of the ground (1s) and excited (2p) state energy levels of the F-centre in sodium chloride in absorption (——) and in emission (---). For convenience the energy of the crystal with the electron in the conduction band is chosen to be the same, and a comparison of similar energy levels for absorption and emission is not meaningful. (Adapted from Fowler 1968.)

values reasonably in accord with the experimental peaks at 2.70 eV and 1.08 eV. Note that the Stokes shift between absorption and emission peak energies arises because of the small outward relaxation of the nearest-neighbour ions. Furthermore the relaxed excited state is determined to be ~0.12 eV below the conduction band compared with the experimental value of 0.09 eV (Chapter 7). The calculated oscillator strength for absorption is $f \simeq 1.0$, corresponding to a radiative lifetime of $\tau_R \simeq 67 \times 10^{-8}$ s: the measured lifetime is about 100×10^{-8} s. (Oscillator strengths and radiative lifetimes are discussed in Chapter 4.) These theoretical results are obtained using an effective dielectric constant $\kappa' = 4.21$ for sodium chloride, midway between high-frequency (2.34) and static (5.9) values.

2.5.3 Continuum models of vacancy aggregate centres

Under appropriate conditions, isolated F-centres may be caused to cluster together in near-neighbour sites to form *aggregate* centres.[†] The simplest aggregate centre involves two nearest-neighbour anion vacancies along a ⟨110⟩ direction, in which may be trapped one, two, or three electrons so creating F_2^+, F_2, or F_2^--centres, respectively. If three anion vacancies cluster in nearest-neighbour sites in the (111) plane[‡] then F_3^+, F_3, and F_3^--centres

[†] A detailed discussion of these centres is given in Chapter 7.

[‡] (We use the usual crystallographic convention: (*hkl*) is one member of the {*hkl*} family of planes and [*hkl*] is a particular one of the ⟨*hkl*⟩ family of directions in a crystal.)

result from the trapping of two, three, and four electrons, respectively. There is no reason why the continuum model should work better for aggregate centres than for the generating F-centre. Nonetheless it has been used, and with considerable success, for F_2^+, F_2, and F_3-centres. Consider the F_2^+-centre which is represented as an H_2^+ molecular ion embedded in a dielectric continuum. The H_2^+ ion is the simplest molecular structure—a one-electron system in which the two nuclei are in fixed positions a distance R apart. The one-electron Hamiltonian is

$$\mathscr{H} = -\frac{\hbar^2}{2m^*}\nabla^2 - \frac{e^2}{4\pi\varepsilon_0\kappa}\left(\frac{1}{r_1}+\frac{1}{r_2}-\frac{1}{R}\right) \tag{2.78}$$

where r_1, r_2 are the distances of the electron from the centres of the vacancies. The simplest (unnormalized) wavefunctions are

$$\sigma_g^+ = |1s_A\rangle + |1s_B\rangle$$

and (2.79)

$$\sigma_u^+ = |1s_A\rangle - |1s_B\rangle$$

where $|1s_{A,B}\rangle$ are hydrogen (1s) wavefunctions on nuclei A and B, respectively. The one-electron orbital states are labelled according to the orbital angular momentum about the molecular axis, $L=\lambda\hbar$, having values given by $\lambda=0$ (σ-state), $\lambda=1$ (π-state), etc. A subscript g or u indicates that the orbital is of even or odd parity, respectively. The energies of these lowest-lying σ-states are given by

$$E_\pm = \frac{\mathscr{H}_{11} \pm \mathscr{H}_{12}}{1 \pm S} \tag{2.80}$$

where S is the overlap integral $\langle 1s_A|1s_B\rangle$, $\mathscr{H}_{11} = \langle 1s_A|\mathscr{H}|1s_A\rangle$, $\mathscr{H}_{12} = \langle 1s_A|\mathscr{H}|1s_B\rangle$ and the positive sign refers to the σ_u^+ states. Such a molecular orbital method gives the binding energy of H_2^+ as 1.8 eV at $R=0.62$ nm (appropriate to potassium chloride), a result in error by about 1 eV. The first series of excited states is obtained by combinations (1s, 2s) and (1s, 2p) of hydrogenic wavefunctions centred on the nuclei A and B.

The dielectric continuum model for F_2^+-centres developed by Herman *et al.* (1956) does not require detailed solution of eigenvalues for each state, it requires only that the effective mass, m^*, and the dielectric constant, κ, be used as adjustable parameters. It is useful to introduce the new coordinates

$$\boldsymbol{r}' = \frac{m^*\boldsymbol{r}}{m\kappa}$$

such that the Schrödinger equation becomes

$$\mathscr{H}'\psi'(\boldsymbol{r}') = E'\psi'(\boldsymbol{r}')$$

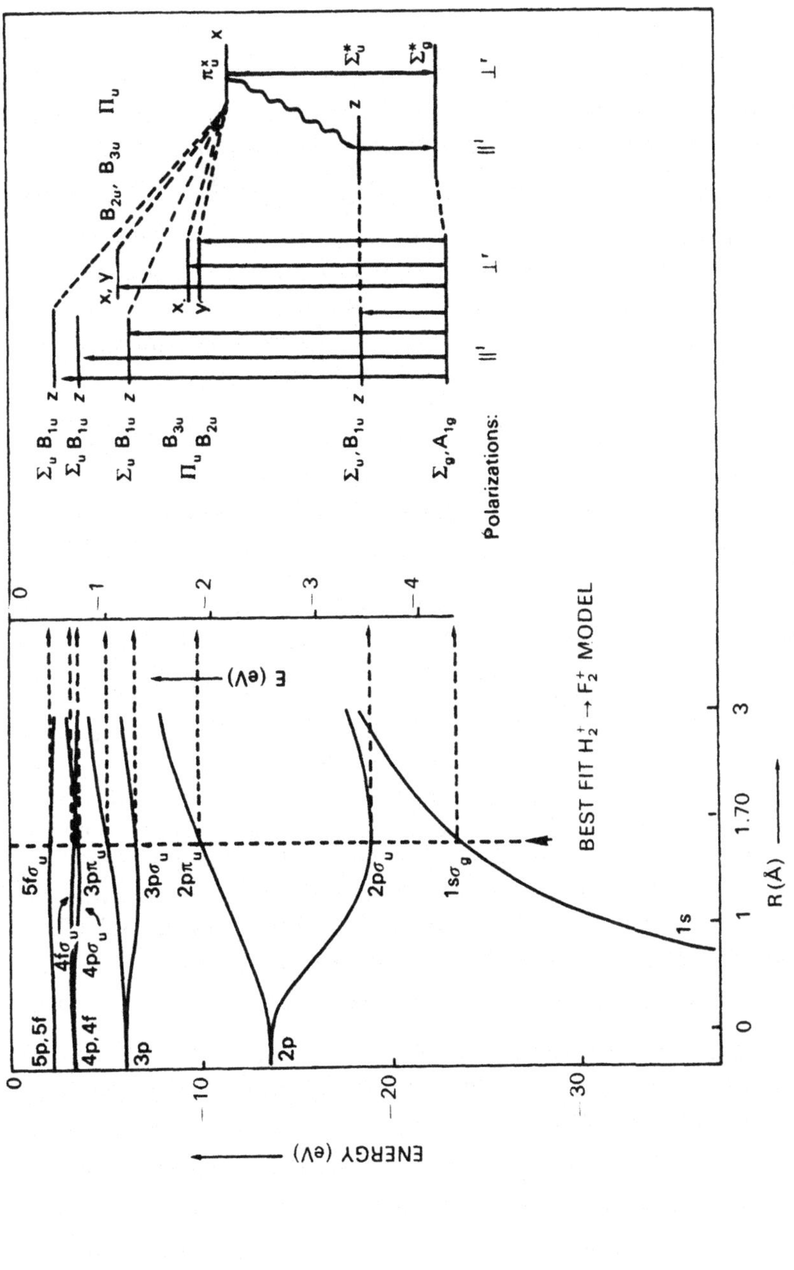

FIG. 2.10. The energy levels of the H_2^+ molecule ion as a function of distance compared with the identified transitions of F_2^+-centres. (Adapted from Aegerter and Lüty 1970.)

which has eigenvalues

$$E' = \frac{m\kappa^{-2}}{m^*} E(H_2^+). \tag{2.81}$$

One therefore obtains eigenvalues E which have the same relationship to R as exist between E and R for the free H_2^+ ion. The energy levels for the H_2^+ centre are compared in Fig. 2.10, with the lowest eight transitions observed by polarization spectroscopy for F_2^+-centres in potassium chloride. The best fit to the experimental data is found for a molecular separation of $R = 0.17$ nm. The value determines both κ and R, the respective values of which are 2.33 and 0.396 nm. Although the agreement is remarkably good the model does not predict the small splitting of the $2p\pi_u$ state nor the precise ordering of excited states determined in excited state absorption measurements.

A similar model has been used for the F_2-centres and with correspondingly good agreement. In this case, however, fewer experimental transitions have been identified. The Hamiltonian for this two-electron system is

$$\mathscr{H} = \frac{-\hbar^2}{2m^*}(\nabla_1^2 + \nabla_2^2) - \frac{e^2}{4\pi\varepsilon_0\kappa}\left(\frac{1}{r_{A1}} + \frac{1}{r_{A2}} + \frac{1}{r_{B1}} + \frac{1}{r_{B2}} - \frac{1}{r_{12}} - \frac{1}{R}\right)$$

where the suffixes 1 and 2 refer to electrons and A and B to the nuclei. The states are now labelled according to the total orbital angular momentum $L = \Lambda\hbar$, where $\Lambda = l_1 + l_2$, measured about the molecular axis. By convention Σ implies $\Lambda = 0$, Π implies $\Lambda = 1$, etc. Assuming $\kappa = 2.22$ and $R = 0.444$ nm, Herman *et al.* (1956) compute that the ${}^1\Sigma_g \rightarrow {}^1\Sigma_u$ energy-level splitting is 1.62 eV whereas the measured ${}^1\Sigma_g \rightarrow {}^1\Pi_u$ transition energy is 2.01 eV. Experimentally the observed transition energies for potassium chloride are 1.55 eV. 2.23 eV, and 2.30 eV. Again the splitting of 0.07 eV in the ${}^1\Pi_u$ state is not predicted theoretically.

Finally we discuss the electronic states of the F_3-centre. The ordering of states has been determined by Silsbee (1965), who adapted the theoretical results of Hirschfelder (1938) for the H_3 molecule using the continuum approximation. For such centres the one-electron orbitals are constructed by taking linear combinations of $\Phi_i = |1s_i\rangle$ orbitals centred on each of the three vacant sites. The ground-state wavefunction, labelled a_1,

$$a_1 = \Phi_1 + \Phi_2 + \Phi_3 \tag{2.82}$$

has a high electron density between the vacancies and at the centre of the triangle. Two other independent linear combinations of these functions are degenerate: these functions

$$e = \begin{matrix} (2\Phi_2 - \Phi_1 - \Phi_3) & (E_y) \\ (\Phi_1 - \Phi_3) & (E_x) \end{matrix} \tag{2.83}$$

have a low charge density between the F-centres and zero charge density at

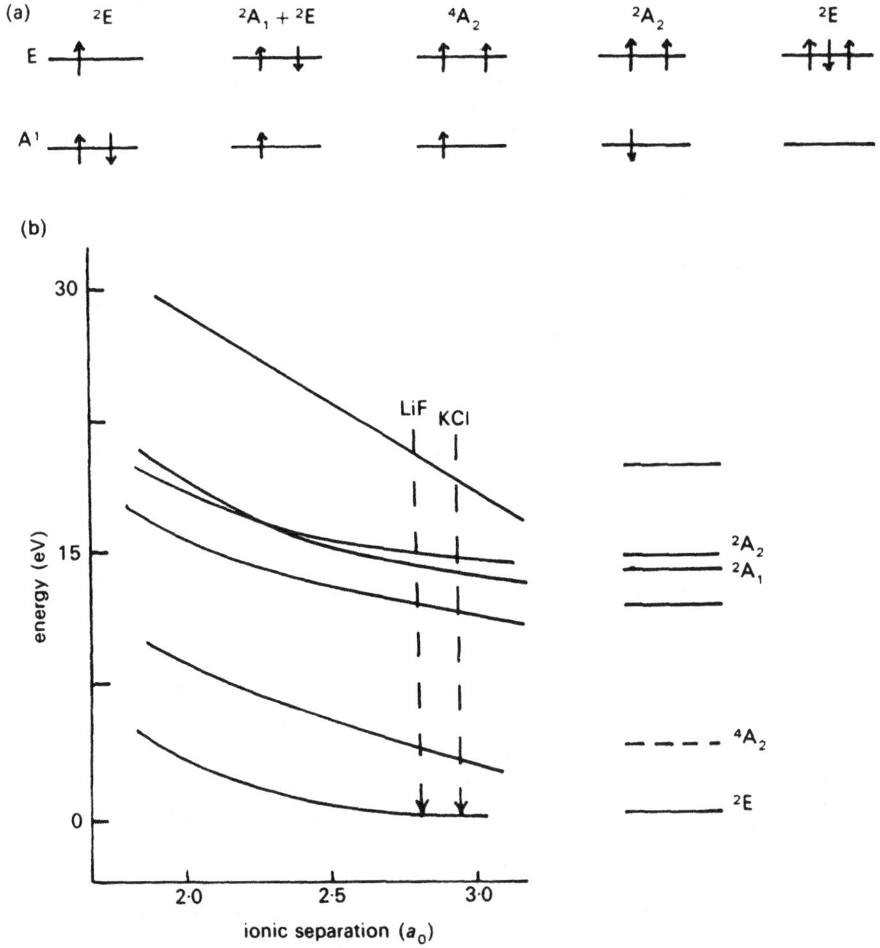

FIG. 2.11. (a) The one-electron molecular configuration orbital used to construct ground and excited state energy levels of F_3-centres in alkali halides and (b) calculated energy levels of H_3 molecules plotted as a function of interionic distance. The vertical dotted lines refer to lithium fluoride. (After Hughes 1966.)

the centre of the triangle. They have a higher energy than the a_1 orbital. A molecular orbital calculation gives an energy separation between the a_1 and e orbitals of $\Delta E \simeq 2$ eV. Invoking the Pauli exclusion principle and filling orbitals according to the ordering of energies the ground-state configuration for the F_3-centre is found to be $(a_1)^2(e)$. The two electrons in the a_1 orbital have their spins aligned antiparallel. Consequently the F_3-centre has a ^2E ground state. By similar arguments we find that the F_3^+ and F_3^--centres have ground states of $^1A_1$ and ^1E symmetry, respectively.

Possible excited configurations of the F_3-centre may be formed from $(a_1)(e)^2$ and $(e)^3$. These are shown in Fig. 2.11a as $^2A_1$, ^2E, $^4A_2$, $^2A_2$, and ^2E. Allowed electric dipole transitions can only be made to spin doublet states. The energies of these states for the H_3 molecule are shown in Fig. 2.11b. Also indicated are the values of the internuclear separation appropriate to lithium fluoride for which $\kappa = 1.96$ and $R = 2.94\ a_H$ (after scaling by κ). Hence we determine $E(^2A_1) - E(^2E) = 3.5$ eV for the $F_3(2)$ transition energy. Experimentally the peak of the $F_3(2)$-band occurs at *c.* 380 nm, which is very close to the predicted value.

For all these aggregate centres the continuum approximation gives a reasonable interpretation of the main spectroscopic results. The justification for the use of the model is this qualitative success. However, accurate calculations require much more sophisticated theoretical techniques (Fowler, 1968, Stoneham, 1985).

2.6 Crystal field states of a single 3d electron

For the transition metal ions of the first series with the outer $3d^n$ electron configuration the interelectron Coulomb interaction, $\mathscr{H}'$, and the crystal field energy, $\mathscr{H}_c$, are comparable. Hence these ions can be treated as intermediate or strong crystal field cases. They are usually treated by the strong crystal field approach as are the ions of the second and third transition metal series, $4d^n$ and $5d^n$, respectively. In this section we consider a single 3d electron in a crystal field; i.e. the electronic configuration of the Ti^{3+} ion. The more complicated case of many d electrons is considered in Chapter 3.

2.6.1 $(3d)^1$ *electronic configuration in an octahedral crystal field*

To illustrate the application of crystal field theory we consider an ion having a single 3d electron outside the closed shell structure situated in the octahedral site, shown in Fig. 2.4(a), a distance a from any one of the six neighbouring ions, each with charge $-Ze$. In this case the electrostatic potential (eqn 2.55) becomes

$$V(x, y, z) = V_x + V_y + V_z \tag{2.84}$$

where

$$V_x = -\frac{Ze}{4\pi\varepsilon_0}\left[\frac{1}{(r^2+a^2-2ax)^{\frac{1}{2}}} + \frac{1}{(r^2+a^2+2ax)^{\frac{1}{2}}}\right] \tag{2.85}$$

with corresponding expressions for V_y and V_z, and $r^2 = x^2 + y^2 + z^2$. The crystal field Hamiltonian due to this octahedral arrangement of six neighbouring ions is $\mathscr{H}_c^{O_h} = -eV$. Assuming $r < a$ and expanding up to terms of sixth degree we obtain after some straightforward but tedious algebra

$$V(x, y, z)=\left[\frac{-1}{4\pi\varepsilon_0}\frac{6Ze}{a}+\frac{35Ze}{4\pi\varepsilon_0 4a^5}\left[(x^4+y^4+z^4)-\frac{3}{5}r^4\right]\right.$$
$$+\frac{21Ze}{4\pi\varepsilon_0 2a^7}\left[(x^6+y^6+z^6)\right. \tag{2.86}$$
$$\left.\left.+\frac{15}{4}(x^2y^4+x^2z^4+y^2x^4+y^2z^4+z^2x^4+z^2y^4)-\frac{15}{14}r^6\right]\right].$$

It is convenient to express this crystal field Hamiltonian in spherical harmonics, eqn (2.57). Inserting the six appropriate sets of value for θ_l, ϕ_l in this expression and putting $a_l=a$ for each we obtain (Sugano *et al.* 1970)

$$\mathscr{H}_c^{O_h}(\boldsymbol{r})=\frac{Ze^2}{4\pi\varepsilon_0}\left[\frac{6}{a}+\frac{7r^4}{2a^5}\left\{C_0^{(4)}(\theta,\phi)+\left(\frac{5}{14}\right)^{\frac{1}{2}}\left(C_4^{(4)}(\theta,\phi)+C_{-4}^{(4)}(\theta,\phi)\right)\right\}\right]$$
$$+\boldsymbol{r}^6 \text{ terms}+\ \ldots \tag{2.87}$$

where $\boldsymbol{r}(r, \theta, \phi)$ is the position of the 3d electron. We must evaluate the effect of $\mathscr{H}_c^{O_h}$ on the 3d electron states, ignoring configuration mixing. In the case of d electrons, with $l=2$, we recall (eqn 2.59) that we do not need terms in $\mathscr{H}_c$ greater than fourth power in r.

The first term in (2.87) is neglected since it contributes the same constant energy to each state. We calculate the matrix elements of the remainder of $\mathscr{H}_c^{O_h}$ within the 3d configuration of states, $|3dm_l\rangle$. The diagonal matrix element for the $|3d0\rangle$ state is obtained with the aid of Table 2.2. It is

$$\langle 3d0|\,\mathscr{H}_c^{O_h}|3d0\rangle=\frac{1}{4\pi\varepsilon_0}\frac{7Ze^2}{2a^5}\langle r^4\rangle_{3d}\cdot\frac{6}{21}=6Dq$$

where

$$D=\frac{1}{4\pi\varepsilon_0}\frac{35Ze^2}{4a^5} \quad \text{and} \quad q=\frac{2}{105}\langle r^4\rangle_{3d}. \tag{2.88}$$

The parameters D and q always occur as a product. Hence Dq can be regarded as a single parameter which characterizes the strength of the octahedral crystal field. Other matrix elements can be similarly calculated, and we list the non-zero matrix elements:

$$\langle 3d0|\,\mathscr{H}_c^{O_h}|3d0\rangle=6Dq$$
$$\langle 3d1|\,\mathscr{H}_c^{O_h}|3d1\rangle=\langle 3d-1|\,\mathscr{H}_c^{O_h}|3d-1\rangle=-4Dq$$
$$\langle 3d2|\,\mathscr{H}_c^{O_h}|3d2\rangle=\langle 3d-2|\,\mathscr{H}_c^{O_h}|3d-2\rangle=Dq$$
$$\langle 3d2|\,\mathscr{H}_c^{O_h}|3d-2\rangle=\langle 3d-2|\,\mathscr{H}_c^{O_h}|3d2\rangle\quad=5Dq. \tag{2.89}$$

The Hamiltonian matrix

$$\begin{array}{c} 3\mathrm{d}2 \\ 3\mathrm{d}1 \\ 3\mathrm{d}0 \\ 3\mathrm{d}-1 \\ 3\mathrm{d}-2 \end{array} \begin{bmatrix} Dq & 0 & 0 & 0 & 5Dq \\ 0 & -4Dq & 0 & 0 & 0 \\ 0 & 0 & 6Dq & 0 & 0 \\ 0 & 0 & 0 & -4Dq & 0 \\ 5Dq & 0 & 0 & 0 & Dq \end{bmatrix} \tag{2.90}$$

is easily diagonalized to obtain the eigenstates and eigenvalues. We see that the $|3\mathrm{d}0\rangle$, $|3\mathrm{d}1\rangle$, $|3\mathrm{d}-1\rangle$ states are exact eigenstates, whereas the $|3\mathrm{d}\pm 2\rangle$ states are mixed. There are two eigenstates with eigenvalue $6Dq$, and three eigenstates with eigenvalue $-4Dq$. The most common form of these eigenstates is

$$\phi_{eu} = |3\mathrm{d}0\rangle = R'_{3\mathrm{d}}(r)\left(\frac{5}{4\pi}\right)^{\frac{1}{2}}\left(\frac{3z^2-r^2}{2r^2}\right)$$

$$\phi_{ev} = \left(\frac{1}{2}\right)^{\frac{1}{2}}(|3\mathrm{d}2\rangle + |3\mathrm{d}-2\rangle) = R'_{3\mathrm{d}}(r)\left(\frac{5}{4\pi}\right)^{\frac{1}{2}} 3^{\frac{1}{2}}\,\frac{x^2-y^2}{2r^2}. \tag{2.91}$$

These are the e crystal field orbitals and have energy $6Dq$.

$$\phi_{t_2\xi} = \left(\frac{i}{2^{\frac{1}{2}}}\right)(|3\mathrm{d}1\rangle + |3\mathrm{d}-1\rangle) = R'_{3\mathrm{d}}(r)\left(\frac{5}{4\pi}\right)^{\frac{1}{2}} 3^{\frac{1}{2}}\left(\frac{yz}{r^2}\right)$$

$$\phi_{t_2\eta} = -\left(\frac{1}{2^{\frac{1}{2}}}\right)(|3\mathrm{d}1\rangle - |3\mathrm{d}-1\rangle) = R'_{3\mathrm{d}}(r)\left(\frac{5}{4\pi}\right)^{\frac{1}{2}} 3^{\frac{1}{2}}\left(\frac{xz}{r^2}\right)$$

$$\phi_{t_2\zeta} = -\left(\frac{i}{2^{\frac{1}{2}}}\right)(|3\mathrm{d}2\rangle - |3\mathrm{d}-2\rangle) = R'_{3\mathrm{d}}(r)\left(\frac{5}{4\pi}\right)^{\frac{1}{2}} 3^{\frac{1}{2}}\left(\frac{xy}{r^2}\right). \tag{2.92}$$

These are the t_2 crystal field orbitals and have energy $-4Dq$.

Thus the octahedral crystal field splits the fivefold degenerate 3d level into a doubly-degenerate level with additional energy $+6Dq$ and a triply-degenerate level with additional energy $-4Dq$, as shown in Fig. 2.12. The general symbol for the crystal field orbital is $\phi_{\Gamma\gamma}$, where Γ refers to either t_2 or e. The significance of these labels will be understood when we discuss the symmetry of the crystal field using group-theoretical methods. The angular properties of the $\phi_{\Gamma\gamma}$ orbitals are illustrated in Fig. 2.13.

The magnitude of the parameter Dq can be measured spectroscopically; the agreement between the experimental value and that calculated using eqn (2.88) is not good. This lack of agreement reflects the crudeness of the point

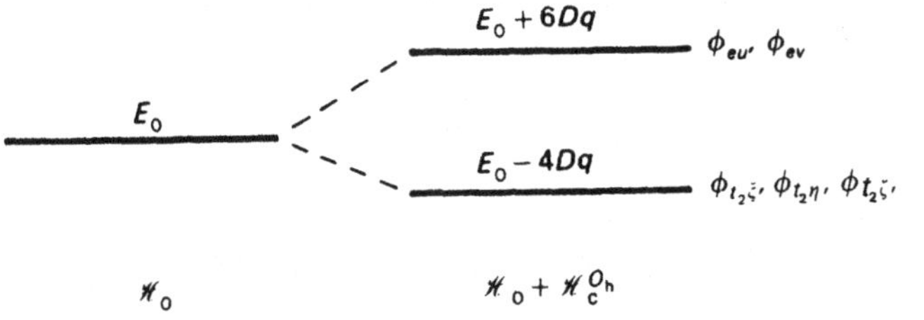

FIG. 2.12. A single outer 3d electron in the central field of the nucleus and of the inner electron core has energy E_0 and has fivefold orbital degeneracy. This level splits in two, separated in energy by $\Delta = 10\,Dq$, in the presence of an octahedral crystal field.

ion model. In a more rigorous model the neighbouring ions would be treated as extended ligands. Nevertheless, the point ion model correctly describes the symmetry of the crystal field. Indeed, Sugano *et al.* (1970) show, on general symmetry arguments, that the octahedral crystal field Hamiltonian has the form

$$\mathscr{H}_c^{O_h}(r) = D(r)\left[C_0^{(4)}(\theta, \phi) + \left(\frac{5}{14}\right)^{\frac{1}{2}} \{C_4^{(4)}(\theta, \phi) + C_{-4}^{(4)}(\theta, \phi)\}\right] + \ldots \tag{2.93}$$

where the integral $\langle D(r)\rangle_{3d}$ is given the value $21Dq$ to conform to our previous analysis. Thus Dq is regarded as a parameter, the value of which is to be determined by experiment.

It is sometimes useful to use the coordinate set shown in Fig. 2.14 in which the Z-axis is parallel to the [111] direction of the crystal and the X, Y, Z-axes are referred to as the *trigonal axes*. Their orientation relative to the cubic axes are shown in this figure. The coordinate transformation between trigonal and cubic axes is

$$\begin{aligned} x &= -\frac{1}{\sqrt{6}}X + \frac{1}{\sqrt{2}}Y + \frac{1}{\sqrt{3}}Z \\ y &= -\frac{1}{\sqrt{6}}X - \frac{1}{\sqrt{2}}Y + \frac{1}{\sqrt{3}}Z \\ z &= \sqrt{\frac{2}{3}}X + \sqrt{\frac{1}{3}}Z. \end{aligned} \tag{2.94}$$

The energy of interaction of an electron at $\mathbf{r}$ with the octahedral crystal field can be expressed in coordinates referred to the trigonal axes (Sugano *et al.* 1970) as

$$\mathscr{H}_c^{O_h} = D'(r)\left[C_0^{(4)}(\theta, \phi) + \left(\frac{10}{7}\right)^{\frac{1}{2}} \{(C_3^{(4)}(\theta, \phi) - C_{-3}^{(4)}(\theta, \phi)\}\right]$$
$$+\text{terms with } k = 6, \text{ etc.} \tag{2.95}$$

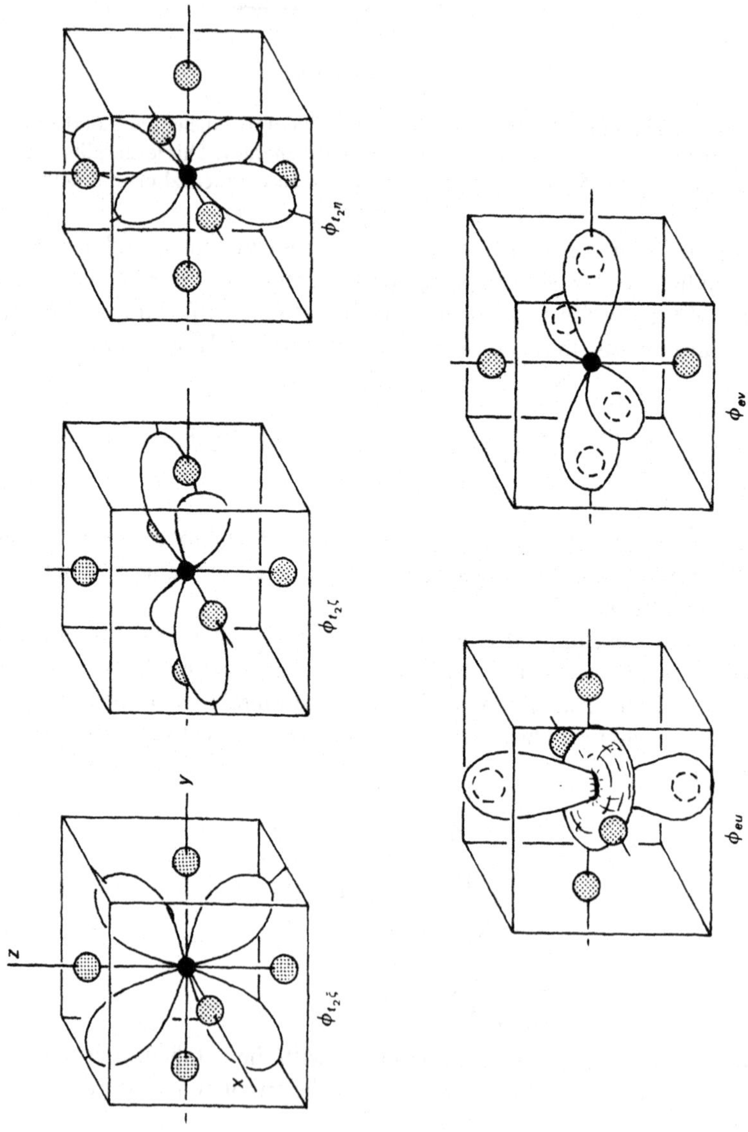

FIG. 2.13. Octahedral crystal field orbitals formed from 3d electron wavefunctions.

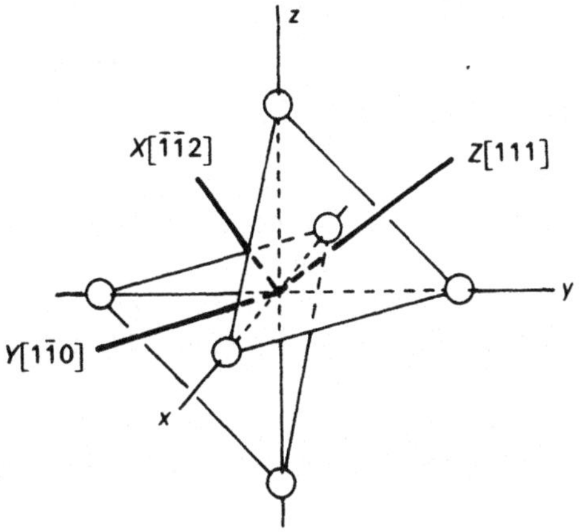

FIG. 2.14. The orientation of the trigonal axes (X, Y, Z) to the cubic axes (x, y, z) and the six equidistant ions causing the octahedral crystal field. The six ions are shown grouped into two triangles whose planes are perpendicular to the Z-axis.

The relationship between $D'(r)$ of eqn (2.95) and the octahedral crystal field parameter Dq is given by $\langle D'(r)\rangle_{3\mathrm{d}} = -14\,Dq$. Useful crystal field orbitals for the t_2 and e states in terms of trigonal coordinates are

$$\phi_{t_2 x_0} = R'_{3\mathrm{d}}(r)\, Y_2^0$$

$$\phi_{t_2 x_+} = R'_{3\mathrm{d}}(r)\left[-\left(\frac{2}{3}\right)^{\frac{1}{2}} Y_2^{-2} - \left(\frac{1}{3}\right)^{\frac{1}{2}} Y_2^1\right]$$

$$\phi_{t_2 x_-} = R'_{3\mathrm{d}}(r)\left[\left(\frac{2}{3}\right)^{\frac{1}{2}} Y_2^2 - \left(\frac{1}{3}\right)^{\frac{1}{2}} Y_2^{-1}\right]$$

$$\phi_{eu_+} = R'_{3\mathrm{d}}(r)\left[-\left(\frac{1}{3}\right)^{\frac{1}{2}} Y_2^{-2} + \left(\frac{2}{3}\right)^{\frac{1}{2}} Y_2^1\right]$$

$$\phi_{eu_-} = R'_{3\mathrm{d}}(r)\left[\left(\frac{1}{3}\right)^{\frac{1}{2}} Y_2^2 + \left(\frac{2}{3}\right)^{\frac{1}{2}} Y_2^{-1}\right]. \qquad (2.96)$$

The general symbol for these orbitals is $\phi_{\Gamma M}$, and the coordinates in eqns (2.95) and (2.96) refer to the trigonal axes. The relationship between $\phi_{\Gamma M}$ referred to the trigonal axes and $\phi_{\Gamma\gamma}$ referred to the cubic axes is written

$$\phi_{\Gamma M} = \sum_{\gamma} \phi_{\Gamma\gamma} \langle \Gamma\gamma | \Gamma M \rangle \qquad (2.97)$$

The numerical coefficients $\langle \Gamma\gamma | \Gamma M \rangle$, which are elements of a unitary matrix, are tabulated in Sugano *et al.* (1970).

2.6.2 The distorted octahedron: lower-symmetry crystal fields

Sites of perfect octahedral symmetry are encountered only in simple crystals; lower-symmetry sites are the more common occurrence. Consider the case in Fig. 2.15 where the octahedral arrangement of Fig. 2.4(a) is distorted, the distances from the origin to the ions along the $\pm z$ cubic axes being equal but different from the distances from the origin to the ions along the $\pm x$, $\pm y$ cubic axes. This arrangement produces an electrostatic field of tetragonal symmetry (indicated by the label D_4). If the appropriate values of a_l, θ_l, ϕ_l are substituted in the point-ion Hamiltonian (eqn 2.57) we find that the crystal field Hamiltonian contains factors $C_0^{(2)}(\theta, \phi)$, $C_0^{(4)}(\theta, \phi)$, $C_4^{(4)}(\theta, \phi)$, and $C_{-4}^{(4)}(\theta, \phi)$, etc. From more general symmetry considerations Sugano *et al.* (1970) have shown that the most general tetragonal crystal field Hamiltonian has the form

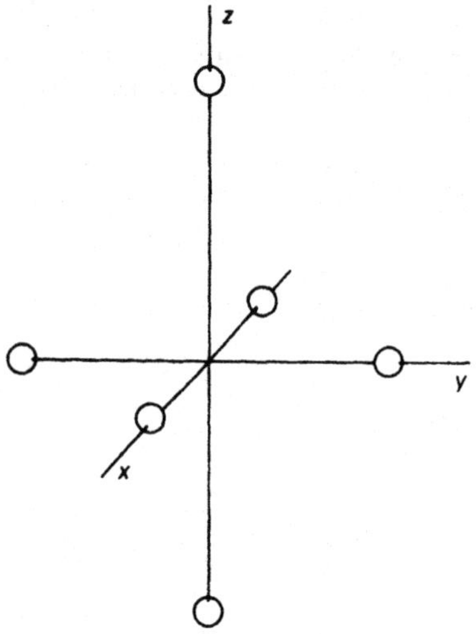

FIG. 2.15. Tetragonally-distorted octahedral arrangement of ions. The distances from the origin to the two ions along the $\pm z$ directions are equal but larger than the distances from the origin to the four equidistant ions along the $\pm x$ and $\pm y$ directions.

$$\begin{aligned}\mathcal{H}_c^{D_4}(r) = {} & A(r)C_0^{(2)}(\theta, \phi) + B(r)C_0^{(4)}(\theta, \phi) \\ & + D(r)\left[C_0^{(4)}(\theta, \phi) + \left(\frac{5}{14}\right)^{\frac{1}{2}}\{C_4^{(4)}(\theta, \phi) + C_{-4}^{(4)}(\theta, \phi)\}\right] \\ & + \text{higher-order terms} \ldots \qquad (2.98)\end{aligned}$$

where the coordinates are referred to the cubic axes. We see that eqn (2.98) contains a term of octahedral symmetry (compare eqn 2.93); the remaining low-order term in $C_0^{(2)}$ and $C_0^{(4)}$ is independent of ϕ and so is invariant under rotation through any angle ϕ about the z-axis. This term contributes a crystal field of axial symmetry. If the distortion from perfect octahedral symmetry is small then it is appropriate to first calculate the eigenstates for octahedral symmetry and then to include the axial field as a perturbation.

Another common distortion of the perfect octahedral arrangement results in a site of trigonal symmetry, as is illustrated in Fig. 2.16; the six ions of an octahedral environment are shown grouped into triangles of ions relative to the trigonal axes (see also in Fig. 2.14). If the two triangles are farther displaced along the $\pm z$ trigonal axis than strict octahedral symmetry requires then the resultant crystal field has trigonal symmetry (indicated by the label D_3). The most general trigonal crystal field Hamiltonian has the form

$$\mathcal{H}_{\mathrm{c}}^{D_3}(r) = A'(r)C_0^{(2)}(\theta,\phi) + B'(r)C_0^{(4)}(\theta,\phi) + D'(r)\left[C_0^{(4)}(\theta,\phi) + \left(\frac{10}{7}\right)^{\frac{1}{2}}\{C_3^{(4)}(\theta,\phi) + C_{-3}^{(4)}(\theta,\phi)\}\right] + \text{higher-order terms}\ldots \quad (2.99)$$

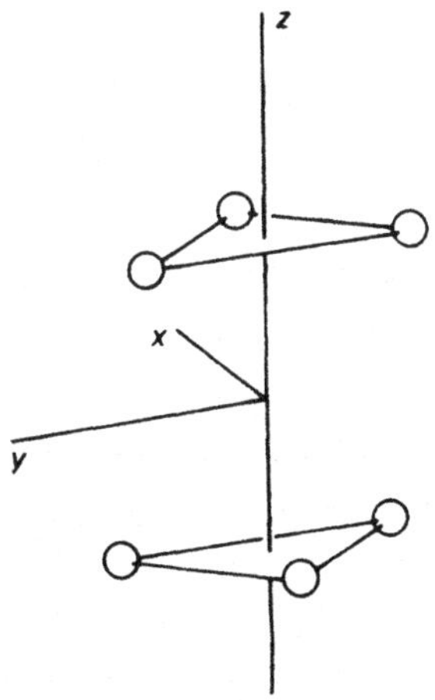

FIG. 2.16. A different view of the octahedral arrangement of ions showing them relative to the trigonal axes. If the two triangles of ions are displaced through equal distances along the $\pm z$ directions the resultant arrangement has trigonal symmetry.

where the coordinates are referred to the trigonal axes (Sugano *et al.* 1970). This formula contains an octahedral crystal field component (compare eqn 2.95) and an axial crystal field along the trigonal z-axis.

The effect of an axial crystal field on the 3d octahedral crystal field levels can be calculated using Table 2.2. In tetragonal symmetry the u, v and ξ, η, ζ functions are still eigenstates of the tetragonal crystal field Hamiltonian but there is an additional energy and a splitting due to the axial component of the crystal field. The splittings and eigenstates are shown in Fig. 2.17 where $\langle A\rangle = \langle A(r)\rangle_{3d}$ and $\langle B\rangle = \langle B(r)\rangle_{3d}$. The effect of a trigonally distorted octahedral crystal field is calculated similarly. The octahedral component splits the 3d states into e and t_2 levels (Fig. 2.12). The effect of the trigonal field is then introduced. Using octahedral basis functions (eqn 2.96) we find that the x_0 orbital is an eigenstate of the trigonal field. However, the axial field mixes both the u_+ and x_+ functions and also the u_- and x_- functions, and in consequence the e level shifts but does not split, whereas the t_2 level splits in two (Sugano *et al.* 1970).

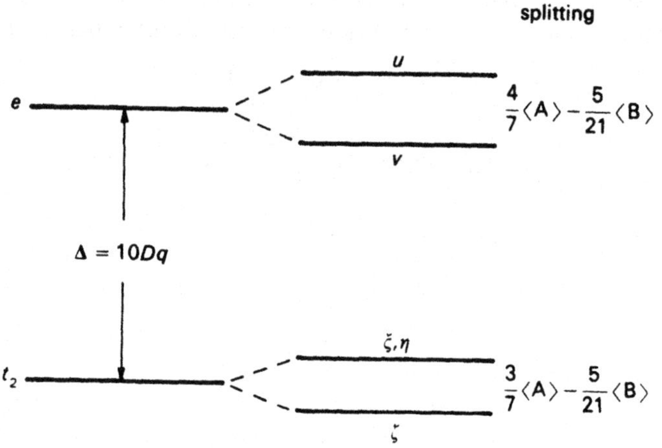

FIG. 2.17. Splitting of the octahedral 3d levels when the crystal field is reduced from the octahedral to tetragonal.

2.6.3 $(3d)^1$ *Electronic configuration in a tetrahedral crystal field*

We now develop a formula for the crystal field energy for a single electron in a field of tetrahedral (T_d) symmetry. As Fig. 2.4(b) shows such a crystal field is due to four point charges, $-Ze$, placed at the vertices of a regular tetrahedron. Inserting the appropriate sets of values of θ_l, ϕ_l and putting $a_l = \sqrt{3}a$ we obtain

$$\mathscr{H}_c^{T_d}(r) = \frac{Ze^2}{4\pi\varepsilon_0}\left[\frac{4}{\sqrt{3}a} - i\frac{\sqrt{40}}{3}\frac{r^3}{(\sqrt{3}a)^4}\left(C_2^{(3)}(\theta,\phi) - C_{-2}^{(3)}(\theta,\phi)\right)\right]$$

$$-\frac{Ze^2}{4\pi\varepsilon_0}\left[\frac{14}{9}\frac{r^4}{(\sqrt{3}a)^5}\left\{C_0^{(4)}(\theta,\phi) + \left(\frac{5}{14}\right)^{\frac{1}{2}}\left(C_4^{(4)}(\theta,\phi) + C_{-4}^{(4)}(\theta,\phi)\right)\right\}\right]$$

$$+\text{higher-order terms}\ldots \tag{2.100}$$

As in the octahedral case we neglect the constant first term. Equation (2.100) contains an odd-parity component to the crystal field energy formula which arises because the tetrahedral arrangement of ions in Fig. 2.4(b) lacks inversion symmetry. Since there are no matrix elements of this odd-parity component within the same (nl) configuration it makes no contribution to the crystal field splitting for a 3d electron, and only the term in r^4 needs to be taken into consideration. The small admixture of odd-parity wavefunction into the even-parity 3d wavefunction may, however, significantly affect the strength of the radiative transitions involving 3d electrons.

The remaining even-parity term (in r^4) has the same functional form as the octahedral crystal field term (compare 2.87). Since the distance from the origin to one of the neighbouring ions in the octahedral case is a while it is $\sqrt{3}a$ in the tetrahedral case (see Fig. 2.4) we see that when the distance from the electronic centre to the neighbouring ions is the same the octahedral crystal field energy is larger than the tetrahedral field energy by a factor of $\frac{9}{4}$ and is of opposite sign. Hence all calculations of crystal field splittings carried out for the octahedral crystal field case can be used for the case of the tetrahedral crystal field case with the appropriate change in value of the parameter, Dq—reversing its sign and reducing its value from the octahedral case.

Finally, the cubic crystal field shown in Fig. 2.4(c) is made up of two tetrahedral components in an arrangement possessing inversion symmetry. The odd-parity terms of the two tetrahedral fields cancel while the even-parity terms add together. Hence the cubic crystal field interaction energy is an even-parity function of $\boldsymbol{r}$, the term in r^4 being of the same functional form as in the tetrahedral field but twice as large. The relationship between the strengths of these three crystal fields is

$$Dq\ (\text{octahedral}) = -\tfrac{9}{4}\ Dq\ (\text{tetrahedral}) = -\tfrac{9}{8}\ Dq\ (\text{cubic})$$

The calculation of the splitting of the multi-electron $(3\text{d})^n$ states by a crystal field is more complicated than in the case of a single 3d electron. It is then helpful to take advantage of the symmetry of the crystal field using the procedures of formal group theory, and it will repay us next to develop an understanding of those aspects of group theory appropriate to our analysis of crystal symmetry.

3

Symmetry and group representation theory

SYMMETRY is a common everyday feature in the physical world. Our modern cities are often laid down on a rectilinear grid, and the buildings thereon possess external features related to one another in a regular manner. If we can catch and observe a snowflake we will find that it displays sixfold rotational symmetry, whereas a ball-bearing looks much the same no matter how it is rotated. The property of any object which shows how its appearance behaves under translations, reflections, rotations, etc., is said to be its symmetry. However, it is not simply the external forms of everyday objects which display symmetry. The physical properties of quantum systems depend as much on spatial arrangements of the components as on their chemical composition. Crystals have a high degree of internal symmetry which may be exhibited in their external shapes. The symmetry properties of crystals are described in terms of a number of operators—inversion, rotation, reflection, and translation operators, all being required in the more general case. Translational symmetry requires that the energy eigenfunctions must be classified by the so-called Bloch waves, useful in the description of electron transport behaviour of crystals. Finite molecules do not possess such periodic structure, and we need only consider local operators e.g. inversion, reflection, and rotation, when discussing their symmetry. Isolated active centres are also sensitive only to this restricted set of operators; we are concerned only with the local symmetry of the centre and its environment.

3.1. Fundamentals of group theory

3.1.1 Symmetry and symmetry operators

In mathematical terms a symmetry operator involves a linear transformation of the coordinates. To see the meaning of this consider the cube in Fig. 3.1. We show two planes ABCD and EFGH, each of which divides the cube into two equal halves. In addition each of these planes acts as a mirror plane in that each point on one side of the plane is the mirror image of a corresponding point on the other side of the plane. An axis of symmetry is an axis which takes the cube into an identical position on rotation through some suitable angle about this axis. The angle of rotation is given by $2\pi/n$ radians where n is an integer for an n-fold rotation axis. This is illustrated by the fourfold rotation axis, C_4, in Fig. 3.1; such an operator rotates the cube through 90°. Inversion through the origin of coordinates takes a point with coordinates (x, y, z) to the

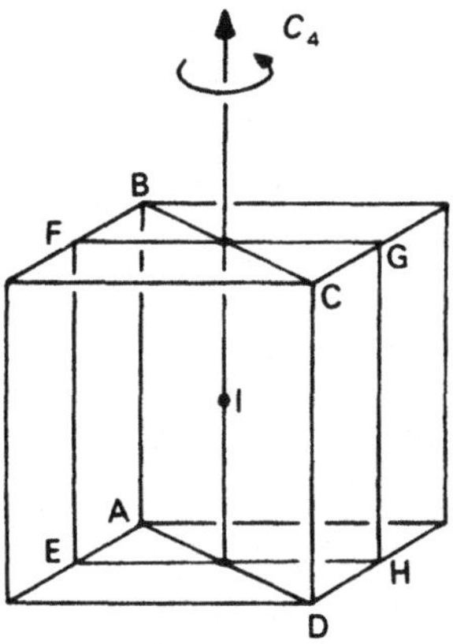

FIG. 3.1. Mirror planes, rotation axis, and inversion centre (I) of a cube.

point $(-x, -y, -z)$. The centre, I, of the cube in Fig. 3.1 is an inversion centre since every point on the cube face, on inversion through I, becomes another point on the cube face. The cube has inversion symmetry through its centre. Mathematical functions can also exhibit symmetry properties. The function

$$\mathrm{g}(x) = a_0 + a_2 x^2 + a_4 x^4 + \ldots \tag{3.1}$$

is an *even* function, whereas

$$\mathrm{u}(x) = a_1 x + a_3 x^3 + a_5 x^5 + \ldots \tag{3.2}$$

is an *odd* function.

3.1.2 Group theory and quantum mechanics I

The classification of the eigenstate of the free-ion Hamiltonian $\mathscr{H}_0$ (eqn 2.29) in terms of L, S, M_L, M_S comes about because the free-ion Hamiltonian commutes with the $\boldsymbol{L}$ and $\boldsymbol{S}$ operators, that is, it commutes with the operator which rotates the coordinates through any angle about any axis. When the ion is in a crystal field the Hamiltonian (eqn 2.53) is invariant under a restricted set of rotation operators. In this section we investigate how the energy levels and wave functions of an electronic centre in a crystalline solid are influenced by the symmetry properties of its Hamiltonian.

As a simple example consider an electron moving in the electrostatic field originating on four equal point charges $-Q$ in the xy-plane, each a distance a from the origin, as shown in Fig. 3.2. The Hamiltonian is

$$\mathscr{H} = \frac{\boldsymbol{p}^2}{2m} + \frac{Qe}{4\pi\varepsilon_0}\left[\frac{1}{[(a-x)^2+y^2+z^2]^{\frac{1}{2}}} + \frac{1}{[(a+x)^2+y^2+z^2]^{\frac{1}{2}}} + \frac{1}{[(a-y)^2+x^2+z^2]^{\frac{1}{2}}} + \frac{1}{[(a+y)^2+x^2+y^2]^{\frac{1}{2}}}\right]. \tag{3.3}$$

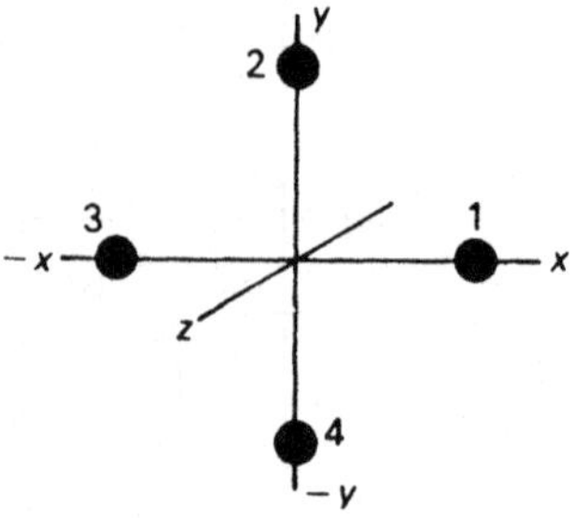

FIG. 3.2. Arrangement of four equal point charges creating an electrostatic crystal field in which an electron moves.

We see that this Hamiltonian is unchanged if we make a transformation of basic functions: $x \to y$, $y \to -x$, $z \to z$. It is similarly unchanged under the transformation $x \to -y$, $y \to x$, $z \to z$. These are rotations of the basis functions x, y, z through $\pm\frac{1}{2}\pi$ about the [001] axis. We label such *rotation-of-function operators* $R^{(F)}(\pm\frac{1}{2}\pi)$, and we see that these operators commute with $\mathscr{H}$, i.e. $[R^{(F)}(\pm\frac{1}{2}\pi), \mathscr{H}]=0$. It follows that if ψ is an eigenstate of $\mathscr{H}$ then $R^{(F)}(\pm\frac{1}{2}\pi)\psi$ is also an eigenstate of $\mathscr{H}$ with the same energy.

The reason why the Hamiltonian is invariant under the rotations of basic function through $\pm\frac{1}{2}\pi$ about the [001] axis is that the arrangement of point charges is invariant under the interchange of ions $1\to2$, $2\to3$, $3\to4$, $4\to1$ and under the interchange $1\to4$, $2\to1$, $3\to2$, $4\to3$. That is, the arrangement of ions is invariant under the rotation of axes through $\pm\frac{1}{2}\pi$ about the [001] axis. We label these *rotation-of-axes operators* $R^{(A)}(\pm\frac{1}{2}\pi)$. The arrangement of ions is also invariant under the interchange $1\to3$, $2\to4$, $5\to6$, that is, under *inversion of axes*. It is a simple matter to check that $\mathscr{H}$ is invariant under the inversion of functions operator, which changes $\boldsymbol{r}\to-\boldsymbol{r}$.

We now generalize these ideas. Consider a Hamiltonian describing an electronic system which is interacting with an arrangement of static charges. Associated with each transformation-of-axes operator which leaves the arrangement of ions invariant is a transformation of functions operator which leaves $\mathscr{H}$ invariant. It is easier to recognize and discuss the symmetry of an electronic system interacting with the arrangement of static ions in terms of the transformation-of-axes operators, so we will particularly concern ourselves with these operators. It is instructive now to examine the relationship between the rotation-of-axes operator and the analogous rotation-of-functions operator. We adopt the more usual notation, C_n(axis), to indicate a rotation through $2\pi/n$ about some particular axis.

Consider the electronic orbital $x\exp(-r/a)$ shown in projection on the xy-plane in Fig. 3.3(a). Under $C_4^{(A)}(001)$ the axes are rotated into the arrangement shown in Fig. 3.3(b). In terms of these axes the orbital is now described by the

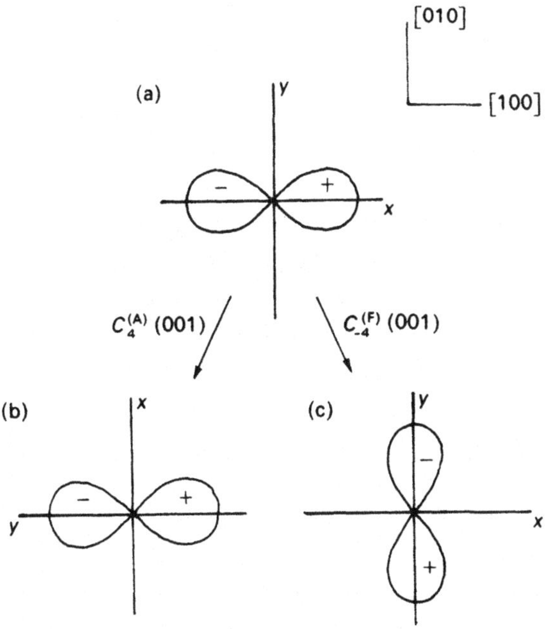

FIG. 3.3. The relationship between a rotation of axes and a rotation of function.

function $-y\exp(-r/a)$. Let us consider the inverse rotation, $C_{-4}^{(F)}(001)$, applied to the basic functions. This changes functions $x \to -y$, $y \to x$, $z \to z$, and when applied to the original orbital it changes it to $-y\exp(-r/a)$, as shown in Fig. 3.3(*c*). Both transformations yield the same result: $x\exp(-r/a) \to -y\exp(-r/a)$. We see that carrying out a rotation-of-axes transformation is equivalent to carrying out the inverse rotation-of-functions transformation.

In Fig. 3.4 we show some of the rotations of axes which leave invariant the octahedral arrangement of ions shown in the figure. This is called an octahedral arrangement. These rotations also leave invariant the cube and the octahedron shown in the figure. The six C_4 rotations are through $\pm\frac{1}{2}\pi$ about each of the $\langle 001 \rangle$-type directions. The six rotations of π about the $\langle 110 \rangle$-type directions are labelled C_2' to distinguish them from the π rotations about the $\langle 100 \rangle$-type directions. There are also eight C_3 rotations about the $\langle 111 \rangle$-type directions. These 23 rotations plus the identity transformation, E (which we can think of as a rotation through 2π) constitute a mathematical group—the *octahedral group* of rotations. This group is labelled O.

These rotations which are about axes through the origin are examples of *point transformations* in which a point (the origin) is fixed. As noted above there are three kinds of point transformation:

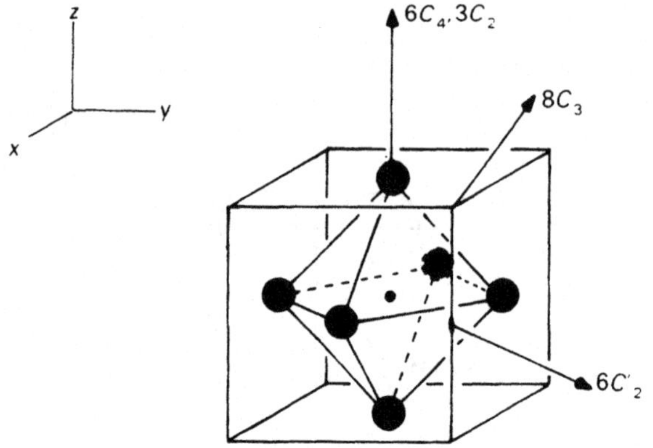

FIG. 3.4. The arrangement of six equally-charged ions, each in the centre of a cube face, is called *an octahedral arrangement*. (We obtain the octahedron by drawing lines between adjacent ions, as shown in the figure.) The rotations which leave invariant the cube, octahedron, and octahedral arrangement of ions are indicated. The six C_4 rotations are about each of the six ⟨100⟩-type directions. Only one C_4 direction is shown in the figure—that along the $+z$-direction. The three C_2 rotations are about [100], [010], [001] directions—the cubic axes. The eight C_3 rotations are about the eight ⟨111⟩-type directions. The six C_2' rotations are about ⟨110⟩-type directions, the primes serving to distinguish these from the C_2 rotations along the three cubic axes.

1. rotation about an axis through the origin,
2. reflection in a plane containing the origin,
3. inversion i $(\boldsymbol{r} \rightarrow -\boldsymbol{r})$.

We see that in addition to the group of 24 rotations already listed there are a number of reflection transformations which leave the octahedron invariant, and the octahedron is also invariant under inversion. Each of these symmetry reflections can, however, also be achieved by applying both a symmetry rotation and an inversion to the octahedron. The group of rotations, O, plus the inversion, i, covers all the point transformations which leave the octahedron invariant. This larger group is labelled O_h and one writes $O_h = O \times i$. O_h contains 48 elements. These are the 24 rotation operators of the O group, and the 24 rotation-plus-inversion operators.

The symmetry properties of the arrangement of ions about the electronic centre and hence the symmetry properties of the Hamiltonian of the electronic centre can best be exploited by applying the arguments and analyses of group theory, and the time spent in acquiring an understanding of the formal methods of group theory and in learning how they are applied to our quantum mechanics problems is well worth while. One very surprising result: despite the very large numbers of molecules and solids which exist in nature, there is only a small number of symmetry types. Once we have deduced the symmetry properties of the system then a brief consultation of a short character table reveals an amount of detail about its quantum mechanical behaviour. By symmetry considerations alone we can:

1. Deduce angular dependences of the system's electronic and vibrational wavefunctions
2. State the possible degeneracies of the energy levels of the system
3. Deduce whether or not static perturbations remove these degeneracies
4. Deduce the selection rules which govern transitions between states.

Thus group theory reveals much about the behaviour of a physical system even before we solve the equations of motion for the system. The amount of labour needed to complete quantum mechanical calculations is thereby reduced enormously. In the next section we introduce some basic definitions in group theory, stating without proof some fundamental theorems, and using as examples groups of point transformation operators. Our aim is to demonstrate how and why group-theoretical ideas facilitate the quantum mechanical analysis of electronic systems in crystals. For a formal and rigorous discussion of group theory, especially as it applies to solid state physics, the reader can consult, for example, the texts by Tinkham (1964), Heine (1960), and Elliott and Dawber (1979).

3.1.3 Properties of groups and sets

A set of entities is only a collection if no relationship among the entities is implied. The elements of a set which bear a relationship with each other constitute a *group*. Any set may be divided into a number of subsets each comprising a fraction of the elements in the set.

By a group is meant a set of elements

$$E, A, B, C, \ldots, R, \ldots$$

which can be 'multiplied' together under the following rules:

1. The set is closed under group multiplication. If A and B are elements in the set, then the product AB is also a member of the set. Note that A and B are an ordered pair since BA does not necessarily yield the same element as AB.
2. The associative law holds: $A(BC)=(AB)C$.
3. There is an unit element E: $ER=RE=R$ for every element in the set.
4. For every element A there is an inverse element A^{-1} which is also an element in the set. The inverse element has the property $AA^{-1}=A^{-1}A=E$.

We are mainly concerned here with groups of point transformations applied to axes or functions. We assume that the group has a finite number of elements. This number is called the *order* of the group and is labelled h.

As an example of (1) we consider the octahedral group of rotations (O) and apply two rotation transformations $C_4(001)\,C_4(100)$, where the order means that $C_4(100)$ is applied first. This product operator is equivalent to a single rotation transformation $C_3(111)$, as Fig. 3.5 shows. Rule (4) says that for each rotation-of-axes operator in the group, the inverse rotation of axes is also in the group. We recall now that carrying out a rotation-of-axes transformation is equivalent to carrying out the inverse rotation to the functions. Hence we see that there is a one-to-one correspondence between the group of rotation-of-axes operators which leave the arrangement of ions invariant and the group of rotation-of-functions operators which leave invariant the Hamiltonian of the electronic centre. If a rotation C_n applied to the axes leaves the arrangement of ions invariant, the same rotation C_n applied to functions leaves the Hamiltonian invariant.

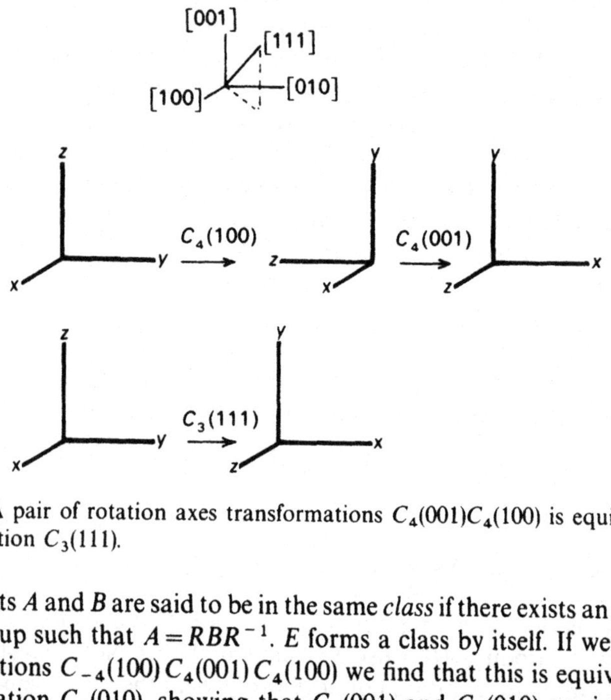

FIG. 3.5. A pair of rotation axes transformations $C_4(001)C_4(100)$ is equivalent to a single rotation $C_3(111)$.

Elements A and B are said to be in the same *class* if there exists an element R of the group such that $A = RBR^{-1}$. E forms a class by itself. If we apply the three rotations $C_{-4}(100)\,C_4(001)\,C_4(100)$ we find that this is equivalent to a single rotation $C_4(010)$, showing that $C_4(001)$ and $C_4(010)$ are in the same class. It is not difficult to see that in the octahedral group of rotations the six C_4 elements form a single class, the eight C_3 elements form a class, the six C_2' elements form a class, and the three C_2 elements form a class. Together with the E element, which forms a class by itself, there are five classes in the octahedral group.

3.1.4 Group representations.

For our purposes we define a *representation* of a group as a set of square matrices in a one-to-one correspondence with the original group. The square matrix corresponding to a group element R is written $D(R)$. $D(E)$ is the unit matrix. The group multiplication appropriate to the group of representation matrices is ordinary matrix multiplication.

If we have one such set of matrices, $D(R)$, then we can generate a second set $D'(R)$, by applying the same similarity transformation to each member $D(R)$:

$$D'(R)=U^{-1}D(R)U. \tag{3.4}$$

Since a similarity transformation leaves matrix equations unchanged the set $D'(R)$ is a group in an one-to-one correspondence with the group $D(R)$.

An unitary matrix U has the property $U^{-1}=U^{\dagger}=\tilde{U}^{*}$ where †, ~, * indicate adjoint, transpose, and complex conjugate, respectively. In terms of elements of the matrix this property can be written

$$(U^{-1})_{ij}=(U^{\dagger})_{ij}=(U^{*})_{ji}. \tag{3.5}$$

A transformation of one set of orthonormal basis functions spanning a function space into another set of orthonormal basis functions spanning the same space is an unitary transformation and is represented by an unitary matrix.

All groups of representation matrices $D'(R)$ which can be obtained from another group $D(R)$ by means of a similarity transformation are said to be *equivalent*. We do not distinguish between equivalent representations: We are concerned with enumerating the inequivalent representations.

One method of creating a representation of a group of operators which is particularly relevant for our purposes is as follows. We find a space of functions which is closed under the group of operators. Let ϕ_i be one of a set of orthonormal basis functions for this space. If we operate ϕ_i with R, one of the operators of the group, we obtain another function also in the space. We can write

$$R\phi_i=\sum_j \phi_j D_{ji}(R) \tag{3.6}$$

where $D_{ji}(R)$ are the components of a square matrix. The set of matrices $D(R)$ is a representation of the group of operators.

As an example consider $C_4(001)$ which transforms the basic functions $(x\ y\ z)$ into $(y\ -x\ z)$. This can be written

$$C_4(001)(x\ y\ z)=(y\ -x\ z)=(x\ y\ z)\begin{bmatrix} 0 & -1 & 0 \\ 1 & 0 & 0 \\ 0 & 0 & 1 \end{bmatrix}. \tag{3.7}$$

We can regard x, y, z as basis functions for a three-dimensional function space which is invariant under the O group of operators. The representation matrix $D(C_4(001))$ is the 3×3 matrix shown in eqn (3.7). Similarly, the representation matrix $D(C_3(111))$ using the same basis functions is

$$\begin{bmatrix} 0 & 0 & 1 \\ 1 & 0 & 0 \\ 0 & 1 & 0 \end{bmatrix}.$$

In another example we consider the five-dimensional space with the spherical harmonics, $Y_2^m(\theta, \phi)$ as functions; this space is closed under O. With these basis functions we can generate a representation in 5×5 matrices. Ordering the basis functions as Y_2^2, Y_2^1, Y_2^0, Y_2^{-1}, Y_2^{-2} we find that the representation matrix for $C_3(111)$ is

$$D(C_3(111)) = \begin{bmatrix} -\frac{1}{4} & \frac{i}{2} & \sqrt{\frac{3}{8}} & -\frac{i}{2} & -\frac{1}{4} \\ -\frac{1}{2} & \frac{i}{2} & 0 & \frac{i}{2} & \frac{1}{2} \\ -\sqrt{\frac{3}{8}} & 0 & -\frac{1}{2} & 0 & -\sqrt{\frac{3}{8}} \\ -\frac{1}{2} & -\frac{i}{2} & 0 & -\frac{i}{2} & \frac{1}{2} \\ -\frac{1}{4} & -\frac{i}{2} & \sqrt{\frac{3}{8}} & \frac{i}{2} & -\frac{1}{4} \end{bmatrix} \tag{3.8}$$

Referring again to eqn (3.6) we can generate another representation by choosing a different set of orthonormal basis functions ψ_i where

$$\psi_i = \sum_j \phi_j U_{ji} \tag{3.9}$$

and we also have the inverse transformation

$$\phi_k = \sum_l \psi_l (U^{-1})_{lk}. \tag{3.10}$$

The new representation matrices are $D'(R)$ where

$$R\psi_i = \sum_l \psi_l D'_{li}(R). \tag{3.11}$$

To see that the group $D'(R)$ is equivalent to $D(R)$ we write

$$\begin{aligned} R\psi_i &= R\sum_j \phi_j U_{ji} \\ &= \sum_j R\phi_j U_{ji} \\ &= \sum_{j,k} \phi_k D_{kj}(R) U_{ji} \\ &= \sum_{j,k,l} \psi_l (U^{-1})_{lk} D_{kj}(R) U_{ji} \\ &= \sum_l \psi_l \left(\sum_{j,k} (U^{-1})_{lk} D_{kj}(R) U_{ji} \right). \end{aligned} \tag{3.12}$$

Comparing eqns (3.12) and (3.11) we see that

$$D'_{li}(R) = \sum_{j,k} (U^{-1})_{lk} D_{kj}(R) U_{ji},$$

or in matrix form

$$D'(R) = U^{-1} D(R) U \tag{3.13}$$

showing that $D'(R)$ and $D(R)$ are equivalent.

Since we will be concerned only with operators which preserve the norm of a function and since we shall generally work with orthonormal basis functions the representation matrices we deal with will be unitary.

If a representation has the form

$$\begin{pmatrix} 1 & & & 0 \\ & 1 & & \\ & & 1 & \\ & & & \ddots \\ 0 & & & & 1 \end{pmatrix}, \left(\begin{array}{c|c} D^{(1)}(A) & 0 \\ \hline 0 & D^{(2)}(A) \end{array}\right), \ldots,$$

$$\left(\begin{array}{c|c} D^{(1)}(R) & 0 \\ \hline 0 & D^{(2)}(R) \end{array}\right), \ldots,$$

where all the $D^{(1)}$ matrices have the same dimension (and consequently the $D^{(2)}$ matrices have the same dimension) the representation is said to be *reducible* into $D^{(1)}$ and $D^{(2)}$ representations. Sometimes a representation may not appear to have the above form but by means of a similarity transformation it can be converted into such a form. This representation is also regarded as reducible, since we regard as equivalent those representations whose matrices are related to each other by means of a similarity transformation. It may happen that the smaller matrix, $D^{(1)}(R)$, can be further reduced by means of another similarity transformation. If this cannot be done the $D^{(1)}$ representation is said to be *irreducible*.

As an example we can choose $\phi_{eu}, \phi_{ev}, \phi_{t_2\xi}, \phi_{t_2\eta}, \phi_{t_2\zeta}$ (defined in eqns (2.91) and (2.92)) as the basis functions for the five-dimensional space discussed in the previous example, and we can form a set of representation matrices $D'(R)$ for the operators in the group O. With the basis functions ordered as above we find, for example,

$$D'(C_3(111)) = \begin{bmatrix} -\frac{1}{2} & -\frac{\sqrt{3}}{2} & 0 & 0 & 0 \\ \frac{\sqrt{3}}{2} & -\frac{1}{2} & 0 & 0 & 0 \\ 0 & 0 & 0 & 0 & 1 \\ 0 & 0 & 1 & 0 & 0 \\ 0 & 0 & 0 & 1 & 0 \end{bmatrix}, \tag{3.14}$$

i.e. consisting of two diagonal submatrices. This means that the ϕ_{eu}, ϕ_{ev} subspace is closed under $C_3(111)$ as is the subspace $\phi_{t_2\xi}, \phi_{t_2\eta}, \phi_{t_2\zeta}$. Using the same five basis functions the representation matrix for each of the 24 operators of the O group has the diagonal form shown in eqn (3.14). Hence the original five-dimensional representation is reducible.

This should not be surprising since, as we saw in the previous chapter, these five functions are the angular functions of the orbital wavefunctions for the d-electron in an octahedral crystal field, and we saw that the ϕ_{eu}, ϕ_{ev} wavefunctions differ in energy from the $\phi_{t_2\xi}$, $\phi_{t_2\eta}$, $\phi_{t_2\zeta}$ wavefunctions. Now each of the O operators commutes with the Hamiltonian for the electron in the octahedral field, and so there are no matrix elements of the O operators between states of different energy.

The relationship between the Y_2^m and the $\phi_{\Gamma\gamma}$ basis functions can be written

$$\phi_{\Gamma\gamma}=\sum_m Y_2^m \langle 2m|\Gamma\gamma\rangle \tag{3.15}$$

where the $\langle 2m|\Gamma\gamma\rangle$ coefficients are the components of an unitary matrix U, and the components of U can be obtained from eqns (2.91) and (2.92). The U^{-1} matrix can then be written down, and it is a straightforward matter to show that

$$U^{-1}D(C_3(111))\,U=D'(C_3(111)) \tag{3.16}$$

where the D and D' matrices are given in eqns (3.8) and (3.14).

We shall see that for groups of finite order the number of irreducible representations is finite. The symbols $\Gamma_1, \Gamma_2, \ldots$ are sometimes used to label the different irreducible representations. We denote by $D^{\Gamma}(R)_{\mu\nu}$ the $\mu\nu$ component of the representation matrix for the operator R in the Γ irreducible representation of the group of operators. We state below, without proof, some properties of representations. The proofs are given in the texts listed earlier.

1. *The great orthogonality theorem*

$$\sum_R D^{\Gamma}(R)^*_{\mu\nu} D^{\Gamma'}(R)_{\mu'\nu'} = \frac{h}{l_\Gamma}\,\delta_{\Gamma\Gamma'}\,\delta_{\mu\mu'}\,\delta_{\nu\nu'} \tag{3.17}$$

where l_Γ is the dimensionality of the Γ irreducible representation and h is the order of the group.

2. *Burnside's theorem.* If there are p inequivalent irreducible representations then

$$\sum_\Gamma l_\Gamma^2 = h \tag{3.18}$$

where the summation is over all p representations.

3. *The number of inequivalent irreducible representations equals the number of classes.* For example, the O group has five classes, and so it has five inequivalent irreducible representations. Generally the various irreducible representations are labelled Γ_1 to Γ_5 but in the case of the octahedral group it is more usual to label them A_1, A_2, E, T_1, T_2.

The character of a representation, D, whether reducible or irreducible, is the set of h numbers which are the traces of the representation matrices

$$\chi(E),\ \chi(A),\ \chi(B),\ \ldots,\ \chi(R),\ \ldots$$

where

$$\chi(R)=\sum_{\mu} D(R)_{\mu\mu}=\operatorname{tr} D(R). \tag{3.19}$$

Since the representation matrices are unitary none of the character elements, $\chi(R)$, is larger than $\chi(E)$, and since the trace of a matrix is invariant under a similarity transformation all equivalent representations have the same character. Further, elements in the same class, A and B say, are related by $B=RAR^{-1}$ where R is an element in the group. The corresponding relationship among the representation matrices is $D(B)=D(R)\,D(A)\,D(R^{-1})$. Since we are dealing only with representation matrices which are unitary we see that the representations of elements in the same class are related by a similarity transformation and so have identical traces. Hence we see that all group elements which form a class have identical character elements.

If we put $\nu=\mu$ and $\nu'=\mu'$ in the great orthogonality theorem, eqn (3.17), and sum over values of μ, μ', we obtain

$$\sum_{R} \chi^{\Gamma}(R)\chi^{\Gamma'}(R)^{*}=h\delta_{\Gamma\Gamma'}. \tag{3.20}$$

We can interpret this equation as follows. We regard the $\chi(R)$ elements as constituting the elements of an h-dimensional vector. Then eqn (3.20) can be viewed as the orthogonality property of these vectors. It follows that two different irreducible representations cannot have the same character: *the character of a representation is unique.*

We can use this property of the character to determine the decomposition of a reducible representation. Suppose a representation, D, is reducible. Each $D(R)$ can be converted, by a similarity transformation, to the diagonal form

$$\left(\begin{array}{c:c:c} D^{\Gamma_1}(R) & 0 & 0 \\ \hdashline 0 & D^{\Gamma_2}(R) & 0 \\ \hdashline 0 & 0 & \\ & & \end{array}\right)$$

where Γ_1, Γ_2, . . . refer to irreducible representations. Since the trace of a matrix is unaffected by a similarity transformation this means that

$$\operatorname{tr} D(R)=\operatorname{tr} D^{\Gamma_1}(R)+\operatorname{tr} D^{\Gamma_2}(R)+\ \ldots \tag{3.21}$$

or

$$\chi(R)=\sum_{\Gamma} a_{\Gamma}\,\chi^{\Gamma}(R) \tag{3.22}$$

where a_{Γ} is the number of times that the Γ irreducible representation appears in D. If we multiply $\chi(R)$ by $\chi^{\Gamma'}(R)^{*}$ and sum over all R we find

$$\sum_R \chi(R)\chi^{\Gamma'}(R)^* = \sum_\Gamma \sum_R a_\Gamma \chi^\Gamma(R)\chi^{\Gamma'}(R)^*$$

$$= \sum_\Gamma a_\Gamma h\, \delta_{\Gamma\Gamma'} \tag{3.23}$$

making use of eqn (3.20). Hence we see that

$$a_\Gamma = \frac{1}{h} \sum_R \chi(R)\chi^\Gamma(R)^*. \tag{3.24}$$

Since the character of an irreducible representation is unique, a_Γ is unique. There is only one decomposition of a reducible representation, and the number of times that the irreducible representation Γ appears in the decomposition is given by a_Γ above.

A *character table* displays the characters of the different irreducible representations of a given group. Since group elements in the same class have the same character element the elements are grouped into classes. The character table for the O group is shown in Table 3.1. There are five classes, hence there are five inequivalent irreducible representations which are usually labelled A_1, A_2, E, T_1, T_2. In general labels A and B are used for one-dimensional representations, E for two-dimensional representations, and T for three-dimensional representations. A_1 is the invariant representation in which each character element has the value unity. All groups possess the invariant representation. In the extreme right-hand column, Table 3.1 assigns to each irreducible representation basis functions (unnormalized) for a space which is closed under all the operators of the group O, and these can be used as basis functions to create the matrices of the irreducible representation. If the orthonormal set ϕ_i is one such set of basis functions then

$$R\phi_i = \sum_j \phi_j D^\Gamma_{ji}(R) \tag{3.25}$$

Table 3.1

Character table of the O group†

	E	$8C_3$	$3C_2$	$6C_2'$	$6C_4$	
A_1	1	1	1	1	1	$x^2+y^2+z^2$
A_2	1	1	1	-1	-1	xyz
E	2	-1	2	0	0	$(x^2-y^2, 3z^2-r^2)$
T_1	3	0	-1	-1	1	$(x, y, z); (R_x, R_y, R_z)$
T_2	3	0	-1	1	-1	(yz, zx, xy)

† In the right-hand column are listed some basis functions (unnormalized) for a space which is closed under all the operations of the O group, and these can be used as basis functions to create the matrices of the various irreducible representations. R_x is the infinitesimal rotation about the x-axis.

where the $D^{\Gamma}(R)$ are the matrices for the Γ irreducible representation. We say that the functions ϕ_i belong to (or transform according to) the Γ irreducible representation. And we say that the function ϕ_i transforms according to the ith column of the Γ irreducible representation. These functions are more usefully written $\phi(\Gamma, i)$.

As an example of how a character table is created we shall show how the components of the character table of the O group are determined. If l_i is the dimensionality of the ith irreducible representation then by Burnside's theorem we have $\sum_{i=1}^{5} l_i^2 = 24$. Since $l_1 = 1$ we are left with $l_2^2 + l_3^2 + l_4^2 + l_5^2 = 23$. The only solution for this is $l_2 = 1$, $l_3 = 2$, $l_4 = l_5 = 3$, in ascending order of dimensionality. We consider the character elements for the other one-dimensional representation which is labelled A_2. We use the orthogonality property, eqn (3.20), remembering that $\chi^{A_2}(E) = 1$ and that $\chi(R)$ for other group elements is not larger than $\chi(E)$. We obtain

$$1 + 8\chi^{A_2}(C_3) + 3\chi^{A_2}(C_2) + 6\chi^{A_2}(C_2') + 6\chi^{A_2}(C_4) = 0 \tag{3.26}$$

which gives for the character elements $\chi^{A_2}(R)$ the set 1, 1, 1, -1, -1, and this is the only solution. We next consider the character χ^{E}. This is orthogonal to χ^{A_1} and χ^{A_2}. Since χ^{A_1} and χ^{A_2} differ only in the character elements of the last two classes, we find $\chi^{E}(C_2') = -\chi^{E}(C_4)$. And $\chi^{E}(E) = 2$. We can now try out some possible values for the character elements, $\chi^{E}(R)$. We eventually find that the set 2, -1, 2, 0, 0 is orthogonal to χ^{A_1} and χ^{A_2} and correctly normalized according to eqn (3.20). By the uniqueness property of the characters this must be the set of character elements for χ^{E}. The same approach using the orthogonality with the χ^{A_1}, χ^{A_2}, χ^{E} characters can be used to locate values of the character elements of the two three-dimensional representations. The character table of the O group is given in Table 3.1. Character tables of all the common groups of point transformations have been worked out and are listed in the textbooks cited in Section 3.1.2.

From Table 3.1 we see that the set of functions x, y, z transforms according to the T_1 representation. These functions, if properly normalized, can be classified as $\phi(T_1, i)$ where $i = 1, 2, 3$. The set of functions $x(5x^2 - 3r^2)$, $y(5y^2 - 3r^2)$, $z(5z^2 - 3r^2)$ also transforms according to the T_1 representation. These two sets transform identically under all the operations of the O group but they form distinct function spaces. In order to distinguish the different function spaces which belong to the same irreducible representation an additional label such as α, β, ... may be used, in which case the functions can be classified $\phi(\alpha, \Gamma, i)$.

In Fig. 3.6 we show the symmetry rotation operators of the tetragonal group D_4. These are the operators which leave invariant the tetragonal prism as well as the arrangement of six ions shown in Fig. 3.6. This is a distortion of the octahedral arrangement shown in Fig. 3.4; the separation from the origin

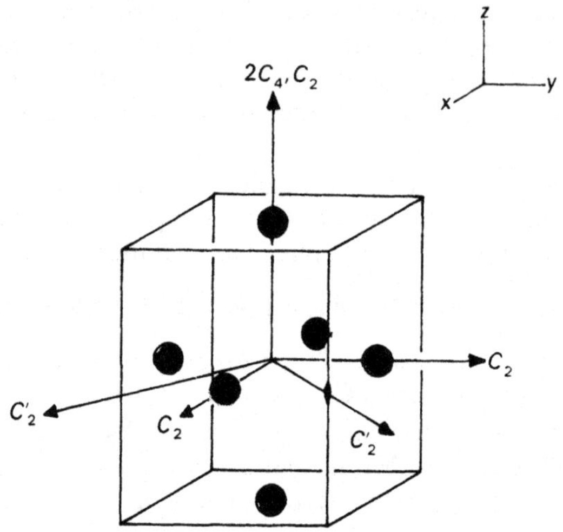

FIG. 3.6. The symmetry rotations of the D_4 group.

of the two ions along the z-axis are equal but differ from the equal separations of the four ions along the x- and y-axes.

We notice that this set of operators is a reduced set of eight of the operators of the O group; the set of D_4 rotations is a subgroup of the O group. Further, whereas in the O group $C_2(001)$ is in the same class with $C_2(100)$ and $C_2(010)$, in the D_4 group $C_2(001)$ forms a class by itself. There are five classes: E; $C_2(001)$; $C_2(100)$, $C_2(010)$; $C_2'(110)$, $C_2'(1\bar{1}0)$; $C_4(001)$, $C_4(00\bar{1})$, and so there are five irreducible representations. From Burnside's theorem we have $\sum_i l_i^2 = 8$, whose only solution is the set 1, 1, 1, 1, 2. There are four one-dimensional and one two-dimensional irreducible representations. The character table is given in Table 3.2 which also shows some function spaces belonging to the different irreducible representations.

Table 3.2

Character table of the D_4 group

	E	C_2^z	$2C_2^{x,y}$	$2C_2'$	$2C_4$	
A_1	1	1	1	1	1	(x^2+y^2); z^2
A_2	1	1	-1	-1	1	z; R_z
B_1	1	1	1	-1	-1	x^2-y^2
B_2	1	1	-1	1	-1	xy
E	2	-2	0	0	0	(x, y) (xz, yz); (R_x, R_y)

We now consider the symmetry rotation operators of the trigonal group, D_3, which we can again usefully relate to a distortion of the octahedral arrangement of ions seen in Fig. 3.4. We show this octahedral arrangement of six ions again in Fig. 3.7 along with some of the symmetry operators and emphasizing that we can regard the ions as forming two triangles in parallel planes perpendicular to the [111] direction. If this arrangement is distorted by displacing the upper and lower triangles through equal distances along the [111] and $[\bar{1}\bar{1}\bar{1}]$ directions, respectively, the arrangement has trigonal symmetry (D_3) with the symmetry operators shown in the figure: C_3 and C_3^2 along [111], and C_2 along $[1\bar{1}0]$, $[\bar{1}01]$, $[01\bar{1}]$. These, along with the identity operator E, form the six rotation operators of the D_3 group, and are a subgroup of O. There are three classes: E; $2C_3$; $3C_2'$. The D_3 character table is given in Table 3.3. The D_3 character table is identical with that of another group C_{3v}, which we will discuss later. Table 3.3 shows the irreducible representations and character elements for both groups, as well as function spaces belonging to the different irreducible representations.

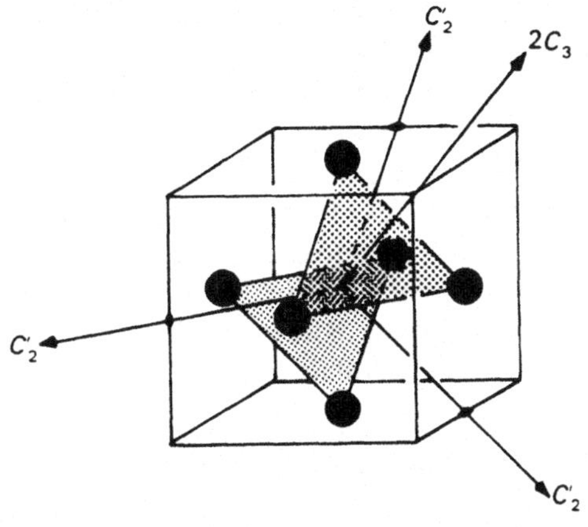

FIG. 3.7. The octahedral arrangement of ions showing some symmetry rotation operators and emphasizing the threefold symmetry about the [111] axis. If the upper and lower triangles of ions are displaced through equal distances along the [111] and $[\bar{1}\bar{1}\bar{1}]$ directions, respectively, the arrangement remains invariant under the operators shown. The arrangement now has D_3 trigonal symmetry and can be regarded as a trigonally-distorted octahedral arrangement, such as was discussed in Chapter 2 and illustrated in Fig. 2.17.

3.1.5 Group theory and quantum mechanics II

We now relate the group-theoretical concepts we have developed to our basic quantum mechanical problem of solving for the eigenstates and eigenvalues of the Hamiltonian for the electronic centre in a crystal. Consider the group $\mathcal{G}$ of

Table 3.3
Character table of the D_3 and C_{3v} groups

D_3	E	$2C_3$	$3C_2'$	D_3	
C_{3v}	E	$2C_3$	$3\sigma_v$		C_{3v}
A_1	1	1	1	$(x^2+y^2); (z^2)$	$(x^2+y^2); (z^2); (z)$
A_2	1	1	-1	$R_z; z$	R_z
E	2	-1	0	$(x, y); (xz, yz); (x^2-y^2, xy)$ (R_x, R_y)	$(x, y); (xz, yz); (x^2-y^2, xy)$ (R_x, R_y)

operators each of which commutes with $\mathscr{H}$. If $\mathscr{G}$ includes all the symmetry operators which commute with $\mathscr{H}$ it is called *the symmetry group of the Hamiltonian.* If ψ is an eigenstate of $\mathscr{H}$ with energy E and if R is one of the operators of the group $\mathscr{G}$, then $R\psi$ is also an eigenstate of $\mathscr{H}$ with the same energy E. The function space containing all the eigenstates of $\mathscr{H}$ of the same energy is closed under $\mathscr{G}$, and we can use these eigenstates as basis functions to form a representation of $\mathscr{G}$. We now ask whether this representation is reducible or irreducible. If it is reducible it means that the space can be separated into at least two subspaces each of which is separately closed under $\mathscr{G}$. There is no symmetry element connecting these spaces, yet they are associated with the same energy. It seems unlikely that the two spaces of wavefunctions which are unrelated to each other by any symmetry property of the system should have the same energy. We feel, therefore, that the original space of wavefunctions cannot be associated with a reducible representation. We adopt as a fundamental postulate that

eigenstates of the Hamiltonian with the same energy belong to an irreducible representation of the symmetry group of $\mathscr{G}$ and eigenstates which belong to inequivalent irreducible representations have different energies.

This means that we can label the energy levels of the system according to the irreducible representations of the symmetry group of the Hamiltonian to which the space of wavefunctions of each energy level belongs.

As an illustration of these ideas consider the wavefunctions for the 3d electron in the octahedral crystal field discussed in Section 2.6.1. In the absence of an octahedral crystal field the 3d level has fivefold orbital degeneracy. In the presence of an octahedral crystal field the wavefunctions for each energy level form a function space which belongs to an irreducible representation of the O group. The largest dimension for an irreducible representation of the O group is 3, hence the largest orbital degeneracy of a level is 3. As a result, in the presence of an octahedral field the fivefold degenerate 3d space is expected to split up into smaller subspaces each of which belongs to an irreducible representation of the O group, and the wavefunctions in the different subspaces have different energies. Our calculations in Section 2.6.1 showed that the 3d energy level splits into two distinct energy levels—one two-dimensional, one three-dimensional. The wavefunctions for the two-dimensional level are given in eqn (2.91), and we see from Table 3.1 that these belong to the E irreducible representation. We use the symbol E to characterize that level. Similarly, the wavefunctions of the three-dimensional level belong to the T_2 irreducible representation; hence the T_2 label characterizes this level. Lower-case letters, e and t_2, are used to characterize the one-electron orbital functions belonging to the E and T_2 irreducible representations, respectively.

When calculating energy levels one may come across 'accidental' degeneracies, where two spaces of wavefunctions each belonging to different irreducible representations have exactly the same energy. The origin of the 'accidental' degeneracy generally lies in some symmetry property of $\mathcal{H}$ having been overlooked, and this symmetry property relates the two spaces of wavefunctions.

3.1.6 Product spaces and product representations.

We consider the set of orthonormal functions $\phi(\Gamma_1, \gamma_1)$ which belong to the Γ_1 irreducible representation of some group G, and we consider a second set $\phi(\Gamma_2, \gamma_2)$ which belongs to the Γ_2 irreducible representation of G. Representation matrices are defined from the equations

$$R\phi(\Gamma_1, \gamma_1') = \sum_{\gamma_1} \phi(\Gamma_1, \gamma_1) D^{\Gamma_1}_{\gamma_1\gamma_1'}(R),$$

$$R\phi(\Gamma_2, \gamma_2') = \sum_{\gamma_2} \phi(\Gamma_2, \gamma_2) D^{\Gamma_2}_{\gamma_2\gamma_2'}(R). \tag{3.27}$$

The product functions $\phi(\Gamma_1, \gamma_1)\, \phi(\Gamma_2, \gamma_2)$ also form a set of basis functions for a representation of G which we write as $D^{\Gamma_1 \times \Gamma_2}$, defined by

$$R\phi(\Gamma_1, \gamma_1')\phi(\Gamma_2, \gamma_2') = \sum_{\gamma_1, \gamma_2} \phi(\Gamma_1, \gamma_1)\phi(\Gamma_2, \gamma_2) D^{\Gamma_1}_{\gamma_1\gamma_1'}(R) D^{\Gamma_2}_{\gamma_2\gamma_2'}(R)$$

$$= \sum_{\gamma_1, \gamma_2} \phi(\Gamma_1, \gamma_1)\phi(\Gamma_2, \gamma_2) D^{\Gamma_1 \times \Gamma_2}_{\gamma_1\gamma_2, \gamma_1'\gamma_2'}(R). \tag{3.28}$$

If the Γ_1 and Γ_2 irreducible representations are of dimensionality l_{Γ_1} and l_{Γ_2}, respectively, the product representation has dimensionality $l_{\Gamma_1} \times l_{\Gamma_2}$.

The character of the product representation can be calculated:

$$\chi^{\Gamma_1 \times \Gamma_2}(R) = \sum_{\gamma_1, \gamma_2} D^{\Gamma_1 \times \Gamma_2}_{\gamma_1\gamma_2, \gamma_1\gamma_2}(R)$$

$$= \sum_{\gamma_1} D^{\Gamma_1}_{\gamma_1\gamma_1}(R) \sum_{\gamma_2} D^{\Gamma_2}_{\gamma_2\gamma_2}(R)$$

$$= \chi^{\Gamma_1}(R) \,.\, \chi^{\Gamma_2}(R). \tag{3.29}$$

In general the $\Gamma_1 \times \Gamma_2$ representation is reducible. Its decomposition is expressed by the symbolic formula

$$D^{\Gamma_1 \times \Gamma_2}(R) = \sum_j a_j D^{\Gamma_j}(R) \tag{3.30}$$

where a_j is the number of times that the Γ_j irreducible representation appears in the decomposition. This decomposition can be expressed in another symbolic formula,

$$\Gamma_1 \times \Gamma_2 = \sum_j a_j \Gamma_j. \tag{3.31}$$

The decomposition formula in terms of characters is

$$\chi^{\Gamma_1 \times \Gamma_2}(R) = \chi^{\Gamma_1}(R) \times \chi^{\Gamma_2}(R) = \sum_j a_j \chi^{\Gamma_j}(R). \tag{3.32}$$

The a_j values are unique and are given by

$$a_j = \frac{1}{h} \sum_R \chi^{\Gamma_1 \times \Gamma_2}(R) \chi^{\Gamma_j}(R)^*. \tag{3.33}$$

As an example, consider the sets of functions x_1, y_1, z_1, and x_2, y_2, z_2 each of which belongs to the T_1 representation of the O group. The product functions form a nine-dimensional space and belong to a nine-dimensional representation. Since the largest irreducible representation of the O group is of dimension 3, the product representation is reducible. We obtain its decomposition from a study of its character. From Table 3.1 we see that the character $\chi^{T_1 \times T_2}$ which equals $\chi^{T_1} \times \chi^{T_2}$ is

$$\chi^{T_1 \times T_2} = 9 \quad 0 \quad 1 \quad 1 \quad 1 \tag{3.34}$$

which by inspection or through the use of eqn (3.33) we see is $\chi^{A_1} + \chi^{E} + \chi^{T_1} + \chi^{T_2}$. We write this decomposition as

$$T_1 \times T_1 = A_1 + E + T_1 + T_2. \tag{3.35}$$

We can similarly derive the following decompositions:

$$\begin{array}{llll} A_1 \times A_1 = A_1, & A_1 \times A_2 = A_2, & A_1 \times E = E, & A_1 \times T_1 = T_1 \\ A_1 \times T_2 = T_2, & A_2 \times A_2 = A_1, & A_2 \times E = E, & A_2 \times T_1 = T_2 \\ A_2 \times T_2 = T_1, & E \times E = A_1 + A_2 + E, & & E \times T_1 = T_1 + T_2, \\ E \times T_2 = T_1 + T_2, & T_1 \times T_2 = A_2 + E + T_1 + T_2, & & \\ & T_2 \times T_2 = A_1 + E + T_1 + T_2. & & \end{array} \tag{3.36}$$

Since the product representations are, in general, reducible it follows that the simple product functions are, in general, not the basis functions for irreducible representations. The basis functions are linear combinations of the simple products. If the product function space $\phi(\Gamma_1, \gamma_1)\phi(\Gamma_2, \gamma_2)$ contains a subspace which belongs to the Γ irreducible representation, then the function of the product space which transforms as the γ column of the Γ irreducible representation is written as

$$\phi(\Gamma, \gamma) = \sum_{\gamma_1, \gamma_2} \phi(\Gamma_1, \gamma_1)\phi(\Gamma_2, \gamma_2)\langle\Gamma_1\Gamma_2\gamma_1\gamma_2|\Gamma\gamma\rangle \tag{3.37}$$

where $\langle\Gamma_1\Gamma_2\gamma_1\gamma_2|\Gamma\gamma\rangle$ are *Clebsch–Gordan coefficients* and are the elements of an unitary matrix. The inverse formula is

$$\phi(\Gamma_1, \gamma_1)\phi(\Gamma_2, \gamma_2) = \sum_{\Gamma, \gamma} \phi(\Gamma, \gamma)\langle\Gamma\gamma|\Gamma_1\Gamma_2\gamma_1\gamma_2\rangle \tag{3.38}$$

where $\langle\Gamma\gamma|\Gamma_1\Gamma_2\gamma_1\gamma_2\rangle = \langle\Gamma_1\Gamma_2\gamma_1\gamma_2|\Gamma\gamma\rangle^*$. The calculation of Clebsch–Gordan coefficients is tedious but has been carried out for a number of groups. Sugano *et al.* (1970) tabulate the coefficients for the O group in the cubic basis and in the trigonal basis. The phases are chosen so that $\langle\Gamma_1\Gamma_2\gamma_1\gamma_2|\Gamma\gamma\rangle = \langle\Gamma_2\Gamma_1\gamma_2\gamma_1|\Gamma\gamma\rangle$ when $\Gamma_1 \neq \Gamma_2$. Tables of values of some Clebsch–Gordan coefficients for the O group in the cubic and trigonal bases are given in Appendix 3B.

3.1.7 Theorems on matrix elements

1. *Two functions which belong to different irreducible representations or which transform according to different columns of the same irreducible representations are orthogonal.*

Since the value of an integral is unchanged if the coordinates are changed by application of one of the operators, R, of the group, we have

$$\begin{aligned}\langle\phi(\alpha_1, \Gamma_1, \gamma_1)|\phi(\alpha_2, \Gamma_2, \gamma_2)\rangle &= R\langle\phi(\alpha_1, \Gamma_1, \gamma_1)|\phi(\alpha_2, \Gamma_2, \gamma_2)\rangle \\ &= \langle R\phi(\alpha_1, \Gamma_1, \gamma_1)|R\phi(\alpha_2, \Gamma_2, \gamma_2)\rangle \\ &= \sum_{\gamma_1', \gamma_2'} D^{\Gamma_1}_{\gamma_1'\gamma_1}(R)^* D^{\Gamma_2}_{\gamma_2'\gamma_2}(R)\phi(\alpha_1, \Gamma_1, \gamma_1')|\phi(\alpha_2, \Gamma_2, \gamma_2').\end{aligned} \tag{3.39}$$

Since this relationship holds for each of the h operators of the group, we have

$$\begin{aligned}\langle\phi(\alpha_1, \Gamma_1, \gamma_1)|\phi(\alpha_2, \Gamma_2, \gamma_2)\rangle = \frac{1}{h}\sum_{\gamma_1', \gamma_2'}\sum_R D^{\Gamma_1}_{\gamma_1'\gamma_1}(R)^* D^{\Gamma_2}_{\gamma_2'\gamma_2}(R) \\ \times \langle\phi(\alpha_1, \Gamma_1, \gamma_1')|\phi(\alpha_2, \Gamma_2, \gamma_2')\rangle.\end{aligned} \tag{3.40}$$

Using the great orthogonality theorem (3.17) this becomes

$$\sum_{\gamma_1'\gamma_2'} \frac{1}{l_{\Gamma_1}} \delta_{\Gamma_1\Gamma_2}\, \delta_{\gamma_1'\gamma_2'}\, \delta_{\gamma_1\gamma_2}\langle\phi(\alpha_1, \Gamma_1, \gamma_1')|\phi(\alpha_2, \Gamma_2, \gamma_2')\rangle \tag{3.41}$$

where l_Γ is the dimensionality of the Γ representation. The integral can therefore be written

$$\langle\phi(\alpha_1, \Gamma_1, \gamma_1)|\phi(\alpha_2, \Gamma_2, \gamma_2)\rangle = \delta_{\Gamma_1\Gamma_2}\,\delta_{\gamma_1\gamma_2}P(\alpha_1, \Gamma_1; \alpha_2, \Gamma_2) \tag{3.42}$$

where

$$P(\alpha_1, \Gamma_1; \alpha_2, \Gamma_2) = \frac{1}{l_\Gamma}\sum_\gamma \langle\phi(\alpha_1, \Gamma_1, \gamma)|\phi(\alpha_2, \Gamma_2, \gamma)\rangle. \tag{3.43}$$

Equation (3.42) tells us that the original integral is zero unless $\Gamma_1 = \Gamma_2$ and $\gamma_1 = \gamma_2$, which proves the theorem. In addition, since P in eqn (3.43) is the average over all such integrals for different γ it is independent of γ. Hence the value of the original integral is independent of the particular γ chosen.

2. *The Wigner–Eckart theorem* Consider the integral $\langle\phi(\alpha, \Gamma, \gamma)|X(\bar{\Gamma}, \bar{\gamma})|\phi(\alpha', \Gamma', \gamma')\rangle$ where $X(\bar{\Gamma}, \bar{\gamma})$ transforms as the $\bar{\gamma}$ row of the $\bar{\Gamma}$ irreducible representation. $X(\Gamma, \gamma)$ need not be normalized. To evaluate this integral we first express the product function $X(\bar{\Gamma}, \bar{\gamma})\phi(\alpha', \Gamma', \gamma')$ as a linear combination of single functions belonging to irreducible representations:

$$X(\bar{\Gamma}, \bar{\gamma})\phi(\alpha', \Gamma', \gamma') = \sum_{\Gamma'', \gamma''} C(X)\phi(\alpha'', \Gamma'', \gamma'')\langle\Gamma''\gamma''|\bar{\Gamma}\Gamma'\bar{\gamma}\gamma'\rangle \tag{3.44}$$

where $C(X)$ is a normalization factor. The integral then becomes

$$\sum_{\Gamma''\gamma''} C(X)\langle\phi(\alpha, \Gamma, \gamma)|\phi(\alpha'', \Gamma'', \gamma'')\rangle\langle\Gamma''\gamma''|\bar{\Gamma}\Gamma'\bar{\gamma}\gamma'\rangle. \tag{3.45}$$

We know from Theorem (i) that this expression is zero unless $\Gamma''\gamma'' = \Gamma\gamma$, and when this condition is satisfied the value of the integral does not depend on the particular γ. It depends on the values of α and α''. Now α'' depends on the particular functions $X(\bar{\Gamma})$ and $\phi(\alpha', \Gamma')$, and $C(X)$ depends only on the magnitude of the X function. Hence $C(X)\langle\phi(\alpha, \Gamma, \gamma)|\phi(\alpha'', \Gamma'', \gamma'')\rangle$ depends on α, α', Γ, Γ', $\bar{\Gamma}$ and not on the γ, γ'', or γ' values. We can write

$$C(X)\langle\phi(\alpha, \Gamma, \gamma)|\phi(\alpha', \Gamma', \gamma')\rangle = \frac{1}{(l_\Gamma)^{\frac{1}{2}}}\langle\phi(\alpha, \Gamma)\|X(\bar{\Gamma})\|\phi(\alpha', \Gamma')\rangle. \tag{3.46}$$

where the integral on the right-hand side is known as the *reduced matrix element*. The original integral thus becomes

$$\langle\phi(\alpha, \Gamma, \gamma)|X(\bar{\Gamma}, \bar{\gamma})|\phi(\alpha', \Gamma', \gamma')\rangle = \frac{1}{(l_\Gamma)^{\frac{1}{2}}}\langle\phi(\alpha, \Gamma)\|X(\bar{\Gamma})\|\phi(\alpha', \Gamma')\rangle\langle\Gamma\gamma|\bar{\Gamma}\Gamma'\bar{\gamma}\gamma'\rangle \tag{3.47}$$

and this is known as the *Wigner–Eckart theorem*. It shows that the matrix elements of $X(\bar{\Gamma}, \bar{\gamma})$ between the different states of the Γ and Γ' manifolds of states are in the ratio given by the appropriate Clebsch–Gordan coefficients. Once one such matrix element is evaluated the reduced matrix element can be obtained, and the other matrix elements are found by reference to the tables of Clebsch–Gordan coefficients.

Because of the Clebsch–Gordan coefficient in eqn (3.47) we see the integral $\langle\phi(\alpha, \Gamma, \gamma)|X(\bar{\Gamma}, \bar{\gamma})|\phi(\alpha', \Gamma', \gamma')\rangle$ is zero unless the product representation $\bar{\Gamma}\times\Gamma'$ contains Γ. If the function X belongs to the invariant representation, that is, if it is unchanged under all the operations of the group, then the function $X\phi(\alpha'\Gamma'\gamma')$ transforms according to the γ' row of the Γ' representation, and the integral will be zero unless $\Gamma\gamma = \Gamma'\gamma'$. This is just a restatement of Theorem 1. The electrostatic crystal field Hamiltonian of our electronic system is one such invariant function. Hence we see that there are no matrix elements between functions belonging to different irreducible representations; the only non-zero matrix elements are between functions which transform according to the same row of the same irreducible representation.

One way of seeking eigenstates of $\mathscr{H}$ is first to seek functions which belong to irreducible representations of $\mathscr{G}$. Say one such set of functions is $\phi(\alpha, \Gamma, \gamma)$. These are usually eigenstates of a similar but simpler Hamiltonian. In general these will not be eigenstates of $\mathscr{H}$ because non-zero matrix elements of $\mathscr{H}$ may exist connecting the set $\phi(\alpha, \Gamma, \gamma)$ and any other set of functions $\phi(\beta, \Gamma, \gamma)$ which belong to the same representation Γ. If, however, the separation in energy between the α and β functions is large the amount of mixing of the α and β functions through $\mathscr{H}$ may be small and the $\phi(\alpha, \Gamma, \gamma)$ set may to very good approximation be regarded as eigenstates of $\mathscr{H}$.

3.1.8 Reduction in symmetry and splitting of energy levels.

We now discuss an important consequence of our fundamental postulate that the eigenstates of a Hamiltonian belong to irreducible representations of the symmetry group of the Hamiltonian, and that eigenstates belonging to different irreducible representations have different energies.

Consider an ion in an octahedral crystal field. If we neglect spin and confine our attention to orbital functions the energy levels are labelled according to the irreducible representations of the O group, i.e. A_1, A_2, E, T_1, T_2. Since the T_1 and T_2 representations are three-dimensional the T_1 and T_2 levels have threefold orbital degeneracy. We now consider the same ion in a crystal field of tetragonal symmetry, D_4. The levels are labelled according to the irreducible representations of D_4, i.e. A_1, A_2, B_1, B_2, E; the maximum orbital degeneracy of a level is now twofold, and all of these levels will have different energies. In general we can see that, in comparison with the case of high symmetry, an ion in a crystal field of low symmetry will have a greater number of energy levels of smaller degeneracy.

We can try to see how these two sets of levels are related by assuming that the reduction in symmetry is due to a small lower symmetry perturbation on the higher symmetry potential. Let us assume that a small axial crystal field distortion reduces the symmetry of the crystal field Hamiltonian from octahedral, $\mathscr{H}_c^O$ to tetragonal, $\mathscr{H}_c^{D_4}$. The wavefunctions of a given energy level associated with, say, a T_1 irreducible representation of O form a function space which is invariant under all the rotations of the O group. The rotation operators of the D_4 group are a subset of the O operators, hence the T_1 space of functions is also closed under all the rotations of the D_4 group, and these functions can be used as basis functions for a three-dimensional representation of the D_4 group. This is a reducible representation since the maximum dimension of the irreducible representations of D_4 is 2. The three-dimensional T_1 space of functions breaks up into smaller spaces of functions each of which belongs to an irreducible representation of D_4. To the approximation that the axial component of crystal field is very weak we can ignore the mixing of states from the different octahedral crystal field levels by the axial crystal field and therefore the wavefunctions of $\mathscr{H}_c^O$ associated with the T_1 irreducible representation become the wavefunctions of $\mathscr{H}_c^{D_4}$ and are associated with more than one irreducible representation of D_4. According to our fundamental postulate, the wavefunctions associated with these irreducible wavefunctions of D_4 have distinct energies. Hence in the presence of the weak axial crystal field the T_1 energy level splits up.

The way in which an octahedral level splits up in the presence of a perturbation which lowers the symmetry to tetragonal can be determined from the decomposition of the irreducible representations of the octahedral group into a number of different irreducible representations of the tetragonal group. This decomposition can be ascertained from the characters of the representations using the decomposition formulae (3.22) and (3.24).

The 24 character elements of the T_1 octahedral representation grouped into the five different classes are given in Table 3.4. The subset of eight of these character elements corresponding to the eight rotations of the D_4 group is also given, and these are grouped into the five different classes of D_4. This set of elements constitutes the character of a three-dimensional representation of D_4. We call this the D representation, and it is reducible.

The decomposition of D into a sum of irreducible representations of D_4 can be performed with the aid of the D_4 character table (Table 3.2) and eqns (3.22) and (3.24), but in a simple case such as we are now analysing the decomposition can be determined by inspection. In Table 3.4 the characters of the A_2 and E irreducible representations of D_4 are also given, and we can see that $\chi^D = \chi^{A_2} + \chi^E$. Hence we see that when the crystal field symmetry is reduced from O to D_4 the octahedral T_1 level splits up into two levels associated with A_2 and E irreducible representations of D_4, and by our fundamental postulate these have different energy levels. The splitting of all energy levels when a crystal field is reduced from O to D_4 is shown schematically in Fig. 3.8.

We can similarly analyse the reduction in crystal field symmetry from O to D_3 due to the addition of a weak axial crystal field. The splittings of the

Table 3.4

O	E	$8C_3$	$3C_2$	$6C_2'$	$6C_4$
T_1	3	0	-1	-1	1
D_4	E	C_2^z	C_2^x, C_2^y	$2C_2'$	$2C_4$
D	3	-1	-1	-1	1
A_2	1	1	-1	-1	1
E	2	-2	0	0	0

D_3	E	$2C_3$	$3C_2'$
D	3	0	-1
A_2	1	1	-1
E	2	-1	0

octahedral energy levels can again be determined from the character Tables 3.1 and 3.3. The T_1 irreducible representation of the O group becomes the D reducible representation of D_3, as Table 3.4 shows. And this is reducible into A_2 and E irreducible representations of D_3. The splittings of all octahedral orbital states when a trigonal crystal field is applied are shown schematically in Fig. 3.8.

In discussing the splitting of octahedral levels we assume that the perturbing axial field is small, and as a result the octahedral eigenstates become the tetragonal or trigonal eigenstates. We now extend the analysis to the case where the axial field is not small. The symmetry arguments are still valid; the levels will be classified according to the irreducible representations of the lower symmetry group. Let us seek the eigenstates of the tetragonal A_1 and B_1 levels derived from the octahedral E level (Fig. 3.8). The two-dimensional space of wavefunctions associated with the octahedral E level can be resolved into two orthogonal functions each of which is closed under all the D_4 rotations, and these transform as the A_1 and B_1 irreducible representations of D_4. These need not be eigenstates of $\mathscr{H}_c^{D_4}$ since, as we showed in our discussion on the Wigner–Eckart theorem, there can be non-zero matrix elements of $\mathscr{H}_c^{D_4}$ between functions belonging to the same irreducible representations of D_4. This may cause a mixing of the tetragonal A_1 function

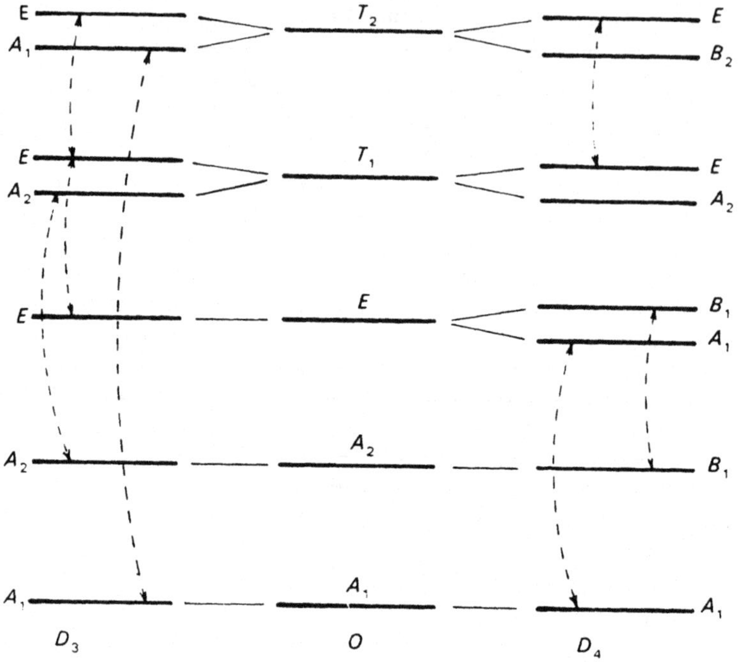

FIG. 3.8. Schematic illustration of the splitting of octahedral crystal field levels, based on orbital wavefunctions, when the crystal field symmetry is reduced from octahedral to tetragonal and from octahedral to trigonal. The broken arrows connecting the levels of the same irreducible representation indicate the states which can be mixed by the low-symmetry axial crystal field which causes the reduction in symmetry.

derived from the excited octahedral E level with the tetragonal A_1 function derived from the ground level. This mixing is indicated in Fig. 3.8 by the broken arrows connecting the A_1 levels. Similarly, mixing of tetragonal E functions and of tetragonal B_1 functions can occur, and these are indicated in Fig. 3.8. When these mixings are taken into account the wavefunctions in the tetragonal crystal field can be calculated. The mixing of states for D_3 symmetry is also shown in Fig. 3.8.

We should emphasize that so far we have considered only the separation in energy of orbital functions; spin functions have not been taken into account. We shall extend our analysis to spin functions in a later section.

A classic experiment was performed by Schawlow *et al.* (1961) which illustrates the splitting of levels when the symmetry of the crystal field is lowered. When Cr^{3+} ions are substituted for Mg^{2+} ions in magnesium oxide, many go into sites of perfect octahedral symmetry. This is illustrated in Fig. 3.9. The ground-state orbital belongs to the octahedral A_2 representation, and the first excited orbital state belongs to the E representation. A

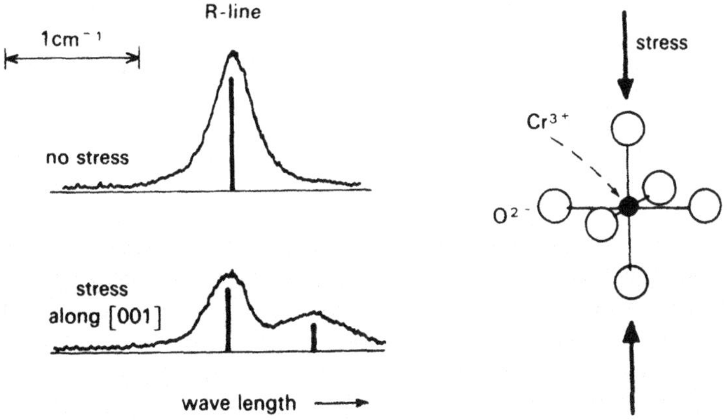

FIG. 3.9. Splitting of the R-line of Cr^{3+} in octahedral sites in magnesium oxide at 77 K under uniaxial stress (11 kg mm^{-2}) along the [001] direction. (After Schawlow *et al.* 1961.)

single sharp luminescence line (the R-line) caused by the $E \rightarrow A_2$ transition is found. These workers subjected the doped magnesium oxide crystal to an uniaxial stress along the [001]-direction which strained the environment of O^{2-} ions surrounding the Cr^{3+} ion and reduced the crystal field symmetry at the site of the Cr^{3+} ion from octahedral to tetragonal. The excited E level splits in two as expected, and this is seen as a splitting of the R-line, as shown in Fig. 3.9. A similar splitting of the R-line occurs under a stress along [110] or [111] directions, the [111] stress reducing the symmetry from octahedral to trigonal. The splitting of the excited E level of Cr^{3+} in a trigonal crystal field is due to a combination of the trigonal crystal field and spin–orbit coupling. (The trigonal crystal field on its own will not split the excited E level, as Fig. 3.8 shows.) We have not yet discussed the effect of spin–orbit coupling on the splittings of levels.

3.1.9 The full rotation group

This is the group of all rotations through any angle about any axis. This group has an infinite number of elements. All rotations through the same angle about any axis are in the same class. There is an infinite number of classes and hence an infinite number of irreducible representations.

It is shown in the textbooks on group theory (e.g. Heine (1960), Tinkham (1964)) and in textbooks on quantum mechanics that the operator which rotates a function through θ about an axis $\boldsymbol{t}$ is $\exp(-i\boldsymbol{J}.\boldsymbol{t}\theta/\hbar)$ where $\boldsymbol{J}$ is the general angular momentum operator. If we are dealing only with orbital functions we put $\boldsymbol{J}=\boldsymbol{L}$; if we are dealing only with spin functions we put $\boldsymbol{J}=\boldsymbol{S}$.

In order to apply this operator we express it as a Taylor expansion, and products of J_i operators ($i = x, y, z$) act on the functions.

Let us for the present confine our attention to orbital functions. The simple one-electron Hamiltonian operator

$$\mathscr{H}_0(\boldsymbol{r}) = \frac{\boldsymbol{p}^2}{2m} + \frac{1}{4\pi\varepsilon_0}\frac{Ze^2}{r} \qquad (3.48)$$

has the property that $\mathscr{H}_0(\boldsymbol{r})$ commutes with $\boldsymbol{l}$ and with all powers of $\boldsymbol{l}$ and $\mathscr{H}_0(\boldsymbol{r})$ is consequently invariant under the rotation operator $\exp(-\mathrm{i}\boldsymbol{l}.\boldsymbol{t}\,\theta/\hbar)$ for all angles θ about any axis $\boldsymbol{t}$. The eigenstates of this Hamiltonian are the set $R_{nl}(r)Y_l^m(\theta, \phi)$ where $R_{nl}(r)$ is invariant under all rotations. Let us consider the application of the rotation operator to the $Y_l^m(\theta, \phi)$ functions. A function space closed under $\boldsymbol{l}$ is also closed under the rotation operator $\exp(-\mathrm{i}\boldsymbol{l}.\boldsymbol{t}\,\theta/\hbar)$. Such a space is one spanned by the orthonormal set of $2l+1$ functions $|l, m\rangle = Y_l^m(\theta, \phi)$ for a specific value of l, and there is no subspace which is closed under $\boldsymbol{l}$. To see this we note that the space closed under $\boldsymbol{l}$ will be closed under $l_\pm = l_x \pm \mathrm{i}l_y$ and repeated applications of these operators to one of the $|l, m\rangle$ functions generates a full set of $2l+1$ functions for a specific value of l. These functions, then, form the basis functions for a space invariant under the rotation operator, we can use them to generate representation matrices, and these are irreducible representations of the full rotation group. We write the resultant representation matrix for a rotation through θ about the $\boldsymbol{t}$ axis obtained by using the set $|l, m\rangle$ as $D^l(\theta_t)$. These representations have dimensions 1, 3, 5, 7, . . . The irreducible representation derived from the $l = 0$ function is labelled s, those derived from $l = 1, 2, 3, \ldots$ are labelled $p, d, f, \ldots$, respectively. No other irreducible representations can be formed from orbital functions.

The character elements of the l irreducible representation are easily obtained. We calculate the character element for the rotation through θ about the z-axis, and this is also the character element for rotation through θ about any axis. The operator for rotation through θ about z is $\exp(-\mathrm{i}l_z\theta/\hbar)$ and since $l_z|l, m\rangle = m\hbar|l, m\rangle$ the representation matrix for this operator is

$$D^l(\theta_z) = \begin{pmatrix} \exp(\mathrm{i}l\theta) & & \mathbf{0} \\ & \exp(\mathrm{i}(l-1)\theta) & \\ \mathbf{0} & & \ddots\ \exp(-\mathrm{i}l\theta) \end{pmatrix}. \qquad (3.49)$$

Hence the character element $\chi^l(\theta)$ is $\sum_{m=-l}^{l} \exp(im\theta)$ and when this summation is carried out we obtain

$$\chi^l(\theta) = \frac{\sin(l+\frac{1}{2})\theta}{\sin\frac{1}{2}\theta}. \qquad (3.50)$$

Some of the common character elements are listed here for later use:

$$\begin{aligned}
\chi^l(E) &= \ 2l+1 \\
\chi^l(\pi) &= \ (-1)^l \\
\chi^l(\tfrac{1}{2}\pi) &= \ 1 \text{ for } l=0, 1, 4, 5, 8, 9, \ldots \\
&= -1 \text{ for } l=2, 3, 6, 7, \ldots \\
\chi^l(\tfrac{2}{3}\pi) &= \ 1 \text{ for } l=0, 3, 6, \ldots \\
&= \ 0 \text{ for } l=1, 4, 7, \ldots \\
&= -1 \text{ for } l=2, 5, 8, \ldots.
\end{aligned} \tag{3.51}$$

We consider now the two-electron Hamiltonian

$$\begin{aligned}
\mathscr{H}(\boldsymbol{r}_1, \boldsymbol{r}_2) &= \mathscr{H}_0(\boldsymbol{r}_1) + \mathscr{H}_0(\boldsymbol{r}_2) + \mathscr{H}'(\boldsymbol{r}_1, \boldsymbol{r}_2) \\
&= \mathscr{H}_0(\boldsymbol{r}_1, \boldsymbol{r}_2) + \mathscr{H}'(\boldsymbol{r}_1, \boldsymbol{r}_2)
\end{aligned} \tag{3.52}$$

where the forms of $\mathscr{H}_0(\boldsymbol{r}_1)$ and $\mathscr{H}_0(\boldsymbol{r}_2)$ are given by eqn (3.48). These are invariant under all rotations applied *independently* to the spaces of electrons 1 and 2, respectively. Hence, $\mathscr{H}_0(\boldsymbol{r}_1, \boldsymbol{r}_2)$ is invariant under all rotations applied independently to the spaces of electrons 1 and 2.

If $\mathscr{H}'(\boldsymbol{r}_1, \boldsymbol{r}_2)$ refers to the Coulomb interaction between the electrons it has the form

$$\mathscr{H}'(\boldsymbol{r}_1, \boldsymbol{r}_2) = \frac{1}{4\pi\varepsilon_0} \frac{e^2}{|\boldsymbol{r}_1 - \boldsymbol{r}_2|} \tag{3.53}$$

and is invariant under all rotations applied *simultaneously* to the spaces of electrons 1 and 2. Hence $\mathscr{H}(\boldsymbol{r}_1, \boldsymbol{r}_2)$ given by eqn (3.52) is invariant under all rotations applied simultaneously to the spaces of electrons 1 and 2.

We know from the discussion on product functions and product spaces that simple products of functions $Y_{l_1}^{m_1}(\theta, \phi) = |l_1\, m_1\rangle$ and $Y_{l_2}^{m_2}(\theta, \phi) = |l_2 m_2\rangle$ can act as basis functions for representations of the group of all rotations applied simultaneously to $\boldsymbol{r}_1$ and $\boldsymbol{r}_2$. These product functions $|l_1, m_1\rangle|l_2, m_2\rangle$ are in general basis functions for a reducible representation. There are more suitable basis functions, made up of linear combinations of the simple product functions, and with these as basis functions all the product representation matrices reduce to a similar diagonal form, consisting of smaller irreducible matrices along the diagonal. This is written symbolically

$$D^{l_1}(\theta) \times D^{l_2}(\theta) = \sum_L a_L D^L(\theta) \tag{3.54}$$

where the a_L are integers, and L can have integer values 0, 1, 2, . . . (The reason for choosing L as the symbol for this integer will become clear afterwards.) We now seek these more suitable basis functions.

The operator which simultaneously rotates the functions of $\boldsymbol{r}_1$ and $\boldsymbol{r}_2$ through the same angle θ about the axis $\boldsymbol{t}$ is

$$\begin{aligned}&\exp(-\mathrm{i}\boldsymbol{l}_1 . \boldsymbol{t}\theta/\hbar)\exp(-\mathrm{i}\boldsymbol{l}_2 . \boldsymbol{t}\theta/\hbar)\\ &= \exp(-\mathrm{i}\boldsymbol{L} . \boldsymbol{t}\theta/\hbar)\end{aligned} \tag{3.55}$$

where $\boldsymbol{L}=\boldsymbol{l}_1+\boldsymbol{l}_2$ is the total orbital angular momentum vector. A space made up of product functions of $\boldsymbol{r}_1$ and $\boldsymbol{r}_2$ (a product space) which is closed under $\boldsymbol{L}$ is closed under all the rotation operators (3.55) and is associated with a representation of the full rotation group. Consider such a product which belongs to the irreducible representation associated with a specific integer L. (We have seen earlier that the irreducible representations of the full rotation group associated with orbital functions are characterized by integers 0, 1, 2, . . . with dimensionality 1, 3, 5, . . . , respectively.) The space associated with integer L has dimensionality $2L+1$. We write the basis functions for this product function space as $|L, M\rangle$ where M ranges over the $2L+1$ integer values between $-L$ and $+L$. These basis functions $|L, M\rangle$ are chosen to transform under rotations in an identical manner to the spherical harmonics $|l, m\rangle$ where $L=l$ and $M=m$. That is, $|L, M\rangle$ belongs to the same row of the same irreducible representation as the spherical harmonic $|l, m\rangle$. The $|L, M\rangle$ functions are made up of linear combinations of the product functions $|l_1, m_1\rangle|l_2, m_2\rangle$, and the relationship is written

$$|L, M\rangle = \sum_{m_1, m_2} |l_1, m_1\rangle|l_2, m_2\rangle\langle l_1 l_2 m_1 m_2|LM\rangle \tag{3.56}$$

where the factors $\langle l_1 l_2 m_1 m_2|LM\rangle$ are the Clebsch–Gordan coefficients for the full rotation group. The specific L values which occur in (3.55) can be found from the relationship between the characters

$$\chi^{l_1}(\theta)\times\chi^{l_2}(\theta)=\sum_L a_L\chi^L(\theta). \tag{3.57}$$

We say that l_1 and l_2 combine to give specific values of L.

Inserting (3.50) into (3.57) we obtain

$$\begin{aligned}\sin(2l_1+1)\theta/2\times\sin(2l_2+1)\theta/2=\sin\theta/2\times(&a_0\sin\theta/2+a_1\sin 3\theta/2\\ &+a_2\sin 5\theta/2+\,\ldots)\end{aligned} \tag{3.58}$$

which can be used to find the specific values of L.

Let us consider the combination of two one-electron s states ($l_1=l_2=0$). We find that eqn (3.58) is satisfied only for values $a_0=1$ and all other coefficients are zero, showing that $l_1=0$ and $l_2=0$ combine to give only $L=0$. We can express this symbolically as

$$\mathrm{s}\times\mathrm{s}=\mathrm{S}. \tag{3.59}$$

In a similar way for an s and a p state we obtain $a_1 = 1$ and all other coefficients are zero:

$$s \times p = P. \tag{3.60}$$

For two p states we have

$$\sin^2 \tfrac{3}{2}\theta = \sin \tfrac{1}{2}\theta(a_0 \sin \tfrac{1}{2}\theta + a_1 \sin \tfrac{3}{2}\theta + a_2 \sin \tfrac{5}{2}\theta + \ldots). \tag{3.61}$$

To find the solution we consider the case of very small angle θ so that (3.61) becomes

$$(\tfrac{3}{2})^2 = a_0(\tfrac{1}{2})^2 + a_1(\tfrac{1}{2})(\tfrac{3}{2}) + a_2(\tfrac{1}{2})(\tfrac{5}{2}) + \ldots$$

or

$$3^2 = a_0 + 3a_1 + 5a_2 + \ldots. \tag{3.62}$$

Since the a's are integers the only solutions are either $a_4 = 1$, others $= 0$, or $a_0 = a_1 = a_2 = 1$, others $= 0$. Putting $\theta = \pi/3$ in eqn (3.61) we obtain

$$1 = a_0/4 + a_1/2 + a_2/4 + a_3/4 - a_4/2 + \cdots \tag{3.63}$$

which is consistent with the second solution only. Hence we have

$$p \times p = S + P + D. \tag{3.64}$$

We can similarly find that

$$d \times d = S + P + D + F + G. \tag{3.65}$$

The general pattern which emerges is, symbolically,

$$l_1 \times l_2 = \sum_i L_i$$

where

$$|l_1 - l_2| < L_i < l_1 + l_2 \tag{3.66}$$

and this is the basis for the 'vector model' of the many-electron atom illustrated for the two-electron case in Fig. 3.10.

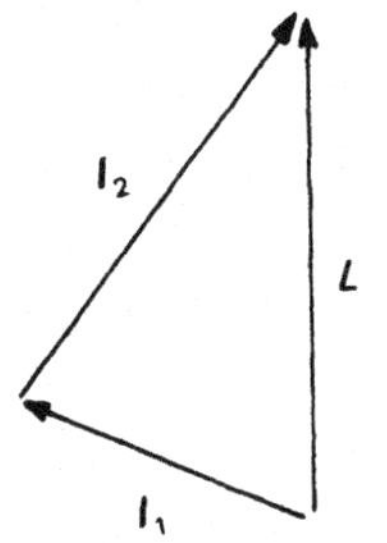

FIG. 3.10. The coupling of two one-electron orbital angular momenta to form the total orbital angular momentum is represented by this vector relationship. This *vector model* should be regarded as a symbolic relationship between quantum numbers rather than as a relationship between physical quantities.

The eigenstates of $\mathcal{H}(\boldsymbol{r}_1, \boldsymbol{r}_2)$ (eqn 3.52) transform according to irreducible representations of the full rotation group applied simultaneously to both electron functions. Starting with the one-electron eigenstates of $\mathcal{H}_0(\boldsymbol{r}_1)$ and $\mathcal{H}_0(\boldsymbol{r}_2)$, $|l_1, m_1\rangle$ and $|l_2, m_2\rangle$, respectively, we have constructed functions $|L, M\rangle$ which transform according to irreducible representations of the full rotation group. Are they eigenstates of $\mathcal{H}(\boldsymbol{r}_1, \boldsymbol{r}_2)$? We know that $\mathcal{H}(\boldsymbol{r}_1, \boldsymbol{r}_2)$ will not mix states $|L, M\rangle$ of different $|L, M\rangle$ values. Hence $\mathcal{H}(\boldsymbol{r}_1, \boldsymbol{r}_2)$ will not mix $|L, M\rangle$ states derived from the same pair of one-electron $|l, m\rangle$ states. On the other hand $\mathcal{H}(\boldsymbol{r}_1, \boldsymbol{r}_2)$ may mix states of the same L, M values derived from different pairs of one-electron states. We illustrate this in Fig. 3.11. We show the $|L, M\rangle$ states derived from the nd, n′d pair of one-electron states and from the np, n′d pair of one-electron states. (We assume n$\neq$n′ so as to avoid complications in the lower level due to the Pauli principle.) The $|L, M\rangle$ states

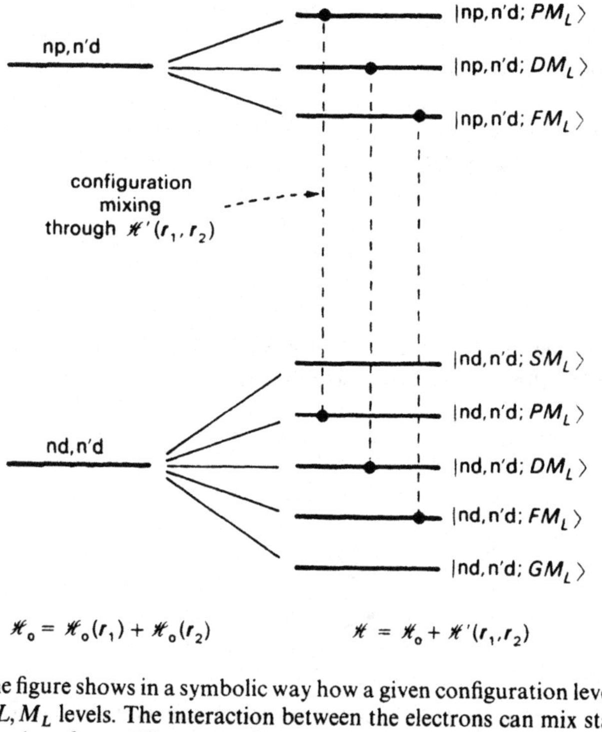

FIG. 3.11. The figure shows in a symbolic way how a given configuration level splits up into distinct L, M_L levels. The interaction between the electrons can mix states of the same L, M_L values from different configurations. The value of n' is assumed to be different from the value of n so that complications due to the Pauli principle do not arise.

which may be mixed by the $H'(\mathbf{r}_1, \mathbf{r}_2)$ component of $H(\mathbf{r}_1, \mathbf{r}_2)$ are connected by the broken lines in Fig. 3.11. This is the configuration mixing discussed in Section 2.3.4, p. 40. If this mixing can be neglected then the $|L, M\rangle$ states defined by eqn (3.56) are eigenstates of $H(\mathbf{r}_1, \mathbf{r}_2)$. And the states with different L values have distinct energies.

The relationship between $|L, M\rangle$ and the $|l_1, m_1\rangle|l_2, m_2\rangle$ product states can be written

$$|L, M\rangle = \sum_{m_1, m_2} (-1)^{l_1 - l_2 + M}(2L+1)^{-\frac{1}{2}} \begin{pmatrix} l_1 & l_2 & L \\ m_1 & m_2 & -M \end{pmatrix} |l_1, m_1\rangle|l_2, m_2\rangle \tag{3.67}$$

where $\begin{pmatrix} l_1 & l_2 & L \\ m_1 & m_2 & -M \end{pmatrix}$ are the Wigner 3j-symbols.

So far we have confined our attention to orbital states and we obtained irreducible representations of the full rotation group of odd dimensions. If we consider more general functions, i.e. spin or combined spin-plus-orbital functions, we obtain irreducible representations of the full rotation group which are of even dimensions. The operators for rotation of spin functions are $\exp(-i\mathbf{S}.\mathbf{t}\theta/\hbar)$ where $\mathbf{S}$ is the spin angular momentum operator. Spin spaces closed under $\mathbf{S}$ can be used to form representations. The set of spin functions $|S, M_S\rangle$ span such a closed space, and these can be used to form irreducible representations of the full rotation group. For half-odd integer values of S these are even-dimensional irreducible representations. The formula (3.50) for the character element in the l-irreducible representation is also valid for the S irreducible representation:

$$\chi^S(\theta) = \frac{\sin(S+\frac{1}{2})\theta}{\sin \theta/2}. \tag{3.68}$$

There are, however, some peculiarities about the rotations of spin functions in the case where S has half-odd integer value. For a rotation through $2\pi + \theta$ we find

$$\chi^S(2\pi+\theta) = \frac{\sin(S+\frac{1}{2})(2\pi+\theta)}{\sin(2\pi+\theta)/2} = (-1)^{2S}\chi^S(\theta). \tag{3.69}$$

We appear to have a double-valued character since a rotation through 2π reverses the sign of such characters rather than leaving them unchanged. A rotation through 4π, however, leaves all characters unchanged. This is a peculiarity of spin functions; when S has half-odd integer values the functions must be rotated through 4π to reproduce themselves.

To include rotations of spin functions in the group-theoretical analysis Bethe (1929) introduced the idea of a *double group* which involves rotations through $2\pi + \theta$ as well as through θ. There is a double group associated with the full rotation group and with the crystal groups discussed previously. For example, if we are considering rotations through $\frac{2}{3}\pi$ about some axis we must also consider rotations through $2\pi+\frac{2}{3}\pi$ about the same axis. For convenience of notation we write a rotation through 2π as R. In this larger group we regard R^2 as the identity element, E. For every group element C_n, say, we have an additional element RC_n which rotates functions through $2\pi+2\pi/n$.

Having generalized our analysis to take into account spin functions as well as orbital functions, we can generalize eqn (3.56) to include the coupling of two general angular momenta, j_1 and j_2. We can use the $(2j_1+1)(2j_2+1)$ product functions $|j_1, m_1\rangle\,|j_2, m_2\rangle$ for specific values of j_1 and j_2 to form a representation of the full rotation group. This representation is, in general, reducible, and with a similar analysis to that used for orbital functions we find

$$D^{j_1}(\theta)\times D^{j_2}(\theta)=\sum_{J=|j_1-j_2|}^{j_1+j_2} D^J(\theta). \tag{3.70}$$

If we write the basis functions for the D^J irreducible representations as $|J, M\rangle$ we have

$$|J, M\rangle=\sum_{m_1,m_2}|j_1, m_1\rangle|j_2, m_2\rangle\langle j_1 j_2 m_1 m_2|JM\rangle \tag{3.71}$$

where $\langle j_1 j_2 m_1 m_2|JM\rangle$ are the Clebsch–Gordan coefficients for the full rotation group.

3.1.10 *The octahedral double group, O′*

For every group element C_n in the octahedral group O we have an additional element RC_n. There are then $2\times 24=48$ elements in the octahedral double group O'. If we are dealing with irreducible representations formed from spin functions such that J is a half odd-integer, we find that, in general, $\chi(\theta)$ may not equal $\chi(R\theta)$ and so the elements $C(\theta)$ and $RC(\theta)$ belong, in general, to separate classes. We observe that for the full rotation group $\chi(C_2)=\chi(RC_2)$ always, so that C_2 and RC_2 might belong to the same class. It can be shown (Heine 1960) that rotations C_2 and RC_2 about a particular axis are in the same class if and only if there is another twofold axis at right angles to this particular C_2 axis. Hence for the octahedral group both C_2 and RC_2 are in the same class, as are C_2' and RC_2'. Similarly, the rotations C_3 and RC_3^2 about the same axis are in the same class, while the pair C_3^2 and RC_3 about the same axis are in a separate class. C_4 and RC_4^3 are in the same class, while C_4^3 and RC_4 together are in a separate class. There are eight separate classes in the O' group of 48 elements, and consequently there are eight inequivalent irreducible representations. Five of these can be formed from purely orbital functions and are the A_1, A_2, E, T_1, T_2 irreducible representations of the O group found previously. From Burnside's theorem we have $\sum_{i=1}^{8} l_i^2=48$ while we know $\sum_{i=1}^{5} l_i^2=24$. We see therefore that $l_6^2+l_7^2+l_8^2=24$ which gives $l_6=l_7=2$ and $l_8=4$. The characters for these three irreducible representations can be found by applying the rules cited previously, and the full character table for the O' group is shown in Table 3.5. The three additional representations Γ_6, Γ_7, Γ_8 are called the *doubled-valued representations.* Double groups for other point symmetry groups can similarly be found.

Table 3.5

Character table of the O' group

O'	$E=R^2$	R	$4C_3$ $4RC_3^2$	$4C_3^2$ $4RC_3$	$3C_2$ $3RC_2$	$6C_2'$ $6RC_2'$	$3C_4$ $3RC_4^3$	$3C_4^3$ $3RC_4$
$A_1=\Gamma_1$	1	1	1	1	1	1	1	1
$A_2=\Gamma_2$	1	1	1	1	1	-1	-1	-1
$E=\Gamma_3$	2	2	-1	-1	2	0	0	0
$T_1=\Gamma_4$	3	3	0	0	-1	-1	1	1
$T_2=\Gamma_5$	3	3	0	0	-1	1	-1	-1
Γ_6	2	-2	1	-1	0	0	$\sqrt{2}$	$-\sqrt{2}$
Γ_7	2	-2	1	-1	0	0	$-\sqrt{2}$	$\sqrt{2}$
Γ_8	4	-4	-1	1	0	0	0	0

3.1.11 Other point symmetry groups

So far we have concentrated on a small number of groups of rotations. There are other groups of point symmetry operators which are of interest and we will discuss them briefly. For example, the Hamiltonian of a centre in a solid may be invariant under reflections in various planes through the centre. A reflection is denoted by σ. The Hamiltonian may be invariant under inversion (i) which takes $\boldsymbol{r}$ into $-\boldsymbol{r}$. We now consider some of these other point symmetry groups.

(a) *The inversion group*: *i*. This group has two elements, E and i, and when i acts on $f(\boldsymbol{r})$ it converts it into $f(-\boldsymbol{r})$. These two elements are in separate classes, and there are two irreducible representations. Table 3.6 shows the character table. The labels g (*gerade*) and u (*ungerade*) are used as subscripts to distinguish the even-parity and odd-parity functions.

Table 3.6

Character table of the i group

i	E	i	
g	1	1	all even-parity functions, ϕ_g
u	1	-1	all odd-parity functions, ϕ_u

From the theorems on matrix elements we see that $\langle\phi_u|\phi_g\rangle=0$. We also consider the matrix element $\langle\phi_i|\xi_j|\phi_k\rangle$ where the subscripts i, j, k are either u or g, and we see that if ξ_j is an even-parity function there can only be non-zero matrix elements between functions of the same parity, while if ξ_j is an odd-parity function there can only be non-zero matrix elements between functions of opposite parity.

(b) *The full octahedral group*: O_h. This is the group of all possible rotations and reflections which take the cube or octahedron into itself. It is the largest of the point symmetry groups and contains 48 elements. As noted in Section 3.1.2, any symmetry operator of this group can also be achieved by an operator of the O group followed by an inversion. So we can write

$$O_h = O \times i. \tag{3.72}$$

We can classify basis functions for the O_h group by how they transform under O and how they transform under i. For example, the set of functions x, y, z are odd-parity functions (u) and they transform according to the T_1 irreducible representation of O. They are classified as T_{1u} functions. Similarly, the set xy, yz, zx belong to the T_{2g} irreducible representation of O_h.

Since the one-electron 3d functions are of even parity, the $(3d)^n$ states of transition metal ions also are of even parity. When the transition metal ion is placed in an octahedral field the crystal field states formed from 3d functions are of even parity, all are distinguished by a g subscript, and, since the electric dipole operator is an odd-parity function, electric dipole transitions are forbidden between these states. Odd-parity crystal field states formed from $(3d)^{n-1}p$ functions are 50 000–100 000 cm^{-1} above the $(3d)^n$ level. If, in addition to the octahedral crystal field, there is a weak odd-parity component of the crystal field energy, this can mix some odd-parity $(3d)^{n-1}p$ states into $(3d)^n$ states. This mixing has a negligible effect on the energy levels but can have a profound effect on the optical transitions between the low-lying levels since it now permits electric dipole transitions to occur between these states.

(c) *The tetrahedral group*: T. This is the group of all proper rotations which transform a regular tetrahedron into itself. It has twelve elements shown in Fig. 3.12.

(d) *The full tetrahedral group*: T_d. This is the group of all symmetry operators (rotations and reflections) which transform a regular tetrahedron into itself. T_d does not contain inversion symmetry.

(e) *The group* C_n. There is one n-fold symmetry axis. For example, group C_4 has elements E, C_4, C_4^2, C_4^3. These form four distinct classes, and there are four one-dimensional irreducible representations.

(f) *The group* C_{nv}. This includes the group C_n and n vertical reflection planes. These are illustrated in the case of C_{3v} in Fig. 3.13. C_{nv} does not include inversion symmetry. The character table of C_{3v} is given in Table 3.3.

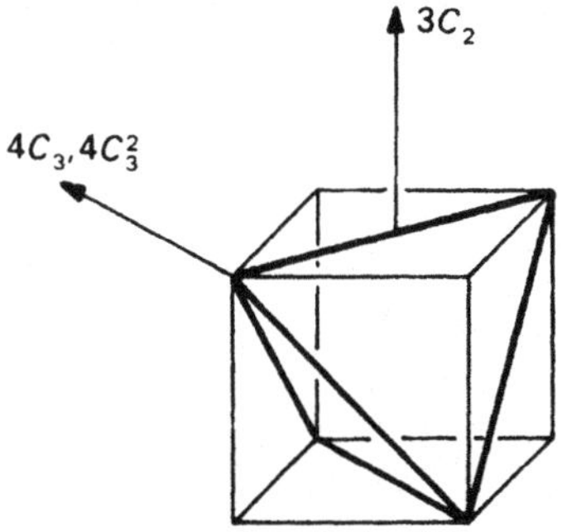

FIG. 3.12. The rotation operators of the tetrahedral group, T.

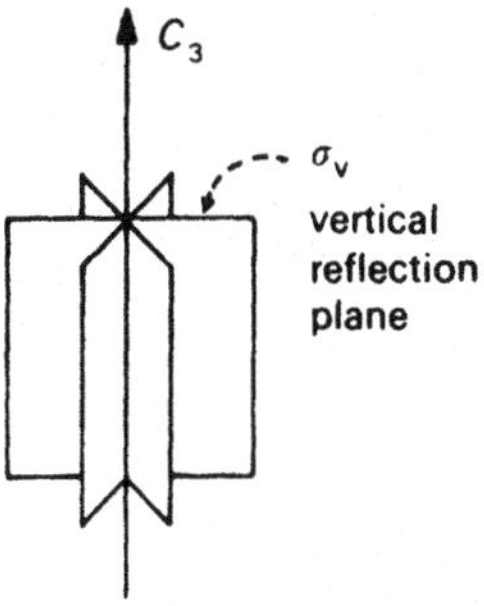

FIG. 3.13. C_3 axis and vertical reflection planes of C_{3v}.

(g) *The group* C_{nh}. This contains a horizontal reflection plane as well as the group C_n. This is illustrated in Fig. 3.14. It is clear that $C_{(2n)h}$ includes inversion symmetry: $C_{(2n)h} = C_{2n} \times i$. On the other hand, C_{nh} where n is an odd integer does not include inversion symmetry.

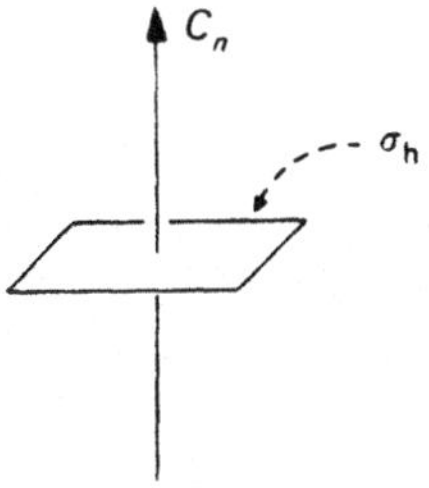

FIG. 3.14. C_n axis and horizontal reflection plane of C_{nh}.

(h) *The group* D_n. These groups include C_n and have in addition n twofold axes perpendicular to the C_n axis, as illustrated in Fig. 3.15 for the case of D_3. The character table of D_3 is given in Table 3.3.

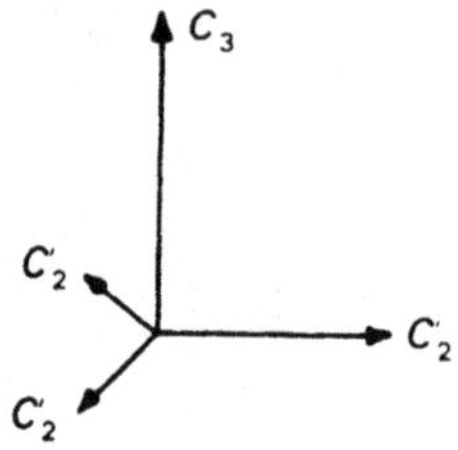

FIG. 3.15. The rotation operators of D_3.

The character tables for the double-group D_3' and C_{3v}' are given in Table 3.7. We see that there are two one-dimensional irreducible representations which are grouped together as $2\bar{A}$. We notice that these are complex conjugates of each other. The wavefunctions which would correspond to these two irreducible representations have the same energy. This means that there is some symmetry element which has not so far been taken into account in our treatment up to now. This is the *time-reversal symmetry*. If the Hamiltonian has time-reversal symmetry all electronic energy levels containing an odd number of electrons will be at least doubly degenerate. This is *Kramers' theorem* (Sugano *et al.* 1970; Tinkham 1964).

Table 3.7
Character table of the D_3 and C_{3v} double groups

D_3'	$E=R^2$	R	RC_3^2, C_3	RC_3, C_3^2	$3C_2'$	$3RC_2'$
C_{3v}'	$E=R^2$	R	RC_3^2, C_3	RC_3, C_3^2	$3\sigma_v$	$3R\sigma_v$
A_1	1	1	1	1	1	1
A_2	1	1	1	1	−1	−1
E	2	2	−1	−1	0	0
$2\bar{A}$ --- {	1	−1	−1	1	i	−i
	1	−1	−1	1	−i	i
$\bar{E}$	2	−2	1	−1	0	0

3.1.12 Decomposition of irreducible representations of the full rotation group

The crystal rotation groups discussed in the previous section are subgroups of the full rotation group. Consider such a crystal rotation group, G, whose elements are indicated by R. We can form a set of representation matrices for G by using the $(2J+1)$-fold function space which belongs to the J irreducible representation of the full rotation group. We write these matrices as $D^J(R)$. In general this representation is reducible and we write

$$D^J(R) = \sum_\Gamma a_\Gamma D^\Gamma(R) \tag{3.73}$$

where Γ refers to the irreducible representations of G. The values of the a_Γ coefficients are readily obtained from the character decomposition formula

$$\chi^J(R) = \sum_\Gamma a_\Gamma \chi^\Gamma(R). \tag{3.74}$$

The character elements $\chi^J(R)$ are obtained from eqns (3.50) or (3.68). And the character elements $\chi^\Gamma(R)$ are obtained from the character table of the group G. Consider the case where G is the octahedral group. The decompositions of the various J representations into irreducible representations of O (or O') are given in Table 3.8.

We apply this decomposition of the J representations to an analysis of the splitting of free-ion levels in a crystal field, and in particular to the splitting of rare-earth ion levels in an octahedral crystal field. As discussed earlier (Section 2.4, p. 47), the crystal field energy term for rare-earth ions is smaller than either the Coulomb interaction among the 4f electrons or the spin–orbit coupling. If the crystal field energy is neglected the rare-earth ion levels are the free ion levels and are characterized by J values. If now the octahedral crystal field term is taken into consideration the symmetry is reduced to octahedral and the free-ion eigenstates for a given J level are the basis functions for a representation of the octahedral group. This representation is, generally, reducible, and the space of eigenstates can be separated into subspaces each of which transforms as an irreducible representation of the octahedral group and each has a distinct energy. The free-ion J level, which is $(2J+1)$-fold degenerate, splits up into sublevels and the number of sublevels is given by the number of different irreducible representations of the octahedral group into which the J representation decomposes (Table 3.8). The splittings of $J=3$ and $J=\frac{5}{2}$ free-ion levels are shown schematically in Fig. 3.16. Whereas group theory tells us the degree of removal of the $(2J+1)$-fold degeneracy of the free-ion level, it does not give the magnitude of the splitting nor say which crystal field level is higher in energy. To obtain this information one must carry out specific crystal field calculations, such as was done for the case of the splitting of the d electron level in an octahedral field, as carried out in Section 2.6.

As long as the crystal field splitting is very much less than the separation between adjacent J levels the crystal field eigenstates can be formed from linear combinations of the free-ion eigenstates belonging to a specific J value. The zero-order approximation is often adequate for rare-earth ions. If the crystal field splitting is not small the crystal field term can mix together zero-order functions belonging to the same crystal field representation but coming from different J levels (J-mixing). Although the crystal field eigenstates have become more complicated the symmetry properties are unchanged; the crystal field labelling is valid for all strengths of crystal field.

Table 3.8

Decomposition of D^J representations into irreducible representations of O'

	$E=R^2$	R	$4C_3$ $4RC_3^2$	$4C_3^2$ $4RC_3$	$3C_2$ $3RC_2$	$6C_2'$ $6RC_2'$	$3C_4$ $3RC_4^3$	$3C_4^3$ $3RC_4$	Irreducible representations of O' into which the D^J representations decompose
D^0	1	1	1	1	1	1	1	1	A_1
$D^{\frac{1}{2}}$	2	−2	1	−1	0	0	$\sqrt{2}$	$-\sqrt{2}$	Γ_6
D^1	3	3	0	0	−1	−1	1	1	T_1
$D^{\frac{3}{2}}$	4	−4	−1	1	0	0	0	0	Γ_8
D^2	5	5	−1	−1	1	1	−1	−1	$E+T_2$
$D^{\frac{5}{2}}$	6	−6	0	0	0	0	$-\sqrt{2}$	$\sqrt{2}$	$\Gamma_7+\Gamma_8$
D^3	7	7	1	1	−1	−1	−1	−1	$A_2+T_1+T_2$
$D^{\frac{7}{2}}$	8	−8	1	−1	0	0	0	0	$\Gamma_6+\Gamma_7+\Gamma_8$
D^4	9	9	0	0	1	1	1	1	$A_1+E+T_1+T_2$
$D^{\frac{9}{2}}$	10	−10	−1	1	0	0	$\sqrt{2}$	$-\sqrt{2}$	$\Gamma_6+2\Gamma_8$
D^5	11	11	−1	−1	−1	−1	1	1	$E+2T_1+T_2$
$D^{\frac{11}{2}}$	12	−12	0	0	0	0	0	0	$\Gamma_6+\Gamma_7+2\Gamma_8$
D^6	13	13	1	1	1	1	−1	−1	$A_1+A_2+E+T_1+2T_2$

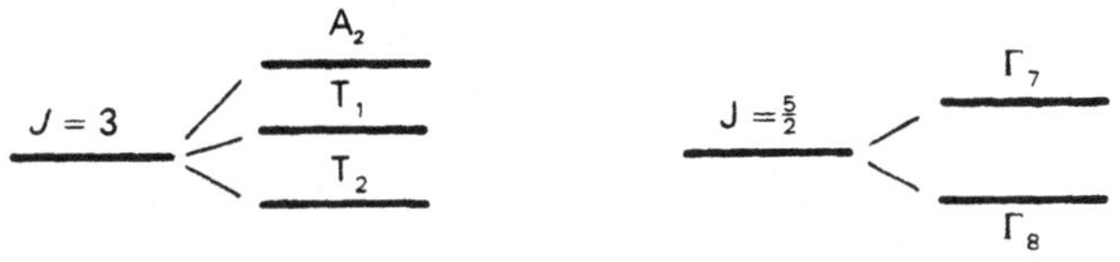

FIG. 3.16. Schematic representation of the splitting of free-ion J levels in a crystal field of octahedral symmetry.

The difference of approach to crystal field calculations of the rare earth ions and transition metal ions arises from the relative sizes of various terms in the Hamiltonian. For rare-earth ions there is a definite ordering

$$\mathscr{H}'(\text{Coulomb}) > \mathscr{H}_{\text{so}} > \mathscr{H}_{\text{c}}$$

and one approaches crystal field calculations by first calculating the eigenstates of $\mathscr{H}' + \mathscr{H}_{\text{so}}$—these are the free-ion eigenstates—and then calculating the effect of $\mathscr{H}_{\text{c}}$. This is *the weak field scheme.*

For the transition metal ions $\mathscr{H}_{\text{c}} > \mathscr{H}_{\text{so}}$. We could use the weak field approach but the J-mixing would be so great that the original free ion states would lose their identities. The J eigenstates are not the most suitable starting functions. One generally can use either the *intermediate* or *strong field schemes* (see Section 2.4, pp. 47–8). In practice the eigenstates of the Hamiltonian—neglecting spin–orbit coupling—are calculated by diagonalizing the matrix for $\mathscr{H}' + \mathscr{H}_{\text{c}}$, and the crystal field term, $\mathscr{H}_{\text{c}}$, refers to the dominant high-symmetry field—octahedral or tetrahedral. Smaller low-symmetry crystal field terms and the spin–orbit coupling term generally are treated afterwards by perturbation theory.

The analysis of the splitting of rare-earth ion J levels in octahedral fields can be extended to the splitting in fields of other symmetry. Table 3.9 shows the splitting of integer J levels when crystal fields of different symmetry are applied. One sees that for a symmetry lower than D_3 the $(2J+1)$-fold symmetry is fully removed.

Table 3.10 gives the splitting of half-odd integer J levels under various symmetry crystal fields. We see that for all symmetries lower than octahedral all degeneracies, except for the twofold Kramers degeneracy, is removed. The Kramers degeneracy can be removed by the application of a magnetic field which destroys the invariance of the Hamiltonian under time reversal.

3.1.13 *Splitting due to spin–orbit coupling*

We can use the same group-theoretical techniques to investigate the splitting of levels due to spin–orbit interaction. We do this first for free ions, and then for ions in a crystal field.

Table 3.9

Number of crystal field levels into which a given integer J free-ion level splits when the ion is placed in crystal fields of different symmetry. For a symmetry lower than D_3 the $(2J+1)$-fold free-ion degeneracy is fully removed.

D^J	degeneracy $(2J+1)$	0	T	D_6	D_4	D_3	C_6	C_4	C_3	D_2	C_2
D^0	1	1	1	1	1	1	1	1	1	1	1
D^1	3	1	1	2	2	2	2	2	2	3	3
D^2	5	2	2	3	4	3	3	4	3	5	5
D^3	7	3	3	5	5	5	5	5	5	7	7
D^4	9	4	4	6	7	6	6	7	6	9	9
D^5	11	4	4	7	8	7	7	8	7	11	11
D^6	13	6	6	9	10	9	9	10	9	13	13
D^7	15	6	6	10	11	10	10	11	10	15	15
D^8	17	7	7	11	13	11	11	13	11	17	17

Table 3.10

Number of crystal field levels into which a half-odd integer J level splits when the ion is placed in different crystal fields. For all symmetries lower than octahedral the degeneracies are removed except for the twofold Kramers degeneracy.

J	degeneracy $(2J+1)$	0	T	D_6	Lower symmetries
$\frac{1}{2}$	2	1	1	1	
$\frac{3}{2}$	4	1	1	2	splitting similar to that of D_6
$\frac{5}{2}$	6	2	2	3	
$\frac{7}{3}$	8	3	3	4	
$\frac{9}{2}$	10	3	3	5	
$\frac{11}{2}$	12	4	4	6	
$\frac{13}{2}$	14	5	5	7	
$\frac{15}{2}$	16	5	5	8	
$\frac{17}{2}$	18	6	6	9	

For the free ion case, if spin–orbit coupling is neglected, the eigenstates are products of orbital and spin states: $|LSM_LM_S\rangle = |L, M_L\rangle|S, M_S\rangle$. The orbital and spin functions transform according to the L and S irreducible representations of the full rotation group with representation matrices D^L and D^S, respectively. This results from the invariance of the Hamiltonian under all rotations applied *separately* to orbital and spin functions. If spin–orbit interaction is now taken into account the Hamiltonian is invariant under all rotations applied *simultaneously* to the orbital and space functions. The product states $|LSM_LM_S\rangle$ for a given LS term form basis functions for a representation $D^L \times D^S$ of the full rotation group. This is generally reducible, so we write

$$D^L \times D^S = \sum_J a_J D^J \tag{3.75}$$

or

$$\chi^L \times \chi^S = \sum_J a_J \chi^J. \tag{3.76}$$

A given LS level splits up into as many levels as there are J values in the above decomposition. This decomposition has been calculated previously (eqn 3.70) and we obtain

$$\begin{aligned} D^L \times D^S &= D^{L+S} + D^{L+S-1} + \ldots + D^{|L-S|} \\ &= \sum_{J=|L-S|}^{L+S} D^J \end{aligned} \tag{3.77}$$

which is a familiar vector model result.

In the case of an ion in a crystal field we find, in the absence of spin–orbit coupling, that the orbital functions are classified according to irreducible representations (Γ) of the point symmetry group. The spin functions are unaffected by the crystal field and are classified according to irreducible representations of the full rotation group, S. The eigenstate is a product of orbital and spin states, labelled $^{2S+1}\Gamma$.

If spin–orbit coupling is now taken into consideration the Hamiltonian is invariant under the point symmetry group applied *simultaneously* to orbital and spin functions. The product states $^{2S+1}\Gamma$ for given S form basis functions for a representation of the point symmetry group: this is generally reducible,

$$D^S \times D^\Gamma = \sum_{\Gamma'} a_{\Gamma'} D^{\Gamma'} \tag{3.78}$$

and the decomposition can be determined from the character elements. The level is split up into as many levels as there are irreducible representations, Γ', in the above decomposition.

As an example we consider the splitting due to spin–orbit coupling of the $^{2S+1}\Gamma = {}^4\mathrm{T}_2$ level of a transition metal ion in an octahedral crystal field. The product state functions belong to the representation $D^{3/2} \times D^{T_2}$ and from Table 3.5 we find the decomposition

$$D^{3/2} \times D^{\mathrm{T}_2}\!: \;\; \Gamma^8 \times \mathrm{T}_2 = \Gamma_6 + \Gamma_7 + 2\Gamma_8 \tag{3.79}$$

showing that $^4\mathrm{T}_2$ splits into four levels. Similarly $^2\mathrm{E}$ belongs to the product representation $D^{\frac{1}{2}} \times D^{\mathrm{E}}$ and we have

$$D^{\frac{1}{2}} \times D^{\mathrm{E}}\!: \;\; \Gamma_6 \times \mathrm{E} = \Gamma_8. \tag{3.80}$$

For the $^4A_2$ level we have

$$D^{3/2} \times D^{A_2}: \ \Gamma_8 \times A_2 = \Gamma_8 \tag{3.81}$$

showing that ^2E and $^4A_2$ are unsplit by spin–orbit coupling.

The R-line of Cr^{3+} in perfect octahedral sites is a transition between the ground $^4A_2$ state and the first excited ^2E state. As our symmetry arguments show these two levels are unsplit by spin–orbit coupling and the transition is characterized by a single sharp line.

We now analyse the case where the crystal field symmetry at the site of a Cr^{3+} ion is reduced to trigonal (D_3 or C_{3v})—a common distortion in chromium-doped oxides. We analyse first the effect of this distortion on the octahedral ^2E level. The orbital state which belongs to the E irreducible representation of O also belongs to the E irreducible representation of D_3 or C_{3v} as we can see by comparing the character elements of these two irreducible representations (Tables 3.1 and 3.3, or Tables 3.5 and 3.7). The orbital state is unsplit by the trigonal distortion—as was stated earlier in Section 3.1.8. The spin state belongs to $D^{\frac{1}{2}}$ of the full rotation group and to the $\bar{E}$ irreducible representation of the D_3 or C_{3v} groups. The combined spin and orbital function belongs to the $\bar{E} \times E$ representation, and from the D_3 or C_{3v} character table we find that this decomposes as $\bar{E} \times E = \bar{E} + 2\bar{A}$, showing that the ^2E level splits due to the combined action of spin–orbit coupling and trigonal crystal field into two levels labelled $2\bar{A}$ and $\bar{E}$.

We next examine the effects of spin–orbit coupling and trigonal crystal field on the $^4A_2$ ground state. The spin state which belongs to $D^{3/2}$ of the full rotation group is found to belong to the reducible representation $(2\bar{A} + \bar{E})$ of D_3 or C_{3v}, as an examination of the relevant character elements shows. The orbital A_2 state of the octahedral group also belongs to the A_2 irreducible representation of C_{3v}. The $^4A_2$ state then belongs to the $(2\bar{A} + \bar{E}) \times A_2$ reducible representation of D_3 or C_{3v} and this decomposes as

$$(2\bar{A} + \bar{E}) \times A_2 = 2\bar{A} + \bar{E} \tag{3.82}$$

showing that the $^4A_2$ level splits in two. The trigonal crystal field at the site of the Cr^{3+} ion in ruby can, to good approximation, be assigned the symmetry C_{3v} (see Chapter 9) and the upper and lower levels of the R transition are each split in two. The ^2E splitting is 29 cm^{-1}, while the $^4A_2$ splitting is 0.38 cm^{-1}. The luminescence is characterised by two sharp R-lines (R_1 and R_2) separated by 29 cm^{-1}. Each of these lines has fine structure due to the ground state splitting. The splittings of the $^{2S+1}\Gamma$ levels in a trigonal crystal field are shown schematically in Fig. 3.17.

3.2 Multi-(3d)-electron systems.

We now apply some of the group theoretical ideas of the last section to the calculation of wavefunctions and energy levels of multi-(3d)-electron systems.

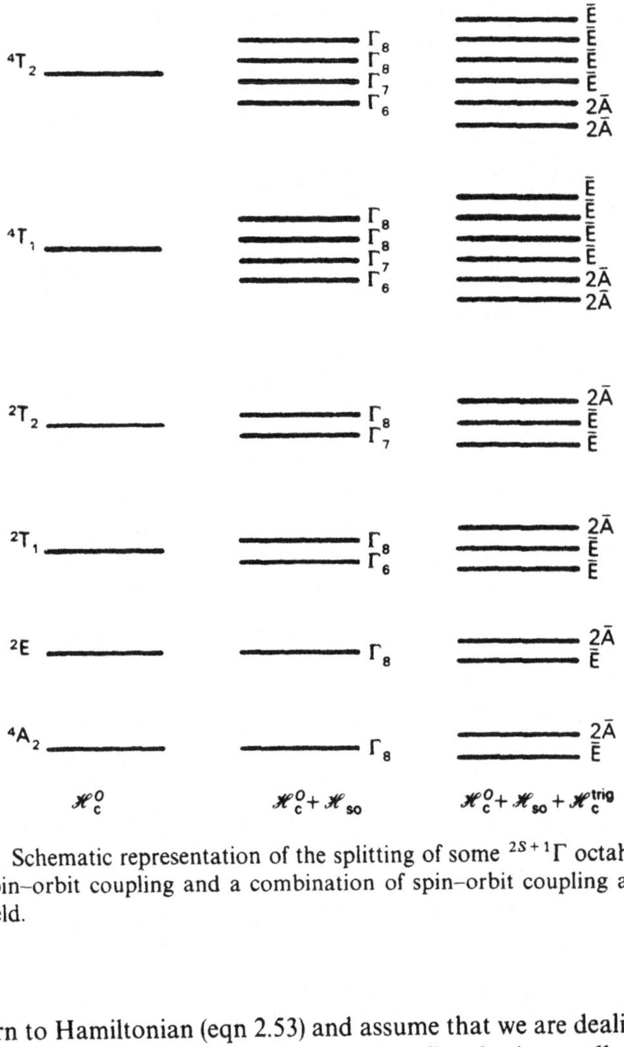

FIG. 3.17. Schematic representation of the splitting of some $^{2S+1}\Gamma$ octahedral levels due to spin–orbit coupling and a combination of spin–orbit coupling and trigonal crystal field.

We return to Hamiltonian (eqn 2.53) and assume that we are dealing with an octahedral crystal field. Since spin–orbit coupling is the smallest term we neglect it initially; it can be taken into account afterwards by perturbation theory. The resultant Hamiltonian is purely orbital and is

$$\mathscr{H} = \sum_i \left(\frac{\boldsymbol{p}_i^2}{2m} + V_{CF}(\boldsymbol{r}_i) + \mathscr{H}_c^O(\boldsymbol{r}_i) \right) + \sum_{i>j} \frac{e^2}{4\pi\varepsilon_0 |\boldsymbol{r}_i - \boldsymbol{r}_j|}$$

$$= \sum_i \mathscr{H}_0^O(\boldsymbol{r}_i) + \sum_{i>j} \mathscr{H}'(\boldsymbol{r}_i, \boldsymbol{r}_j). \tag{3.83}$$

$\mathscr{H}_0^O(\boldsymbol{r}_i)$ is the Hamiltonian for a single electron interacting with the central field of the ion core and with the octahedral crystal field. The wavefunctions derived from 3d-electron orbitals have already been found (Section 2.6). These are the $\phi_{t_2\xi}$, $\phi_{t_2\eta}$, $\phi_{t_2\zeta}$ orbitals which belong to the T_2 irreducible representation of O and which for convenience we now write simply as ξ, η, ζ, and the ϕ_{eu}, ϕ_{ev} orbitals which belong to the E irreducible representation and which for convenience we write as u, v. The u, v functions are 10 Dq higher in energy than the ξ, η, ζ functions. We seek the wavefunctions of the multi-electron Hamiltonian (3.83) in terms of products of these one-electron orbitals (strong crystal field basis functions).

3.2.1 Two-(3d)-electron wavefunctions.

In this case the Hamiltonian is

$$\begin{aligned}\mathscr{H}(\boldsymbol{r}_1,\boldsymbol{r}_2) &= \mathscr{H}_0^{\mathrm{O}}(\boldsymbol{r}_1)+\mathscr{H}_0^{\mathrm{O}}(\boldsymbol{r}_2)+\mathscr{H}'(\boldsymbol{r}_1,\boldsymbol{r}_2)\\ &= \mathscr{H}_0^{\mathrm{O}}(\boldsymbol{r}_1,\boldsymbol{r}_2)+\mathscr{H}'(\boldsymbol{r}_1,\boldsymbol{r}_2)\end{aligned} \tag{3.84}$$

$\mathscr{H}_0^{\mathrm{O}}(\boldsymbol{r}_1,\boldsymbol{r}_2)$, which is the sum of two one-electron crystal field Hamiltonians, is invariant under all the rotation operators of the group O applied independently to $\boldsymbol{r}_1$ and $\boldsymbol{r}_2$. On the other hand, $\mathscr{H}'(\boldsymbol{r}_1,\boldsymbol{r}_2)$ is invariant under all rotations of O applied simultaneously to $\boldsymbol{r}_1$ and $\boldsymbol{r}_2$. This is also the symmetry of $\mathscr{H}(\boldsymbol{r}_1,\boldsymbol{r}_2)$. The wavefunctions will transform according to irreducible representations of the O group. With the aid of eqn (3.37) we can write down functions formed from products of the one-electron crystal field orbitals which transform according to irreducible representations of O. We start with two e-orbitals. Since $\mathrm{E}\times\mathrm{E}=\mathrm{A}_1+\mathrm{A}_2+\mathrm{E}$ these product functions will belong to the A_1, A_2, and E irreducible representations. From eqn (3.37) and the table of Clebsch–Gordan coefficients for octahedral basis functions (Appendix 3B) we obtain the product functions

$$\begin{aligned}|\mathrm{e}^2,\mathrm{A}_1\rangle &= \frac{1}{\sqrt{2}}[u(\boldsymbol{r}_1)u(\boldsymbol{r}_2)+v(\boldsymbol{r}_1)v(\boldsymbol{r}_2)]\\ |\mathrm{e}^2,\mathrm{A}_2\rangle &= \frac{1}{\sqrt{2}}[u(\boldsymbol{r}_1)v(\boldsymbol{r}_2)-v(\boldsymbol{r}_1)u(\boldsymbol{r}_2)]\\ |\mathrm{e}^2,\mathrm{E}u\rangle &= \frac{1}{\sqrt{2}}[u(\boldsymbol{r}_1)u(\boldsymbol{r}_2)-v(\boldsymbol{r}_1)v(\boldsymbol{r}_2)]\\ |\mathrm{e}^2,\mathrm{E}v\rangle &= \frac{1}{\sqrt{2}}[u(\boldsymbol{r}_1)v(\boldsymbol{r}_2)+v(\boldsymbol{r}_1)u(\boldsymbol{r}_2)].\end{aligned} \tag{3.85}$$

These are the e^2 orbitals. No other product states can be formed from two e-orbitals.

Since $\mathscr{H}'(\boldsymbol{r}_1,\boldsymbol{r}_2)$ can only mix product states which transform according to the same row of the same irreducible representation this Coulomb interaction term will not mix these states (3.85) together. It may, however, mix in states formed from different two-electron crystal field configurations, for example, states formed from t_2e orbitals. To the extent that this mixing is small and can be neglected the above product states (3.85) are approximate orbital wavefunctions.

The $|e^2, A_1\rangle$ state defined in (3.85) is symmetric under interchange of $\boldsymbol{r}_1$ and $\boldsymbol{r}_2$. In accordance with the Pauli principle this orbital function must be multiplied by an antisymmetric spin function,

$$|S=0, M_S=0\rangle = \frac{1}{\sqrt{2}}[\alpha(\sigma_1)\beta(\sigma_2) - \beta(\sigma_1)\alpha(\sigma_2)]$$

so that the full spin-plus-orbital wavefunction is antisymmetric under the interchange of both spin and orbital coordinates of the two electrons. The full wavefunction is

$$|e^2, {}^1A_1, M_S=0\rangle = \frac{1}{2}[u(\boldsymbol{r}_1)u(\boldsymbol{r}_2) + v(\boldsymbol{r}_1)v(\boldsymbol{r}_2)]\,[\alpha(\sigma_1)\beta(\sigma_2) - \beta(\sigma_1)\alpha(\sigma_2)]$$

$$= \frac{1}{\sqrt{2}}(|u^+u^-| + |v^+v^-|) \tag{3.86}$$

where $|u^+u^-|$ is the *Slater determinant*

$$\frac{1}{\sqrt{2}}\begin{vmatrix} u(\boldsymbol{r}_1)\alpha(\sigma_1) & u(\boldsymbol{r}_2)\alpha(\sigma_2) \\ u(\boldsymbol{r}_1)\beta(\sigma_1) & u(\boldsymbol{r}_2)\beta(\boldsymbol{r}_2) \end{vmatrix}. \tag{3.87}$$

In a similar way we see that the $|e^2, A_2\rangle$ orbital function is antisymmetric under the interchange of $\boldsymbol{r}_1$ and $\boldsymbol{r}_2$. Hence we multiply this by the symmetric spin function to satisfy the Pauli principle. There are three such symmetric spin functions, $|S, M_s\rangle = |1, 1\rangle, |1, 0\rangle, |1, -1\rangle$. We find

$$|e^2, {}^3A_2, M_S=1\rangle = \frac{1}{\sqrt{2}}[u(\boldsymbol{r}_1)v(\boldsymbol{r}_2) - v(\boldsymbol{r}_1)u(\boldsymbol{r}_2)]\alpha(\sigma_1)\alpha(\sigma_2)$$

$$= |u^+v^+|. \tag{3.88}$$

Similarly we find

$$|e^2, {}^3A_2, M_S=0\rangle = \frac{1}{2}[u(\boldsymbol{r}_1)v(\boldsymbol{r}_2) - v(\boldsymbol{r}_1)u(\boldsymbol{r}_2)]\,[\alpha(\sigma_1)\beta(\sigma_2) + \beta(\sigma_1)\alpha(\sigma_2)]$$

$$= \frac{1}{\sqrt{2}}(|u^+v^-| + |u^-v^+|)$$

$$|e^2, {}^3A_2, M_S=-1\rangle = |u^-v^-|. \tag{3.89}$$

The orbital functions $|e^2, Eu\rangle$ and $|e^2, Ev\rangle$ are both symmetric under interchange of $\boldsymbol{r}_1$ and $\boldsymbol{r}_2$, hence they must be multiplied by an antisymmetric

$|S=0, M_S=0\rangle$ spin function. The full wavefunctions are then

$$|e^2, {}^1Eu, M_S=0\rangle = \frac{1}{\sqrt{2}}(|u^+u^-| - |v^+v^-|)$$

$$|e^2, {}^1Ev, M_S=0\rangle = \frac{1}{\sqrt{2}}(|u^+v^-| - |u^-v^+|). \quad (3.90)$$

We see that from two one-electron e orbitals we have found approximate wavefunctions of the type ${}^1A_1$, ${}^1E$, ${}^3A_2$. These differ in energy because of the Coulomb interaction term $\mathscr{H}'(\boldsymbol{r}_1, \boldsymbol{r}_2)$.

We proceed in the same way with two t_2 orbitals. Since $T_2 \times T_2 = A_1 + E + T_1 + T_2$ there are four sets of product orbital functions. The orbital functions made up of product t_2 orbitals and which belong to these four irreducible representations can be written down from eqn (3.37) and the table of Clebsch–Gordan coefficients (Appendix 3B). One finds that the A_1, E, and T_2 product orbital functions are symmetric under the interchange of $\boldsymbol{r}_1$ and $\boldsymbol{r}_2$. Hence they will be associated with antisymmetric spin singlet states. The T_1 product orbital function is antisymmetric under the interchange of $\boldsymbol{r}_1$ and $\boldsymbol{r}_2$, hence it will be multiplied by a symmetric spin triplet function. To the extent that mixing from other crystal field configurations can be neglected these products of spin and orbital functions are approximate wavefunctions of the Hamiltonian. These wavefunctions are given in Table 3.11.

We now consider the t_2e configuration. Since $E \times T_2 = T_1 + T_2$ there are two sets of product orbital functions. From eqn (3.37) and the table of Clebsch–Gordan coefficients we find, for example,

$$|t_2e, T_1\gamma\rangle = \zeta(\boldsymbol{r}_1)v(\boldsymbol{r}_2) \quad (3.91)$$

which is neither symmetric nor antisymmetric. There is, however, another $T_1\gamma$ function, $v(\boldsymbol{r}_1)\zeta(\boldsymbol{r}_2)$, and since these two functions transform according to the same row of the same irreducible representation there can be matrix elements of $\mathscr{H}'(\boldsymbol{r}_1, \boldsymbol{r}_2)$ between them, and individually they are not eigenstates of $\mathscr{H}(\boldsymbol{r}_1, \boldsymbol{r}_2)$. The symmetric and antisymmetric combinations of these two product functions, however, are eigenstates of $\mathscr{H}(\boldsymbol{r}_1, \boldsymbol{r}_2)$—to the extent that we neglect mixing from other crystal field configurations—and they have different energy values. The symmetric combination is symmetric under interchange of $\boldsymbol{r}_1$ and $\boldsymbol{r}_2$ and must be multiplied by the antisymmetric spin singlet. This gives us

$$|t_2e, {}^1T_1\gamma, M_S=0\rangle = \frac{1}{2}[\zeta(\boldsymbol{r}_1)v(\boldsymbol{r}_2) + v(\boldsymbol{r}_1)\zeta(\boldsymbol{r}_2)]\,[\alpha(\sigma_1)\beta(\sigma_2) - \beta(\sigma_1)\alpha(\sigma_2)]$$

$$= \frac{1}{\sqrt{2}}(|\zeta^+v^-| - |\zeta^-v^+|). \quad (3.92)$$

Table 3.11

Approximate octahedral crystal field wavefunctions formed from t_2^2 product orbitals (from Sugano *et al.* 1970).

$$|t_2^2\,{}^1A_1\rangle = \frac{1}{\sqrt{3}}(|\xi^+\xi^-| + |\eta^+\eta^-| + |\zeta^+\zeta^-|)$$

$$|t_2^2\,{}^1Eu\rangle = \frac{1}{\sqrt{6}}(-|\xi^+\xi^-| - |\eta^+\eta^-| + 2|\zeta^+\zeta^-|)$$

$$|t_2^2\,{}^1Ev\rangle = \frac{1}{\sqrt{2}}(|\xi^+\xi^-| - |\eta^+\eta^-|)$$

$$|t_2^2\,{}^1T_2\zeta\rangle^* = \frac{1}{\sqrt{2}}(|\xi^+\eta^-| - |\xi^-\eta^+|)$$

$$|t_2^2\,{}^3T_1\alpha, M_s = 1\rangle^{\dagger\ddagger} = |\eta^+\zeta^+|$$

* The ξ and η product functions can be obtained by the appropriate cyclic permutation of the ξ, η, ζ one-electron orbitals.

† The β and γ product functions can be obtained by the appropriate cyclic permutation of the ξ, η, ζ one-electron orbitals.

‡ The $M_s = 0, -1$ functions can be obtained by operating on this function with the S_- operator.

The antisymmetric combination must similarly be multiplied by the symmetrical spin triplet. The resultant $M_S = 1$ wavefunction is listed on Table 3.12. We have thus found that ${}^3T_1$ and ${}^1T_1$ wavefunctions can be formed from the t_2e configuration. Similarly we find that ${}^3T_2$ and ${}^1T_2$ wavefunctions can be formed from the t_2e configuration, and some of these are listed in Table 3.12.

Table 3.12

Some approximate octahedral crystal field wavefunctions formed from t_2e product orbitals (from Sugano *et al.* 1970).

$$|t_2e\,{}^1T_1\gamma\rangle = \frac{1}{\sqrt{2}}(|\zeta^+v^-| - |\zeta^-v^+|)$$

$$|t_2e\,{}^3T_1\gamma, M_S = 1\rangle = |\zeta^+v^+|$$

$$|t_2e\,{}^1T_2\zeta\rangle = \frac{1}{\sqrt{2}}(|\zeta^+u^-| - |\zeta^-u^+|$$

$$|t_2e\,{}^3T_2\zeta, M_S = 1\rangle = |\zeta^+u^+|$$

The Coulomb interaction energy between the 3d electrons is comparable to the crystal field energy so we cannot assume that the mixing of functions derived from different crystal field configurations is small. This mixing must be taken into account if we wish to obtain accurate energy levels. We can use the approximate wavefunctions (strong field functions) to diagonalize the Hamiltonian (3.83), the off-diagonal matrix elements being caused by the Coulomb interaction $\mathscr{H}'(\boldsymbol{r}_1, \boldsymbol{r}_2)$. Since $\mathscr{H}'$ is invariant under all group O operators applied simultaneously to $\boldsymbol{r}_1$ and $\boldsymbol{r}_2$ the only non-zero off-diagonal matrix elements are those between functions which transform as the same row of the same irreducible representation. There are only matrices of small dimension to be diagonalized.

For example, there are two $^1A_1$ levels, one from t_2^2 and one from e^2 orbitals, and there will be matrix elements of $\mathscr{H}'$ between them. These matrix elements have been evaluated by Tanabe and Sugano (1954) in terms of the Racah parameters A, B, C. The full Hamiltonian matrix for $\mathscr{H}_0^0 + \mathscr{H}'(\boldsymbol{r}_1, \boldsymbol{r}_2)$ is

$$^1A_1(^1G, {}^1S)$$

$$\begin{array}{cc} t_2^2 & e^2 \\ \end{array}$$

$$\begin{pmatrix} A+10B+5C & \sqrt{6}(2B+C) \\ \sqrt{6}(2B+C) & 20Dq+A+8B+4C \end{pmatrix}.$$

We notice that the Racah parameter A enters once in the diagonal terms and does not affect the energy difference between the levels. (It is similarly found not to contribute to the energy difference between crystal field levels of any $(3d)^n$ system, and is not included in the tabulated matrix elements of the Coulomb interaction.) The above matrix can be diagonalized to find the energy of the two levels in terms of Dq, B, C. If Dq is put equal to zero the two levels belong to the free-ion ^1G and ^1S levels, and this is shown above the matrix. The full set of matrices is given in Table 3.13 for the d^2 configuration.

A useful method of displaying the variation of energy with the Dq, B, C parameters was used by Tanabe and Sugano (1954) who took advantage of the fact that the ratio $C/B = \gamma$ should be almost constant through the transition metal series. If a value of γ is chosen this leaves only one adjustable Racah parameter, B, as well as Dq. Tanabe and Sugano then plotted E/B against Dq/B where E is the energy of the level. All energies were plotted relative to the lowest level.

3.2.2 *Energy levels of* $(3d)^n$ *electron systems in an octahedral field.*

The full set of matrices of the Coulomb interaction for the $(3d)^n$ configuration, for $n = 2$ to 8, have been calculated by Tanabe and Sugano (1954) using strong crystal field basis function, and they are tabulated in Griffith (1961) and in Sugano *et al.* (1970). To get the full matrix elements for $\mathscr{H}_0^0 + \mathscr{H}'$ one adds the

Table 3.13

Matrix elements of $H_o^0 + \mathscr{H}'(\mathbf{r}_1, \mathbf{r}_2)$ for the d^2 configuration. (From Sugano *et al.* 1970.)

$^1A_1(^1G, ^1S)$

t_2^2	e^2
$10B+5C$	$\sqrt{6}(2B+C)$
$\sqrt{6}(2B+C)$	$20Dq+8B+4C$

$^1E(^1D, ^1G)$

t_2^2	e^2
$B+2C$	$-2\sqrt{3}B$
$-2\sqrt{3}B$	$20Dq+2C$

$^1T_2(^1D, ^1G)$

t_2^2	t_2e
$B+2C$	$2\sqrt{3}B$
$2\sqrt{3}B$	$10Dq+2C$

$^3T_1(^3F, ^3P)$

t_2^2	t_2e
$-5B$	$6B$
$6B$	$10Dq+4B$

$t_2e\ ^1T_1(^1G)$	$10Dq+4B+2C$
$t_2e\ ^3T_2(^3F)$	$10Dq-8B$
$e^2\ ^3A_2(^3F)$	$20Dq-8B$

appropriate diagonal crystal field matrices elements (multiples of $10Dq$), as has been done in Table 3.13. The resulting matrices must now be diagonalized to get the energy levels in terms of the positive parameters Dq, B, C. As long as one is dealing with pure d states the $(10-n)$-electron system can be regarded as an n-hole system, except for a constant energy term which can be ignored (Sugano *et al.* 1970). The Coulomb energy of mutual interaction between the n positive charges is the same as that between n negative charges, hence the free ion energy levels will be the same for the n-electron system and for the $(10-n)$-electron system. But the energy of interaction of n positive charges with the crystal field will be opposite in sign to that of n negative charges. Hence to calculate the crystal field energies for the $(10-n)$-electron system we carry out the diagonalization procedure for the n-electron system but use a negative value for Dq. For example, to calculate the crystal field levels for Co^{2+} $(n=7)$ in an octahedral crystal field we use the matrices of the $n=3$ system but use a negative value of Dq.

Plots of E/B against Dq/B are given for each $(3d)^n$ configuration from $n=2$ to $n=8$ in Figs. 3.18–3.24, and these are calculated by diagonalization of the relevant matrices. We observe the similarity between the $(10-n)$- and n-electron systems. If Figs. 3.20 $(n=3)$ and 3.21 $(n=7)$ are compared one finds

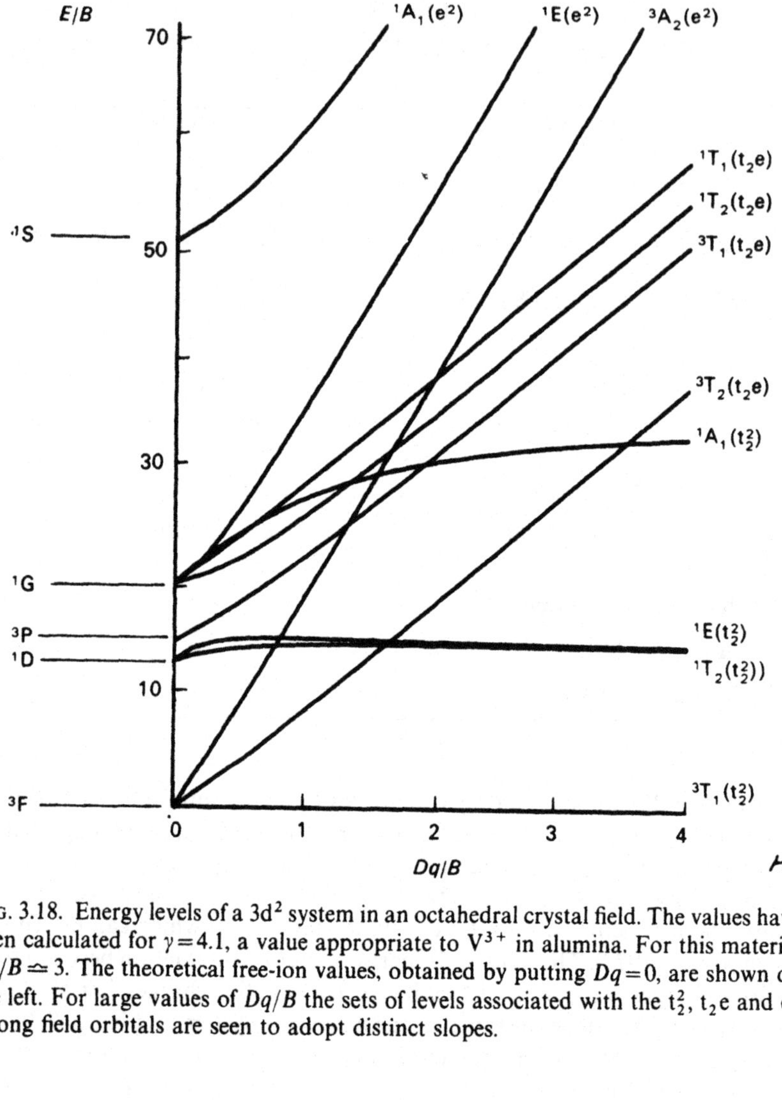

FIG. 3.18. Energy levels of a $3d^2$ system in an octahedral crystal field. The values have been calculated for $\gamma = 4.1$, a value appropriate to V^{3+} in alumina. For this material $Dq/B \simeq 3$. The theoretical free-ion values, obtained by putting $Dq = 0$, are shown on the left. For large values of Dq/B the sets of levels associated with the t_2^2, t_2e and e^2 strong field orbitals are seen to adopt distinct slopes.

that both have the same free-ion levels, one finds that the ground free-ion state ^4F is split into three by the crystal field in both cases, but the order of the splitting is reversed in the two cases, being $^4A_2$, $^4T_2$, $^4T_1$ in ascending order for $(3d)^3$ and $^4T_1$, $^4T_2$, $^4A_2$ in ascending order for $(3d)^7$.

For the case of a $(3d)^5$-electron system the values of n and $(10-n)$ are equal, hence, since one can go from the n- to the $(10-n)$-electron system by reversing

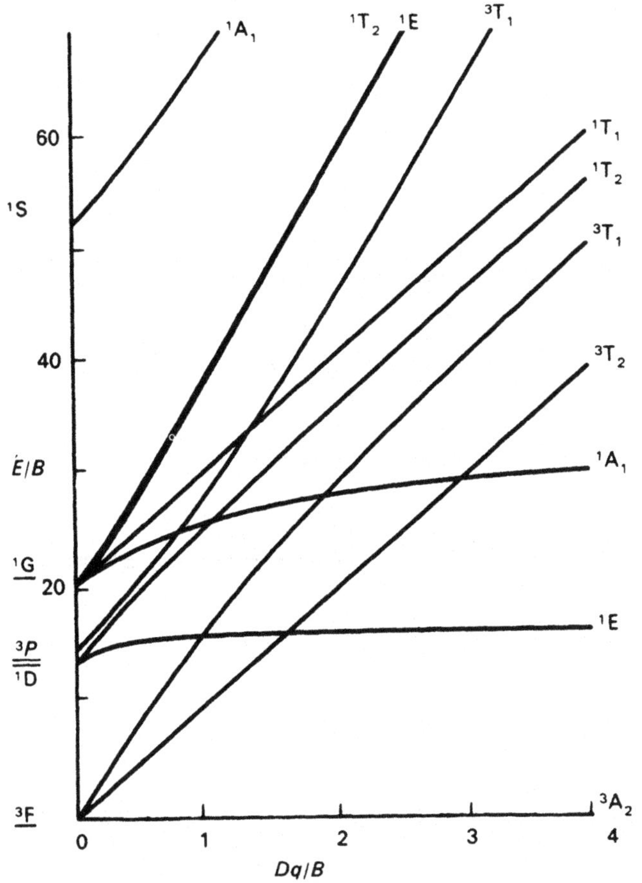

FIG. 3.19. Energy levels of a $3d^8$ system in an octahedral crystal field. The values have been calculated for $\gamma = 4.4$, a value appropriate to Ni^{2+} in magnesium oxide. For this material $Dq/B = 0.9$. The theoretical free-ion values, obtained by putting $Dq = 0$, are shown on the left. The order of the free-ion levels for the $3d^8$ system is identical to the order for the $3d^2$ system (Fig. 3.18), as explained in the text. The reversed order of the crystal field splitting in the $3d^2$ and $3d^8$ systems is obscured by the mixing of identical crystal field states from different free-ion levels.

the sign of *Dq* the splitting pattern in the $(3d)^5$-electron system must be independent of the sign of *Dq*. The splittings, then, do not vary linearly with *Dq*, as is evident in Fig. 3.24.

3.2.3 Energy levels of $(3d)^n$ *electron systems in a tetrahedral field*

We have seen in Section 2.6.3 that for a transition metal ion in a tetrahedral crystal field the even-parity component of the crystal field can again be

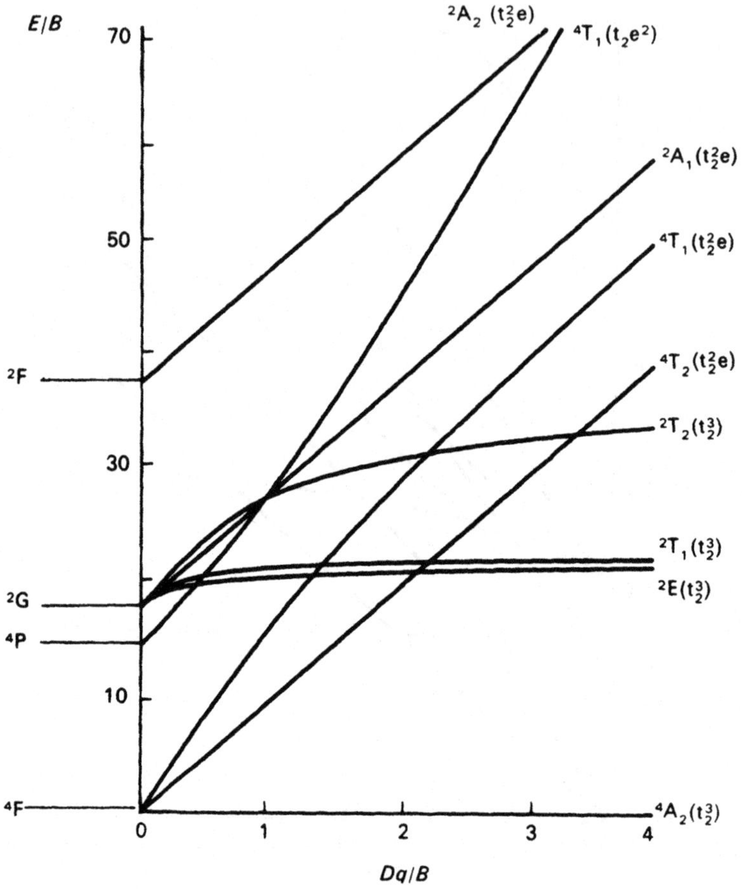

FIG. 3.20. Energy levels of a $3d^3$ system in an octahedral crystal field. The values have been calculated for $\gamma = 4.8$, a value appropriate for Cr^{3+} in aluminium oxide. For this material $Dq/B = 2.8$. The theoretical free-ion levels, obtained by putting $Dq = 0$, are shown on the left.

characterized by a parameter Dq, but this now has a negative value. Thus the octahedral crystal field matrices can be used with Dq regarded as a negative quantity. For example, the crystal field splitting of a $(3d)^7$ system in a tetrahedral crystal field is the same as in an octahedral crystal field but with negative Dq, and, from the analysis of the previous section, we know that this has the same pattern as the splitting of a $(3d)^3$ system in an octahedral field. Thus the splitting of the levels of Co^{2+} $(n = 7)$ in a tetrahedral field is similar to

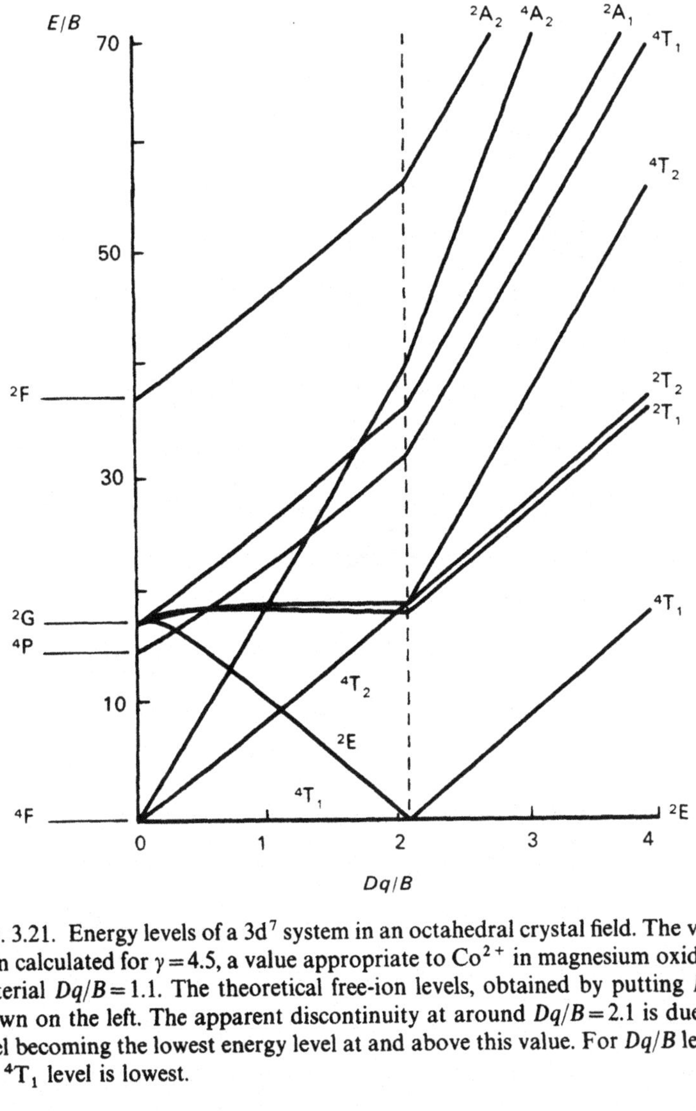

FIG. 3.21. Energy levels of a $3d^7$ system in an octahedral crystal field. The values have been calculated for $\gamma = 4.5$, a value appropriate to Co^{2+} in magnesium oxide. For this material $Dq/B = 1.1$. The theoretical free-ion levels, obtained by putting $Dq = 0$, are shown on the left. The apparent discontinuity at around $Dq/B = 2.1$ is due to the ^2E level becoming the lowest energy level at and above this value. For Dq/B less than 2.1 the $^4T_1$ level is lowest.

the splitting of Cr^{3+} ($n = 3$) in an octahedral field. In all cases the Dq value used in the figure should be regarded as the magnitude of the appropriate octahedral or tetrahedral crystal field parameter. For the case of a $(3d)^5$-electron system, for which the octahedral splitting pattern is independent of the sign of Dq, the splitting patterns are identical for octahedral and tetrahedral crystal fields.

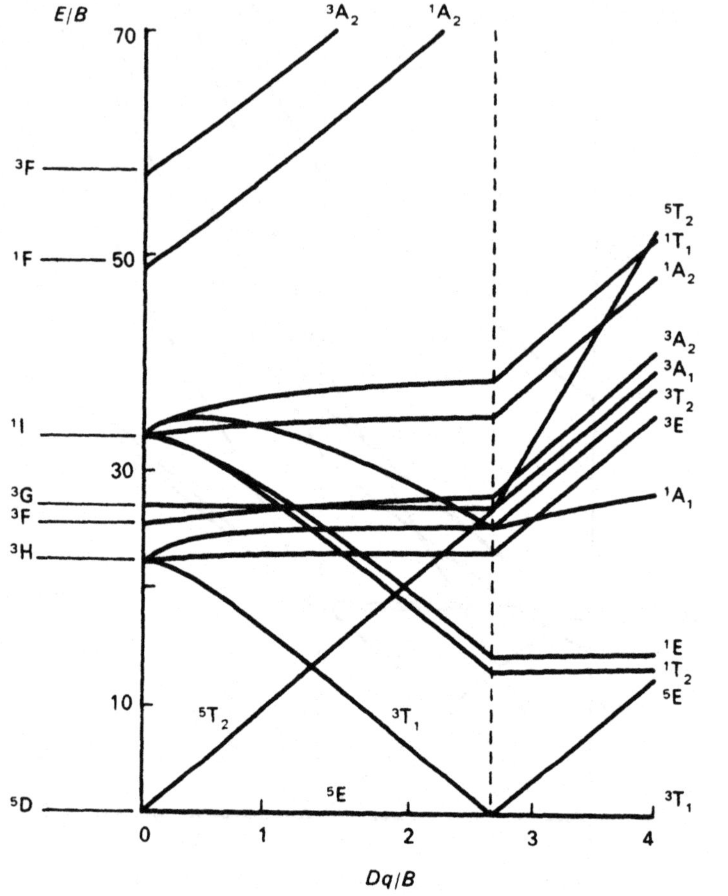

FIG. 3.22. Energy levels of a 3d^4 system in an octahedral crystal field. The values are calculated for $\gamma = 4.61$. The theoretical free ion values, obtained by putting $Dq = 0$, are shown on the left.

The splitting of (3d)^n-electron systems in a cubic field is similar to the splitting in a tetragonal field but the crystal field is twice as strong.

3.2.4 *Lower-symmetry fields and spin–orbit coupling*

In many cases the crystal field is not one of perfect octahedral or tetragonal symmetry, and additional lower-symmetry crystal fields may occur. These may cause a splitting of the octahedral or tetrahedral levels. If the lower symmetry is known one can use group theory to determine if splitting occurs, but one must evaluate matrix elements of the low-symmetry crystal field energy if one is to estimate the splittings quantitatively. By means of the

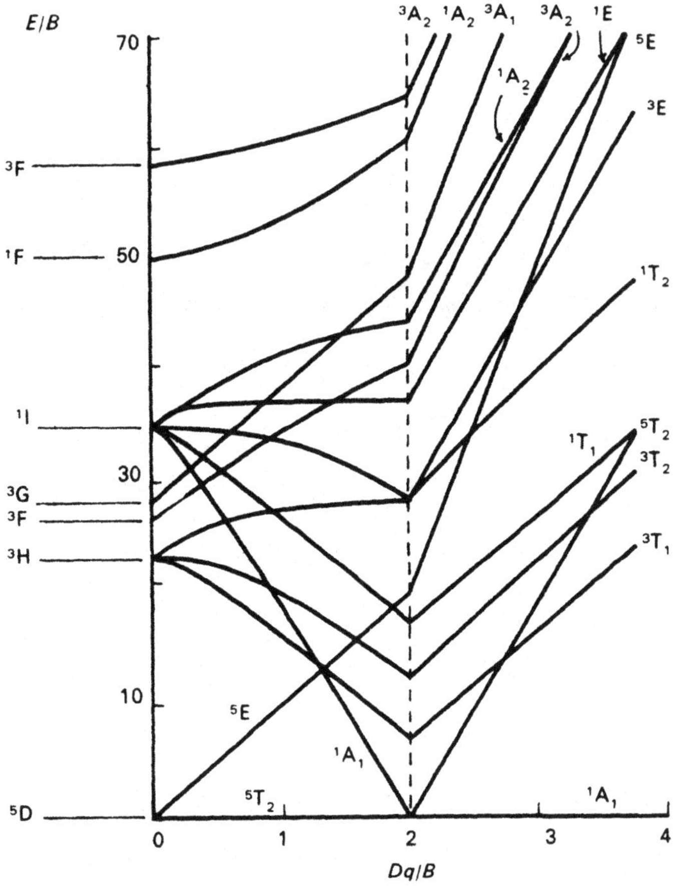

FIG. 3.23. Energy levels of a $3d^6$ system in an octahedral crystal field. The values are calculated for $\gamma = 4.81$. The theoretical free ion values, obtained by putting $Dq = 0$, are shown on the left.

Wigner–Eckart theorem one can estimate the splittings and shifts of the energy levels in terms of one or more reduced matrix elements of the lower-symmetry field, and these reduced matrix elements can then be regarded as the parameters which classify the strength of the lower-symmetry field.

The effects of lower symmetry fields are often comparable in size to the effects of spin–orbit coupling, and one may need to consider both lower symmetry crystal fields and spin–orbit coupling together. Matrix elements of $\mathscr{H}_{so} = \sum_i \xi(r_i)\boldsymbol{l}_i \cdot \boldsymbol{s}_i$ must be evaluated between $|S\Gamma M_s\gamma\rangle$ states, where $\Gamma\gamma$ characterize the orbital eigenstates with respect to the symmetry of the crystal field, while SM_s characterize the spin eigenstate. Expressions for the matrix

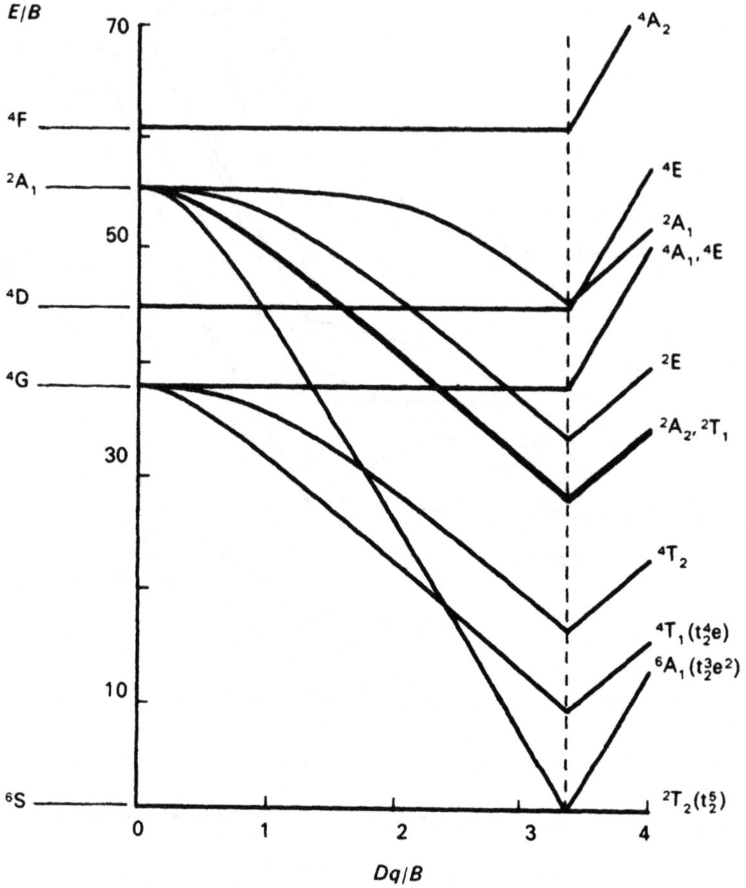

FIG. 3.24. Energy levels of a $3d^5$ system in an octahedral crystal field. The values have been calculated for $\gamma = 5.55$, a value appropriate to Mn^{2+} in manganese fluoride. The theoretical free-ion values are shown on the left.

element $\langle S\Gamma M_s\gamma|\mathscr{H}_{so}|S'\Gamma' M_s'\gamma'\rangle$ have been worked out by Sugano *et al.* (1970) in terms of reduced matrix elements and these are tabulated in their textbook.

Appendix 3A. Clebsch–Gordan coefficients for basis functions of the full rotation group

Linear combinations of products of angular momentum eigenstates, $|j_1 m_1\rangle$, $|j_2 m_2\rangle$, for a specific set of values of j_1 and j_2, can be chosen which are also eigenstates of total angular momentum characterized by the quantum numbers, J, M: $|j_1 j_2 JM\rangle$. These are given by the Wigner formula

$$|j_1 j_2 JM\rangle = \sum_{m_1, m_2} |j_1 m_1\rangle |j_2 m_2\rangle \langle j_1 j_2 m_1 m_2 | j_1 j_2 JM\rangle \tag{3A.1}$$

where $\langle j_1 j_2 m_1 m_2 | j_1 j_2 JM\rangle$ are the Clebsch–Gordan coefficients. They are real and satisfy the relationship $\langle j_1 j_2 m_1 m_2 | j_1 j_2 JM\rangle = \langle j_1 j_2 JM | j_1 j_2 m_1 m_2\rangle$. The choice of phases for these coefficients are such that

$$\langle j_2 j_1 m_2 m_1 | j_2 j_1 JM\rangle = (-1)^{j_1 + j_2 - J} \langle j_1 j_2 m_1 m_2 | j_1 j_2 JM\rangle. \tag{3A.2}$$

Equation (3A.1) is sometimes written in abbreviated form

$$|JM\rangle = \sum_{m_1, m_2} |m_1\rangle |m_2\rangle \langle m_1 m_2 | JM\rangle \tag{3A.3}$$

where a specific set of j_1, j_2 values is understood. Values of these coefficients for some low values of j_1 and j_2 are listed in Table 3A.1.

Table 3A.1

Values of Clebsch–Gordan coefficients $\langle j_1 j_2 m_1 m_2 | j_1 j_2 JM\rangle$ for some specific values of j_1, j_2

$j_1 = \frac{1}{2}, j_2 = \frac{1}{2}$

	J	1	1	1	0
m_1	m_2 M	1	0	-1	0
$\frac{1}{2}$	$\frac{1}{2}$	1			
$\frac{1}{2}$	$-\frac{1}{2}$		$\sqrt{\frac{1}{2}}$		$\sqrt{\frac{1}{2}}$
$-\frac{1}{2}$	$\frac{1}{2}$		$\sqrt{\frac{1}{2}}$		$-\sqrt{\frac{1}{2}}$
$-\frac{1}{2}$	$-\frac{1}{2}$			1	

$j_1 = 1, j_2 = \frac{1}{2}$

	J	$\frac{3}{2}$	$\frac{3}{2}$	$\frac{3}{2}$	$\frac{3}{2}$	$\frac{1}{2}$	$\frac{1}{2}$
m_1	m_2 M	$\frac{3}{2}$	$\frac{1}{2}$	$-\frac{1}{2}$	$-\frac{3}{2}$	$\frac{1}{2}$	$-\frac{1}{2}$
1	$\frac{1}{2}$	1					
1	$-\frac{1}{2}$		$\sqrt{\frac{1}{3}}$			$\sqrt{\frac{2}{3}}$	
0	$\frac{1}{2}$		$\sqrt{\frac{2}{3}}$			$-\sqrt{\frac{1}{3}}$	
0	$-\frac{1}{2}$			$\sqrt{\frac{2}{3}}$			$\sqrt{\frac{1}{3}}$
-1	$\frac{1}{2}$			$\sqrt{\frac{1}{3}}$			$-\sqrt{\frac{2}{3}}$
-1	$-\frac{1}{2}$				1		

$j_1 = 1, j_2 = 1$

		J	2	2	2	2	2	1	1	1	0
m_1	m_2	M	2	1	0	-1	-2	1	0	-1	0
1	1		1								
1	0			$\sqrt{\frac{1}{2}}$				$\sqrt{\frac{1}{2}}$			
0	1			$\sqrt{\frac{1}{2}}$				$-\sqrt{\frac{1}{2}}$			
1	-1				$\sqrt{\frac{1}{6}}$				$\sqrt{\frac{1}{2}}$		$\sqrt{\frac{1}{3}}$
0	0				$\sqrt{\frac{2}{3}}$				0		$-\sqrt{\frac{1}{3}}$
-1	1				$\sqrt{\frac{1}{6}}$				$-\sqrt{\frac{1}{2}}$		$\sqrt{\frac{1}{3}}$
0	-1					$\sqrt{\frac{1}{2}}$				$\sqrt{\frac{1}{2}}$	
-1	0					$\sqrt{\frac{1}{2}}$				$-\sqrt{\frac{1}{2}}$	
-1	-1						1				

$j_1 = \frac{3}{2}, j_2 = \frac{1}{2}$

		J	2	2	2	2	2	1	1	1
m_1	m_2	M	2	1	0	-1	-2	1	0	-1
$\frac{3}{2}$	$\frac{1}{2}$		1							
$\frac{3}{2}$	$-\frac{1}{2}$			$\sqrt{\frac{1}{4}}$				$\sqrt{\frac{3}{4}}$		
$\frac{1}{2}$	$\frac{1}{2}$			$\sqrt{\frac{3}{4}}$				$-\sqrt{\frac{1}{4}}$		
$\frac{1}{2}$	$-\frac{1}{2}$				$\sqrt{\frac{1}{2}}$				$\sqrt{\frac{1}{2}}$	
$-\frac{1}{2}$	$\frac{1}{2}$				$\sqrt{\frac{1}{2}}$				$-\sqrt{\frac{1}{2}}$	
$-\frac{1}{2}$	$-\frac{1}{2}$					$\sqrt{\frac{3}{4}}$				$\sqrt{\frac{1}{4}}$
$-\frac{3}{2}$	$\frac{1}{2}$					$\sqrt{\frac{1}{4}}$				$-\sqrt{\frac{3}{4}}$
$-\frac{3}{2}$	$-\frac{1}{2}$						1			

$j_1 = \frac{3}{2}, j_2 = 1$

		J	$\frac{5}{2}$	$\frac{5}{2}$	$\frac{5}{2}$	$\frac{5}{2}$	$\frac{5}{2}$	$\frac{5}{2}$	$\frac{3}{2}$	$\frac{3}{2}$	$\frac{3}{2}$	$\frac{3}{2}$	$\frac{1}{2}$	$\frac{1}{2}$
m_1	m_2	M	$\frac{5}{2}$	$\frac{3}{2}$	$\frac{1}{2}$	$-\frac{1}{2}$	$-\frac{3}{2}$	$-\frac{5}{2}$	$\frac{3}{2}$	$\frac{1}{2}$	$-\frac{1}{2}$	$-\frac{3}{2}$	$\frac{1}{2}$	$-\frac{1}{2}$
$\frac{3}{2}$	1		1											
$\frac{3}{2}$	0			$\sqrt{\frac{2}{5}}$					$\sqrt{\frac{3}{5}}$					
$\frac{1}{2}$	1			$\sqrt{\frac{3}{5}}$					$-\sqrt{\frac{2}{5}}$					
$\frac{3}{2}$	-1				$\sqrt{\frac{1}{10}}$					$\sqrt{\frac{2}{5}}$			$\sqrt{\frac{1}{2}}$	
$\frac{1}{2}$	0				$\sqrt{\frac{3}{5}}$					$\sqrt{\frac{1}{15}}$			$-\sqrt{\frac{1}{3}}$	
$-\frac{1}{2}$	1				$\sqrt{\frac{3}{10}}$					$-\sqrt{\frac{8}{15}}$			$\sqrt{\frac{1}{6}}$	
$\frac{1}{2}$	-1					$\sqrt{\frac{3}{10}}$					$\sqrt{\frac{8}{15}}$			$\sqrt{\frac{1}{6}}$
$-\frac{1}{2}$	0					$\sqrt{\frac{3}{5}}$					$-\sqrt{\frac{1}{15}}$			$-\sqrt{\frac{1}{3}}$
$-\frac{3}{2}$	1					$\sqrt{\frac{1}{10}}$					$-\sqrt{\frac{2}{5}}$			$\sqrt{\frac{1}{2}}$
$-\frac{1}{2}$	-1						$\sqrt{\frac{3}{5}}$					$\sqrt{\frac{2}{5}}$		
$-\frac{3}{2}$	0						$\sqrt{\frac{2}{5}}$					$-\sqrt{\frac{3}{5}}$		
$-\frac{3}{2}$	-1							1						

$j_1 = 2, j_2 = \frac{1}{2}$

		J	$\frac{5}{2}$	$\frac{5}{2}$	$\frac{5}{2}$	$\frac{5}{2}$	$\frac{5}{2}$	$\frac{5}{2}$	$\frac{3}{2}$	$\frac{3}{2}$	$\frac{3}{2}$	$\frac{3}{2}$
m_1	m_2	M	$\frac{5}{2}$	$\frac{3}{2}$	$\frac{1}{2}$	$-\frac{1}{2}$	$-\frac{3}{2}$	$-\frac{5}{2}$	$\frac{3}{2}$	$\frac{1}{2}$	$-\frac{1}{2}$	$-\frac{3}{2}$
2	$\frac{1}{2}$		1									
2	$-\frac{1}{2}$			$\sqrt{\frac{1}{5}}$					$\sqrt{\frac{4}{5}}$			
1	$\frac{1}{2}$			$\sqrt{\frac{4}{5}}$					$-\sqrt{\frac{1}{5}}$			
1	$-\frac{1}{2}$				$\sqrt{\frac{2}{5}}$					$\sqrt{\frac{3}{5}}$		
0	$\frac{1}{2}$				$\sqrt{\frac{3}{5}}$					$-\sqrt{\frac{2}{5}}$		
0	$-\frac{1}{2}$					$\sqrt{\frac{3}{5}}$					$\sqrt{\frac{2}{5}}$	
-1	$\frac{1}{2}$					$\sqrt{\frac{2}{5}}$					$-\sqrt{\frac{3}{5}}$	
-1	$-\frac{1}{2}$						$\sqrt{\frac{4}{5}}$					$\sqrt{\frac{1}{5}}$
-2	$\frac{1}{2}$						$\sqrt{\frac{1}{5}}$					$-\sqrt{\frac{4}{5}}$
-2	$-\frac{1}{2}$							1				

$j_1 = 2, j_2 = 1$

	J	3	3	3	3	3	3	3	2	2	2	2	2	1	1	1
m_1	m_2 M	3	2	1	0	-1	-2	-3	2	1	0	-1	-2	1	0	-1
2	1	1														
2	0		$\sqrt{\frac{1}{3}}$						$\sqrt{\frac{2}{3}}$							
1	1		$\sqrt{\frac{2}{3}}$						$-\sqrt{\frac{1}{3}}$							
2	-1			$\sqrt{\frac{1}{15}}$						$\sqrt{\frac{1}{3}}$				$\sqrt{\frac{3}{5}}$		
1	0			$\sqrt{\frac{8}{15}}$						$\sqrt{\frac{1}{6}}$				$-\sqrt{\frac{3}{10}}$		
0	1			$\sqrt{\frac{6}{15}}$						$-\sqrt{\frac{1}{2}}$				$\sqrt{\frac{1}{10}}$		
1	-1				$\sqrt{\frac{1}{5}}$						$\sqrt{\frac{1}{2}}$				$\sqrt{\frac{3}{10}}$	
0	0				$\sqrt{\frac{3}{5}}$						0				$-\sqrt{\frac{2}{5}}$	
-1	1				$\sqrt{\frac{1}{5}}$						$-\sqrt{\frac{1}{2}}$				$\sqrt{\frac{3}{10}}$	
0	-1					$\sqrt{\frac{6}{15}}$						$\sqrt{\frac{1}{2}}$				$\sqrt{\frac{1}{10}}$
-1	0					$\sqrt{\frac{8}{15}}$						$-\sqrt{\frac{1}{6}}$				$-\sqrt{\frac{3}{10}}$
-2	1					$\sqrt{\frac{1}{15}}$						$-\sqrt{\frac{1}{3}}$				$\sqrt{\frac{3}{5}}$
-1	-1						$\sqrt{\frac{2}{3}}$						$\sqrt{\frac{1}{3}}$			
-2	0						$\sqrt{\frac{1}{3}}$						$-\sqrt{\frac{2}{3}}$			
-2	-1							1								

Appendix 3B. Clebsch–Gordan coefficients for octahedral basis functions

Linear combinations of products of octahedral basis functions (belonging to the Γ_1 and Γ_2 irreducible representations) can be chosen which belong to the Γ irreducible representation. Using the cubic basis functions this is written

$$\phi(\Gamma, \gamma) = \sum_{\gamma_1, \gamma_2} \phi(\Gamma_1, \gamma_1)\, \phi(\Gamma_2, \gamma_2) \langle \Gamma_1 \Gamma_2 \gamma_1 \gamma_2 | \Gamma \gamma \rangle \tag{3B.1}$$

with the inverse relationship

$$\phi(\Gamma_1, \gamma_1)\, \phi(\Gamma_2, \gamma_2) = \sum_{\Gamma, \gamma} \phi(\Gamma, \gamma) \langle \Gamma \gamma | \Gamma_1 \Gamma_2 \gamma_1 \gamma_2 \rangle. \tag{3B.2}$$

For trigonal basis functions we have

$$\phi(\Gamma, M) = \sum_{M_1, M_2} \phi(\Gamma_1, M_1)\, \phi(\Gamma_2, M_2) \langle \Gamma_1 \Gamma_2 M_1 M_2 | \Gamma M \rangle. \tag{3B.3}$$

The $\langle \Gamma_1 \Gamma_2 \gamma_1 \gamma_2 | \Gamma \gamma \rangle$ and $\langle \Gamma_1 \Gamma_2 M_1 M_2 | \Gamma M \rangle$ factors are the Clebsch–Gordan coefficients and are the elements of an unitary matrix. With cubic basis functions the Clebsch–Gordan coefficients are real, while for the trigonal basis functions we have

$$\langle \Gamma_1 \Gamma_2 M_1 M_2 | \Gamma M \rangle = \langle \Gamma M | \Gamma_1 \Gamma_2 M_1 M_2 \rangle^*. \tag{3B.4}$$

Values of the Clebsch–Gordan coefficients (taken from Sugano *et al.* 1970) are given in Tables 3B.1 and 3B.2.

Table 3B.1

Clebsch–Gordan coefficients, $\langle \Gamma_1 \Gamma_2 \gamma_1 \gamma_2 | \Gamma \gamma \rangle$, with cubic bases ($\langle \Gamma_1 \Gamma_2 \gamma_1 \gamma_2 | \Gamma \gamma \rangle = \langle \Gamma_2 \Gamma_1 \gamma_2 \gamma_1 | \Gamma \gamma \rangle$ when $\Gamma_1 \neq \Gamma_2$)

$A_2 \times A_2$

		Γ	A_1
γ_1	γ_2	γ	e_1
e_2	e_2		-1

$A_2 \times T_2$

		Γ	T_1		
γ_1	γ_2	γ	α	β	γ
	ξ		-1	0	0
e_2	η		0	-1	0
	ζ		0	0	-1

$A_2 \times E$

γ_1	γ_2	Γ / γ: E u	E v
e_2	u	0	-1
	v	1	0

$E \times E$

γ_1	γ_2	Γ / γ: A_1 e_1	A_2 e_2	E u	E v
u	u	$\frac{1}{\sqrt{2}}$	0	$-\frac{1}{\sqrt{2}}$	0
	v	0	$\frac{1}{\sqrt{2}}$	0	$\frac{1}{\sqrt{2}}$
v	u	0	$-\frac{1}{\sqrt{2}}$	0	$\frac{1}{\sqrt{2}}$
	v	$\frac{1}{\sqrt{2}}$	0	$\frac{1}{\sqrt{2}}$	0

$A_2 \times T_1$

γ_1	γ_2	Γ / γ: T_2 ξ	T_2 η	T_2 ζ
	α	1	0	0
e_2	β	0	1	0
	γ	0	0	1

$E \times T_1$

γ_1	γ_2	Γ / γ: T_1 α	T_1 β	T_1 γ	T_2 ξ	T_2 η	T_2 ζ
	α	$-\frac{1}{2}$	0	0	$\frac{\sqrt{3}}{2}$	0	0
u	β	0	$-\frac{1}{2}$	0	0	$-\frac{\sqrt{3}}{2}$	0
	γ	0	0	1	0	0	0
	α	$\frac{\sqrt{3}}{2}$	0	0	$\frac{1}{2}$	0	0
v	β	0	$-\frac{\sqrt{3}}{2}$	0	0	$\frac{1}{2}$	0
	γ	0	0	0	0	0	-1

$E \times T_2$

γ_1	γ_2	Γ / γ: T_1 α	T_1 β	T_1 γ	T_2 ξ	T_2 η	T_2 ζ
	ξ	$-\frac{\sqrt{3}}{2}$	0	0	$-\frac{1}{2}$	0	0
u	η	0	$\frac{\sqrt{3}}{2}$	0	0	$-\frac{1}{2}$	0
	ζ	0	0	0	0	0	1
	ξ	$-\frac{1}{2}$	0	0	$\frac{\sqrt{3}}{2}$	0	0
v	η	0	$-\frac{1}{2}$	0	0	$-\frac{\sqrt{3}}{2}$	0
	ζ	0	0	1	0	0	0

$T_1 \times T_1$

Γ		A_1	E		T_1			T_2		
γ_1	γ_2 γ	e_1	u	v	α	β	γ	ξ	η	ζ
	α	$-\frac{1}{\sqrt{3}}$	$\frac{1}{\sqrt{6}}$	$-\frac{1}{\sqrt{2}}$	0	0	0	0	0	0
α	β	0	0	0	0	0	$-\frac{1}{\sqrt{2}}$	0	0	$-\frac{1}{\sqrt{2}}$
	γ	0	0	0	0	$\frac{1}{\sqrt{2}}$	0	0	$-\frac{1}{\sqrt{2}}$	0
	α	0	0	0	0	0	$\frac{1}{\sqrt{2}}$	0	0	$-\frac{1}{\sqrt{2}}$
β	β	$-\frac{1}{\sqrt{3}}$	$\frac{1}{\sqrt{6}}$	$\frac{1}{\sqrt{2}}$	0	0	0	0	0	0
	γ	0	0	0	$-\frac{1}{\sqrt{2}}$	0	0	$-\frac{1}{\sqrt{2}}$	0	0
	α	0	0	0	0	$-\frac{1}{\sqrt{2}}$	0	0	$-\frac{1}{\sqrt{2}}$	0
γ	β	0	0	0	$\frac{1}{\sqrt{2}}$	0	0	$-\frac{1}{\sqrt{2}}$	0	0
	γ	$-\frac{1}{\sqrt{3}}$	$-\frac{2}{\sqrt{6}}$	0	0	0	0	0	0	0

$T_1 \times T_2$

Γ		A_2	E		T_1			T_2		
γ_1	γ_2 γ	e_2	u	v	α	β	γ	ξ	η	ζ
	ξ	$-\frac{1}{\sqrt{3}}$	$-\frac{1}{\sqrt{2}}$	$-\frac{1}{\sqrt{6}}$	0	0	0	0	0	0
α	η	0	0	0	0	0	$\frac{1}{\sqrt{2}}$	0	0	$-\frac{1}{\sqrt{2}}$
	ζ	0	0	0	0	$\frac{1}{\sqrt{2}}$	0	0	$\frac{1}{\sqrt{2}}$	0
	ξ	0	0	0	0	0	$\frac{1}{\sqrt{2}}$	0	0	$\frac{1}{\sqrt{2}}$
β	η	$-\frac{1}{\sqrt{3}}$	$\frac{1}{\sqrt{2}}$	$-\frac{1}{\sqrt{6}}$	0	0	0	0	0	0
	ζ	0	0	0	$\frac{1}{\sqrt{2}}$	0	0	$-\frac{1}{\sqrt{2}}$	0	0
	ξ	0	0	0	0	$\frac{1}{\sqrt{2}}$	0	0	$-\frac{1}{\sqrt{2}}$	0
γ	η	0	0	0	$\frac{1}{\sqrt{2}}$	0	0	$\frac{1}{\sqrt{2}}$	0	0
	ζ	$-\frac{1}{\sqrt{3}}$	0	$\frac{2}{\sqrt{6}}$	0	0	0	0	0	0

$T_2 \times T_2$

γ_1	γ_2	Γ / γ	A_1 e_1	E u	E v	T_1 α	T_1 β	T_1 γ	T_2 ξ	T_2 η	T_2 ζ
ξ	ξ		$\frac{1}{\sqrt{3}}$	$-\frac{1}{\sqrt{6}}$	$\frac{1}{\sqrt{2}}$	0	0	0	0	0	0
	η		0	0	0	0	0	$\frac{1}{\sqrt{2}}$	0	0	$\frac{1}{\sqrt{2}}$
	ζ		0	0	0	0	$-\frac{1}{\sqrt{2}}$	0	0	$\frac{1}{\sqrt{2}}$	0
η	ξ		0	0	0	0	0	$-\frac{1}{\sqrt{2}}$	0	0	$\frac{1}{\sqrt{2}}$
	η		$\frac{1}{\sqrt{3}}$	$-\frac{1}{\sqrt{6}}$	$-\frac{1}{\sqrt{2}}$	0	0	0	0	0	0
	ζ		0	0	0	$\frac{1}{\sqrt{2}}$	0	0	$\frac{1}{\sqrt{2}}$	0	0
ζ	ξ		0	0	0	0	$\frac{1}{\sqrt{2}}$	0	0	$\frac{1}{\sqrt{2}}$	0
	η		0	0	0	$-\frac{1}{\sqrt{2}}$	0	0	$\frac{1}{\sqrt{2}}$	0	0
	ζ		$\frac{1}{\sqrt{3}}$	$\frac{2}{\sqrt{6}}$	0	0	0	0	0	0	0

Table 3B.2

Clebsch–Gordan Coefficients with Trigonal Bases

$\langle \Gamma_1 \Gamma_2 M_1 M_2 | \Gamma M \rangle = \langle \Gamma M | \Gamma_1 \Gamma_2 M_1 M_2 \rangle^*$

$(\langle \Gamma_1 \Gamma_2 M_1 M_2 | \Gamma M \rangle = \langle \Gamma_2 \Gamma_1 M_2 M_1 | \Gamma M \rangle$ when $\Gamma_1 \neq \Gamma_2)$

$A_2 \times E$

M_1	M_2	Γ / M: E u_+	E u_-
e_2	u_+	$-i$	0
	u_-	0	i

$A_2 \times T_2$

M_1	M_2	Γ / M: T_1 a_+	T_1 a_-	T_1 a_0
e_2	x_+	-1	0	0
	x_-	0	-1	0
	x_0	0	0	-1

$A_2 \times T_1$

M_1	M_2	Γ / M: T_2 x_+	T_2 x_-	T_2 x_0
e_2	a_+	1	0	0
	a_-	0	1	0
	a_0	0	0	1

$E \times E$

M_1	M_2	Γ / M: A_1 e_1	A_2 e_2	E u_+	E u_-
u_+	u_+	0	0	0	-1
	u_-	$-\frac{1}{\sqrt{2}}$	$-\frac{i}{\sqrt{2}}$	0	0
u_-	u_+	$-\frac{1}{\sqrt{2}}$	$\frac{i}{\sqrt{2}}$	0	0
	u_-	0	0	1	0

$E \times T_1$

Γ		T_1			T_2		
$M_1\ M_2$	M	a_+	a_-	a_0	x_+	x_-	x_0
$u_+\ a_+$		0	$\frac{1}{\sqrt{2}}$	0	0	$\frac{i}{\sqrt{2}}$	0
$u_+\ a_-$		0	0	$-\frac{1}{\sqrt{2}}$	0	0	$-\frac{i}{\sqrt{2}}$
$u_+\ a_0$		$\frac{1}{\sqrt{2}}$	0	0	$\frac{i}{\sqrt{2}}$	0	0
$u_-\ a_+$		0	0	$-\frac{1}{\sqrt{2}}$	0	0	$\frac{i}{\sqrt{2}}$
$u_-\ a_-$		$-\frac{1}{\sqrt{2}}$	0	0	$\frac{i}{\sqrt{2}}$	0	0
$u_-\ a_0$		0	$\frac{1}{\sqrt{2}}$	0	0	$-\frac{i}{\sqrt{2}}$	0

$E \times T_2$

Γ		T_1			T_2		
$M_1\ M_2$	M	a_+	a_-	a_0	x_+	x_-	x_0
$u_+\ x_+$		0	$-\frac{i}{\sqrt{2}}$	0	0	$\frac{1}{\sqrt{2}}$	0
$u_+\ x_-$		0	0	$\frac{i}{\sqrt{2}}$	0	0	$-\frac{1}{\sqrt{2}}$
$u_+\ x_0$		$-\frac{i}{\sqrt{2}}$	0	0	$\frac{1}{\sqrt{2}}$	0	0
$u_-\ x_+$		0	0	$-\frac{i}{\sqrt{2}}$	0	0	$-\frac{1}{\sqrt{2}}$
$u_-\ x_-$		$-\frac{i}{\sqrt{2}}$	0	0	$-\frac{1}{\sqrt{2}}$	0	0
$u_-\ x_0$		0	$\frac{i}{\sqrt{2}}$	0	0	$\frac{1}{\sqrt{2}}$	0

$T_1 \times T_1$

Γ		A_1	E		T_1			T_2		
M_1	M_2 M	e_1	u_+	u_-	a_+	a_-	a_0	x_+	x_-	x_0
	a_+	0	0	$-\frac{1}{\sqrt{3}}$	0	0	0	0	$-\frac{\sqrt{2}}{\sqrt{3}}$	0
a_+	a_-	$\frac{1}{\sqrt{3}}$	0	0	0	0	$\frac{i}{\sqrt{2}}$	0	0	$-\frac{1}{\sqrt{6}}$
	a_0	0	$-\frac{1}{\sqrt{3}}$	0	$\frac{i}{\sqrt{2}}$	0	0	$\frac{1}{\sqrt{6}}$	0	0
	a_+	$\frac{1}{\sqrt{3}}$	0	0	0	0	$-\frac{i}{\sqrt{2}}$	0	0	$-\frac{1}{\sqrt{6}}$
a_-	a_-	0	$\frac{1}{\sqrt{3}}$	0	0	0	0	$\frac{\sqrt{2}}{\sqrt{3}}$	0	0
	a_0	0	0	$-\frac{1}{\sqrt{3}}$	0	$-\frac{i}{\sqrt{2}}$	0	0	$\frac{1}{\sqrt{6}}$	0
	a_+	0	$-\frac{1}{\sqrt{3}}$	0	$-\frac{i}{\sqrt{2}}$	0	0	$\frac{1}{\sqrt{6}}$	0	0
a_0	a_-	0	0	$-\frac{1}{\sqrt{3}}$	0	$\frac{i}{\sqrt{2}}$	0	0	$\frac{1}{\sqrt{6}}$	0
	a_0	$-\frac{1}{\sqrt{3}}$	0	0	0	0	0	0	0	$-\frac{\sqrt{2}}{\sqrt{3}}$

$T_1 \times T_2$

Γ		A_2	E		T_1			T_2		
M_1	M_2 M	e_2	u_+	u_-	a_+	a_-	a_0	x_+	x_-	x_0
	x_+	0	0	$-\frac{i}{\sqrt{3}}$	0	$\sqrt{\frac{2}{3}}$	0	0	0	0
a_+	x_-	$\frac{1}{\sqrt{3}}$	0	0	0	0	$\frac{1}{\sqrt{6}}$	0	0	$\frac{i}{\sqrt{2}}$
	x_0	0	$\frac{i}{\sqrt{3}}$	0	$-\frac{1}{\sqrt{6}}$	0	0	$\frac{i}{\sqrt{2}}$	0	0
	x_+	$\frac{1}{\sqrt{3}}$	0	0	0	0	$\frac{1}{\sqrt{6}}$	0	0	$-\frac{i}{\sqrt{2}}$
a_-	x_-	0	$-\frac{i}{\sqrt{3}}$	0	$-\sqrt{\frac{2}{3}}$	0	0	0	0	0
	x_0	0	0	$-\frac{i}{\sqrt{3}}$	0	$-\frac{1}{\sqrt{6}}$	0	0	$-\frac{i}{\sqrt{2}}$	0
	x_+	0	$\frac{i}{\sqrt{3}}$	0	$-\frac{1}{\sqrt{6}}$	0	0	$-\frac{i}{\sqrt{2}}$	0	0
a_0	x_-	0	0	$-\frac{i}{\sqrt{3}}$	0	$-\frac{1}{\sqrt{6}}$	0	0	$\frac{i}{\sqrt{2}}$	0
	x_0	$-\frac{1}{\sqrt{3}}$	0	0	0	0	$\sqrt{\frac{2}{3}}$	0	0	0

$T_2 \times T_2$

		Γ	A_1	E		T_1			T_2		
M_1	M_2	M	e_1	u_+	u_-	a_+	a_-	a_0	x_+	x_-	x_0
	x_+		0	0	$\frac{1}{\sqrt{3}}$	0	0	0	0	$\sqrt{\frac{2}{3}}$	0
x_+	x_-		$-\frac{1}{\sqrt{3}}$	0	0	0	0	$-\frac{i}{\sqrt{2}}$	0	0	$\frac{1}{\sqrt{6}}$
	x_0		0	$\frac{1}{\sqrt{3}}$	0	$-\frac{i}{\sqrt{2}}$	0	0	$-\frac{1}{\sqrt{6}}$	0	0
	x_+		$-\frac{1}{\sqrt{3}}$	0	0	0	0	$\frac{i}{\sqrt{2}}$	0	0	$\frac{1}{\sqrt{6}}$
x_-	x_-		0	$-\frac{1}{\sqrt{3}}$	0	0	0	0	$-\sqrt{\frac{2}{3}}$	0	0
	x_0		0	0	$\frac{1}{\sqrt{3}}$	0	$\frac{i}{\sqrt{2}}$	0	0	$-\frac{1}{\sqrt{6}}$	0
	x_+		0	$\frac{1}{\sqrt{3}}$	0	$\frac{i}{\sqrt{2}}$	0	0	$-\frac{1}{\sqrt{6}}$	0	0
x_0	x_-		0	0	$\frac{1}{\sqrt{3}}$	0	$-\frac{i}{\sqrt{2}}$	0	0	$-\frac{1}{\sqrt{6}}$	0
	x_0		$\frac{1}{\sqrt{3}}$	0	0	0	0	0	0	0	$\sqrt{\frac{2}{3}}$

4
Radiative transition rates and selection rules

ALTHOUGH the discussion in the preceding two chapters concentrated on the stationary states of an electronic system, the need to consider perturbations which mix stationary states and thereby excite transitions between these states was noted. In this chapter we apply time-dependent perturbation theory to estimate the rates of such transitions, first by developing a general formula for the transition probability between initial $|i\rangle$ and final $|f\rangle$ states; if the initial state $|i\rangle$ is lower in energy than $|f\rangle$ an absorption transition is involved, whereas if $|f\rangle$ is the lower state emission takes place. We then consider in some detail how such transitions are caused by interactions between the electronic system and the radiation field.

4.1 Time-dependent perturbation theory

In the absence of perturbations the quantum system in described by a Hamiltonian $\mathscr{H}$ with eigenstates and eigenvalues $|n\rangle$ and E_n, respectively, where n refers to the set of labels which characterizes the eigenstate. The system is now perturbed by an interaction $\mathscr{H}_1$ which we speak of as the perturbation; we are interested in the action of $\mathscr{H}_1$ in causing transitions out of a state $|i\rangle$. Because of the perturbation the Hamiltonian is now $\mathscr{H}+\mathscr{H}_1$ and we seek new eigenstates which we can write as linear combinations of the original stationary states

$$\Psi=\sum_n c_n(t)|n\rangle \exp(-\mathrm{i}E_n t/\hbar) \tag{4.1}$$

and Ψ satisfies the Schrödinger equation

$$(\mathscr{H}+\mathscr{H}_1)\sum_n c_n(t)|n\rangle \exp(-\mathrm{i}E_n t/\hbar)$$
$$=\mathrm{i}\hbar\frac{\partial}{\partial t}\sum_n c_n(t)|n\rangle \exp(-\mathrm{i}E_n t/\hbar) \tag{4.2}$$

Since $\mathscr{H}|n\rangle=E_n|n\rangle$, eqn (4.2) reduces to

$$\sum_n c_n(t)\mathscr{H}_1|n\rangle \exp(-\mathrm{i}E_n t/\hbar)=\mathrm{i}\hbar\sum_n \frac{\mathrm{d}c_n}{\mathrm{d}t}|n\rangle \exp(-\mathrm{i}E_n t/\hbar). \tag{4.3}$$

Multiplying from the left by $\langle f|$ and integrating both sides of eqn (4.3) with respect to spatial coordinates one obtains by using the orthogonality property of the $|n\rangle$ states

$$\frac{\mathrm{d}c_f}{\mathrm{d}t} = -\frac{\mathrm{i}}{\hbar}\sum_n c_n(t)\langle f|\mathscr{H}_1|n\rangle \exp(\mathrm{i}\omega_{fn}t) \tag{4.4}$$

where $\hbar\omega_{fn} = E_f - E_n$. Equation (4.4) is a set of coupled equations. Let us assume that at $t=0$ the system is in an initial state $|n\rangle = |i\rangle$: $c_i(0)=1$, $c_n(0)=0$ for $n \neq i$. We further assume that the perturbation $\mathscr{H}_1$ is weak so that for a period after $t=0$ we can replace $c_n(t)$ by $c_n(0)$ on the right-hand side of eqn (4.4) which then becomes

$$\frac{\mathrm{d}c_f}{\mathrm{d}t} = -\frac{\mathrm{i}}{\hbar}\langle f|\mathscr{H}_1|i\rangle \exp(\mathrm{i}\omega_{fi}t). \tag{4.5}$$

The perturbations which are of interest are real and oscillatory in time with angular velocity ω and have the form

$$\mathscr{H}_1 = V\exp(-\mathrm{i}\omega t) + V^* \exp(\mathrm{i}\omega t) \tag{4.6}$$

where V is a function of spatial coordinates only. Inserting eqn (4.6) into eqn (4.5) gives

$$\frac{\mathrm{d}c_f(t)}{\mathrm{d}t} = -\frac{\mathrm{i}}{\hbar}[V_{fi}\exp \mathrm{i}(\omega_{fi}-\omega)t + V^*_{fi}\exp \mathrm{i}(\omega_{fi}+\omega)t] \tag{4.7}$$

where $V_{fi} = \langle f|V|i\rangle$ and $V^*_{fi} = \langle f|V^*|i\rangle$. Integrating over time we find

$$c_f(t) = -\frac{\mathrm{i}}{\hbar}\left[V_{fi}\left(\frac{\exp \mathrm{i}(\omega_{fi}-\omega)t - 1}{\mathrm{i}(\omega_{fi}-\omega)}\right) + V^*_{fi}\left(\frac{\exp \mathrm{i}(\omega_{fi}+\omega)t - 1}{\mathrm{i}(\omega_{fi}+\omega)}\right)\right]. \tag{4.8}$$

The value of $|c_f(t)|^2$ is the probability that the electronic system, initially in state i, will be found in state f after time t. The value of $c_f(t)$ will only be significant when one of the denominators in eqn (4.8) becomes zero. We consider two cases separately.

1. $\omega_{fi} - \omega = 0$. ω is a positive quantity, the angular velocity of the oscillatory perturbation. We require $\omega = \omega_{fi} = (E_f - E_i)/\hbar$, i.e.

$$E_f - E_i = \hbar\omega. \tag{4.9}$$

State f is higher in energy than state i; this is an *absorption* process.

2. $\omega_{fi} + \omega = 0$. In this case state f is lower in energy than state i by an amount $\hbar\omega$ and we are dealing with an *emission* process.

Let us consider the absorption process in more detail. We have

$$c_f(t) = -\frac{V_{fi}}{\hbar}\left[\frac{\exp i(\omega_{fi}-\omega)t-1}{\omega_{fi}-\omega}\right]$$

$$= -\frac{V_{fi}}{\hbar}\left[\frac{\exp i(\omega_{fi}-\omega)t/2\,\{\exp i(\omega_{fi}-\omega)t/2-\exp -i(\omega_{fi}-\omega)t/2\}}{(\omega_{fi}-\omega)}\right]. \tag{4.10}$$

This gives

$$|c_f(t)|^2 = 4\left|\frac{V_{fi}}{\hbar}\right|^2 \frac{\sin^2\frac{1}{2}(\omega_{fi}-\omega)t}{(\omega_{fi}-\omega)^2}. \tag{4.11}$$

Figure 4.1, which plots $\sin^2[(\omega_{fi}-\omega)t/2]/(\omega_{fi}-\omega)^2$, has a maximum when $\omega_{fi}=\omega$, showing that the absorption transition probability is a maximum when $E_f - E_i = \hbar\omega$.

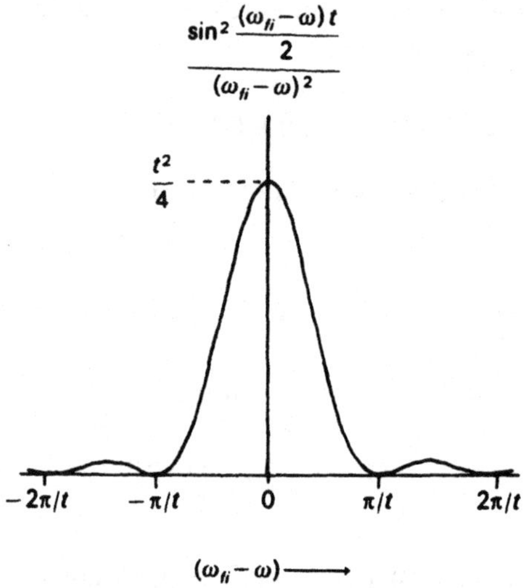

FIG. 4.1. A plot of $[\sin^2(\omega_{fi}-\omega)t/2]/(\omega_{fi}-\omega)^2$ against $(\omega_{fi}-\omega)$. This has a maximum when $\omega=\omega_{fi}$, that is when $\hbar\omega = E_f - E_i$.

In the radiative absorptive process induced by interaction with the radiation field the electronic energy difference is made up by extracting a photon of energy $\hbar\omega$ from the radiation field. Conservation of energy requires $E_f - E_i = \hbar\omega$. This is a resonance condition. Figure 4.1 shows also, however, that the absorption process can occur for frequencies *near* resonance, this discrepancy in energy being within the uncertainty in energy allowed by the Heisenberg uncertainty principle.

If the perturbation is due to a large number of oscillatory modes, as is the case when the electronic system interacts with the radiation field, there is a quasicontinuum of values of ω and the perturbation has the form

$$H_1 = \sum_{\omega} V^{(\omega)} \exp(-i\omega t) + \text{complex conjugate}. \tag{4.12}$$

The probability of a transition from i to f in a time t due to this perturbation is found by summing over the probabilities for all the modes individually:

$$|c_f(t)|^2 = \sum_{\omega} \left| \frac{V_{fi}^{(\omega)}}{\hbar} \right|^2 \frac{\sin^2 \frac{1}{2}(\omega_{fi} - \omega)t}{[\frac{1}{2}(\omega_{fi} - \omega)]^2}. \tag{4.13}$$

If the density of modes is large we can replace the summation by an integration. The density of modes—the number of distinct modes per unit angular velocity—is written as $\rho(\omega)$. The probability can be written

$$|c_f(t)|^2 = \int \left| \frac{V_{fi}^{(\omega)}}{\hbar} \right|^2 \frac{\sin^2 \frac{1}{2}(\omega_{fi} - \omega)t}{[\frac{1}{2}(\omega_{fi} - \omega)]^2} \rho(\omega) \mathrm{d}\omega. \tag{4.14}$$

As Fig. 4.1 shows, only a narrow band of modes, those with $\omega \approx \omega_{fi}$, contributes to the absorption probability. Assuming that the perturbation $V(\omega)$ can be regarded as constant over this range of modes, and that $\rho(\omega)$ varies sufficiently slowly that $\rho(\omega)$ may be replaced by $\rho(\omega_{fi})$ over this range of modes, then we find that

$$\begin{aligned} |c_f(t)|^2 &= \left| \frac{V_{fi}^{(\omega)}}{\hbar} \right|^2 \rho(\omega_{fi}) \int_0^{\infty} \frac{\sin^2 \frac{1}{2}(\omega_{fi} - \omega)t}{[\frac{1}{2}(\omega_{fi} - \omega)]^2} \mathrm{d}\omega \\ &= \left| \frac{V_{fi}^{(\omega)}}{\hbar} \right|^2 \rho(\omega_{fi}) 2t \int_{-\infty}^{+\infty} \frac{\sin^2 X}{X^2} \mathrm{d}X. \end{aligned} \tag{4.15}$$

We have assumed that the $\sin^2 X/X^2$ function has fallen to zero at $X = \omega_{fi} t/2$ so that the upper limit to the integral can be replaced by $+\infty$. The value of the integral is π. Therefore eqn (4.15) becomes

$$|c_f(t)|^2 = \frac{2\pi}{\hbar^2} t \, |V_{fi}^{(\omega)}|^2 \rho(\omega = \omega_{fi}). \tag{4.16}$$

This formula, derived for an absorption process when the state $|f\rangle$ is higher in energy than state $|i\rangle$, was obtained by considering the first term in eqn (4.8). For the emission process, where the final state, f, is lower in energy than the initial state, i, ω_{fi} is negative and the second term in eqn (4.8), becomes effective.

Consider now the processes of absorption and emission between the ground state, a, and an excited state, b, which involve transition probabilities determined by $|V_{ba}^{(\omega)}|^2$ and $|V_{ab}^{(\omega)*}|^2$, respectively. Since $|V_{ba}^{(\omega)}| = |V_{ab}^{(\omega)*}|$ the transition probabilities for absorption and emission are equal. We define the transition probability, W_{if}, as the probability of a transition $|i\rangle \rightarrow |f\rangle$ per unit time. We can write

$$W_{if} = \frac{2\pi}{\hbar^2} |V_{fi}^{(\omega)}|^2 \rho(\omega) \tag{4.17}$$

where $\hbar\omega = \pm(E_f - E_i)$. The positive sign implies an absorption transition and the negative sign an emission transition. This equation is known as *Fermi's Golden Rule.*

4.2 The radiation field

The perturbation responsible for optical transitions in atomic centres is the interaction between the atomic centre and the electromagnetic field. In this section we review the properties of the electromagnetic field, defined by stating the values of E and B at all positions and at all times. These field values are given by Maxwell's equations in terms of charge density and current density. The electromagnetic field will be assumed to be contained in a cube of linear dimension L and volume V, and this space is assumed to be filled with a linear isotropic non-magnetic material. If there are no charges other than the bound charges in the material and no conventional current except for any vibratory motion of the bound charges then Maxwell's equations have a particularly simple form:

$$\begin{aligned} \nabla . \boldsymbol{E} &= 0 \\ \nabla . \boldsymbol{B} &= 0 \\ \nabla \times \boldsymbol{E} &= -\frac{\partial B}{\partial t} \\ \nabla \times \boldsymbol{B} &= \mu_0 \varepsilon_0 \kappa \frac{\partial E}{\partial t} \end{aligned} \tag{4.18}$$

where κ is the dielectric constant of the material. $\boldsymbol{E}$ and $\boldsymbol{B}$ expressed in terms of a vector potential $\boldsymbol{A}$ and a scalar potential ϕ are

$$\boldsymbol{E} = -\nabla\phi - \frac{\partial \boldsymbol{A}}{\partial t} \tag{4.19a}$$

$$\boldsymbol{B} = \nabla \times \boldsymbol{A}. \tag{4.19b}$$

$\boldsymbol{A}$ and ϕ are not uniquely defined by these equations. To see this we consider $\boldsymbol{A}' = \boldsymbol{A} + \nabla f$, f being any scalar function of space and time. Since $\nabla \times \nabla f = 0$, $\nabla \times \boldsymbol{A}' = \nabla \times \boldsymbol{A}$ and so there is an indeterminacy in the definition of $\boldsymbol{A}$. Since $\boldsymbol{A}$ and ϕ are both involved in the definition of the electric field strength, $\boldsymbol{E}$, there is also an indeterminacy in ϕ. Consider $\phi'=\phi-(\partial f/\partial t)$. Then

$$-\nabla\phi'-\frac{\partial A'}{\partial t}=-\nabla\phi+\frac{\partial}{\partial t}\nabla f-\frac{\partial A}{\partial t}-\frac{\partial}{\partial t}\nabla f$$

$$=-\nabla\phi-\frac{\partial A}{\partial t}$$

$$=E. \tag{4.20}$$

So ϕ' and A' are equally good potentials where

$$\phi'=\phi-\frac{\partial f}{\partial t}$$

$$A'=A+\nabla f. \tag{4.21}$$

These are called *gauge transformations.*

To define A and ϕ without indeterminacy we need to introduce an additional condition. The *Coulomb gauge*, which requires that

$$\nabla \cdot A=0 \tag{4.22}$$

is convenient for discussion of radiation problems. Since $\nabla \cdot E=0$ we find from eqn (4.19a) that $0=-\nabla^2\phi-(\partial/\partial t)\nabla \cdot A$, giving

$$\nabla^2\phi=0 \quad \text{(Poisson's equation)}. \tag{4.23}$$

In the Coulomb gauge ϕ is a function of the distribution of charge, but for a linear isotropic material the bound charge density is zero, and in the absence of any free charges we can take $\phi=0$. We now have

$$E=-\frac{\partial A}{\partial t} \tag{4.24a}$$

and

$$B=\nabla\times A. \tag{4.24b}$$

Inserting these into the fourth Maxwell equation we derive the following differential equation for A:

$$\nabla^2 A=\mu_0\varepsilon_0\kappa\frac{\partial^2 A}{\partial t^2}=\left(\frac{n}{c}\right)^2\frac{\partial^2 A}{\partial t^2} \tag{4.25}$$

where $\mu_0\varepsilon_0=c^{-2}$, c is the velocity of light *in vacuo*, and n is the refractive index ($n=\kappa^{\frac{1}{2}}$).

The solutions of this differential equation which concern us are the travelling electromagnetic waves:

$$A_k^0\exp \mathrm{i}(k \cdot r-\omega t) \tag{4.26}$$

where k and ω are related by

$$\frac{\omega}{k}=\frac{c}{n}=v \tag{4.27}$$

and where v is the velocity of the wave and $\boldsymbol{k}$ is the wavevector which points in the direction of propagation of the wave. The wavelength is

$$\lambda=\frac{2\pi}{k} \tag{4.28}$$

where k is the magnitude of the wavevector. We impose *periodic boundary conditions* on these travelling wave solutions:

$$\exp \mathrm{i}\boldsymbol{k}.(\boldsymbol{r}+L\hat{\boldsymbol{x}})=\exp \mathrm{i}\boldsymbol{k}.\boldsymbol{r}, \text{ etc.,} \tag{4.29}$$

where $\hat{\boldsymbol{x}}$ indicates a unit vector in the x direction. From this we find that the k values are restricted to $\boldsymbol{k}=k_x\hat{\boldsymbol{x}}+k_y\hat{\boldsymbol{y}}+k_z\hat{\boldsymbol{z}}$, where

$$k_x=\frac{2\pi n_x}{L}, \quad k_y=\frac{2\pi n_y}{L}, \quad k_z=\frac{2\pi n_z}{L}, \tag{4.30}$$

n_x, n_y, n_z being positive or negative integers. For a $\boldsymbol{k}$ value satisfying these conditions, the value of angular velocity is

$$\omega=\omega_k=(ck/n). \tag{4.31}$$

The solutions are characterized by the values of the wavevector, $\boldsymbol{k}$.

Using eqn (4.30) the number of different k values whose magnitude lies between k and $k+\mathrm{d}k$ can be written as

$$4\pi\left(\frac{L}{2\pi}\right)^3 k^2\,\mathrm{d}k=\frac{V}{2\pi^2}k^2\,\mathrm{d}k. \tag{4.32}$$

Substituting from eqn (4.31) into eqn (4.32) defines the number of solutions with angular velocity between ω and $\omega+\mathrm{d}\omega$, i.e. the *density of* $\boldsymbol{k}$*-modes*, $\rho_k(\omega)$, as

$$\rho_k(\omega)=\frac{V}{2\pi^2 v^3}\omega^2. \tag{4.33}$$

For each $\boldsymbol{k}$-mode there are two distinct polarization modes, i.e. two distinct directions of the $\boldsymbol{E}$ vector of the radiation.

The general solution for $\boldsymbol{A}(\boldsymbol{r}, t)$ can now be written

$$\boldsymbol{A}(\boldsymbol{r}, t)=\sum_{\substack{\text{all}\\ \text{modes}}}[\boldsymbol{A}_k^{(0)}\exp \mathrm{i}(\boldsymbol{k}.\boldsymbol{r}-\omega t)+\boldsymbol{A}_k^{(0)*}\exp -\mathrm{i}(\boldsymbol{k}.\boldsymbol{r}-\omega t)] \tag{4.34}$$

where the summation is over all $\boldsymbol{k}$-modes and all polarization modes. The requirement that the coefficients in front of the two oscillating terms be complex conjugates of each other follows from the requirement that $\boldsymbol{E}$ and $\boldsymbol{B}$, which are calculated from $\boldsymbol{A}$ (eqn 4.24), are real. The expressions for $\boldsymbol{E}$ and $\boldsymbol{B}$ are

$$\boldsymbol{E}(\boldsymbol{r}, t) = \sum_{\text{modes}} [(\boldsymbol{E}_k^{(0)} \exp \mathrm{i}(\boldsymbol{k}.\boldsymbol{r} - \omega t) + \boldsymbol{E}_k^{(0)*} \exp - \mathrm{i}(\boldsymbol{k}.\boldsymbol{r} - \omega t)] \quad (4.35)$$

$$\boldsymbol{B}(\boldsymbol{r}, t) = \sum_{\text{modes}} [\boldsymbol{B}_k^{(0)} \exp \mathrm{i}(\boldsymbol{k}.\boldsymbol{r} - \omega t) + \boldsymbol{B}_k^{(0)*} \exp - \mathrm{i}(\boldsymbol{k}.\boldsymbol{r} - \omega t)] \quad (4.36)$$

where

$$\boldsymbol{E}_k^{(0)} = \mathrm{i}\omega_k \boldsymbol{A}_k^{(0)}, \qquad \boldsymbol{B}_k^{(0)} = \frac{\boldsymbol{k} \times \boldsymbol{E}_k^{(0)}}{\omega_k}. \quad (4.37)$$

The magnitudes of $\boldsymbol{E}_k^{(0)}$ and $\boldsymbol{B}_k^{(0)}$ are related by

$$|\boldsymbol{B}_k^{(0)}| = \frac{|\boldsymbol{E}_k^{(0)}|}{v}. \quad (4.38)$$

$\boldsymbol{E}_k^{(0)}$ and $\boldsymbol{A}_k^{(0)}$ are parallel to each other. Since $\nabla . \boldsymbol{A} = \nabla . \boldsymbol{E} = 0$ we see that for each mode $\boldsymbol{k}.\boldsymbol{A}_k^{(0)} = 0$, so that $\boldsymbol{A}_k$ and $\boldsymbol{E}_k$ are transverse to the direction of propagation of the $\boldsymbol{k}$-mode, and from eqn (4.37) $\boldsymbol{B}_k$ is transverse to both $\boldsymbol{k}$ and $\boldsymbol{E}_k$.

The electromagnetic energy contained in a k-mode is $2\varepsilon_0 \kappa |\boldsymbol{E}_k^{(0)}|^2 V$. Hence if $\boldsymbol{E}_k^{(0)}$ is a slowly-varying function of k we can express the energy density per unit angular velocity, per unit volume, as

$$u(\omega) = 4\varepsilon_0 \kappa |\boldsymbol{E}_k^{(0)}|^2 \rho_k(\omega) \quad (4.39)$$

where a factor of 2 is included for the two polarization modes, assumed to have the same value of $|\boldsymbol{E}_k^{(0)}|$. The energy density can also be expressed in terms of the number of photons in each allowed mode. If n_ω is the number of photons in a mode of angular velocity ω with a particular polarization which are introduced into the cube V, then the energy introduced at this frequency and polarization is $n_\omega \hbar\omega$. Hence the energy per unit angular velocity and per unit volume in the introduced radiation is

$$u(\omega) = 2n_\omega \hbar\omega \, \rho_k^{(\omega)}/V \quad (4.40)$$

where the factor of 2 takes into account the two senses of polarization. This introduced radiation is in addition to the ever-present *zero-point radiation.*

4.3 Interaction between the electronic centre and the radiation field

We now apply the time-dependent perturbation formula (eqn 4.17) to the interaction between the electronic centre and the radiation field. This enables us to calculate such experimental variables as the lifetime and absorption coefficient of the centre.

4.3.1 The electromagnetic interaction Hamiltonian

The Hamiltonian for the electronic centre in a solid (eqn 2.53) must now be modified to include the interaction of the centre with the electromagnetic field. This requires the replacement of $\boldsymbol{p}_i$ by $\boldsymbol{p}_i + e\boldsymbol{A}$ in the kinetic energy term $\sum_i \boldsymbol{p}_i^2/2m$. Also, the interaction energy of the magnetic moment of the spinning electron with the magnetic field must be added. The extra terms in the Hamiltonian due to these interactions are

$$\sum_i \frac{(\boldsymbol{p}_i + e\boldsymbol{A}(\boldsymbol{r}_i, t))^2}{2m} - \sum_i \frac{\boldsymbol{p}_i^2}{2m} + \frac{e}{m}\sum_i \boldsymbol{s}_i \cdot \boldsymbol{B}(\boldsymbol{r}_i, t) \tag{4.41}$$

where $\boldsymbol{r}_i$ is the position of the ith electron. This perturbation term can be written

$$\mathscr{H}_1 = \sum_i \frac{e}{2m}(\boldsymbol{p}_i \cdot \boldsymbol{A} + \boldsymbol{A} \cdot \boldsymbol{p}_i + 2\boldsymbol{s}_i \cdot \boldsymbol{B} + e^2\boldsymbol{A}^2). \tag{4.42}$$

The $e^2\boldsymbol{A}^2$ term allows interactions between photons through their electrostatic coupling to the atomic centre; for normal light intensities such interactions are negligible. Hence we ignore the $e^2\boldsymbol{A}^2$ term and write $\boldsymbol{p}_i \cdot \boldsymbol{A} = \boldsymbol{A} \cdot \boldsymbol{p}_i - i\hbar\nabla \cdot \boldsymbol{A} = \boldsymbol{A} \cdot \boldsymbol{p}_i$ since we are in the Coulomb gauge. The perturbation is then

$$\mathscr{H}_1 = \frac{e}{m}\sum_i (\boldsymbol{A} \cdot \boldsymbol{p}_i + \boldsymbol{s}_i \cdot \boldsymbol{B}). \tag{4.43}$$

Using eqns (4.34), (4.35), (4.36) and (4.37) this becomes

$$H_1 = \frac{e}{\mathrm{i}m}\sum_{\text{modes}}\sum_i \exp \mathrm{i}(\boldsymbol{k} \cdot \boldsymbol{r}_i - \omega_k t)\left(\frac{\boldsymbol{p}_i \cdot \boldsymbol{E}_k^{(0)}}{\omega_k} + \mathrm{i}\,\boldsymbol{s}_i \cdot \boldsymbol{B}_k^{(0)}\right)$$
$$+ \text{complex conjugate.} \tag{4.44}$$

This is of the form we assumed previously (eqn 4.12) with $V^{(\omega)} = V^{(\omega_k)}$ given by

$$V^{(\omega_k)} = \frac{e}{\mathrm{i}m}\sum_i \exp(\mathrm{i}\,\boldsymbol{k} \cdot \boldsymbol{r}_i)\left(\frac{\boldsymbol{p}_i \cdot \boldsymbol{E}_k^{(0)}}{\omega_k} + \mathrm{i}\boldsymbol{s}_i \cdot \boldsymbol{B}_k^{(0)}\right). \tag{4.45}$$

If one adopts maximum possible values for the expectation values of $\boldsymbol{p}_i$ and $\boldsymbol{s}_i$ then in the optical region the effect of the $\boldsymbol{p}_i \cdot \boldsymbol{E}_k^{(0)}/\omega_k$ term is much larger than that of the $\boldsymbol{s}_i \cdot \boldsymbol{B}_k^{(0)}$ term. And since in the optical region $\boldsymbol{k} \cdot \boldsymbol{r}_i \ll$ unity, $\exp(\mathrm{i}\boldsymbol{k} \cdot \boldsymbol{r})$ can be approximated by $1 + \mathrm{i}\boldsymbol{k} \cdot \boldsymbol{r}$. Retaining only first- and second-order terms $V^{(\omega_k)}$ can be written as

$$V^{(\omega_k)} \simeq \frac{e}{\mathrm{i}m}\sum_i \left(\frac{\boldsymbol{p}_i \cdot \boldsymbol{E}_k^{(0)}}{\omega_k} + \mathrm{i}\frac{(\boldsymbol{k} \cdot \boldsymbol{r}_i)(\boldsymbol{p}_i \cdot \boldsymbol{E}_k^{(0)})}{\omega_k} + \mathrm{i}\,\boldsymbol{s}_i \cdot \boldsymbol{B}_k^{(0)}\right). \tag{4.46}$$

The second term on the right-hand side of eqn (4.46) can be expanded by

writing
$$(\boldsymbol{k}.\boldsymbol{r}_i)(\boldsymbol{E}_k^{(0)}.\boldsymbol{p}_i)=\tfrac{1}{2}[(\boldsymbol{k}.\boldsymbol{r}_i)(\boldsymbol{E}_k^{(0)}.\boldsymbol{p}_i)+(\boldsymbol{E}_k^{(0)}.\boldsymbol{r}_i)(\boldsymbol{k}.\boldsymbol{p}_i)]$$
$$+\tfrac{1}{2}[(\boldsymbol{k}.\boldsymbol{r}_i)(\boldsymbol{E}_k^{(0)}.\boldsymbol{p}_i)-(\boldsymbol{E}_k^{(0)}.\boldsymbol{r}_i)(\boldsymbol{k}.\boldsymbol{p}_i)]. \qquad (4.47)$$

Using the vector relationship

$$(\boldsymbol{a}\times\boldsymbol{b}).(\boldsymbol{c}\times\boldsymbol{d})=(\boldsymbol{a}.\boldsymbol{c})(\boldsymbol{b}.\boldsymbol{d})-(\boldsymbol{a}.\boldsymbol{d})(\boldsymbol{b}.\boldsymbol{c}) \qquad (4.48)$$

and taking care to maintain the order of the $\boldsymbol{r}_i$ and $\boldsymbol{p}_i$ operators the second term on the right in eqn (4.47) becomes with the aid of eqn (4.37)

$$\tfrac{1}{2}(\boldsymbol{k}\times\boldsymbol{E}_k^{(0)}).(\boldsymbol{r}_i\times\boldsymbol{p}_i)=\tfrac{1}{2}\omega_k\boldsymbol{l}_i.\boldsymbol{B}_k^{(0)}. \qquad (4.49)$$

The first term on the right-hand side of eqn (4.47) can be written, using $\boldsymbol{p}_i=m\dot{\boldsymbol{r}}_i$ as

$$\frac{m}{2}[(\boldsymbol{k}.\boldsymbol{r}_i)(\boldsymbol{E}_k^{(0)}.\dot{\boldsymbol{r}}_i)+(\boldsymbol{k}.\dot{\boldsymbol{r}}_i)(\boldsymbol{E}_k^{(0)}.\boldsymbol{r}_i)]$$
$$=\frac{m}{2}\frac{\mathrm{d}}{\mathrm{d}t}[(\boldsymbol{k}.\boldsymbol{r}_i)(\boldsymbol{E}_k^{(0)}.\boldsymbol{r}_i)] \qquad (4.50)$$

Now the only spatial operator in the first term in the expression for $V^{(\omega_k)}$ (eqn 4.46) is $\boldsymbol{p}_i$. Hence the matrix element of $\boldsymbol{p}_i$ between electronic states $|i\rangle$ and $|f\rangle$, obtained using $\boldsymbol{p}_i$ as $m\dot{\boldsymbol{r}}_i$ and the operator relationship

$$\dot{\boldsymbol{r}}=-\frac{i}{\hbar}[\boldsymbol{r},\boldsymbol{H}] \qquad (4.51)$$

is

$$\langle f|\boldsymbol{p}_i|i\rangle=im\omega_{fi}\langle f|\boldsymbol{r}_i|i\rangle \qquad (4.52)$$

where $\hbar\omega_{fi}=E_f-E_i$. Since the conservation of energy requires that $\omega_{fi}=\omega_k$ we can replace the operator $\boldsymbol{p}_i$ in the first part of $V^{(\omega_k)}$ (eqn 4.46) by $im\omega_k\boldsymbol{r}_i$. In a similar way the operator shown in eqn (4.50), which is multiplied by $e/im\omega_k$ in the expression for $V^{(\omega_k)}$, can be replaced by

$$\frac{im\omega_{fi}}{2}(\boldsymbol{k}.\boldsymbol{r}_i)(\boldsymbol{E}_k^{(0)}.\boldsymbol{r}_i)=\frac{im\omega_{fi}}{2}\boldsymbol{r}_i\boldsymbol{r}_i:\boldsymbol{k}\boldsymbol{E}_k^{(0)} \qquad (4.53)$$

which we write as a contraction of two second-rank tensors. With these substitutions the expression for $V^{(\omega_k)}$ becomes

$$V^{(\omega_k)}\simeq\sum_i[e\boldsymbol{r}_i.\boldsymbol{E}_k^{(0)}+\frac{e}{2m}(\boldsymbol{l}_i+2\boldsymbol{s}_i).\boldsymbol{B}_k^{(0)}+\tfrac{1}{2}e\boldsymbol{r}_i\boldsymbol{r}_i:\boldsymbol{k}\boldsymbol{E}_k^{(0)}]. \qquad (4.54)$$

Since this is real it is the appropriate form for both emission and absorption transitions. The first term in this $V^{(\omega_k)}$ formula contains $\sum_i e\boldsymbol{r}_i$ and is the *electric dipole* (ED) term, the second containing $e/2m\sum_i(\boldsymbol{l}_i+2\boldsymbol{s}_i)$ is the *magnetic dipole* (MD) term, while the third term containing $e\boldsymbol{r}_i\boldsymbol{r}_i$ is the *electric quadrupole* (EQ) term. The relative strengths of the three allowed transitions are in the approximate ratio ED:MD:EQ$\simeq(ea_0)^2:(\mu_B/c)^2:(ea_0^2\pi/\lambda)^2$, where a_0 and μ_B are the Bohr radius and Bohr magnetron, respectively. This ratio is approximately $1:10^{-5}:10^{-6}$.

Let us now use the electric dipole term in eqn (4.54) to calculate the interaction of the electronic centre with the radiation field. The value of $|\boldsymbol{E}_k^{(0)}|$ is the same for each k-mode within the bandwidth allowed by the uncertainty principle. However, $\boldsymbol{r}_i.\boldsymbol{E}_k^{(0)}$ can vary with the polarization mode, so that the appropriate form of Fermi's Golden Rule is

$$W_{if}(\text{ED})=\frac{2\pi}{\hbar^2}\sum_{\text{pol}}|\langle f|\sum_i e\boldsymbol{r}_i.\boldsymbol{E}_k^{(0)}|i\rangle|^2\rho_k(\omega) \tag{4.55}$$

where the summation is over the two polarization modes, and ω is the centre frequency of the transition. In the case of radiation in a cavity where all the $\boldsymbol{k}$-modes contain some radiation the appropriate form of $\rho_k(\omega)$ is that given by eqn (4.33), whereas for a directed beam of radiation the form of $\rho_k(\omega)$ would be different. Writing $\boldsymbol{E}_k^{(0)}=E_k^{(0)}\hat{\varepsilon}_k$, where $\hat{\varepsilon}_k$ is a unit polarization vector, and using eqn (4.39) for the energy density in the radiation field we can write $W_{if}(\text{ED})$ as

$$W_{if}(\text{ED})=\frac{\pi}{2\varepsilon_0\kappa\hbar^2}\sum_{\text{pol}}|\langle f|\sum_i e\boldsymbol{r}_i.\hat{\varepsilon}_k|i\rangle|^2\,4\varepsilon_0\kappa|E_k^{(0)}|^2\,\rho_k(\omega)$$

$$=\frac{\pi}{2\varepsilon_0\kappa\hbar^2}\sum_{\text{pol}}|\langle f|\sum_i e\boldsymbol{r}_i.\hat{\varepsilon}_k|i\rangle|^2u(\omega). \tag{4.56}$$

This result, that the transition probability is proportional to the energy density of the radiation, holds also for magnetic dipole and electric quadrupole transitions.

In the case of randomly polarized radiation we have

$$\sum_{\text{pol}}|\langle f|\sum_i e\boldsymbol{r}_i.\hat{\varepsilon}_k|i\rangle|^2=\frac{2}{3}|\langle f|\sum_i e\boldsymbol{r}_i|i\rangle|^2. \tag{4.57}$$

On the other hand, if the radiation has all $\boldsymbol{E}_k$ vectors pointing along one direction, the z-direction, for example, only this polarization mode need be taken into consideration, and

$$\sum_{\text{pol}}|\langle f|\sum_i e\boldsymbol{r}_i.\hat{\varepsilon}_k|i\rangle|^2=|\langle f|\sum_i ez_i|i\rangle|^2. \tag{4.58}$$

Since most electronic energy levels are degenerate we can label the individual states of level i by i_n and the individual states of level f by f_m, the degeneracies of the levels being g_i and g_f, respectively. Now if level i is below level f the electric dipole transition probability for an absorption transition from i to f is calculated by summing over the transitions to the different final states and by averaging over the transitions from the initial states, hence

$$W_{if}(\text{ED}) = \frac{\pi}{2\varepsilon_0 \kappa \hbar^2} \frac{1}{g_i} \sum_{i_n f_m} \sum_{\text{pol}} |\langle f | \Sigma e r_i . \hat{\varepsilon}_E | i \rangle|^2 u(\omega). \tag{4.59}$$

The magnetic dipole and electric quadrupole transition probabilities have the same form as eqn (4.59), the electric dipole operator being replaced by the appropriate term from $V^{(\omega_\kappa)}$ (eqn 4.54). If level i is above level f an emission process takes place, the electric dipole transition probability of which is also given by eqn (4.59).

In the succeeding sections we will apply the transition probability formulae to the case of radiative transitions between the ground level (a) and the excited level (b) of an electronic centre. The degeneracy in the ground level is g_a, and the individual ground states are labelled $|a_n\rangle$. In the excited level the degeneracy is g_b and the individual excited states are labelled $|b_m\rangle$. The probability of an absorption transition from a to b, W_{ab}, is given by eqn (4.59) in which i and f are replaced by a and b, while the probability of an emission transition from b to a, W_{ba}, is also given by eqn (4.59) but with i and f replaced by b and a, respectively. For the same energy density of the radiation field our quantum mechanical analysis predicts that the absorption and emission probabilities are related by

$$g_a W_{ab} = g_b W_{ba}. \tag{4.60}$$

4.3.2 Local field correction in solids

In the formulae developed so far we have been careful to include the refractive index ($n = c/v$) and the dielectric constant ($\kappa = n^2$) to allow for the case where the electronic centre is in a dielectric material. There is another effect of the material which must be taken into account arising from the fact that the E and B fields in the above equations refer to macroscopic or average fields in the material. However, because the radiation polarizes the neighbouring atomic environment the local field and macroscopic field may not be the same. For non-magnetic materials the local magnetic field is the same as the macroscopic magnetic field. However, the local electric field $\boldsymbol{E}_{\text{loc}}$, is in general different from the macroscopic electric field, $\boldsymbol{E}$, and a multiplicative correction factor, $(\boldsymbol{E}_{\text{loc}}/\boldsymbol{E})^2$, must be applied to the transitions. There is no simple general formula for this correction factor. In crystals of high local symmetry, e.g. the octahedral site of cations in rock-salt-structured ionic crystals, the correction factor is

$$\left(\frac{E_{\text{loc}}}{E}\right)^2 = \left(\frac{n^2+2}{3}\right)^2 \tag{4.61}$$

where n is the refractive index of the material. Although not strictly applicable to lower-symmetry situations, eqn (4.61) is nevertheless often used when the local symmetry is not octahedral.

4.4 Spontaneous and stimulated optical transitions

4.4.1 Einstein A and B coefficients

In his analysis of the interaction of radiation and matter Einstein (1905, 1906) assumed the probability of a transition stimulated by the electromagnetic field to be proportional to the energy density of the radiation field at the transition frequency, $u(\omega)$. In Section 4.3.1 this result was derived by quantum mechanical analysis. Following Einstein we write the transition probability for absorption between states a and b as

$$W_{ab} = B_{ab} u(\omega). \tag{4.62}$$

B_{ab} is called the Einstein coefficient for a *stimulated* absorption transition from $|a\rangle$ to $|b\rangle$. Similarly the emission transition probability from b to a is

$$W_{ba} = B_{ba} u(\omega). \tag{4.63}$$

B_{ba} is the Einstein coefficient for a stimulated emission transition from b to a. From eqn (4.60) we find

$$g_a B_{ab} = g_b B_{ba}. \tag{4.60$'$}$$

If there is a large number of electronic centres interacting with radiation through stimulated absorption and emission transitions, then from eqn (4.60) we deduce that the equilibrium populations in the two levels, N_a and N_b, should be in the ratio $N_a/N_b = g_a/g_b$. However, the correct ratio for a system in thermal equilibrium is given by the Boltzmann factor, i.e.

$$\frac{N_a}{N_b} = \frac{g_a}{g_b} \exp[(E_b - E_a)/kT] \tag{4.64}$$

so that in general the equilibrium population in the excited level, is lower than that in ground level. In order to obtain the correct population ratio Einstein proposed an additional emission process, one not stimulated by the radiation field, which he called *spontaneous emission.* The Einstein coefficient for spontaneous emission, A_{ba}, is the transition probability for this spontaneous process from b to a. To find the relationship between A_{ba} and the B coefficients we consider a collection of electronic centres in a cavity filled with the radiation, the whole system being in thermal equilibrium at temperature T. The energy density in the radiation field in this case is given by the blackbody radiation formula

$$u(\omega) = \frac{(\hbar\omega^3/\pi^2 v^3)}{\exp(\hbar\omega/kT) - 1}. \tag{4.65}$$

The variations in level populations are given by the rate equations

$$\frac{dN_a}{dt} = -N_a B_{ab} u(\omega) + N_b B_{ba} u(\omega) + N_b A_{ba},$$

$$\frac{dN_b}{dt}=-\frac{dN_a}{dt} \tag{4.66}$$

where $\hbar\omega=E_b-E_a$. In equilibrium, when $dN_a/dt=0$, we obtain from eqns (4.60′) and (4.66)

$$u(\omega)=\frac{A_{ba}/B_{ba}}{\dfrac{g_b}{g_a}\dfrac{N_a}{N_b}-1}. \tag{4.67}$$

Comparing eqn (4.67) with the blackbody radiation formula (eqn 4.65) we obtain

$$\frac{N_a}{N_b}=\frac{g_a}{g_b}\exp\left(\frac{\hbar\omega}{kT}\right) \tag{4.68}$$

in agreement with eqn (4.64), and

$$\frac{A_{ba}}{B_{ba}}=\frac{\hbar\omega^3}{\pi^2 v^3}. \tag{4.69}$$

Using this last equation and eqn (4.65) we can write the full emission transition probability, $A_{ba}+B_{ba}u(\omega)$, as

$$W_{ba}=A_{ba}\left[1+\frac{1}{\exp(\hbar\omega/kT)-1}\right]=A_{ba}[1+n_\omega(T)] \tag{4.70}$$

where $n_\omega(T)$ is the photon occupancy in the electromagnetic mode at angular velocity ω, being the equilibrium number of photons in a mode at angular velocity ω in the cavity radiation (blackbody radiation) at temperature T.

From eqns (4.69) and (4.65) we can write

$$A_{ba}=B_{ba}\left[\frac{u(\omega)}{n_\omega(T)}\right]=B_{ba}2\rho_k(\omega)\hbar\omega_k \tag{4.71}$$

showing that the spontaneous transition probability is numerically equal to the stimulated transition probability stimulated by one photon in each electromagnetic mode. We can likewise write the transition probability due to stimulated absorption as

$$W_{ab}=B_{ab}\,u(\omega)=\left(\frac{g_b}{g_a}\right)A_{ba}n_\omega(T). \tag{4.72}$$

The absorption and emission transition probabilities are illustrated in Fig. 4.2.

We derived the formula for the spontaneous transition probability from an analysis of the interaction of the electronic system with blackbody radiation. Equations (4.70) and (4.72) are valid, however, for any value of the photon occupancy n_ω, as long as it represents the occupancy of an electromagnetic

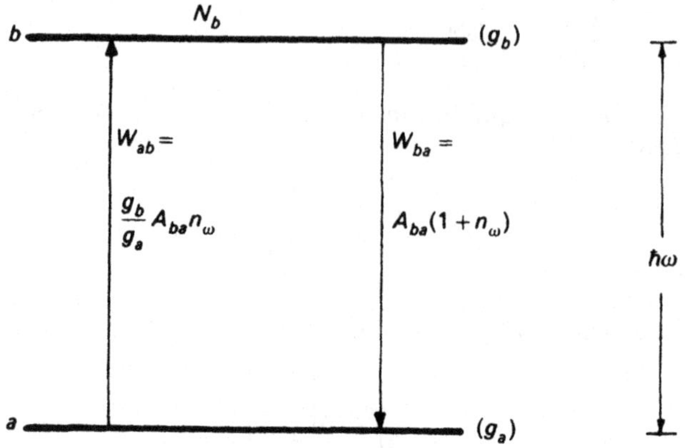

FIG. 4.2. Emission and absorption transition probabilities between levels a and b in terms of the spontaneous transitions probability A_{ba}. The photon occupancy in the modes at the transition frequency is $n(\omega)$, and the radiation is assumed to be randomly polarized. g_a and g_b are the statistical weights of levels a and b, respectively.

mode at the resonance frequency. In a quantum treatment of the radiation field we find that even at $T=0$, where $n_\omega=0$ for all ω, there is nevertheless some electromagnetic radiation present. This is the so-called zero-point radiation energy, which can stimulate an emission process, and is responsible for the 'spontaneous' emission process described above. Since the zero-point radiation cannot be removed it cannot cause an absorption process, and there is no spontaneous absorption process.

4.4.2 Calculation of the spontaneous transition probability

We have seen that the spontaneous transition probability is numerically equal to the stimulated transition probability, where the stimulation is due to radiation with one photon per mode. Consider the electric dipole process for which $A_{ba}(\mathrm{ED})$ is given by eqn (4.59), with $u(\omega)=2\hbar\omega_k\rho_k(\omega)$ and i, f being replaced by b, a. Using eqn (4.57) and the local field correction factor, eqn (4.61), we obtain

$$A_{ba}(\mathrm{ED})=\frac{\pi}{\varepsilon_0\kappa\hbar^2}\frac{1}{g_b}\sum_{a_n,b_m}|\langle a_n|\sum_i er_i|b_m\rangle|^2\left(\frac{E_{\mathrm{loc}}}{E}\right)^2\frac{\hbar\omega^3}{3\pi^2 v^3}. \quad (4.73)$$

Writing $v=c/n$, where n is the refractive index, and writing the electric dipole moment operator $(\Sigma_i\, er_i)$, as μ_e, the spontaneous transition probability can be written

$$A(\mathrm{ED})=\frac{1}{4\pi\varepsilon_0}\frac{4n\omega^3}{3\hbar c^3}\left(\frac{E_{\mathrm{loc}}}{E}\right)^2\frac{1}{g_b}\sum_{a_n,b_m}|\langle a_n|\mu_e|b_m\rangle|^2 \quad (4.74)$$

and we have omitted the subscript in the symbol for the spontaneous transition probability. For an allowed electric dipole transition we expect $\langle \mu_e \rangle = e\langle r \rangle$ where $\langle r \rangle \simeq 10^{-10}$ m. Taking $n=1.7$ we calculate $A(\text{ED}) \simeq 10^8\,\text{s}^{-1}$ for a transition in the visible spectrum.

To obtain the spontaneous emission probability for magnetic dipole process we replace the electric dipole term $\sum_i e\boldsymbol{r}_i \,.\, \boldsymbol{E}_k^{(0)}$ by the magnetic dipole term $\sum_i (e/2m)(\boldsymbol{l}_i + 2\boldsymbol{s}_i) \,.\, \boldsymbol{B}_k^{(0)}$ (see eqn 4.54). Recalling from eqn (4.38) that

$$|\boldsymbol{B}_k^{(0)}| = \frac{|\boldsymbol{E}_k^{(0)}|}{v} = n\frac{|\boldsymbol{E}_k^{(0)}|}{c} \tag{4.38}$$

we obtain

$$\begin{aligned} A(\text{MD}) &= \frac{1}{4\pi\varepsilon_0}\frac{4n^3\omega^3}{3\hbar c^5}\frac{1}{g_b}\sum_{a_n, b_m} |\langle a_n | \mu_m | b_m \rangle|^2 \\ &= \frac{\mu_0}{4\pi}\frac{4n^3\omega^3}{3\hbar c^3}\frac{1}{g_b}\sum_{a_n, b_m} |\langle a_n | \mu_m | b_m \rangle|^2 \end{aligned} \tag{4.75}$$

where μ_m, the magnetic dipole operator, is $\sum_i (e/2m)(\boldsymbol{l}_i + 2\boldsymbol{s}_i)$. There is no local field correction to be made for magnetic fields in non-magnetic materials. Thus for an allowed magnetic dipole transition $A(\text{MD}) \simeq 10^3\,\text{s}^{-1}$.

It is useful at this point to define a spontaneous transition probability for emission at polarization $\hat{\varepsilon}$. If the particular radiation field in a volume V contains only photons of polarization $\hat{\varepsilon}$ with density of k-modes $\rho_k(\omega)$ and if there is a single photon per mode in this radiation field we can derive a formula for the transition probability for spontaneous emission into the volume V of a photon of polarization $\hat{\varepsilon}$ stimulated by this radiation field. This formula is

$$A(\text{ED}, \hat{\varepsilon}) = \frac{\pi}{\varepsilon_0 \kappa \hbar^2 V}\left(\frac{E_{\text{loc}}}{E}\right)^2 \frac{1}{g_b}\sum_{a_n, b_m} |\langle a_n | \mu_e \,.\, \hat{\varepsilon} | b_m \rangle|^2 \rho_k(\omega)\hbar\omega. \tag{4.76}$$

In discussing transition probabilities it was assumed that levels a and b are very sharply defined in energy; hence the transition was assumed to be sharp. However, as is discussed in subsequent sections, all transitions have a finite width which can be described by a lineshape function, $g(\omega)$ or $g(\nu)$, normalized so that

$$\int g(\omega)\mathrm{d}\omega = \int g(\nu)\mathrm{d}\nu = 1; \quad g(\nu) = 2\pi g(\omega). \tag{4.77}$$

Consequently the spontaneous transition probability per unit angular velocity, $A(\omega)$, is defined by

$$A(\omega) = Ag(\omega). \tag{4.78}$$

Applying this to eqn (4.76) we define $A(\text{ED}, \hat{\varepsilon})g(\omega)$ as the transition probability for emission of a photon of polarization ε per unit angular velocity at ω

stimulated by a radiation field containing one photon of polarization ε in each k-mode. Since the density of k-modes is $\rho_k(\omega)$ we can define $A(\text{ED}, \omega, \hat{\varepsilon}) = A(\text{ED}, \hat{\varepsilon})g(\omega)/\rho_k(\omega)$ as the spontaneous transition probability for emission into the volume V of a photon in the mode $(\hat{\varepsilon}, \omega)$. This is

$$A(\text{ED}, \omega, \hat{\varepsilon}) = \frac{\pi\omega}{\varepsilon_0 \kappa \hbar V}\left(\frac{E_{\text{loc}}}{E}\right)^2 \frac{1}{g_b} \sum_{a_n, b_m} |\langle a_n | \mu_e . \hat{\varepsilon} | b_m \rangle|^2 g(\omega). \qquad (4.79)$$

Analogous formulae to eqns (4.76) and (4.79) can be written for the magnetic dipole process.

The quantity $A(\omega, \hat{\varepsilon})$ can be used to calculate the transition rate for atoms interacting with a beam of polarized radiation as it traverses an absorbing medium. In this case, by analogy with eqn (4.70), we find that $W_{ba}(\omega, \hat{\varepsilon}) = A(\omega, \hat{\varepsilon})(1 + n(\omega, \hat{\varepsilon}))$ is the probability per second that an atom in level b will emit one $(\omega, \hat{\varepsilon})$ photon into volume V when $n(\omega, \hat{\varepsilon})$ photons of this $(\omega, \hat{\varepsilon})$ mode have been introduced into this volume.

Similarly, in analogy with eqn (4.72), we write $W_{ab}(\omega, \hat{\varepsilon})$ $(g_b/g_a)A(\omega, \hat{\varepsilon})n(\omega, \hat{\varepsilon})$ as the probability that a $(\omega, \hat{\varepsilon})$ photon will be absorbed in one second by an atom in level a from radiation introduced into volume V, when there are $n(\omega, \hat{\varepsilon})$ photons of mode $(\omega, \hat{\varepsilon})$ in the introduced radiation.

4.4.3 *The absorption coefficient*

We consider a parallel monochromatic beam of radiation linearly polarized in the $\hat{\varepsilon}$-direction of intensity $I(\omega, \hat{\varepsilon})$ travelling through a block of material of thickness l (Fig. 4.3). We can express the intensity density as a flux of photons:

$$I(\omega, \hat{\varepsilon}) = N(\omega, \hat{\varepsilon})\hbar\omega v \qquad (4.80)$$

where $N(\omega, \hat{\varepsilon})$ is the number of photons of mode $(\omega, \hat{\varepsilon})$ per unit volume in the beam and v and is the velocity of the beam $(v = c/n)$. As the flux of photons moves through the material it interacts with the electronic centres in the material, and the emission and absorption processes which can change the density of photons in the beam are shown in Fig. 4.4. The number of photons in the mode $(\omega, \hat{\varepsilon})$ is assumed to be much greater than unity so that the spontaneous emission process is negligible compared with the stimulated emission process, and this is taken into account in Fig. 4.4. N_a and N_b are the number densities of electronic centres in the ground and excited levels, respectively. We assume that the beam is weak enough that the values of N_a and N_b are not significantly altered because of the interaction with the beam; these quantities can be regarded as constant throughout the material.

We confine our attention to a volume V of the radiation as it traverses the material. There are $N_a V$ and $N_b V$ atoms in the ground and excited states which are interacting with this radiation and causing a variation in the number of photons $n(\omega, \hat{\varepsilon})$ in this volume of the radiation. $n(\omega, \hat{\varepsilon}) = N(\omega, \hat{\varepsilon}) V$.

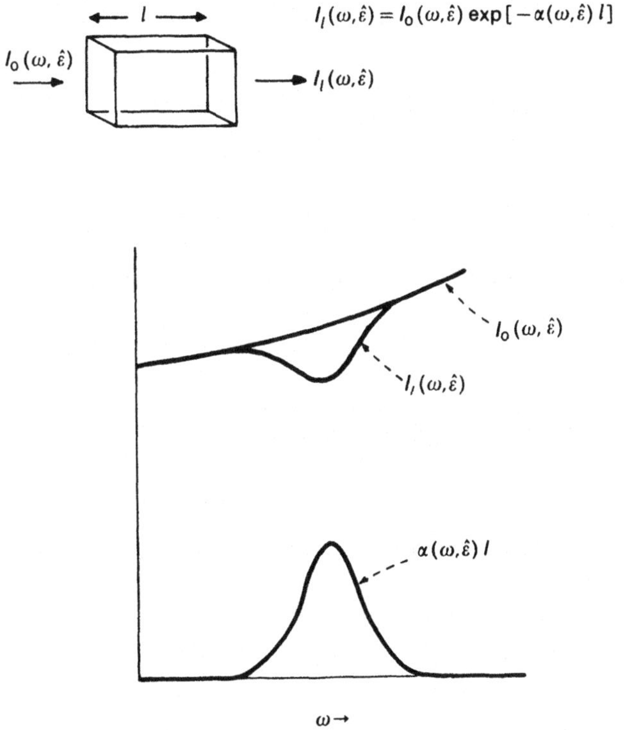

FIG. 4.3. The polarized beam is reduced in intensity on passing through a block of material. The absorption coefficient is obtained from the transmission curve using the formula $\alpha(\omega,\hat{\varepsilon}) . l = \ln[I_0(\omega,\hat{\varepsilon})/I_l(\omega, \hat{\varepsilon})]$.

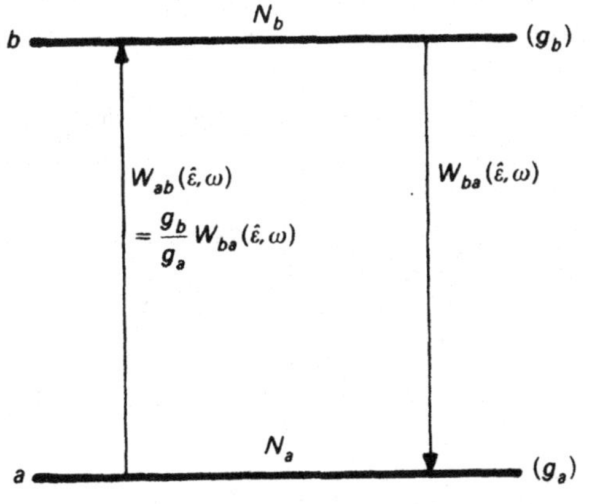

FIG. 4.4. Transition probabilities in mode ω, $\hat{\varepsilon}$ stimulated by the beam $I(\omega, \hat{\varepsilon})$ passing through the material. The density of photons in mode ω, $\hat{\varepsilon}$ in the beam is assumed to be much larger than unity so that spontaneous transitions can be neglected.

The variation in $n(\omega, \hat{\varepsilon})$ is given by (Fig. 4.4)

$$\frac{dn(\omega, \hat{\varepsilon})}{dt} = N_b V W_{ba}(\omega, \hat{\varepsilon}) - N_a V W_{ab}(\omega, \hat{\varepsilon})$$

$$= \left[N_b - \frac{g_b}{g_a} N_a \right] V A(\omega, \hat{\varepsilon}) n(\omega, \hat{\varepsilon})$$

$$= \Phi(\omega, \hat{\varepsilon}) n(\omega, \hat{\varepsilon}). \tag{4.81}$$

Since the density of photons $N(\omega, \hat{\varepsilon})$ scales as $n(\omega, \hat{\varepsilon})$ then

$$\frac{dN(\omega, \hat{\varepsilon})}{dt} = \Phi(\omega, \hat{\varepsilon}) N(\omega, \hat{\varepsilon}). \tag{4.81a}$$

This equation can be integrated to give the variation in time of $N(\omega, \hat{\varepsilon})$ in the beam as it traverses the material:

$$N(\omega, \hat{\varepsilon})_t = N(\omega, \hat{\varepsilon})_0 \exp(\Phi(\omega, \hat{\varepsilon}) t). \tag{4.82}$$

During the time t the beam has travelled a distance $l = vt$ through the material, and we can insert $t = l/v = ln/c$ in eqn (4.82). Since the intensity density $I(\omega, \hat{\varepsilon})$ in the beam is proportional to $N(\omega, \hat{\varepsilon})$ we obtain an expression for the variation in the intensity with distance through the material:

$$I_l(\omega, \hat{\varepsilon}) = I_0(\omega, \hat{\varepsilon}) \exp(-\alpha(\omega, \hat{\varepsilon}) l) \tag{4.83}$$

where the *absorption coefficient* $\alpha(\omega, \hat{\varepsilon})$ is

$$\alpha(\omega, \hat{\varepsilon}) = \frac{-\Phi(\omega, \hat{\varepsilon})}{v} = A(\omega, \hat{\varepsilon}) \left[N_a \frac{g_b}{g_a} - N_b \right] \frac{V}{v}. \tag{4.84}$$

In deriving this formula we have assumed that N_a and N_b have uniform values throughout the material and that these values are constant in time. If, however, the beam is very intense the transitions may change the values of N_a and N_b. The initial part of a very intense beam could cause such strong absorption that the population in the excited level increases until $N_a/g_a = N_b/g_b$. After that the absorption is zero, as eqn (4.86) shows, and the remainder of the beam is no longer attenuated. Such *self-induced transparency* effects have been observed and studied.

If the radiation interacts with the atoms through an electric dipole process we can insert the formula for $A(\mathrm{ED}, \omega, \hat{\varepsilon})$ (eqn 4.79) into eqn (4.84) and we obtain

$$\alpha(\mathrm{ED}, \omega, \hat{\varepsilon}) = \left(N_a \frac{g_b}{g_a} - N_b \right) \frac{\pi\omega}{\varepsilon_0 \kappa h v} \left[\frac{E_{\mathrm{loc}}}{E} \right]^2 \frac{1}{g_b} \sum_{a_n, b_m} |\langle a_n | \mu_e \cdot \hat{\varepsilon}_E | b_m \rangle|^2 g(\omega). \tag{4.85}$$

By integrating over all frequencies we obtain the *absorption strength at polarization* $\hat{\varepsilon}$:

$$\int \alpha(\mathrm{ED}, \omega, \hat{\varepsilon}_E)\mathrm{d}\omega = \left(N_a \frac{g_b}{g_a} - N_b\right) \frac{\pi\omega}{\varepsilon_0 \hbar c n} \left[\frac{E_{\mathrm{loc}}}{E}\right]^2 \frac{1}{g_b} \sum_{a_n, b_m} |a_n| \mu_e \cdot \hat{\varepsilon}_E | b_m \rangle|^2. \tag{4.86}$$

The equivalent expression for the magnetic dipole absorption strength is

$$\int \alpha(\mathrm{MD}, \omega, \hat{\varepsilon}_B)\mathrm{d}\omega = \left(N_a \frac{g_b}{g_a} - N_b\right) \frac{\pi\omega n}{\varepsilon_0 \hbar c^3} \frac{1}{g_b} \sum_{a_n, b_m} |\langle a_n | \mu_m \cdot \hat{\varepsilon}_B | b_m \rangle|^2. \tag{4.86$'$}$$

In these equations the polarization vectors $\hat{\varepsilon}_E$ and $\hat{\varepsilon}_B$ point along the directions of the *electric* and *magnetic* fields, respectively, of the radiation.

In the case of atoms in high-symmetry sites in optically isotropic materials the absorption coefficient should be independent of polarization and $|\langle \mu . \hat{\varepsilon} \rangle|^2 = |\langle \mu \rangle|^2/3$. The isotropic absorption coefficient valid for both electric dipole and magnetic dipole transitions can now be expressed in terms of the spontaneous transition probability (A_{ba}) by using eqn (4.73). We obtain

$$\alpha(\omega) = \left(N_a \frac{g_b}{g_a} - N_b\right) A_{ba} \frac{\pi^2 c^2}{n^2 \omega^2} g(\omega). \tag{4.87}$$

The absorption strength in this isotropic case is

$$\int \alpha(\omega)\mathrm{d}\omega = \left(N_a \frac{g_b}{g_a} - N_b\right) A_{ba} \frac{\lambda^2}{4n^2} \tag{4.86$''$}$$

where λ is the vacuum wavelength at the centre of the transition.

Under normal conditions and where the light beam is not very intense we have $N_b \ll N_a$ and N_b can be ignored. In this case the isotropic absorption strength can be written

$$\int \alpha(\omega)\mathrm{d}\omega \simeq N_a \frac{g_b}{g_a} A_{ba} \frac{\lambda^2}{4n^2}. \tag{4.86$'''$}$$

This formula is often used to calculate the transition probability in terms of the experimentally measured absorption strength. In applying this formula we must take care with the units. $\alpha(\omega)$ is the absorption coefficient at a particular value of the angular velocity, ω. It could equally be labelled according to the frequency of the radiation, $\alpha(\nu)$, and $\alpha(\nu) = \alpha(\omega)$. It is more usual to express the integral in the formula for the absorption strength of the transition in units of frequency and to regard the absorption coefficient as a function of frequency. We then have

$$\int \alpha(\nu)\mathrm{d}\nu = \frac{1}{2\pi} \int \alpha(\omega)\mathrm{d}\omega \simeq N_a \frac{g_b}{g_a} A_{ba} \frac{\lambda^2}{8\pi n^2}. \tag{4.88}$$

The formula for $\alpha(\nu)=\alpha(\omega)$ for the case where the N_b term is not omitted is, from eqns (4.87) and (4.77),

$$\alpha(\nu)=\left(N_a\frac{g_b}{g_a}-N_b\right)A_{ba}\frac{\lambda^2}{8\pi n^2}g(\nu). \tag{4.89}$$

This formula is used as the starting point of our discussion on laser action (Chapter 11).

4.4.4 *The lifetime and natural linewidth of transitions*

Equation (4.66) gives the rate at which levels change population in the presence of radiation. If there is no introduced radiation, $n(\omega)=0$, we obtain

$$\frac{dN_b}{dt}=-N_bA_{ba}$$

which on integration yields

$$N_b(t)=N_b(0)\exp(-A_{ba}t). \tag{4.90}$$

An intense short pulse of radiation at $t=0$ creates a population $N_b(0)$ in the excited state. The intensity of emission—the energy radiated per second—can be written as $I(\omega)_t=A_{ba}N_b(t)\hbar\omega$, hence at times $t>0$ we expect $I(\omega)_t=I(\omega)_0\exp(-A_{ba}t)$. The exponential term is usually written as $\exp(-t/\tau_R)$, where τ_R is the radiative decay time. Thus we have

$$\tau_R^{-1}=A_{ba}. \tag{4.91}$$

The *expectation value of t* in the excited state, denoted by $\langle t\rangle$, is the average time that an atom spends in the excited state; $\langle t\rangle$ is calculated from

$$\langle t\rangle=\frac{1}{N_b(0)}\int_{-\infty}^{\infty}N_b(t)dt=\int_{0}^{\infty}\exp(-t/\tau_R)dt$$

$$=\tau_R=(A_{ba})^{-1}. \tag{4.92}$$

Observed values of τ_R range from $\sim10^{-8}$ s for allowed electric dipole transitions on free atoms and molecules, (e.g. the 2p→1s transition in atomic hydrogen), to $\geqslant 1$ ms for triplet–singlet transitions in some organic molecules. The exponential decrease in intensity can be measured experimentally by suddenly exciting a sample with a short pulse of radiation, and then observing the decay of emitted intensity. Figure 4.5 shows the experimentally measured intensity decay patterns from Cr^{3+} and Nd^{3+} ions in a single crystal of yttrium aluminium garnet (YAG) doped with neodymium and chromium after pulsed excitation at 77 K. Plotting the logarithm of the intensity as a function of time shows (a) that the decay of the Cr^{3+} luminescence follows a single exponential with a decay time of 5.3 ms and (b) that the decay of the

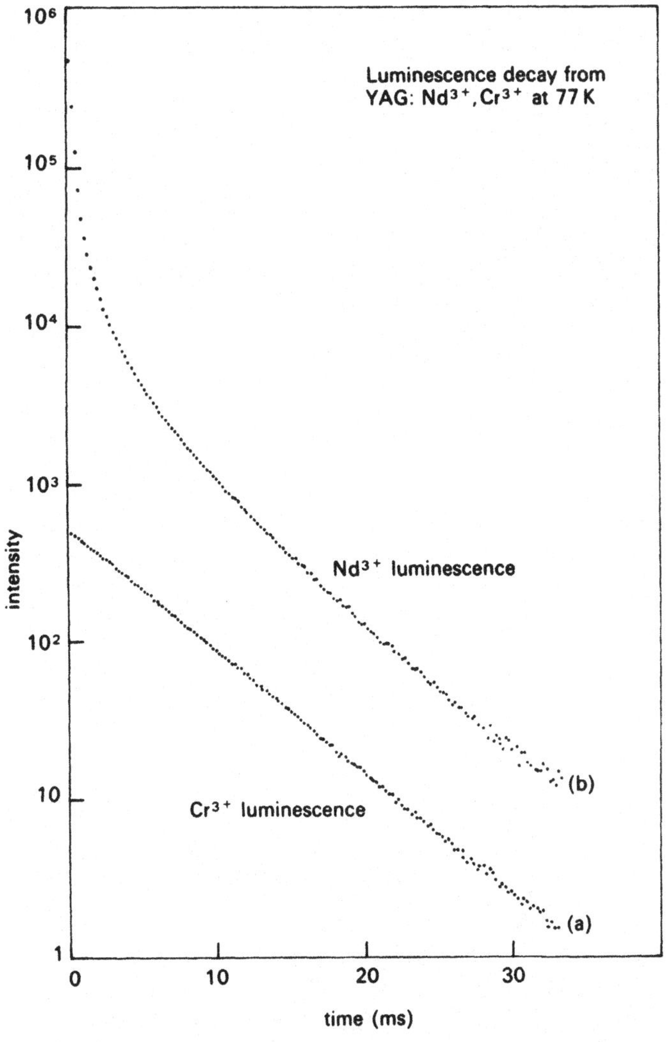

FIG. 4.5. Decay patterns of Nd^{3+} and Cr^{3+} luminescence in YAG: Nd, Cr measured at 77 K after pulsed excitation. The Cr^{3+} luminescence has a single exponential decay. The Nd^{3+} luminescence decay, however, is non-exponential because of excitation transfer from excited Cr^{3+} ions.

Nd^{3+} luminescence is more complex. The fast initial decay is at the intrinsic radiative decay rate of the excited Nd^{3+} ions, while the later slower decay has an asymptotic decay time of 5.3 ms. The non-exponential decay pattern of the Nd^{3+} ions is interpreted as follows. Some Nd^{3+} ions are raised to an excited state by the excitation light pulse, and these decay at the intrinsic decay rate of

Nd^{3+} ions. Other Nd^{3+} ions are raised to an excited state by excitation transfer from nearby excited ions, this transfer occurs while there are excited Cr^{3+} ions available, hence these Nd^{3+} ions decay in step with the Cr^{3+} ions and exhibit the decay pattern of the Cr^{3+} ions. (Excitation transfer is analysed in detail in Chapter 10.)

The linewidth of the transition is another important experimental quantity. To discuss this we consider once more the two-level system of ground state a and excited state b separated in energy by $E_b - E_a = \hbar\omega_{ba}$. Applying the perturbation analysis (Section 4.1) to this system we write the rate equations (eqn 4.4)

$$\frac{dc_a}{dt} = -\frac{i}{\hbar} c_b(t) \langle a|\mathscr{H}_1|b\rangle \exp(i\omega_{ab}t) \tag{4.93a}$$

$$\frac{dc_b}{dt} = -\frac{i}{\hbar} c_a(t) \langle b|\mathscr{H}_1|a\rangle \exp(i\omega_{ba}t) \tag{4.93b}$$

where we have assumed that the diagonal matrix elements of the perturbation, $\langle a|\mathscr{H}_1|a\rangle$ and $\langle b|\mathscr{H}_1|b\rangle$, are zero. Since $\mathscr{H}_1$ is of the general form

$$\mathscr{H}_1 = V^{(\omega)} \exp(-i\omega t) + V^{(\omega)*} \exp(i\omega t) \tag{4.6$'$}$$

and since we are interested in the effect of radiation near resonance $(\omega \sim \omega_{ba})$ the rate equations become

$$\frac{dc_a(t)}{dt} = -\frac{i}{\hbar} c_b(t) (V^{(\omega)*})_{ab} \exp i(\omega_{ab} + \omega)t \tag{4.94a}$$

$$\frac{dc_b(t)}{dt} = -\frac{i}{\hbar} c_a(t) (V^{(\omega)})_{ba} \exp i(\omega_{ba} - \omega)t \tag{4.94b}$$

and the initial conditions are $c_b(0) = 1$, $c_a(0) = 0$.

Assuming that the probability of finding the atom in the excited state varies as $\exp(-t/\tau_R)$ we consider $c_b(t) = \exp(-t/2\tau_R)$ as a solution of eqn (4.94). Substituting for $c_b(t)$ in eqn (4.94a) gives

$$\frac{dc_a}{dt} = -\frac{i}{\hbar} (V^{(\omega)*})_{ab} \exp[i(\omega_{ab} + \omega)t - t/2\tau_R] \tag{4.95}$$

which has a solution

$$c_a(t) = \frac{(V^{(\omega)*})_{ab}}{\hbar} \frac{\exp[i(\omega_{ab} + \omega)t - t/2\tau_R] - 1}{(\omega_{ab} + \omega) + i/2\tau_R}. \tag{4.96}$$

Since $|c_a(t)|^2$ is the probability that the atom has reached the ground state this quantity also represents the probability that the atom has emitted a photon at angular velocity ω. After very long observation times,

$$|c_a(\infty)|^2 = \left|\frac{(V^{(\omega)*})_{ab}}{\hbar}\right|^2 \frac{1}{(\omega_{ab}+\omega)^2 + \left(\dfrac{1}{2\tau_R}\right)^2}$$

$$= \left|\frac{(V^{(\omega)*})_{ab}}{\hbar}\right|^2 \frac{1}{(\omega_{ba}-\omega)^2 + \left(\dfrac{1}{2\tau_R}\right)^2} \tag{4.97}$$

gives the distribution of frequencies in the transition. If $V^{(\omega)*}$ varies sufficiently slowly with ω so as to be regarded as constant over the frequency range of the transition, then the frequency distribution in the transition has a Lorentzian lineshape with full width at half maximum, $\Delta\omega$, given by

$$\Delta\omega = \frac{1}{\tau_R} = A_{ba} \tag{4.98}$$

as is illustrated in Fig. 4.6. This identification of the linewidth with the spontaneous transition probability (or the radiative decay rate) is related to the uncertainty principle. The time available to measure the energy of the excited state is Δt which we equate with the expectation value of t in the excited state: $\Delta t = \tau_R$. The width in energy of the transition is $\Delta E = \hbar\Delta\omega$. Hence $\Delta E \,.\, \Delta t = \hbar\Delta\omega\, \tau_R = \hbar$ follows from eqn (4.98). Thus the finite lifetime of the excited state leads to a Lorentzian distribution of emission frequencies.

In this analysis the excited state was considered to decay only by a radiative decay. Since additional decay processes may occur one adds all individual

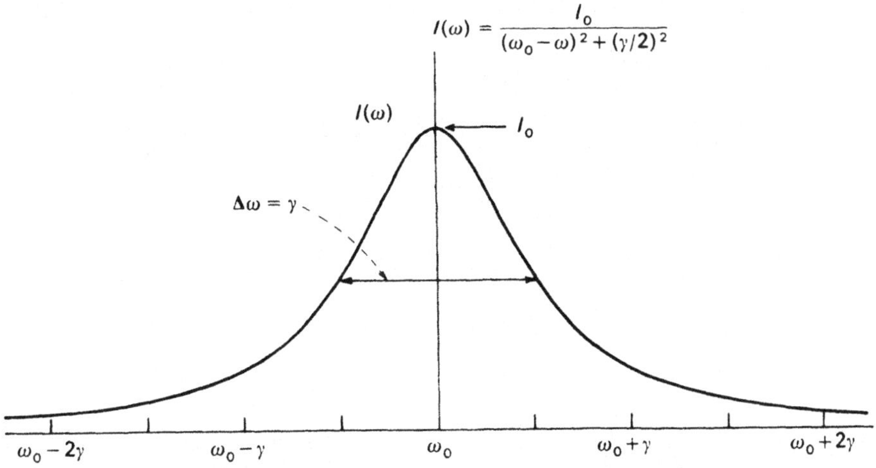

FIG. 4.6. Lorentzian lineshape of a transition after lifetime $\tau = 1/\gamma$ in the excited state. FWHM $= \Delta\omega = \gamma = 1/\tau$.

decay rates to obtain the total decay rate, $1/\tau$, and it is this decay time, $\tau \leqslant \tau_R$ which determines the width of the transition: $\Delta\omega = 1/\tau$. If both upper and lower levels of the radiative transition have finite lifetimes then the energy width of the transition is the sum of the uncertainties for the two levels. Broadening by such lifetime processes is usually known as *natural*, *homogeneous*, or *lifetime broadening* and is characterized by a Lorentzian lineshape.

Although the above result relating the finite width of the transition to the lifetime was deduced using purely quantum effects arising from the uncertainty principle, it also follows from a simple classical argument. Consider a collection of atoms excited at time $t=0$. For time $t>0$ the emitted radiation has an associated electric field at some nearby point given by

$$\boldsymbol{E}(t) = \boldsymbol{E}_0 \exp(\mathrm{i}\omega_0 t) \exp(-t/2\tau_R) \tag{4.99}$$

whereas for $t<0$, $\boldsymbol{E}(t)$ is zero. Figure 4.7 represents the imaginary part of $\boldsymbol{E}(t)$. This electric field oscillates at ω_0, the central frequency of the transition on the atoms. Since the intensity varies as $|\boldsymbol{E}(t)|^2$ the intensity decreases exponentially in time with decay time τ_R. The distribution of frequencies in the electric field oscillations is obtained by Fourier analysing $\boldsymbol{E}(t)$ into a frequency

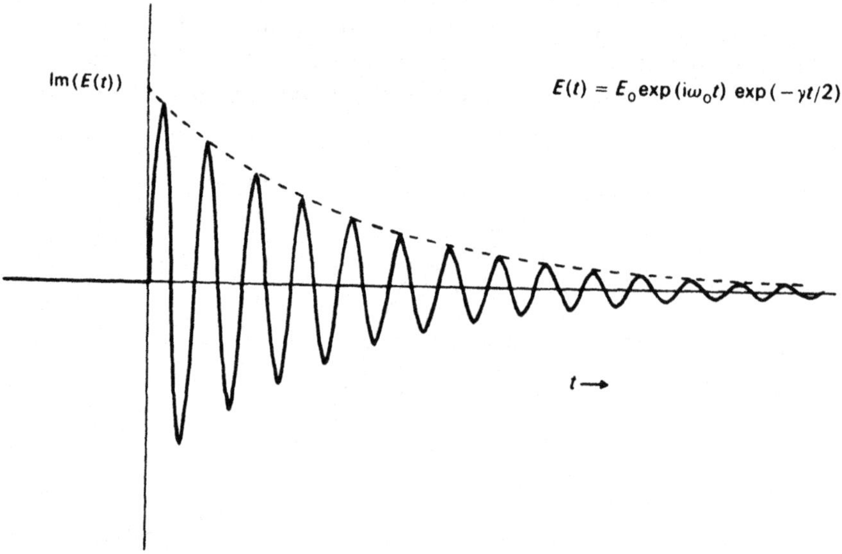

FIG. 4.7. The effect of radiation damping on the electric field amplitude oscillating at frequency ω_0 is to cause an exponential decay $(\exp - t/2\tau_R)$ which produces the natural broadening of a spectroscopic transition.

spectrum, i.e.

$$\boldsymbol{E}(\omega)=\int_{-\infty}^{\infty}\boldsymbol{E}(t)\exp(-\mathrm{i}\omega t)\,\mathrm{d}t. \tag{4.100}$$

Substitution of $\boldsymbol{E}(t)$ from eqn (4.99) into eqn (4.100) yields

$$\begin{aligned}\boldsymbol{E}(\omega)&=\int_{0}^{\infty}\boldsymbol{E}_0\exp\mathrm{i}(\omega_0-\omega)t\exp(-t/2\tau_{\mathrm{R}})\,\mathrm{d}t\\&=\frac{\boldsymbol{E}_0}{-\mathrm{i}(\omega_0-\omega)+\left(\dfrac{1}{2\tau_{\mathrm{R}}}\right)}\\&=\frac{\boldsymbol{E}_0}{\sqrt{(\omega_0-\omega)^2+\left(\dfrac{1}{2\tau_{\mathrm{R}}}\right)^2}}\exp(\mathrm{i}\phi(\omega))\end{aligned} \tag{4.101}$$

where $\phi(\omega)$ is a constant phase factor. Since $I(t)\sim|\boldsymbol{E}(t)|^2$ we obtain the intensity distribution of frequencies given in eqn (4.97): i.e.

$$I(\omega)=\frac{I_0}{(\omega_0-\omega)^2+\left(\dfrac{1}{2\tau_{\mathrm{R}}}\right)^2} \tag{4.102}$$

with linewidth $\Delta\omega=1/\tau_{\mathrm{R}}$.

Typically in an allowed electric dipole transition $\tau_{\mathrm{R}}\simeq10^{-8}$ s so that $\Delta E\simeq0.5\times10^{-3}\,\mathrm{cm}^{-1}$ giving a linewidth in the visible spectrum of $\simeq2\times10^{-3}$ nm. The homogeneous width is the minimum width which we might expect to measure experimentally. The several other mechanisms contributing to the observed width of spectra from atoms, molecules and solids include

1. Doppler broadening in gases due to the thermal motion of atoms and molecules
2. Unresolved structure, such as fine structure or hyperfine structure
3. Static distortions in the crystalline environment of optically active centres in solids
4. Dynamic distortions of the crystalline environment associated with lattice vibrations.

All these contributions to linewidth are discussed in subsequent chapters.

4.5 Selection rules and oscillator strengths in optical spectroscopy

4.5.1 Oscillator strengths for optical transitions

The strengths of optical transitions are often discussed by analogy with the emission by a classical radiating dipole. Consider an electron executing simple harmonic motion described by $x = x_0 \cos \omega_0 t$. Classical analysis gives the energy radiated per second by such an oscillator as

$$P(\hat{x}) = \frac{e^2 \omega_0^4 x_0^2}{6\pi\varepsilon_0 c^3} \tag{4.103}$$

and this is all emitted as $\hat{x}$-polarized radiation. The total energy of the oscillator is $E = m\omega_0^2 x_0^2$. The amplitude of the oscillation gradually diminishes because of the energy loss through radiation; we assume this effect to be small enough that the frequency is unchanged. Since both $P(\hat{x})$ and E vary as x_0^2 we can write

$$P(\hat{x}) = \frac{e^2 \omega_0^2}{6\pi\varepsilon_0 m c^3} E = \gamma E. \tag{4.104}$$

For an electron oscillating in the x-, y-, and z-directions the energy becomes $E = m\omega^2 (x_0^2 + y_0^2 + z_0^2) = 3m\omega_0^2 x_0^2$ and the power radiated can again be written

$$P = \gamma E. \tag{4.105}$$

If there are f electrons in the centre the power radiated by the centre can be written

$$P = \gamma f E \tag{4.106a}$$

where f is the classical *oscillator strength*.

We now apply similar formulae to an isolated quantum system in the excited state b which can undergo spontaneous transitions to the ground state a. The radiated power is $P = \hbar\omega A_{ba}$, where A_{ba} is given by eqn (4.73) for electric dipole processes. By analogy with the classical formula $E = 3m\omega_0^2 x_0^2$ we write for the quantum system $E = 3\hbar\omega$ and this is the energy above the ground state. Hence from eqn (4.106a) we deduce

$$\hbar\omega\, A_{ba} = \gamma f 3\hbar\omega \tag{4.106b}$$

from which we calculate the oscillator strength to be

$$f = \frac{A_{ba}}{3\gamma} = \frac{2m\omega_0}{3\hbar e^2} \frac{1}{g_b} \sum_{a_n, b_m} |\langle a_n | \mu_e | b_m \rangle|^2 \tag{4.107}$$

since for a free electronic centre the index of refraction $n = 1$ and there is no local field correction. This formula introduces a quantum mechanical oscillator strength. Sometimes this is written as

$$f = \frac{2m\omega_0}{3\hbar e^2} \frac{1}{g_b} S_{ba}(\mathrm{ED}) \tag{4.107'}$$

where $S_{ba}(\mathrm{ED}) = \sum_{a_n, b_m} |\langle a_n|\mu_e|b_m\rangle|^2$ is referred to as the strength of the electric dipole transition.

Following Bethe and Salpeter (1957) we define the quantum mechanical oscillator strength for an electric dipole process from initial state i to final state f as

$$f_{if}(\mathrm{ED}) = \frac{2m\omega_{if}}{3\hbar e^2}\frac{1}{g_i}|\langle f|\mu_e|i\rangle|^2 \tag{4.108}$$

where $\omega_{if} = (E_f - E_i)/\hbar$. Thus f_{if} is positive for absorption and negative for emission transitions. From this definition the sum rule

$$\sum_f f_{if} = Z \tag{4.109}$$

follows for any Z-electron system (Bethe and Salpeter 1957). The state i can be the ground state or an excited state, and the summation is over all possible final states, f. If i is an excited state some f_{if} values are negative. It is more useful for us, however, always to regard the oscillator strength as a positive quantity, so we replace ω_{if} in eqn (4.107) by $|\omega_{if}|$. And in the case where the initial and final levels are degenerate we use the form (4.107) for the oscillator strength. For an allowed electric dipole transition the matrix element of μ_e has a magnitude $\simeq ea_0$, where a_0 is the Bohr radius. This gives a value for f of the order of unity. If the transition occurs by magnetic dipole process we can similarly define a quantum mechanical oscillator strength, $f(\mathrm{MD})$, by replacing the operator μ_e in eqn (4.107) by μ_m/c.

Using eqn (4.73) for the spontaneous transition rate in an electric dipole process we can express a relationship between $A(\mathrm{ED})$ and $f(\mathrm{ED})$:

$$A(\mathrm{ED}) = \frac{1}{\tau_R} = \frac{1}{4\pi\varepsilon_0}\frac{2\omega^2 e^2}{mc^3}[(E_{\mathrm{loc}}/E)^2 n]\, f(\mathrm{ED}). \tag{4.110}$$

By inserting numerical values for the physical constants and the local field correction in terms of the refractive index (eqn 4.61) we deduce

$$f(\mathrm{ED})\tau_R = 1.5\times 10^4 \frac{\lambda_0^2}{[1/3(n^2+2)]^2 n} \tag{4.111}$$

where λ_0 is the wavelength *in vacuo* and n is the refractive index. The analogous relationship for a magnetic dipole process is

$$f(\mathrm{MD})\tau_R = 1.5\times 10^4 \frac{\lambda_0^2}{n^3}. \tag{4.112}$$

For $\lambda = 0.5\ \mu$m and $n = 1$ we obtain $f(\mathrm{ED})\tau_R \simeq 4\times 10^{-9}$. Thus an allowed electric dipole transition in the visible on a free atom has a radiative time of 4×10^{-9} s whereas for an electronic centre in a solid for which $n \simeq 1.5$,

$\tau_R \simeq 10^{-8}$ s for an allowed electric dipole transition. The measured radiative decay times for dopant ions in solids are generally much longer than this, indicating that the oscillator strengths are significantly less than unity. These transitions are either weakly allowed electric dipole processes or magnetic dipole processes.

4.5.2 Selection rules and the Wigner–Eckart theorem

The probability of a radiative transition between states a and b is proportional to the square of the matrix element $\langle b|\boldsymbol{\mu}.\hat{\varepsilon}|a\rangle$, where $\boldsymbol{\mu}.\hat{\varepsilon}$ is the appropriate operator. For electric dipole transitions this is $\boldsymbol{\mu}_e.\hat{\varepsilon}_E$ where $\boldsymbol{\mu}_e=\sum_i e\boldsymbol{r}_i$ and $\hat{\varepsilon}_E$ is the unit electric polarization vector parallel to the $\boldsymbol{E}$ field of the radiation. For magnetic dipole transitions the appropriate operator is $\boldsymbol{\mu}_m.\hat{\varepsilon}_B$ where $\boldsymbol{\mu}_m=\sum_i \frac{e}{2m}(\boldsymbol{l}_i+2\boldsymbol{s}_i)$ and $\hat{\varepsilon}_B$ is the unit vector along the direction of the $\boldsymbol{B}$ field of the radiation. Whether or not a transition is allowed and, if allowed, how strong it is depends on the value of this matrix element. By analysing this matrix element we learn the *selection rules* governing the transition.

To evaluate the matrix element we use the theorems on matrix elements particularly the Wigner–Eckart theorem (eqn 3.47) which gives the formula for the evaluation of $\langle a(\Gamma,\lambda)|\mu(\bar{\Gamma},\bar{\lambda})|b(\Gamma',\lambda')\rangle$, where $|a(\Gamma,\lambda)\rangle$ indicates that this function belongs to the λ row of the Γ-irreducible representation of the symmetry group of the electronic centre. The symmetry properties of the electronic centre affect the value of the matrix element. In the case where the electronic centre has inversion symmetry as well as a particular rotational symmetry it is usual to examine the matrix element in respect of the inversion symmetry separately from its considerations in respect of rotational symmetry. From the Wigner–Eckart theorem we find that the value of the matrix element depends on the Clebsch–Gordan coefficient $\langle\Gamma\gamma|\bar{\Gamma}\Gamma'\,\bar{\gamma}\gamma'\rangle$, and in particular we have the rule that the matrix element is zero unless the product representation $\bar{\Gamma}\times\Gamma'$ contains the Γ-irreducible representation, or in another form, unless the product representation $\Gamma\times\bar{\Gamma}\times\Gamma'$ contains the identity representation, A_1 or Γ_1.

Let us first consider a situation when the electronic centre has inversion symmetry. The a and b wavefunctions are either even (g) or odd (u). The electric dipole operator $\sum_i e\boldsymbol{r}_i$ is an odd-parity function. Since $\Gamma_u\times\Gamma_u=\Gamma_g$ and $\Gamma_u\times\Gamma_g=\Gamma_u$ it follows from the above rule that the electric dipole matrix element is zero unless the a and b wavefunctions are of opposite parity. This is the *Laporte selection rule*. The magnetic dipole operator is of even parity, and since $\Gamma_g\times\Gamma_u=\Gamma_u$ and $\Gamma_g\times\Gamma_g=\Gamma_g$ we see that the magnetic dipole matrix element is zero unless the a and b states are of similar parity.

We now examine the matrix element from the viewpoint of rotational symmetry, and as a first example we take the case of free atoms where the wavefunctions are classified according to the irreducible representations of the full rotation group (J values). We observe that the components of the electric dipole and magnetic dipole operators belong to the $J=1$ irreducible representation. The matrix element in question can be written $\langle a(JM)|\mu(\bar{J}=1, \bar{M})|b(J'M')\rangle$. Since $D^1 \times D^{J'} = D^{J'+1} + D^{J'} + D^{J'-1}$ (except when $J'=0$ or $\frac{1}{2}$) we have the selection rule $\Delta J = J - J' = 0, \pm 1$. Since $D^1 \times D^0 = D^1$, and $D^1 \times D^{\frac{1}{2}} = D^{\frac{1}{2}} + D^{\frac{3}{2}}$, we see that $J=0 \leftrightarrow J'=0$ is forbidden, and only $0 \rightarrow 1$ and $\frac{1}{2} \rightarrow \frac{1}{2}, \frac{3}{2}$ transitions are allowed for $J=0$ and $J=\frac{1}{2}$ states, respectively.

Whether or not a transition on the free atom is allowed *in a particular polarization* can also be determined by examining the value of the Clebsch–Gordan coefficient $\langle JM|\bar{J}=1J'\bar{M}M'\rangle$. The polarization of the radiation is usually defined with reference to some direction (normally labelled the $\hat{z}$-direction) defined by the crystal symmetry or by an externally applied force (a stress field or static $\boldsymbol{E}$, $\boldsymbol{B}$ fields). We assume that the $\hat{x}$- and $\hat{y}$-directions are equivalent. The definitions of the different senses of linear polarization are given in Fig. 4.8. π-linearly polarized radiation travels in a direction perpendicular to $\hat{z}$ and the direction of its electric field, $\hat{\varepsilon}_E$, is parallel to $\hat{z}$. The direction of the magnetic field, $\hat{\varepsilon}_B$, is perpendicular to $\hat{z}$ (Fig. 4.8(a)), say along the $\hat{x}$-direction. The electric dipole operator for a π transition is then

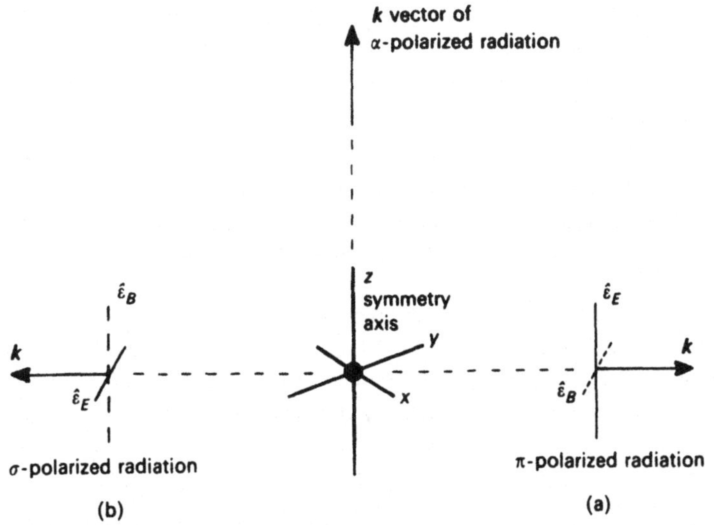

FIG. 4.8. Definitions of light beams linearly polarized with respect to a local symmetry axis. For π- and σ-polarizations the $\boldsymbol{k}$ vector can point in any direction in the xy-plane. For α-polarization $\boldsymbol{k}$ points along the symmetry axis and the $\hat{\varepsilon}_E$ and $\hat{\varepsilon}_B$ vectors are in the xy-plane.

$\sum_j e\boldsymbol{r}_j.\hat{z}=\sum_j ez_j$, and the magnetic dipole operator is

$$\frac{e}{2m}\sum_j(\boldsymbol{l}_j+2\boldsymbol{s}_j).\hat{x}=\frac{e}{2m}\sum_j\{(\boldsymbol{l}_j)_x+2(\boldsymbol{s}_j)_x\}.$$

For σ-linearly polarized radiation (Fig. 4.8(b)) the direction of propagation is perpendicular to $\hat{z}$, and the directions of its electric and magnetic fields are $\hat{y}$(or $\hat{x}$) and $\hat{z}$, respectively. The electric dipole and magnetic dipole operators for a σ transition are then $\sum_j ey_j\left(\text{or }\sum_j ex_j\right)$ and $\frac{e}{2m}\sum_j\{(\boldsymbol{l}_j)_z+2(\boldsymbol{s}_j)_z\}$, respectively.

Now consider the form of the operators for circularly polarized radiation. We define right circularly polarized (RCP) radiation as having electric and magnetic polarization vectors which rotate clockwise when viewed from behind along the direction of propagation (Fig. 4.9). In the formalism used here the electric and magnetic dipole operators for absorption transitions induced by RCP light propagating in the $\hat{z}$-direction are

$$e\sum_j\frac{x_j+\mathrm{i}y_j}{\sqrt{2}}\quad\text{and}\quad\frac{e}{2m}\sum_j\left(\frac{(\boldsymbol{l}_j)_x+2(\boldsymbol{s}_j)_x}{\sqrt{2}}+\mathrm{i}\frac{(\boldsymbol{l}_j)_y+2(\boldsymbol{s}_j)_y}{\sqrt{2}}\right)\tag{4.113}$$

where j labels the individual electrons. For the emission process the operators are the complex conjugate of the above. Accordingly for left circularly polarized (LCP) radiation, where the sense of rotation is anticlockwise, the operators are the complex conjugates of the RCP operators. (The operators for circularly polarized radiation are derived in Appendix 4A.)

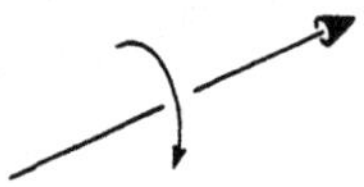

FIG. 4.9. Definition of RCP light beam used in this text. The senses of rotation of the E and B vectors are given by the right-hand screw convention.

To illustrate the method of calculating the selection rules for electric dipole transitions we consider the D_1 and D_2 absorption lines on atomic sodium, i.e. the $^2S_{\frac{1}{2}}\rightarrow{}^2P_{\frac{1}{2}}$, $^2P_{\frac{3}{2}}$ transitions. Figure 4.10 shows the Zeeman splitting of the appropriate energy levels caused by a static magnetic field along the $\hat{z}$-

FIG. 4.10. Zeeman splittings (drawn to scale) of the $^2P_{3/2}$, $^2P_{1/2}$, and $^2S_{1/2}$ levels of a sodium atom when a magnetic field is applied along the z-axis. The magnetic field removes the M_J degeneracy of each J state. The relative magnitudes of the electric dipole matrix elements for the absorption of linearly-polarized radiation are given for each transition, as are the predicted intensity patterns in various polarizations. Also shown are some experimentally-observed absorption spectra.

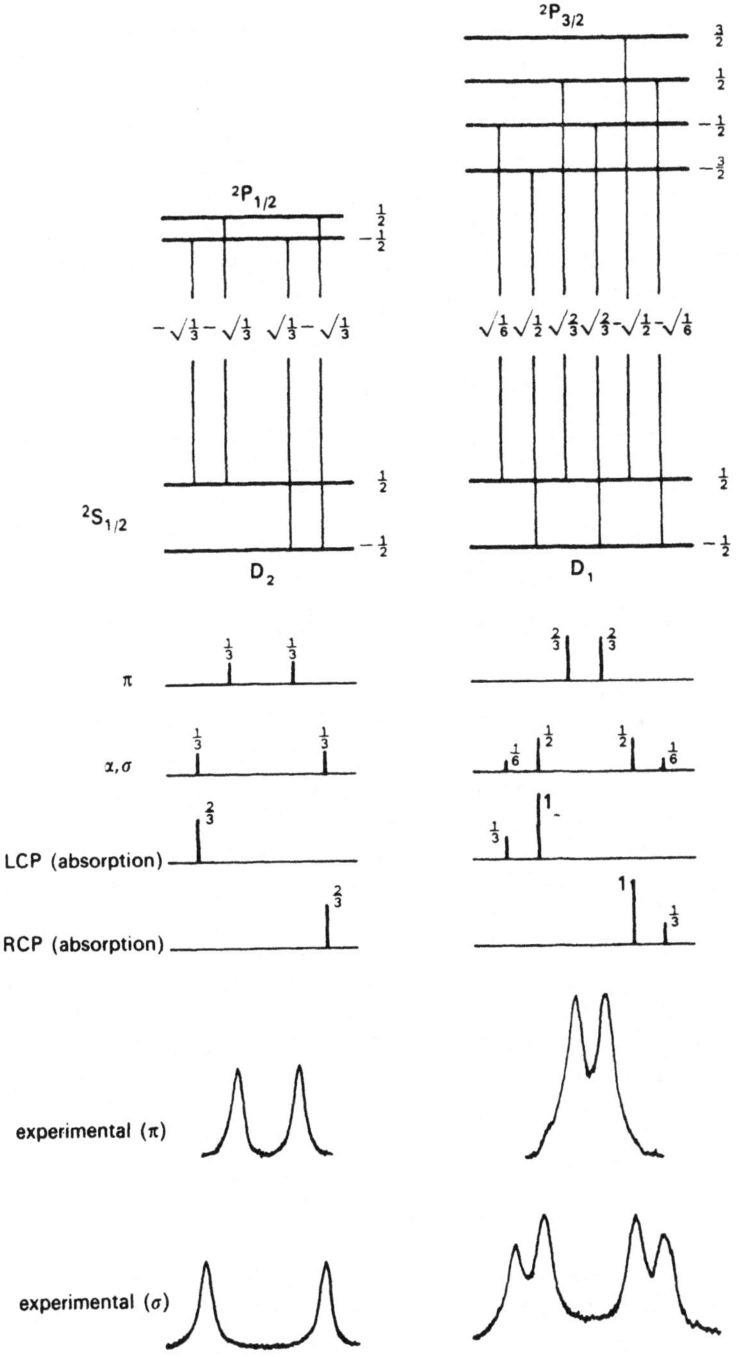

²P3/2
²P1/2
²S1/2
D2
D1
π
α, σ
LCP (absorption)
RCP (absorption)
experimental (π)
experimental (σ)

direction under the approximation that the Zeeman splitting is much less than the spin–orbit splitting between the $^2P_{\frac{1}{2}}$ and $^2P_{\frac{3}{2}}$ levels. In Fig. 4.10 the wavefunctions are labelled by J, M values. Consider first a π-polarized transition for which the operator, ez or $\sqrt{(4\pi/3)}\,erY_1^0(\theta, \phi)$, transforms as the $M=0$ row of the $J=1$ irreducible representation of the full rotation group. For the D_1 transitions, i.e. $(^2S_{\frac{1}{2}} \to {}^2P_{\frac{1}{2}})$, the Clebsch–Gordan coefficient appropriate to the $|^2S_{\frac{1}{2}}M'\rangle \to |^2P_{\frac{1}{2}}M\rangle$ π-polarized transition is $\langle \frac{1}{2}M | 1\frac{1}{2}\,0\,M' \rangle$. Table 3A.1 of Clebsch–Gordan coefficients for $j_1=1, j_2=\frac{1}{2}$ in Appendix 3A shows this coefficient to be zero unless $\Delta M = M - M' = 0$. Similarly, for σ-polarized transitions the operator is

$$ex = \sqrt{\frac{4\pi}{3}}\,er\frac{1}{\sqrt{2}}(Y_1^{-1} - Y_1^1),$$

so that the appropriate Clebsch–Gordan coefficients for the $|^2S_{\frac{1}{2}}M'\rangle \to |^2P_{\frac{1}{2}}M\rangle$ transitions are $\langle \frac{1}{2}M | 1\frac{1}{2}1M' \rangle$ and $\langle \frac{1}{2}M | 1\frac{1}{2}-1M' \rangle$. The first coefficient is non-zero only for $\Delta M = M - M' = 1$, the second for $\Delta M = M - M' = -1$. Thus for σ polarization we have the selection rule $\Delta M = \pm 1$. For RCP transitions the operator is

$$e\frac{x+iy}{\sqrt{2}} = -\sqrt{\frac{4\pi}{3}}\,Y_1^1$$

and the appropriate Clebsch–Gordan coefficient is $\langle \frac{1}{2}M | 1\frac{1}{2}1M' \rangle$. For the D_2 transitions $^2S_{\frac{1}{2}} \to {}^2P_{\frac{3}{2}}$ the Clebsch–Gordan coefficients (Table 3A.1) involve $J' = \frac{3}{2}$ instead of $J' = \frac{1}{2}$. The intensities of the transitions between the different M, M' states vary as the squares of the relevant Clebsch–Gordan coefficients. Figure 4.10 shows the allowed $^2S_{\frac{1}{2}}M' \to {}^2P_{\frac{1}{2}}M$, $^2P_{\frac{3}{2}}M$ transitions with the relevant coefficients and the theoretical intensities of the transitions for linear and circular polarizations. The experimentally observed splittings in the π- and σ-polarizations are also shown.

It is illustrative to carry out the direct integration of some matrix elements involved in these selection rules. For a one-electron atom in the absence of spin–orbit coupling the wavefunctions are $R'_{nl}(r)\,Y_l^{m_l}(\theta, \phi)$. The components of the electric dipole moment operator in spherical harmonies are

$$ex = er\sqrt{\frac{4\pi}{3}}\,\frac{Y_1^{-1} - Y_1^1}{\sqrt{2}}$$

$$ez = er\sqrt{\frac{4\pi}{3}}\,Y_1^0.$$

The matrix element for an electric dipole transition between one-electron states involves an integral of the type

$$\langle R'_{nl}(r)\,Y_l^{m_l} | r\,Y_1^{\bar{m}} | R'_{n'l'}(r)\,Y_{l'}^{m'_l} \rangle = \langle R'_{nl} | r | R'_{n'l'} \rangle \langle Y_l^{m_l} | Y_1^{\bar{m}} | Y_{l'}^{m'_l} \rangle.$$

The radial integral depends on the values of n and l, and it does not automatically become zero for particular values of n and l. Consequently there is no selection rule on the quantum number n.

The angular integral is non-zero only under the conditions (see eqn 2.59)

1. $m_l - m_l' = \bar{m}$
2. $1 + l + l' =$ even integer
3. $|l - l'| \leqslant 1 \leqslant l + l'$.

From (2) and (3) we find $l - l' = \Delta l = \pm 1$. This is related to the Laporte selection rule. The components of the electric dipole moment, formed from $l = 1$ spherical harmonics, are odd-parity functions. Hence the relevant matrix element will be zero unless the two wavefunctions have opposite parity (Laporte rule). As was shown in Chapter 2 (eqn 2.12) the one-electron orbital wavefunction characterized by quantum number l has parity $(-1)^l$. Consequently the matrix element will be zero unless l and l' differ by an odd integer. This is condition (2). Condition (3) shows that this odd integer can only be ± 1. Condition (1) indicates that for π-polarized transitions, involving dipole functions with $\bar{m} = 0$, the change in m is given by $\Delta m = m_l - m_{l'} = 0$. In contrast σ-polarized transitions require that $\Delta m = \pm 1$. Although it is a simple matter to evaluate the matrix elements and obtain selection rules for transitions on one-electron atoms, for more complicated multi-electron systems we must use the formal methods based on the Wigner–Eckart theorem.

It is instructive to regard the radiating atom classically. For example, the π-polarized component of the $2p \rightarrow 1s$ transition on a one-electron atom involves the selection rule $\Delta m = 0$ and couples the $|2p0\rangle$ and $|1s0\rangle$ states. During such transitions the atom exists in a non-stationary state consisting of an admixture of $|1s0\rangle$ and $|2p\,0\rangle$ eigenstates. Such a superposition state has an electric charge density which oscillates along the $+z$ or $-z$ directions outwards from the inner core. In consequence the atom has an electronic dipole moment which oscillates at frequency $\omega = (E_{2p} - E_{1s})/\hbar$, just like a classical radiating dipole. Such an oscillating dipole does not emit along the z-direction (Section 2.2). Quantum mechanically the probability that the atom undergoes a transition from a $|2p0\rangle$ state to the $|1s0\rangle$ state, emitting a photon in the z-direction, is also zero. Similarly, superposition states involving admixtures of $|2p+1\rangle$ and $|2p-1\rangle$ states into the $|1s0\rangle$ state result in the emisssion of helical or circularly-polarized radiation, depending on the amount of mixing and the phase relationship assumed to exist between the mixed $|2p+1\rangle$ and $|2p-1\rangle$ states. These radiations can be related to the analogous quantum transition probabilities.

In addition to the exact selection rules illustrated above, certain approximate selection rules can be very useful. For example, in the approximation of negligible spin–orbit coupling, where the free atom wavefunction is a product of orbital and spin functions and the energy level is characterized by LS values, the orbital and spin matrix elements may be examined separately. The electric dipole operator is a purely orbital operator belonging to the $L=1$ irreducible representation, and evaluation of the matrix element $\langle a(LS)|\mu(L=1)|b(L'S')\rangle$ using the Wigner–Eckart theorem leads to the selection rule $\Delta S=0$, $\Delta L=0, \pm 1$, with $L=0 \rightarrow L=0$ forbidden. Transitions between states of different total spin which violate the $\Delta S=0$ selection rule are said to be 'spin-forbidden'. However, in general this is not an exact selection rule; in the presence of spin–orbit coupling spin-forbidden usually means weakly allowed, being perhaps one to two orders of magnitude weaker than corresponding 'spin-allowed' ($\Delta S=0$) transitions.

The case of radiative transitions on an electronic centre in a solid is treated in an analogous manner to the free-atom case, although the symmetry is now considerably lower. There are many different site symmetries encountered in solids, e.g. $O_h = O \times i$, T_d, $D_{3d} = D_3 \times i$, C_{3v}, C_{2v}, etc. Since many of these contain inversion symmetry (i), we examine the implications of inversion symmetry first. The presence of inversion symmetry leads to the Laporte selection rule, that electric dipole transitions can only occur between states of opposite parity. This selection rule implies that in general the sharp radiative transitions between low-lying levels of rare-earth impurities and between low-lying levels of transition metal impurities in ionic crystals should be electric dipole-forbidden when the ions occupy sites of inversion symmetry. Nevertheless, transitions on rare-earth ions frequently occur by electric dipole processes even when the ions occupy sites of *apparent* inversion symmetry. This occurs because some perturbation close to the rare-earth impurity destroys the inversion symmetry at this site. This perturbation may be too weak to shift the energy levels significantly, and the energy levels may still be classified according to the higher symmetry classification. After examining the implications of inversion symmetry, if it occurs, the transition matrix elements are analysed from the viewpoint of the other symmetry properties.

As a final example we consider the F_3-centre in an alkali halide, a defect which has C_{3v} symmetry. The character table for C_{3v} symmetry, Table 3.3, shows there to be three distinct irreducible representations, A_1, A_2, and E. Table 3.3 also gives the spaces of functions which belong to these irreducible representation. The π and σ electric dipole operators vary as z and x, y which transform as A_1 and E irreducible representations, while the π and σ magnetic dipole operators transform as R_x, $R_y(E)$ and $R_z(A_2)$, respectively. The probability of a π-polarized electric dipole transition between the A_1 and E states, for example, involves the matrix element $\langle a(E)|\mu(A_1)|b(A_1)\rangle$. Since $A_1 \times A_1 = A_1$, this matrix element is zero and the electric dipole transitions between these states is forbidden. We can investigate other possible transitions in the same way, using the easily derived relationships

$$\begin{array}{lll} A_1 \times A_1 = A_1 & A_2 \times A_1 = A_2 & E \times A_1 = E \\ A_1 \times A_2 = A_2 & A_2 \times A_2 = A_1 & E \times A_2 = E \\ A_1 \times E = E & A_2 \times E = E & E \times E = A_1 + A_2 + E. \end{array}$$

We can then draw a selection rule table (Table 4.1).

Now the ground state of the F_3-system places two electrons in the lowest a_1 state and one electron in the e state, i.e. the ground state has the configuration $(a_1)^2\,(e)$. The product orbital in the ground state thus belongs to the $A_1 \times A_1 \times E = E$ irreducible representation. The excited state may be constructed by promoting one of the two a_1 electrons into an e state giving a configuration $(a_1)(e)^2$. Since $A_1 \times E \times E = A_1 + A_2 + E$ there are three distinct excited orbitals with this $(a_1)(e)^2$ configuration, belonging to the A_1, A_2, and E irreducible representations. As Table 4.1 shows, there are distinct polarization features in the absorption transitions from the ground state to the three excited states.

Table 4.1

Final state	(Initial state) A_1	A_2	E
A_1	ED(π)	MD(σ)	ED(σ) MD(π)
A_2	MD(σ)	ED(π)	ED(σ) MD(π)
E	ED(σ) MD(π)	ED(σ) MD(π)	ED(π), ED(σ) MD(π), MD(σ)

Appendix 4A. Electric and magnetic dipole operators for RCP and LCP radiation

We compare the formulae for the interaction of an atomic centre with a linearly polarized beam and with a circularly polarized beam. This will enable us to obtain the dipole operator for circularly polarized radiation, analogous to that for linear polarization derived in Section 4.3. A beam of linearly polarized ($\hat{x}$-direction) radiation travelling in the $\hat{z}$-direction contains an electric field $E = E_0 \hat{x} \cos \omega t = E_0 \hat{x}(\exp(\mathrm{i}\omega t) + \exp(\mathrm{i}\omega t))/2$. We can write the electric dipole interaction term as $\mathcal{H}_1 = -\boldsymbol{p}.\boldsymbol{E} = \sum e\boldsymbol{r}.\boldsymbol{E} = \sum \boldsymbol{E}_0/2$ $e\boldsymbol{x}_j(\exp(-\mathrm{i}\omega t) + \exp(\mathrm{i}\omega t))$. The first and second terms induce absorption and

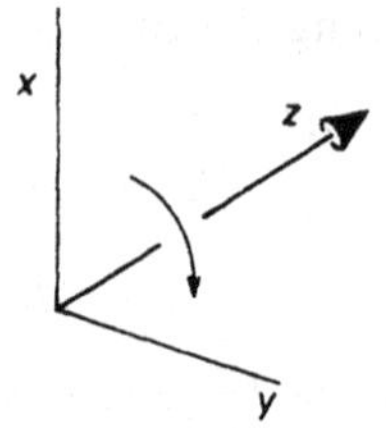

FIG. 4A.1. In a RCP beam travelling in the z-direction the electric and magnetic vectors rotate in a clockwise sense from x to y.

emission transitions, respectively, and the electric dipole operator, $\sum e\boldsymbol{x}_j$, is the same for both transitions.

A beam of RCP radiation travelling in the $\hat{z}$-direction (Fig. 4A.1) contains an electric field

$$\begin{aligned}E &= \hat{x}\frac{E_0}{\sqrt{2}}\cos\omega t + \hat{y}\frac{E_0}{\sqrt{2}}\cos(\omega t - \pi/2)\\ &= \frac{E_0}{2\sqrt{2}}\hat{x}(\exp(\mathrm{i}\omega t)+\exp(-\mathrm{i}\omega t)) + \frac{E_0}{2\sqrt{2}}\hat{y}(-\mathrm{i}\exp(\mathrm{i}\omega t)+\mathrm{i}\exp(-\mathrm{i}\omega t))\\ &= \frac{E_0}{2}\frac{\hat{x}+\mathrm{i}\hat{y}}{\sqrt{2}}\exp(-\mathrm{i}\omega t) + \frac{E_0}{2}\frac{\hat{x}-\mathrm{i}\hat{y}}{\sqrt{2}}\exp(\mathrm{i}\omega t).\end{aligned} \tag{4A.1}$$

We again write the electric dipole interaction term as

$$\mathscr{H}_1 = -\boldsymbol{p}\cdot\boldsymbol{E} = \sum_j \frac{E_0}{2}e\frac{x_j+\mathrm{i}y_j}{\sqrt{2}}\exp(-\mathrm{i}\omega t) + \sum_j \frac{E_0}{2}e\frac{x_j-\mathrm{i}y_j}{2\sqrt{2}}\exp(\mathrm{i}\omega t)$$

which is of the form $V\exp(-\mathrm{i}\omega t) - V^*\exp(\mathrm{i}\omega t)$. Hence for the absorption process, where the first term is effective, the electric dipole operator is $\sum_j e(x_j+\mathrm{i}y_j)/\sqrt{2}$. For the emission process, where the second term is effective, the electric dipole operator is $\sum_j e(x_j-\mathrm{i}y_j)/\sqrt{2}$. For LCP radiation the electric field varies as $\hat{x}E_0/\sqrt{2}\cos\omega t + \hat{y}E_0/\sqrt{2}\cos(\omega t+\pi/2)$ which leads to electric dipole operators which are the complex conjugate of those for RCP radiation.

A similar analysis can be made for magnetic dipole transitions, when the phase relationships between $\boldsymbol{B}_x$ and $\boldsymbol{B}_y$ are identical to those between $\boldsymbol{E}_x$ and $\boldsymbol{E}_y$ for RCP and LCP beams. We obtain the expression for the magnetic dipole operators given in the text.

5

Electronic centres in a vibrating crystalline environment

5.1 General considerations

An optical centre in a crystal participates in the vibrational motion of that crystal, and this motion affects the optical properties of the centre. It is now appropriate to consider how lattice vibrations can modify such properties. The Hamiltonian describing the vibrating lattice is

$$\mathscr{H}_{\text{lattice}} = \sum_l \frac{\boldsymbol{P}_l^2}{2M_l} + V_1(\boldsymbol{R}_l) \tag{5.1}$$

where $\boldsymbol{P}_l$ and $\boldsymbol{R}_l$ are the momentum and position of the lth ion and $V_1(\boldsymbol{R}_l)$ is the interionic potential energy. If $\boldsymbol{R}_l(0)$ is the equilibrium position of the lth ion then we can write

$$\boldsymbol{R}_l = \boldsymbol{R}_l(0) + \boldsymbol{q}_l \tag{5.2}$$

where $\boldsymbol{q}_l$ is the displacement of the lth ion from its equilibrium position. The interionic potential energy is then

$$V_1(\boldsymbol{R}_l) = V_1(\boldsymbol{R}_l(0)) + V_1(\boldsymbol{q}_l) \tag{5.3}$$

where $V_1(\boldsymbol{R}_l(0))$ is a constant energy term, and $V_1(\boldsymbol{q}_l)$ is the additional potential energy due to the displacement of the ions from their equilibrium positions. V_1 has a simple form if the additional potential energy is harmonic in the ionic displacements, particularly if only nearest-neighbour forces between the ions are considered. In that case, and ignoring end effects at the edges of the lattice, we have

$$V_1(\boldsymbol{q}_l) = \tfrac{1}{2}K\sum_l (\boldsymbol{q}_l - \boldsymbol{q}_{l-1})^2 \tag{5.4}$$

where K is the spring constant. Neglecting the constant term $V_1(R_l(0))$, the resulting dynamic lattice Hamiltonian is

$$\mathscr{H}_{\substack{\text{dynamic}\\ \text{lattice}}} = \sum_l \left[\frac{\boldsymbol{P}_l^2}{2M_l} + \tfrac{1}{2}K(\boldsymbol{q}_l - \boldsymbol{q}_{l-1})^2\right]. \tag{5.5}$$

Before attempting to solve for the eigenstates and eigenvalues of this Hamiltonian we review the quantum analysis of the simplest oscillating system, the linear harmonic oscillator. Then the simplest example of a dynamic lattice is examined—a monatomic line of atoms bound together by nearest-neighbour harmonic forces. We find that the vibrational modes of this lattice can be analysed as a collection of linear harmonic oscillators. The energy quanta of these vibrational modes are referred to as *phonons*. This analysis is then extended to the case of a three-dimensional crystal lattice. Finally we analyse the effects of the vibrating crystalline environment on the spectroscopic properties of an optically active centre which is an element of the lattice

5.2 Lattice vibrations

5.2.1 The linear harmonic oscillator

The linear harmonic oscillator consists of a single mass M moving in the x-direction under the action of a harmonic force $F=-KX$, K being the spring constant and X the position of the mass. A classical analysis shows that the mass oscillates sinusoidally about $X=0$ with angular frequency $\omega_0=2\pi\nu_0=(K/M)^{\frac{1}{2}}$. The Hamiltonian for this system is

$$\begin{aligned}\mathscr{H}_{\text{HO}}&=\frac{P_x^2}{2M}+\frac{1}{2}KX^2\\&=\frac{P_x^2}{2M}+\frac{1}{2}M\omega_0^2X^2\end{aligned} \tag{5.6}$$

where $P_x=-\text{i}\hbar\partial/\partial X$. The operators X and P_x satisfy the commutation relationship

$$[X, P_x]=\text{i}\hbar. \tag{5.7}$$

It is instructive to derive the eigenstates and eigenvalues of $\mathscr{H}_{\text{HO}}$ using an operator technique since this can be generalized into a method for solving the harmonic crystal lattice problem. We define lowering and raising operators a and $a^\dagger$:

$$\begin{aligned}a&=\frac{1}{(2M\hbar\omega_0)^{\frac{1}{2}}}[M\omega_0X+\text{i}P_x]\\a^\dagger&=\frac{1}{(2M\hbar\omega_0)^{\frac{1}{2}}}[M\omega_0X-\text{i}P_x].\end{aligned} \tag{5.8}$$

From these definitions and using eqn (5.7) we find

$$[a, a^\dagger]=1. \tag{5.9}$$

$\mathscr{H}_{\text{HO}}$ can be written in terms of these operators:

$$\mathscr{H}_{\text{HO}}=\frac{\hbar\omega_0}{2}(aa^\dagger+a^\dagger a)=\hbar\omega_0\left(a^\dagger a+\frac{1}{2}\right). \tag{5.10}$$

We now define state functions $|n\rangle$, where n is an integer, as follows:

$$\boldsymbol{a}|n\rangle = n^{\frac{1}{2}}|n-1\rangle$$

$$\boldsymbol{a}^{\dagger}|n\rangle = (n+1)^{\frac{1}{2}}|n+1\rangle. \tag{5.11}$$

It can be shown that these definitions are self-consistent. Thus eqn (5.11) shows that the effect of operating on the state function $|n\rangle$ with $\boldsymbol{a}^{\dagger}$ is to change this function to another representing the state $|n+1\rangle$, which, as we see below, is higher in energy than $|n\rangle$ by one quantum $\hbar\omega_0$. $\boldsymbol{a}^{\dagger}$ is called a *raising* or *creation* operator. Similarly $\boldsymbol{a}$ is a *lowering* or *annihilation* operator. From eqn (5.11) we see that

$$\boldsymbol{a}^{\dagger}\boldsymbol{a}|n\rangle = n|n\rangle. \tag{5.12}$$

Hence the functions $|n\rangle$ are eigenstates of $\mathscr{H}$:

$$\mathscr{H}_{\mathrm{HO}}|n\rangle = \hbar\omega_0(n+\tfrac{1}{2})|n\rangle \tag{5.13}$$

and the energy of state $|n\rangle$, E_n, is $\hbar\omega_0(n+\frac{1}{2})$. Except for the factor of $\frac{1}{2}$ this result was postulated by Planck when he used the linear harmonic oscillator model of the atom to analyse the interaction of the atom with radiation. The quantum analysis shows that the linear harmonic oscillator has a minimum energy $\hbar\omega_0/2$, the *zero-point vibrational energy*, required for consistency with the uncertainty principle.

This formalism is very useful for calculating properties of the harmonic oscillator. For example, to calculate the expectation value of X^2 when the harmonic oscillator is in state n, we first express X in terms of $\boldsymbol{a}$ and $\boldsymbol{a}^{\dagger}$:

$$X = \left(\frac{\hbar}{2M\omega_0}\right)^{\frac{1}{2}}(\boldsymbol{a}+\boldsymbol{a}^{\dagger}). \tag{5.14}$$

Hence the expectation value of X^2 is

$$\langle X^2\rangle = \frac{\hbar}{2M\omega_0}\langle n|(\boldsymbol{a}+\boldsymbol{a}^{\dagger})(\boldsymbol{a}+\boldsymbol{a}^{\dagger})|n\rangle = \frac{\hbar}{M\omega_0}(n+\tfrac{1}{2}). \tag{5.15}$$

This operator formalism is easily generalized. In fact, any quantum system satisfying a Hamiltonian of the form $\mathscr{H} = AP^2 + BQ^2$, P and Q being operators which satisfy $[Q, P] = i\hbar$, can be analysed in the same way after first defining the appropriate $\boldsymbol{a}$, $\boldsymbol{a}^{\dagger}$ operators and state functions $|n\rangle$.

5.2.2 *The linear monatomic chain*

The simplest lattice is a one-dimensional chain of $N+1$ identical atoms bound together by nearest-neighbour harmonic forces. N is assumed to be a large even integer. The equilibrium positions of the atoms are $X_l = la$, l being an integer between 0 and N which serves as a label for the atom, and a being the

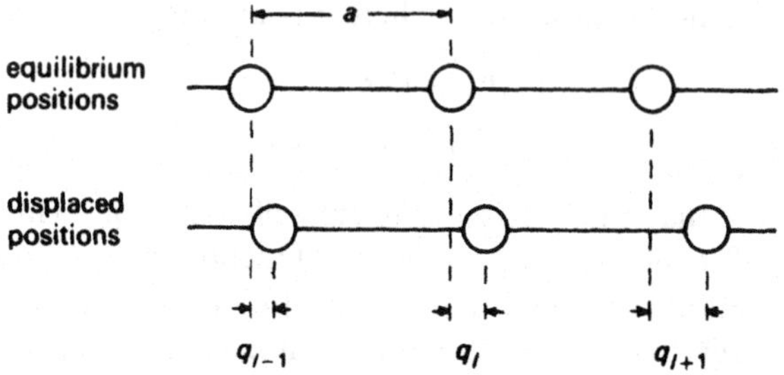

FIG. 5.1. The longitudinal displacement of the lth atom is q_l. The extension of the separation between atoms l and $l-1$ is $(q_l - q_{l-1})$.

equilibrium separation between adjacent atoms (Fig. 5.1). Ignoring end-effects, which are relatively unimportant if N is large, the dynamic lattice Hamiltonian is

$$\mathscr{H} = \sum_l \frac{P_l^2}{2M} + \frac{1}{2} K \sum_l (q_l - q_{l-1})^2 \tag{5.5'}$$

which is the one-dimensional analogue of the Hamiltonian for the three-dimensional simple lattice (eqn 5.5). The operators q_l and P_l are related by

$$[q_l, P_{l'}] = i\hbar\delta_{ll'}. \tag{5.16}$$

The Hamiltonian (5.5′) can be converted to the correct form, $AP^2 + BQ^2$, by transforming P_l and q_l to new 'coordinates' and 'momenta'. Since q_l is a function defined only at discrete points la, we can make the expansion

$$q_l = N^{-\frac{1}{2}} \sum_{n=-\frac{1}{2}N}^{\frac{1}{2}N} Q_n \exp(-2\pi i n l/N) = N^{-\frac{1}{2}} \sum_{k_n} Q_{k_n} \exp(-ik_n la) \tag{5.17}$$

where n is an integer with values between $-\frac{1}{2}N$ and $+\frac{1}{2}N$. It is more convenient to use $k_n = 2\pi n/Na$ rather than n to label the coefficients in the expansion. A similar expansion for P_l is

$$P_l = N^{-\frac{1}{2}} \sum_{n=-\frac{1}{2}N}^{\frac{1}{2}N} P_n \exp(2\pi i n l/N) = N^{-\frac{1}{2}} \sum_{k_n} P_{k_n} \exp(ik_n la) \tag{5.18}$$

The coefficients Q_{k_n} and P_{k_n} in eqns (5.17) and (5.18) are operators defined by

$$Q_{k_n} = N^{-\frac{1}{2}} \sum_{l=0}^{N} q_l \exp(ik_n la) \tag{5.19}$$

and

$$P_{k_n} = N^{-\frac{1}{2}} \sum_{l=0}^{N} P_l \exp(-ik_n la). \tag{5.20}$$

From these definitions of Q_{k_n} and P_{k_n} and making use of eqn (5.16) we find

$$[Q_{k_n}, P_{k_{n'}}] = i\hbar\delta_{nn'} \tag{5.21}$$

However, Q_{k_n} and P_{k_n} are not Hermitian operators. From their definitions we see that

$$Q^{\dagger}_{k_n} = Q_{k_{-n}}, \qquad P^{\dagger}_{k_n} = P_{k_{-n}}. \tag{5.22}$$

After some mathematical manipulation the Hamiltonian for the dynamic lattice, eqn (5.5′), can be written as a sum of Hamiltonians:

$$\mathscr{H} = \sum_{k_n} \left[\frac{P_{k_n} P^{\dagger}_{k_n}}{2M} + \tfrac{1}{2} G(k_n) Q_{k_n} Q^{\dagger}_{k_n} \right] \tag{5.23}$$

$$= \sum_{k_n} \mathscr{H}_{k_n} \tag{5.24}$$

where

$$G(k_n) = 4K \sin^2 \tfrac{1}{2}(k_n a) \tag{5.25}$$

and each individual Hamiltonian is of the required form, viz.

$$\mathscr{H}_{k_n} = \frac{P_{k_n} P^{\dagger}_{k_n}}{2M} + \tfrac{1}{2} G(k_n) Q_{k_n} Q^{\dagger}_{k_n}, \tag{5.26}$$

Q_{k_n} and P_{k_n} having the correct commutation relationship. The fact that the Hamiltonian (5.26) contains $PP^{\dagger}$ instead of PP is not inconsistent with the linear harmonic oscillator analysis since, as P_x and X are Hermitian, the original linear harmonic oscillator Hamiltonian could have been written as

$$\mathscr{H}_{\text{HO}} = \frac{P_x P^{\dagger}_x}{2M} + \tfrac{1}{2} K X X^{\dagger}. \tag{5.27}$$

Comparing eqns (5.26) and (5.27), and from the definition of Q_{k_n}, we see that the Hamiltonian $\mathscr{H}_{k_n}$ describes a coupled mode of oscillation with normal coordinate Q_{k_n}, the frequency of this oscillation being

$$\omega_{k_n} = \left(\frac{G(k_n)}{M} \right)^{\frac{1}{2}} = 2 \left(\frac{K}{M} \right)^{\frac{1}{2}} \left| \sin \frac{k_n a}{2} \right|. \tag{5.28}$$

The k_n values are in the range $-\pi/a < 0 < +\pi/a$, so that the frequencies cover a quasicontinuum of discrete values, as indicated by the dispersion relationship shown in Fig. 5.2. This dispersion relationship is identical to that obtained from a classical analysis of the vibrations of a linear monatomic chain. For convenience we replace the index k_n by k, remembering that k covers a quasi-continuum of discrete values.

We may proceed with the quantum analysis of the vibrations of the linear monatomic chain by analogy with that used for the linear harmonic oscillator.

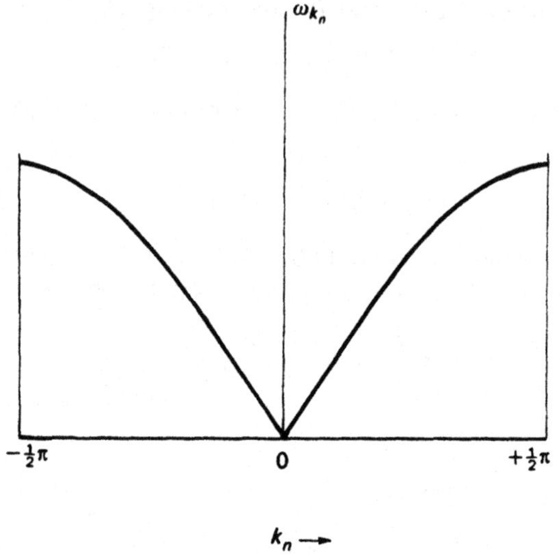

FIG. 5.2. Variation of ω_k with k for the vibrational modes of the one-dimensional line of atoms.

For each Hamiltonian, $\mathscr{H}_k$, operators $\boldsymbol{a}_k$ and $\boldsymbol{a}_k^\dagger$ are defined as in eqn (5.8), replacing X and P_x by the appropriate collective motion coordinates. It is not immediately apparent, however, whether X should be replaced by Q_k or by $Q_k^\dagger$ and whether P_x should be replaced by P_k or $P_k^\dagger$. The following definitions are adopted:

$$\begin{aligned} \boldsymbol{a}_k &= \left(\frac{1}{2M\hbar\omega_k}\right)^{\frac{1}{2}} (M\omega_k Q_k^\dagger + \mathrm{i}P_k) \\ \boldsymbol{a}_k^\dagger &= \left(\frac{1}{2M\hbar\omega_k}\right)^{\frac{1}{2}} = (M\omega_k Q_k - \mathrm{i}P_k^\dagger) \end{aligned} \tag{5.29}$$

from which, using eqn (5.22), it follows that

$$[a_k, a_{k'}^\dagger] = \delta_{kk'}. \tag{5.30}$$

State functions $|n_k\rangle$ are also defined such that

$$\begin{aligned} \boldsymbol{a}_k|n_k\rangle &= n_k^{\frac{1}{2}}|n_k - 1\rangle \\ \boldsymbol{a}_k^\dagger|n_k\rangle &= (n_k + 1)^{\frac{1}{2}}|n_k + 1\rangle \end{aligned} \tag{5.31}$$

where n_k is a positive integer or zero. Next we evaluate $(\boldsymbol{a}_k\boldsymbol{a}_k^\dagger + \boldsymbol{a}_k^\dagger\boldsymbol{a}_k)\hbar\omega_k/2$, obtaining the value

$$\tfrac{1}{2}(a_k a_k^\dagger + a_k^\dagger a_k)\hbar\omega_k = \mathscr{H}_k - \frac{\mathrm{i}}{\hbar^2\omega_k}(P_k Q_k + Q_k P_k - P_k^\dagger Q_k^\dagger - Q_k^\dagger P_k^\dagger)$$

where $\mathscr{H}_k$ is given by eqn (5.26). Because of the last term on the right this is not quite analogous to eqn (5.10). However, after summing over all k values we obtain

$$\mathscr{H} = \sum_k \mathscr{H}_k = \sum_k \frac{\hbar\omega_k}{2}(a_k a_k^\dagger + a_k^\dagger a_k). \tag{5.32}$$

This can be written

$$\mathscr{H} = \sum_k \mathscr{H}_k^{\mathrm{HO}} \tag{5.33}$$

where

$$\mathscr{H}_k^{\mathrm{HO}} = \frac{\hbar\omega_k}{2}(a_k a_k^\dagger + a_k^\dagger a_k). \tag{5.34}$$

We observe that $\mathscr{H}_k^{\mathrm{HO}}$ is not the same as $\mathscr{H}_k$; $\mathscr{H}_k^{\mathrm{HO}}$ is *not* the Hamiltonian for one of the vibrational modes of the lattice. However, $\mathscr{H}_k^{\mathrm{HO}}$ has the form of the Hamiltonian for the linear harmonic oscillator (eqn 5.10) and so, in analogy with the case of the linear harmonic oscillator, we see that the state functions $|n_k\rangle$ are eigenstates of $\mathscr{H}_k^{\mathrm{HO}}$:

$$\mathscr{H}_k^{\mathrm{HO}}|n_k\rangle = (n_k + \tfrac{1}{2})\hbar\omega_k|n_k\rangle \tag{5.35}$$

and n_k gives the number of quanta of energy in the collective mode of vibrations k above the zero-point energy. The average value of n_k at temperature T is given by the Bose–Einstein factor $[\exp(\hbar\omega_k/kT)-1]^{-1}$. These quanta of vibrational energy are *phonons.* The eigenstate of the Hamiltonian $\mathscr{H}$ is the product state $|n_{k1}\rangle|n_{k2}\rangle|n_{k3}\rangle \ldots$ which we can write as $|n_{k1}, n_{k2}, n_{k3}, \ldots\rangle$. A convenient shorthand for this product state is $|n_1, n_2, n_3, \ldots\rangle$ or simply $|n\rangle$. The energy of the eigenstate $|n\rangle$ is $\Sigma(n_k+\tfrac{1}{2})\hbar\omega_k$. This procedure of quantizing the energy of the normal modes of vibration is known as *second quantization.*

The normal coordinates, Q_k, are composed of displacements of the individual atoms, q_l, multiplied by appropriate phase factors (eqn 5.19). Conversely we can describe the displacement, q_l, of the lth atom from its equilibrium position in terms of the normal coordinates (eqn 5.17). From eqn (5.29) we find

$$Q_k = \left(\frac{\hbar}{2M\omega_k}\right)^{\frac{1}{2}}(a_k^\dagger + a_{-k}) \tag{5.36}$$

and using eqn (5.19) we obtain, after some algebra,

$$q_l = \sum_k q_l^{(k)} = \sum_k \left(\frac{\hbar}{2M\omega_k N}\right)^{\frac{1}{2}} (\boldsymbol{a}_k^\dagger + \boldsymbol{a}_{-k}) \exp(-ikla)$$

or, by a rearrangement of terms,

$$q_l = \sum_k \left(\frac{\hbar}{2M\omega_k N}\right)^{\frac{1}{2}} [\boldsymbol{a}_k^\dagger \exp(-ikla) + \boldsymbol{a}_k \exp(ikla)]. \tag{5.37}$$

This is a very useful formula since it relates the displacement of the lth atom to the creation and annihilation operators for phonons.

A lattice vibration k introduces a periodic distortion of the environment, in particular it introduces a periodic variation in the separation between adjacent atoms. In the long wavelength limit (vibrational wavelength $\gg a$) the local 'strain' in the vicinity of the lth atom caused by the kth mode can be written as

$$\varepsilon_k(l) = \frac{q_{l+1}^{(k)} - q_l^{(k)}}{a} = -ik\left(\frac{\hbar}{2MN\omega_k}\right)^{\frac{1}{2}} (\boldsymbol{a}_k^\dagger \exp(-ikla) - \boldsymbol{a}_k \exp(ikla)). \tag{5.38}$$

Since we will be dealing with the effect of strain upon a specific atom or ion the atom or ion in question can be taken to have the label $l=0$, so that

$$\varepsilon_k = -ik\left(\frac{\hbar}{2MN\omega_k}\right)^{\frac{1}{2}} (\boldsymbol{a}_k^\dagger - \boldsymbol{a}_k) = -i\left(\frac{\hbar\omega_k}{2MNv_k^2}\right)^{\frac{1}{2}} (\boldsymbol{a}_k^\dagger - \boldsymbol{a}_k) \tag{5.39}$$

a result valid only in the long wavelength limit. v_k is the velocity of the kth mode.

So far we have considered only a simple monatomic harmonic lattice. Real crystals are three-dimensional; they generally contain more than one type of atom, the forces acting between which are not solely nearest-neighbour forces nor are they purely harmonic. If we neglect the anharmonic effects the quantization of such a system can proceed along identical lines as for the monatomic linear lattice, but the algebra is much more complicated. The normal modes of vibration are also more complicated. There are different branches to the dispersion curves of ω versus $\boldsymbol{k}$ as Fig. 5.3 shows, and the dispersion relationship differs for different propagation directions in the crystal, each branch of the dispersion curve being separable into *acoustical* and *optical* modes of vibration. Finally there are transverse and longitudinal vibrations. For long-wavelength acoustical modes there is generally a small phase difference between the motions of adjacent atoms or ions, while for optical modes the motions of adjacent atoms or ions can be in antiphase. In the case of ionic materials transverse optical modes cause a large oscillating

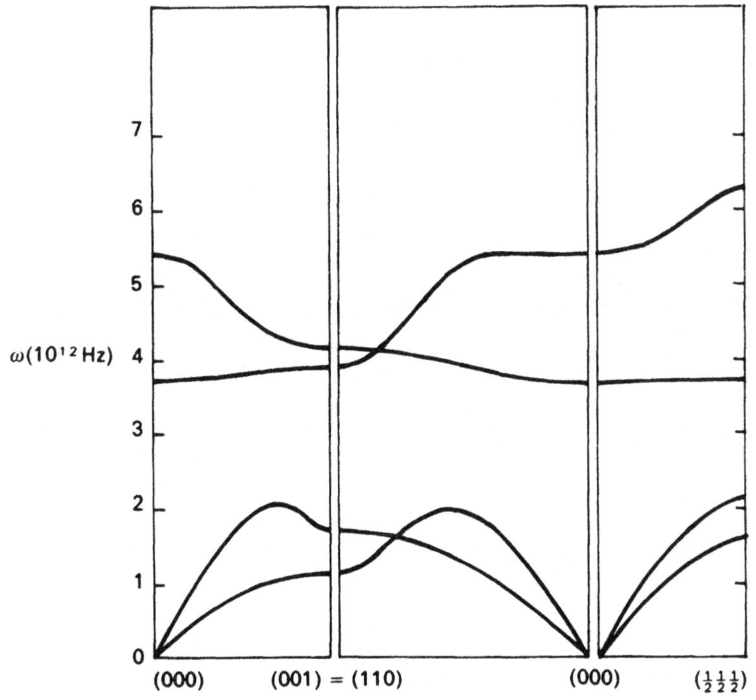

FIG. 5.3. Dispersion curves for sodium iodide based on a simple shell model (Woods *et al.* 1963). The measured values are in general agreement with the calculations. The dispersion curves in the acoustical branches show a similarity to the curve derived for the one-dimensional line of atoms (Fig. 5.2).

dipole moment which can interact strongly with electromagnetic radiation. Exact phase matching between this vibrational mode and the radiation requires that $k \simeq 0$ for this interacting vibrational mode. The frequency at which this strong interaction occurs is in the infrared and it is known as the *reststrahl* frequency. For our purposes eqns (5.37) and (5.39) with their minimum formalism are generally adequate to describe displacements and strains in real crystals associated with lattice vibrations of long wavelength.

When the vibrational wavelength becomes comparable with the interionic or interatomic distances the distortions of the crystal environment about the electronic centre are more complicated. It is then useful to classify the distortions according to irreducible representations of the equilibrium symmetry of the electronic centre and its surroundings. For example, Fig. 5 4 shows the relevant normal mode distortions of an octahedral complex, in which the central ion is surrounded by six neighbouring ions equidistant along $\pm x$, $\pm y$, $\pm z$-directions. The symmetry classifications of these distor-

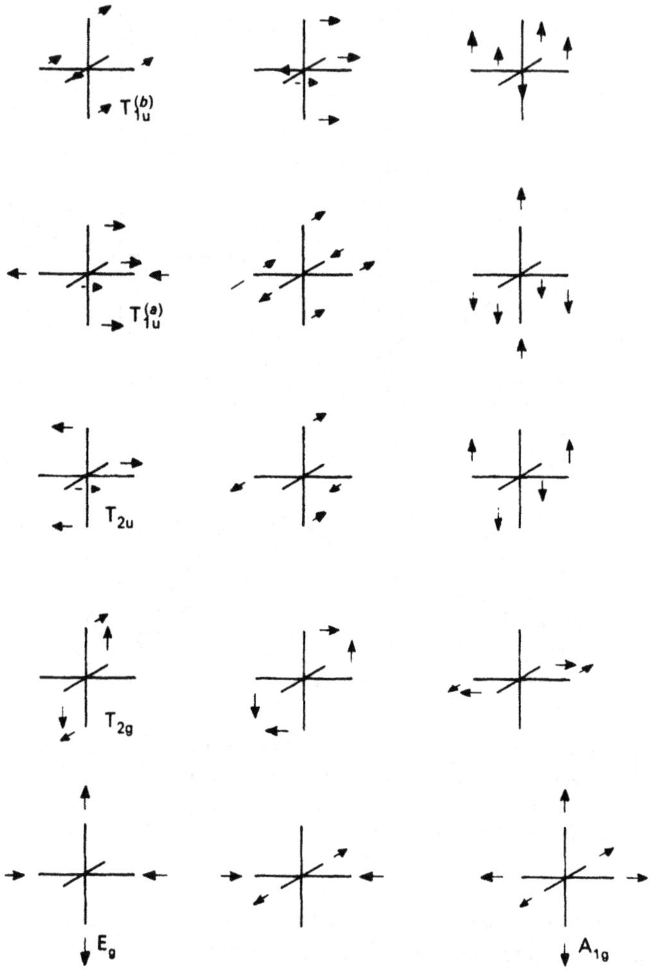

FIG. 5.4. Normal modes of distortion of the complex consisting of an electronic centre with six neighbouring ions equidistant along the $\pm x$, $\pm y$, $\pm z$-directions.

tions are also shown. Note that distortions of T_{1u} and T_{2u} symmetries destroy the inversion symmetry at the site of the electronic centre. Such odd-parity distortions have important consequences for the spectroscopic properties of the electronic centre.

5.3 The coupled electron–lattice system

We now investigate the situation in which the optically-active centre is an element in a dynamic lattice. For convenience we will assume that the centre in question is an ion. The Hamiltonian describing this system can be written

$$\mathscr{H} = \mathscr{H}_{\mathrm{FI}}(\boldsymbol{r}_i) + \mathscr{H}_{\mathrm{c}}(\boldsymbol{r}_i, \boldsymbol{R}_l) + V_{\mathrm{I}}(\boldsymbol{R}_l) + \sum_l \frac{\boldsymbol{P}_l^2}{2M_l} \tag{5.40}$$

where $\mathscr{H}_{\mathrm{FI}}$ is the free-ion Hamiltonian, eqn (2.35), $\mathscr{H}_{\mathrm{c}}(\boldsymbol{r}_i, \boldsymbol{R}_l)$ is the crystal field Hamiltonian, eqn (2.56), $V_{\mathrm{I}}(\boldsymbol{R}_l)$ is the interion potential energy, and the last term is the kinetic energy of the lattice ions. The free ion and crystal field Hamiltonians have been discussed in Sections 2.3 and 2.4, respectively. $\boldsymbol{r}_i$ and $\boldsymbol{R}_l$ are the variables of the system. The difficulty with the above Hamiltonian is that the term $\mathscr{H}_{\mathrm{c}}(\boldsymbol{r}_i, \boldsymbol{R}_l)$, which contains both electronic and ionic coordinates, couples the electronic motion with the vibrations of the lattice ions. The extent to which the electronic and ionic variables may be decoupled was first considered by Born and Oppenheimer (1927) in their analysis of the coupling of electronic and nuclear motions of diatomic molecules. We shall generally follow their approach.

Let us first omit the ionic kinetic energy term and consider the Hamiltonian with the static lattice

$$\begin{aligned}\mathscr{H}_0 &= \mathscr{H}_{\mathrm{FI}}(\boldsymbol{r}_i) + \mathscr{H}_{\mathrm{c}}(\boldsymbol{r}_i, \boldsymbol{R}_l) + V_{\mathrm{I}}(\boldsymbol{R}_l)\\ &= \mathscr{H}_{\mathrm{e}}(\boldsymbol{r}_i, \boldsymbol{R}_l) + V_{\mathrm{I}}(\boldsymbol{R}_l)\end{aligned} \tag{5.41}$$

in which $\boldsymbol{R}_l$ is regarded as a *parameter* rather than a variable; $\boldsymbol{r}_i$ is the variable. $\mathscr{H}_{\mathrm{e}}(\boldsymbol{r}_i, \boldsymbol{R}_l)$ is the electronic part of this Hamiltonian and it was considered in detail in Section 2.4. We write the eigenfunctions of $\mathscr{H}_0$ as $\psi_a(\boldsymbol{r}_i, \boldsymbol{R}_l)$:

$$\mathscr{H}_0 \psi_a(\boldsymbol{r}_i, \boldsymbol{R}_l) = E^{(a)}(\boldsymbol{R}_l)\psi_a(\boldsymbol{r}_i, \boldsymbol{R}_l). \tag{5.42}$$

The subscript a labels the particular electronic state. We indicate in eqn (5.42) that the energy depends *parametrically* on the set of $\boldsymbol{R}_l$ values chosen. $V_{\mathrm{I}}(\boldsymbol{R}_l)$ is included in $\mathscr{H}_0$ because the interionic energy depends on the coupling between the central ion (which is part of the lattice) and the remainder of the lattice. The energy eigenvalue $E^{(a)}(\boldsymbol{R}_l)$ is then

$$\begin{aligned}E^{(a)}(\boldsymbol{R}_l) &= \langle \psi_a | \mathscr{H}_{\mathrm{e}}(\boldsymbol{r}_i, \boldsymbol{R}_l) | \psi_a \rangle + \langle \psi_a | V_{\mathrm{I}}(\boldsymbol{R}_l) | \psi_a \rangle\\ &= \mathscr{H}_{\mathrm{e}}^{(a)}(\boldsymbol{R}_l) + V_{\mathrm{I}}^{(a)}(\boldsymbol{R}_l).\end{aligned} \tag{5.43}$$

We note that $\mathscr{H}_{\mathrm{e}}^{(a)}(\boldsymbol{R}_l)$ is the electronic energy of the ion in the static crystal, as discussed in previous chapters. To this is added the interion potential energy $V_{\mathrm{I}}^{(a)}(\boldsymbol{R}_l)$; the label a indicates that the interion potential energy depends on the electronic state a of the centre.

We now return to the Hamiltonian which includes the dynamic lattice, eqn (5.40): $\mathscr{H}_0 + \sum_l (\boldsymbol{P}_l^2/2M_l)$ and we seek eigenfunctions of the form $\psi_a(\boldsymbol{r}_i, \boldsymbol{R}_l)\chi_a(\boldsymbol{R}_l)$ where ψ_a is an eigenstate of $\mathscr{H}_0$ and $\chi_a(\boldsymbol{R}_l)$ is a function of the *variables* $\boldsymbol{R}_l$. The Schrödinger equation is

$$\left[\mathscr{H}_0+\sum_l \frac{\boldsymbol{P}_l^2}{2M_l}\right]\psi_a(\boldsymbol{r}_i, \boldsymbol{R}_l)\chi_a(\boldsymbol{R}_l)=E\psi_a(\boldsymbol{r}_i, \boldsymbol{R}_l)\chi_a(\boldsymbol{R}_l). \tag{5.44}$$

Writing $\boldsymbol{P}_l$ as $-i\hbar\nabla_l$ we see that

$$\frac{\boldsymbol{P}_l^2}{2M_l}\psi_a\chi_a=-\frac{\hbar^2}{2M_l}[\chi_a\nabla_l^2\psi_a+2(\nabla_l\chi_a)\cdot(\nabla_l\psi_a)+\psi_a\nabla_l^2\chi_a]. \tag{5.45}$$

Assuming only a weak parametric dependence of $\psi_a(\boldsymbol{r}_i, \boldsymbol{R}_l)$ on $\boldsymbol{R}_l$ the first two terms in eqn (5.45) are small in comparison with the last term; they may possibly be neglected. If these terms are neglected it means that the electronic centre is not changed out of its electronic state by the variations in $\boldsymbol{R}_l$. Rather does the electronic state 'adjust' to the changing $\boldsymbol{R}_l$ values. On physical grounds we might expect such gradual adjustments of the electronic state to be a reasonable assumption because the electronic motion is so much faster than the ionic motion. The approximation in which the first two terms in eqn (5.45) are neglected was first made by Born and Oppenheimer and is known as the *Born–Oppenheimer approximation* or as the *adiabatic approximation*, in view of the implication that the electronic state adjusts adiabatically to the slowly varying ionic positions.

The effects of the terms $\nabla_l^2\psi_a$ and $\nabla_l\chi_a\cdot\nabla_l\psi_a$, which have been neglected, can be included using perturbation theory. These terms mix the state $\psi_a\chi_a$ with other states $\psi_b\chi_b$, but mixing will be small provided that the energy separation between these states is large. The Born–Oppenheimer approximation should be valid for non-degenerate electronic states. If there is electronic degeneracy present then it may not be possible to express the eigenstate of the Hamiltonian (eqn 5.40) as a simple product of the electronic and vibrational functions. One consequence of the coupling of electronic and lattice states when degeneracy occurs is the *Jahn–Teller distortion*. In the analysis which follows we adopt the Born–Oppenheimer approximation.

If we adopt the Born–Oppenheimer approximation, the ion-plus-lattice eigenstates are of the form $\psi_a(\boldsymbol{r}_i, \boldsymbol{R}_l)\chi_a(\boldsymbol{R}_l)$ and these are known as Born–Oppenheimer states. The Schrödinger equation (5.44) becomes

$$[\mathscr{H}_0\psi_a(\boldsymbol{r}_i, \boldsymbol{R}_l)]\chi_a(\boldsymbol{R}_l)+\left[\sum_l \frac{-\hbar^2}{2M_l}\nabla_l^2\chi_a(\boldsymbol{R}_l)\right]\psi_a(\boldsymbol{r}_i, \boldsymbol{R}_l)$$
$$=E\psi_a(\boldsymbol{r}_i, \boldsymbol{R}_l)\,\chi_a(\boldsymbol{R}_l) \tag{5.46}$$

which reduces to

$$\left[\sum_l \frac{-\hbar^2}{2M_l}\nabla_l^2+E^{(a)}(\boldsymbol{R}_l)\right]\chi_a(\boldsymbol{R}_l)=E\chi_a(\boldsymbol{R}_l) \tag{5.47}$$

where $E^{(a)}(\boldsymbol{R}_l)$ has already been given (eqn 5.43). This is the potential energy in which the ions move.

Since $\boldsymbol{R}_l$ oscillates about its average value we write

$$\boldsymbol{R}_l = \boldsymbol{R}_l^{(a)}(0) + q_l^{(a)} \tag{5.48}$$

where $\boldsymbol{R}_l^{(a)}(0)$ is the average position of the lth ion when the system is in electronic state a and $q_l^{(a)}$ is the displacement of the ion from its average value. The electronic energy term and the interion potential term can then be written

$$\mathscr{H}_{\mathrm{e}}^{(a)}(\boldsymbol{R}_l) = \mathscr{H}_{\mathrm{e}}^{(a)}(\boldsymbol{R}_l^{(a)}(0)) + V_{\mathrm{e}}^{(a)}(\boldsymbol{q}_l^{(a)}) \tag{5.49}$$

and

$$V_{\mathrm{I}}^{(a)}(\boldsymbol{R}_l) = V_{\mathrm{I}}^{(a)}(\boldsymbol{R}_l^{(a)}(0)) + V_{\mathrm{I}}^{(a)}(\boldsymbol{q}_l^{(a)}). \tag{5.50}$$

The average or rigid lattice energy is

$$E_0^{(a)} = \mathscr{H}_{\mathrm{e}}^{(a)}(\boldsymbol{R}_l^{(a)}(0)) + V_{\mathrm{I}}^{(a)}(\boldsymbol{R}_l^{(a)}(0)). \tag{5.51}$$

The ionic potential energy $E^{(a)}(\boldsymbol{R}_l)$ in eqn (5.43) becomes, from eqns (5.49) and (5.50)

$$\begin{aligned} E^{(a)}(\boldsymbol{R}_l) &= E_0^{(a)} + V_{\mathrm{e}}^{(a)}(\boldsymbol{q}_l^{(a)}) + V_{\mathrm{I}}^{(a)}(\boldsymbol{q}_l^{(a)}) \\ &= E_0^{(a)} + V^{(a)}(\boldsymbol{q}_l^{(a)}). \end{aligned} \tag{5.52}$$

The term $V_{\mathrm{e}}^{(a)}(\boldsymbol{q}_l^{(a)})$ gives the effect of lattice distortion on the electronic energy, and $V_{\mathrm{I}}^{(a)}(\boldsymbol{q}_l^{(a)})$ gives the effect of the lattice distortion on the interionic potential energy.

The Schrödinger equation for the lattice state, eqn (5.47), can now be written

$$\left[\sum_l \frac{\boldsymbol{P}_l^2}{2M_l} + V^{(a)}(\boldsymbol{q}_l^{(a)})\right]\chi^{(a)}(\boldsymbol{q}_l^{(a)}) = (E - E_0^{(a)})\chi^{(a)}(\boldsymbol{q}_l^{(a)}). \tag{5.53}$$

The difficulty encountered in solving this equation is that $V^{(a)}(\boldsymbol{q}_l^{(a)})$ is not separable into a sum of terms each of which is a function only of an individual $\boldsymbol{q}_l^{(a)}$ variable; the dynamic lattice is a coupled system. As we did in the case of the monatomic line of atoms we change from a description in terms of displacements of the individual ions, $\boldsymbol{q}_l^{(a)}$, to a description in terms of normal coordinates, Q_k, of the complex of ions. We write the lattice state as $\chi_a(Q_k)$. If $V^{(a)}$ is harmonic in the normal coordinates then the lattice eigenstates are products of linear harmonic oscillator functions, one for each normal mode k:

$$\chi_a = \prod_k |n_k\rangle. \tag{5.54}$$

The total energy E (eqn 5.53) is now given by

$$E = E_0^{(a)} + \sum_k \hbar\omega_k^{(a)}(n_k + \tfrac{1}{2}). \tag{5.55}$$

$\omega_k^{(a)}$ is the angular velocity (usually termed the frequency) of mode k, and the superscript indicates that the frequencies can depend on the electronic state of the system. n_k gives the number of quanta of vibrational energy (phonons) in mode k above the zero-point energy. At temperature T the average value of n_k is given by the Bose–Einstein factor:

$$\langle n_k \rangle = \frac{1}{\exp(\hbar\omega_k/kT)-1}. \tag{5.56}$$

The electronic part of the Born–Oppenheimer state, $\psi_a(\boldsymbol{r}_i, \boldsymbol{R}_l)$, is an eigenstate of $\mathscr{H}_0$ (eqn 5.41) with energy $E^{(a)}(\boldsymbol{R}_l)$ given by eqn (5.52), and the $\boldsymbol{R}_l$ are a set of variable parameters. In practice, in the calculation of ψ_a one replaces these by a set of fixed $\boldsymbol{R}_l$ values, $\boldsymbol{R}_l(0)$, and the ion-plus-lattice system in now described by 'crude Born–Oppenheimer states':

$$\psi_a(\boldsymbol{r}_i, \boldsymbol{R}_l(0))\chi_a(\boldsymbol{R}_l). \tag{5.57}$$

Returning to the lattice state, $\chi_a(Q_k)$, there are many normal modes of vibration, k, which must be taken into account. This can complicate the analysis of transitions between various ion-plus-lattice states. Consequently it is very helpful to make the simplifying assumption that we can confine our attention to one representative mode only. It is usual to choose as this vibrational mode the *breathing mode* in which the ionic environment pulsates about the optically active ion. The distance from this central ion to the first shell of neighbouring ions is labelled Q, it is the *configurational coordinate* and is the variable of the lattice state. In this *single configurational coordinate model* Q oscillates about its equilibrium value, $Q_0^{(a)}$, and the crude Born–Oppenheimer wavefunction can be written as $\psi_a(\boldsymbol{r}_i, Q_0^{(a)})\chi_a(Q)$. The ionic potential energy (eqn 5.52) can be written as

$$E^{(a)}(Q) = E_0^{(a)} + V^{(a)}(Q). \tag{5.58}$$

Figure 5.5 shows the ionic potential energy $E^{(a)}(Q)$ in the case where the system is in electronic state a, and $V^{(a)}(Q)$ is drawn in the form of a Morse potential. In the harmonic approximation this is replaced by a harmonic potential, shown by the broken curve in the figure. In the harmonic approximation the lattice state $\chi_a(n)$ can be written as $|n\rangle$, where n is the number of vibrational quanta above the zero-point energy. Some of the vibrational states are drawn in Fig. 5.5. When the system changes to electronic state b the equilibrium positions of the ions are $\boldsymbol{R}_l^{(b)}(0)$ and Q oscillates about a new equilibrium value, $Q_0^{(b)}$. The ionic potential energy $E^{(b)}(Q) = E_0^{(b)} + V^{(b)}(Q)$ is drawn in the figure and $V^{(b)}(Q)$ is shown as a Morse potential. In the harmonic approximation this is replaced by a harmonic potential, shown by the broken curve in the figure; the lattice state $\chi_b(m)$ is now a harmonic oscillator state and can be written as $|m\rangle$ where m is the number of vibrational quanta above the zero-point energy. We can allow for the breathing mode

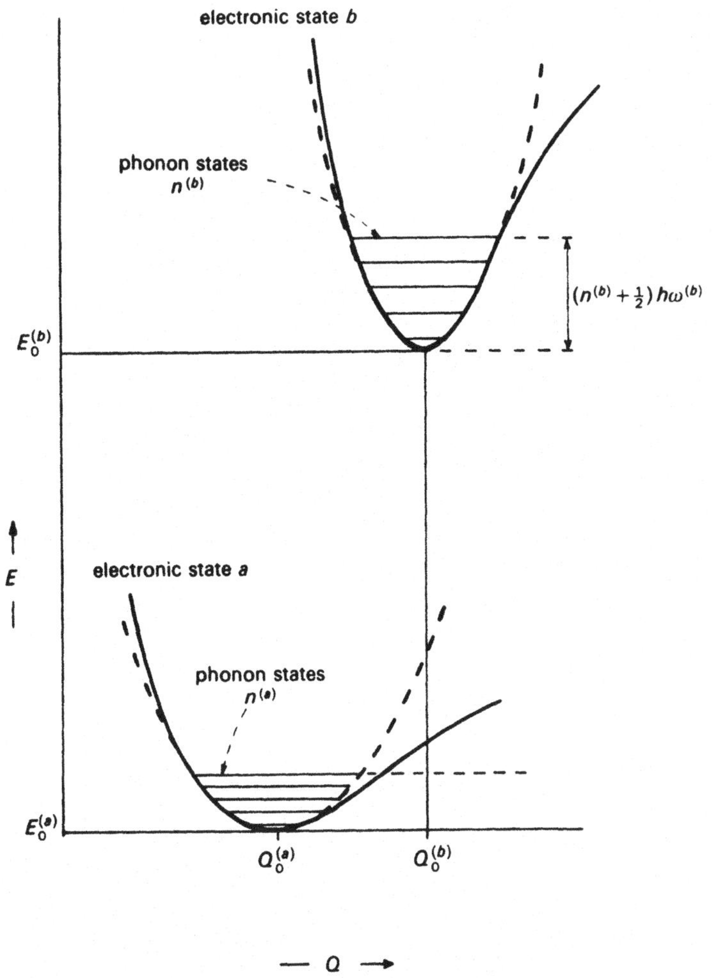

FIG. 5.5. The ionic potential energy curves in the case of the 'breathing-mode' vibration when the electronic centre is in states a and b. The broken curves are the approximate harmonic oscillator potentials.

frequencies $\omega^{(a)}$ and $\omega^{(b)}$ to be different for the two electronic states. Fig. 5.5 is known as a *configurational coordinate diagram.* (Williams 1951; Curie 1963.)

5.4 Radiative transitions using the configurational coordinate model

We consider a radiative transition between different electronic states a and b on an optically-active ion in a vibrating lattice. We use the single configurational coordinate model in the harmonic approximation where the ground

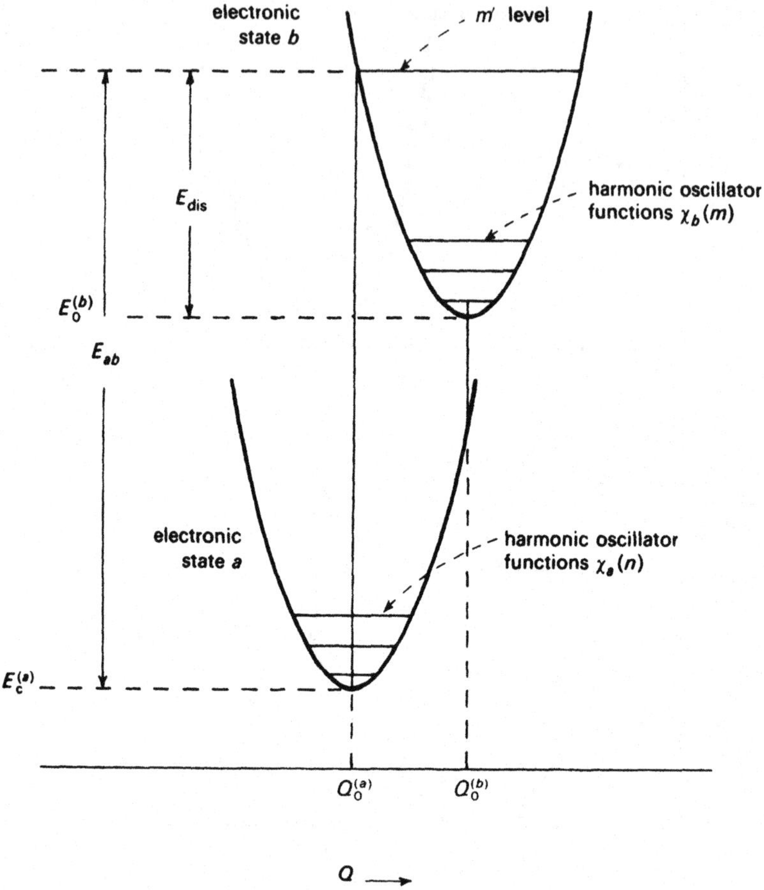

FIG. 5.6. Configurational coordinate diagram in the harmonic approximation for electronic states a and b. The vibrational frequencies in the two electronic states are assumed to be the same. In the classical Condon transition the peak of the absorption energy is represented by the length of the vertical line from the bottom of the ground-state parabola to the point of intersection with the excited-state parabola.

and excited state wavefunctions are $\psi_a(\boldsymbol{r}_i, \boldsymbol{R}_l)\chi_a(n)$ and $\psi_b(\boldsymbol{r}_i, \boldsymbol{R}_l)\chi_b(m)$, respectively. The ionic potential energy curves in the ground (a) and excited (b) states are shown in Fig. 5.6. For simplicity in the analysis we assume that the vibrational frequencies are the same in the a and b states, but we allow the average values of the configurational coordinate, Q, to be different in the two cases. This difference in the average values of Q arises because of the difference in coupling between the optically active ion and the lattice (the electron–lattice coupling) in states a and b. The larger the difference in coupling the larger is $Q_0^{(b)} - Q_0^{(a)}$.

The ionic potential energy in the ground state is $E^{(a)}(Q)=E_0^{(a)}+\frac{1}{2}M\omega^2(Q-Q_0^{(a)})^2$, where M is some effective ionic mass and ω is the vibrational frequency. It is convenient to treat $E_0^{(a)}$ as the zero of energy, so that the ionic potential in the ground state is

$$E^{(a)}(Q)=\tfrac{1}{2}M\omega^2(Q-Q_0^{(a)})^2. \tag{5.59}$$

With this same zero of energy the ionic potential in the excited state is

$$\begin{aligned} E^{(b)}(Q) &= E_{ab}-\tfrac{1}{2}M\omega^2(Q_0^{(b)}-Q_0^{(a)})^2+\tfrac{1}{2}M\omega^2(Q-Q_0^{(b)})^2 \\ &= E_{ab}+\tfrac{1}{2}M\omega^2(Q-Q_0^{(a)})^2-M\omega^2(Q_0^{(b)}-Q_0^{(a)})(Q-Q_0^{(a)}) \end{aligned} \tag{5.60}$$

where E_{ab} is defined in Fig. 5.6. Since

$$E^{(b)}(Q)-E^{(a)}(Q)=E_{ab}-M\omega^2(Q_0^{(b)}-Q_0^{(a)})(Q-Q_0^{(a)}) \tag{5.61}$$

is linear in $(Q-Q_0^{(a)})$, this is referred to as the *linear coupling case.*

The excited state ionic potential (5.60) can be written as

$$E_b^{(b)}(Q)=E_{ab}+\tfrac{1}{2}M\omega^2(Q-Q_0^{(a)})^2-A\hbar\omega\left(\frac{M\omega}{\hbar}\right)^{\frac{1}{2}}(Q-Q_0^{(a)}) \tag{5.62}$$

where the dimensionless constant $A=(M\omega/\hbar)^{\frac{1}{2}}(Q_0^{(b)}-Q_0^{(a)})$ characterizes the difference in electron–lattice coupling between electronic states a and b. It is usual to characterize the difference in electron–lattice coupling by another dimensionless constant, the Huang–Rhys parameter, S, defined as

$$S=\frac{A^2}{2}=\frac{1}{2}\frac{M\omega^2}{\hbar\omega}(Q_0^{(b)}-Q_0^{(a)})^2=\frac{E_{\text{dis}}}{\hbar\omega} \tag{5.63}$$

where E_{dis} is defined in Fig. 5.6. If the vertical line from $Q=Q_0^{(a)}$ on the configurational coordinate diagram intersects the upper parabola at the vibrational level m' (Fig. 5.6) then

$$E_{\text{dis}}=S\hbar\omega=(m'+\tfrac{1}{2})\hbar\omega. \tag{5.64}$$

The absorption and emission transitions between electronic states a and b can be analysed using this configurational coordinate diagram. The shapes of the spectra are found to depend strongly on the difference in electron–lattice coupling between the two states.

5.4.1 *Absorption transitions*

By extension of the techniques developed in Chapter 4 to discuss radiative transition rates it is apparent that the probability of an absorption transition from electronic–vibrational state a, n to electronic–vibrational state b, m is proportional to the square of the matrix element

$$\langle\psi_b(r_i, Q)\chi_b(m)|\mu|\psi_a(r_i, Q)\chi_a(n)\rangle \tag{5.65}$$

where the integration is over electronic and vibrational coordinates and μ is the appropriate electronic dipole operator. To simplify this matrix element we use crude Born–Oppenheimer functions, replacing the variable Q in ψ_a and ψ_b by some average value. At very low temperatures the initial state is $\psi_a\chi_a(0)$ and the harmonic oscillator function $\chi_a(0)$ has maximum amplitude at $Q=Q_0^{(a)}$. Hence we use this value of Q as the average in the electronic wavefunctions. The matrix element (5.65) is then given approximately by

$$\langle\psi_b(r_i, Q_0^{(a)})|\mu|\psi_a(r_i, Q_0^{(a)})\rangle\langle\chi_b(m)|\chi_a(n)\rangle. \tag{5.66}$$

Converting the matrix element (5.65) into the form (5.66) is known as the *Condon approximation.* The transition probability then is

$$W_{an-bm}=P_{ab}|\langle\chi_b(m)|\chi_a(n)\rangle|^2 \tag{5.67}$$

where P_{ab} is the purely electronic transition probability and is the same for all vibrational states n, m. The relative values of W_{an-bm} depend on the squares of the vibrational overlap integrals $\langle\chi_b(m)|\chi_a(n)\rangle$. $\chi_a(n)$ and $\chi_b(m)$ are each a similar set of harmonic oscillator functions but are defined with respect to different zeros of the variables Q. As a result the overlap integrals are generally not zero.

The shape of the absorption band is given by

$$I_{ab}(E)=I_0\,\mathrm{Av}_n\sum_m|\langle\chi_b(m)|\chi_a(n)\rangle|^2\delta(E_{b,m}-E_{a,n}-E) \tag{5.68}$$

where Av_n indicates a thermal average over the initial vibrational state n, and where $E_{a,n}$ is the energy of the nth vibrational state when the electronic centre is in state a. This band will consist of a series of delta-functions at various energies differing by $\hbar\omega$. The overlap integral can be expressed in closed form (Keil 1965)

$$\langle\chi_b(m)|\chi_a(n)\rangle=\exp(-A^2/4)(n!/m!)^{\frac{1}{2}}(-A/\sqrt{2})^{m-n}L_n^{m-n}(A^2/2) \tag{5.69}$$

where L_n^{m-n} are associated Laguerre polynomials of which $L_0^m(x)=1$.

At $T=0$ K only the $n=0$ vibrational state is occupied, so we require values of $|\langle\chi_b(m)|\chi_a(0)\rangle|^2$. This is called the zero-temperature Franck–Condon factor, $F_m(0)$, and

$$F_m(0)=|\langle\chi_b(m)|\chi_a(0)\rangle|^2=\frac{\exp(-A^2/2)(A^2/2)^m}{m!} = \frac{\exp(-S)S^m}{m!}. \tag{5.70}$$

At $T=0$ K the absorption band shape function is

$$\begin{aligned} I_{ab}(E) &= I_0 \sum_m \frac{\exp(-S)S^m}{m!} \delta(E_{bm}-E_{a0}-E) \\ &= I_0 \sum_m \frac{\exp(-S)S^m}{m!} \delta(E_0+m\hbar\omega-E) \end{aligned} \tag{5.71}$$

where $E_0=E_{b0}-E_{a0}$ is the energy of the transition between the zero vibrational levels of both initial and final states; this is the energy of the *zero-phonon transition.*

Since $\sum_m |\langle\chi_b(m)|\chi_a(n)\rangle|^2=1$ the intensity of the full band is I_0 and is independent of the value of S. This also means that the intensity of the absorption band is independent of temperature. The zero-phonon line has the intensity $I_0 \exp(-S)$, and if $S=0$ all the intensity is contained in the zero-phonon line. $S=0$ means there is no lateral displacement of the harmonic oscillator parabolae in Fig. 5.6. In this case $Q_0^{(a)}=Q_0^{(b)}$, and $\chi_a(n)$ and $\chi_b(m)$ are identical sets of harmonic oscillator wavefunctions. Hence by the orthogonality of the $\chi_a(n)$ functions we see $\langle\chi_b(m)|\chi_a(0)\rangle=\delta_{m0}$. As S increases the intensity in the zero-phonon line decreases and this is compensated for by the appearance of *vibrational sidebands* which are observed at energies $m\hbar\omega$ above the zero-phonon line. We can say that the sideband intensity is *borrowed* from the zero-phonon line. At sufficiently large values of S the zero-phonon line does not appear.

The predicted bandshapes for different values of S (eqn 5.71) are plotted in Fig. 5.7. The total transition probability is independent of S so all of these should have the same integrated intensity; but for clarity they are drawn to have the same maximum intensity. The zero-phonon line ($0\rightarrow0$) is a transition between purely electronic states and should appear as a sharp line. The sideband transitions, $0\rightarrow m$ ($m>0$), occurring at higher energies, involve the creation of m phonons in the excited state. Since there is a wide spectrum of lattice vibrational frequencies rather than a single breathing-mode vibrational frequency, we expect that in practice the sideband features above the zero-phonon line will appear as broad bands. In Fig. 5.8(a) we show the low temperature absorption band shape predicted for the case $S=7$ where $\hbar\omega\simeq 250$ cm^{-1} and where the $0\rightarrow m$ sideband feature at energy $m\hbar\omega$ above the zero-phonon line is assumed to have a width of amount $m\hbar\omega$. We notice that the higher-order sidebands lose their separate identities and we obtain a smooth bandshape at the high-energy side of the absorption peak. In Fig. 5.8(b) we show the observed $^4A_2\rightarrow{}^4T_2$ absorption band of Cr^{3+} in aluminium oxide (ruby). The experimental bandshape is quite similar to the theoretical bandshape for the chosen set of parameters.

Two aspects of Fig. 5.7 are worth noting. The first is that for $S\geqslant1$ the maximum absorption occurs at an energy $(S-\frac{1}{2})\hbar\omega$ above the zero-phonon

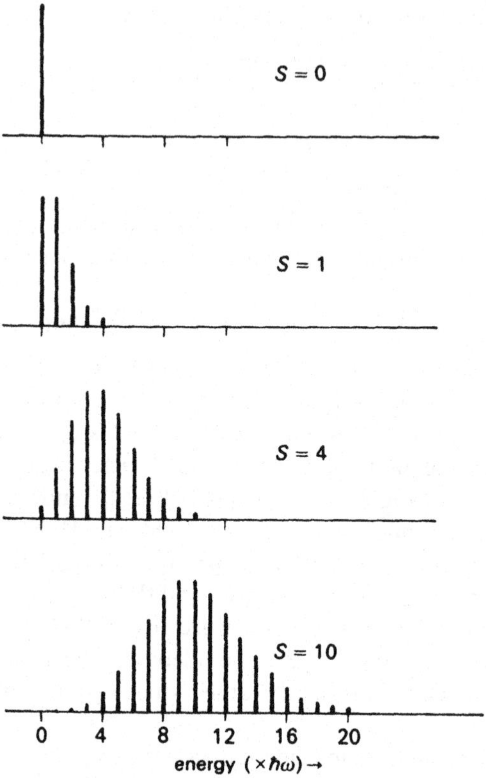

FIG. 5.7. The relative intensities in the different electronic–vibrational lines change with the strength of the coupling (S) in accordance with eqn (5.71). The envelope of the individual intensities gives the predicted bandshapes.

line. This means that the most likely final state vibrational level is at an energy $S\hbar\omega$ above the bottom of the excited state parabola. But from eqn (5.64) we have $S\hbar\omega = E_{\text{dis}}$, where E_{dis} is shown in Fig. 5.6. Hence the most likely final vibrational level is the m' level (Fig. 5.9), which touches the excited state parabola at $Q = Q_0^{(a)}$. This means that the most probable absorption transition is indicated on the configurational coordinate model by the *vertical line* from the centre of the ground-state parabola to where this vertical line touches the excited-state parabola. We can understand this result by examining the overlap integrals $\langle \chi_b(m) | \chi_a(0) \rangle$. We first note that in the electronic ground state, a, the vibrational wavefunction $\chi_a(0)$ has its maximum amplitude at $Q_0^{(a)}$. Hence we expect that the overlap integral will be a maximum for the state $\chi_b(m)$ with maximum amplitude at $Q = Q_0^{(a)}$. The amplitudes of some harmonic oscillator wavefunctions of $\chi_b(m)$ with $m = m'+1$, m', $m'-1$ are shown schematically in Fig. 5.9. The maximum occurs for $m = m'$, the vibrational

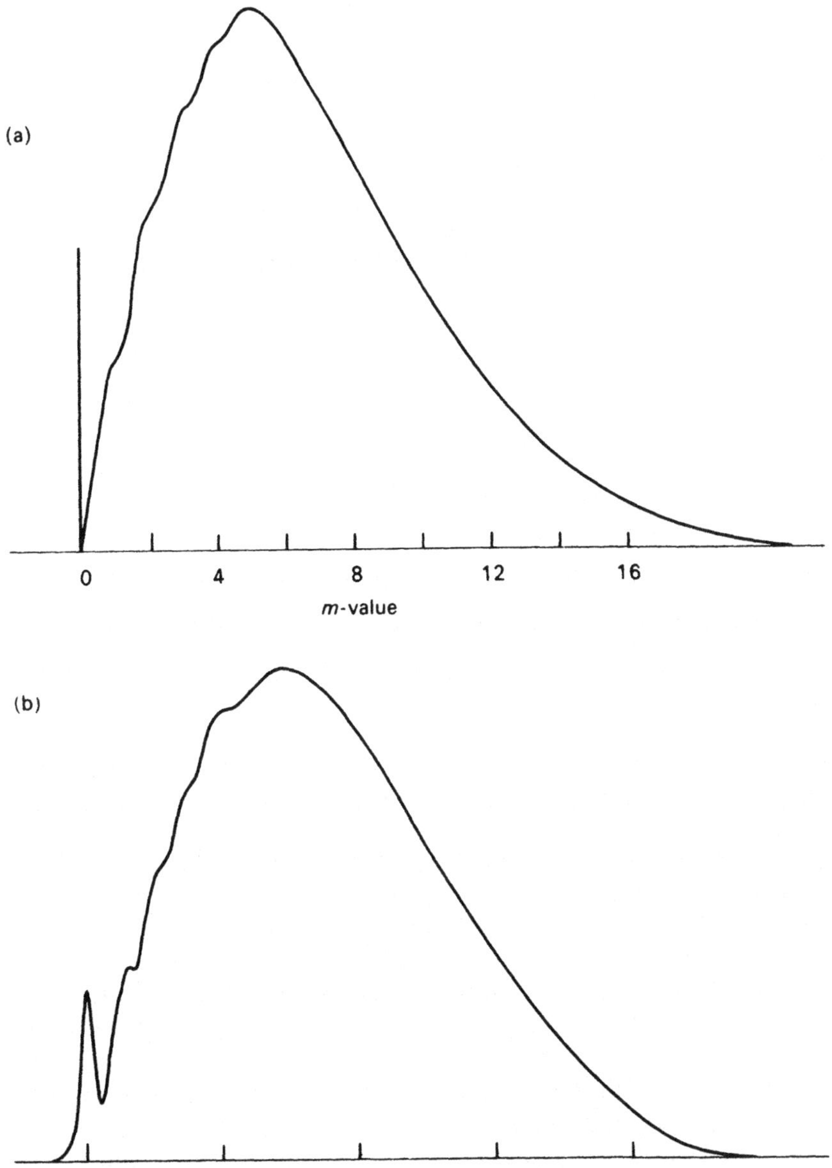

FIG. 5.8. (a) Low-temperature absorption bandshape predicted for the case of $S=7$ and $\hbar\omega = 250\ \text{cm}^{-1}$ where the sideband feature at energy $m\hbar\omega$ above the zero-phonon line is assumed to have a width of amount $m\hbar\omega$. (b) Observed $^4A_2 \rightarrow {^4T_2}$ absorption band of Cr^{3+} in aluminium oxide (ruby) at 77 K.

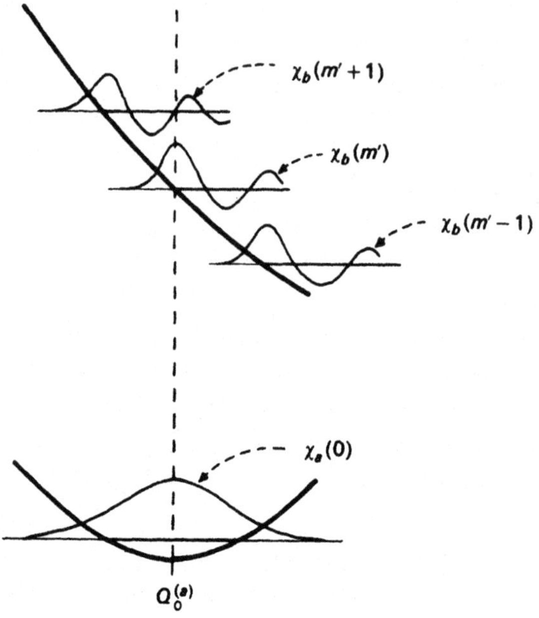

FIG. 5.9. $\chi_a(0)$ and $\chi_b(m')$ have maximum amplitudes at the same value of the configurational coordinate, $Q_0^{(a)}$.

state whose energy level crosses the b parabola at $Q = Q_0^{(a)}$. This is the quantum basis for the semi-classical model which envisages the absorption maximum being represented by a vertical line from the bottom of the ground-state parabola until the excited-state parabola is reached (the classical Franck–Condon principle).

The second aspect to be noted in Fig. 5.7 is that the bandshape is not symmetrical for small S but becomes more symmetrical for larger values of S. The change in bandshape as a function of the Huang–Rhys parameter, S, is illustrated in Fig. 5.10, which examines transitions for both large S and small S. In each case the peak of the absorption band occurs at an energy given by the vertical line from the bottom of the ground-state parabola to the excited-state parabola. The width of the transition is approximately the energy spread between the vertical lines from the points where the zero vibrational level meets the ground-state parabola (i.e. the dotted vertical lines in Fig. 5.10). For large values of S the vertical lines intersect the excited-state parabola at a region of approximately constant curvature, and the resultant absorption band is quite symmetrical. For small S values the vertical lines intersect the excited-state parabola in a region where the curvature is changing rapidly and the resultant absorption band has an asymmetry. The envelope of the

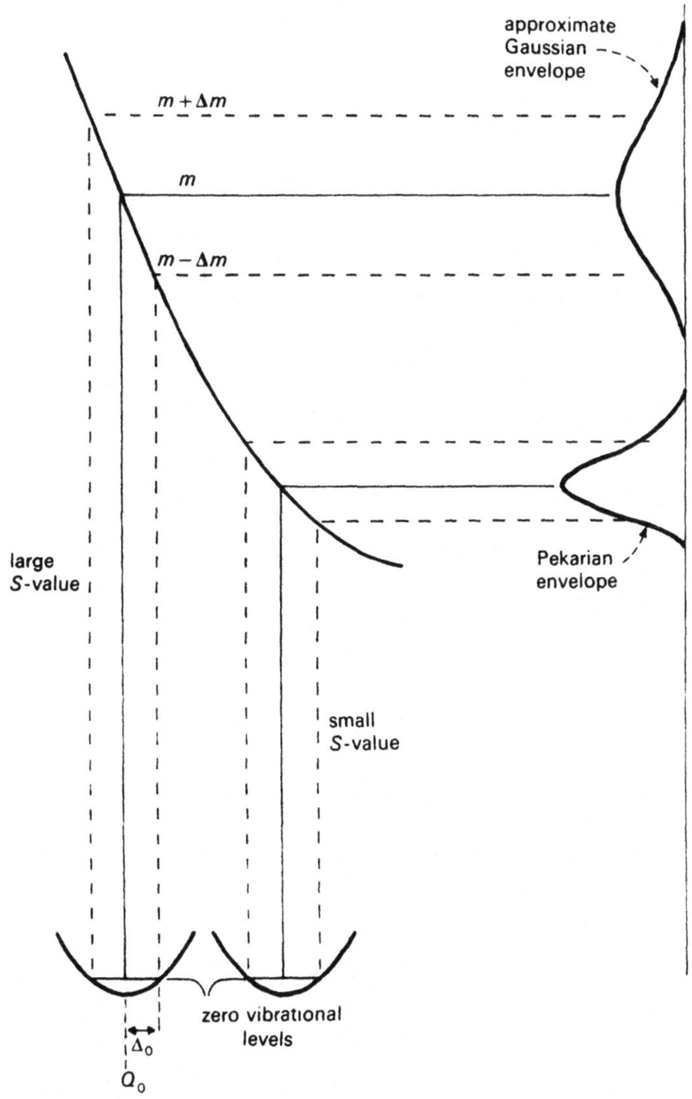

FIG. 5.10. The width and shape of the absorption band depends on the configurational coordinate displacement between the ground-state and excited-state parabolae.

absorption transitions (eqn 5.71) is a Pekarian curve. However, for large values of S it becomes Gaussian-like and the bandwidth increases.

We now use Fig. 5.10 to make an approximate calculation of the bandwidth. The width is taken as the spread in energy of the excited-state parabola between the intercepts with the dotted vertical lines. The zero vibrational level involves a variation $\pm\Delta_0$ about Q_0. We see that

$$\tfrac{1}{2}M\omega^2(Q_0^{(b)}-Q_0^{(a)})^2=(m'+\tfrac{1}{2})\hbar\omega \tag{5.72}$$

$$\tfrac{1}{2}M\omega^2(Q_0^{(b)}-Q_0^{(a)}-\Delta_0)^2=(m'-\Delta m+\tfrac{1}{2})\hbar\omega \tag{5.73}$$

$$\tfrac{1}{2}M\omega^2(Q_0^{(b)}-Q_0^{(a)}+\Delta_0)^2=(m'+\Delta m'+\tfrac{1}{2})\hbar\omega. \tag{5.74}$$

Assuming that $\Delta_0 \ll Q_0^{(b)}-Q_0^{(a)}$ we can write

$$M\omega^2(Q_0^{(b)}-Q_0^{(a)})\Delta_0 \simeq \Delta m\hbar\omega \simeq \Delta m'\hbar\omega, \tag{5.75}$$

therefore the bandwidth at $T=0$ is $\Gamma(0)$ where

$$\Gamma(0)\simeq(\Delta m+\Delta m')\hbar\omega \simeq 2M\omega^2(Q_0^{(b)}-Q_0^{(a)})\Delta_0. \tag{5.76}$$

From the definition of the Huang–Rhys parameter we have

$$\tfrac{1}{2}M\omega^2(Q_0^{(b)}-Q_0^{(a)})^2=S\hbar\omega \tag{5.77}$$

and from the energy of the zero-vibrational level we have

$$\tfrac{1}{2}M\omega^2\Delta_0^2=\tfrac{1}{2}\hbar\omega. \tag{5.78}$$

Hence we find

$$\Gamma(0)\simeq 2\hbar\omega(2S)^{\frac{1}{2}}. \tag{5.79}$$

An exact calculation of the second moment of a Pekarian-shaped bandwidth gives

$$\Gamma(0)=2.36\hbar\omega S^{\frac{1}{2}}. \tag{5.80}$$

An expression for the absorption bandshape at temperatures above $T=0$ is derived by carrying out the thermal average over the initial vibrational states in eqn (5.68). Keil (1965) shows that the absorption bandshape as a function of temperatures is given by

$$\begin{aligned} I_{ab}(E)&=I_0\sum_{p=-\infty}^{\infty}\exp\left(-S\coth\frac{\hbar\omega}{2kT}+\frac{p\hbar\omega}{2kT}\right)I_p\left(S\operatorname{csch}\frac{\hbar\omega}{2kT}\right)\delta(E_0+p\hbar\omega-E)\\ &=I_0\sum_{p=-\infty}^{\infty}\exp(-S(1+2n))\left(\frac{1+n}{n}\right)^{p/2}I_p(2S\sqrt{n(n+1)})\delta(E_0+p\hbar\omega-E) \end{aligned} \tag{5.81}$$

where $n=[\exp(\hbar\omega/kT)-1]^{-1}$ is the mean thermal occupancy in the vibrational mode, and $I_p(x)$ is the modified Bessel function. The $\exp(-S)$ term at $T=0$ (eqn 5.71) has become $\exp[-S(1+2n)]$. The term $S(1+2n)$ is sometimes called the effective Huang–Rhys parameter at high temperatures. We see that at $T>0$ there can be components of the band with negative values of p, that is, absorption sidebands at energies lower than the zero-phonon line. These are called the *anti-Stokes sidebands.* Figure 5.11 shows how tempera-

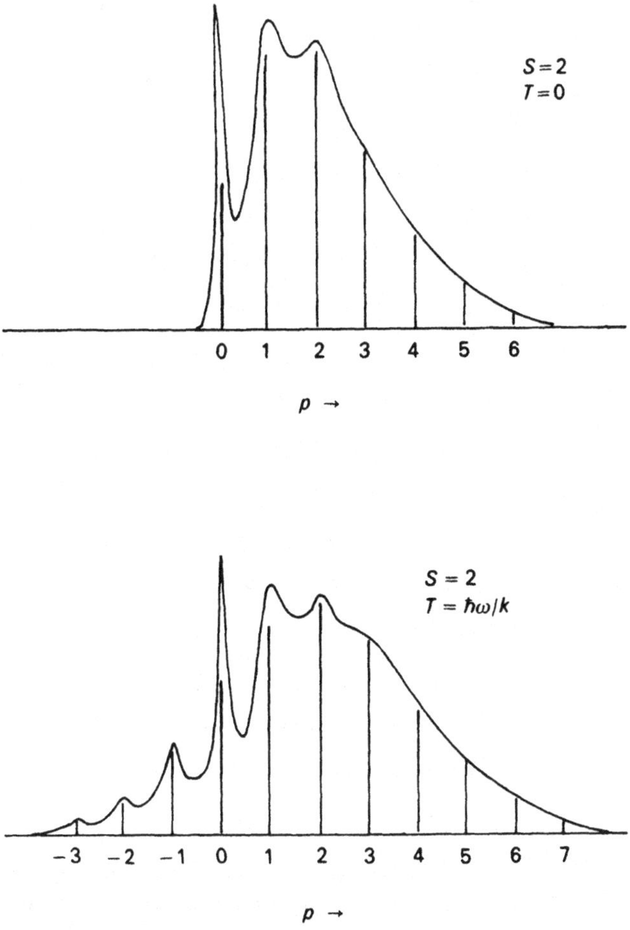

FIG. 5.11. At $T=0$ the sidebands appear only on one side of the zero-phonon line. These are the Stokes sidebands. When the temperature is raised additional sidebands (anti-Stokes sidebands) appear on the other side of the zero-phonon line.

ture affects the absorption band. Although the lineshape changes, the total absorption intensity is independent of the temperature. According to eqn (5.81) the intensity in the zero-phonon line (at $p=0$) is

$$I_0 \exp\left(-S \coth\frac{\hbar\omega}{2kT}\right) = I_0 \exp(-S(1+2n)) \tag{5.82}$$

which decreases with increasing temperature. When S is much less than unity, $S \ll 1$, the zero-phonon line intensity becomes $I_0(1 - S\coth(\hbar\omega/2kT))$, showing that the sideband intensity increases with temperature as $I_0 S \coth(\hbar\omega/2kT)$.

Equation (5.76) shows that at $T=0$ the bandwidth is $\Gamma(0) \simeq 2M\omega^2(Q_0^{(b)} - Q_0^{(a)})\Delta_0$. The temperature dependence of the bandwidth can be taken into account by writing $\Gamma(T) \simeq 2M\omega^2(Q_0^{(b)} - Q_0^{(a)})(\mathrm{Av}_n \Delta_n^2)^{\frac{1}{2}}$ where Δ_n is the amplitude of the breathing mode oscillating in the nth vibrational level. Since $\frac{1}{2}M\omega^2\Delta_n^2 \simeq \hbar\omega(n+\frac{1}{2})$ we find

$$\Delta_n^2 = \Delta_0^2(1+2n) \tag{5.83}$$

therefore

$$\mathrm{Av}_n \Delta_n^2 = \sum_n \frac{\Delta_n^2 \exp(-n\hbar\omega/kT)}{\sum_m \exp(-m\hbar\omega/kT)} \tag{5.84}$$

which can be evaluated to yield

$$\mathrm{Av}_n \Delta_n^2 = \Delta_0^2 \coth \frac{\hbar\omega}{2kT}. \tag{5.85}$$

Hence

$$\Gamma(T) \simeq \Gamma(0)\sqrt{\coth \frac{\hbar\omega}{2kT}}. \tag{5.86}$$

This is also the result of an exact calculation based on eqn (5.81). Although the width of the band increases with increasing temperatures, the centre of gravity of the band is expected to remain constant.

5.4.2 *Emission transitions*

When at low temperature the electronic–vibrational system is raised to some higher vibrational level, $\chi_b(m)$, associated with an excited electronic state b, it decays quickly by multiphonon emission to the ground vibrational level, $\chi_b(0)$. Subsequently radiative decay returns the system to vibrational level, $\chi_a(n)$, of the ground electronic state a. The relative probability of returning to this vibrational state is given by the square of the overlap integral $\langle \chi_a(n)|\chi_b(0)\rangle$. The shape of the emission band at temperature $T=0$ is

$$I_{ba}(E) = I_0 \sum_n \frac{\exp(-S)S^n}{n!} \delta(E_0 - n\hbar\omega - E). \tag{5.87}$$

Figure 5.12 compares the absorption and emission transitions on the configurational coordinate model. The emission band occurs at lower energy, and the shift in energy between the emission and absorption band peaks is known as the *Stokes shift*. If the excited-state and ground-state parabolae are identical, the emission and absorption bandshapes are mirror images of each other and the Stokes shift has magnitude $(2S-1)\hbar\omega$. When the parabolae are not identical this mirror symmetry does not occur.

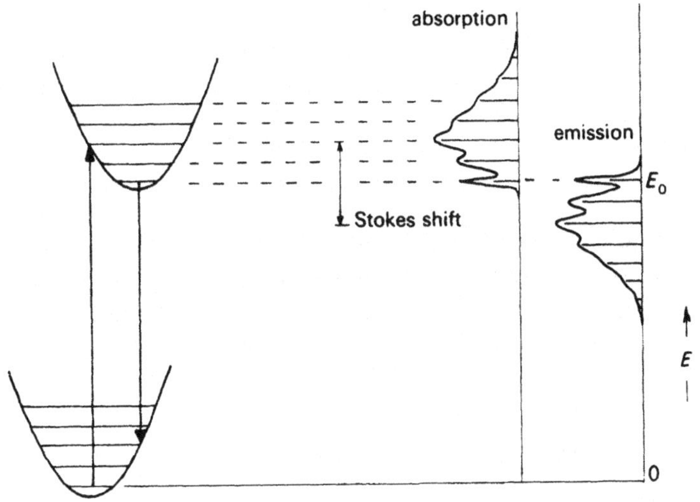

FIG. 5.12. A comparison of the low-temperature absorption and emission transitions on the configurational coordinate diagram. The band peaks occur at the energies given by the lengths of the vertical arrows, hence the emission band is at lower energy. The shift in energy between the absorption and emission band peaks is known as the Stokes shift.

5.4.3 Vibronic coupling and electronic degeneracy

The treatment of vibronic coupling has so far neglected possible electronic degeneracies of the states involved in optical transitions. Indeed this omission is implicit in the use of Born–Oppenheimer product functions to represent the vibronic states of the system. In general, such orbital degeneracies may not persist when the electronic system occupies a highly symmetric environment. This follows from the theorem due to Jahn and Teller (1937), the consequences of which have been reviewed by Sturge (1967), Englman (1972), and Ham (1972) among others. The origin of the effect is that where orbital degeneracy exists there are, in addition to the usual terms quadratic in the displacements, terms which are linear in the displacements of the neighbouring ions. Jahn–Teller coupling affects the properties of non-linear molecules and localized systems of electrons such as impurity ions, defects, and molecular centres in solids, which may have incipient instability against symmetry-lowering distortions which reduce the degeneracy so that, at most, only Kramers degeneracy remains. In some instances the optical or magnetic resonance spectrum immediately reveals the reduction in symmetry associated with the presence of the Jahn–Teller interaction. However, there are also certain subtle property changes such as different responses to uniaxial stress or magnetic field by optical zero-phonon lines and broad bands (Silsbee 1965).

The general features of the Jahn–Teller interactions are illustrated by reference to the F_3-centre in the alkali halides (Silsbee 1965; Hughes 1966)—an aggregate of three F-centres arranged in nearest-neighbour anion sites on {111} planes. Allowed electric dipole transitions occur between an orbital singlet ground state, A, and an orbital doublet state, E; these states transform as irreducible representations of the C_{3v} group. The orbital doublet state undergoes a Jahn–Teller distortion. Figure 5.13a shows two components, Q_1 and Q_2, of an E-mode of distortion of the F_3-centre, which transform as the basis vectors $|E_x\rangle$ and $|E_y\rangle$ of the representation E (see Table 3.3). (In reality it is the cooperative motion of all the ions in the neighbourhood of the centre which produces this mode.) The Hamiltonian representing the dependence of

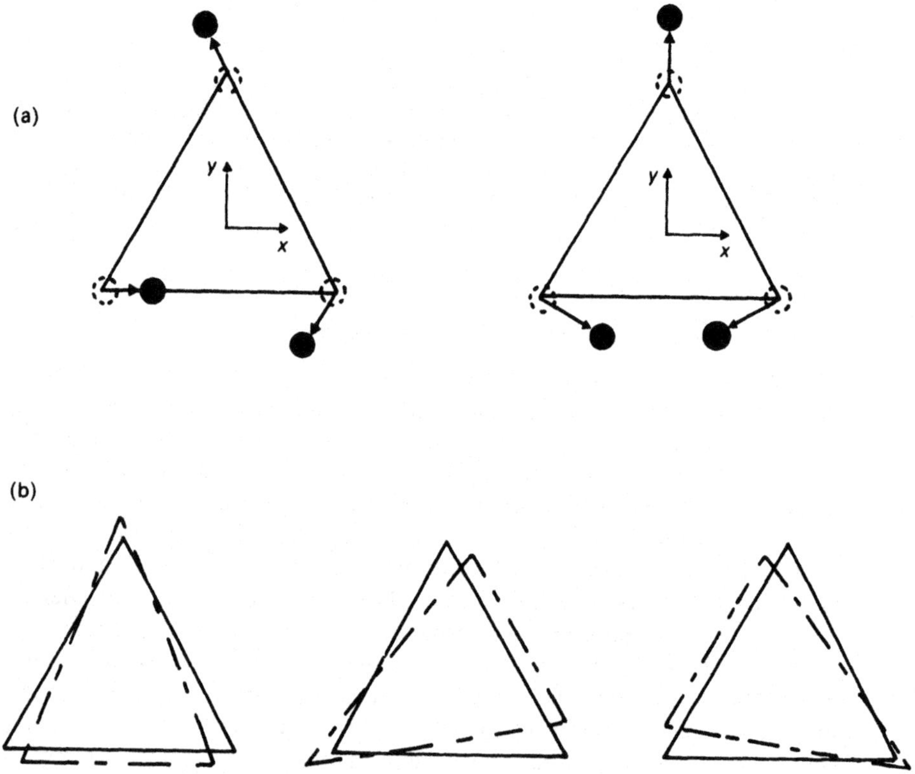

FIG. 5.13. The Jahn–Teller distortion of F_3-centres in alkali halides. Three F-centres form an aggregate in nearest-neighbour anion sites on the (111) plane. The directions shown are $x||[1\bar{1}0]$ and $y||[11\bar{2}]$. (a) Shows the two modes of distortion of the equilateral triangle corresponding to two components of an E-mode of vibration of the F_3-centre. (b) Shows three equivalent distortions of the equilateral triangle which might be stabilized by additional non-linear terms in the Hamiltonian.

the electronic energy on the normal modes Q_1 and Q_2 may then be written as the sum of two terms: the elastic energy in the modes *quadratic* in Q_1 and Q_2, and the interaction between the electronic states and the modes, the leading term of which is *linear* in Q_1 and Q_2, hence

$$\mathscr{H} = \tfrac{1}{2}M\omega^2(Q_1^2+Q_2^2)+q_1Q_1+q_2Q_2 \tag{5.88}$$

where M is an effective 'molecular' mass, ω is the mode frequency, and q_1, q_2 are electronic terms describing the sensitivity of the electronic states to Q_1 and Q_2 distortions. Symmetry arguments alone show that the electronic operator q_1 and q_2 must transform as Q_1 and Q_2, respectively. In consequence the only finite matrix elements of the Jahn–Teller Hamiltonian, $\mathscr{H}_{JT}=q_1Q_1+q_2Q_2$, between states of E symmetry are

$$\langle E_x|\mathscr{H}_{JT}|E_x\rangle = -\langle E_y|\mathscr{H}_{JT}|E_y\rangle = \alpha Q_2$$

and

$$\langle E_x|\mathscr{H}_{JT}|E_y\rangle = \langle E_y|\mathscr{H}_{JT}|E_x\rangle = \alpha Q_1.$$

In consequence the matrix of the Hamiltonian $\mathscr{H}$ is

$$\tfrac{1}{2}M\omega^2(Q_1^2+Q_2^2)I+\alpha\begin{bmatrix} Q_2 & Q_1 \\ Q_1 & -Q_2 \end{bmatrix} \tag{5.89}$$

in which I is a 2×2 unit matrix. The associated energy eigenvalues as functions of the Q's provide the energy surfaces on which we seek the minima corresponding to stable solutions. Making the parametric substitution

$$|E_x\rangle = Q_1 \equiv \rho\cos\theta$$
$$|E_y\rangle = Q_2 \equiv \rho\sin\theta \tag{5.90}$$

where ρ measures the absolute magnitude of the distortion of the equilateral triangle, we find that the Hamiltonian, $\mathscr{H}$, has eigenvalues

$$E = \tfrac{1}{2}M\omega^2\rho^2 \pm \alpha\rho. \tag{5.91}$$

A planar section through the E versus ρ surfaces (Fig. 5.14) shows two branches (eqn 5.91) which have equivalent stable minima at $\rho=\pm\alpha/M\omega^2$, depressed in energy relative to the symmetric ($\rho=0$) configuration by the Jahn–Teller energy, $E_{JT}=\alpha^2/2M\omega^2$. Obviously twofold *electronic* degeneracy has been removed at the expense of producing a twofold degenerate *vibronic* state. This is a general feature of the Jahn–Teller theorem. Note that the eigenvalues (eqn 5.91) do not depend on the angle θ, implying only a dynamic effect in which rotation about $\rho=0$ would be unrestricted; the presence of a Jahn–Teller effect may not then be immediately apparent.

The assumption of terms only linear in Q_1 and Q_2 is not necessarily a good approximation; usually quadratic interaction terms and anharmonic potential energy terms must be included. These terms tend to stabilize particular

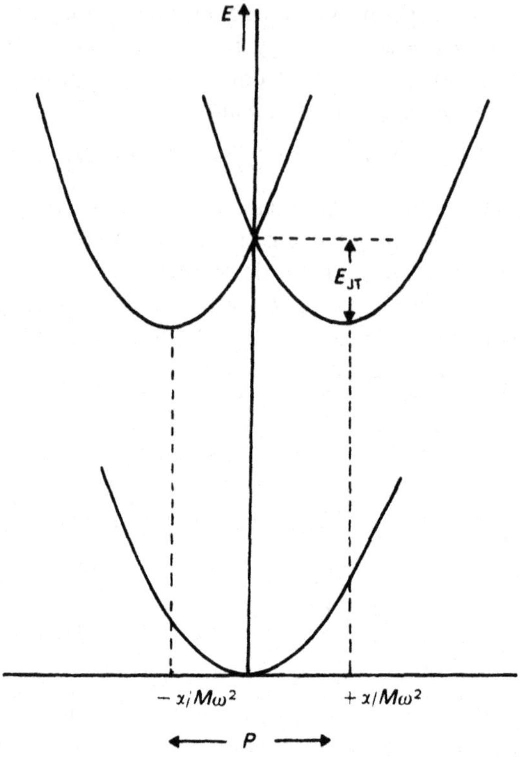

FIG. 5.14. A planar section through the energy versus displacement surfaces of an E electronic state of the F_3-centre, showing the removal of orbital degeneracy by distortions associated with the Jahn–Teller effect.

distortions. In the case of the F_3-centres there are three equivalent distorted configurations (Fig. 5.13(b)) from an equilateral triangle to an isosceles triangle. If these higher-order terms are strong enough, i.e. $E_{JT} \gg \hbar\omega$, we have the static Jahn–Teller effect and the twofold *electronic* degeneracy is converted into a threefold *configurational* degeneracy. Tunnelling from one of the singlet orbital electronic configurations to another lifts this threefold degeneracy into a singlet and a doublet. The singlet–doûblet energy separation is larger the lower the barrier against tunnelling. If the barrier is large then the distortion will be 'frozen-in' and the symmetry of the F_3-centre reduced to C_{1h} from C_{3v}. The experimental results (Silsbee 1965; Huges 1966) do not favour such a static effect: they suggest rather that the barrier is low enough to allow tunnelling to leave a E vibronic state lowest by at least 30 cm^{-1}. This is the *dynamic* Jahn–Teller effect ($E_{JT} \sim \hbar\omega$).

To an extent the division into *static* and *dynamic* effects is arbitrary since distortions which are stabilized at low temperature may be averaged out by thermally-activated reorientation at high temperature. For example, the ESR spectrum of Cu^{2+} in silver chloride is anisotropic below 90 K due to a static effect, whereas between 90 K and room temperature both isotropic and anisotropic spectra are observed. However, in magnesium oxide the Cu^{2+} ESR spectrum is isotropic down to 1.6 K due to tunnelling between axes of distortion.

Ham (1965, 1972) showed that there are consequences of the Jahn–Teller effect other than the lowering of local symmetry at a defect. For the F_3-centre, in the absence of a Jahn–Teller interaction, there is an off-diagonal matrix element of orbital angular momentum perpendicular to the plane of the defect given by

$$L_0 = \langle \Psi_1 | L_z | \Psi_2 \rangle. \tag{5.92}$$

The Born–Oppenheimer states of the system after the Jahn–Teller distortion, $\psi_1\chi_1$ and $\psi_2\chi_2$, are centred about $Q = \pm\alpha/M\omega^2$ and the orbital momentum is

$$\begin{aligned} L &= \langle \psi_1\chi_1 | L_z | \psi_2\chi_2 \rangle \\ &= \langle \psi_1 | L_z | \psi_2 \rangle \langle \chi_1 | \chi_2 \rangle \\ &= \langle \chi_1 | \chi_2 \rangle L_0. \end{aligned} \tag{5.93}$$

Only off-diagonal elements are affected, diagonal elements of the vibrational overlap factor being unity. Evaluating the overlap integral $\langle \chi_1 | \chi_2 \rangle$ between a Jahn–Teller distorted state and an undistorted state leads to

$$L = L_0 \exp(-\beta\, E_{JT}/\hbar\omega) \tag{5.94}$$

where β depends upon the particular electronic system, showing that the angular momentum is reduced by the Ham factor $\exp(-\beta\, E_{JT}/\hbar\omega)$. The reduction in off-diagonal operators (e.g. orbital Zeeman interaction, spin–orbit coupling, or uniaxial stress) by the Ham effect reduces the effectiveness of external perturbations on optical zero-phonon lines more than on broad bands. This is because the overlap integral $\langle \chi_b(m) | \chi_a(0) \rangle$ for the broad band contains large contributions close to $\rho = 0$ where both $\chi_a(0)$ and $\chi(m)$ are large whereas $\langle \chi_b(0) | \chi_a(0) \rangle$ is small because $\chi_a(0)$ is small for displacements towards $\rho = |\alpha/M\omega^2|$. We have discussed only the case of an electronic E state coupled to E vibrational modes. In octahedral or tetrahedral symmetry orbital triplet states are possible which may couple to tetragonal distortions (E-modes) or trigonal distortions (T-modes), as discussed by Sturge (1967) and Ham (1972). Of course, coupling to A modes will also be present; no special effects then occur because there is no lowering of symmetry.

5.4.4 *Observed bandshapes*

It is now appropriate to illustrate the basic theoretical ideas with a few well-characterized experimental examples. Figure 5.15 shows the luminescence

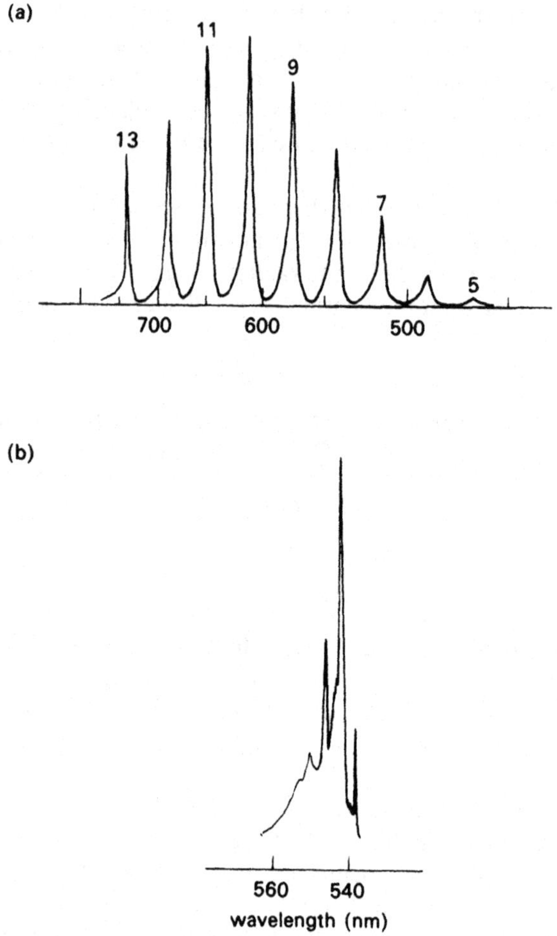

FIG. 5.15. (a) Luminescence spectrum (electronic–vibrational transitions) of O_2^- in potassium bromide at liquid nitrogen temperatures. The electronic system is strongly coupled to the molecular vibrational mode in the O_2^- ion at around 1000 cm^{-1} (Rebane 1974). (b) Detail of the $n = 8$ electronic–vibrational transition at liquid helium temperatures. The sharp electronic–vibrational line is accompanied by a weak sideband due to the coupling with the potassium bromide vibrational models of the host lattice (Rebane and Rebane 1975).

spectrum of O^{2-} in potassium bromide at 77 K (Rebane and Rebane 1975). O_2^- is an interesting spectroscopic species since the optically-active electronic levels are strongly coupled ($S \simeq 10$) to the intramolecular vibrational mode with energy $E \simeq 1000$ cm^{-1}, and weakly coupled ($S \simeq 1$) to the phonon spectrum of the potassium bromide crystal in which the maximum vibrational frequency is only $\simeq 200$ cm^{-1}. Hence the luminescence spectrum at 77 K shows a series of equally spaced lines (Fig. 5.15(a)), each of which is accompanied at low temperature ($T \simeq 4.2$ K) by a weak vibrational sideband characteristic of the host lattice (Fig. 5.15(b)).

The optical spectra of F-centres are associated with $^2A_{1g} \rightarrow {}^2T_{1u}$ transitions at sites with octahedral symmetry. The T_{1u} electronic state may couple to symmetric (A_1), tetragonal (E), and trigonal (T) distortions. In the alkali halides and alkaline earth halides such centres are characterized by strong electron–phonon coupling ($S \sim 20$) to symmetric modes. Hence the absorption and emission bands are broad and almost Gaussian in shape, there being a strong Stokes shift between them (Chapter 7). Essentially effects due to the Jahn–Teller distortion are smeared out by coupling to symmetric modes. The bandshapes are well described by the configurational coordinate model after allowance is made for coupling to a band of vibrational frequencies rather than to a single breathing-mode vibration, the effect of which is to broaden the spectrum to look like the envelope encompassing the spectrum of sharp sidebands shown in Fig. 5.7. Typical absorption and emission bands for the F-centre in potassium bromide are shown in Fig. 5.16. As temperature is raised these bands broaden. The increase in width is accurately described by eqn (5.86) in which the representative frequency is $\omega = 2.96 \times 10^{12}\ \mathrm{s}^{-1}$ (Konitzer and Markham 1960). The centre of gravity of the band also shifts with

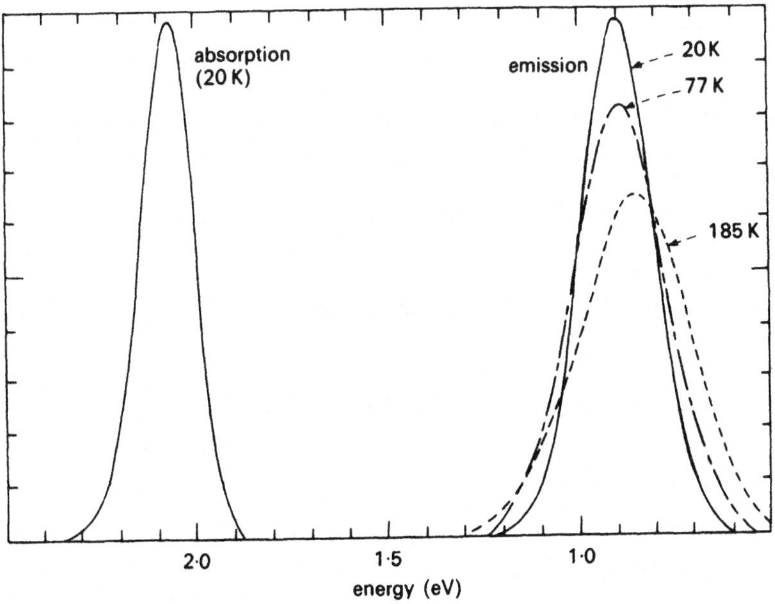

FIG. 5.16. Absorption and luminescence spectra of the F-centre in potassium bromide. (After Gebhardt and Kuhnert 1964.)

temperature, due to differences in the vibrational frequencies of the ground and excited electronic states. In contrast to the F-centre in potassium bromide we find for the F^+-centre in calcium oxide that the coupling to symmetric modes is weak ($S \sim 1$), and the Jahn–Teller coupling is dominant ($S_E \sim S_T \sim 2.5$). The absorption and emission bands then have different shapes (Chapter 7).

Another optical centre which we discuss at length in subsequent chapters is the Cr^{3+} ion, much studied in ionic solids such as magnesium oxide and alumina. For the present purposes the splitting of the octahedral levels is ignored and the ^2E level is assumed to be the lowest excited level. The wavefunctions of the $^4A_2$ ground state and the ^2E excited state are formed from the t_2^3 strong field orbitals, whereas the $^4T_2$ state wavefunction is formed from the t_2^2e strong field orbitals. Since the form of the t_2 and e orbitals are different (see Fig. 2.14), the electron–lattice coupling of the Cr^{3+} ion in the $^4A_2$ and ^2E states should be very similar but different from that when the Cr^{3+} ion is in the $^4T_2$ state. The resulting configurational coordinate diagram for these three Cr^{3+} states is illustrated in Fig. 5.17, where the $^4T_2$ state is assumed to be higher in energy than ^2E. That the $^4A_2 \rightarrow {}^4T_2$ absorption transition is a broad band, as indicated in the figure, is anticipated from the large lateral displacement of the two parabolae. (The observed $^4A_2 \rightarrow {}^4T_2$ band for Cr^{3+} in ruby is shown in Fig. 5.8.) In contrast, the similar coupling to the lattice of $^4A_2$ and ^2E states, results in harmonic oscillator parabolae which have a small relative lateral displacement. The transition between these states is characterized by a small value of S, and a strong zero-phonon line is observed together with a weak vibronic sideband which appears on the high energy side of the sharp line in absorption, and on the low-energy side in emission (Fig. 5.17). This transition is discussed below using an analysis suitable for weak electron–lattice coupling.

At low temperatures there is no emission from the $^4T_2$ state. Ions optically pumped into the $^4T_2$ state tend to decay back to the ^2E state non-radiatively, and then to decay radiatively between the ^2E state and the ground state. Because of the rapid non-radiative processes between $^4T_2$ and ^2E there is a thermalization of excited ion population among the various electronic and vibrational levels of the ^2E and $^4T_2$ states. If the temperature is high enough this may lead to an equilibrium population in the $^4T_2$ state so that radiative decay occurs from $^4T_2$. When this occurs the $^4T_2 \rightarrow {}^4A_2$ emission appears on the low-energy side of the $^4A_2 \rightarrow {}^4T_2$ absorption transition, the Stokes shift between absorption and emission bands being some 2000–3000 cm^{-1}.

An interesting feature of transition-metal ion spectra is that they exhibit both broad bands (associated with strong coupling, $S > 1$) and sharp transitions dominated by zero-phonon lines (associated with weak coupling, $S < 1$). The two transitions represented in Fig. 5.17 illustrate this point. Transition metal ions are reasonably strongly coupled to their crystalline environments. However, it is the *difference in the coupling* when the ion is in

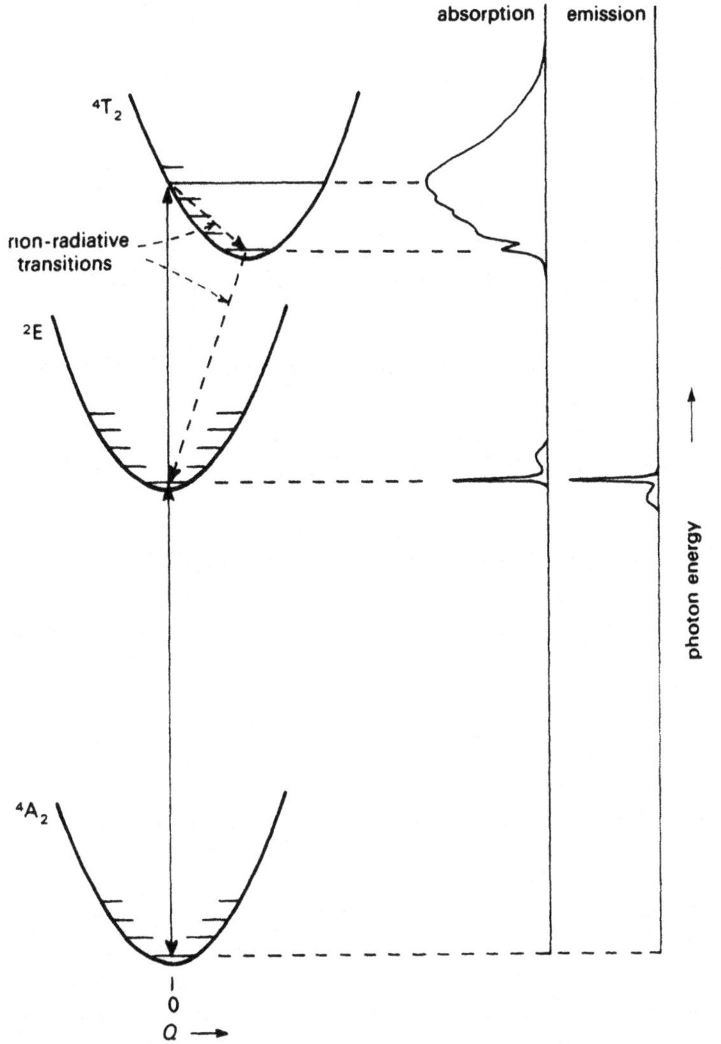

FIG. 5.17. Configurational coordinate diagram for Cr^{3+} in octahedral site of high crystal field strength, where the $^4T_2$ parabola is higher in energy than the ^2E parabola. Absorption occurs from the $^4A_2$ ground state to both ^2E and $^4T_2$ states. At low temperatures the decay of the $^4T_2$ state is dominated by non-radiative decay to the ^2E state. Emission occurs from the ^2E state.

the initial and final states which determines the shape of the transition between these states.

Finally we comment on the electronic transitions on rare-earth ions in solids. The interaction between rare-earth ions of the $(4f)^n$ configuration and the crystalline environment is weak, hence all transitions between $(4f)^n$ levels are characterized by small S values. The transitions are dominated by zero-phonon lines, attended by very weak one-phonon sidebands. These are best analysed using the weak-coupling approximation to be discussed in Section 5.5.

5.4.5 Deficiencies of the configurational coordinate model

The representation of the full spectrum of lattice vibrations by a single breathing-mode vibration leads to an unrealistic description of the sidebands accompanying the zero-phonon line. For large S transitions the overlap of sidebands effectively washes out details of the vibrational spectrum and the configurational coordinate model gives a reasonably good description of the relatively smooth bandshapes observed (e.g. Fig. 5.8). On the other hand, when the Huang–Rhys parameter is small ($S<1$) the observed spectrum appears as a zero-phonon line with a one-phonon sideband. The one-phonon sideband should reflect the rich detail of the full spectrum of lattice vibrations. Thus the use of a single 'representative' mode to describe this full spectrum is crude.

On the configurational coordinate model the breathing-mode vibration can be considered to *frequency modulate* the pure electronic transition, and the frequency-modulated sideband thereby produced has the same dipole nature as the pure electronic transition. This is not an unrealistic point of view if the pure electronic transition is an allowed electric dipole transition, but it is not adequate if the pure electronic transition is only weakly allowed. If, for example, the electronic centre is in a site of inversion symmetry and if the initial and final electronic states have the same parity then the pure electronic transition occurs by a magnetic dipole process and may be very weak. Lattice vibrations, however, can cause odd-symmetry distortions of the crystalline environment of the electronic centre. These destroy the inversion symmetry and allow electric dipole transitions to occur in the electronic centre. The stronger electric dipole transition thereby allowed appears in the sidebands only; odd-symmetry distortions caused by lattice vibrations cannot affect the zero-phonon line. As a result the observed ratio of sideband to zero-phonon line may be much larger that that predicted by the configurational coordinate model. We shall discuss this point again in the following sections.

5.5 The electron–lattice interaction in the weak coupling limit

If the coupling between the optically-active ion and the lattice is weak it is convenient to use a simpler separation of the electronic and lattice motions. We return to the Hamiltonian

$$\mathcal{H}=\mathcal{H}_{\mathrm{FI}}+\mathcal{H}_{\mathrm{c}}(\boldsymbol{r}_i, \boldsymbol{R}_l)+V_{\mathrm{l}}(\boldsymbol{R}_l)+\sum_l \frac{\boldsymbol{P}_l^2}{2M_l} \tag{5.40}$$

and consider the case where the coupling is sufficiently weak that the equilibrium values of $\boldsymbol{R}_l$ do not depend on the electronic state. We write these values as $\boldsymbol{R}_l(0)$. This implies that $V_1(\boldsymbol{R}_l)$ is independent of the electronic state. The crystal field term, $\mathscr{H}_c(\boldsymbol{r}_i, \boldsymbol{R}_l)$, can be expanded as

$$\mathscr{H}_c(\boldsymbol{r}_i, \boldsymbol{R}_l) = \mathscr{H}_c(\boldsymbol{r}_i, \boldsymbol{R}_l(0)) + V_c(\boldsymbol{r}_i, \boldsymbol{q}_l). \tag{5.95}$$

We can also write

$$V_1(\boldsymbol{R}_l) = V_1(\boldsymbol{R}_l(0)) + V_1(\boldsymbol{q}_l). \tag{5.96}$$

The various terms in the Hamiltonian can then be gathered as

$$\mathscr{H} = [\mathscr{H}_{\mathrm{FI}} + \mathscr{H}_c(\boldsymbol{r}_i, \boldsymbol{R}_l(0))] + \left[V_1(\boldsymbol{R}_l(0)) + V_1(\boldsymbol{q}_l) + \sum_l \frac{\boldsymbol{P}_l^2}{2M_l} \right] + V_c(\boldsymbol{r}_i, \boldsymbol{q}_l). \tag{5.97}$$

The terms gathered in the first square bracket constitute the Hamiltonian of the optically active ion in the static crystal field, $\mathscr{H}_e(\boldsymbol{r}_i, \boldsymbol{R}_l(0))$, whose eigenstates are $\psi_a(\boldsymbol{r}_i, \boldsymbol{R}_l(0))$. The $V_1(\boldsymbol{R}_l(0))$ term in the second square bracket has the same value for all electronic states and can be omitted from further consideration. The remaining terms in the square bracket constitute the dynamic lattice Hamiltonian which, in the harmonic approximation, has eigenstates in the form of products of linear harmonic oscillator functions (eqn 5.54). These eigenstates are independent of the electronic state of the optically-active ion. The lattice state can be characterized by giving the number of phonons in each lattice mode. The remaining $V_c(\boldsymbol{r}_i, \boldsymbol{q}_l)$ term, which can also be written as $V_c(\boldsymbol{r}_i, Q_k)$, couples the electronic and dynamic lattice systems and can be termed the *electron–phonon coupling*. If this is neglected the wave function is a product of independent electronic and lattice states, $\psi_a(\boldsymbol{r}_i, \boldsymbol{R}_l(0))\Pi_k|n_k\rangle$, (this is the crude Born–Oppenheimer function), which it is convenient to write this as the ket $|a, n\rangle$. Perturbation theory can be used to determine how these wavefunctions are perturbed by $V_c(\boldsymbol{r}_i, Q_k)$. However, we must first derive an appropriate form of this coupling term, $V_c(\boldsymbol{r}_i, Q_k)$.

A vibrational mode k distorts the environment of the electronic centre by introducing an oscillating crystal field of symmetry Γ, which changes the electronic energy. The one-dimensional model (Fig. 5.18) shows that the separation between neighbouring ions is changed by the lattice vibrations by the amount $q_{l+1} - q_l$. In the long-wavelength approximation this can be written as $(\partial q/\partial x)a$ or εa, where ε is the strain introduced by the lattice vibrations and a is the equilibrium separation between ions. The operator for the strain ε_k due to mode k is given by eqn (5.39). Assuming that the change in crystal field energy of the electronic centre induced by lattice vibrations can be expressed as a power series in the strain, the electron–phonon coupling can be written as

$$V_c(\boldsymbol{r}_i, Q_k) = \sum_{k,\Gamma} V^{k,\Gamma}(\boldsymbol{r}_i)\varepsilon_k + \text{higher powers in } \varepsilon_k. \tag{5.98}$$

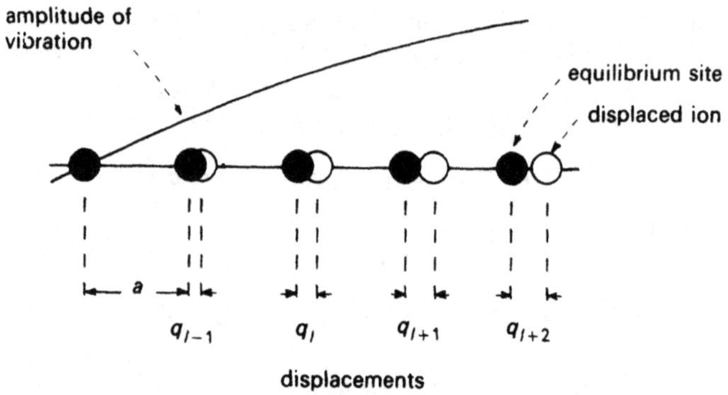

FIG. 5.18. One-dimensional model of the displacements of the ions due to a lattice vibration. The change in crystal field at a particular ion depends on the change in the separation of the neighbouring ions, $q_l - q_{l-1}$, and in the long wavelength limit we can write $q_l - q_{l-1} = (\partial q/\partial x)a = \varepsilon a$. ε is the strain introduced by the lattice vibration, and a is the average separation between adjacent ions.

The lattice mode k introduces a distortion which can be regarded as made up of separate distortions of different symmetries, Γ. $V^{k,\Gamma}(\boldsymbol{r}_i)$ is the electronic energy due to a distortion of symmetry Γ caused by unit strain of mode k. In the long wavelength limit it is independent of the particular mode k. For the case of weak coupling only the linear term in eqn (5.98) is necessary. The operator for strain, ε_k, is given by eqn (5.39) as

$$\varepsilon_k = -\mathrm{i}\left(\frac{\hbar\omega_k}{2MNv_k^2}\right)^{\frac{1}{2}}(\boldsymbol{a}_k^\dagger - \boldsymbol{a}_k)$$

where $\boldsymbol{a}_k^\dagger$ and $\boldsymbol{a}_k$ are phonon creation and annihilation operators. The electron–phonon interaction perturbs the wavefunction $|b, n\rangle$, changing it to $\psi(b, n)$, where $\psi(b, n)$ is given by

$$\begin{aligned}\psi(b,n) &= C_b\left[\,|b,n\rangle + \sum_{t,k,\Gamma,n'} \frac{\langle t,n'|\,V^{k,\Gamma}\varepsilon_k|b,n\rangle}{E_{b,n}-E_{t,n'}}|t,n'\rangle\right] \\ &= C_b\left[\,|b,n\rangle + \sum_{t,k,\Gamma,n'} \frac{\langle t|\,V^{k,\Gamma}|b\rangle}{E_{b,n}-E_{t,n'}}\langle n'|\varepsilon_k|n\rangle|t,n'\rangle\right]\end{aligned} \tag{5.99}$$

where t represents those electronic states of the centre mixed in by the interaction, n' represents other lattice states, and C_b is a normalization constant. It is convenient to write the electronic matrix element as $V_{t,b}^{k,\Gamma}$. A similar expression can be obtained for the perturbation of wavefunction $|a, n\rangle$.

5.5.1 Radiative transitions in the weak coupling limit

We now analyse how these perturbations affect the radiative transitions between electronic states b and a in the particular case that both initial and final electronic states are of similar and definite parity; we assume them to be of even parity. The electronic states are then labelled $|a_g\rangle$ and $|b_g\rangle$. Distortions of even parity, Γ_g, mix in even-parity electronic states; $|t_g\rangle$, odd-parity distortions, Γ_u, mix in odd-parity electronic states, $|t_u\rangle$. Let us restrict $V_c(r_i, Q_k)$ to terms linear in the strain, ε_k. Since the strain ε_k contains the operator $(a_k^\dagger - a_k)$, the lattice state $|n'\rangle$ in eqn (5.99) differs from $|n\rangle$ only by having one more or one less phonon i.e. $|n'\rangle = |n+1\rangle$ or $|n-1\rangle$. We rewrite $\psi(b, n)$ more explicitly as follows:

$$\psi(b,n) = C_b\Bigg[|b_g, n\rangle$$

$$+\sum_{t_g,k,\Gamma_g}\frac{V_{t,b}^{k,\Gamma_g}}{E_{b_g}-E_{t_g}-\hbar\omega_k}(-\mathrm{i})\left(\frac{\hbar\omega_k}{2MNv_k^2}\right)^{\frac{1}{2}}(1+n_k)^{\frac{1}{2}}|t_g, n+1_k\rangle$$

$$+\sum_{t_g,k,\Gamma_g}\frac{V_{t,b}^{k,\Gamma_g}}{E_{b_g}-E_{t_g}+\hbar\omega_k}(\mathrm{i})\left(\frac{\hbar\omega_k}{2MNv_k^2}\right)^{\frac{1}{2}}(n_k)^{\frac{1}{2}}|t_g, n-1_k\rangle$$

$$+\sum_{t_u,k,\Gamma_u}\frac{V_{t,b}^{k,\Gamma_u}}{E_{b_g}-E_{t_u}-\hbar\omega_k}(-\mathrm{i})\left(\frac{\hbar\omega_k}{2MNv_k^2}\right)^{\frac{1}{2}}(1+n_k)^{\frac{1}{2}}|t_u, n+1_k\rangle$$

$$+\sum_{t_u,k,\Gamma_u}\frac{V_{t,b}^{k,\Gamma_u}}{E_{b_g}-E_{t_t}+\hbar\omega_k}(\mathrm{i})\left(\frac{\hbar\omega_k}{2MNv_k^2}\right)^{\frac{1}{2}}(n_k)^{\frac{1}{2}}|t_u, n-1_k\rangle\Bigg] \quad (5.100)$$

and similarly for the perturbed ground state, $\psi(a, n)$. Now consider the emission transition between the states $\psi(b, n)$ and $\psi(a, n')$, where the final vibrational state n' differs from the initial vibrational state n. The matrix element for this transition is $\langle\psi(a, n')|\mu|\psi(b, n)\rangle$ where μ is the appropriate dipole operator. Figure 5.19 distinguishes between the several emission processes involved in this transition, the different features of which are:

1. The zero-phonon magnetic dipole transition at energy E_0 between $\psi(b, n)$ and $\psi(a, n)$, the matrix element of which is $(C_a)^*C_b\langle a_g, n|\mu_m|b_g, n\rangle = (C_a)^*C_b\langle a_g|\mu_m|b_g\rangle$, where μ_m is the magnetic dipole operator. Because of the parity selection rule (Chapter 4) electric dipole transitions are forbidden between the even-parity states involved in this matrix element.

2. The magnetic dipole sideband transition at energy $E_0 - \hbar\omega_k$ between states $\psi(b, n)$ and $\psi(a, n+1_k)$ induced by phonons of A_{1g} symmetry. The

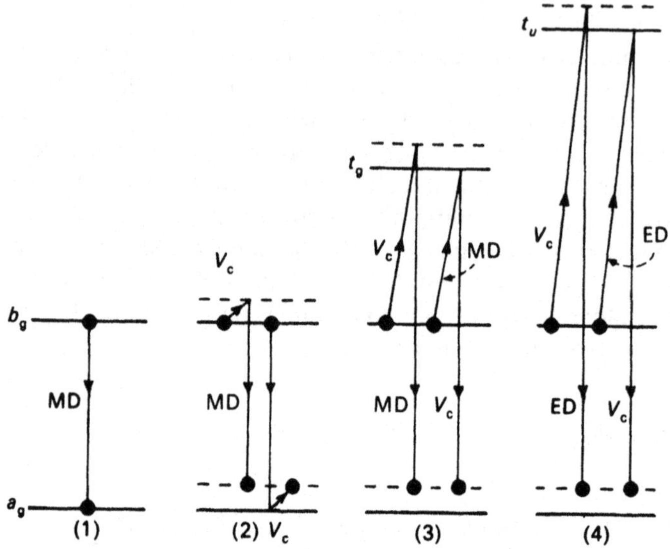

FIG. 5.19. Representations of the various radiative transitions between states *b* and *a*. The broken levels correspond to higher vibrational levels. Process (i) is the zero-phonon transition. The other processes are Stokes sideband transitions ending on higher vibrational levels. Process (iv) is the vibronic transition.

matrix element for this transition is

$$\mathrm{i}(C_a)^* C_b \langle a_g|\mu_{\mathrm{m}}|b_g\rangle \left(\frac{\hbar\omega_k}{2MNv_k^2}\right)^{\frac{1}{2}} \left[\frac{V^{k,A_{1g}}_{b_g,b_g} - (V^{k,A_{1g}}_{a_g,a_g})^*}{\hbar\omega_k}\right](1+n_k)^{\frac{1}{2}} \tag{5.101}$$

which depends on the *difference* in coupling to the A_{1g} modes in the excited and ground states. There is an analogous anti-Stokes transition at energy $E_0+\hbar\omega_k$, the matrix element of which is identical to eqn (5.101) except that $(1+n_k)^{\frac{1}{2}}$ is replaced by $(n_k)^{\frac{1}{2}}$, and in the denominator $\hbar\omega_k$ is replaced by $-\hbar\omega_k$.

3. The Stokes sideband magnetic dipole process at energy $E_0-\hbar\omega_k$ caused by lattice distortions of even symmetry which mix in intermediate states t_g. This has a matrix element

$$(C_a)^* C_b \sum_{t_g,\Gamma_g} (-\mathrm{i})\left(\frac{\hbar\omega_k}{2MNv_k^2}\right)^{\frac{1}{2}} \left[\frac{\langle a_g|\mu_{\mathrm{m}}|t_g\rangle V^{k,\Gamma_g}_{t_g,b_g}}{E_{b_g}-E_{t_g}-\hbar\omega_k} - \frac{\langle t_g|\mu_{\mathrm{m}}|b_g\rangle (V^{k,\Gamma_g}_{t_g,a_g})^*}{E_{a_g}-E_{t_g}-\hbar\omega_k}\right](1+n_k)^{\frac{1}{2}} \quad \text{plus converse.} \tag{5.102}$$

The analogous anti-Stokes sideband process at energy $E_0+\hbar\omega_k$ has an identical matrix element to eqn (5.102) except that $(1+n_k)^{\frac{1}{2}}$ is replaced by $(n_k)^{\frac{1}{2}}$, and in the energy denominators $\hbar\omega_k$ replaced by $-\hbar\omega_k$.

4. The Stokes sideband electric dipole transition at energy $E_0-\hbar\omega_k$ caused by lattice distortions of odd parity which mix in intermediate states of odd parity, t_u. The matrix element is

$$(C_a)^*C_b\sum_{t_u,\Gamma_u}(-i)\left(\frac{\hbar\omega_k}{2MNv_k^2}\right)^{\frac{1}{2}}\left[\frac{\langle a_g|\mu_e|t_u\rangle V^{k,\Gamma_u}_{t_u,b_g}}{E_{b_g}-E_{t_u}-\hbar\omega_k}\right.$$
$$\left.-\frac{\langle t_u|\mu_e|b_g\rangle(V^{k,\Gamma_u}_{t_u,a_g})^*}{E_{a_g}-E_{t_u}-\hbar\omega_k}\right](1+n_k)^{\frac{1}{2}}. \tag{5.103}$$

The analogous anti-Stokes sideband transition at energy $E_0+\hbar\omega_k$ has an identical matrix element to eqn (5.103) except that $(1+n_k)^{\frac{1}{2}}$ is replaced by $(n_k)^{\frac{1}{2}}$, and in the energy denominators $\hbar\omega_k$ is replaced by $-\hbar\omega_k$.

Sideband transition (4) is an electric dipole process in contrast to the zero-phonon line and other sideband transitions which proceed by a magnetic dipole process. Since the electric dipole transition probability is four to five orders of magnitude larger than the magnetic dipole transition probability process, (4) may be the dominant sideband mechanism in many cases.

We now consider as a specific example the optical transitions on transition metal ions in octahedral symmetry sites where the transitions are between states of the same $(3d)^n$ configuration and therefore between states of the same parity. (The optical transitions in rare-earth ions between states of the same $(4f)^n$ configuration are likewise between states of the same parity.) Process (4) involves the mixing of odd-parity $(3d)^{n-1}5p$ states into the even-parity $(3d)^n$ states through odd-parity vibrations. These odd- parity $(3d)^{n-1}5p$ states are some 50 000 cm^{-1} above the even-parity $(3d)^n$ states, giving energy denominators in eqn (5.102) of 30 000–50 000 cm^{-1}. On the other hand the energy denominators in magnetic dipole process (2) is only $\hbar\omega \sim 300$–500 cm^{-1}: in the case of sidebands this is the distance from the zero-phonon line. Hence the energy denominator squared which appears in the formula for the same probability for magnetic dipole process (2) is about 10^4 times smaller than the analogous case of electric dipole process (4). However, since the electric dipole mechanism is more efficient than the magnetic dipole mechanism by 10^4–10^5 we see that process (4) may dominate the sideband process. We reserve the name 'vibronic' for this vibrationally-induced electric dipole sideband process.

Because of the large energy denominators the amount of mixing of odd-parity wavefunction with the even-parity wavefunctions is very small, and it can be neglected when calculating the normalization constant C_b in eqn (5.99). If the magnetic dipole matrix elements $\langle a_g|\mu_m|t_g\rangle$, $\langle a_g|\mu_m|b_g\rangle$, and

$\langle t_g|\mu_m|b_g\rangle$ have the same value *the total transition probability for processes* (1)+(2)+(3) *remains constant independent of temperature.* As temperature is raised the equilibrium value of n_k increases and as a result the magnetic dipole sideband processes increase relative to the magnetic dipole zero-phonon line. When both Stokes and anti-Stokes sidebands are considered the ratio of this sideband to zero-phonon line intensity varies as $1+2n_k=\coth(\hbar\omega_k/2kT)$.

The vibronic process (4) is in addition to these constant magnetic dipole processes, and for the transition metal ions, may dominate the transition process. As eqn (5.103) shows the probability for this Stokes sideband varies as $1+n_k$ while the anti-Stokes sideband probability varies as n_k, hence *the vibronic process increases with temperature as* $1+2n_k=\coth(\hbar\omega_k/2kT)$.

We have concentrated on the sidebands due to a single mode k. If we take into account the full spectrum of lattice vibrations, whose density of modes is $\rho(\omega)$, we see that the sideband shape will bear a resemblance to this density of modes. The sideband shape due to process (2), whose matrix element square varies as ω^{-1}, will, in the long wavelength limit, have a shape which varies as $\rho(\omega)/\omega$. Similarly, the sideband shape due to process (3) should vary as $\omega\rho(\omega)$ in the long-wavelength limit. A more detailed analysis of the distortions due to odd-parity vibrations (Stokowski *et al.* 1966) shows that the vibronic sideband shape in the long wavelength limit should vary as $\omega^3\rho(\omega)$. Irrespective of which process dominates we expect that the sideband shapes will reflect the peaks in the density of vibrational modes.

Figure 5.20 compares the shape of the $^2E\rightarrow{}^4A_2$ emission sideband transition of Cr^{3+} ions in octahedral sites in magnesium oxide with the known density of vibrational modes. The zero-phonon line of this transition occurs through the magnetic dipole process. A number of interesting observations can be made about this sideband spectrum. First, there is an overall similarity of shape between the sideband spectrum and the density of vibrational modes although there is a difference in the positions of the peaks in the two spectra. This is due to the Cr^{3+} ion being a defect in the magnesium oxide lattice and as a result the lattice vibrations in the vicinity of the Cr^{3+} ion are modified from those of the perfect magnesium oxide lattice (Sangster 1972). Next, the sideband is clearly a one-phonon process. There is little evidence of higher-order sidebands, which justifies treating this transition in the weak-coupling limit. According to the configurational coordinate model the absence of second- and higher-order sidebands suggests that the Huang–Rhys parameter (S) is less than unity. We take $S\sim0.3$. Hence at 77 K the zero-phonon line should be about three times as intense as the magnetic dipole sideband. Experimental measurements show, however, that the sideband is about four times more intense than the zero-phonon line. The probable explanation for such relative intensities is that the sideband is vibronically induced and is an order of magnitude stronger than the magnetic dipole sideband process.

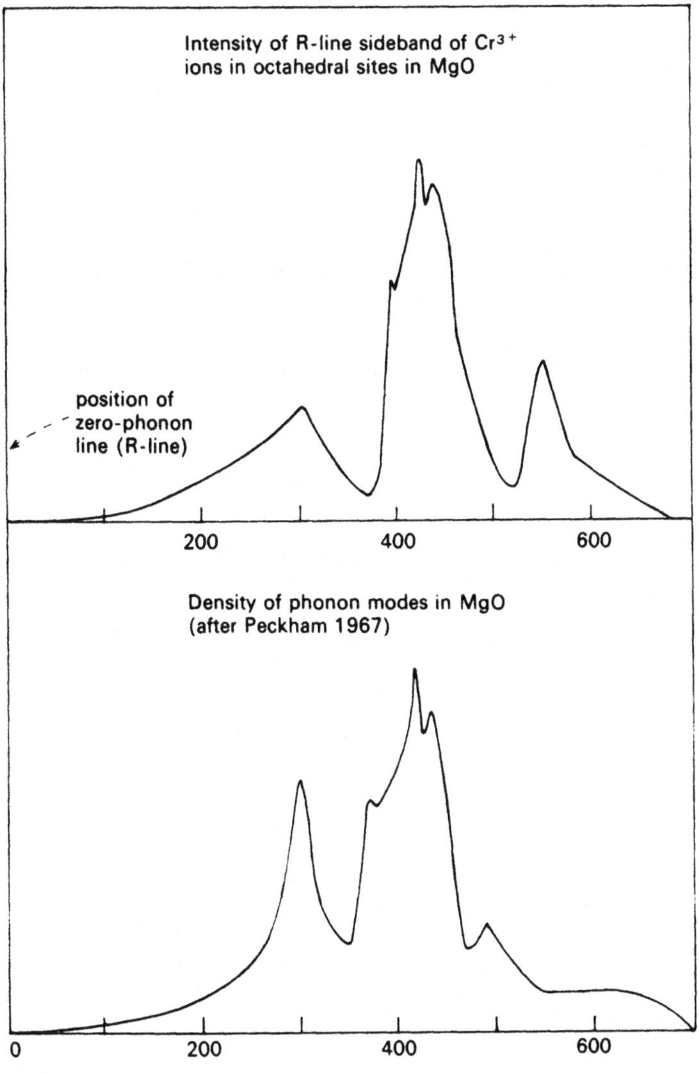

Fig. 5.20. Comparison of the vibrational sideband accompanying the R-line of $MgO:Cr^{3+}$ with the density of phonon modes in magnesium oxide as measured by neutron scattering (Peckham 1967).

5.6 Vibronic processes in broad bands

The electric dipole-induced vibronic process was considered above when discussing transitions between electronic states of similar parity in the limit of weak electron–lattice coupling. We now examine vibronic processes which occur in broadband transitions between electronic states of similar parity. In Section 5.4 the configurational coordinate model was used to analyse such broadband transitions. This model only considers A_{1g} vibrations of the ionic complex, and these modulate the crystal field experienced by the optically-active ion. Such modulations do not change the parity of the initial or final electronic states, hence on the configurational coordinate model the full transition is a magnetic dipole process. It is necessary to modify the configurational coordinate approach to allow mixing of odd-parity electronic states into even-parity states. We cannot use the electron–phonon coupling terms, $V_c(r_i, Q_k)$, developed in Section 5.5 since this approach is valid only in the weak coupling limit. Instead we consider the terms neglected in the Schrödinger equation for the Born–Oppenheimer states $\psi_a \chi_a$, which are (from eqn (5.45))

$$\sum_l \frac{-\hbar^2}{2M_l} [\chi_a \nabla_l^2 \psi_a + 2(\nabla_l \chi_a).(\nabla_l \psi_a)]. \tag{5.104}$$

The first term is the smaller in view of the weaker dependence of ψ_a on R_l. The remaining term is written $\mathscr{H}_{NA}\psi_a\chi_a$ and is termed the non-adiabatic Hamiltonian:

$$\mathscr{H}_{NA}\psi_a\chi_a = \sum_l \frac{-\hbar^2}{M_l} (\nabla_l \chi_a).(\nabla_l \psi_a). \tag{5.105'}$$

This can mix other electronic states into ψ_a.

If the non-adiabatic Hamiltonian is expressed in terms of normal co-ordinates we have

$$\mathscr{H}_{NA}\psi_a\chi_a = \sum_k \frac{-\hbar^2}{M} \left(\frac{\partial \chi_a}{\partial Q_k}\right)\left(\frac{\partial \psi_a}{\partial Q_k}\right) \tag{5.105'}$$

where M is a representative effective mass.

It is a complicated matter to take formal account of the non-adiabatic Hamiltonian; the methods used by various workers have been discussed by Auzel (1978). We simplify matters by assuming a single configurational model and taking χ_a to be a harmonic oscillator state describing the oscillations of Q about the equilibrium value of $Q_0^{(a)}$. The harmonic oscillator wavefunctions are

$$\chi_a(n) = \frac{N}{(2^n n!)^{\frac{1}{2}}} H_n(x) \exp(-x^2/2)$$

where $x = \left(\frac{M\omega}{\hbar}\right)^{\frac{1}{2}} (Q - Q_0^{(a)})$, $H_n(x)$ is the nth-order Hermite polynomial, n is the vibrational quantum number, and N is a normalizing constant. From the properties of the Hermite polynomials, $\mathrm{d}H_n/\mathrm{d}x = 2nH_{n-1}$ and

$H_{n+1} = 2xH_n - 2nH_{n-1}$ we see that

$$\frac{d\chi_a(n)}{dQ} = \left(\frac{M\omega}{2\hbar}\right)^{\frac{1}{2}} [n^{\frac{1}{2}}\chi_a(n-1) - (n+1)^{\frac{1}{2}}\chi_a(n+1)]. \tag{5.106}$$

Other electronic states ψ_t can be mixed into ψ_a through $\mathscr{H}_{\mathrm{NA}}$ and the matrix element is written

$$\frac{-\hbar^2}{M}\langle\psi_t\left|\frac{d}{dQ}\right|\psi_a\rangle = R_{ta}. \tag{5.107}$$

How do these terms modify the configurational coordinate model for the transition between $\psi_{b_g}\chi_b(m)$ and $\psi_{a_g}\chi_a(n)$, where the g subscripts indicate that both electronic states are of even parity? In the absence of $\mathscr{H}_{\mathrm{NA}}$ the matrix element for this magnetic dipole transition is

$$\langle\psi_{a_g}\chi_a(n)|\mu_{\mathrm{m}}|\psi_{b_g}\chi_b(m)\rangle. \tag{5.108}$$

$\mathscr{H}_{\mathrm{NA}}$ can mix in odd-parity states ψ_{t_u} which allows this transition to proceed by the electronic dipole process with the matrix element

$$\sum_{t_u,\chi_t(p)} \frac{\langle\psi_{a_g}\chi_a(n)|\mu_e|\psi_{t_u}\chi_t(p)\rangle\langle\psi_{t_u}\chi_t(p)|\mathscr{H}_{\mathrm{NA}}|\psi_{b_g}\chi_b(m)\rangle}{E(\psi_{b_g}\chi_b(m)) - E(\psi_{t_u}\chi_t(p))}$$

$$\text{plus converse.} \tag{5.109}$$

Here p is the vibrational quantum number in the lattice state χ_t. If we assume that the energy denominators are large we can neglect their dependence on p and m and write them as $\Delta E_{b,t}$. The matrix element is now

$$\sum_{t_u,\chi_t(p)} \left(\frac{M\omega}{2\hbar}\right)^{\frac{1}{2}} \frac{\langle\psi_{a_g}|\mu_e|\psi_{t_u}\rangle R_{t_u b_g}}{\Delta E_{b,t}} [\langle x_a(n)|\chi_t(p)\rangle$$
$$\times\langle\chi_t(p)|(m^{\frac{1}{2}}\chi_b(m-1) - (m+1)^{\frac{1}{2}}\chi_b(m+1))\rangle]$$

$$\text{plus converse.} \tag{5.110}$$

Summing over $\chi_t(p)$ and invoking closure this becomes

$$\sum_{t_u} \frac{\langle\psi_{a_g}|\mu_e|\psi_{t_u}\rangle R_{t_u b_g}}{\Delta E_{b,t}} [\langle\chi_a(n)|\chi_b(m-1)\rangle m^{\frac{1}{2}} - \langle\chi_a(n)|\chi_b(m+1)\rangle(m+1)^{\frac{1}{2}}]$$

$$\text{plus converse.} \tag{5.111}$$

The transition may terminate on many different levels of $\chi_a(n)$ thereby giving the transition its width. To calculate the total intensity we square the matrix element and sum over all the n values, giving

$$I = \left|\sum_{t_u} \frac{\langle\psi_{a_g}|\mu_e|\psi_{t_u}\rangle R_{t_u b_g}}{\Delta E_{b,t}}\right|^2 \left(\frac{M\omega}{2\hbar}\right)(1+2m). \tag{5.112}$$

The equilibrium value of m is $[\exp(\hbar\omega/kT)-1]^{-1}$ so that $1+2m=\coth(\hbar\omega/2kT)$. Thus the vibronic process occurs in the broadband transitions with a rate which increases with temperature as $\coth(\hbar\omega/2kT)$, just as was found in the weak coupling case. Particularly if the odd and even states are well separated in energy the mixing of these states is very small and hardly affects the magnetic dipole process. The vibronic process occurs in addition to the magnetic dipole process, and whereas the magnetic dipole process has constant total intensity the intensity of the vibronic process increases with temperature as $\coth(\hbar\omega/2kT)$. Holmes and McClure (1957) made accurate measurements of the oscillator strength of the broadband absorption in the blue of Ni^{2+} ions in $NiSO_4.7H_2O$, a transition which occurs between $(3d)^8$ configurational states. They showed that the relative oscillator strength increased with temperature as $\coth(\hbar\omega/2kT)$ where $\hbar\omega/k=500$ K, showing that the absorption is dominated by a vibronic process.

5.7 Phonon-induced relaxation processes

So far this discussion has centred on radiative processes in which the emission or absorption of radiation was accompanied by the emission or absorption of vibrational energy. We now discuss transitions in which only vibrational energy is released or absorbed. Some of the possible phonon relaxation processes are illustrated in Fig. 5.21.

5.7.1 *Direct relaxation between adjacent levels*

The direct process (Fig. 5.21(i)) involves relaxation between electronic states $|b\rangle$ and $|a\rangle$ of an ion whose energy separation (E_b-E_a) is within the range of phonon energies of the lattice; relaxation from $|b\rangle$ to $|a\rangle$ releases a phonon of energy $\hbar\omega_k=E_b-E_a$. Since they are close in energy, being separated only by a phonon energy, $\hbar\omega_k$, these states may be considered to have very similar electron–phonon coupling. Hence the electron–phonon interaction $V_c(r_i, Q_k)$, eqn (5.98), is used to describe this coupling. In the absence of the electron–phonon interaction the electron–lattice state is described by the wavefunction $|b, n_k\rangle=|b\rangle|n_k\rangle$, where $|b\rangle$ is assumed to be independent of the vibrating lattice coordinates, and n_k is the phonon occupancy in mode k. (We are only interested in the k-mode since this is the mode involved in the direct process.) The electron–phonon coupling term is given by eqn (5.98). Using a simple scalar description of the strain ε_k and taking advantage of the fact that $V^{k,\Gamma}$ is independent of the mode k in the long wavelength limit, the electron–phonon interaction can be written as

$$V_c=\sum_k V_1\varepsilon_k+\sum_{k,k'} V_2\varepsilon_k\varepsilon_{k'}+\ldots \tag{5.113}$$

where the strain is given by eqn (5.39). The transition probability $W_{ba}^{\rm dir}$ for a

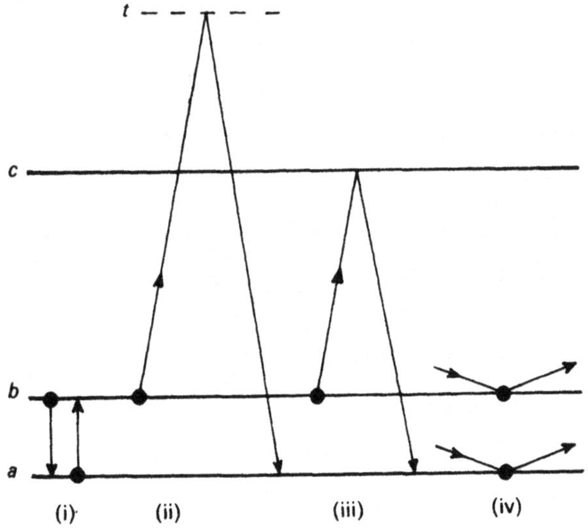

FIG. 5.21. Phonon-induced relaxation processes between states a and b: (i) is the one-phonon direct process relaxation, (ii) is a two-phonon Raman relaxation process from b to a proceeding through a virtual intermediate state t, (iii) is a two-phonon Orbach process from b to a proceeding through a real electronic state c, (iv) shows intrinsic Raman broadening mechanisms in a and b.

direct process relaxation from $|b, n_k\rangle$ to $|a, n_k+1\rangle$ is

$$W_{ba}^{\mathrm{dir}}=\frac{2\pi}{\hbar^2}|\langle a|V_1|b\rangle|^2|\langle n_k+1|a_k^{\dagger}|n_k\rangle|^2\,\rho(\omega_k)\frac{\hbar\omega_k}{2MNv_k^2}. \quad (5.114)$$

$\rho(\omega_k)$ is the density of phonon modes in frequency space assumed from a simple Debye spectrum to be

$$\rho(\omega_k)=\frac{3V\omega_k^2}{2\pi^2 v_k^3} \quad (5.115)$$

where V is the volume of the crystal. In the long-wavelength region v_k is independent of k and we can drop the subscript. More correctly a separate density of modes should be used for longitudinal (l) and transverse (t) waves. These differ only in their having different velocities and in there being two senses of polarization for each transverse mode:

$$\rho_{\mathrm{l}}(\omega_k)=\frac{V\omega_k^2}{2\pi^2 v_{\mathrm{l}}^3},\quad \rho_{\mathrm{t}}(\omega_k)=\frac{2V\omega_k^2}{2\pi^2 v_{\mathrm{t}}^3}. \quad (5.116)$$

The transition probability at temperature T is then

$$W_{ba}^{\mathrm{dir}}=|\langle a|V_1|b\rangle|^2\frac{\omega_k^3}{2\pi\hbar\rho}\left(\frac{1}{v_{\mathrm{l}}^5}+\frac{2}{v_{\mathrm{t}}^5}\right)(n_k+1)=W_{ba}^{\mathrm{dir}}(0)(n_k+1) \quad (5.117)$$

where $W_{ba}^{\text{dir}}(0)$ is the direct process b to a transition at $T=0$ and the mass density of the material is $\rho = MN/V$. Since the electronic matrix element is approximately independent of the phonon variables W_{ba}^{dir} should vary as $(\hbar\omega_k)^3$, that is, as the cube of the energy separation between levels b and a. The fifth power in the lattice wave velocity can be important; harder crystals have larger phonon velocities, and a factor of 3 in velocity introduces a factor of over two orders of magnitude in relaxation rate.

A similar calculation for the *upward* transition probability at temperature T gives

$$W_{ab}^{\text{dir}} = W_{ba}^{\text{dir}}(0)n_k. \tag{5.118}$$

For these relaxation processes we can write the rate equations for the populations N_a and N_b in the two levels, where $N_a + N_b$ is constant, as

$$\frac{\mathrm{d}N_a}{\mathrm{d}t} = -N_a W_{ab}^{\text{dir}} + N_b W_{ba}^{\text{dir}}$$

$$\frac{\mathrm{d}N_b}{\mathrm{d}t} = -\frac{\mathrm{d}N_a}{\mathrm{d}t}. \tag{5.119}$$

In equilibrium, when $\mathrm{d}N_a/\mathrm{d}t = \mathrm{d}N_b/\mathrm{d}t = 0$, we obtain

$$\frac{N_a^{(0)}}{N_b^{(0)}} = \frac{W_{ba}^{\text{dir}}}{W_{ab}^{\text{dir}}} = \frac{n+1}{n} = \exp\left(\frac{\hbar\omega}{kT}\right), \tag{5.120}$$

the superscript indicating equilibrium population values. If at time $t=0$ the populations are $N_a(0)$ and $N_b(0)$, and if these differ from the equilibrium values, the rate equations show that the populations approach equilibrium as

$$N_a(t) - N_a^{(0)} = (N_a(0) - N_a^{(0)})\exp(-ft)$$

$$N_b(t) - N_b^{(0)} = -(N_a(0) - N_a^{(0)})\exp(-ft) \tag{5.121}$$

where

$$f = W_{ba}^{\text{dir}} + W_{ab}^{\text{dir}} = W_{ba}^{\text{dir}}(0)(1+2n) = W_{ba}^{\text{dir}}(0)\coth(\hbar\omega/2kT). \tag{5.122}$$

$1/f$ is the spin–lattice relaxation time T_1 for the direct process.

If the levels a and b are components of a Zeeman splitting due to a magnetic field B then there is a field dependence to the relaxation rate: $1/T_1 \propto B_0^3$. If, however, the centre has an odd number of electrons and the levels a and b are degenerate in energy in the absence of B_0 (i.e. Kramers degeneracy) then there are no matrix elements of V_c between $|a\rangle$ and $|b\rangle$. Generally, however, the Zeeman field mixes some other states into $|a\rangle$ and $|b\rangle$. The resultant spin–lattice relaxation is a weak process, the rate of which varies as a higher power of the magnetic field B_0.

If the populations are disturbed in a and b and allowed to relax to equilibrium a high density of phonons in a narrow band at frequency ω (where

$\hbar\omega = E_b - E_a$) is created. Unless these phonons can rapidly release their energies to some local thermal reservoir a *phonon bottleneck* may develop.

5.7.2 *Raman and Orbach relaxation processes*

In addition to the one-phonon direct process, two-phonon relaxation processes also exist. They are illustrated in Fig. 5.21(ii) and (iii). In the *Raman process* (ii) a high-energy phonon k_1 is absorbed and a higher-energy phonon k_2 is created, the energy difference being obtained from the electronic system when it changes from b to a. This second-order process, involving phonon operators $\boldsymbol{a}^\dagger_{k_2}\,\boldsymbol{a}_{k_1}$ and $\boldsymbol{a}_{k_1}\,\boldsymbol{a}^\dagger_{k_2}$, requires an operator quadratic in ε. Such an operator can come from $\sum_k V_1\varepsilon_k$ in second-order and from $\sum_{k,k'} V_2\varepsilon_k\varepsilon_{k'}$ in first-order. The transition probability is

$$W^{\text{Raman}}_{ba} = \frac{2\pi}{\hbar^2}\int\!\!\int |M|^2\rho(\omega_1)\rho(\omega_2)\delta\left(\omega_2 - \omega_1 - \frac{E_b - E_a}{\hbar}\right)\mathrm{d}\omega_1\,\mathrm{d}\omega_2 \tag{5.123}$$

where the matrix element, M, is given by

$$\begin{aligned} M &= 2\left[\sum_t 2\frac{\langle a|V_1|t\rangle\langle t|V_1|b\rangle}{\Delta_t} + \langle a|V_2|b\rangle\right]\frac{\hbar(\omega_1\omega_2)^{\frac{1}{2}}}{2MNv^2}(n_1(n_2+1))^{\frac{1}{2}} \\ &= 2\langle a|\tilde{V}|b\rangle\frac{\hbar(\omega_1\omega_2)^{\frac{1}{2}}}{2MNv^2}(n_1(n_2+1))^{\frac{1}{2}}. \end{aligned} \tag{5.124}$$

Δ_t is an energy separation such as $E_b - E_t \pm \hbar\omega_{1,2}$. If Δ_t is assumed to be much larger than the range of phonon energies then all the energy denominators may be given the same value of Δ_t. We assume a density of states $\rho(\omega) = 3V\omega^2/2\pi^2v^3$, where v is some average over the longitudinal and transverse velocities. The integration over phonon modes involves the expression

$$\int\!\!\int \frac{9V^2\omega_1^3\omega_2^3}{(2\pi^2v^3)^2}\,\frac{\exp(\hbar\omega_2/kT)}{\left(\exp\left(\frac{\hbar\omega_1}{kT}\right)-1\right)\left(\exp\left(\frac{\hbar\omega_2}{kT}\right)-1\right)} \times\delta\left(\omega_1 - \omega_2 - \frac{E_b - E_a}{\hbar}\right)\mathrm{d}\omega_1\,\mathrm{d}\omega_2. \tag{5.125}$$

Because they have a higher density of states the higher-frequency modes are more effective. If $\hbar\omega_1$ and $\hbar\omega_2$ are much larger than $E_b - E_a$, we can assume $\omega_1 = \omega_2$ thus greatly simplifying this integral, which becomes

$$\int_0^{\omega_0} \frac{9V^2}{(2\pi^2v^3)^2}\,\frac{\omega^6\exp(\hbar\omega/kT)}{\left(\exp\left(\frac{\hbar\omega}{kT}\right)-1\right)^2}\,\mathrm{d}\omega = \frac{9V^2}{(2\pi^2v^3)^2}\left(\frac{kT}{\hbar}\right)^7\int_0^{x_0}\frac{x^6e^x}{(\exp(x)-1)^2}\,\mathrm{d}x \tag{5.126}$$

where $x=\hbar\omega/kT$, $x_0=\hbar\omega_0/kT$ and ω_0 is the Debye cutoff frequency. Writing $V/MN=1/\rho$, where ρ is the mass density, we obtain

$$W_{ba}^{\text{Raman}}=\frac{9|\langle a|\tilde{V}|b\rangle|^2}{2\pi^3 v^{10}\rho^2}\left(\frac{kT}{\hbar}\right)^7\int_0^{x_0}\frac{x^6\exp(x)}{(\exp(x)-1)^2}\,\mathrm{d}x. \tag{5.127}$$

At very low temperatures, where x_0 can be set equal to infinity, the Raman relaxation rate varies as T^7. In the approximation adopted in deriving eqn (5.125), that E_b-E_a is much less than the maximum phonon energy, we see that $W_{ab}^{\text{Raman}}=W_{ba}^{\text{Raman}}$. Solving the rate equations for populations N_a and N_b (eqn 5.119) gives a Raman relaxation rate equal to $2W_{ba}^{\text{Raman}}$.

In discussing the Raman process we assume that the intermediate electronic states t have energies E_t such that $|E_t-E_a|$ and $|E_t-E_b|$ are greater than the phonon energies. If these energy differences are within the range of phonon energies some of the energy denominators in eqn (5.124) will be zero and this theory is no longer applicable. It is best to treat this resonant process, the Orbach process, separately; we can then maintain our simple formulation of the Raman process for those intermediate states above the range of phonon energies.

Orbach relaxation is the resonant two-phonon process shown in Fig. 5.21(iii). We label the intermediate electronic state c, assume that $|E_c-E_b|$ and $|E_c-E_a|$ are less than the maximum phonon energies, and write W_1 and W_2 for the relaxation rates due to the direct process at 0 K between c and b and between c and a, respectively (Fig. 5.22). The upward and downward direct process transition rates at temperature T are given in the figure in terms of W_1, W_2, and the phonon occupancy, n. The rate equations involving populations N_a and N_b governed by the Orbach process are

$$\begin{aligned}\frac{\mathrm{d}N_a}{\mathrm{d}t}&=-N_aW_{ac}+N_cW_{ca}\\ \frac{\mathrm{d}N_b}{\mathrm{d}t}&=-N_bW_{bc}+N_cW_{cb}.\end{aligned} \tag{5.128}$$

These can be solved and the population difference is found to approach equilibrium according to

$$(N_a-N_b)-(N_a^{(0)}-N_b^{(0)})=A_1\exp(-\alpha_1 t)+A_2\exp(-\alpha_2 t) \tag{5.129}$$

where α_1 and α_2 are (Harris and Yngvesson 1968)

$$\alpha_{1,2}=\frac{W_{cb}+W_{bc}+W_{ca}+W_{ac}\pm[(W_{cb}+W_{bc}-W_{ca}-W_{ac})^2+4W_{ca}W_{cb}]^{\frac{1}{2}}}{2}. \tag{5.130}$$

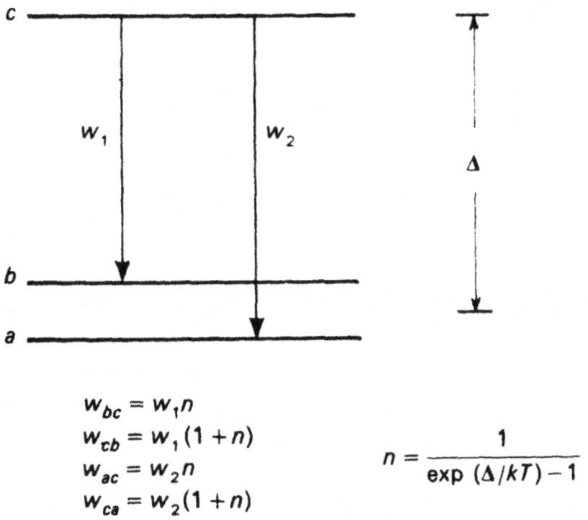

FIG. 5.22. W_1 and W_2 are the zero-temperature relaxation rates between c and b and between c and a, respectively. The higher temperature transitions rates are given in terms of W_1 and W_2. Δ is assumed to be within the range of phonon energies, and n is the occupancy at temperature T for phonons of energy Δ.

The slower of these (α_2) is the observable decay rate. If $W_1 = W_2$

$$\alpha_2 = \frac{W_1}{\exp(\Delta/kT) - 1} \simeq W_1 \exp(-\Delta/kT) \quad \text{when} \quad \Delta \gg kT, \tag{5.131}$$

whereas if $W_1 \gg W_2$ or $W_2 \gg W_1$, then, for $\Delta \gg kT$, we find

$$\alpha_2 = \frac{2W_1 W_2}{W_1 + W_2} \exp(-\Delta/kT) \tag{5.132}$$

showing that the slower of the rates, W_1 or W_2, determines the decay rate.

The Orbach process is important in the case of relaxation between the Zeeman split levels of a Kramers ion since the direct and Raman process may not be very effective here. In the case of a Kramers ion state c will also consist of a degenerate Kramers doublet split by the magnetic field (Fig. 5.23). Symmetry arguments show that $W_1 = W'_1$ and $W_2 = W'_2$. If the energy separation Δ is much larger than the Zeeman splittings and if $W_2 \gg W_1$ (which is usually the case because W_1 involves a spin-flip and W_2 does not) the formula for the relaxation rate is

$$\alpha_2 = \frac{4W_1 W_2}{W_1 + W_2} \exp(-\Delta/kT) \simeq 4W_1 \exp(-\Delta/kT) \tag{5.133}$$

showing that the process proceeds by the slower of the two rates.

An example of this relaxation process is found in the adjustment of population in the Zeeman-split levels of the $\bar{E}(^2E)$ luminescent state of Cr^{3+}

in aluminium oxide (Geschwind *et al.* 1965). The splitting of the $\bar{E}$ level shows up as a splitting of the R_1 luminescence line and the adjustment of population can be monitored optically. The $2\bar{A}(^2E)$ level lies 29 cm^{-1} above $\bar{E}(^2E)$, and this upper level is split by the magnetic field. The energy level structure is similar to that in Fig. 5.23; the a, b levels correspond to the Zeeman-split levels of $\bar{E}(^2E)$, the c, c' levels are the Zeeman split $2\bar{A}$ levels, and $\Delta = 29$ cm^{-1}. The measured spin–lattice relaxation times, T_1, plotted in Fig. 5.24, vary as $1/T_1 = C\exp(-\Delta/kT)$, showing that the spin–lattice relaxation in the $\bar{E}(^2E)$ state occurs by Orbach relaxation via the higher $2\bar{A}(^2E)$ state.

All these mechanisms which induce transitions from b to a cause lifetime broadening of the b level. *Intrinsic* broadening mechanisms also exist which broaden a level without causing transitions out of that level. These are considered in the next section.

5.7.3 *Intrinsic Raman broadening processes*

This process is illustrated in Fig. 5.21(iv); the electron–lattice state $|b, n_{k_1}, n_{k_2}\rangle$ undergoes a transition to state $|b, n_{k_1}+1, n_{k_2}-1\rangle$ induced by the electron–phonon coupling, the k_1 and k_2 phonons having equal energy. From the treatment of the Raman process in Section 5.7.2 the relaxation rate of the intrinsic Raman process can, by analogy with eqn (5.127), be written as

$$W_{bb}^{\text{Raman}} = \frac{9|\langle b|\tilde{V}|b\rangle|^2}{2\pi^3 v^{10}\rho}\left(\frac{kT}{\hbar}\right)^7 \int_0^{x_0} \frac{x^6 e^x}{(e^x-1)^2}\,dx. \tag{5.134}$$

This process gives a finite lifetime, $(W_{bb}^{\text{Raman}})^{-1}$, to the electron–lattice state $|b, n\rangle$ before it relaxes to another state $|b, n'\rangle$, with the same energy. Consequently this gives each state $|b, n\rangle$ an intrinsic width, $\Gamma = \hbar W_{bb}^{\text{Raman}}$, which will broaden zero-phonon transitions out of and into the state $|b, n\rangle$.

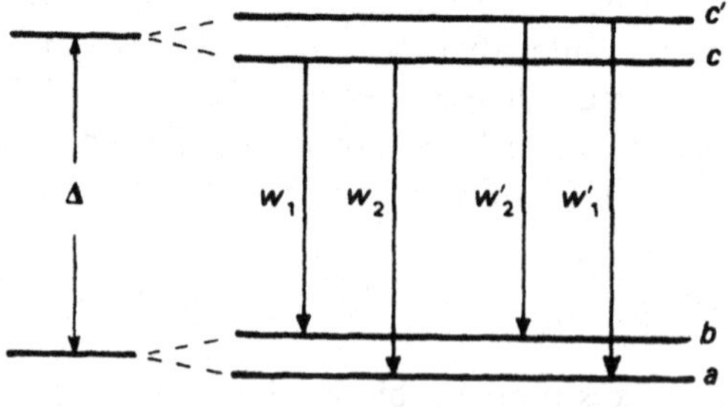

FIG. 5.23. Zeeman splitting of the levels of a hypothetical Kramers ion. Δ is assumed to be within the range of phonon energies.

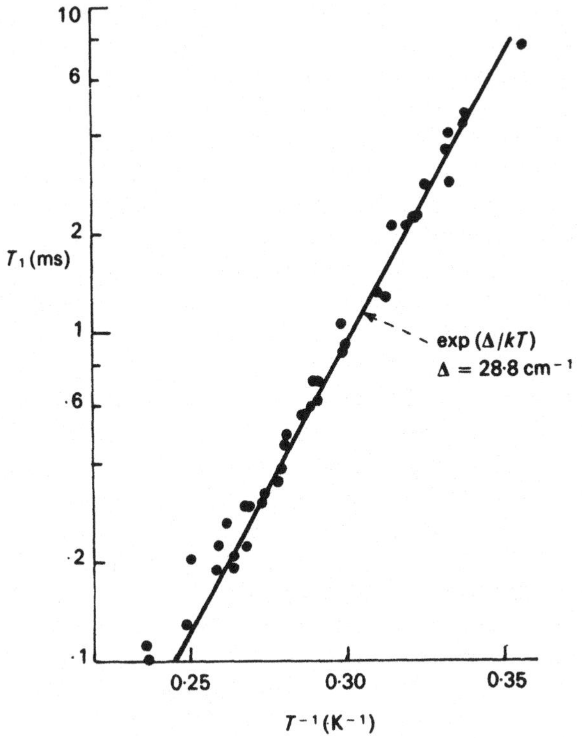

FIG. 5.24. Measured spin–lattice relaxation times in the $\bar{E}(^2E)$ luminescent level of $Al_2O_3:Cr^{3+}$. The measured values follow the predictions of an Orbach process proceeding through the $2\bar{A}(^2E)$ level 28.8 cm^{-1} higher in energy. (After Geschwind *et al.* 1965.)

5.7.4 Linewidths of zero-phonon transitions

The relaxation mechanisms in different levels, analysed in the previous sections, contribute to the linewidths of zero-phonon transitions between these levels. This is *homogeneous or lifetime broadening*, described in Section 4.4. There is an additional *inhomogeneous broadening* due to strains and defects in the host solid. Such strains and defects result in the sites occupied by individual ions being slightly different, and since the energies of the electronic states of the ions depend on their environments the ions in a given state possess a spread of energies. We say that the electronic levels of the solid as a whole are strain-broadened, such levels giving rise to strain-broadened transitions. For a random strain distribution the observed strain-broadened line has a Gaussian shape. Lifetime broadening mechanisms, on the other hand, cause a Lorentzian lineshape. The final lineshape when both mechanisms are operative is a complicated convolution of Gaussian and Lorentzian lineshapes known as a *Voigt profile*. Care must be exercised in deducing from the observed lineshape the contributions from homogeneous and inhomogeneous broadening mechanisms (Wertheim *et al.* 1974).

Strain broadening of zero-phonon lines, even in the best materials, is commonly in excess of 0.1 cm^{-1} and this can mask weak homogeneous broadening processes. In favourable cases the technique of fluorescence line narrowing (FLN) can be used to significantly reduce the influence of strain broadening and permit homogeneous linewidths orders of magnitude smaller than the inhomogeneous linewidth to be measured. (The FLN technique is described in Chapter 6.)

Finally, we comment on the absence of the Doppler broadening effects in optical transitions in solids. Even at the absolute zero of temperature the optically-active ions participate in the vibrational motion of the solid. This motion might be expected to affect the frequency of emission (or absorption) of the ion. The situation can be examined classically by treating the emitting ion as a classical continuously-emitting oscillator with natural frequency, ω_0. (This is the frequency of the zero-phonon line.) The classical oscillator vibrates sinusoidally, the displacement of the oscillator in the x-direction due to the kth mode of vibration is $x = x_0 \sin \omega_k t$, hence the velocity of the oscillator in the x-direction is $v(t) = \omega_k x_0 \cos \omega_k t$. The radiation emitted by the oscillator in the x-direction is frequency shifted by the Doppler effect and $\omega(t)$ is given by

$$\omega(t) = \omega_0\left(1 + \frac{v(t)}{c}\right) = \omega_0\left(1 + \frac{\omega_k x_0}{c}\cos \omega_k t\right). \tag{5.135}$$

The amplitude function of the output signal from the oscillator is $A \sin \theta(t)$ where $d\theta(t)/dt$ is the instantaneous frequency, and therefore we have

$$\frac{d\theta(t)}{dt} = \omega_0\left(1 + \frac{\omega_k x_0}{c}\cos \omega_k t\right). \tag{5.136}$$

On integration eqn (5.136) gives

$$\theta(t) = \omega_0 t + \frac{\omega_0 x_0}{c}\sin \omega_k t. \tag{5.137}$$

Fourier analysis of the output signal, $\sin \theta(t)$, where $\theta(t)$ is given by eqn (5.137), yields the expansion

$$\sin \theta(t) = \sin \omega_0 t \cos\left(\frac{\omega_0 x_0}{c}\sin \omega_k t\right) + \cos \omega_0 t \sin\left(\frac{\omega_0 x_0}{c}\sin \omega_k t\right), \tag{5.138}$$

which can be further expanded in terms of Bessel functions. In fact it is much simpler to take advantage of the fact that for lattice vibrations the modulation index $\omega_0 x_0/c$ is much less than unity. Hence eqn (5.138) can be further simplified to obtain

$$\sin\theta(t)=\sin\omega_0 t+\cos\omega_0 t\,\frac{\omega_0 x_0}{c}\sin\omega_k t$$

$$=\sin\omega_0 t+\frac{\omega_0 x_0}{2c}[\sin(\omega_0+\omega_k)t-\sin(\omega_0-\omega_k)t]. \qquad (5.139)$$

Thus the output consists of the main emission line at frequency ω_0 accompanied by a weak sideband on either side of this line but shifted from it by the frequency of the lattice vibration, ω_k. This is just a weak frequency modulation of the natural oscillator frequency ω_0 by the lattice vibration. *The main emission line is not broadened by this frequency modulation process.*

The intensity of each sideband is $(\omega_0 x_0/2c)^2$. Consequently the intensity in the main line at the central frequency ω_0 is reduced from unity to $1-2(\omega_0 x_0/2c)^2=1-\omega_0^2 x_0^2/2c^2$. (A correct Fourier analysis shows that the fractional intensity at the central frequency is $[J_0(\omega_0 x_0/c)]^2$, where J_0 is the Bessel function of order zero, and, for small values of $\omega_0 x_0/c$, this agrees with our simple intensity formula.) The fractional intensity in the main line can be approximated by $\exp(-\omega_0^2 x_0^2/2c^2)$ in the present instance. We can take the full spectrum of lattice vibrations into account by replacing $x_0^2/2$ by $\langle u^2\rangle/3$, where $\langle u^2\rangle$ is the mean-square lattice displacement. The fractional intensity in the main line is then $\exp(-\omega_0^2\langle u^2\rangle/3c^2)$ which is just the Debye–Waller factor familiar to us from the theory of the Mössbauer effect. This factor gives the fraction of the gamma radiation emitted (or absorbed) by gamma-active nuclei in a vibrating solid which appears in the sharp Mössbauer line. Our Doppler analysis of the emissions from a vibrating oscillator is just the classical analogue of the Mössbauer effect. Whereas for nuclear gamma rays ($\omega_0 \simeq 10^{20}\,\mathrm{s}^{-1}$) the fractional intensity in the sharp zero-phonon line, as calculated from the Debye–Waller factor, can be much less than unity, for optical transitions ($\omega_0 \simeq 3\times10^{15}\,\mathrm{s}^{-1}$) essentially all the radiative transitions occur in the sharp zero-phonon lines and the sidebands are too weak to be detected. Hence optical zero-phonon lines from ions in solids are unaffected, as regards intensity or broadening, by the Doppler modulation.

5.7.5 *Experimental studies of line broadening*

McCumber and Sturge (1963) first systematically analysed the temperature-dependent homogeneous broadening of zero-phonon lines in their investigation of the temperature dependence of the R-lines in ruby, $Al_2O_3{:}Cr^{3+}$. They showed that the temperature-dependent width of the R_1-line was due to Raman relaxation (eqn 5.134). In these early experiments accurate measurements of homogeneous linewidths smaller than the inhomogeneous linewidths ($\gtrsim 0.1\,\mathrm{cm}^{-1}$) could not be made. Later in a very ingeneous experiment Muramoto *et al.* (1974) eliminated strain effects and accurately measured the R_1 linewidth at very low temperatures, where the homogeneous broadening is

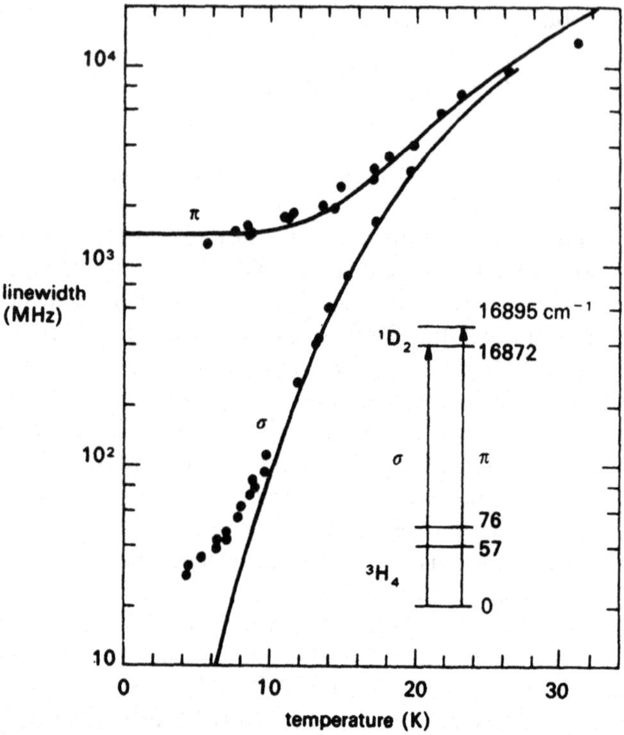

FIG. 5.25. Homogeneous linewidths of the $^3H_4 \rightarrow {}^1D_2$ transition of Pr^{3+} in lanthanum fluoride at low temperatures. The homogeneous linewidth of the σ transition becomes considerably narrower than the inhomogeneous broadening at very low temperatures. The measurements were made by Erickson (1975) using narrowband laser excitation.

dominated by the Raman process. Homogeneous linewidths much lower than the inhomogeneous width can be measured using FLN techniques (Ericson 1975). Figure 5.25 shows the measured temperature dependence of the linewidth of the $^2H_4 \rightarrow {}^1D_2$ transition of Pr^{3+} in lanthanum fluoride. The limiting width of the π-transition reflects the direct process relaxation from the upper $^1D_2$ level to the lower $^1D_2$ level. The solid curves are theoretical estimates of the linewidth derived from the direct and Raman broadening mechanisms. The deviation of the width at low temperatures from the theoretical curve is due to the presence of hyperfine splitting.

5.7.6 *Phonon-induced line shifts*

In addition to inducing transitions between levels the electron–phonon coupling ($V_c = V_1\varepsilon + V_2\varepsilon^2 + \dots$) contributes to the energy of the electron–lattice state, $|b, n\rangle$. We will use perturbation theory to estimate this contribution. Since our concern is with the diagonal matrix elements of V_c and since ε is linear in the a_k and $a_k^\dagger$ operators, $V_1\varepsilon$ contributes only in second-order whereas $V_2\varepsilon^2$ contributes in first-order. The additional energy of $|b, n\rangle$ is

$$\langle b, n|V_2\varepsilon^2|b, n\rangle + \sum_{t,n'} \frac{\langle b, n|V_1\varepsilon|t, n'\rangle \langle t, n'|V_1\varepsilon|b, n\rangle}{E_{b,n} - E_{t,n'}}$$

$$= \sum_k \langle b|V_2|b\rangle \langle n_k|a_k^\dagger a_k + a_k a_k^\dagger|n_k\rangle \frac{\hbar\omega_k}{2MNv_k^2}$$

$$+ \sum_{k,t} |\langle b|V_1|t\rangle|^2 \left[\frac{\langle n_k|a_k^\dagger a_k|n_k\rangle}{E_b - E_t + \hbar\omega_k} + \frac{\langle n_k|a_k a_k^\dagger|n_k\rangle}{E_b - E_t - \hbar\omega_k}\right] \frac{\hbar\omega_k}{2MNv_k^2}. \quad (5.140)$$

The summation over t is over all electronic states including state b. Distinguishing the case that $t = b$ from other states the additional energy term is

$$\sum_k \frac{|\langle b|V_1|b\rangle|^2}{\hbar\omega_k} \langle n_k|a_k^\dagger a_k - a_k a_k^\dagger\ n_k\rangle \frac{\hbar\omega_k}{2MNv_k^2}$$

$$+ \sum_k \left[\langle b|V_2|b\rangle + \sum_t{}' \frac{|\langle b|V_1|t\rangle|^2}{E_b - E_t}\right] \langle n_k|a_k^\dagger a_k + a_k a_k^\dagger|n_k\rangle \frac{\hbar\omega_k}{2MNv_k^2}. \quad (5.141)$$

The prime in the summation over t means that $t = b$ is excluded from the summation. In addition, it is assumed that when $t \neq b$, $|E_b - E_t| \gg \hbar\omega_k$ so that the $\hbar\omega_k$ term can be omitted from the energy denominators. Since $a_k^\dagger a_k - a_k a_k^\dagger = -1$ and since $a_k^\dagger a_k + a_k a_k^\dagger = 1 + 2n_k$ the energy contribution to state $|b, n\rangle$ becomes

$$\Delta E_b = \sum_k \left[-\frac{|\langle b|V_1|b\rangle|^2}{\hbar\omega_k} + \sum_t{}' \frac{|\langle b|V_1|t\rangle|^2}{E_b - E_t} + \langle b|V_2|b\rangle\right] \frac{\hbar\omega_k}{2MNv_k^2}$$

$$+ \sum_k \left[\sum_t{}' \frac{|\langle b|V_1|t\rangle|^2}{E_b - E_t} + \langle b|V_2|b\rangle\right] \frac{\hbar\omega_k}{2MNv_k^2} \cdot 2n_k. \quad (5.142)$$

The first term in eqn (5.142), $\Delta E_b(0)$, is temperature-independent. The second term, $\Delta E_b(T)$, is temperature-dependent; it can be evaluated assuming a Debye spectrum of lattice modes of density $\rho(\omega)$ and we obtain

$$\Delta E_b(T) = \Delta V_b \int_0^{\omega_0} 2\rho(\omega) \frac{\hbar\omega}{2MNv^2} \cdot \frac{d\omega}{\exp(\hbar\omega/kT) - 1}$$

$$= \Delta V_b \int_0^{\omega_0} \frac{3\hbar\omega^3}{2\pi^2\rho v^5} \cdot \frac{d\omega}{\exp(\hbar\omega/kT) - 1}$$

$$= C_b \int_0^{\omega_0} \frac{\omega^3 d\omega}{\exp(\hbar\omega/kT) - 1} \quad (5.143)$$

where ΔV_b is the sum of the two electronic matrix elements in the second term in eqn (5.142), ρ is the mass density of the material, and C_b is a temperature-independent constant. Writing $x=\hbar\omega/kT$ in eqn (5.143) gives

$$\Delta E_b(T)=C_b\left(\frac{kT}{\hbar}\right)^4\int_0^{x_0}\frac{x^3\,\mathrm{d}x}{\mathrm{e}^x-1} \tag{5.144}$$

where $x_0=\hbar\omega_0/kT$, ω_0 being the Debye cutoff frequency. The energy of the transition between $|b\rangle$ and $|a\rangle$ is expected to vary with temperature as

$$\Delta E_{ba}(T)=(C_b-C_a)\left(\frac{kT}{\hbar}\right)^4\int_0^{x_0}\frac{x^3\,\mathrm{d}x}{\mathrm{e}^x-1}. \tag{5.145}$$

This formula, first derived by McCumber and Sturge (1963), was tested against the measured values of the temperature-dependent shifts of the R-lines of Cr^{3+} in ruby and magnesium oxide. These lines shift to lower energies with increasing temperature, and the temperature dependences well accounted for by eqn (5.140). In the case of the R-line in MgO:Cr^{3+} the shift to lower energies in going from 0 K to room temperature is about 20 cm^{-1}.

The temperature-independent term in eqn (5.142) which we write as $\Delta E_b(0)$ is due to zero-point vibrations, and this leads to a temperature-independent line shift $\Delta E_{ba}(0)=\Delta E_b(0)-\Delta E_a(0)$. There is an additional electronic matrix element, $-|\langle b|V_1|b\rangle|^2/\hbar\omega_k$, in $\Delta E_b(0)$ which is not present in the expression for $\Delta E_b(T)$. For the sharp R-line transition of Cr^{3+} in magnesium oxide the value of $\langle b|V_1|b\rangle$ is known from static stress experiments, and we can estimate that this term makes a negligible contribution to the temperature-independent line shift $\Delta E_{ba}(0)$. The shifts $\Delta E_{ba}(T)$ and $\Delta E_{ba}(0)$ are then related through

$$\frac{\Delta E_{ba}(T)}{\Delta E_{ba}(0)}=\langle 2n\rangle \tag{5.146}$$

where $\langle n\rangle$ is the phonon thermal occupancy averaged over all modes. From the measured values of $\Delta E_{ba}(T)$ for the R-line of Cr^{3+} in magnesium oxide we deduce that $\Delta E_{ba}(0)\sim 50\ \mathrm{cm}^{-1}$ for this transition.

The zero-point vibration of the Cr^{3+} ion and its near neighbours in magnesium oxide shift the R-line at absolute zero some 50 cm^{-1} away from the position it would have in the 'stationary' lattice. Chromium has four relatively abundant isotopes with masses expressed in atomic mass units of 50 (4.3%), 52 (83.8%), 53 (9.6%), and 54 (2.4%). Since a lighter isotope should have a larger vibrational amplitude its temperature-independent line shift should be larger than that of heavier isotopes, and the R-lines of the different isotopes are shifted relative to each other (Fig. 5.26(a)).

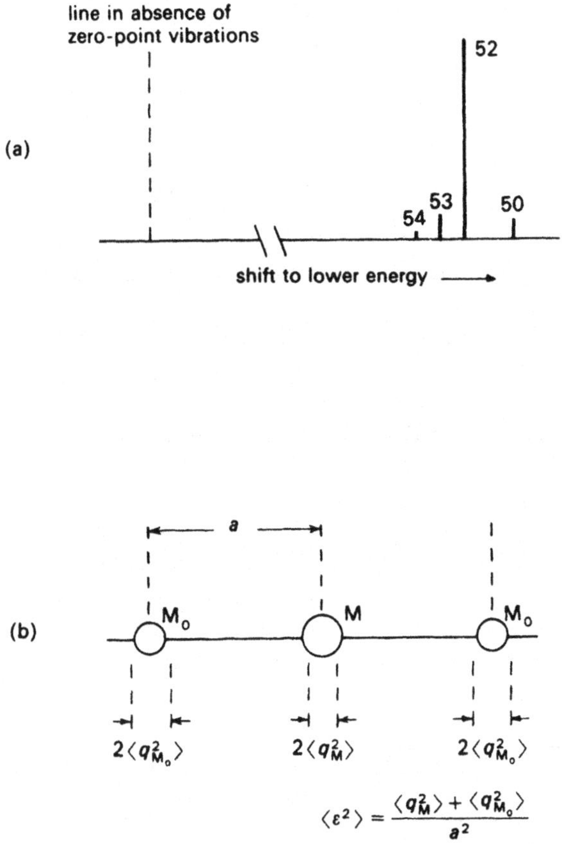

FIG. 5.26. (a) Because of zero-point vibrations the zero-phonon lines from different chromium isotopes are shifted by different amounts. (b) Simple model to derive the expression for the mean-square strain at the optically-active ion.

We can make an order of magnitude estimate of this effect as follows. From eqn (5.140) we see that the line shift, ΔE, is proportional to the mean-square strain, $\langle \varepsilon^2 \rangle$. Hence the zero temperature shift, $\Delta E(0)$, is proportional to $\langle \varepsilon^2 \rangle_0$, the mean-square strain due to zero-point lattice vibrations. Using the fact that the high-frequency modes are the most effective we adopt the simple picture (Fig. 5.26(b)) that

$$\langle \varepsilon^2 \rangle_0 = \frac{\langle q_M^2 \rangle + \langle q_{M_0}^2 \rangle^2}{a^2} \tag{5.147}$$

where $\langle q_M^2 \rangle$ and $\langle q_{M_0}^2 \rangle$ are the mean-square displacements at 0 K of the impurity ion, M, and of the neighbouring ion, M_0, respectively, and a is the average separation between these ions. It can be argued that for impurity ions of mass M in a crystal the mean-square displacement scales as $M^{-\frac{1}{2}}$. Based on this we estimate the differential isotope shift, when the mass M changes by one atomic mass unit, $\delta(\Delta E(0))$, as

$$\frac{\delta(\Delta E(0))}{\Delta E(0)} \simeq \frac{1}{2M} \frac{\langle q_{M_0}^2 \rangle}{\langle q_M^2 \rangle + \langle q_{M_0}^2 \rangle}. \tag{5.148}$$

If we take the $\langle q_{M_0}^2 \rangle \sim \frac{1}{2} \langle q_M^2 \rangle$ (the optically-active ion is the more massive) we obtain $(6M)^{-1}$ for the ratio of differential isotope shift to zero-temperature shift.

In the case of the R-line of $MgO:Cr^{3+}$ we have $\Delta E(0) = 50\ cm^{-1}$, $M \sim 50$, hence we expect a differential isotope shift of around $0.17\ cm^{-1}$. Such a differential isotope shift was observed in the $^2E \rightarrow {}^4A_2$ zero-phonon line of Cr^{3+} ions in magnesium oxide and Al_2O_3 when samples of very low strain were examined at liquid helium temperatures (Imbusch *et al.* 1964). By using narrowband laser techniques to overcome inhomogeneous broadening, Jessop and Szabo (1980) observed the isotope shift in the R_1-line of $Al_2O_3:Cr^{3+}$ with much greater clarity (Fig. 5.27). In similarly high-resolution measurements Pelletier-Allard and Pelletier (1984) succeeded in making accurate measurements of the thermal shift $\Delta E(T)$ and differential isotope shift for two optical lines of Nd^{3+} in lanthanum trichloride. From their thermal shift one can estimate $\Delta E(0) \sim 9\ cm^{-1}$, and using $M = 144$ for neodymium the simple formula predicts a differential isotope shift of $0.010\ cm^{-1}$, close to the measured values, $0.0065\ cm^{-1}$, and $0.0058\ cm^{-1}$, for the two lines.

In this analysis of the temperature-dependent shift and broadening of zero-phonon lines we employed perturbation theory and the effects were calculated to lowest order in the ion–phonon coupling. This approach is only valid for weak coupling, but the results tend to be used in cases of stronger coupling where the calculations may not be strictly applicable. A non-perturbative theory of shift and broadening has been presented by Hsu and Skinner (1984*a*, *b*) which is more generally applicable. Their calculations involve a lengthy diagrammatic analysis but their results for shift and broadening are presented in the form of integrals which can be performed analytically or numerically.

5.8 Non-radiative transitions involving multiphonon emission

Up to this point we have examined only those non-radiative transitions between adjacent levels whose energy separation is within the range of phonon energies. In the direct process such a transition results in the generation of a single phonon. However, non-radiative transitions may also

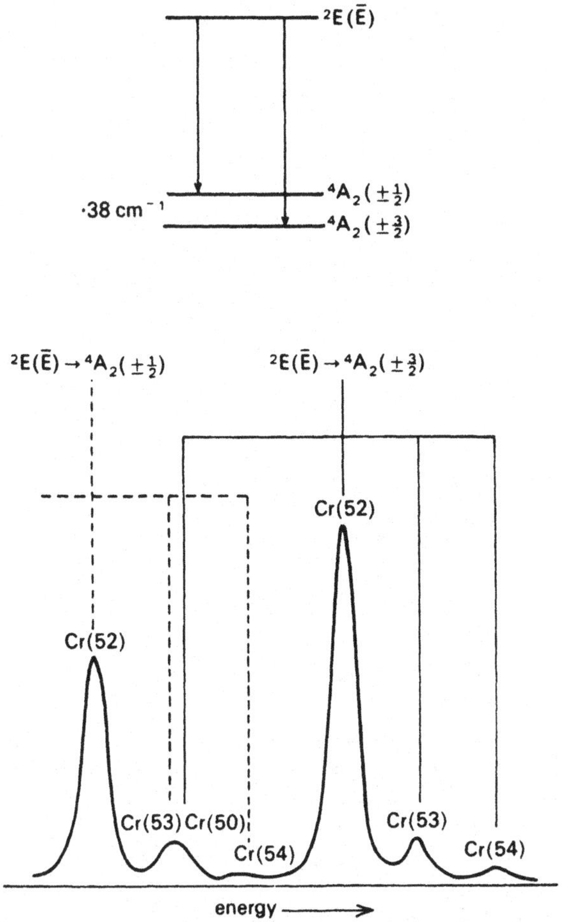

FIG. 5.27. Fine structure in the R_1-line of ruby obtained by resonance fluorescence excitation using a very narrow bandwidth dye laser (after Jessop and Szabo 1980). The two transitions to the split levels of the ground state are well resolved and each transition contains structure due to different chromium isotopes. The measured differential isotope shift of 0.12 cm^{-1} is close to the calculated value.

take place between levels which have a much larger energy separation than the highest-energy phonons of the lattice. Consider for example the simple energy-level diagram for Mn^{2+} ions in an octahedral field (Fig. 5.28). Luminescence is observed from the first excited $^4T_1$ state but not from higher energy states. When a Mn^{2+} ion is raised to one of these higher excited states there are two decay modes available to it: radiative decay to a lower state, W_r, and non-radiative decay to a lower state, W_{nr}. First we examine the non-radiative process in the weak coupling regime. If the energy gap to the next

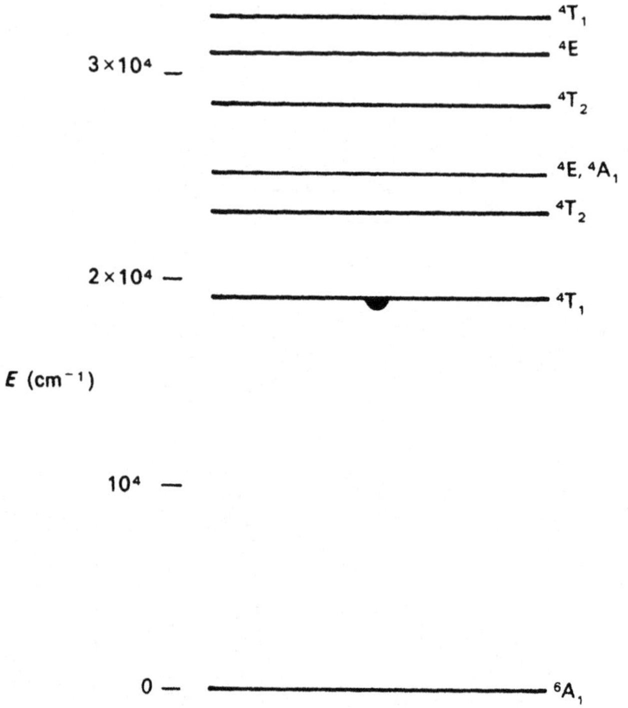

FIG. 5.28. Simple energy level diagram for Mn^{2+} ions in an octahedral crystal field (for example, Mn^{2+} in potassium magnesium fluoride). The pendant semicircle on the first excited level indicates that luminescence occurs from this state. Luminescence does not occur from the higher energy-level states.

lower level is larger than the maximum energy of a phonon, the non-radiative process will involve the generation of several phonons. This is a higher-order process which will involve the use of the electron–phonon coupling to a higher order of perturbation theory. In general the higher the order of perturbation theory required in the calculation of transition rates the smaller is the transition probability. Hence as the energy gap to the next lower level increases the non-radiative decay process involves a larger number of phonons and consequently its transition probability becomes smaller. The radiative decay probability, on the other hand, increases as the cube of the energy separation between initial and final states (eqn 4.74). Hence we anticipate that de-excitation transitions between levels well separated in energy will be mainly radiative, whereas de-excitation transitions between levels separated by a small energy will mainly be non-radiative. These ideas are nicely illustrated by the Mn^{2+} system (Fig. 5.28). The higher excited levels are relatively close together, being separated by not more than about

3000 cm^{-1}, and efficient non-radiative decay processes operate between these levels. A Mn^{2+} ion raised to one of these higher excited levels drops to the next lowest excited level by non-radiative decay. The lowest excited level, however, is roughly 18 000 cm^{-1} above the ground state, and non-radiative decay across this gap would involve the emission of some 30–40 phonons. Such a very high-order process is not efficient. As a result the $^4T_1$ level decays radiatively, despite the fact that in a strictly static environment this transition is both spin- and parity-forbidden. (The observed radiative decay time for this $^4T_1 \rightarrow {}^6A_1$ transition can be $\sim$50 ms.) Dopant rare-earth ions with their abundance of luminescent levels and wide range of energy gaps give very clear evidence of the dependence of the non-radiative transitions rate on the size of the energy gap to the next lower level. A glance ahead at the Dieke diagram (Fig. 8.1), which shows the energy levels of the trivalent rare-earth ions in lanthanum trichloride, shows luminescent levels only where the gap to the next lower level is reasonably large.

Since non-radiative processes compete with the radiative processes they may greatly affect the luminescence efficiency. In the non-radiative process the change of electronic state to one of lower energy is accompanied by the release of vibrational energy. To effect a non-radiative transition the electron–lattice interaction which couples the two electronic states must allow the generation of phonons. It is convenient to discuss separately the effect of non-radiative transitions on the luminescence transitions on F-centres and on dopant ions.

5.8.1 Non-radiative transitions on F*-centres*

The general features of the optical transitions of F-centres are analysed in detail in Chapters 2 and 7. They are also discussed in terms of the configurational coordinate model in Section 5.4. This model, shown again in Fig. 5.29, is a useful starting point for discussion of non-radiative transitions. In this figure X is the 'crossover' point between the a and b parabolae. At low temperatures the absorption is centred on the AB transition ($\psi_a\chi_a(0) \rightarrow \psi_b\chi_b(m)$) which terminates in a higher vibrational state. For a small configurational coordinate offset (Fig. 5.29(a)) the point X is higher in energy than point B. Excitation of the centre to the $\psi_b\chi_b(m)$ state (point B) is followed by rapid relaxation involving phonon emission until the state $\psi_b\chi_b(0)$ is reached (point C). Radiative emission then centres on transition CD.

In Fig. 5.29(b) the configurational displacement is larger and the point X is lower in energy than the point B. Once raised to point B, the centre again de-excites downwards through the $\chi_b(m)$ states but only until the $\psi_b\chi_b(m')$ state is reached which coincides with the point X, and which is degenerate in energy with a very highly excited vibrational level of the ground electronic state, $\psi_a\chi_a(n')$. The wavefunctions $\chi_a(n')$ and $\chi_b(m')$ have maximum amplitudes at the same value of the configurational coordinate, Q_x, and we expect strong mixing of these wavefunctions to occur. Thus the excited centre passing

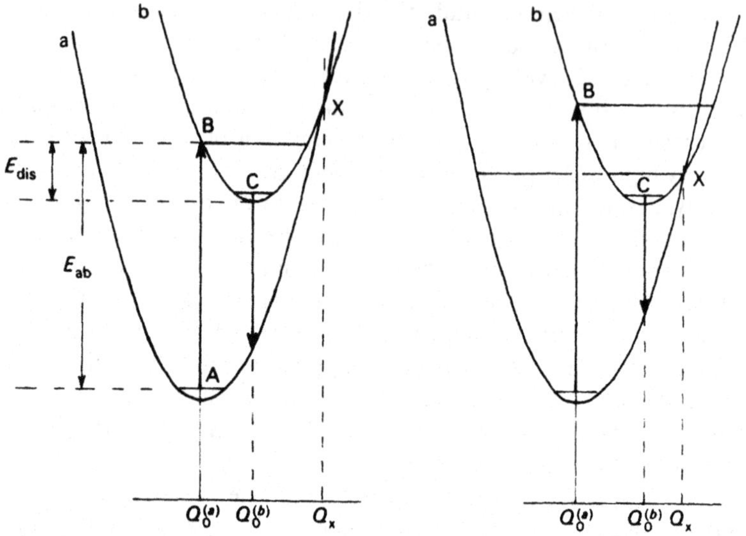

FIG. 5.29. Configurational coordinate diagrams for (a) the case where the crossover point (X) of parabolae *a* and *b* is at a higher energy than the point of peak absorption (B), and (b) the case where X is below B. Except for the configurational coordinate shift the *a* and *b* parabolae are assumed similar.

downwards from the point B reaches the level corresponding to the crossing point X and becomes strongly coupled to the higher vibrational states of the ground electronic state $\psi_a \chi_a(n')$. Since n' is a very large number, non-radiative relaxation from $\chi_a(n')$ to $\chi_a(n'-1)$ is much stronger than from $\chi_b(m')$ to $\chi_b(m'-1)$. Hence the system in the state corresponding to point X relaxes to a higher vibrational state of the electronic ground state, and then will stay in this $\chi_a(n)$ set, decaying by multiphonon emission to the ground state. No luminescence occurs. A simple criterion for determining whether or not there is luminescent de-excitation (Dexter *et al.* 1955), is that the crossover point X should be above the point B.

This condition can be expressed in terms of measurable parameters with the aid of Fig. 5.29, in which the vibrational frequencies of ground and excited states are assumed equal. Points B and X have the same energy when the point $Q_0^{(b)}$ is midway between $Q_0^{(a)}$ and Q_x: i.e. $(Q_x - Q_0^{(a)}) = 2(Q_0^{(b)} - Q_0^{(a)})$. From the figure we observe

$$E_{\text{dis}} = E_B - E_0^{(b)} = \tfrac{1}{2} M\omega^2 (Q_0^{(b)} - Q_0^{(a)})^2 \quad (5.149)$$

from which we obtain

$$E_{ab} = \tfrac{1}{2} M\omega^2 (Q_x - Q_0^{(a)})^2 - \tfrac{1}{2}\hbar\omega$$
$$\simeq \tfrac{1}{2} M\omega^2 (Q_x - Q_0^{(a)})^2 \simeq 4E_{\text{dis}}. \quad (5.150)$$

When the point X is above B luminescence occurs. In this case point $Q_0^{(b)}$ is closer to $Q_0^{(a)}$ than to Q_x, and $E_{ab} > 4E_{\text{dis}}$. Hence the criterion for luminescence to occur is that

$$\Lambda = \frac{E_{\text{dis}}}{E_{ab}} < \frac{1}{4}. \tag{5.151}$$

In the case of the F-centre E_{dis} is, to a good approximation, half the Stokes shift and E_{ab} is the peak of the absorption transition. In cases where luminescence is not observed E_{dis} can be obtained from the equation $E_{\text{dis}} \simeq S\hbar\omega$ where the Huang–Rhys parameter, S, can be determined from the shape of the absorption band. This easy-to-apply criterion seems to reflect very accurately the luminescence behaviour of F-centres (Bartram and Stoneham 1975, 1985). In fact, for all the lithium halides and for sodium bromide and sodium chloride $\Lambda > 0.25$, so that luminescence is not observed.

5.8.2 Non-radiative transitions on dopant ions

Systematic studies of non-radiative decay of rare-earth ions were carried out independently by Moos and collaborators (Moos 1970) and by Weber (1973). Moos (1970) measured the quantum efficiency of a level $[W_r(W_r + W_{nr})^{-1}]$ and the decay time, τ, of the level, where $1/\tau = W_r + W_{nr}$ (Fig. 5.30). From these two measurements the radiative probability (W_r) and the non-radiative probability (W_{nr}) can separately be found. Weber calculated W_r from the measured absorption data. The difference between the calculated W_r and the observed decay rate $1/\tau$ was attributed to non-radiative decay. Note that each method requires detection of some luminescence from an excited state in order to obtain a value for W_{nr} from that state.

Figure 5.31 shows the measured values of W_r from various levels of different trivalent rare-earth ions in five different host materials (Riseberg and Weber 1975). The values of W_{nr} are plotted against the energy gap to the next lower level. The Moos–Weber analysis indicates that in each host material there is a

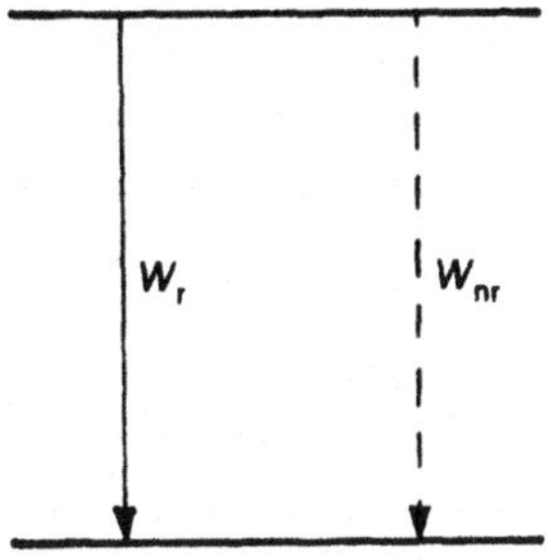

FIG. 5.30. Radiative and non-radiative decay processes can occur from an excited state with transition probabilities W_r and W_{nr}, respectively.

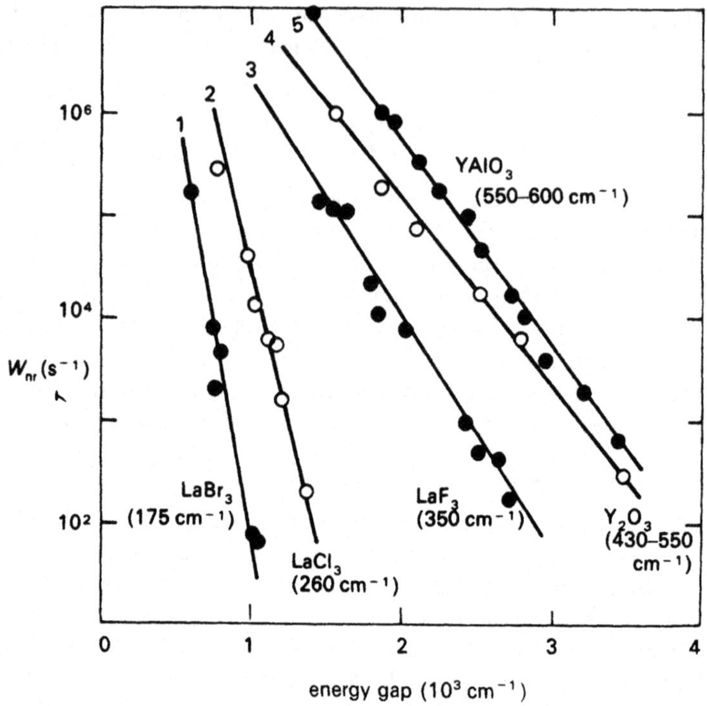

FIG. 5.31. Measured values of W_{nr} form various levels of different rare-earth ions in five different host materials (Riseberg and Weber 1975). These are plotted against energy gap to the next lower level. For each host material the effective phonon energies are given.

clear exponential dependence on energy gap; this is clear from Fig. 5.31 which shows that values of log W_{nr} for each crystal fall on a straight line. These lines, numbered 1 to 5, characterize the non-radiative transitions in the five host materials. The 'effective' phonon energies indicated for the materials in Fig. 5.31 are those phonons expected to participate in non-radiative decay and are assumed to be the highest-energy phonons which have a reasonable density of phonon states for that material. Dividing the energy gap by the effective phonon energy gives the number of phonons (p) involved in the transitions. The straight lines in Fig. 5.31 are plotted against p in Fig. 5.32 for these five host crystals. (In this plot the straight lines are continued to higher W_{nr} values, beyond the region of measurements, to show that the lines appear to converge at $p \approx 3$. The shaded area in the figure represents the range of typical values of radiative decay rate for trivalent rare-earth ions. These plots show that for rare-earth ions in lanthanum bromide and lanthanum chloride (lines 1 and 2) the breakeven point between radiative and non-radiative processes

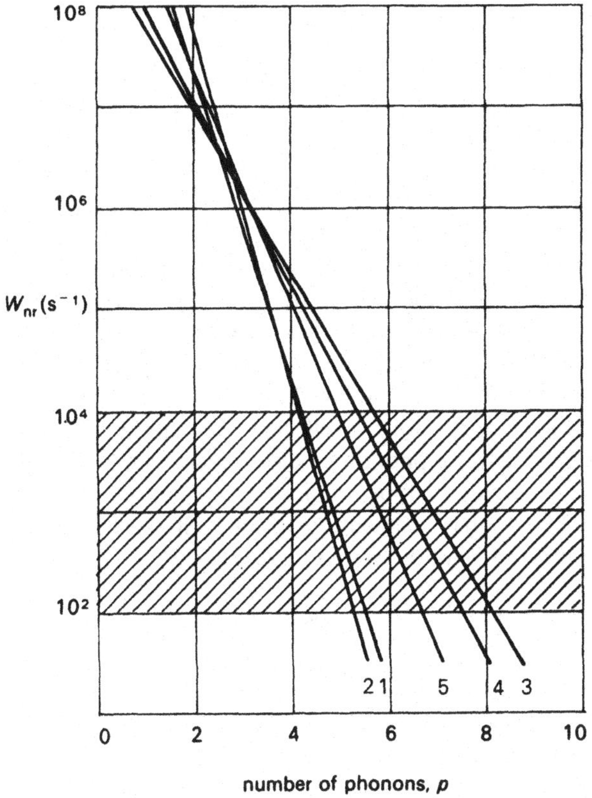

FIG. 5.32. Plot of the straight lines of Fig. 5.31 against p, the number of phonons involved in the non-radiative transition. The lines are extrapolated to larger W_{nr} values beyond the range of measurements in order to show the convergence of the lines for p between 2 and 3. We observe from these plots that W_{nr} decreases by a factor of 10–40 when p increases by unity. The shaded area represents the range of typical values of the radiative decay rate for the dopant trivalent rare-earth ions in crystal hosts.

occurs at transitions involving about five phonons. Non-radiative decay involving the generation of more than five phonons is weaker than the radiative process. For these host materials the non-radiative process dominates for transitions across energy gaps of fewer than five phonons. The breakeven point for the other materials is at around six phonons. The dependence of the measured W_{nr} values on energy gap to the next lowest level (ΔE) and on a number of effective phonons (p) is written

$$W_{nr} = A\exp(-\alpha\Delta E) = A\exp(-\beta p). \qquad (5.152)$$

It is instructive to test these experimental results theoretically. For rare-earth ions with weak coupling to the lattice it is reasonable to use the electron–phonon coupling (eqn 5.92) which we write in scalar form

$$V_c = V_1\varepsilon + V_2\varepsilon^2 \ldots + V_s\varepsilon^s + \ldots. \tag{5.153}$$

The strain, ε, is linear in the phonon creation and annihilation operators. Further, for the case of weak coupling the set of lattice eigenstates is the same irrespective of the state of the ion. Consider a process changing the electronic state from ψ_b to ψ_a with the release of p phonons. Since p phonons are created the operator responsible for the transition must contain ε^p. This can occur in a number of ways: through the first term, $V_1\varepsilon$, operating by pth-order perturbation theory; through the pth term, $V_p\varepsilon^p$, operating by the first-order perturbation theory; or processes intermediate between these two which effectively involves an ε^p operator. For large p values the calculations of this transition probability is very involved.

The creation of p phonons is effected through the ε^p operator which contains the product of p phonon creation operators, $a_1^\dagger\, a_2^\dagger\, a_3^\dagger \ldots a_p^\dagger$ where the subscripts represent the types of phonons created. (These may all be of the same type.) The matrix element governing this transition contains the following product of phonon matrix elements:

$$\langle n_1+1|a_1^\dagger|n_1\rangle\langle n_2+1|a_2^\dagger|n_2\rangle \ldots \langle n_p+1|a_p^\dagger|n_p\rangle$$
$$= (1+n_1)^{\frac{1}{2}}(1+n_2)^{\frac{1}{2}} \ldots (1+n_p)^{\frac{1}{2}}. \tag{5.154}$$

Assuming that all the p phonons involved in the transition are of approximately the same energy then the square of this matrix element is written as $(1+n_{\rm eff})^p$, where $n_{\rm eff}$ is the occupancy of the effective phonon modes:

$$n_{\rm eff} = (\exp(\hbar\omega_{\rm eff}/kT) - 1)^{-1}. \tag{5.155}$$

This gives the temperature dependence of the non-radiative relaxation rate

$$W_{\rm nr}^{(p)}(T) = W_{\rm nr}^{(p)}(0)(1+n_{\rm eff})^p. \tag{5.156}$$

Figure 5.33 is a plot of $(1+n_{\rm eff})^p$ for a process across a gap of 1200 cm^{-1} using three different processes: four phonons each of energy 300 cm^{-1}, five phonons each of energy 240 cm^{-1}, six phonons each of energy 200 cm^{-1}, and seven phonons each of energy 171 cm^{-1}. The temperature dependences are quite distinct so that one should be able to infer from the observed temperature dependence of $W_{\rm nr}$ the number of phonons involved (p) and hence the energy of the effective phonons ($\Delta E/p$). In general the variation with temperature of non-radiative transitions in rare-earth ions is accurately described by eqn (5.156). The calculation of the absolute value of $W_{\rm nr}$ is, however, a very difficult task.

We have seen how many different perturbation terms can contribute to the decay process. In that regard the analysis of Hagston and Lowther (1973) is of interest. They showed on the basis of a point charge model that the squares of

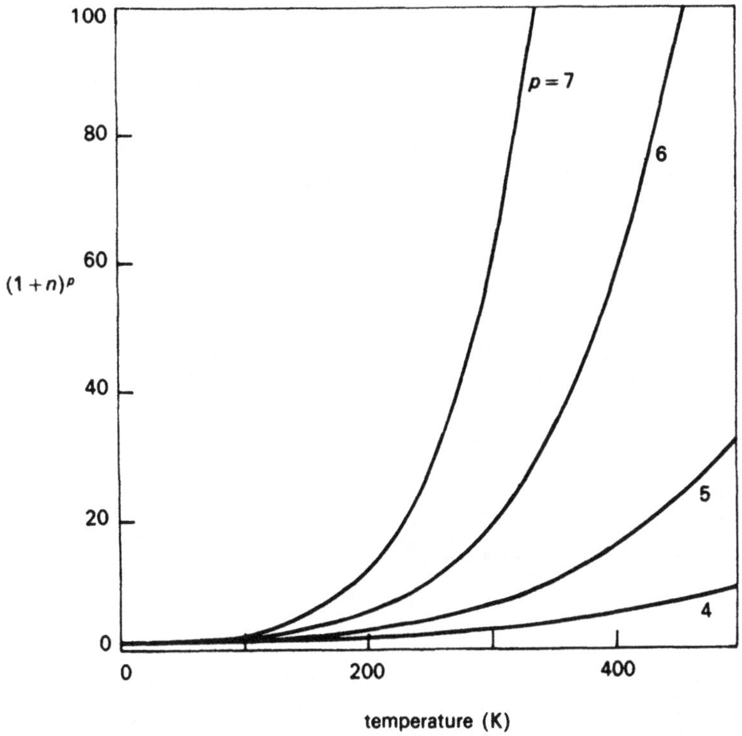

FIG. 5.33. Plot of $(1+n)^p$ for the case of a gap of 1200 cm^{-1} using four different decay modes: four phonons each of 300 cm^{-1}, five phonons each of 240 cm^{-1}, six phonons each of 200 cm^{-1}, and seven phonons each of 171 cm^{-1}.

the electronic matrix elements of the various V_s terms of the electron–phonon coupling are increasing in size as s increases. Their result is

$$2 \leqslant \left| \frac{\langle a|V_r|b\rangle}{\langle a|V_{r-1}|b\rangle} \right|^2 \leqslant 4 \tag{5.157}$$

showing that the ratio is independent of r. Hagston and Lowther compared the contribution to the transition probability arising from the pth term ($V_p \varepsilon^p$) in first-order perturbation theory with the contribution of the first term ($V_1 \varepsilon$) in pth-order perturbation theory and showed that the contribution from the $V_p \varepsilon^p$ term is the larger. Also, contributions from intermediate terms are collectively smaller than that from the $V_p \varepsilon^p$ term. Hence we can adopt the following formula for the non-radiative transition probability from $|b, m\rangle$ to

$|a, n\rangle$ as

$$W_{\rm nr}^{(p)} = \frac{2\pi}{\hbar}|\langle a|V_p|b\rangle|^2\,|\langle n|\varepsilon^p|m\rangle|^2\,\delta\!\left(E_b - E_a - \sum_{i=1}^{p}\hbar\omega_i\right) \tag{5.158}$$

where the phonon state n differs from m in having an additional p phonons.

We can estimate the ratio, R, of $W_{\rm nr}^{(p)}$ to $W_{\rm nr}^{(p-1)}$ as

$$R = \frac{W_{\rm nr}^{(p)}}{W_{\rm nr}^{(p-1)}} \simeq \left|\frac{\langle a|V_p|b\rangle}{\langle a|V_{p-1}|b\rangle}\right|^2 \langle\varepsilon^2\rangle. \tag{5.159}$$

$\langle\varepsilon^2\rangle$ can be evaluated using a Debye density of states, $\rho(\omega) = 3V\omega^2/2\pi^2 v^3$, and we obtain at low temperatures,

$$\langle\varepsilon^2\rangle \simeq \frac{\hbar\omega_0^4}{50\rho v^5}. \tag{5.160}$$

Taking $\omega_0 = 10^{14}\,\mathrm{s}^{-1}$ as the Debye cutoff frequency, $v \simeq 5\times 10^3\,\mathrm{ms}^{-1}$, and the mass density $\rho = 4\times 10^3\,\mathrm{kg\,m}^{-3}$ we estimate $\langle\varepsilon^2\rangle \simeq 0.016$. Taking the ratio of the matrix elements $|\langle V_p\rangle/\langle V_{p-1}\rangle|^2 \simeq 3$ gives the result

$$R \simeq 0.05 \tag{5.161}$$

which is in reasonable agreement with the experimental measurements plotted in Fig. 5.32. An important point here is that this ratio is independent of p. Hence we can write

$$\frac{W_{\rm nr}^{(p)}}{W_{\rm nr}^{(0)}} = R^p = \exp(-p\ln(1/R)) \tag{5.162}$$

which gives the exponential gap law, since $p = \Delta E/\hbar\omega$. Writing it in the form $\exp(-\beta p)$ we find $\beta = \ln(1/R) \approx 3$.

We return to Fig. 5.32 which plots the measured values of $W_{\rm nr}$ against p, and we observe that by extending the linear plots to lower p values, beyond the region of measurement, there is a tendency for the plots to converge. This is seen much more clearly in Fig. 5.34 which shows the extended plots of $W_{\rm nr}$ against p for a large number of rare-earth ions in a large number of host materials (Schuurmans and Van Dijk 1984). The point of convergence is close to $p = 2$. This would seem to indicate that one can write a general formula for $W_{\rm nr}$, applicable to all rare-earth ions in any host material, of the form $W_{\rm nr} = A\exp[-\alpha(\Delta E - 2\hbar\omega)]$ where A varies only over an order of magnitude for a wide range of crystals and glasses, and, with one or two exceptions, α has a value of $4.5(\pm 1)\times 10^3$ cm. This can also be written in the form $A\exp[-\beta(p-2)]$ and β is in the range 1–5 (Schuurmans and Van Dijk 1984).

Since for transition metal ions the interaction between the ion and the crystalline environment is not weak we cannot use the electron–phonon coupling term V_c as we did for the rare-earth ions. Instead we return to the

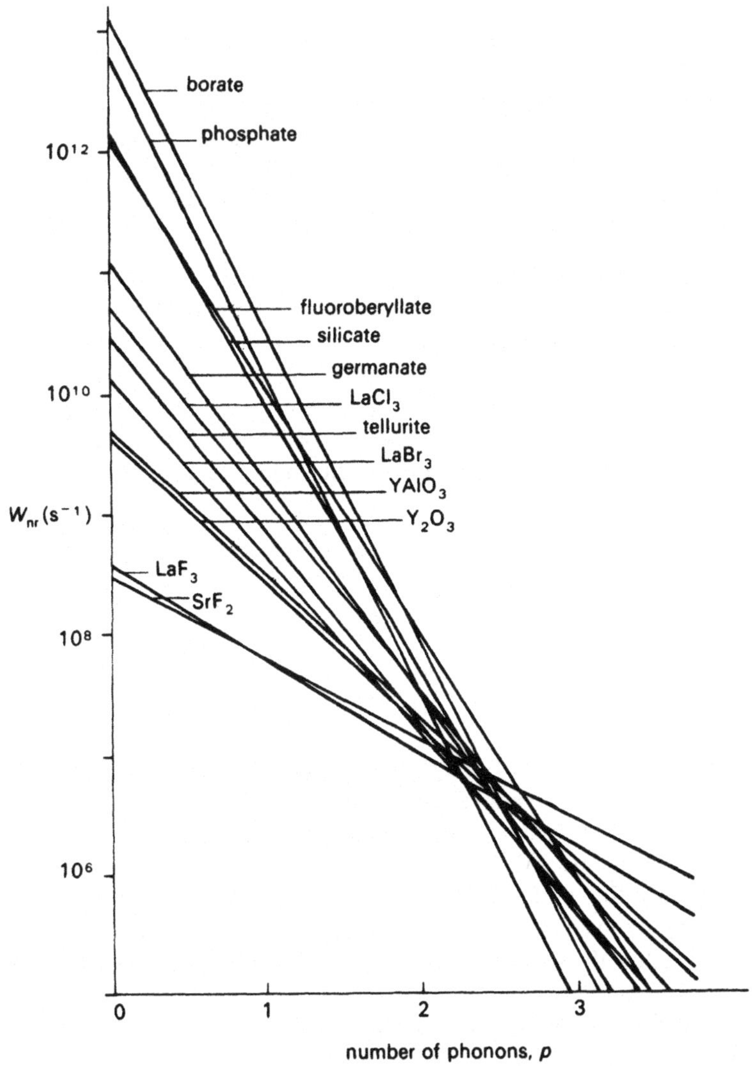

FIG. 5.34. The exponential dependences of W_{nr} from various levels of different rare-earth ions in a number of different crystalline and glass host materials. The experimental measurements were made in the region of p values between 3 and 5, the lines are drawn through the experimental points, and the lines are continued backwards to $p=0$. (After Schuurmans and Van Dijk 1984.)

non-adiabatic terms omitted in the Born–Oppenheimer approximation. The non-adiabatic Hamiltonian written in terms of normal coordinates is

$$\mathscr{H}_{\mathrm{NA}}\psi_a(\mathbf{r}, Q_k)\chi_a(Q_k)=\sum_k \frac{-\hbar^2}{M}\left(\frac{\partial\psi_a}{\partial Q_k}\right)\left(\frac{\partial\chi_a}{\partial Q_k}\right). \tag{5.105}$$

The probability of a non-radiative transition between $\psi_b\chi_b(m)$ and $\psi_a\chi_a(n)$ is

$$W_{\mathrm{nr}}=\frac{2\pi}{\hbar}\sum_{\chi_a(n)}\mathrm{Av}_m\left|\sum_k\frac{-\hbar^2}{M}\left\langle\psi_a\left|\frac{\partial}{\partial Q_k}\right|\psi_b\right\rangle\left\langle\chi_a(n)\left|\frac{\partial}{\partial Q_k}\right|\chi_b(m)\right\rangle\right|^2\delta(E_{a,n}-E_{b,m}) \tag{5.163}$$

where Av_m denotes a thermal average over all initial states $\chi_b(m)$ and we sum over all final states $\chi_a(n)$ subject to energy conservation. We write $\chi_a(n)=\Pi_k|n_k\rangle$ and $\chi_b(m)=\Pi_k|m_k\rangle$ where $|n_k\rangle$ and $|m_k\rangle$ are, respectively, linear harmonic oscillator states associated with the ground and excited electronic states and with n_k and m_k phonons of mode k in each state. The parabolae for $|n_k\rangle$ and $|m_k\rangle$ are centred on different values of the configurational coordinate Q_k. The non-radiative transition probability now becomes

$$W_{\mathrm{nr}}=\frac{2\pi}{\hbar}\sum_n\mathrm{Av}_m\left|\sum_k R_{ab}^{(k)}\left\langle n_k\left|\frac{\partial}{\partial Q_k}\right|m_k\right\rangle\prod_{\substack{k'\neq k\\ k''\neq k}}\langle n_{k'}|m_{k''}\rangle\right|^2\delta(E_{a,n}-E_{b,m}) \tag{5.164}$$

where

$$R_{ab}^{(k)}=-\frac{\hbar^2}{M}\left\langle\psi_a\left|\frac{\partial}{\partial Q_k}\right|\psi_b\right\rangle.$$

The evaluation of eqn (5.164) is quite complicated and has been carried out by a number of workers using a variety of mathematical techniques. These calculations are discussed by Auzel (1978). One interpretation of eqn (5.164) is that mode k, which is involved in the electronic matrix element, permits the electronic transition and so is called a 'promoting' mode. The other modes k', k'' which are involved in the overlap integrals are called the 'accepting' modes in that they allow the energy to be accepted as a higher vibrational state of the ground electronic state. This formulation of the non-radiative decay process has a number of points of similarity with the formulation of the process of radiative emission into vibrational sidebands, as was demonstrated in the work of Miyakawa and Dexter (1970) and stressed by Auzel (1978).

We now simplify matters by adopting a single configurational coordinate model (only one mode k) with identical frequencies, ω, in the ground and excited electronic states. If in addition we restrict our attention to $T=0$, eqn (5.164) becomes

$$W_{\mathrm{nr}}=\frac{2\pi}{\hbar}R_{ab}^2\left|\left\langle p\left|\frac{\partial}{\partial Q}\right|0\right\rangle\right|^2 \tag{5.165}$$

where the energy gap between the zero-vibrational levels of ψ_a and ψ_b is $\Delta E = p\hbar\omega$. Using eqn (5.106) we obtain

$$W_{nr} = \frac{\pi M\omega}{\hbar^2} R_{ab}^2 |\langle p|1\rangle|^2. \tag{5.166}$$

From eqn (5.69) the overlap integral squared is

$$|\langle p|1\rangle|^2 = \frac{\exp(-S)S^{p-1}}{p!} L_1^{p-1}(S) \simeq \frac{\exp(-S)S^{p-1}}{(p-1)!} \tag{5.167}$$

where we have used $L_1^{p-1}(S) = p - S \simeq p$ since $S \ll p$ for the situations of experimental interest to us. We note that this is also the value of the zero-temperature Franck–Condon factor, $F_{p-1}(0) = |\langle p-1|0\rangle|^2$. Using Stirling's approximation, $m! = m^m(2\pi m)^{\frac{1}{2}}\exp(-m)$, the non-radiative decay probability can be written as

$$W_{nr} \simeq \left(\frac{\pi}{2(p-1)}\right)^{\frac{1}{2}} \frac{M\omega}{\hbar^2} R_{ab}^2 \exp(-S)\exp(-\beta(p-1)) \tag{5.168}$$

where $\beta = (\ln(p/S) - 1)$. We observe that this formula contains an approximate exponential energy gap law. For rare-earth ions with $S \simeq 0.05$ and $p \simeq 5$ we obtain $\beta \simeq 3.5$ which is close to the observed values. Since it contains $\ln p$ we see that β is not a true constant. Further, the factor in front of the exponential also contains p. Schuurmans and Van Dijk (1984) argue that for the range of values of S and p where measurements of W_{nr} are made for rare-earth ions the Franck–Condon factor (eqn 5.167) is more properly represented by $A\exp[-\beta(p-1-\delta)]$ where A is a constant independent of S and p, and δ is close to unity. This is consistent with the observation made on the data in Fig. 5.34. The evaluation of eqns (5.163) and (5.164) for general values of S and p leads to much more complicated expressions. Huang and Rhys (1950) derive an expression involving a sum of Franck–Condon factors, F_{p+2}, F_{p+1}, F_p, F_{p-1}, F_{p-2}. For small S and low temperatures the F_{p-2} term would seem to be dominant which leads to the exponential gap law of the form $\exp[-\beta(p-2)]$. Miyakawa and Dexter's (1970) calculation of W_{nr} for small S and $T=0$ gives

$$W_{nr} = R^2\left(1 - \frac{p}{S}\right)^2 \frac{S^p}{p!} \simeq R^2 \frac{e^2}{(2\pi p)^{\frac{1}{2}}} \exp(-\beta(p-2)) \tag{5.169}$$

where R is an electronic matrix element, $\beta = \ln((p/S) - 1)$, and we have used Stirling's approximation to obtain the exponential factor.

For transition metal ions the value of S can vary by more than an order of magnitude about unity. Because of the strong coupling the non-radiative process competes with the radiative process over much larger gaps (to $10\,000\ \mathrm{cm}^{-1}$ in some cases) and consequently the measured non-radiative process generally involves large values of p (>20 in some cases). In the single configurational coordinate model with the same value of the frequency in the ground and excited electronic states we can take the general formula for W_{nr} as R^2F_p where R is an electronic matrix element and F_p is the Franck–Condon factor. We take $p = \Delta E/\hbar\omega$ where $\hbar\omega$ is the energy of the 'effective' phonon involved in the non-radiative process. For general temperatures the Franck–Condon factor becomes (Rebane 1970)

$$F_p(T) = \exp(-S(1+2m))\left(\frac{1+m}{m}\right)^{p/2} I_p(2S\sqrt{m(m+1)}) \tag{5.170}$$

where $I_p(x)$ is the modified Bessel function and m is the mean thermal occupancy in the vibrational mode

$$m = \frac{1}{\exp(\hbar\omega/kT)-1}. \tag{5.171}$$

We therefore have for the general non-radiative transition rate

$$W_{nr} = R^2\exp(-S(1+2m))\left(\frac{1+m}{m}\right)^{p/2} I_p(2S\sqrt{m(m+1)}). \tag{5.172}$$

The temperature dependence is complicated since m appears in the various factors of the formula. We can rewrite this formula using a series expansion for the Bessel function:

$$\begin{aligned} W_{nr} &= R^2\exp(-S(1+2m))\frac{S^p}{p!}\sum_{l=0}^{\infty}\frac{p!}{l!(l+p)!}(1+m)^{l+p}(mS^2)^l \\ &= R^2\frac{\exp(-S(1+2m))S^p}{p!}(1+m)^p\left(1+\frac{S^2m(m+1)}{p+1}+\ldots\right). \end{aligned} \tag{5.173}$$

For small S and low temperatures the temperature dependence reduces to $(1+m)^p$ which was already found for rare-earth ions. In addition to the more complicated temperature dependence for transition metal ions, the energy separation between electronic states a and b may vary with temperature and this can change the order, p, of the multiphonon process. This can strongly affect the non-radiative decay rate through the $p!$ term in the denominator.

For his study of the non-radiative decay from the $^4T_2$ excited state of Co^{2+} in potassium magnesium fluoride, Sturge (1973) modified eqn (5.172) to take account of the continuous nature of the phonon spectrum. He was able to account for the temperature dependence of the observed non-radiative decay rate over nearly four orders of magnitude (Fig. 5.35), but his attempt to calculate the absolute value of the rate using the harmonic approximation (which is the basis of our treatment) was unsuccessful. He pointed out that the

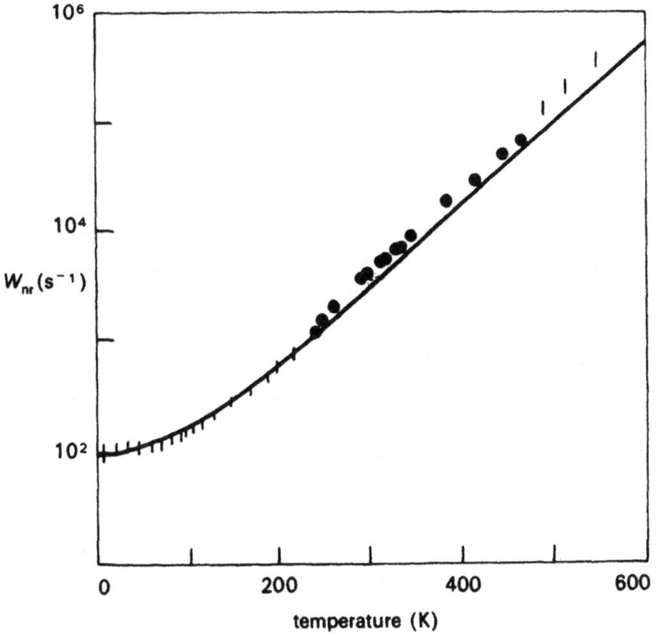

FIG. 5.35. Temperature dependence of the non-radiative decay rate from the $^4T_2$ state of Co^{2+} in $KMgF_3$. The curve shows the theoretical calculations, the points and vertical lines show the experimental points. (After Sturge 1973.)

value of the overlap integral which appears in the formula for W_{nr} can be drastically altered if anharmonicities are present.

In a series of papers Struck and Fonger (1970, 1975, 1976) developed the single configurational coordinate model in the harmonic approximation in a computational form which is very suitable for testing the dependence of the theoretical estimates on the values of the various parameters of the model. They and other workers have used these model calculations to analyse experimental measurements of non-radiative decay rates and quantum efficiencies, for example, the luminescence of Cr^{3+} and Eu^{3+} (Fonger and Struck 1970; Struck and Fonger 1976).

6
Experimental techniques

THIS book is concerned with the optical properties of inorganic solids which have a measurable forbidden energy gap between the valence and conduction bands. The bandgap may be small, as in some semiconductors, or large, as in materials such as diamond and aluminium oxide. Both impurity ions and defects may have energy levels in the gap, and transitions may be excited between these gap levels or between band states and the gap levels. Many experimental techniques have been used to investigate such transitions, those of longest standing being optical absorption spectroscopy and photoluminescence spectroscopy. In crystalline materials absorption and luminescence transitions may be strongly polarized due to the electronic centre occupying a low-symmetry site in the solid or to the effects of an external perturbation (e.g. magnetic field, electric field, or uniaxial stress). Analysis of the luminescence decay after pulsed excitation may indicate the dipole nature of the radiative process (Section 2), and the occurrence of dynamic processes such as excitation transfer between centres (Section 10.1).

In the past four decades there have been many developments in solid state optical spectroscopy. Measurements of the Stark and Zeeman effects and the stress splitting of zero-phonon lines came to the fore in the 1960s. Almost coincidentally modulation spectroscopy gave new insights into the band structure subtleties of metals, semiconductors, and insulators. Optically detected spin resonance gave a new dimension to the understanding of interactions in excited electronic configurations. However, the most dramatic developments awaited the invention of the tunable dye laser, which has led to measurements with unprecedentedly high resolution in both the frequency and time domains. The underlying principles of such experimental techniques are surveyed in this chapter.

6.1 Optical absorption spectroscopy

In optical absorption spectroscopy, electromagnetic radiation in the near ultraviolet, visible, or near infrared regions is used to excite transitions between the electronic states. Experimental measurements are presented as graphs of absorbed intensity versus photon energy ($E = h\nu$), frequency (ν) or wavelength (λ). Whereas atoms in a low-pressure gas discharge exhibit very sharp lines (Corney 1977), electronic centres in a solid display a variety of different bandshapes as described in Chapter 5, so that in general the absorption intensity is a strong function of the photon wavelength (or energy). Optical spectroscopists may quote bandpeaks and widths in energy units E (the energy of the optical photon in eV), in wavenumbers $\bar{\nu}$ (the reciprocal of the wavelength in cm^{-1}), in frequency units (ν or ω), or wavelength (in nanometres (nm) or micrometres (μm)). The following approximate relationships exist among these quantities: $1\ cm^{-1} = 1.24 \times 10^{-4}$ eV; $1\ eV = 8066\ cm^{-1}$; and $E(eV) = 1.24/\lambda(\mu m)$.

6.1.1 *Experimental considerations*

Typically one aims to measure $\alpha(\nu)$ over the wavelength range 185–3000 nm. The first essential is a broadband source; none exist which cover the entire range. In commercial spectrophotometers, deuterium, hydrogen, xenon, and tungsten lamps are commonly used. Their outputs cover different spectral regions. One may also use tunable dye lasers; each dye is tunable over a small spectral region. Although the most desirable experimental format plots $\alpha(\nu, \varepsilon)$ as a function of ν (or $\bar{\nu}$), this is only one of the options available in commercial spectrophotometers. Very often the spectrometer output is given in terms of the specimen transmission, $T = I(\nu, \hat{\varepsilon})/I_0(\nu, \hat{\varepsilon})$ expressed as a percentage, or the *optical density*, $OD = \log_{10}(1/T)$. These experimental parameters are related to the absorption coefficient $\alpha(\nu, \varepsilon)$ by

$$OD = \log_{10}(1/T) = \alpha(\nu, \hat{\varepsilon})l/2.303. \tag{6.1}$$

In the literature optical density is often referred to as the *absorbance*. (We recall that the relationship between the transmitted intensity and the absorption coefficient was derived in Chapter 4. It was shown there that α is a strong function of radiation frequency ν, and sometimes polarization $\hat{\varepsilon}$, hence the designation $\alpha(\nu, \hat{\varepsilon})$.)

Accurate measurements of these quantities are best made using a double-beam spectrophotometer (Fig. 6.1). The exciting beam, after passage through a grating monochromator, is divided into sample and reference channels by a beam-splitting chopper. After recombination at the phototube, phase-sensitive detection and amplification is used to obtain a signal suitable for output to a pen recorder or for further signal processing. Chopping at a pre-selected frequency permits narrowband amplification of the detected signal. Thus any noise components in the signal are limited to a narrow band centred at the chopping frequency. The choice of single-beam versus double-beam operation is a compromise of sensitivity versus resolution. In absorption measurements the sensitivity is of the order 10^{14} centres cm^{-3}; the accuracy of such a measurement is much enhanced in the double-beam spectrophotometer. However, ensuring a high light throughput in both sample and reference channels requires that a low-resolution monochromator be used. Hence very sharp lines, $\Delta\lambda < 1.0$ nm, are not usually resolved by such methods. Single-beam instruments use a grating monochromator of focal length *c.* 1 m and

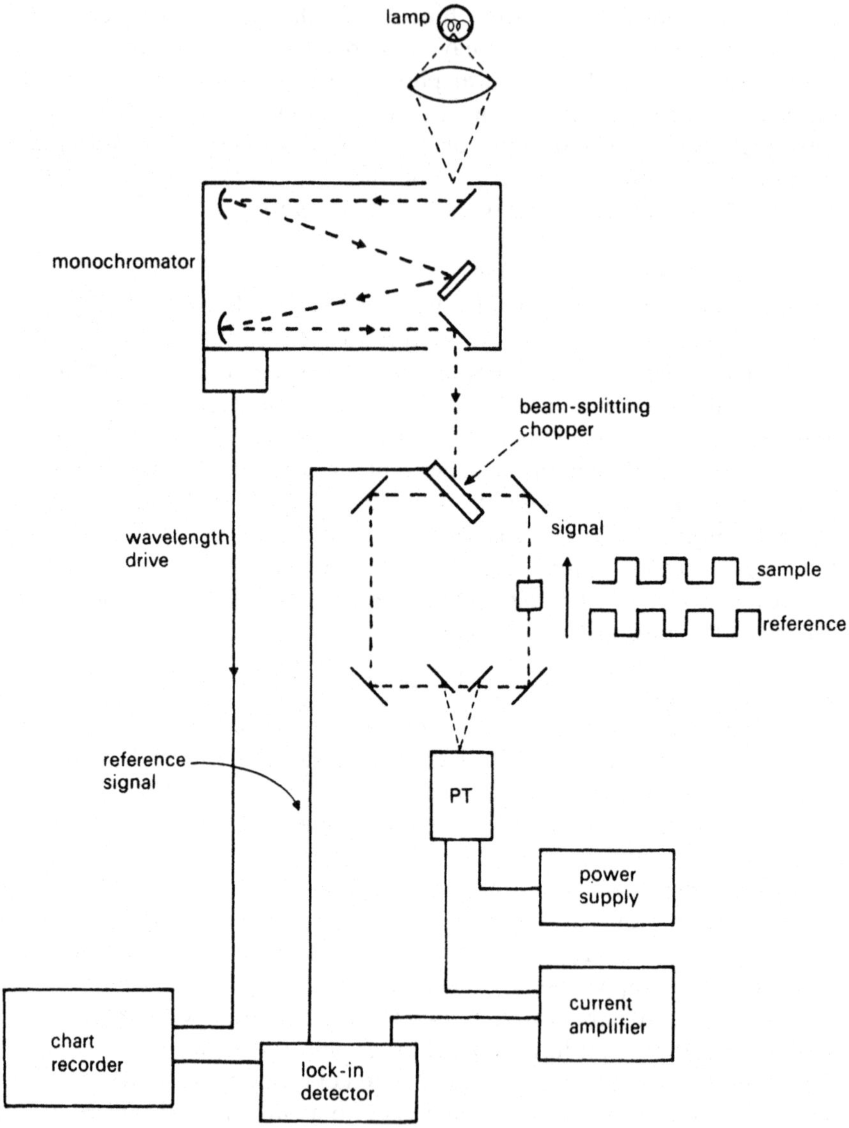

FIG. 6.1. Block diagram of a dual-beam absorption spectrometer.

have a first-order dispersion of 1 nm mm^{-1}, giving a resolution of order 0.1 nm.

6.1.2 Absorption coefficient and oscillator strengths

We have already discussed the theory of stimulated optical transitions (Section 4.4). For a solid containing N non-interacting absorbing centres per unit volume which absorb radiation at frequency ν and polarization $\hat{\varepsilon}$, the attenuation of a beam of intensity $I_0(\nu, \hat{\varepsilon})$ by a solid of thickness l is given by the Beer-Lambert law as

$$I(\nu, \hat{\varepsilon}) = I_0(\nu, \hat{\varepsilon}) \exp(-\alpha(\nu, \hat{\varepsilon}) l). \tag{6.2}$$

Experimentally $I_0(\nu, \hat{\varepsilon})$ represents the transmission of the system in the absence of an absorbing specimen. In practice $I_0(\nu, \hat{\varepsilon})$ and $I(\nu, \hat{\varepsilon})$ are measured and the value of the absorption coefficient $\alpha(\nu, \hat{\varepsilon})$ at a particular frequency is obtained using the formula

$$\alpha(\nu, \hat{\varepsilon}) = \frac{1}{l} \ln \frac{I_0(\nu, \hat{\varepsilon})}{I(\nu, \varepsilon)}. \tag{6.3}$$

$\alpha(\nu, \hat{\varepsilon})$ has the units cm^{-1} or m^{-1}. The variation of the absorption coefficient with frequency is difficult to predict. In general, the absorption transition has a finite width and the absorption strength, $\int \alpha(\nu, \hat{\varepsilon}) \mathrm{d}\nu$, is related to the density of absorbing centres and to the transition probability, as discussed in Section 4.4.3.

The value of the absorption coefficient for a particular polarization depends upon the value of the square of the matrix element of $\mu_e . \hat{\varepsilon}_E$ or of $\mu_m . \hat{\varepsilon}_B$, as eqns (4.85), (4.86) and (4.86a) show. In the case of centres in sites of high symmetry, where the material acts isotropically, the absorption coefficient is independent of polarization and from eqn (4.87)

$$\begin{aligned} \alpha(\nu) = \alpha(\omega) &= \left(N_a \frac{g_b}{g_a} - N_b \right) A_{ba} \frac{\pi^2 c^2}{n^2 \omega^2} g(\omega) \\ &= \left(N_a \frac{g_b}{g_a} - N_b \right) A_{ba} \frac{\pi^2 c^2}{n^2 4\pi^2 \nu^2} \frac{g(\nu)}{2\pi} \\ &= \left(N_a \frac{g_b}{g_a} - N_b \right) A_{ba} \frac{c^2}{8\pi \nu^2} \frac{1}{n^2} g(\nu). \end{aligned} \tag{6.4}$$

N_a and N_b are the population densities in the ground and excited states, g_a and g_b are the statistical weights of the states, $g(\nu)$ is the lineshape function ($\int g(\nu) \mathrm{d}\nu = 1$), and n is the refractive index. In deriving eqn (6.3), N_a and N_b have been assumed to be invariant with time and unaffected by the absorption process. However, if N_a and N_b do change with time then $\alpha(\nu)$ will vary with time. Furthermore, if $I_0(\nu)$ is very intense then absorption will change the values of N_a and N_b and hence modify the rate of energy abstraction from the beam. Indeed, the initial part of a very intense beam may be so strongly absorbed that the excited state population increases until $N_a/g_a = N_b/g_b$, whence according to eqn (6.4) $\alpha(\nu) = 0$. The remainder of the beam is no longer attenuated. This is the phenomenon of *self-induced transparency* (McCall and Hahn 1969).

Returning to the normal condition of weak excitation we can replace N_a by N, the density of centres, and ignore the small N_b value. We then obtain for the *normalized absorption coefficient*

$$k(\nu)=\alpha(\nu)/N=\sigma g(\nu)$$

$$=A_{ba}\frac{c^2}{8\pi\nu^2}\frac{1}{n^2}\frac{g_b}{g_a}g(\nu) \tag{6.5}$$

where σ is the *absorption cross-section per centre*. The absorption strength is related to the cross-section by

$$\int\alpha(\nu)\mathrm{d}\nu=N\sigma. \tag{6.6}$$

From eqn (6.5) we obtain a relationship between cross-section and Einstein spontaneous transition probability:

$$\sigma=A_{ba}\frac{g_b}{g_a}\frac{\lambda_0^2}{8\pi n^2}, \tag{6.7}$$

λ_0 being the vacuum wavelength at the centre of the absorption band. Equation (6.7) applies irrespective of the dipole nature of the transition. In the case where the material absorption is not isotropic the cross-section can be defined by eqn (6.6) using an appropriate average absorption coefficient.

In the absence of an analytic shape function $g(\nu)$, Gaussian or Lorentzian lineshapes are sometimes assumed for broad bands. For Gaussian bands and unpolarized radiation

$$\alpha(\nu)=\alpha(\nu_0)\exp(-k(\nu-\nu_0)^2) \tag{6.8}$$

where $\alpha(\nu_0)$ is the peak absorption coefficient at the centre frequency, ν_0, of the band, and $k=4\ln 2(\Delta\nu)^{-2}$, $\Delta\nu$ being the frequency interval between the half-power points, i.e. the full width at half maximum (FWHM). Since $\int\exp(-\gamma x^2)\mathrm{d}x=\sqrt{\pi/\gamma}$ we obtain

$$\int_{-\infty}^{+\infty}\alpha(\nu)\mathrm{d}\nu=\sqrt{\frac{\pi}{4\ln 2}}\alpha(\nu_0).\Delta\nu. \tag{6.9}$$

Equation (4.109) relates A_{ba} to the emission oscillator strength, f_{ba}. We can use this equation to relate the absorption strength to the absorption oscillator strength ($f_{ab}=f_{ba}g_b/g_a$) and we obtain

$$\int a(\nu)\mathrm{d}\nu=Nf_{ab}(\mathrm{ED})\frac{e^2}{4\varepsilon_0 mc}\left[\left(\frac{E_{\mathrm{loc}}}{E}\right)^2\frac{1}{n}\right] \tag{6.10a}$$

for an electric dipole transition, and

$$\int a(\nu)\mathrm{d}\nu=Nf_{ab}(\mathrm{MD})\frac{e^2}{4\varepsilon_0 mc}n \tag{6.10b}$$

for a magnetic dipole transition. If we ignore the refractive index and local field correction factors and if we use eqn (6.8) we obtain for a Gaussian-shaped absorption band

$$Nf_{ab} = 0.87 \times 10^{17} \alpha(\nu_0) \Delta\nu \tag{6.11}$$

where $\alpha(\nu_0)$ is measured in cm^{-1}, $\Delta\nu$ is measured in eV and N is the number of centres cm^{-3}. Equation (6.11) is often referred to as *Smakula's formula.* For a Lorentzian lineshape the factor 0.87 is replaced by 1.29. As eqn (6.11) shows, to obtain the oscillator strength from the area under the absorption band one needs an independent determination of the concentration density of centres. For impurity ions in solids, N may be determined by chemical assay. However, chemical analysis cannot be used to measure defect concentrations and it is then necessary to resort to other techniques. When defects are paramagnetic their concentration may in principle be measured by electron spin resonance (ESR). Silsbee (1956) used this technique for F-centres in potassium chloride. The difficulties of making spin assays to better than 5–10 per cent are well known. There will be a correspondingly large inaccuracy in the oscillator strength. As discussed in Chapter 2 and Chapter 7, the F-centre behaves optically much like a hydrogen atom embedded in a dielectric continuum. The absorption transitions may be crudely thought of as 1s→2p-like based on hydrogenic wavefunctions, hence this is a fully-allowed electric dipole transition. As we saw in Chapter 4, for an allowed electric dipole transition the oscillator strength is approximately unity. Silsbee's (1956) measurement for F-centres in potassium chloride gave $f_{ab} = 0.56$, whereas for F^+-centres in magnesium oxide Henderson and King (1966) obtained $f_{ab} = 0.7$. These values are very much in keeping with theoretical expectations.

6.2 Luminescence measurements

To study luminescence from ionic materials it is usually possible to optically pump using high-intensity sources into strong absorption bands in the near ultraviolet and blue regions of the spectrum. Typical sources with broad bands in these regions are 200–500 W mercury or xenon arc lamps. The xenon arc lamp is also particularly useful for exciting luminescence in the yellow-red region of the spectrum since xenon does not show interfering sharp line emission in this region. In addition, a variety of lasers may be used including Ar^+, Kr^+, He–Ne, and He–Cd lasers which have emissions at fixed wavelengths. Tunable dye lasers can be selected to closely match the absorption spectra of particular materials.

The light emitted by the solid may be resolved into its component lines/bands using a prism or reflection grating monochromator. For medium resolution a 1 m Czerny–Turner monochromator will give a spectral resolution of about 0.02 nm. Resolution an order of magnitude lower can be achieved using a grating spectrometer with focal length 0.25 m. The light emerging from the monochromator is detected using an appropriate electron multiplier phototube with associated high-voltage power supplies. In the visible region the wavelength range of gallium arsenide phototubes is particularly useful; they operate with good quantum efficiency in the range 280–900 nm. Furthermore they have a sufficiently low dark current that cooling of the phototube is often unnecessary. For measurements in the near infrared a lead sulphide cell, cooled germanium photodetector, or special III–V compound photodiode may be used. Under steady-state optical pumping a steady-state luminescence output is obtained and detected as a photocurrent which is amplified and converted to a voltage signal to be displayed on a pen recorder. Most authors quote the emission output in 'arbitrary units', due to the difficulty in calibrating the detection system accurately. Luminescence detection is inherently more sensitive than absorption measurements, and sensitivities of 10^{11} centres cm^{-3} are routine. This steady-state technique is relatively straightforward. However, if the signals are weak then phase-sensitive detection and/or computer averaging can be employed.

6.2.1 Phase-sensitive detection

Phase-sensitive detection techniques have already been discussed by reference to absorption measurements. In luminescence measurements the excitation intensity is switched on and off at a certain reference frequency, so that the luminescence intensity is modulated at this same frequency. The detection system is then set to record signals at the reference frequency only. This effectively eliminates all noise signals except those closely centred on the modulation frequency. Furthermore, the signal recovery system is also phase-selective, a facility of very great value in discriminating between centres with different radiative decay times. A typical experimental set-up is shown in Fig. 6.2. The pumping light is square-wave modulated (on/off) by a mechanical light-chopper operating in the range 10–10^3 Hz. A reference signal is taken from the chopper to one channel of a lock-in detector. The magnitude and phase of the luminescence signal is then compared with the reference signal. Because of the finite radiative lifetime of the emission, and to some extent also because of the phase changes within the electronics, the luminescence signal is not in phase with the reference signal. Hence to maximize the output from the lock-in detector the phase control of the reference signal is adjusted until input (luminescence) and reference signals to the lock-in detector are in phase. Of course, the phase of the reference signal may also be adjusted so that reference and luminescence signals are in quadrature giving zero output from the lock-in. This method of phase-adjusting may enable one to separate overlapping luminescence bands from different centres. In such experiments the chopping frequency is adjusted so that there is an ap-

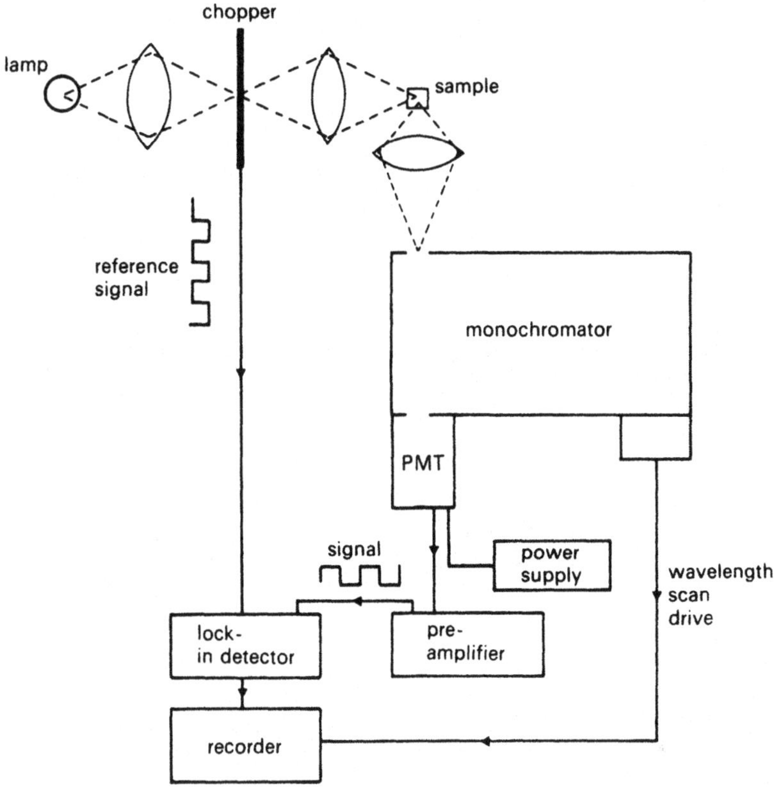

FIG. 6.2. Block diagram of the experimental arrangement for measuring luminescence spectra by lock-in detection techniques.

preciable reduction in the luminescence intensity during the 'off' half-cycle. This effectively puts an upper limit on the rate at which the lock-in system can operate.

As an example of the usefulness of the phase adjustment we review the application of phase-sensitive techniques to the study of the luminescence from MgO:Cr^{3+} (Henry *et al.* 1976). In this material the Cr^{3+} ions can occupy three principal sites, sites of octahedral symmetry (which emit the sharp R-line at 698.1 nm and accompanying vibrational sidebands), sites of tetragonal symmetry (which emit sharp N-lines at 703.5, 703.9 nm, and accompanying vibrational sidebands), and orthorhombic (which emit in a broad band centred at 800 nm). The decay times of the luminescence signals are 11.4 ms, 8.6 ms, and 35 μs for the octahedral, tetragonal, and orthorhombic sites, respectively. The luminescence spectrum of MgO:Cr^{3+} at 77 K under steady state optical excitation is shown in Fig. 6.3(a). Figures

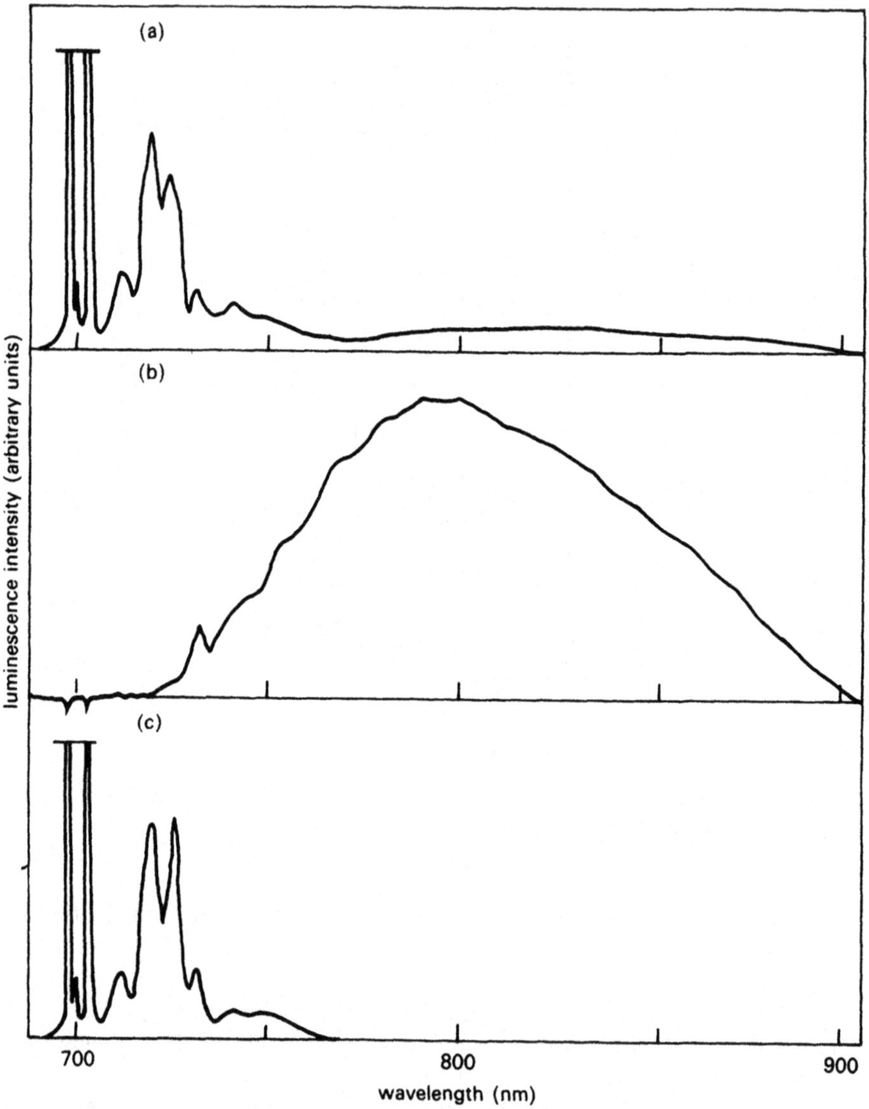

FIG. 6.3. Using the phase control of the lock-in detector to discriminate between overlapping luminescence spectra emitted by a Cr^{3+}-doped magnesium oxide crystal. Spectrum (a) is a composite of slow components (R- and N-lines) and a fast component (800 nm band). On adjusting the phase setting to null the R- and N-line components (b) only the broad band is observed. When the broad band signal is nulled to zero (c) the R- and N-lines and their vibronic sidebands are observed.

6.3(b) and 6.3(c) show the output of the phase-sensitive detector when the excitation beam is chopped at two different phase settings. Because of its much larger decay rate the modulated luminescence signal from the orthorhombic centres is out of phase with the modulated luminescence signals from the octahedral and tetragonal centres. In Fig. 6.3(b) the phase of the reference signal is adjusted to be in approximate quadrature with the modulated luminescence signal from the octahedral and tetragonal sites. Thus there is essentially a null signal from these sites. At this phase setting the modulated luminescence signal from the orthorhombic sites is not in quadrature with the reference signal, so that the output of the phase-sensitive detector measures only the luminescence spectrum of the orthorhombic centres. The weak zero-phonon line at 730 nm in this spectrum was previously undetected, being masked by the much stronger vibrational sidebands of the R- and N-lines. In Fig. 6.3(c) the phase of the reference signal is adjusted to be in quadrature with the modulated luminescence from the orthorhombic centres, and only the R- and N-lines and their vibrational sidebands are observed. Because of the difference in the decay times of the luminescence from the octahedral and tetrahedral centres one can also separately null the signal from one or other of these centres. This allows the overlapping sidebands of the R- and N-lines to be resolved from each other. The phase resolution technique was used to great advantage by Engstrom and Mollenauer (1965) in their study of the luminescence signal from exchange-coupled pairs of Cr^{3+} ions in $Al_2O_3:Cr^{3+}$.

6.2.2 Excitation spectroscopy

In many cases inorganic solids have strong overlapping absorption bands due to non-luminescent centres. These absorptions can completely overwhelm absorption features related to a particular luminescence centre. An example of this is magnesium oxide doped with vanadium which enters the oxide host as both V^{2+} and V^{3+}. V^{2+} is isoelectronic with Cr^{3+}; in magnesium oxide, V^{2+} ions emit luminescence in the near infrared (900 nm). The absorption bands of V^{2+} ions, however, are masked by the very strong absorption bands of V^{3+} and are not detected in absorption measurements. These difficulties are overcome by excitation spectroscopy, in which the intensity of the luminescence output is recorded as a function of the wavelength of an excitation beam. Strong emission at a particular excitation wavelength signals that the emitting centre is absorbing strongly at that wavelength. In this way it is possible to determine the shape and position of the absorption bands which excite the emission process. The experimental apparatus would involve the addition of a low-resolution scanning monochromator in the excitation channel of the luminescence spectrometer. As illustrated in Fig. 6.4 this scanning monochromator is placed immediately after the chopper and light from its exit slit is then focused onto the sample.

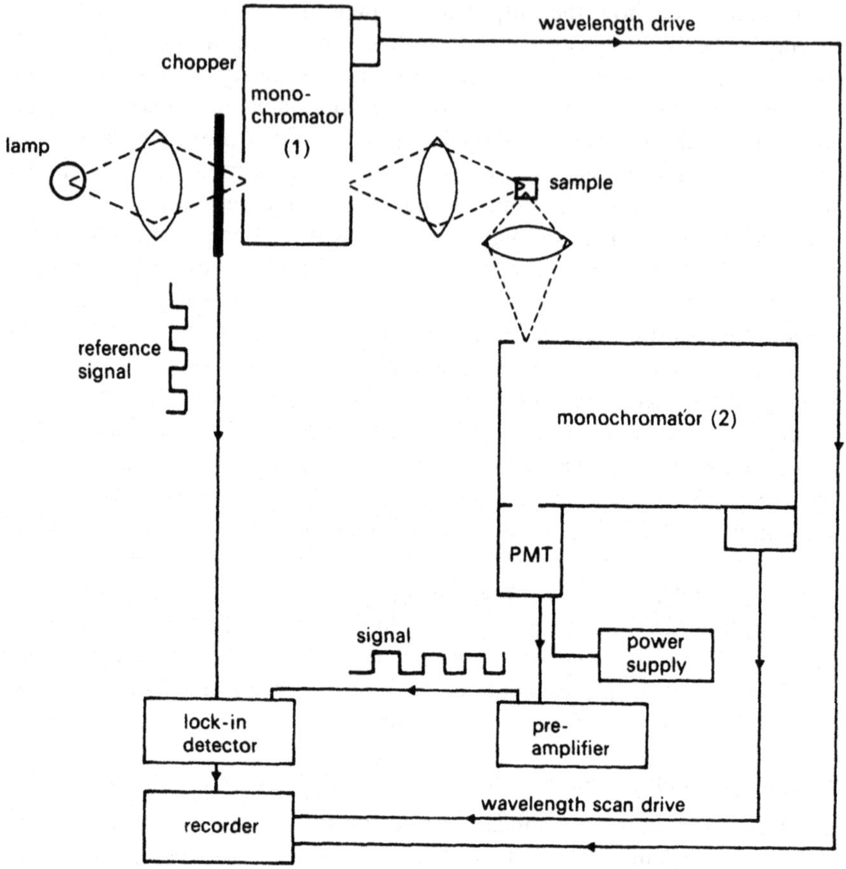

FIG. 6.4. A schematic representation of an apparatus for measuring excitation spectra.

With such an apparatus it is possible to distinguish absorption transitions from several centres whose absorption and luminescence bands partially overlap. The example given in Fig. 6.5 shows the luminescence pattern emitted by F_2-centres in magnesium oxide and the excitation spectrum associated with this emission. Other strong absorption bands which overlap the F_2-absorption bands are strongly discriminated against by the selective detection of the F_2-centre luminescence.

A second example of excitation spectroscopy (Fig. 6.6(a)) makes use of the much studied optical properties of multi-quantum well structures (MQWS) based upon GaAs/$Al_xGa_{1-x}As$ epilayers. In such structures alternate layers of two different semiconductors are grown on top of each other so that the bandgap varies in one dimension with the periodicity of the epitaxial layers

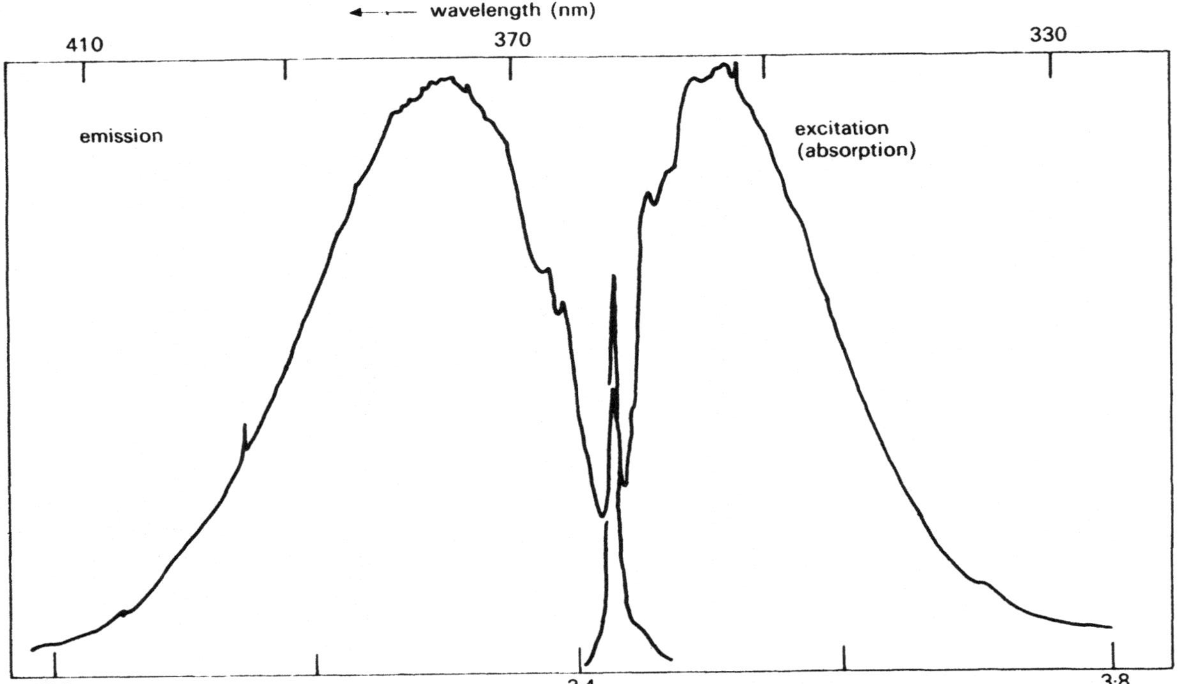

Fig. 6.5. Illustrating the advantage of using excitation spectroscopy. Absorption spectroscopy in the wavelength range 300–400 nm of neutron-irradiated magnesium oxide is difficult, due to large optical density and overlapping bands. However, once the F_2-centre luminescence band has been recorded it is comparatively easy to set the luminescence wavelength to the peak of the band and scan through the excitation dependence of this luminescence using a second monochromator.

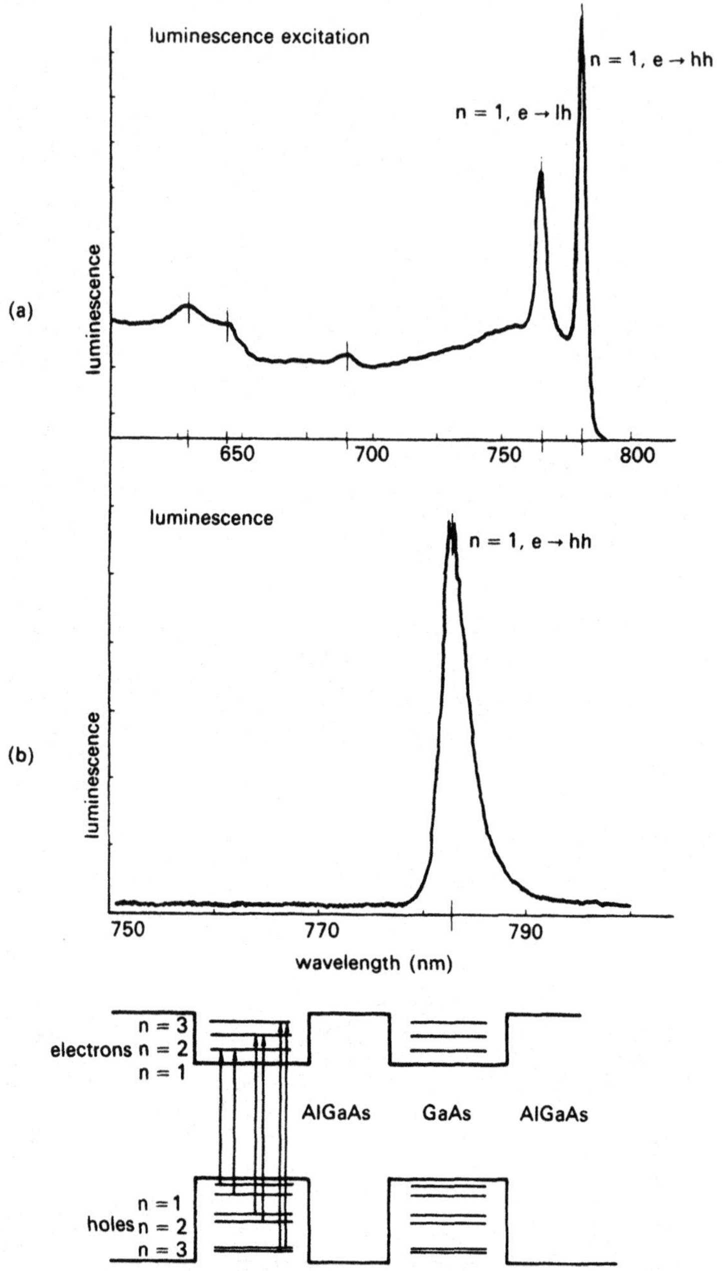

FIG. 6.6. (a) Luminescence excitation spectra of MQWS based on GaAs/$Al_xGa_{1-x}As$ measured at 6 K (after Dawson 1986, unpublished). (b) Schematic energy level structure and optical transitions for a one-dimensional MQW.

(Fig. 6.6(b)). Using the simple approximation of the infinite one-dimensional square-well potential (Section 2.5.1), electrons and holes in the narrower gap material can be assigned energy eigenvalues using a modified form of eqn (2.62). In the case of holes in the valence band, however, the square well potential energy levels are split by spin–orbit interaction into light (lh) and heavy holes (hh). Optical transitions between valence band and conduction band are then determined by the selection rule $\Delta n = 0$. There is, in consequence, a range of different absorption transitions across the bandgap. Due to rapid energy relaxation the electrons recombine via emission from the $n = 1$ level. Measurement of the excitation dependence of the luminescence will reveal absorption transitions between the various electron and hole levels. The example given is for a MQWS of 5.5 nm thick gallium arsenide wells and 12.5 nm thick barriers of $Al_{0.27}Ga_{0.73}As$, the measurements being made at $T = 6$ K. There is a single emission line at $h\nu = 1.602$ eV due to radiative decay from the $n = 1$ level of electrons (1e) in the conduction band into the $n = 1$ level of light holes (1 lh) in the valence band (i.e. the transition 1e $\rightarrow$ 1 lh). The first absorption transition detected by excitation spectroscopy is the 1 lh $\rightarrow$ 1e transition, which occurs at a slightly lower photon energy than the 1 hh $\rightarrow$ 1e transition (Fig. 6.6(a)). The assignment of the other transitions are confirmed by theoretical techniques; they reveal that the $\Delta n = 0$ selection is not rigorously upheld, which is, perhaps, not too surprising. Note that the 1lh $\leftrightarrow$ 1e absorption and emission transitions occur at almost identical photon energies.

6.2.3 *Radiative lifetime measurements*

In order to measure the radiative lifetime of a transition it is necessary to use a sharp intense pulse of excitation in the wavelength region of the appropriate absorption band, together with some means of recording the temporal evolution of the luminescence signal. The simplest procedure is to use a high-speed shutter to select pulses of the exciting radiation. The emission signal is detected using a photomultiplier tube in conjunction with an oscilloscope. The method is quite suitable and accurate for strong luminescence with lifetimes down to a few milliseconds (Henderson *et al.* 1969). Alternative excitation sources include pulsed lasers, flash lamps, or stroboscopes. Modern laser systems produce pulses of duration 0.1–100 ps in a routine, albeit expensive, manner whereas flash lamps and stroboscopes will produce pulses of order 10^{-8} s and 10^{-5} s, respectively.

Consider first luminescence lifetime measurements in the range 10^{-3}–10^{-6} s. A possible experimental arrangement is shown in Fig. 6.7. Usually the luminescence yield following a single excitation pulse is too small for good signal-to-noise throughout the decay period. In consequence, repetitive pulsing techniques are used together with signal averaging to obtain good decay statistics. The pulse reproducibility of the stroboscope is advan-

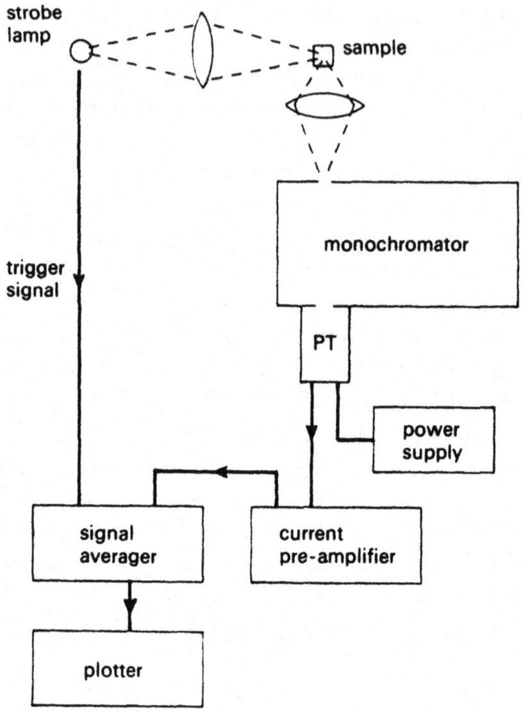

FIG. 6.7. Apparatus for measuring luminescence decay times using a strobe lamp and signal averaging techniques.

tageous in the signal averaging process, in which the output from the detector is sampled at equally-spaced time intervals after each excitation pulse. If the pulse is repeated N times then there is an $N^{\frac{1}{2}}$ improvement in the signal-to-noise ratio. If a multichannel analyser is used as part of the detection system, the excitation pulse may be used to trigger the analyser, and hence the time between pulses need not be constant. Radiative lifetimes in the range 10^{-3}–10^{-6} s may also be measured by phase-sensitive detection. The phase angle between maximum and zero signal is $\phi = \tan^{-1}\omega\tau_r$, where ω is the modulation frequency and τ_r is the luminescence decay time.

An illustration of the data obtainable using the stroboscopic technique is shown in Fig. 6.8. The luminescence signal is the broad band emission from Cr^{3+} ions in orthorhombic symmetry sites in magnesium oxide measured at 77 K (Fig. 6.3(b)). At low Cr^{3+} ion concentrations (*c.* 100 ppm) the radiative lifetime of this luminescence centre is 35 μs. The data in Fig. 6.8 refer to a sample containing 1300 ppm of Cr^{3+} ions. On sampling the emission at times long relative to τ_R there is a component with characteristic decay time of 11.4 ms which is the lifetime of Cr^{3+} ions in octahedral symmetry sites in

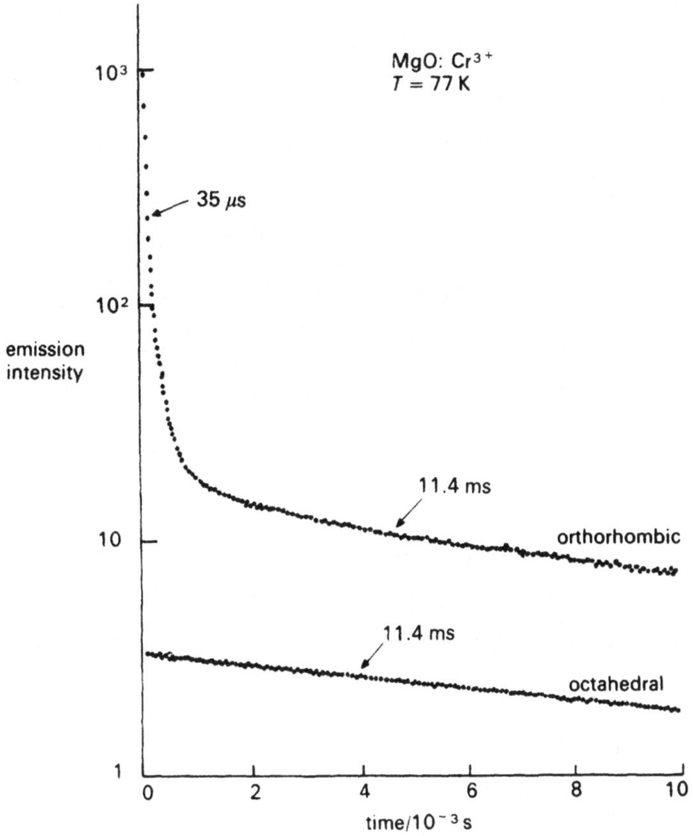

FIG. 6.8. The temporal evolution of the intensity emitted by Cr^{3+} ions in orthorhombic symmetry sites in magnesium oxide, showing two components, one fast ($\tau \simeq 35$ μs) and one slow ($\tau \simeq 11.4$ ms).

magnesium oxide. This result implies that excitation is being transferred from excited Cr^{3+} ions in octahedral sites to Cr^{3+} ions in orthorhombic sites. The dynamics of the transfer process is described in Chapter 10. These data, and those in Fig. 4.5 also, serve to remind us that the evolution of the intensity during the pulse-decay cycle is not necessarily in the form of a single exponential decay determined by an Einstein A_{ba} coefficient. As we discuss in Chapters 5 and 10 measured decay times give information about non-radiative decay and energy transfer between centres as well as radiative decay.

For rather faster decay processes (10^{-10}–10^{-8} s), fast flashlamps may be used to excite the luminescence. Modern gated flashlamps have extremely reproducible pulses, down to 0.8 ns width with repetition rates of up to 50 kHz. The usual gases for such lamps are hydrogen, deuterium, and air. Hydrogen has several advantages, not the least being the continuum output in the ultraviolet and visible ranges, with pulse profiles which are independent of wavelength. The combination of pulse sampling techniques and computer deconvolution of the decaying luminescence enables decay times to be measured down to 200 ps. However, judicious choice of photomultiplier tube and careful design of the photomultiplier dynode chain is necessary to eliminate signal noise. It is usual to use coincidence single-photon counting techniques to obtain good decay data (Ware 1983). The example given in Fig. 6.9 shows how the rise of the luminescence output follows the excitation pulse. The luminescence decay is a single exponential process with decay time of 5.4 ns. The solid line is a computer reconvolution of the luminescence output, calculated using a non-linear least-squares fitting routine.

6.3 Polarized absorption and luminescence

We have already noted in Chapter 4 that the strength of optical transitions may depend strongly on polarization. Here we discuss how the polarization of spectra may be used to give information about the site symmetry of optically-active centres in crystals.

6.3.1 General considerations

The probability of a transition induced by linearly polarized radiation varies as the square of the matrix element of $\boldsymbol{\mu}.\hat{\varepsilon}$, where $\boldsymbol{\mu}.\hat{\varepsilon}$ is $\boldsymbol{\mu}_e.\hat{\varepsilon}_E$ for an electric dipole transition and $\boldsymbol{\mu}_m.\hat{\varepsilon}_B$ for a magnetic dipole transition. The unit polarization vector, $\hat{\varepsilon}$, has direction cosines $\cos\alpha$, $\cos\beta$, $\cos\gamma$ (Fig. 6.10(a)) and the square of the dipole matrix element is then

$$|\langle b|\mu_x\cos\alpha+\mu_y\cos\beta+\mu_z\cos\gamma|a\rangle|^2. \tag{6.12}$$

In the case where the centre occupies a site of octahedral symmetry it can be shown by symmetry arguments that the matrix element squared becomes

$$\langle\mu_x\rangle^2\cos^2\alpha+\langle\mu_y\rangle^2\cos^2\beta+\langle\mu_z\rangle^2\cos^2\gamma \tag{6.13}$$

using $\langle\mu_x\rangle$, $\langle\mu_y\rangle$, and $\langle\mu_z\rangle$ as an obvious shorthand. Because of the octahedral symmetry we have

$$\langle\mu_x\rangle^2=\langle\mu_y\rangle^2=\langle\mu_z\rangle^2 \tag{6.14}$$

hence $|\langle\boldsymbol{\mu}.\hat{\varepsilon}\rangle|^2$ becomes $\langle\mu_x\rangle^2$ and the strength of the transitions is independent of the direction of the polarization of the incident radiation, and,

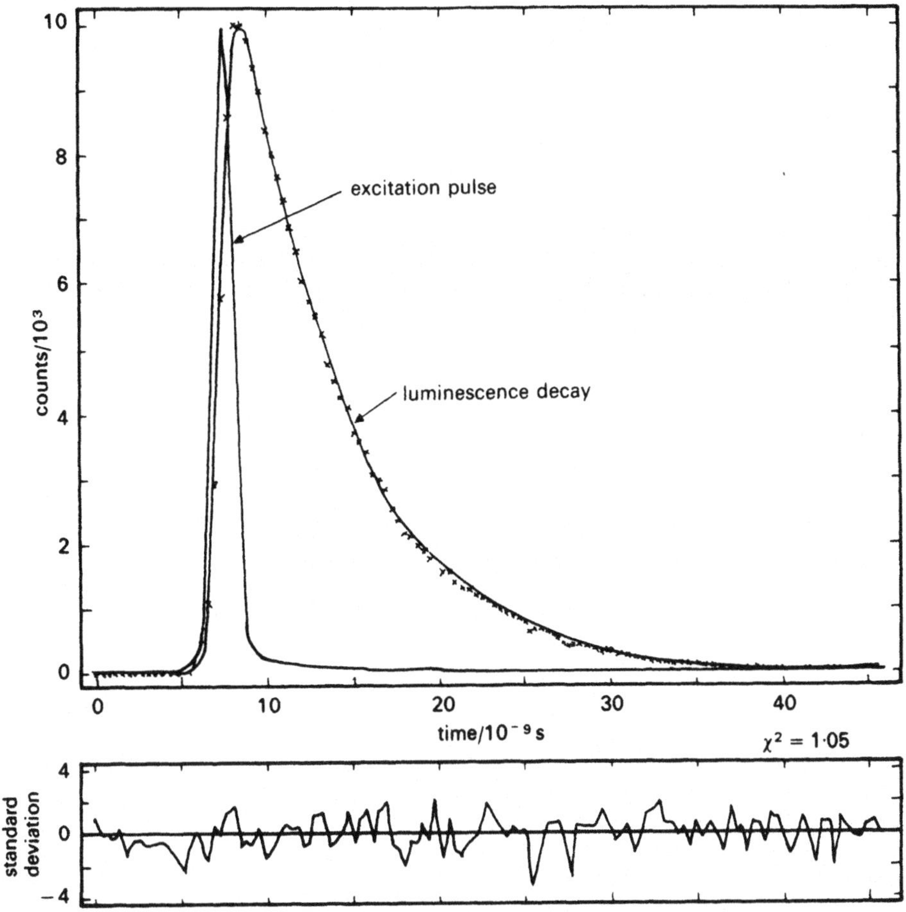

FIG. 6.9. The luminescence from a plastic solar collector containing perylimede. The excitation pulse is centred at 460 nm and the luminescence decay is at 570 nm (×); the computer reconvolved decay (solid line) uses a decay time of $\tau_R = 5.4$ ns. (After Birch and Imhof 1986, unpublished.)

clearly, independent of the direction of propagation, k. The spectroscopic properties of the centre are *isotropic*.

In the case where the centre occupies a site of tetragonal symmetry, the z-axis being the axis of symmetry, the transition probability is again given by eqn (6.13) but now we have

$$\langle \mu_x \rangle^2 = \langle \mu_y \rangle^2 \neq \langle \mu_z \rangle^2 \tag{6.15}$$

and the transition probability for radiation at polarization $\hat{\varepsilon}$ is proportional

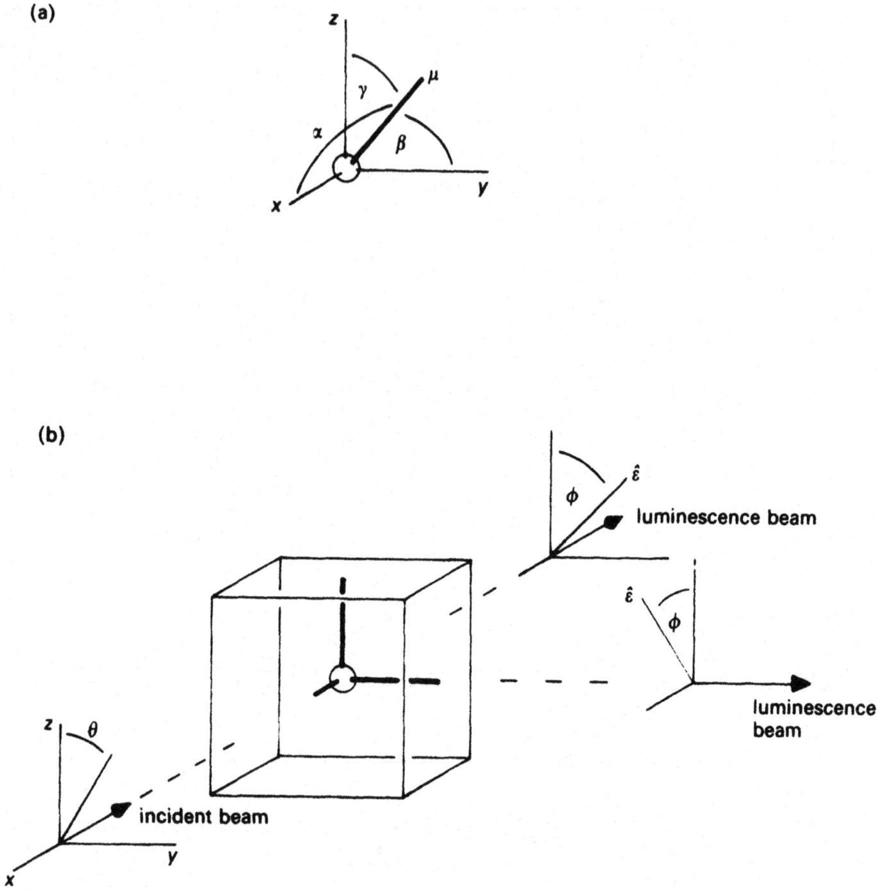

FIG. 6.10. Showing (a) the orientation of the dipole moment, μ, relative to the x, y, and z-axes and (b) the geometrical arrangement used for measuring the polarization of optical absorption/luminescence spectra. Note the two different aspects for detecting the emitted radiation.

to

$$\langle\mu_x\rangle^2(\cos^2\alpha+\cos^2\beta)+\langle\mu_z\rangle^2\cos^2\gamma \tag{6.16}$$

$$=\langle\mu_x\rangle^2+[\langle\mu_z\rangle^2-\langle\mu_x\rangle^2]\cos^2\gamma \tag{6.17}$$

$$=A+B\cos^2\gamma. \tag{6.18}$$

The spectroscopic properties of the centre are now *anisotropic*.

6.3.2 *Application to cubic crystals*

If the host crystal contains only centres in sites of octahedral symmetry then clearly the spectroscopic properties of the crystal are isotropic. However, a host crystal may possess a cubic unit cell in which the sites of the electronic centre are anisotropic, there being a regular distribution of equivalent sites with their anisotropy axes along different equivalent symmetry directions of the unit cell. Polarized absorption/luminescence studies of such crystals can reveal details of the local site anisotropy (Feofilov 1961). Consider first the case of an alkali halide crystal containing F_A-centres. Each F_A-centre is an F-centre with a cation impurity in a nearest neighbour site. The local symmetry is tetragonal, and there are equal densities of F_A-centres with anisotropy axes along each of the [100], [010], and [001] directions. (The tetragonal crystal field of the F_A-centre raises the degeneracy of the s→p(x), p(y), p(z)-like absorption components of the F-centre. Specifically this discussion applies to the excitation of s-p(z) transitions, which may be well separated in wavelength from the two orthogonally polarized transitions.) Figure 6.10(b) shows a typical experimental geometry for such a study. The exciting beam is directed along the [100] direction. Luminescence is observed either along the [100] direction (the 'straight through' geometry) or along the [010] direction. We calculate the absorption strength for electric dipole transitions with light propagating along the [100] direction with its electric vector, $\hat{\varepsilon}$, at some angle θ to the [001] direction. F_A-centres aligned along [100] absorb with strength A (eqn 6.18), while F_A-centres along [001] and [010] absorb with strengths $A+B\cos^2\theta$ and $A+B\sin^2\theta$, respectively. The total absorption strength, $\langle\mu_x\rangle^2+\langle\mu_y\rangle^2+\langle\mu_z\rangle^2=3A+B$, is independent of the angle θ. The crystal absorbs light isotropically. Indeed, we can extend the argument to show that the strength of the absorption is independent of the k, $\hat{\varepsilon}$ directions of the radiation.

Let us suppose that we detect not the absorbed intensity but the emission intensity at some specific k, $\hat{\varepsilon}$ value. We adopt the geometry shown in Fig. 6.10(b) where the absorbed radiation is incident along the [100] direction with polarization vector at an angle θ to the [001] direction, and the emitted radiation is measured in the [100] direction with polarization vector along [001] (i.e. $\phi=0$). In this particular case only those F_A-centres with anisotropy axes along [001] can emit such luminescence. Hence the emitted intensity varies as the number of excited F_A-centres with anisotropy axes along [001] which varies as $A+B\cos^2\theta$. The detected luminesence varies from $A+B$ to A as $\hat{\varepsilon}$ is rotated from being parallel to [001] to being parallel to [010]. A similar analysis applies when the absorption polarization is fixed and the polarization of the luminescence is rotated. The particular angular dependence found is characteristic of centres with anisotropy axes along ⟨100⟩ directions. An example of such a polarized luminescence measurement is given in Chapter 7.

The same experimental geometry can be used to analyse the optical behaviour of Cr^{3+} ions in orthorhombic sites in magnesium oxide. In this case the distortion arises because a cation vacancy occupies one of the nearest-neighbour cation sites of the Cr^{3+} ion along a crystal $\langle 110 \rangle$ axis. There are 48 equivalent orientations of such orthorhombic Cr^{3+} ion–vacancy centres in a cubic crystal. A typical centre will have [101], [010], and $[10\bar{1}]$ axes forming a mutually orthogonal right-handed set of axes (x, y, z), where z is drawn from the Cr^{3+} ion through the vacancy. However since we cannot distinguish optically between positive- and negative-going axes of such centres the number of equivalent sites is effectively reduced to six. Consequently there are six equivalent z-axes, each being equally populated with centres. The relative orientations of the six distortion axes are shown in Fig. 6.11(a). Polarized absorption/luminescence measurements for this system were recorded with the emission polarization fixed with at $\phi = 45°$, and the emitted intensity measured as the absorption polarization angle θ was varied (Henry *et al.* 1965). These authors showed that for this particular geometry the emission intensity varies with angle θ according to

$$I(\theta) = A + B \sin 2\theta \tag{6.19}$$

Figure 6.11b shows the experimental results obtained by Henry *et al.* (1965) for the broadband luminescence spectrum shown in Fig. 6.3(b). The agreement between experiment and theory is excellent, confirming the origin of this luminescence spectrum, as due to the Cr^{3+} ions in orthorhombic symmetry sites.

6.3.3 Non-cubic crystals

Several different possibilities exist in non-cubic crystals. If the local symmetry axis of all centres in the crystal point in the same direction then the crystal as a whole has an axis of symmetry in this direction. An example is aluminium oxide where Al^{3+} ions occupy sites which are distorted from octahedral symmetry by a weak crystal field of trigonal symmetry. Thus the trigonal axis for all optically active species on Al^{3+} ion sites point in this same direction, and consequently the absorption and luminescence spectra of ions in aluminium oxide are naturally polarized, as we discuss in Chapter 9.

6.4 Perturbation spectroscopy

Studies of the effects of magnetic and electric fields on absorption and emission spectra of atoms and molecules have been commonplace for several decades. Similar techniques have also been applied in solids, although experimental analysis may be complicated by the greater width of optical transitions in solids. Even purely electronic transitions, the so-called zero-phonon lines, have considerably larger widths than the transitions of atoms in gases, due largely to inhomogeneous broadening by random lattice strain. In solids one may add uniaxial stress to the list of perturbations which may be used as tools in probing defect structure. Note that the application of a

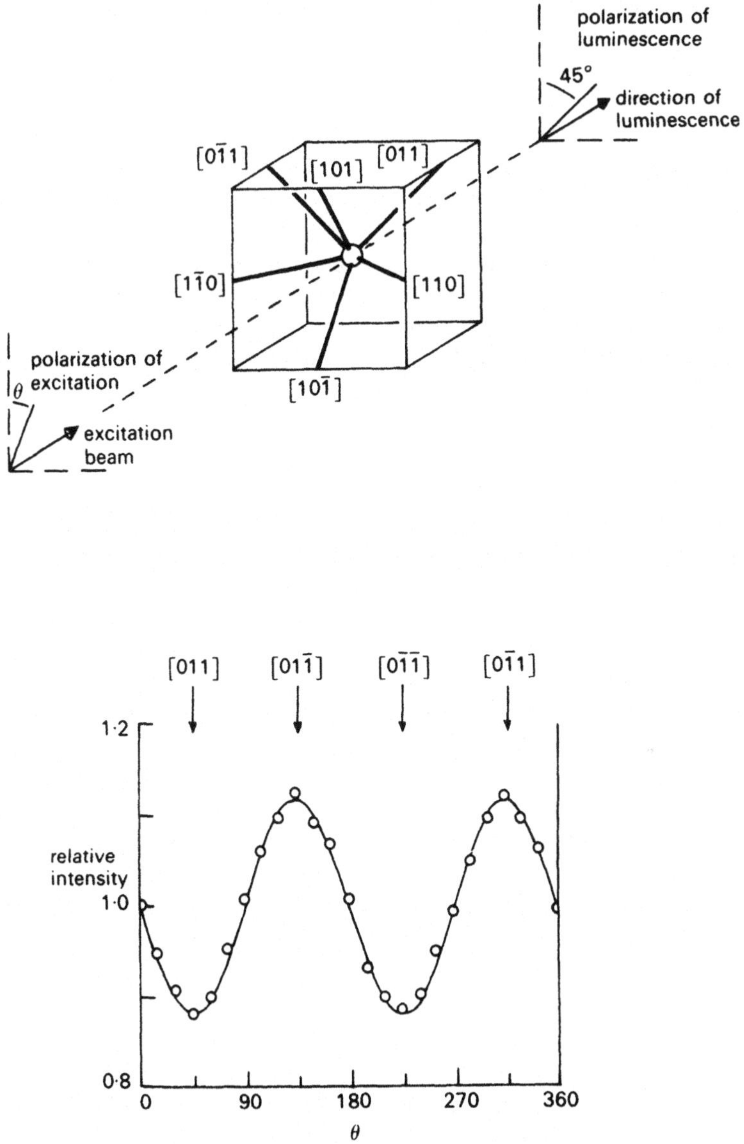

FIG. 6.11. The luminescence polarization of Cr^{3+} ions in orthorhombic symmetry sites in magnesium oxide; (a) shows the six equivalent $\langle 110 \rangle$ orientations and the excitation/emission geometry, and (b) shows the relative broadband emission intensity measured in the $[\bar{1}00]$direction and polarized along the $[011]$ direction of the cube, as a function of the polarization angle of the angle θ of the exciting light. The emission polarization is fixed at $\phi = 45°$, and the orientation of the E-field of the exciting radiation varied from $\theta = 0$ to $\theta = 360°$.

hydrostatic pressure to a solid preserves the symmetry of the centre in the solid but changes the strength of the interaction between the electronic centre and the crystal environment. Hydrostatic pressure causes a shift but not a splitting of transitions. The principles of perturbation spectroscopy are quite general: the application of an external constraint, be it uniaxial stress, static magnetic field, or electric field, may change the energy of an electronic centre and so remove fully or partially the degeneracy of the participating electronic energy levels. Hence the associated spectroscopic lines are shifted and/or split into a number of components. The symmetry of the electronic states and of the defect environment may be deduced from the splitting pattern. In each case, the external perturbation provides a unique symmetry axis in the crystal, and the analyses of the effects of the perturbations have obvious similarities.

6.4.1 Piezospectroscopy

Among the first examples of the use of uniaxial stress in solid state optical spectroscopy was the study by Kaplyanskii (1959) of the effects of stress on the spectra of rare-earth ions in solids. In this experiment the degenerate energy levels of equivalent centres with different crystal orientations were split since these centres responded in different ways to the uniaxial stress. In the experiments reported by Schawlow *et al.* (1961) uniaxial stress reduced the symmetry of the Cr^{3+} ions from octahedral to lower symmetry and caused a splitting of the sharp zero-phonon line (Section 3.1.8). Subsequently Kaplyanskii (1964) published a group-theoretical analysis of the splitting patterns of spectra in cubic crystals. The usual experimental procedure for cubic crystals is to apply stress along $\langle 100 \rangle$, $\langle 110 \rangle$ and $\langle 111 \rangle$ crystal axes and examine the different spectra in π- and σ-polarizations. The quantities to be determined are the number of components for each stress direction, the energy shift of each component, together with its intensity and polarization. The line shifts may be related to the potential energy change at the defect by an operator

$$V = \sum_{ij} A_{ij} \sigma_{ij} = \sum_{kl} B_{kl} \varepsilon_{kl} \tag{6.20}$$

where σ_{ij} and ε_{kl} are the stress and strain tensors, whereas the A_{ij}, B_{kl} are electronic operators linearly related to each other by the elastic constants of the crystal.

The simplest situation in cubic crystals is for anisotropic centres having singly degenerate electronic states. Hence different equivalent directions in the unstressed crystal become inequivalent in the presence of an externally applied uniaxial stress. This type of degeneracy is referred to as *orientational degeneracy*. Application of a uniaxial stress in a particular direction can affect these centres differently and can remove all or part of the orientational degeneracy. If the anisotropic centres have tetragonal symmetry with anisotropy axes parallel to the ⟨100⟩ directions a maximum splitting of sharp optical transitions into three lines can be observed for a general direction of the applied stress. When the stress is applied along the [111] direction no splitting is observed, only a shift, since the stress direction makes an equal angle with all anisotropy axes. However, for stress parallel to [100] two components are observed, one due to the centres with anisotropy axes parallel to the stress direction, and one due to the centres with anisotropy axes in the plane perpendicular to the stress direction. In addition, if the transitions on each centre are strongly polarized with respect to the local symmetry axis the two components may be further distinguished by their polarizations. Note that for a cubic crystal under a ⟨110⟩ stress the crystal becomes biaxial and the number of σ-polarized components depends upon the viewing direction. For ⟨100⟩ and ⟨111⟩ stress directions the viewing direction is of no consequence. There are only seven different anisotropic centres with non-degenerate electronic states in cubic crystals (Kaplyanskii 1964).

If the centres occupy sites of high symmetry they may possess degenerate electronic energy levels, and the applied stress removes the *electronic degeneracy* and splits the transition into a number of components. The number of components into which a transition splits under stress may be found by noting the way in which the particular irreducible representations which characterize the electronic levels in the unstressed crystal split up into irreducible representations of the lower symmetry characteristic of the deformed crystal (see Chapter 3). It is sometimes simpler to analyse the splitting pattern by inspection of the angular form of the wavefunction. Such is the case of F-centres in the alkali halides, for which the orbital wavefunctions transform as the A_1 and T_1 representations of the octahedral group. Consider the transition between the $^2A_1$ (ground state) and $^2T_1$ (excited state). Under a [100] stress the crystal has tetragonal symmetry, and the basis function T_1 splits into an orbital doublet (E) and an orbital singlet (A_2). Since the $^2A_1$ ground state is unaffected two components are observed. For a general stress direction, [*hkl*], three components are observed.

In a third category the centres possess *both* orientational and electronic degeneracy, both types of which are removed by uniaxial stress. Only defects with tetragonal or trigonal symmetry having doubly degenerate electronic state (E symmetry) are involved.

The number of components, shifts, and polarizations for each of the three classes are tabulated by Kaplyanskii (1964), Runciman (1965) and Hughes and Runciman (1966). Experimentally the stress is applied at a temperature in the range 1.6–77 K using a system of levers and standard weights or by hydraulic techniques. The stress splitting patterns of zero-phonon lines associated with numerous defects in the alkali halides, alkaline earth fluorides, and alkaline earth oxides have been reported in the literature (Fitchen 1968; Hayes and Stoneham 1974; Hughes and Henderson 1972). In general, the magnitudes of the splittings increase linearly with increasing stress. Analysis of the stress patterns was used to determine the symmetry and orientation of the defects in the various crystals. By plotting the line shifts as a function of stress the stress coupling operators, A_{ij}, defined in eqn (6.20) may be deduced. Examples of such measurements for colour centres are discussed in Chapter 7. The removal of electronic degeneracy associated with the ^2E state of Cr^{3+} ions in octahedral sites in magnesium oxide when the symmetry is reduced by uniaxial stress was discussed in Chapter 3.

6.4.2 *The Zeeman effect*

The Zeeman effect is the splitting of energy levels by a static magnetic field. Examples of the application of the Zeeman effect in optical spectroscopy were described in Chapters 4 and 5. The Zeeman splitting of optical lines by a static magnetic field is related to the removal of the spin degeneracy of the levels involved in the transition. The magnetic properties of these levels can be described by an *effective* or *fictitious spin*, S; in general there are $2S+1$ orientations of the magnetic dipole in a magnetic field. A *spin Hamiltonian*, containing only polynomials in the electronic spin quantum number of the state, is used to calculate by perturbation theory the energy level shifts induced by the magnetic field. The numerous forms of the spin Hamiltonian appropriate to crystals with particular symmetries are discussed in texts on electron spin resonance (see for example Abragam and Bleaney 1970; Orton 1968). This Hamiltonian is a convenient shorthand representation of the experimental results, in which the magnetic splittings are related to the g-values, fine structure, and hyperfine structure parameters of the centre. Frequently the hyperfine splittings in solids are hidden in the optical line-width. In this situation Zeeman studies reveal information only about electronic structure.

As an example consider the optical Zeeman effect of the R-line of $MgO{:}Cr^{3+}$, the zero-phonon $^2E \rightarrow {}^4A_2$ transition on Cr^{3+} ions in octahedral symmetry. In the A_{2g} state ground state the individual spins are coupled to give a total spin $S = \frac{3}{2}$. The first excited state, E_g, is a spin doublet ($S=\frac{1}{2}$). The R-line emission, which involves the transition $^2E_g \rightarrow {}^4A_{2g}$ is both spin- and parity-forbidden (Section 4.5). In consequence the R-line emission is a weak magnetic dipole transition ($f \sim 10^{-6}$), spin–orbit coupling being responsible for relaxing the spin-forbidden nature of the transition. The splitting of the ^2E and $^4A_2$ levels (Fig. 6.12(a)) is calculated using the spin Hamiltonian

$$\mathscr{H} = \mu_B \boldsymbol{B} . \hat{g} . \boldsymbol{S} \tag{6.21}$$

where $\mu_B = 9.27 \times 10^{-24}\,\mathrm{JT}^{-1}$ is the Bohr magneton, $\boldsymbol{S}$ is the fictitious spin

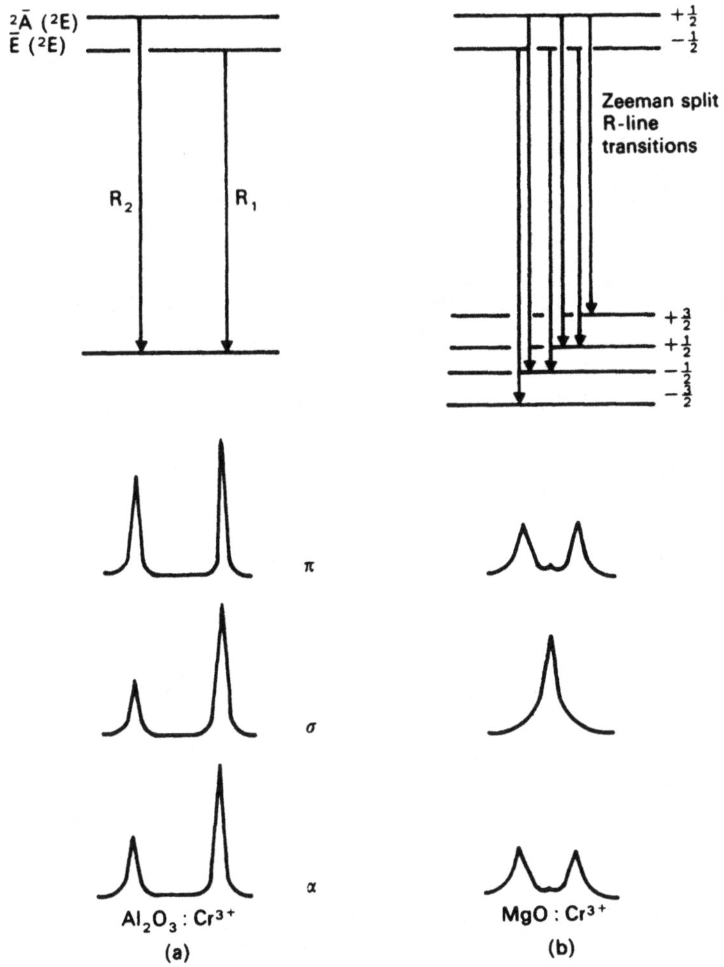

FIG. 6.12. The ^{2}E level of Cr^{3+} ions in aluminium oxide is split into $\bar{E}(^2E)$ and $2\bar{A}(^2E)$ by the trigonal crystal field. The resulting R_1 and R_2 zero-phonon lines when measured in linear polarization at 77 K have the relative intensities in (a). No such crystal field splitting occurs for Cr^{3+} ions in octahedral sites in magnesium oxide. The relative intensities of the linearly polarized Zeeman components observed at 77 K for a magnetic field of about 2 T along a ⟨100⟩ crystal axis are shown in (b).

and the spectroscopic splitting factor, $\hat{g}$, is a tensor of second rank. Since the projection quantum number M_s takes the values of $M_s = +S, S-1 \ldots -S$, we see that in a magnetic field the $^4A_2$ state splits into four levels characterized by $M_s = \pm\frac{3}{2}, \pm\frac{1}{2}$, and the ^{2}E excited state splits into two levels ($M_s = \pm\frac{1}{2}$). In both cases the spin energy levels deviate linearly with increasing magnetic field, i.e.

$$E = M_s \mu_B g(\theta) B \tag{6.22}$$

where θ is the angle between the direction of $\boldsymbol{B}$ and the cubic z-axis and $g(\theta)$ is isotropic for the ground state and slightly dependent on orientation in the ^{2}E excited state. In both ground and excited states $g \sim 2.0$. The transition probabilities of the various Zeeman components are calculated in Chapter 9, and the theoretical intensities for linearly polarized light given in Fig. 9.2. Since in good-quality crystals the R-line width at 77 K is only ~ 0.025 nm, the magnetic splitting of about 0.09 nm T^{-1} is easily resolved with a high-resolution monochromator and modest magnetic field as Fig. 6.12(b) shows. Conventional electromagnets produce magnetic fields of up to 2.5 T; much higher fields ($\sim$ 10.0 T) are achieved in superconducting magnets.

In order to determine the dipole nature of a transition a unique axis in the crystal is necessary so as to define σ-, π-, and α-polarizations. The magnesium oxide crystal is optically isotropic, and it is then necessary to impose a unique symmetry axis by the application of an appropriate external perturbation. The Zeeman field provides a unique symmetry axis for magnesium oxide, whereas in aluminium oxide there is an intrinsic symmetry axis associated with the crystal structure. In the presence of such a symmetry axis, comparison of the π-, σ-, and α-polarization patterns can be used to determine whether the transition is electric or magnetic dipole in character. The principle of this test can be explained by reference to Fig. 4.8, and illustrated by comparison of the polarization patterns of the R-line spectra in magnesium oxide and aluminium oxide. The ^{2}E and $^4A_2$ levels of Cr^{3+} in aluminium oxide are each split into two levels by the trigonal crystal field, the splittings being 29 cm^{-1} and 0.38 cm^{-1}, respectively. This large ^{2}E splitting results in two well-separated zero-phonon lines, R_1 and R_2, from the $\bar{E}(^2E)$ and $2\bar{A}(^2E)$ levels as Fig. 6.12(a) shows; the relative intensities of the R_1- and R_2-lines of ruby in π-, σ-, and α-polarizations are also shown. The senses of these linear polarizations defined with respect to the local symmetry axis are given in Fig. 4.8. In aluminium oxide the symmetry axis of all the Al^{3+} sites point along the optical axis of the crystal. For an electric dipole transition the matrix elements for $\boldsymbol{\mu}_e.\hat{\varepsilon}_E$ are identical for $\hat{\varepsilon}_E$ along $\hat{x}$ and $\hat{y}$ (and for all directions in the xy-plane) but different for $\hat{\varepsilon}_E$ along $\hat{z}$. However, as Fig. 4.8 shows, σ- and α-polarized transitions contain $\hat{\varepsilon}_E$ along $\hat{z}$. Hence for an electric dipole transition we predict that the σ- and α-polarization patterns will be identical but will differ from the π-polarized pattern. Inspection of the relative intensities of the R_1- and R_2-lines in ruby (Fig. 6.12(a)) show that these lines are induced by an electric dipole transition. A similar analysis for magnetic dipole transitions, for which the relevant operator is $\boldsymbol{\mu}_m.\hat{\varepsilon}_B$, shows that for π- and α-polarizations $\hat{\varepsilon}_B$ is in the xy-plane while $\hat{\varepsilon}_B$ points along the z-axis for σ-polarization. In consequence π- and α-polarization patterns are identical but differ from the σ-polarization pattern. Figure 6.12(b) shows the experimental Zeeman pattern of the R-line in magnesium oxide for π-, σ-, and α-polarizations, where these linear polarizations are defined with respect to the direction to the magnetic field applied along a $\langle 100 \rangle$ axis. The similarity of the π and α patterns along with the different σ pattern confirms the magnetic dipole nature of this transition.

6.4.3 Electric field effects

The splitting of spectral lines by an applied electric field is known as the Stark effect. In atomic hydrogen where the 2s and 2p states are very close together the electric field-induced shifts of the 2p → 1s lines are linear in the applied field. In heavier atoms, where spin–orbit splittings are large, the shifts occur to second order in perturbation theory because states of opposite parity mixed by the applied field are well separated in energy. Application of a strong steady electric field to a solid can result in splittings which may be more difficult to detect because of the large linewidths in solids. Nonetheless, both linear and quadratic Stark shifts have been reported in the literature. In favourable circumstances the very high-resolution technique of optical hole-burning has been used in conjunction with the Stark effect to reveal spectral detail not resolved by conventional techniques. Examples are given in Section 7.4.3.

6.5 Method of moments in optical spectroscopy

In many cases the linewidths of optical spectra in solids mask fine structure in the energy level diagram. Although for zero-phonon lines clearly resolved splittings can be obtained by the application of external perturbations, such splittings are not usually resolved for broad bands. Henry *et al.* (1965) showed that in such circumstances it is advantageous to measure the moments of the bandshape. Figure 6.13 shows an asymmetric broad band, identifying the band peak I_0, the centroid E_0, and full width at half maximum (FWHM), Γ. If this is an absorption band then $I(E)$ is a measure of the absorption coefficient, $\alpha(E)$. It is convenient to consider energy intervals $E_j - E_{j-1}$ and to write the area under the band as

$$A = \sum_j (E_j - E_{j-1}) I_j \tag{6.23}$$

where the summation is over the full band. We can also write the area as

$$A = \int I(E) \mathrm{d}E. \tag{6.24}$$

This will be referred to as the *zeroth moment* of the band, M_0. For an absorption band where $I(E) = \alpha(E)$ the zeroth moment is a measure of the density of absorbing centres (eqns 6.10a and 6.10b).

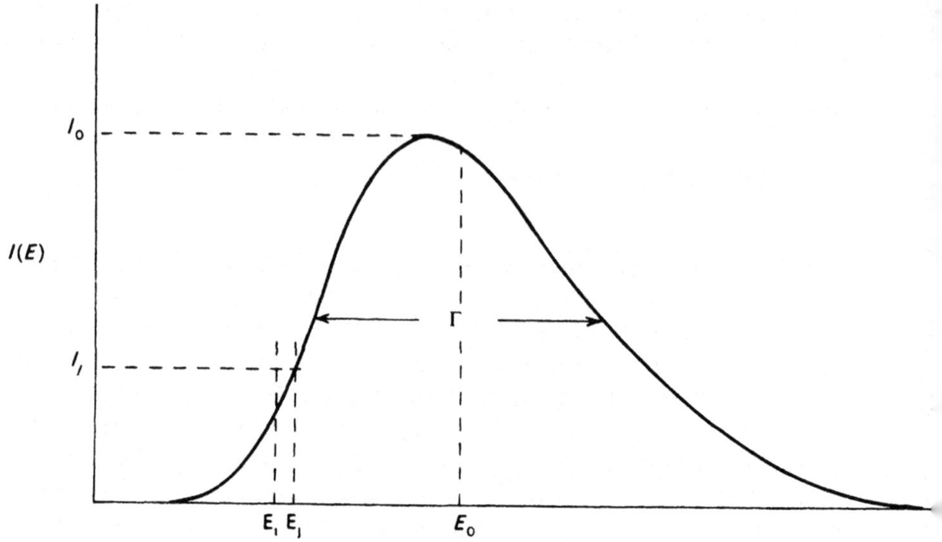

FIG. 6.13. Indicating how the moments of the bandshape of a transition with FWHM Γ and peak energy E_0 may be calculated.

We define the *first moment*, M_1, by

$$M_1 = \frac{1}{M_0} \sum_j E_j (E_j - E_{j-1}) I_j \tag{6.25}$$

or alternatively by

$$M_1 = \frac{1}{M_0} \int I(E) E \, \mathrm{d}E. \tag{6.26}$$

We define the *centroid* of the band as being the first moment: $E_0 = M_1$. It is convenient to define higher moments with respect to the centroid. These are defined as

$$M_n = \frac{1}{M_0} \sum_j (E_j - E_0)^n (E_j - E_{j-1}) I_j \tag{6.27}$$

or

$$M_n = \frac{1}{M_0} \int (E - E_0)^n I(E) \mathrm{d}E. \tag{6.28}$$

For a symmetrical band, E_0 is the position of the band peak and $M_n = 0$ for all odd-integer values of n. If the band has a Gaussian bandshape we find

$$M_0^{(G)} = 1.065\, I_0 \Gamma \tag{6.29}$$

while for a Lorentzian bandshape the numerical constant is 1.57.

6.5.1 *Moments of a vibronically broadened band*

We consider the values of the moments for the case of a transition broadened by a strong coupling with the lattice, such as was described by the single configurational coordinate model in the harmonic approximation in Chapter 5. At $T=0$ the absorption bandshape function is (from eqn 5.71)

$$I_j = I(E_j) = I_0 \sum_m \frac{\exp(-S)S^m}{m!} \delta(E_{zp} + m\hbar\omega - E_j) \tag{6.30}$$

where E_{zp} is the energy of the zero-phonon line and ω is the frequency of the representative vibrational mode. We evaluate the moments making use of the following useful formulae:

$$\sum_{m=0}^{\infty} \frac{\exp(-S)S^m}{m!} = 1,$$

$$\sum_m \frac{\exp(-S)S^m}{m!} m = S,$$

$$\sum_m \frac{\exp(-S)S^m}{m!} m^2 = S(S+1),$$

$$\sum_m \frac{\exp(-S)S^m}{m!} m^3 = S(S^2+3S+1). \tag{6.31}$$

The centroid is obtained from eqn (6.25)

$$E_0 = M_1 = \frac{\sum_j I_j E_j \hbar\omega}{\sum_j I_j \hbar\omega} = \frac{\sum_j I_j E_j}{\sum_j I_j} \tag{6.32}$$

where $E_j - E_{j-1}$ is given by $\hbar\omega$ for all values of j. Using eqn (6.30) this can be written as

$$E_0 = \frac{\sum_m \frac{\exp(-S)S^m}{m!}(E_{zp} + m\hbar\omega)}{\sum_m \frac{\exp(-S)S^m}{m!}}$$

$$= E_{zp} + S\hbar\omega \tag{6.33}$$

making use of eqn (6.31). If we define the centroid relative to the zero-phonon line we get $E_0 = S\hbar\omega$.

The formula for the higher moments (eqn 6.27) defined with respect to the centroid can similarly be written

$$M_n = \sum_m \frac{\exp(-S)S^m}{m!} [(m-S)\hbar\omega]^n \tag{6.34}$$

from which we obtain

$$M_2 = S(\hbar\omega)^2 \tag{6.35}$$

and

$$M_3 = S(\hbar\omega)^3. \tag{6.36}$$

For the emission band we can apply a similar analysis and we obtain the above formulae for M_2 and M_3, but the centroid is given by $E_{zp} - S\hbar\omega$.

In general the absorption and emission bands are not a series of delta-functions, as eqn (6.30) implies; the effect of the other vibrations is to create a single band, as described in Chapter 5. The peaks, I_j, in eqn (6.30), however, give a good description of the band envelope, and eqns (6.33, 35, 36) are expected to apply to the moments of this band. These equations show that accurate measurements of the moments give the concentration of defects (from M_0), the position of the centroid relative to the zero-phonon line ($E_0 = M_1 = S\hbar\omega$), the energy of the representative vibrational mode ($\hbar\omega = M_2/M_1$), and the Huang–Rhys parameter ($S = M_1^2/M_2$).

For large values of S the bandshape is approximately Gaussian and for such a shape we calculate the second moment as

$$M_2 = \frac{\Gamma^2}{8\ln 2}. \tag{6.37}$$

From eqn (6.35) this gives

$$\Gamma^2 = 8\ln 2\, S(\hbar\omega)^2. \tag{6.38}$$

This analysis of the lattice-broadened lineshapes applies only at $T = 0$ K. Carrying out a proper thermal average over the occupied phonon levels in the initial state $|i\rangle$ yields the equation for the half width

$$\Gamma(T)^2 = \Gamma(0)^2 \coth(\hbar\omega/2kT) \tag{6.39}$$

where $\Gamma(0)^2 = 8\ln 2\, S\hbar^2\omega^2$. In consequence we may determine all the important parameters in the configurational coordinate model through measurements of the bandwidth as a function of temperature. When applied properly the method is widely applicable even to bands which show sharp vibronic structure. However, the most detailed application has been to F-centres in alkali halide crystals, as we discuss in Section 7.2.4.

6.5.2 *Application to perturbation spectroscopy*

In eqn (6.30) the optical bandshape was represented as a series of delta-functions. Any external perturbation may shift or split such delta-function lines resulting in a broadening of the optical band. Henry *et al.* (1965) have shown how the method of moments may be applied to experiments involving hydrostatic pressure, uniaxial stress, magnetic field, or electric field. They show that for defects with small spin–orbit coupling the zeroth moment of the band which measures the defect concentration (eqns 6.9, 6.23) is unchanged by the applied field. However, in general the perturbation changes the shape function from $g(E, \hat{\varepsilon})$ to $g(E, \hat{\varepsilon})+\Delta g(E, \hat{\varepsilon})$. The corresponding change in the nth moment is then given by

$$\Delta M_n=(M_0)^{-1}\int(E-E_0)^n\Delta g(E, \hat{\varepsilon})\mathrm{d}E \quad (6.40)$$

where E_0 is the band centroid. It follows that changes in the moments of the bandshape induced by the applied perturbation may introduce changes in the linear and circular polarization of the absorbed or emitted light intensity. We conclude this section with a summary of the principal results obtained using the F-centre as a prototype for such studies.

Very crudely, the F-band in the alkali halides is analogous to an S → P transition on an alkali atom. The excited state is split by spin-orbit coupling into $P_{3/2}$ and $P_{1/2}$-like states, their separation Δ (say) being masked by the large width of the F-band. An applied magnetic field tends to broaden the F-band measured using unpolarized light, due to the removal of angular momentum degeneracies in the ground and excited states. When measured using circularly polarized light propagating along the magnetic field, the induced broadening shows up as a shift in the first moment of the band. Henry *et al.* (1965) showed that the shift observed using $\sigma_\pm$-polarized light is given by

$$\Delta M_1(\pm) = \pm(g_{\mathrm{orb}}\mu_B B+\tfrac{2}{3}\Delta\langle S_z\rangle) \quad (6.41)$$

which is the sum of diamagnetic and paramagnetic terms. The temperature dependence of $\Delta M_1(\pm)$ is given by $\langle S_z\rangle$, the average component of spin along the direction of the light. Since

$$\langle S_z\rangle = -\tfrac{1}{2}\tanh(\mu_B B/kT) \quad (6.42)$$

reflects the Boltzmann distribution in the ground-state spin system, it is obviously the dominant term at low temperature and high field. Computation of higher-order shifts is quite complicated and the reader is referred to the original literature (e.g. Henry *et al.* 1965; Henry and Slichter 1968). Such calculations demonstrated that by measuring changes in the second and third moments of the bands (ΔM_2 and $\Delta M_3(\pm)$) one could evaluate the contributions to the bandwidth (M_2) by cubic (A_{1g}) and non-cubic (E_g, T_{2g}) modes of vibration and by spin–orbit coupling. Furthermore, it was possible to explain the structure in the F-bands of the caesium halides in terms of the very large spin–orbit coupling on the Cs^+ ions in the neighbourhood of the anion vacancy.

Similar analyses were carried out of the effects of applied hydrostatic pressure and uniaxial stress. An hydrostatic pressure causes an electronic perturbation of the F-centre of symmetry A_{1g}, which shifts all levels of the first excited state (P-like) by the same amount. In consequence the F-band is shifted rigidly (i.e. without change of shape) to higher energies, the shift being linear in pressure, p. In other words $\Delta M_0 = 0$, but $\Delta M_1 = -A_1 p$, where A_1 measures the strength of coupling to vibrational modes of A_{1g} symmetry. However, a uniaxial stress, σ_{hkl}, applied parallel to a particular $\langle hkl \rangle$ direction shifts the three differently polarized S–P transitions by different amounts. In general the shift in the band centroid, M_1, is different for light polarized parallel to the stress axis than for light polarized perpendicular to the stress axis. Henry *et al.* (1965) show that the changes in first moments yield for a [100] stress

$$\Delta M_1(\parallel) = \tfrac{1}{3}\sigma_{100}(A_1 + 2A_3) \tag{6.43}$$

$$\Delta M_1(\perp) = \tfrac{1}{3}\sigma_{100}(A_1 - 2A_3), \tag{6.44}$$

and for a [110] stress

$$\begin{aligned} \Delta M_1(\parallel) &= \tfrac{1}{6}\sigma_{110}(2A_1 + A_3 + 3A_5) \\ \Delta M_1(\perp[1\bar{1}0]) &= \tfrac{1}{6}\sigma_{110}(2A_1 + A_3 - 3A_5). \end{aligned} \tag{6.45}$$

In other words, measurements can be made of the coupling constants A_i, for modes of $A_{1g}(A_1)$ $E_g(A_3)$ and $T_{2g}(A_5)$ modes of vibration. A similar analysis of the changes in second, ΔM_2, and third, ΔM_3, moments was used by Schnatterly (1965) to estimate the contributions by cubic and non-cubic modes to the F-band width in the potassium and rubidium halides. Obviously where both stress effects and magnetic circular dichroism are measured there is an over-determination of the parameters which contribute to the width of the F-band. Such effects are discussed at greater length in Chapter 7.

6.6 Optically detected magnetic resonance (ODMR)

In optical absorption spectroscopy, electronic transitions (usually) out of the ground state may result in one of a rich tapestry of possible bandshapes, depending upon the strength of the electron–phonon coupling. Photoluminescence measurements involve transitions which originate on an excited electronic state; the terminal state is (frequently) the electronic ground state. Optical absorption and photoluminescence spectra involving the same pair of electronic states do not necessarily have identical bandshapes (Sections 5.3–5.5). In view of this and also because of overlapping optical bands it can be difficult to assign particular optical absorption and luminescence bands to particular optical centres.

Electron spin resonance (ESR) and its derivative electron nuclear double resonance (ENDOR) may provide a precise model of an atom or defect and its immediate environment in a solid. For observation of ESR an atom or defect must have a non-zero total spin, $S \geqslant \frac{1}{2}$. The traditional application of

ESR is to measurements on the ground state of an electronic spin system (e.g. Abragam and Bleaney 1972; Wertz and Bolton 1972; Orton, 1968). In 1959 Geschwind and his colleagues first reported studies of electron spin resonance (ESR) in optically excited states. Their interest was in measuring the magnetic properties of the excited electronic state and the coupling of the spin system to the phonon spectrum of the crystal. Since the mean lifetimes of excited states of electronic centres tend to fall in the range 10^{-3}–10^{-8} s, it is no trivial matter to measure excited-state ESR using the normal microwave detection techniques pioneered in ground-state studies. Geschwind *et al.* (1959) developed techniques in which the excited-state ESR was detected optically; the absorption of one microwave photon *triggers* the absorption or emission of one optical photon. As we shall see, in favourable cases this method enables one to correlate in a single experiment ESR spectra in the ground state and the excited state with particular optical absorption and luminescence spectra. The technique may be termed optically detected magnetic resonance (ODMR). As conceived and developed by Geschwind and his colleagues (Geschwind *et al.* 1959, 1965) ODMR and the information derived therefrom was of interest in its own right. In the hands of defect spectroscopists, recognizing the difficulty of unambiguously assigning ground-state ESR and optical spectra to particular defect species, ODMR was used for the identification and characterization of optical centres in semiconductors and insulators.

6.6.1 Basic features of ESR and ENDOR

The effect of an applied magnetic field, B, is to remove spin degeneracy according to eqn (6.22) resulting in the splitting of a particular level into $(2S+1)$ levels separated by $g(\theta)\mu_B B$. Transitions may be induced between these levels by an oscillating magnetic field, B_1, of frequency, ν, applied in a direction perpendicular to the static magnetic field. These are magnetic dipole transitions. When a transition occurs with the selection rule $\Delta M_s = \pm 1$ we obtain the resonance condition

$$h\nu = g(\theta)\mu_B B. \tag{6.46}$$

For an experiment at a field $B = 0.34T$ and with $g(\theta) \sim 2.0$ the resonance occurs at a frequency $\nu \sim 9.5 \times 10^9$ Hz in the X-band microwave range. In most spectrometers the microwave frequency is held constant and the static field varied until resonance is observed. Obviously resonant absorption of microwaves occurs only if there are population differences between the $(2S+1)$ spin levels. If the spin system is in thermal equilibrium then the lowest-lying level has the greatest population. Hence ESR transitions, which tend to equalize populations, are predominantly absorptive. Spin–lattice relaxation tends to re-establish thermal equilibrium. so that a steady absorption of microwave energy is observed. The net absorption of microwave power is then proportional to the population difference between the states. For $S=\frac{1}{2}$ there are just two levels, $M_s=\pm\frac{1}{2}$, in a magnetic field, and application of Boltzmann statistics shows that the population difference as a fraction of the total population is given by $\Delta N/N \simeq h\nu/2kT$ in the limit that $h\nu \ll kT$. Thus the sensitivity of the ESR system increases with increasing microwave frequency and decreasing temperature. Roughly, at X-band the sensitivity at $T=4$K is about 10^{10} spins/mT linewidth.

In practice the ESR spectrum is more complex than the single-line spectrum implied by eqn (6.46). This is because crystal field effects for $S>\frac{1}{2}$ and the anisotropy in $g(\theta)$ for low-symmetry sites result in a multiplicity of adjacent electronic levels, each of which is further split by the Zeeman field. The spectrum is then analysed in terms of the general spin Hamiltonian

$$\mathscr{H}=\mu_B \boldsymbol{B}.\hat{\boldsymbol{g}}.\boldsymbol{S}+\boldsymbol{S}.\hat{\boldsymbol{D}}.\boldsymbol{S}+\boldsymbol{I}.\boldsymbol{A}.\boldsymbol{S}+\sum_i \boldsymbol{I}_i.\boldsymbol{A}_i.\boldsymbol{S} \qquad (6.47)$$

where $\boldsymbol{g}$, $\boldsymbol{D}$, $\boldsymbol{A}$, and $\boldsymbol{A}_i$ are all second-rank tensors. The first terms is just the electronic Zeeman interaction and the second term represents the crystal field interaction on the electronic spin $S \geqslant 1$. $\boldsymbol{I}.\boldsymbol{A}.\boldsymbol{S}$ represents the hyperfine coupling between the central nuclear spin, $\boldsymbol{I}$, and the electronic spin, $\boldsymbol{S}$, where $\boldsymbol{A}$ is the hyperfine interaction tensor. Finally, there is a hyperfine interaction between the electron spin and the nuclear magnetic moments of neighbouring atoms (i). This last term gives rise to what is termed *super* or *transferred* hyperfine structure in the ESR spectrum. In the presence of hyperfine interactions the selection rule for ESR becomes $\Delta M_s=\pm 1$, $\Delta M_I=0$ so that nuclear spin is conserved.

Clearly the ESR spectra of defects may be complex due to many overlapping lines. The ESR spectrum of F-centres in potassium chloride consists of a single, broad, and structureless line due to many overlapping hyperfine components. To measure the hyperfine structure in such cases one must resort to Electron Nuclear Double Resonance (ENDOR), a technique which combines the inherent sensitivity of ESR with the high resolution of nuclear magnetic resonance (Wertz and Bolton 1972). In ENDOR a particular ESR transition is partially saturated with microwave power. Simultaneously an intense radiofrequency field is applied, the frequency of which is adjusted until resonance is achieved between the nuclear spin states. The disturbance in the populations of these states is manifest as a change in the ESR signal. The orientation dependence of ENDOR spectra then locates the position of particular nuclei in the lattice. The sureness with which ESR/ENDOR techniques measure defect structure confers upon these techniques the power of microscopy on an atomic scale. The wealth of experimental information enables one not only to identify the structure of the defect but also to measure the displacements of ions in the neighbourhood of the paramagnetic species and to map out the unpaired electron spin over the neighbouring lattice. As we discuss in the next chapter, in alkali halide crystals the F-centre electron spreads out over more than ten neighbouring shells of nuclei.

ESR and ENDOR studies have been used over more than three decades to probe the electronic ground state of an amazing variety of impurities and lattice defects in many different types of solid. Very early in this period workers became interested in the magnetic properties of excited electronic states. To make ESR measurements in such states requires that a sufficient population be maintained in the excited state of interest. Usually such a population cannot be achieved by thermal processes, since in thermal equilibrium the population in a state some 20 000 cm^{-1} above the ground state is negligibly small; one must necessarily appeal to optical excitation. Then the important experimental parameter becomes the lifetime of the excited state, irrespective of whether the state decays radiatively or non-radiatively. For long-lived excited states ($\tau_R \geqslant 1$ s) a substantial excited-state population may be achieved for relatively modest pump power. Conventional ESR spectroscopy may then be used to detect this state. Examples are the metastable spin triplet and quartet states, respectively, of F_2- and F_3-centres in halide crystals. If $\tau_R \leqslant 1$ ms then the steady state population achieved by optical excitation may be too little for the necessary sensitivity of $\sim 10^{11}$ spins to be achieved. In this circumstance, however, we may be able to use the much more sensitive technique of ODMR.

6.6.2 The ODMR experiment

In ODMR measurements a sufficiently large population is created in the excited state by optical pumping, and microwave transitions induced between the Zeeman levels of the excited state are detected by the change in intensity of the absorption or luminescence spectrum. Figure 6.14 is a schematic drawing of an ODMR spectrometer. There are three necessary channels; a microwave channel and channels for optical excitation and detection. The microwave system is relatively simple comprising a klystron or Gunn diode operating at some frequency in the range 8.5–50 GHz, followed by an isolator to protect the microwave source from unwanted reflected signals in the waveguide path. The microwave power is then square-wave modulated at frequencies up to 10 kHz, using a PIN diode. A variable attenuator determines the power incident upon the resonant cavity, although for high-power operation a travelling-wave amplifier might be added to the waveguide system.

The cheapest source for optically exciting the sample is a broadband lamp used in conjunction with optical filters or a low-resolution monochromator. Ion lasers may also be used: they can direct intense radiation onto the sample but have the disadvantage that their sharp lines may not coincide with the absorption spectra of the electronic centre. Alternatively, tunable dye lasers may be used; suitable choice of dyes allows an output from about 300 nm to

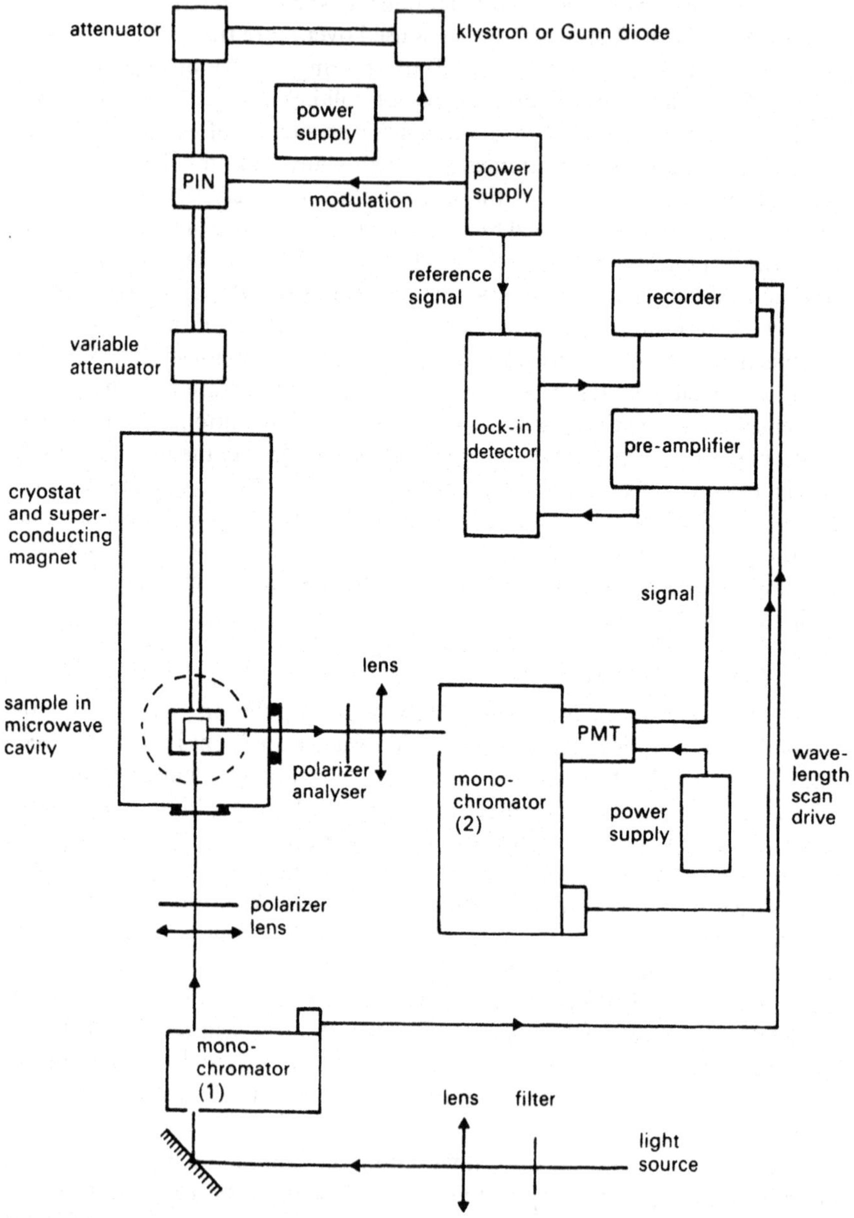

FIG. 6.14. A schematic representation of a spectrometer for measuring optically detected magnetic resonance (ODMR).

the near infrared region. Light from the excitation source is focused onto the sample using suitable lenses. In some cases the radiation is linearly or circularly polarized. The sample is contained in the microwave cavity, which is specifically designed to allow optical access of the sample for light travelling either parallel or perpendicular to the magnetic field direction. In most cases the cavity is submerged in liquid helium to achieve as large a population difference as possible between the Zeeman levels. The magnetic field is provided either by an electromagnet ($B \simeq 0$–2.0 T) or a superconducting solenoid ($B \simeq 0$–6.5 T). Radiation from the sample is focused onto the detection system, which in its simplest form consists of suitable filters, a polarizer, and photomultiplier tube. A high-resolution monochromator may be used instead of the filters to resolve sharp features in the optical spectrum. The signal from the phototube is processed using a phase-sensitive detector, or alternatively using computer data collection with a multichannel analyser or transient recorder. The recorded spectrum is plotted out using a pen recorder as a function of either magnetic field or of photon energy (or wavelength). With such an experimental arrangement one may examine the spectral dependence of the ODMR signal on the wavelength of the optical excitation or on the wavelength of the detected luminescence by use of one of the two scanning monochromators. This is a very important ingredient in the ODMR experiment, the value of which will become apparent in our subsequent discussion.

There are three methods by which one may study ODMR (Geschwind 1972); these employ (i) selective reabsorption, (ii) a high-resolution optical spectrometer, or (iii) the measurement of circular dichroism or circular polarization. The first of these techniques is not general: it is of purely historic interest, since it was used by Geschwind *et al.* (1959) in their pioneering studies on the detection of ODMR in ruby. Optical detection of ESR using high-resolution spectrometers or by monitoring the circular polarization of luminescence is described in detail in Chapter 12, where the ODMR spectra of several systems are analysed. We discuss here only the measurement of the circular polarization of absorption and its use as a probe to the magnetic splittings of the electronic states involved in the optical transitions.

6.6.3 Magnetic circular dichroism (MCD)

Consider that circularly polarized absorption transitions are excited between two Kramers doublets as shown in Fig. 6.15(a). With light propagating along the direction of the magnetic field, the selection rules are such that σ_--polarized light induces absorption transitions in which $\Delta M_s = +1$ and σ_+-polarized light induces absorption transitions in which $\Delta M_s = -1$. As a result of the electronic Zeeman effect the absorption peak splits into two overlapping bands (Fig. 6.15(b)) centred at different frequencies with absorption coefficients $\alpha_\pm(\nu)$ in σ_+ and σ_--polarizations that are different at particular

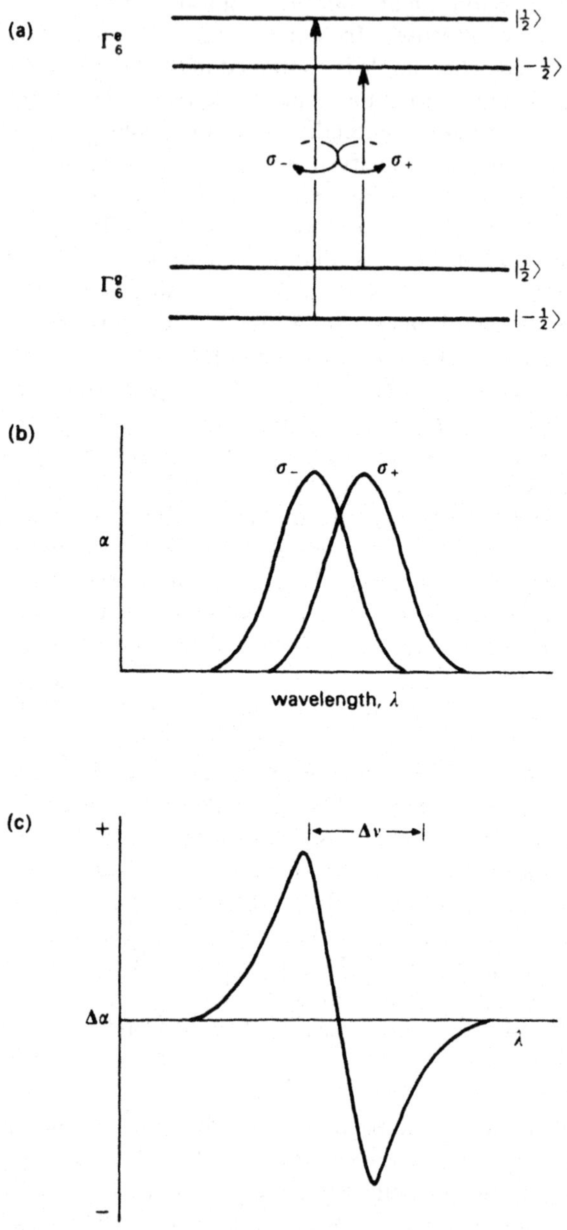

FIG. 6.15. (a) Identifies the splittings of $S=\frac{1}{2}$ ground and excited states in a magnetic field and the hypothetical selection rules for circularly polarized transitions between the levels. In (b) we show the schematic shift of a broad absorption transition between the two polarizations, whereas (c) illustrates the MCD signal.

frequencies. The peaks in the two oppositely polarized bands (Fig. 6.15(b)) are separated in energy by $(g_e+g_g)\mu_B B$, where the g-values refer to the excited (e) and ground (g) states. This energy difference corresponds to a frequency splitting $\Delta\nu=(g_e+g_g)eB/4\pi m$, which for $g_e=g_g=2.0$ and $B=1$ T gives a separation between band peaks of $\simeq$0.04 nm for a band centred at 500 nm. Materials which absorb different amounts of differently polarized light are said to be *dichroic*. In a magnetic field the difference ($\Delta\alpha(\nu)$) in the absorption coefficients for σ_+- and σ_--circularly polarized light is referred to as *magnetic circular dichroism* (MCD).

As Fig. 6.15(c) shows, $\Delta\alpha(\nu)$ is strongly frequency dependent. According to the above definition of $\Delta\alpha(\nu)$,

$$\Delta\alpha(\nu)=\alpha_+(\nu)-\alpha_-(\nu). \tag{6.48}$$

Using eqn (6.2) in the limit of small absorption coefficient leads to

$$\Delta\alpha(\nu)\simeq-\frac{2(I_+(\nu)-I_-(\nu))}{l(I_+(\nu)+I_-(\nu))} \tag{6.49}$$

where l is the sample thickness and $I_+(\nu)$, $I_-(\nu)$ refer to the transmitted intensities of the σ_+- and σ_--circularly polarized light at frequency ν. Experimentally we determine the transmitted intensities, and since it is a simple procedure to take sums and differences of measured quantities electronically it is convenient to define the MCD signal as

$$\Sigma(\nu)=\Delta\alpha(\nu)l=-\Delta I(\nu)/I_{DC}(\nu) \tag{6.50}$$

where $\Delta I(\nu)=I_+(\nu)-I_-(\nu)$ and $I_{DC}=\frac{1}{2}(I_+(\nu)+I_-(\nu))$. For most defect centres a splitting of only 0.04 nm is much less than the width of the absorption band, and may not be resolved in direct Zeeman effect measurements on the broad band. However, this Zeeman structure may be resolved by measuring $\Delta\alpha(\nu)$ as a function of magnetic field, as can be seen from a simple estimate. We approximate the MCD signal, $\Sigma(\nu)$, as the product of the magnetic splitting, $\Delta\nu$, the rate of change of the absorption coefficient with frequency, $\mathrm{d}\alpha(\nu)/\mathrm{d}\nu$, and the sample thickness (l). Hence $\Sigma(\nu)=\Delta\nu\times\mathrm{d}\alpha(\nu)/\mathrm{d}\nu\times l$. For a symmetrical, structureless band we approximate $\mathrm{d}\alpha(\nu)/\mathrm{d}\nu\simeq\alpha(\nu_0)/\Gamma$ so that

$$\Delta\nu=\frac{\Sigma(\nu)}{l}\times\frac{\Gamma}{\alpha(\nu_0)}.$$

In a typical experimental situation $\Sigma(\nu)\simeq l\Delta\alpha(\nu)\simeq10^{-5}$ and $l\alpha(\nu_0)\sim1$, then we estimate $\Delta\nu\simeq10^{-5}\Gamma$. For (say) F-centres in alkali halides, $\Gamma\simeq0.25$ eV $\simeq2000$ cm^{-1}, which gives $\Delta\nu\simeq0.02$ cm^{-1} or 0.05 nm, of the same order as the Zeeman splitting calculated above. Furthermore, although the intensity changes which determine the magnitude of $\Delta\alpha(\nu)$ may be quite small, they may be measured very precisely using lock-in techniques. This is done very efficiently by replacing the circular polarizer in the excitation system by a stress-modulated quarter-wave plate, a device which transmits circularly polarized light the polarization of which is switched between σ_+ and σ_- at the vibration frequency of the plate, usually *c.* 50 kHz. Using this piezo-optic modulator, MCD signals as low as 10^{-6} can be measured (Kemp 1966).

The MCD signal is not only strongly frequency dependent but temperature dependent also. Since at low temperatures the populations, $N_\pm$, of the $M_s = \pm\frac{1}{2}$ levels of the ground state are different it is convenient to specify absorption coefficients, $\gamma_\pm$, for $\sigma_\pm$ light, appropriate to the case that the particular $M_s = \pm\frac{1}{2}$ state is fully occupied. The MCD signal (eqn 6.50) is then given by

$$\Sigma = (n_+ \gamma_+ g(\nu_0 + \tfrac{1}{2}\Delta\nu) - n_- \gamma_- g(\nu_0 - \tfrac{1}{2}\Delta\nu))l$$

where the fractional populations of the ground-state spin levels are $n_\pm = N_\pm/(N_- + N_+)$ and $g(\nu \pm \frac{1}{2}\Delta\nu)$ is the shape function of the band. Since both ground and excited states are Kramers doublets we can write $\gamma_+ = \gamma_- = \gamma_0$ and hence $\Sigma(\nu) = \gamma_0 l g(\Delta\nu)(n_- - n_+)$. By writing $\alpha_0(\nu) = \alpha_+(\nu) + \alpha_-(\nu)$ we determine the MCD signal to be given by

$$\Sigma = \alpha_0(\nu) g(\Delta\nu) l (n_- - n_+) \tag{6.51}$$

since $(n_- + n_+) = 1$. Furthermore, $(n_- - n_+)$ is just the population difference between spin-levels in the ground $s = \frac{1}{2}$ state. When in thermal equilibrium we apply Maxwell–Boltzmann statistics and obtain

$$\Sigma = \alpha_0(\nu) g(\Delta\nu) l \tanh\left[\frac{g\mu_B B}{2kT}\right]. \tag{6.52}$$

In this expression $\alpha_0(\nu)$ and the sample thickness, l, are experimental constants and, in consequence, the MCD signal only varies through the Brillouin function for the $s = \frac{1}{2}$ ground state, i.e. $\tanh(g\mu_B B/2kT)$. This MCD signal is paramagnetic, being strongest at high field and low temperature; measurement of its magnitude probes the ground state magnetization. In order to test eqn (6.52) experimentally, it is best to work at either the positive or negative peak illustrated schematically in Fig. 6.15(c) and so maximize the MCD signal. Having thus obtained a suitable MCD signal, its variation with temperature and magnetic field can then be measured. In order to carry out ODMR, microwave radiation of fixed frequency, ν, is introduced, also while the optical wavelength is kept at the positive or negative peak in Fig. 6.15(c). The magnetic field is then adjusted until the ESR condition, $h\nu = g\mu_B B$, is satisfied. Since ESR transitions tend to equalize the fractional populations n_+ and n_-, resonance is observed as a decrease in $\Sigma(\nu)$ and as the microwave power is increased the MCD gradually tends to zero. This point is reached under conditions of saturated microwave pumping. In certain circumstances the ground-state spin polarization may be used to monitor *excited-state* ESR transitions, because of the selectivity of the transitions induced by circular polarized radiation. Examples involving the optical detection of ground-state and excited-state ESR using this MCD technique are discussed at length in Chapter 12 as are experiments involving the detection of ODMR by observation of the magnetic circular polarization of emission.

6.6.4 Advantages of ODMR spectroscopy

One important aspect of the ODMR method is high sensitivity. The technique is an example of 'trigger detection', one microwave photon in absorption triggering the detection of one optical photon emitted. The enhancement in sensitivity relative to the normal ESR technique is approximately in the ratio of optical to microwave frequency (i.e. $\sim 10^{15}/10^{10} = 10^5$). A simple example will serve to illustrate the high sensitivity of the optical detection technique. The total number of photons incident upon the phototube is fn_4/τ_R where we are monitoring the σ_+ light emitted by level 4 in Fig. 6.15 and f is a geometrical factor which takes into account the fact that the total radiant emission is isotropic. Typically f will be $\sim$0.1 per cent. If the fractional change in population induced by the magnetic resonance is η then the detected signal is just $\Sigma = (f\eta n_4/\tau_R)\varepsilon$, where ε is the efficiency of the phototube, this change being induced by magnetic resonance. This signal must be compared to the shot noise in the detector i.e. $N = (f\varepsilon n_4/\tau_R)^{\frac{1}{2}}$. Hence the signal-to-noise ratio Σ/N is given by

$$\Sigma/N = (f\eta^2 n_4 \varepsilon/\tau_R)^{\frac{1}{2}}$$

for a time constant of 1 s. Hence for $\Sigma/N = 2$ we have

$$n_4 = 4\tau_R/f\eta^2\varepsilon.$$

Although for excited states in thermal equilibrium the intensity of light may change by only $\sim$0.1 per cent, where selective feeding of the spin states results the changes may be very much larger. For ODMR in the spin triplet states of defects in oxides and semiconductors, η may lie between 0.1 and 3, so that if a phototube of $\sim$1 per cent efficiency is used the excited-state populations between $4 \times 10^{+2}\,\tau_R$ and $4 \times 10^{+5}\,\tau_R$ may be detected. Since for such defects lifetimes are observed in the range 3–25 ms, n_4 can range from 10 up to 10^4, far smaller than can be detected by conventional ESR.

The ability to detect very small concentrations and to measure very small magnetic splittings are not the sole attributes of ODMR. An additional advantage is the relatively simple equipment with which one may probe ODMR in molecular, ionic, and semiconductor systems. Furthermore, with the ODMR technique one may gather information on a wide range of important solid state processes including spin–lattice and cross relaxation, spin memory, energy transfer, electron–hole recombination, phonon bottlenecks, and spin coherence effects. Furthermore a major attribute of the

ODMR technique is illustrated in Fig. 6.16 showing the optical characteristics of the ODMR spectrum in calcium oxide. These spectra were measured at 18.7 GHz and 1.6 K with the magnetic field along a crystal ⟨100⟩ direction. A high-pressure xenon discharge lamp and monochromator (M_1 in Fig. 6.14) set at 400 nm was used to excite the fluorescence, which was detected through monochromator M_2 in Fig. 6.14. The spectrum consists of four equally spaced lines due to an $S=1$ state of a centre with tetragonal symmetry. Then with the magnetic field set at the strongest ODMR line the excitation wavelength is scanned using monochromator M_1 (Fig. 6.14) over

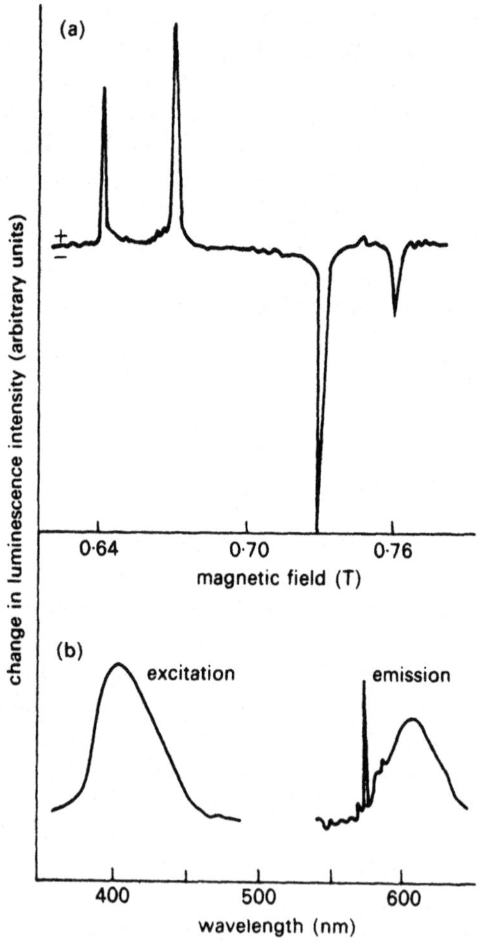

FIG. 6.16. (a) The ODMR spectrum of F-centres in calcium oxide measured at $\nu \simeq 18.7$ GHz and $T=1.6$ K. In (b) is shown how the intensity of the most intense ODMR line depends upon excitation and emission wavelength.

the visible and near ultraviolet region. A single broad structureless excitation peak is observed at 400 nm corresponding to the $^1A_{1g} \rightarrow {}^1T_{1u}$ absorption band of the F-centre (Fig. 6.16(b)). Subsequently, the excitation spectrometer is set at the peak of this excitation band and the same magnetic field, while the detecting monochromator (M_2 in Fig. 6.14), is scanned over the fluorescence spectrum. This spectral dependence (Fig. 6.16(b)) shows a sharp zero-phonon line at a wavelength of 574 nm with an accompanying broad vibronic sideband with peak at 602 nm. In a single experiment a unique and unambiguous relationship has been established between the ESR spectrum, absorption, and fluorescence bands of an intrinsic lattice defect.

6.7 Laser spectroscopy of solids

The discussion of luminescence spectroscopy so far has ignored the nature of the excitation source, broadband lamps and narrowband lasers being of equal use in creating excited-state populations. However, there are techniques specific to laser excitation which offer opportunities and challenges in all areas of optical investigation. The impact of lasers in spectroscopy has been all the more profound since the advent of tunable lasers and ultra-short pulse trains. Although much more expensive, tunable dye lasers are now a very powerful alternative to broadband lamps and monochromators for experiments in the 250–2500 nm wavelength range.

6.7.1 Developments in tunable lasers

Discrete lines from gas lasers (He–Ne, Ar^+, Kr^+) and solid state lasers (ruby, Nd-YAG) have been used for many years for studies of fluorescence spectra. Such studies are only possible when the available laser line overlaps the particular defect absorption spectrum. Despite a wide variety of available lines, and the use of frequency-doubling and Raman-shifting techniques, such lasers do not provide a continuous range of wavelengths. Modern organic dyes provide efficient luminescence bands with widths in the range 20–30 nm. Given the profusion of available dye–solvent mixtures, the luminescence spectra of dyes extends from *c.* 350–1000 nm. These are ideal media for tunable lasers since, by insertion of a dispersive element (e.g. prism, grating, etc.) inside the laser cavity, the dye laser output can be narrowed to a fraction of the overall width of the luminescence band. Further, by adjusting the dispersive element the frequency of the laser can be tuned over most of the luminescence band of the dye. The extent of this narrowing is dictated by the choice of dispersing element. Surprisingly (perhaps) only a fraction of the output power is lost in narrowing the spectral distribution of the laser beam. In other words, such a tuned laser output has the twin benefits of narrow bandwidth and high power. Choice is then dictated by the type of experiment being carried out. A simple prism can reduce the wavelength range over which the laser operates from (say) 50 nm to 2 nm and still enable the laser to maintain *c.* 50 per cent of the output power. A diffraction grating or birefringent filter can be used to reduce the output bandwidth to only 0.1 nm. By use of an external Fabry–Perot etalon dye lasers may be used to produce a bandpass of order 5–10 MHz. The ultimate in linewidths appears to be obtained using interferometer control of a ring dye laser, from which a stable output of 500–600 mW over a band pass of only 120–150 kHz is achievable in the visible region of the spectrum.

There have been very dramatic developments in dye laser technology for spectroscopy in the time domain. In some of the very early experiments on saturation spectroscopy, 30 ns pulses were obtained from dye lasers pumped by a pulsed nitrogen laser. Even shorter pulses (5–10 ps) may be obtained by synchronously pumping a dye laser using a mode-locked Ar^+ ion laser or a frequency-doubled, mode-locked Nd-YAG laser. However, the most dramatic results have been achieved using pulse compression in optical fibres; the pulse widths may then vary between 50–150 fs. This technique is especially important in the near-infrared region around 1.5 μm where the non-linear refractive index and negative group velocity dispersion result in pulse compression and propagation as solitons; i.e. pulse propagation without change of shape (Mollenauer and Stolen 1984). It is one unfortunate aspect of the uncertainty principle that we cannot combine such narrow lines (0.1 MHz) and short pulses (50 fs) in the same system. Nevertheless, these are complementary facilities which enable an ever-widening range of experiments both in the frequency and time domain to be performed with impressive ease.

6.7.2 Scattering experiments

Light-scattering experiments are now a routine feature in many optical laboratories. The first observations of light scattering by small particles were reported by Tyndall (1869). Subsequently theoretical work by Lord Rayleigh (1871) showed both that the scattered intensity varied as the fourth power of the frequency and that the scattering was due to molecules rather than dust particles. Many of the early studies were concerned with the depolarization of the light after being scattered by the dust-free atmosphere. Of course, in the pre-laser era sufficient light intensity could only be achieved by use of strongly condensing lenses to focus light onto the gas cell. Very great care was then necessary to obtain reliable depolarization measurements. Even in the laser era it is still essential to avoid any effects due to parasitic light which often plague light-scattering experiments.

A significant early result from scattering of light by gases was that the scattered light intensity varied with the density of the gas being used as the sample. However, Lord Rayleigh discovered that the intensity scattered per molecule decreased by a factor of order 10 on condensation to the liquid phase. There is a somewhat smaller decrease in going from the liquid phase to the solid. Obviously, then, some scattering experiments become rather difficult in the solid state. The classical experimental geometry for studying Rayleigh scattering is in the 90° orientation for the scattered radiation. This is also the most useful orientation for Rayleigh scattering in solids.

One important feature of the structure of solids is the periodic disturbance of the crystal structure by the propagation of quantised elastic waves, i.e. phonons. Those elastic waves which travel at the velocity of sound, i.e. sonic waves, are essentially thermal density fluctuations in the elastic medium. Brillouin predicted that such fluctuations should give rise to fine structure in the Rayleigh scattered light when the Bragg coherence condition $\lambda_l = 2\lambda_p \sin(\phi/2)$ is obeyed. Here λ_l is the wavelength of light, λ_p is the wavelength of those phonons responsible for scattering the light and ϕ is the scattering angle. Because the scattering centres are in motion the scattered light is frequency shifted by the Doppler effect. It is an easy matter to show that the Doppler shift, $\Delta\nu$, is given by

$$\Delta\nu = \pm\nu_p = \pm 2\nu_l(v/c)\sin(\phi/2) \tag{6.53}$$

where ν_p is the frequency of the density fluctuations in the medium and v is the velocity of sound in the medium. For light in the visible region then that part of the phonon spectrum probed by the Brillouin scattering is in the gigahertz frequency region. In addition, the Brillouin components are completely polarized for 90° scattering. Before the advent of lasers, the study of Brillouin scattering effects in solids was exceedingly difficult. It remains a technique more used in gases than in condensed media. However, Brya *et al.* (1968) have reported a particularly elegant Brillouin scattering study of the phonon bottleneck in $MgO:Ni^{2+}$.

C. V. Raman (1928) was but one of numerous scientists actively engaged in research into light scattering during the decade 1920–30. Much of his work was carried out using sunlight as a source. However, in experiments using monochromatic light he observed in the spectrum of light scattered at 90° by liquid samples new lines at wavelengths not present in the original light. The frequency displacement of these new lines from source frequency was found to be independent of the wavelength of the incident light. This was contrary both to fluorescence excitation and Brillouin scattering (eqn. 6.53); hence was born a new scattering phenomenon for which Raman was awarded the Nobel prize and which now bears his name. The frequency shifts in the Raman spectrum of a particular substance are related to, but not identical to, infrared absorption frequencies. In general infrared transitions occur when there is a change in the electric dipole moment of a centre as a consequence of the local atomic vibrations. The Raman lines occur when a change in polarizability is involved during atomic vibrations. This usually means that infrared transitions occur only between states of opposite parity whereas Raman transitions occur between states of the same parity. Thus the infrared and Raman spectra give complementary information about the vibrational spectra of spectroscopic centres.

Raman scattering measurement have found wide application in condense-matter physics. The first studies of colour centre phenomena were reported by Worlock and Porto (1965) who made measurements of Raman scattering from F-centres in sodium chloride and potassium chloride (Section 7.2). These investigations showed that the first-order scattering is predominantly associated with defect-induced localized modes. However, for F_A(Li)-centres in potassium chloride, Fritz *et al.* (1975) observed three very sharp Raman-active local modes at 47 cm^{-1}, 216 cm^{-1} and 266 cm^{-1} for the ^{7}Li isotope. These results and later polarized absorption/luminescence studies indicated that the Li^+ ion lies in an off-centre position in a $\langle 110 \rangle$ crystal direction relative to the z-axis of the F_A-centre. Detailed polarized Raman spectroscopy resonant and non-resonant with the F_A-centre absorption bands are shown in Fig. 6.17 (Joosen *et al.* 1988). These spectra show that under resonant excitation in the F_{A_1} absorption band each of the three lines due to the sharp localized modes is present in the spectrum. The polarization dependence confirms that the 266 cm^{-1} mode is due to Li^+ ion motion in the mirror plane and parallel to the defect axis. The 216 cm^{-1} mode is the stronger under non-resonant excitation, reflecting the off-axis vibrations of the Li^+ ion vibrating in the mirror plane perpendicular to the z-axis. On the other hand the low-frequency mode is an amplified band mode of the centre which hardly involves the motion of the Li^+ ion. Resonant Raman scattering has also been observed for tne Tl^0(1)-centre (Section 12.3.2) in alkali halides. The spectra show a sharp local mode of the Tl^0-centre and several overtones, superimposed upon the defect-induced first-order spectrum. The strong overtone spectrum is due to non-linear electron–phonon coupling.

6.7.3 Optical holeburning and fluorescence line-narrowing spectroscopies

Very narrow lines are ideal light sources for saturation experiments in optical spectroscopy. Early examples of such experiments were studies of Doppler-free atomic spectra. The resolution is then sufficiently good that all the fine structure in the H_α transition of atomic hydrogen becomes discernible (Hänsch *et al.* 1972). Inhomogeneous broadening arises when atoms are distinguished by the frequency at which they absorb light. The absorption profile is then the sum of separate, independent absorption lines. In atomic spectroscopy the major source of the spectral linewidth is Doppler broadening; the frequency shift is $(\Delta\nu/\nu) = \pm(v_z/c)$ due to an atom moving with velocity component v_z towards (+) or away from (−) the observer. At thermal equilibrium a Gaussian lineshape is observed because of the Maxwell–Boltzmann velocity distribution. In solids, however, atoms oscillate about a mean lattice point in the crystal, and the transition is homogeneously

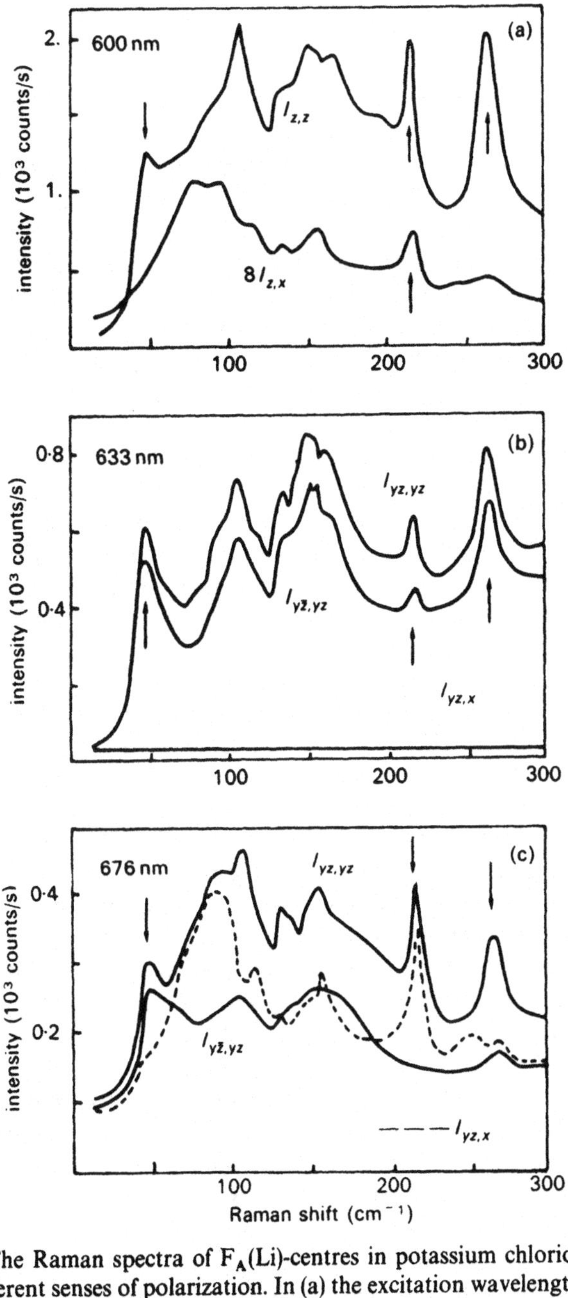

FIG. 6.17. The Raman spectra of F_A(Li)-centres in potassium chloride measured at 10 K for different senses of polarization. In (a) the excitation wavelength $\lambda = 600$ nm is midway between the peaks of the F_{A1}- and F_{A2}-bands, (b) $\lambda = 632.8$ nm is resonant with the F_{A1}-band and (c) $\lambda = 676.4$ nm is non-resonant. (After Joosen *et al.* 1988.)

broadened by the vibronic sidebands (see Chapter 5). However, the distribution of internal strains, to which electronic centres in crystals are sensitive, is a source of inhomogeneous broadening. Because crystals contain imperfections, electronic centres experience crystal fields which vary slightly from site to site in the crystal; in consequence zero-phonon lines may have linewidths of order 0.1–50 cm^{-1}. The contribution of inhomogeneous broadening to the lineshape of broad vibronic bands is negligible. The use of narrow-band laser excitation makes it possible to eliminate inhomogeneous broadening and realize a resolution limited only by the homogeneous width of the transition, which in crystals can vary from kilohertz to gigahertz. This factor of 10^3–10^4 improvement in resolution enables the spectroscopist to carry out high-resolution studies of the physical properties and electronic structure of centres and of the mechanisms responsible for homogeneous broadening. Contributions to homogeneous width come from population dynamics and random modulation of the optical frequency by phonons, nuclear spins, etc.

There are two experimental methods of recovering the homogeneous width of an optical transition, viz. optical holeburning (OHB) and fluorescence line-narrowing (FLN). Both can be understood by considering Fig. 6.18. An inhomogeneously broadened line of width Γ_{inh} is produced by many narrow components of homogeneous width $\Gamma_{hom} \ll \Gamma_{inh}$, each component being centred at a different frequency within the inhomogeneous line profile. If a narrow laser line of frequency ν_L and bandwidth $\Gamma_L < \Gamma_{hom}$ is incident upon an atomic assembly having an inhomogeneously broadened linewidth Γ_{inh}, the resulting absorption of laser radiation depletes only that sub-assembly of excited centres whose energies are within Γ_{hom} of the laser line frequency ν_L. In other words, the absorption signal is determined selectively by some atoms with absorption transitions within the overall line profile but not by others. Consequently there is a distortion of the lineshape in the neighbourhood of ν_L; a 'hole' is burned in the inhomogeneously broadened line. Resolution of the homogeneous width requires that $\Gamma_L < \Gamma_{hom} \ll \Gamma_{inh}$. In holeburning spectroscopy the narrow laser linewidth and high power make it possible to maintain a significant fraction of those atoms with transition frequency, ν_L, in the excited state, where they no longer contribute to the absorption at this frequency. To observe holeburning experimentally requires that ~5 per cent of those centres within the laser bandwidth must be transferred to the excited state, and this depletion is confined to those centres whose transition frequency falls within the bandpass of the pump laser. For a non-degenerate two-level system the population ratio assuming continuous-wave (CW) pumping is $N_2/N_1 = B_{if}u(\omega)/A_{if} = (\lambda^3/4hn^3)u(\omega)$ where $u(\omega)$ is the energy density in the beam per unit angular velocity, n is the refractive index, and λ is the wavelength of the transition. Modern ring lasers will produce linewidths of $\leqslant 1$ MHz, so for a transition in the visible region as little as 5–100 $mW\,cm^{-2}$ of the laser intensity is required to give sufficient saturation.

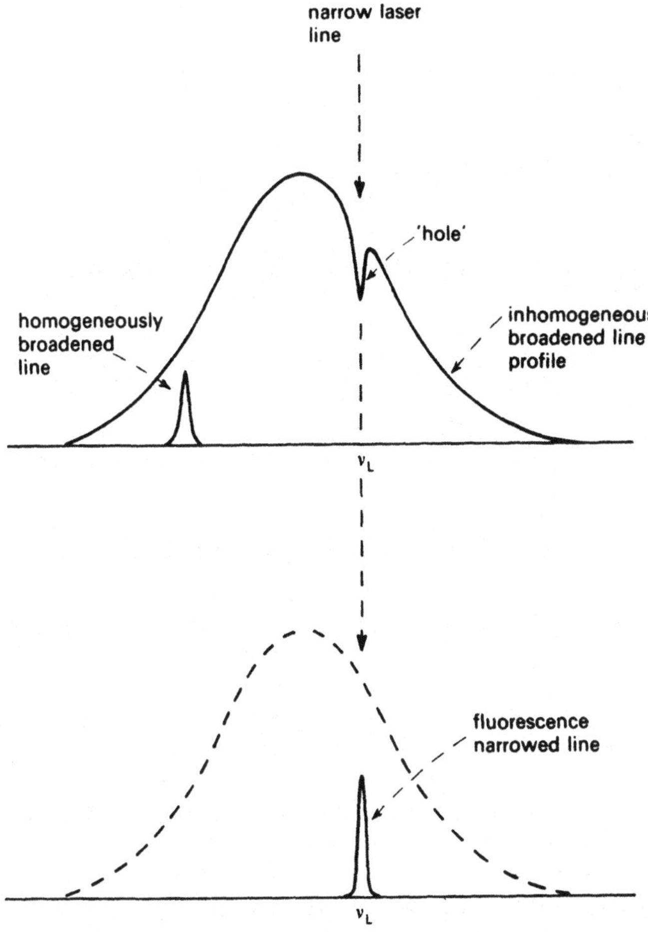

FIG. 6.18. OHB and FLN of an inhomogeneously broadened spectroscopic line.

Under conditions of pulsed excitation the population ratio is

$$N_2/N_1 = B_{if}\, u(\omega)\Delta t$$

where Δt is the laser pulsewidth, which must be shorter than the radiative lifetime. In general a pulsed laser will have a much greater linewidth than a CW ring laser. The linewidth limited by the uncertainty principle for a 10 ns pulse is 100 MHz, so for a simple two-level system, e.g. the F-centre in calcium oxide, with lifetime ~3 ms, we require an intensity some 10^6 higher than the CW case to achieve the appropriate saturation. Hence, to carry out holeburning experiments with pulsed lasers requires rather tight focusing and it becomes more difficult to arrange overlap between pump and probe beam.

In consequence, CW laser systems are to be preferred for saturation experiments in ionic crystals. To burn a hole requires the laser to be focused on the sample for periods of order 10^2–10^3 s, depending upon the specific system. When the laser excitation is switched off the holes recover on some timescale characteristic of the physical process responsible for holeburning, and this to some extent determines the choice of experimental system. For short-lived holes, e.g. R-lines in ruby, the exciting beam is divided unequally using a beam splitter into pump and probe beams. The weak probe beam is passed through an opto-acoustic modulator, which scans it backward and forward over the hole (Seltzer 1981). However, holes observed in some colour centre lines in ionic crystals and in diamond show very long lifetimes. To observe such long-lived holes the sample is irradiated for a short time in the zero-phonon line with a few hundred milliwatts of single-mode dye laser light with a width of a few megahertz. The shape of the hole is then displayed by reducing the laser intensity to a few milliwatts and scanning the laser over the inhomogeneous line profile.

Figure 6.19 shows an example of holeburning in the 601.28 nm line of 0.5 per cent $Pr^{3+}:LaCl_3$ using some 200 mW/cm^2 of single-frequency laser light with a bandwidth of 2 MHz. The zero-phonon line has an inhomogeneous width of 7.5 GHz. The homogeneous width, as measured by this holeburning experiment, is $\Gamma_{hom} = 10$ MHz which corresponds to a lifetime of 10 ns. The weak 'sideholes' in Fig. 6.19 are due to nuclear hyperfine interaction. Such very narrow hole linewidths point to dramatic improvements in spectral resolution in the presence of applied perturbations. Macfarlane *et al.* (1983) have taken advantage of such high resolution to demonstrate that earlier assignments of colour centre symmetry can be incorrect. There have been many reports of holeburning spectroscopy on transition metal ions and rare-earth ions in inorganic materials. For rare-earth ions, holeburning with lifetimes determined by the nuclear spin relaxation processes have been reported to vary from *c.* $10 - 3 \times 10^3$ s. Many measurements are aimed at the mechanisms leading to the homogeneous width of optical transitions. In these cases techniques have been developed for the detection of coherent transients, e.g. photon echo or free induction decay, because the measurements are made on the timescale of the dephasing, and are not affected by spectral diffusion and other such processes. Examples of these measurements in colour centres and rare-earth ions are given by Macfarlane *et al.* (1980, 1983), Genack *et al.* (1980) and by Shelby and Macfarlane (1982).

The complementary technique of FLN can also be understood by reference to Fig. 6.18. A narrow laser line is used to pump within the inhomogeneous linewidth, Γ_{inh}. The laser interacts only with the subset of levels spanning the bandwidth of the laser, Γ_L. These centres re-radiate to some lower-lying level, which may or may not be the ground state. The fluorescence linewidth is then much narrower than the inhomogeneous width, and the fluorescence line-

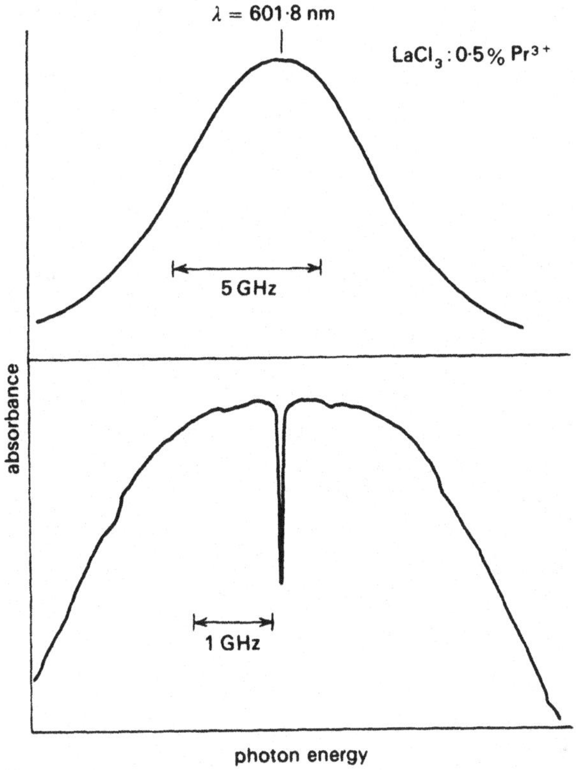

FIG. 6.19. The inhomogeneous line profile of the 601.28 nm absorption line of Pr^{3+} in lanthanum chloride and the effect of hole burning at the line centre using 200 mW/cm^2 laser radiation from a single-mode ring dye laser. The very weak side holes in the lower trace are associated with the presence of hyperfine structure. (After Harley and Macfarlane 1986, unpublished.)

width approaches the homogeneous width. In fact, for centres involved in a resonance fluorescence transition the total FLN lineshape is a convolution of the laser lineshape and twice the homogeneous lineshape (once for the pump bandwidth and once for the fluorescence). The FLN linewidth, Γ, is then usually written as $\Gamma = \Gamma_L + 2\Gamma_h$ (Kushida and Takushi 1975). The situation is slightly more complex for non-resonant fluorescence. Experimentally, FLN requires a little more sophistication than does holeburning spectroscopy. Of course, one still requires a stable, high-resolution laser. Care must be used in extracting the true homogeneous linewidth, especially for non-resonant fluorescence. It is sometimes a problem to extract weak fluorescence in the presence of strong scattered laser light. Furthermore, in cases where the fluorescence linewidth is of order 1 MHz, one must use a Fabry–Perot interferometer with very high finesse. Many of the experimental problems are discussed in the review by Selzer (1981), and in numerous examples given by Yen and Selzer (1981).

The applications of FLN to the study of spectral dynamics are given elsewhere in this text. However, it is useful to give an example of CW FLN, in which the homogeneous width should be measured. We use that well-known testing ground in solid state spectroscopy, the Cr^{3+} ion in aluminium oxide, the material in which FLN in crystals was first demonstrated (Szabo 1970). The fluorescence lifetime is 3.4 ms at 4.2 K. Hence the homogeneous width is of the order 0.3 kHz. However, there is also a direct-phonon relaxation process between the two ^2E levels, $2\bar{A}$ and $\bar{E}$, which are separated in energy by 29 cm^{-1}. This process essentially broadens the homogeneous width to *c.* 130 kHz. In early CW measurements aimed at determining the homogeneous width, values in excess of 100 MHz were reported. The problem here appears to be relaxations due to super-hyperfine interactions with neighbouring aluminium nuclei. It was found, however, that the application of a d.c. magnetic field has the effect of inhibiting relaxation due to local fields at the Cr^{3+} ions due to the ^{27}Al nuclear moments. Figure 6.20 shows FLN in the R_1-line of ruby (Jessop *et al.* 1980), an example of the very considerable narrowing that can be achieved—in this case in a magnetic field of 40 mT.

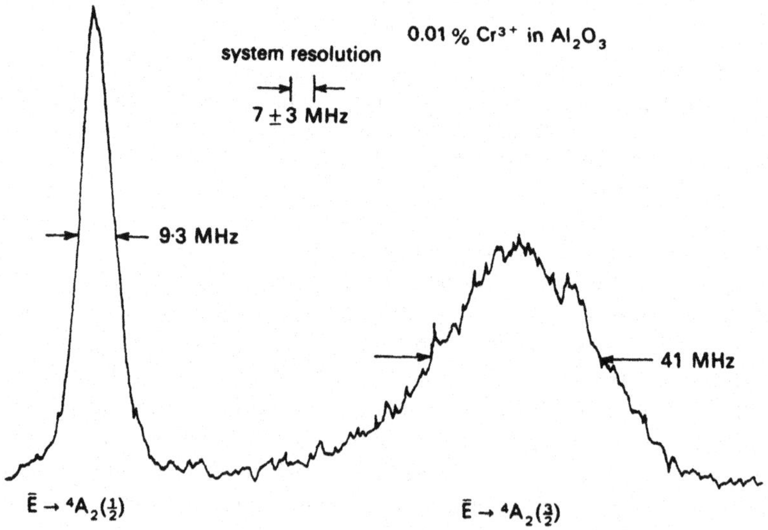

FIG. 6.20. FLN in the R_1-transition of ruby. The separation between the transitions corresponds to the crystal field splitting (0.38 cm^{-1}) in the $^4A_2$ ground state, illustrating the resolution possible with CW FLN techniques. (After Jessop *et al.* 1980.)

6.8 Picosecond spectroscopy of solids

During the past two decades there have been quite remarkable developments in techniques for generating and measuring ultrashort pulses. Pulses in the picosecond time domain are routinely available, and spectroscopy in the femtosecond domain is becoming comparatively familiar. In semiconductors a very wide range of ultrafast phenomena are being studied—electron–hole plasma formation, exciton and biexciton formation dynamics, hot electron effects, phase-conjugate self-defocusing, degenerate four-wave mixing. However, one very general optical phenomenon that may be addressed using ultrashort pulses involves non-radiative decay times in non-resonant fluorescence spectra. Such processes include ionic relaxations around a centre in decaying from an excited state, sometimes including reorientations of anisotropic centres. This is a field of study which is developing very rapidly. We discuss only one example: the measurement of the configurational relaxation time of F_A-centres in potassium chloride (Mollenauer *et al.* 1983*b*). Many picosecond phenomena, especially non-radiative decay processes, are studied by excite-and-probe techniques in which light pulses at wavelength λ_1 are used to excite a phenomenon of interest and then a delayed optical pulse at wavelength λ_2 interrogates a change of some optical property of this phenomenon. Thus ideally we require two sources of picosecond pulses at two different and independently tunable wavelengths which must be synchronized on the picosecond timescale. As we have already discussed, the excited-state relaxation of anion vacancy centres is of great importance in determining whether an excess population can be maintained in the excited state of potential colour centre laser. As illustrations we will discuss some measurements on F_A-centres (Section 7.4.6).

The F_A(Li)-centre in potassium chloride is an interesting example of configurational relaxation because there is a large Stokes shift between absorption and emission and because of the transition from a single potential well to a saddlepoint configuration during the optical pumping cycle. Mollenauer *et al.* (1983*b*) used the experimental system shown in Fig. 6.21 to carry out measurements of the configurational relaxation time. During such de-excitation many phonons are excited in the localized modes coupled to the electronic states which must be dissipated into the continuum of lattice modes. Measurement of the relaxation time constitutes a probe of possible phonon damping. A mode-locked dye laser (Fig. 6.21) producing pulses of 0.7 ps duration at 612 nm was used both to pump the centre in the F_{A2}-absorption band and to provide the timing beam. Such pumping leads to optical gain in the luminescence band and prepares the centres in their relaxed state. The probe beam, collinear with the pump beam, is generated by a CW F_A(Li)-centre laser operating at 2.62 μm. The probe beam and gated pulses from the dye laser are mixed in a non-linear optical crystal (lithium

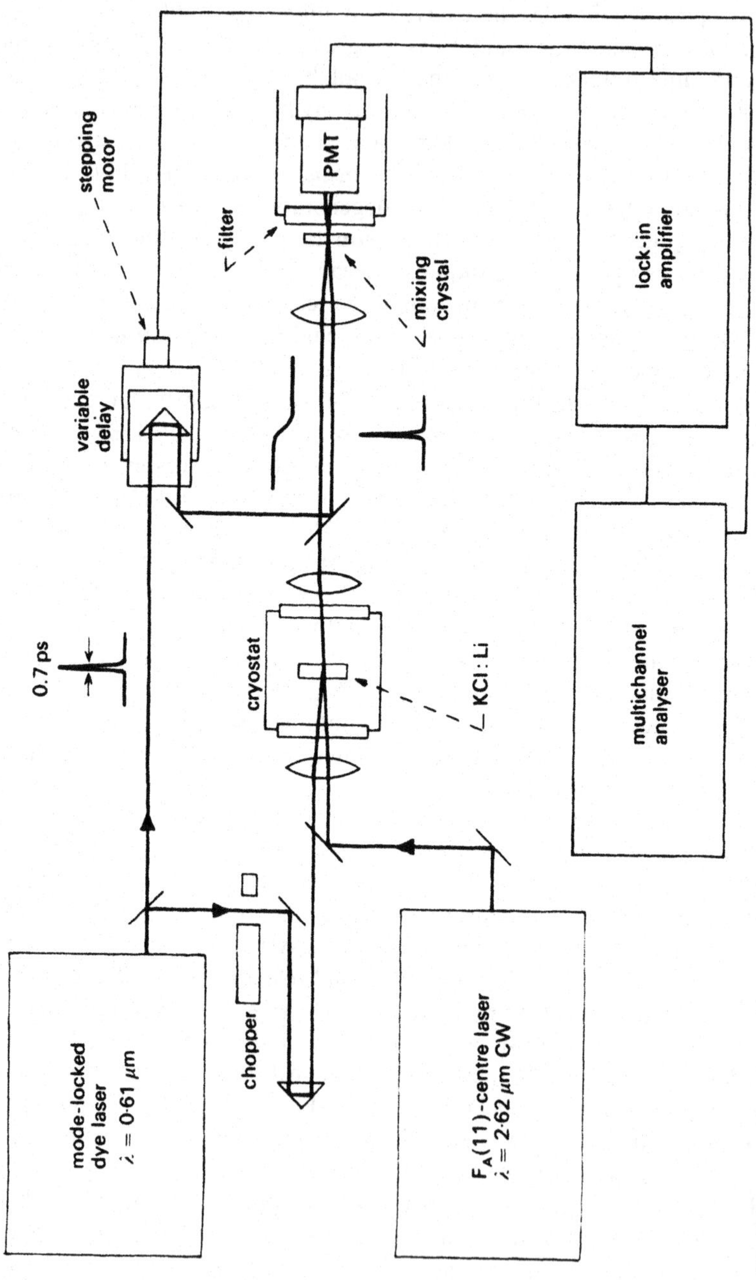

FIG. 6.21. Schematic representation of pulse and probe technique used for measuring vibrational relaxation rates in F_A-centres. (After Mollenauer *et al.* 1983b.)

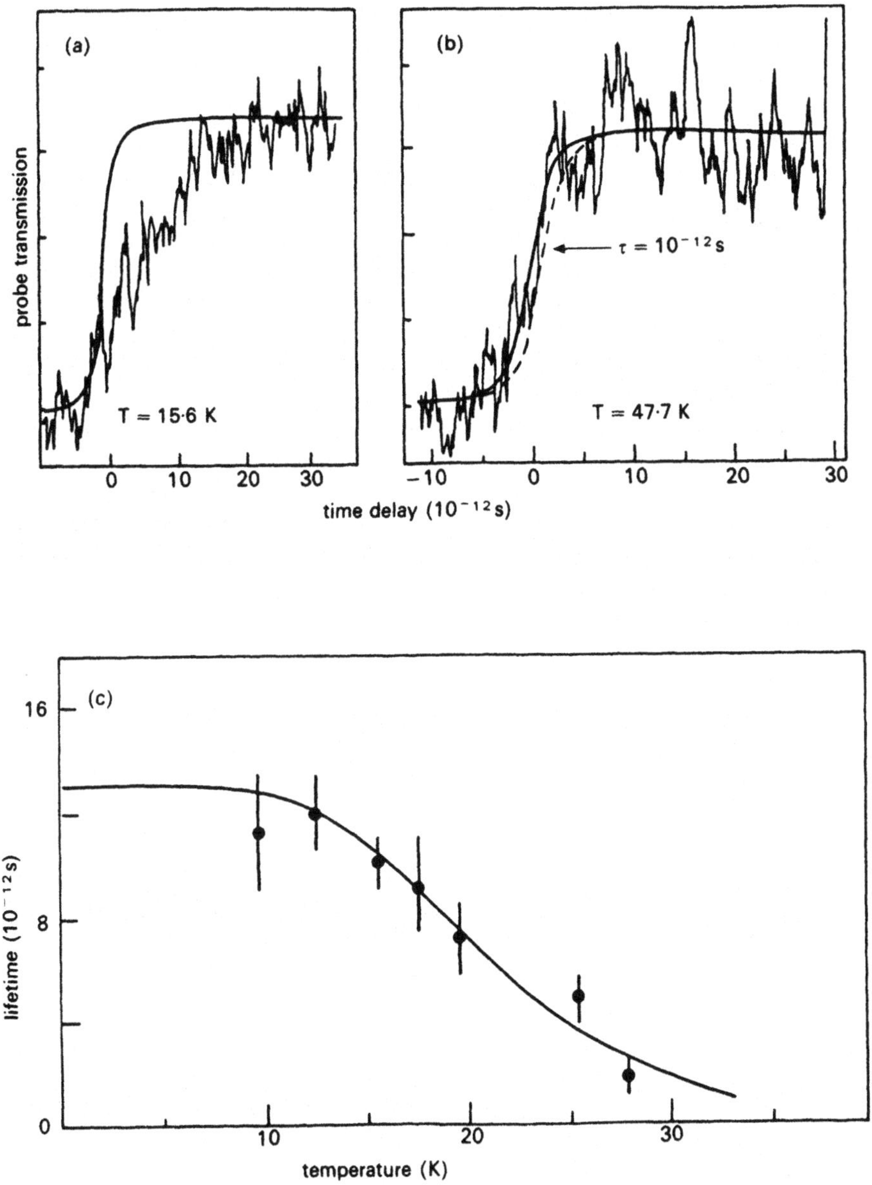

FIG. 6.22. Time resolved measurements of probe beam transmission for F_A(Li)-centres in potassium chloride. The rise of gain is shown at $\lambda = 2.62$ μm for temperatures of (a) 15.6 ± 0.4 K and (b) 47.7 ± 0.2 K. The solid line is the instantaneous response of the system, whereas the dashed line in (b) is the instantaneous response convolved with a 1·0 ps rise time. Measured rise times are 10.1 ps in (a) and <0.5 ps for (c) showing the temperature dependence of the relaxation time, (b). (Mollenauer *et al.* 1983b.)

iodate). A filter allows only the sum frequency at 496 nm to be detected. The photomultiplier tube then measures the rise in intensity of the probe beam which signals the appearance of gain when the F_A(Li)-centres have reached the relaxed excited state. The pump beam is chopped at low frequency to permit phase-sensitive detection. The temporal evolution of F_A(Li)-centre gain (Fig. 6.22) was measured by varying the time delay between pump and gating pulses. Measurements of the temperature dependence of the relaxation times of F_A(Li)- in potassium chloride, (Fig. 6.22) show that the process is very fast, typically of order 10 ps at 4 K. Furthermore configurational relaxation is a multiphonon process which involves mainly the creation of some 20 low-energy phonons of energy $E_p/hc \simeq 47\ \text{cm}^{-1}$. This is just the mode energy which occurs so strongly in the Raman spectrum of F_A(Li) centres (Fig. 6.17). That only *c.* $20 \times 47\ \text{cm}^{-1}/8066\ \text{cm}^{-1} = 0.1$ eV is deposited into the $47\ \text{cm}^{-1}$ mode, whereas 1.6 eV of optical energy is lost to the overall relaxation process, indicates that other higher-energy modes of vibrations must be involved. As Raman studies show (Joosen *et al.* 1988) two high-frequency modes do occur at $216\ \text{cm}^{-1}$ and $266\ \text{cm}^{-1}$ under resonant excitation in the F_{A2}-band. This division of relaxation energy between the various modes reflects the coupling resonance between optically prepared and energy-accepting phonons.

7
Colour centres in ionic crystals

COLOUR centre spectroscopy has played an important role in solid state physics since shortly after World War I. Pohl and his colleagues had observed that in very thin alkali halide crystals the usual reststrahlen absorption was accompanied by additional weak absorption bands. Since these absorption bands were in the visible spectrum, the 'pure' crystals were coloured: sodium chloride yellow and potassium bromide blue. This visible colouration was attributed to the presence of *Farbzentrum*, or colour centres, introduced into the crystal during growth. Such centres became known simply as F-*centres*. After many man-years of research it became clear that the primary source of colouration in alkali halide crystals was isolated anion vacancies each of which had trapped a single electron. This simple defect species is now referred to exclusively as the F-centre.

The term 'colour centre' might well be used to describe any electronic centre present in an ionic crystal. However, in the literature the phrase is usually restricted to those optically-active centres associated with intrinsic lattice constituents. They may be electrons or holes associated with vacancies or interstitials. Impurities are not included, except as they are optically inert. In addition to the alkali halides colour centre phenomena have been studied in a variety of insulators including the alkaline earth halides (e.g. calcium fluoride) and oxides (e.g. magnesium oxide), and a miscellany of such crystals as zinc oxide, aluminium oxide, barium chlorofluoride, sodium nitrate, potassium magnesium fluoride, and many others. The following discussion is concerned largely with the properties of colour centres of which the electronic and atomic structures have been unambiguously determined. More emphasis is placed on the alkali halides, because of the more complete understanding of defects in these materials. However, case histories developed for alkali halide crystals currently provide the basis for interpreting the data garnered on more complex materials.

7.1 Models of colour centres

The method of naming colour centres in insulators developed haphazardly over many years. In recent years spectroscopists have tended to use the system recommended by Sonder and Sibley (1972), which has the merit of maintaining consistency between differently charged polar compounds. We show in Fig. 7.1 a schematic representation of several simple defects in alkali

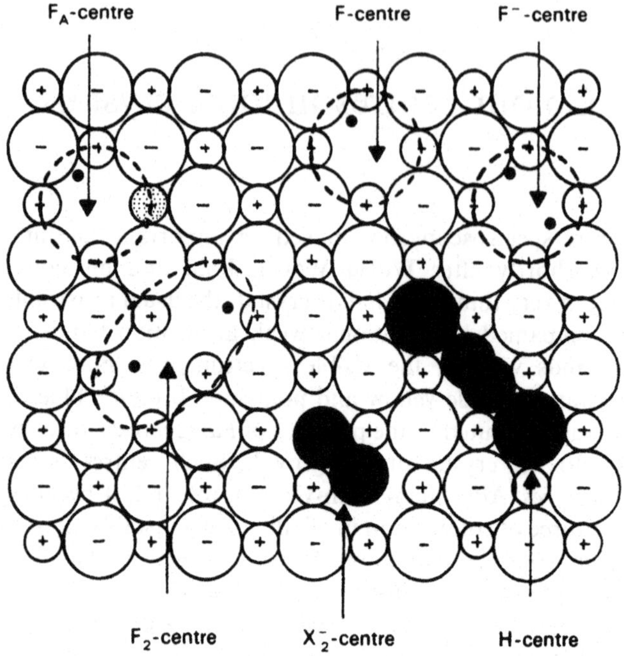

FIG. 7.1. Models of some common colour centres in the alkali halides. (After Henderson 1972.)

halide crystals; these structures were derived by a combination of spectroscopic techniques including paramagnetic resonance, optical absorption and luminescence, magnetic circular dichroism, photoconductivity, etc. Vacant lattice sites (or simply vacancies) are effectively charged relative to the crystal, and electrons or positive holes may be trapped in the field of a vacancy of appropriate charge. Note that bandgap photons may excite electrons from the valence band to the conduction band. Such electrons may then be trapped at anion vacancies so forming F-centres. Without the trapped electron the anion vacancy is referred to as an F^+-centre. However, the anion vacancy may trap two electrons: since this entity is negatively charged it is called the F^--centre. The liberation of an electron from the valence band corresponds to the formation of a *positive hole* in the outer p-shell of the halide ion. These holes may then be trapped at cation vacancies, if such are present in the crystal, resulting in the formation of V-centres. (Isolated V-centres have never been identified in the alkali halides, although their counterpart exists in the alkaline earth oxides.) Usually, however, such holes become self-trapped forming X_2^--centres, where X_0 is a neutral halide atom. The self-trapping arises because the neutral halide atom is unstable in the crystal and to overcome this instability the X^0 atom bonds covalently with a neighbouring X^- ion to produce an X_2^- molecule ion. The H^- centre is a molecular interstitial X_2^- ion occupying a single anion site and loosely bonded to the two adjacent halide ions. Other intrinsic lattice defects e.g. F^-, F_2, F_3, and F_4-centres are also shown in Figs. 7.1 and 7.2: some of these centres are aggregates of F-centres in nearest neighbour anion sites. For example, the F_3-centre is an array of three F-centres in neighbouring sites on the (111) plane (Fig. 7.2). These intrinsic point defects may be modified by nearby impurity ions. The F_A-centre results when a nearest-neighbour cation differs from the host lattice cations. Similarly H_A- and $[X_2]_A^-$-centres have been identified in alkali halides suitably doped with cation impurities. In crystals containing anion impurities, molecule ions of the form $[XY^-]$ may be produced after suitable treatment.

So far we have discussed the structure of defects as they occur in the alkali halides. As we have already noted, similar defects exist in other crystals. Normally, once we have determined the structure it is a relatively simple task to label the defect using the Sondor–Sibley rules (1972). However, it is sometimes the case, especially in the more complex crystal structures, that the experimental evidence is not clear-cut in determining the structure of the defect. When this is the case a good deal of intuition, relying on the evidence from related materials, becomes necessary.

7.2 Single vacancy centres in alkali halides

7.2.1 F^+*-centres*

Anion vacancy centres may be produced in the alkali halide crystals by additive colouration, a technique which involves heating the crystal under a high pressure of alkali metal vapour. Figure 7.3(a) shows the optical absorption spectrum of potassium bromide after additive coloration. The principal features are the F-band and the so-called β-band. The latter absorption band is believed to be due to the creation of an exciton in the neighbourhood of the F-centre. Illumination of the crystal with light in the F-band at temperatures in the range 77–200 K produces changes in the optical absorption spectrum (Fig. 7.3(b)), including the appearance of the broad F^--band , which underlies the F-band, and the α-band. The α-band is associated with exciton trapping at an anion vacancy (F^+-centre). These changes are due to the efficient photoconversion of F-centres into F^--centres and F^+-centres. Note that the F^+-centre does not have an absorption band of the usual type. Nevertheless, ultraviolet bands of the type shown in Fig. 7.3 and labelled α-band have been clearly shown to be due to the F^+-centre. That the band position, $h\nu = 6.14$ eV in potassium bromide, is very close to the first exciton band, suggested the interpretation of the band arising from the formation of an

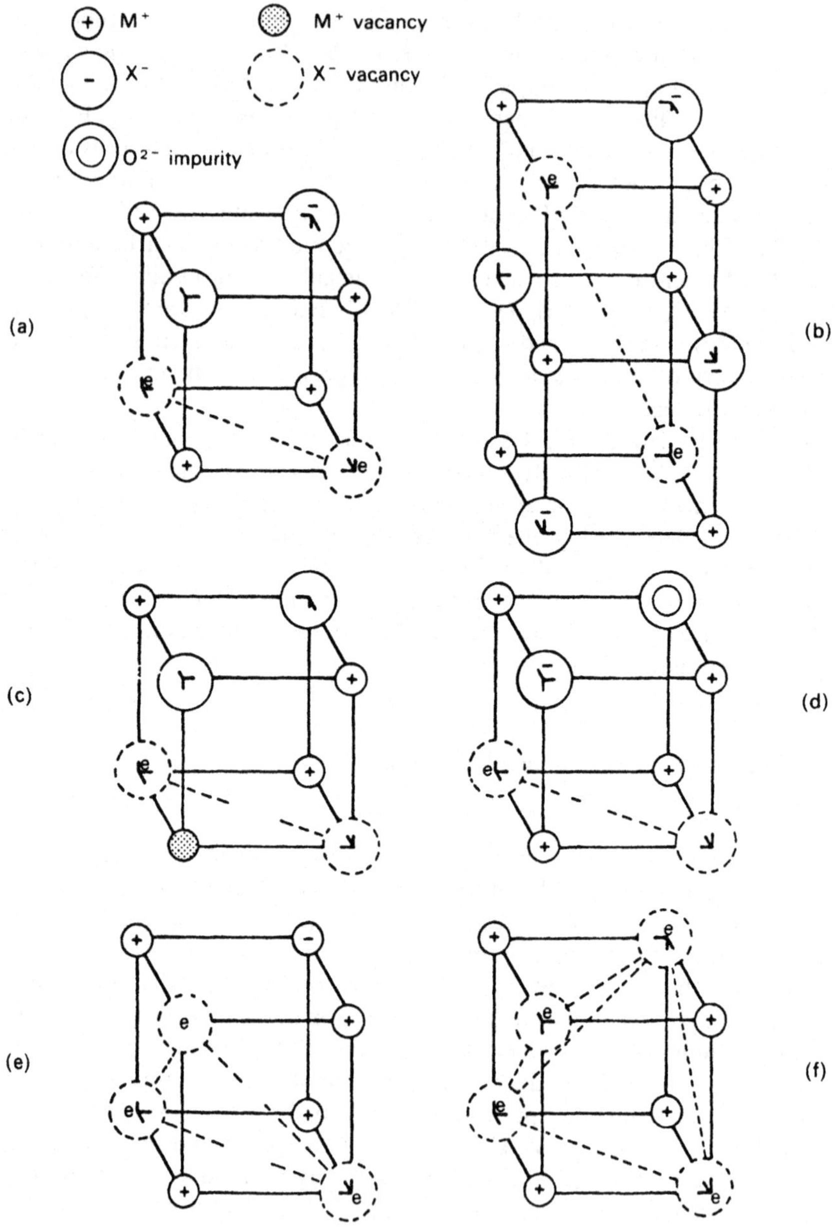

FIG. 7.2. Three-dimensional representations of F_2-, F_3-, and F_4-centres in alkali halides. In this illustration the electrons are indicated e in the vacancies, although the associated charge distribution is shared between vacancies. In (a) and (b) are shown two different arrangements of the anion vacancies in the F_2-centres, whereas (c) and (d) are stabilized versions of the F_2^+-centre. The structures of F_3- and F_4-centres are illustrated in (e) and (f) respectively.

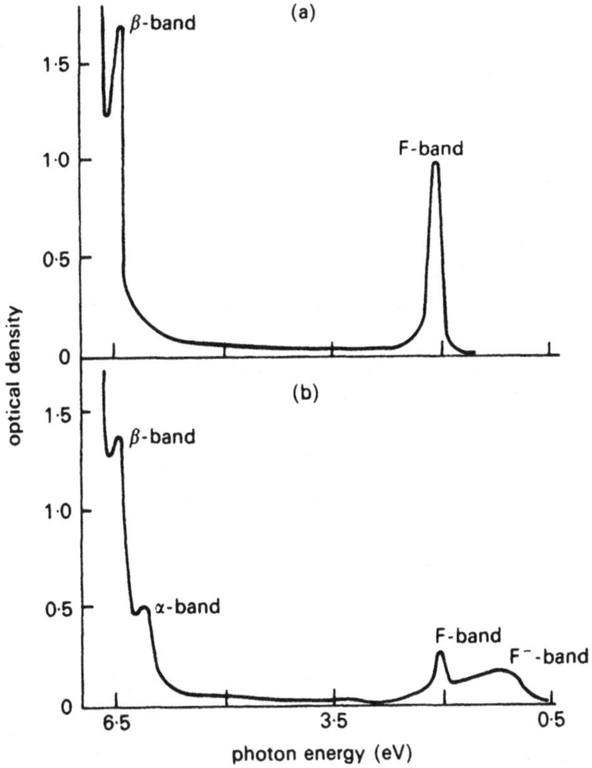

FIG. 7.3. The optical absorption spectrum of additively coloured potassium bromide measured at 10 K (a) before and (b) after bleaching with F-band light at 125 K. (After Crandall 1965.)

exciton near to an anion vacancy. Obviously measurement of the relative intensities of F- and F^--bands together with that of the α- and β-bands constitutes a very useful method of following the photoconversion dynamics in additively coloured crystals.

7.2.2 ESR of F*-centres*

The F-centre is the most widely studied defect; it has been unequivocally established that the F-centre is an electron trapped in the field of a negative ion vacancy. Although a vast body of experimental evidence supports this assignment, it is the ESR/ENDOR data that most directly point to the de Boer model of the F-centre shown in Fig. 7.1. Since a single unpaired electron is involved the F-centre is paramagnetic, a property long known from direct susceptibility measurements. ESR/ENDOR spectroscopy both confirms the model and discriminates against other models (Henderson and Garrison 1973; Seidel and Wolf 1968; Spaeth 1976).

In the alkali halides all the important isotopes possess a nuclear spin. The ESR spectra of F-centres are rich in hyperfine structure, measurements of which show that although the electron is strongly bound to the vacancy, it nevertheless overlaps onto several shells of nuclei. Determination of the number and intensity of the components, and the angular dependence of the energy splittings between the components, completely specify the electronic and ionic structure of the centre. The magnetic resonance spectra are interpreted using the spin Hamiltonian

$$\mathscr{H} = \mu_B g \boldsymbol{B}.\boldsymbol{S} + \sum_i \sum_j (\mu_N g_N \boldsymbol{B}.\boldsymbol{I}_i + \boldsymbol{I}_i.\tilde{\boldsymbol{a}}.\boldsymbol{S}) \tag{7.1}$$

where $\boldsymbol{B}$ is the static magnetic field, $\boldsymbol{S}$ and $\boldsymbol{I}$ are the electron and nuclear spins, and the hyperfine tensor, $\tilde{\boldsymbol{a}}$, is diagonal in the principal axis system. The summation is carried out over the n_i nuclei in the j shells of identical ions centred on the F-centre. The spectroscopic splitting factor, g, is isotropic for F-centres and very close to the free-electron value $g \simeq 2.00$. For example, in sodium fluoride the six nearest-neighbour sites of the F-centre are occupied by the 100% abundant nucleus $^{23}Na(I=\frac{3}{2})$. Thus the ESR spectrum should contain $2n_i I_i + 1 = 19$ components due to this first shell of nuclei, the values of M_I being $+9$, $+8, \ldots, -9$ for B along a $\langle 111 \rangle$ direction. The intensities of these components are in the ratios 1:6:21:56 . . . 580:546:456 . . . 1. Each of these lines is split into 13 lines by the isotropic hyperfine interaction of the unpaired electron with the 12 nearest-neighbour ^{19}F nuclei $(I=\frac{1}{2})$. Additional splittings arise from the anisotropic hyperfine interaction and the ESR spectrum may be very complex. Since the hyperfine structure is very different for the various alkali halides, so too are the resulting ESR spectra. Sometimes (e.g. in sodium fluoride) resolution is good enough that the first and second shell hyperfine constants may be measured with reasonable accuracy but in other cases (e.g. potassium chloride) a broad isotropic ESR line is observed, the envelope of which is made up of many closely-spaced hyperfine lines. Consequently the higher resolution offered by ENDOR is essential to unravel the wealth of information to be obtained from the hyperfine structure. Given such resolution it is possible to describe the hyperfine structure in terms of isotropic, a, and anisotropic, b, contributions to the interaction. The ENDOR frequencies are approximately given by

$$h\nu_N = -g_N \mu_N B + m_s[a + b(3\cos^2\theta - 1)] \tag{7.2}$$

where the selection rule is $\Delta m_I = \pm 1$, $\Delta m_s = 0$. The anisotropic constant, b, represents in part the dipolar interaction between electron and nuclear spins. It is this term that identifies the particular shell of nuclei involved in the spectrum.

The significance of the isotropic or contact hyperfine constant, a, is that it measures the delocalization of the ground state wavefunction. The isotropic constant is related to the charge density at the nucleus $|\psi(0)|^2$ through the Fermi contact expression

$$a=\frac{8\pi}{3}\, g\mu_B g_N \mu_N G|\psi(0)|^2 \tag{7.3}$$

where G is a numerical factor which varies with nuclear species. In Table 7.1 we list the appropriate values of *a*, *b*, and $|\psi(0)|^2$ for the first eight shells of ions around the F-centre in potassium bromide. Notably, $|\psi(0)|^2$ decreases as we go out from the centre of the vacancy. Typically 63 per cent of the electron charge density lies inside the vacancy, 30 per cent inside the second shell, and 6 per cent inside the third shell. The remainder of the charge distribution is associated with the more remote shells. By use of ENDOR, data are obtained on a large number of shells of nuclei overlapped by the F-centre wavefunction. The point of this brief excursion into ESR/ENDOR spectroscopy is to emphasize that we now have a very detailed description of the F-centre wavefunction in the electronic ground state. It is the most compelling experimental description of the F-centre, and as such serves as a model for rigorous theoretical tests of ground-state properties.

7.2.3 *Optical absorption by* F-*centres*

As we have already seen, the principal optical absorption characteristic of the F-centre is a 1s→2p-like transition (Section 2.5) which gives rise to broad absorption bands in the visible region of the spectrum. The F absorption band in potassium bromide is shown in Fig. 7.3. A simple theory (Section 2.5) predicts that the photon energy at the F-band peak scales vary crudely as the inverse square of the lattice parameter and accounts qualitatively for the temperature dependence of the bandshape and the relative position of the

Table 7.1

The ENDOR parameters for shells I to VIII in the potassium bromide F-centre. (After Seidel and Wolf 1968.)

Shell	Nucleus	a/h (MHz)	b/h (MHz)	$\|\psi(0)\|^2$
I	^{39}K	18.3	0.74	1
II	^{81}Br	42.8	2.77	0.20
III	^{39}K	0.27	0.022	0.015
IV	^{81}Br	5.70	0.41	0.027
V	^{39}K	0.16	0.02	0.009
VI	^{81}Br	0.83	0.086	0.004
VIII	^{81}Br	0.53	0.06	0.003

emission band. This model also implies that other states exist within the bandgap. Alternatively, we might compare the F-centre to an 'inside-out' hydrogen atom, so that there will be an infinite series of levels merging into the conduction band. Examination of the spectra at shorter wavelengths than the F absorption band peaks show a number of other features. The β-band (Fig. 7.3(a)), observed on the side of the ultraviolet edge, always has the same intensity relative to the F-band; it has been attributed to excitons recombining in the neighbourhood of F-centres. Also apparent in the optical absorption spectrum of additively coloured potassium bromide (Fig. 7.3), semblances of a band, labelled the K-band, are shown on the high-energy wing of the F-band. The K-band is especially well resolved in rubidium chloride, in which crystal it is asymmetric in shape and insensitive to temperature (Smith and Spinolo 1965). Other absorption bands may also be observed; the so-called L-bands are thought to be directly associated with F-centres, but they are not well understood. A theoretical reconstruction of the F- and K-bands in rubidium chloride demonstrates the asymmetry of the K-band by assuming transitions from the 1s-like ground state of the F-centre to an infinite number of bound p-like states associated with the coulombic tail of the F-centre potential. The oscillator strength of the K-band is of order $f \simeq 0.1$, in comparison with the F-band for which $f \simeq 1$.

7.2.4 *Electron–phonon interaction at* F*-centres*

In the discussion of the electron–phonon interaction in Chapter 5 it was shown that where strong vibronic coupling exists broad structureless bands are the norm, and that large shifts between absorption and emission are expected. The temperature dependence of the second moment of the band, $M_2(T)$, (see Section 6.5.1), is given by

$$M_2(T) = S(\hbar\omega)^2 \coth(\hbar\omega/2kT) \tag{7.4}$$

and the Stokes shift by

$$\Delta E = (2S - 1)\hbar\omega \tag{7.5}$$

where $\hbar\omega$ is the mean energy of the lattice phonons coupled to the electronic states. These equations assume linear coupling, i.e. identical phonon frequencies in the ground and excited state. The observed bandwidths at temperature T fit eqn (7.4) exceptionally well for most of the alkali halides, implying that the effect of all modes is *qualitatively* well represented by a single mode. Least-squares fit of the experimental second moment to eqn (7.4) yields values of the Huang–Rhys parameter, S, and the effective phonon energy, $\hbar\omega$. These are shown in Table 7.2. Huang–Rhys parameters lie between 28 (for sodium fluoride) and 61 (for lithium chloride), clearly precluding the presence of vibronic structure at low temperature. The

Table 7.2

Absorption and emission bands of F-centres in alkali halides. (Adapted from Fowler 1968.)

Crystal	FWHM at 0 K Absorption Γ_0(eV)	$\hbar\omega$(eV)	S	Peak energy (eV) Absorption	Emission
LiF	0.688	0.0455	41	5.083	—
LiCl	0.382	0.0207	61	3.256	—
LiBr	0.319	0.0185	54	2.769	—
NaF	0.373	0.0369	28	3.707	1.665
NaCl	0.270	0.0177	42	2.746	0.975
NaBr	0.388	0.0247	44	2.345	—
NaI	0.276	0.0176	44	2.063	—
KF	0.269	0.0149	59	2.873	1.66
KCl	0.195	0.0150	31	2.295	1.215
KBr	0.160	0.0095	51	1.059	0.916
KI	0.146	0.0084	54	1.874	0.827
RbF	0.192	0.0161	26	2.409	1.328
RbCl	0.153	0.0112	34	2.034	1.328
RbBr	0.155	0.0137	23	1.853	0.87
RbI	0.134	0.0089	41	1.706	0.81

effective phonon energies, when compared with the phonon spectrum for alkali halide crystals, suggest that the F-band width is determined by coupling to acoustical band phonons for alkali halides with smaller atomic masses (lithium fluoride, sodium fluoride, sodium chloride) and to optical phonons for those with heavier masses (potassium iodide, potassium bromide).

The strong electron–phonon coupling also suggests large Stokes shifts between the peaks in absorption and emission. Fig. 5.16 shows the absorption and emission spectra for F-centres in potassium bromide; the Stokes shift is large ($\simeq 1.4$ eV) as it is for all investigated F-centre systems. The position of the emission bands might be expected to follow a similar dependence on lattice constant as that shown for the absorption bands. A rough correlation is obtained, but the fit is not nearly as good as in absorption. The emission band narrows and shifts to higher energy with decreasing temperature. Analysis of the emission second-moment measurements yields frequencies which are in general different from those observed in absorption, which implies that the linear coupling approximation is unsatisfactory. The essential information which the bandshape analysis yields is the single average frequency for the modes interacting with the centre. If the F-centre interacts primarily with modes well separated from the centre, then those modes will be longitudinal optical modes of the lattice. On the other hand, if localized modes predominate then mainly nearest neighbours are involved which should have much lower frequencies than typical lattice phonons. Just which situation occurs can sometimes be obtained from Raman scattering experiments. Worlock and Porto (1965) showed that the major part of the scattering at the F-centres in sodium chloride (Fig. 7.4) is from modes with frequencies centred at $\hbar\omega = 145\ \text{cm}^{-1}$ rather than at the longitudinal optical frequency of $\hbar\omega = 270\ \text{cm}^{-1}$. Hence the vibrational interaction is characteristic of the lattice close to F-centres, i.e., localized modes. In potassium iodide, as illustrated in Fig. 7.5, one observes a very sharp Raman line at a frequency $\hbar\omega = 68\ \text{cm}^{-1}$ which lies between the acoustical and optical branches of the phonon spectrum. This is a very good example of a gap mode (Buisson *et al.* 1975), which can be explained by assuming a 70 per cent reduction in force constant acting between the electron and neighbouring cations, compared with the corresponding perfect lattice force constant. Also shown is a sharp

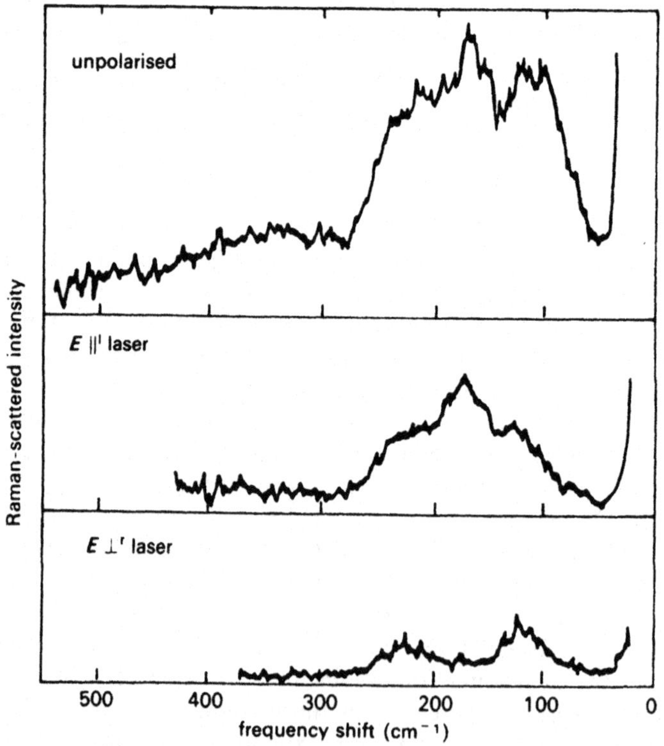

FIG. 7.4. Raman spectra of F-centres in sodium chloride for different polarizations of the detection radiation. The shifts are measured relative to the 514.5 nm line from an Ar^+ laser. (After Worlock and Porto 1965.)

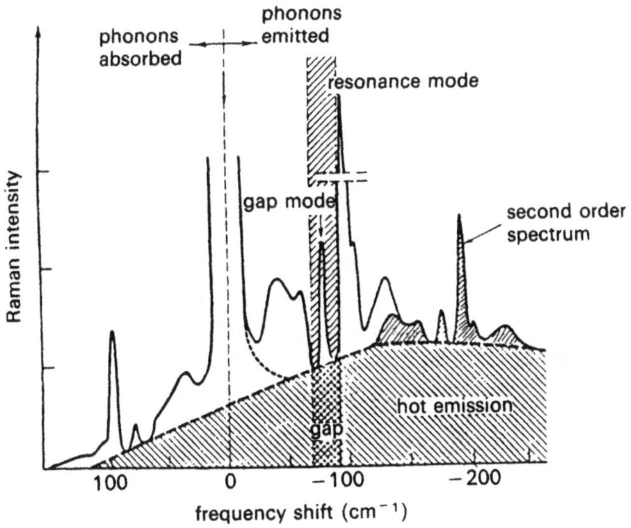

FIG. 7.5. Showing the appearance of a gap mode in the Raman spectrum of potassium iodide containing F-centres. (After Buisson *et al.* 1975.)

resonance mode at 96 cm^{-1} and the second-order Raman scattering of the host lattice.

7.2.5 *Excited states of the* F-*centre*

It is now appropriate to discuss the nature of the lowest-lying excited states of F-centres. We have rather loosely described the state reached in absorption as 2p-like. If this is a reasonable description then the return to the ground state, a 2p→1s transition, should occur via an allowed electric dipole transition with oscillator strength $f \simeq 1.0$ and radiative lifetime $\tau_R \simeq 10^{-9}$s. Clearly, measurements of the quantum efficiency and radiative lifetime should yield information on the nature of the excited states. The radiative lifetime in eqn (4.91) is just the reciprocal of the Einstein coefficient given by eqn (4.73), i.e.,

$$A_{if} = \frac{8\pi^2 n}{3\varepsilon_0 \hbar \lambda_0^3}\left(\frac{E_{\text{loc}}}{E}\right)^2 |\langle f|er|i\rangle|^2$$

where $|f\rangle$ and $|i\rangle$ are the final and initial state wavefunctions and λ_0 is the wavelength at the peak of the emission band. Because of the strong electron–phonon interaction the state reached in absorption may differ from that on which the emission originates, due in large measure to charge distribution in the excited state differing from that in the ground state. As shown in Chapters 2 and 5 this results in the centre of ionic vibration being different in the two states, so that the matrix elements involved in the absorption and emission transitions, ⟨2p|*er*|1s⟩ and ⟨1s|*er*|2p⟩′, respectively, and corresponding oscillator strengths may be quite different. There is a large Stokes shift between absorption and emissions bands for F-centres (Table 7.2).

The quantum efficiency, η, measured at low temperature ($T < 30$ K) is unity for most alkali halides. However, at some temperature, characteristic of the particular material, η decreases exponentially with increasing temperatures. The lifetime of the luminescence, τ, is constant at low temperatures and then decreases with increasing temperature. Data for potassium chloride, potassium fluoride, and sodium fluoride are shown in Fig. 7.6. A related effect is the onset of a photocurrent at temperatures above 90 K. To explain this Swank and Brown (1963) proposed a three-channel model of decay of the excited state involving radiative (τ_R) and non-radiative decay (τ_{NR}) as well as thermal ionization of the F-centre. The total decay rate is given by

$$\frac{1}{\tau} = \frac{1}{\tau_R} + \frac{1}{\tau_{NR}} + \frac{1}{\tau_0}\exp(-E_i/kT),$$

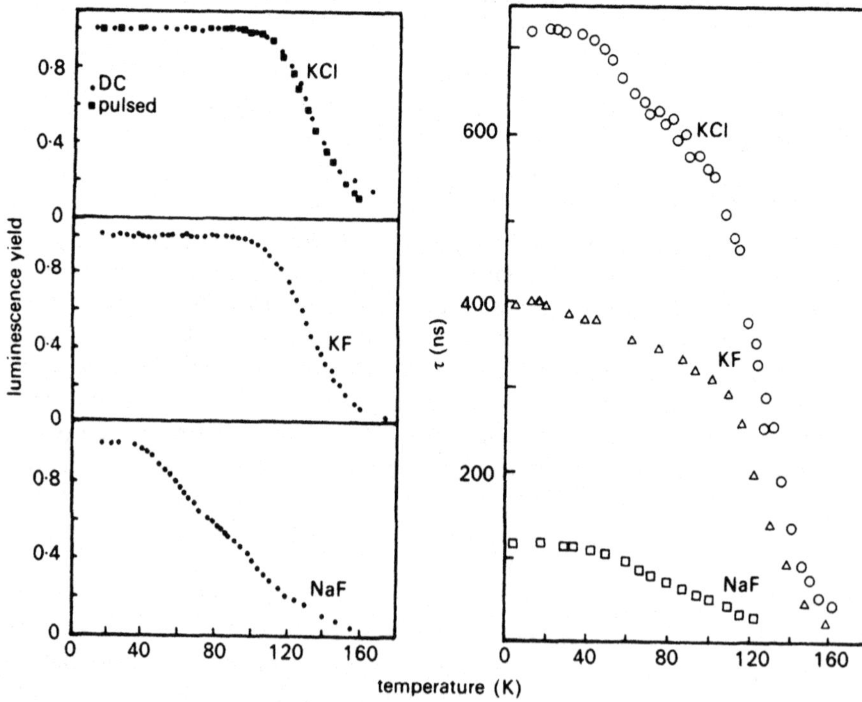

FIG. 7.6. Temperature dependence of the luminescence yield, η, and the decay time, τ, for several alkali halides. (After Stiles *et al.* 1969.)

E_i being the ionization energy of the F-centre in its excited state. Since non-radiative decay to the ground state can be neglected at low temperature this equation implies that thermal ionization, and hence photoconductivity, competes with radiative decay in the 2p-state de-excitation process. Hence we may write

$$\tau = \tau_R / [1 + (\tau_R/\tau_0) \exp(-E_i/kT)] \tag{7.6}$$

$$\eta = \tau/\tau_R = [1 + (\tau_R/\tau_0) \exp(-E_i/kT)]^{-1}. \tag{7.7}$$

The experimental data for the quantum efficiency, $\eta = \tau/\tau_R$, fit eqn (7.7) rather well, whereas eqn (7.6) is followed quite well at high temperatures ($T \geqslant 60$ K), and may be used to obtain τ_R, E_i and $1/\tau_0$. For potassium chloride $\tau_R = 0.57\ \mu$s, $E_i = 150$ meV, and $\tau_0 = 10^{-12}$ s; values for other alkali halides are given in Table 7.3. The value of τ_R, thus obtained, used with eqn (7.6) leads to values of τ in the low-temperature regime that are somewhat longer than the observed low-temperature value. In consequence, the Swank and Brown model requires modification at low temperatures. This breakdown in the model is very clear in Fig. 7.6, which shows that τ_R continues to decrease at temperatures for which η is constant. This feature, together with the fact that τ_R is too long by a factor of order 10^2, requires explanation.

Two models have been proposed to explain the longevity of the excited state. The basis of Fowler's model (1968) is a very diffuse wavefunction in the excited state after relaxation, whereas Bogan and Fitchen (1970) stress the possibility of admixtures of the 2s and 2p states. Local electric fields associated with longitudinal optical phonons of odd parity will mix the $|2s\rangle$ and $|2p\rangle$ functions. In potassium chloride two states of mixed s–p character are produced, the lowest of which, with 60 per cent $|2s\rangle$-like and 40 per cent $|2p\rangle$-like character, has a rather low transition probability to the $|1s\rangle$-like ground state.

Table 7.3

Parameters required to model the properties of the relaxed excited state of F-centres in alkali halides. (After Stiles *et al.* 1970.)

	τ_R (ns)	τ_0 (ps)	E_i (meV)	Δ (meV)	α^2
NaF	110	—	55	17	0.1
KF	400	0.5	131	17	0.3
KCl	710	1.0	148	18	0.5

7.2.6 Uniaxial stress, Stark and Zeeman spectroscopy

So far the discussion has centred on straightforward bandshape features, peak positions, and lifetimes. The large widths of the F-bands due to strong electron–phonon coupling mask a number of important features, such as the symmetry of the phonons which predominate in the vibronic coupling, the magnitude of spin–orbit interaction, and the extent of excited-state mixing. Some measure of these phenomena may be obtained by measurements of bandshapes in the presence of uniaxial stress, electric field, or magnetic field. In each case the splittings induced, and their associated polarizations, may be quite small relative to the bandwidth. Fortunately techniques involving lock-in detection enable polarization changes as small as 10^{-2} per cent to be measured with good sensitivity (Section 6.3).

In the case of piezospectroscopy, stress-induced splittings of the F-band are quite small relative to the F-band width of *c.* 0.2 eV. Since the $|1s\rangle$ ground state is unsplit by the stress, any splittings will be due to the effect of stress on the 2p-like excited state. The stress is usually applied parallel to the $\langle 100\rangle$, $\langle 110\rangle$, and $\langle 111\rangle$ crystallographic directions in turn, and measurements made of changes in the moments and polarization of the spectra. For a $\langle 100\rangle$ stress the largest effect is on the p_x level; p_y and p_z remain degenerate. As shown in Fig. 7.7, with light propagating in the [001] direction only dipoles associated with $s\rightarrow p_x$ and $s\rightarrow p_y$ are allowed for electric dipole transitions. Hence under zero stress there is no linear dichroism. However, for finite stress the two transitions are not degenerate and there is an induced linear

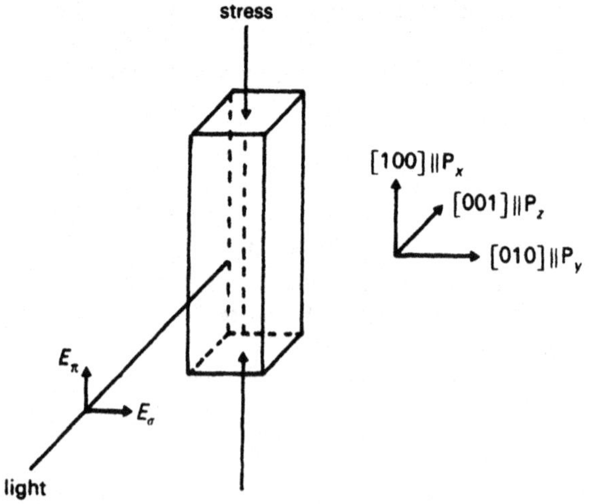

FIG. 7.7. Showing the measurement geometry for uniaxial stress or electric field induced linear dichroism of the F-centre absorption.

polarization

$$P = \frac{I(\nu, \pi) - I(\nu, \sigma)}{I(\nu, \pi) + I(\nu, \sigma)} \tag{7.8}$$

where ν is the transition frequency and the senses of π- and σ-polarisations, are indicated in Fig. 7.7. For the alkali halide F-centres the linear dichroism is linearly proportional to the applied stress, as is the change in first moment (i.e. peak position) for a given sense of polarization. Measurement of the linear variations enables accurate determinations to be made of the various coupling coefficients, the importance of which is their relationship to the interaction between the electronic states and the long-wavelength phonon modes. Such phonon modes produce distortions of particular symmetries. For example, in cubic crystals the phonon modes produce distortions which transform as the irreducible representations A_{1g}, E_g, and T_{2g} of the O_h group (Fig. 5.4). For F-centres in potassium chloride the stress coupling coefficients listed in Table 7.4 show clearly that coupling to breathing-mode phonons (A_{1g}) is predominant, although coupling to tetragonal modes (E_g) and trigonal modes (T_{2g}) is significant.

Stress measurements on the F-centre luminescence band also show a linear polarization which is directly proportional to the magnitude of the applied stress. However, this polarization is independent of incident excitation polarization, emission wavelength, measurement temperature, and defect concentration. This dramatic contrast with the absorption data implies that compact 2p-like states are reached in a Franck–Condon absorption transition. If it were otherwise then there would be no contribution to the moments of the bandshape directly identifiable with the different phonon modes. In emission the independence of the polarization from all experimental variables suggests that the relaxed excited state is spatially diffuse,

Table 7.4

Stress coupling coefficients for coupling to distortions of A_{1g}, E_g, and T_{2g} symmetry at F-centres in alkali halides (units 10^4 eV kg^{-1} mm^{-2}), measured for the absorption band. (After Schnatterley 1965.)

	$A_1(A_{1g})$	$A_2(E_g)$	$A_3(T_{2g})$
NaCl	16.8 ± 0.5	2.4 ± 0.4	6.7 ± 0.8
KCl	13.6 ± 0.5	6.1 ± 0.9	8.8 ± 1.2
KBr	14.6 ± 0.5	4.4 ± 0.6	6.3 ± 0.9
KI	17.6 ± 0.5	6.2 ± 0.9	9.8 ± 1.4
RbCl	13.9 ± 0.5	7.0 ± 1.0	5.8 ± 0.8

and in consequence insensitive to symmetry-changing distortions associated with lattice phonons or to externally applied stress. The linear polarization may arise from stress-induced admixture of higher-lying d-states into the electronic ground state.

Further information on the structure of the excited states of the F-centre comes from measurements of the optical properties of the defect in the presence of an electric field. In principle, for centres with inversion symmetry there should be no linear Stark effect unless there are states of opposite parity which are close enough in energy for significant mixing to be induced by the E-field. In the case of the F-centre the 2p and 2s levels, which are almost degenerate, are strongly admixed by the field (Fig. 7.8), inducing a change in the F-band oscillator strength, and a repulsion between the admixed $|2s'\rangle$ and $|2p'\rangle$ states. Hence we expect both a linear dichroism and an increase in the second moment of the bandshape. Results in the F- and K-band regions in potassium chloride are shown in Fig. 7.9 at $T = 77\,K$ and $250\,K$. The changes in absorption coefficient were attributed by Chiarotti *et al.* (1969), to

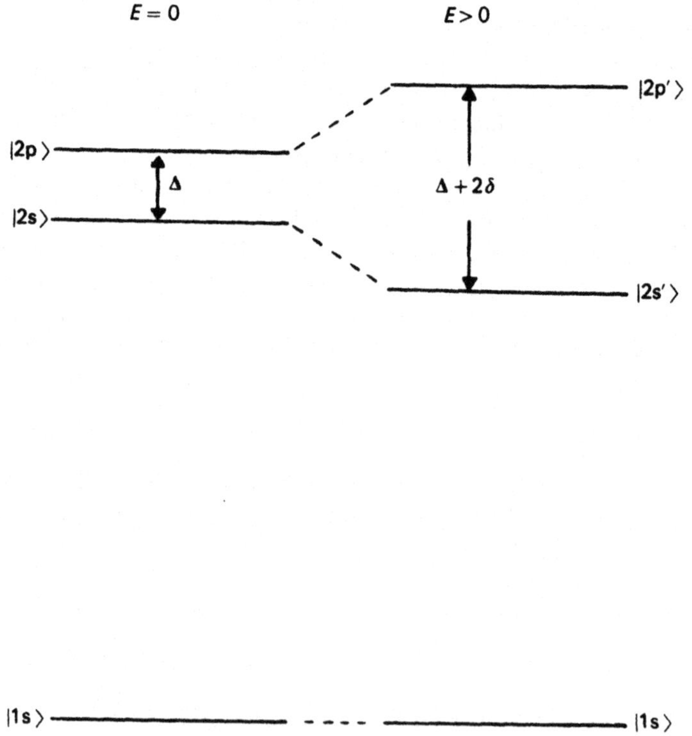

FIG. 7.8. The energy levels of F-centres in the presence of an electric field, showing the effect of s–p admixture which leads both to dichroism and broadening of the F-absorption band. Δ is the spin–orbit splitting and 2δ measures the additional splitting in the electric field, E.

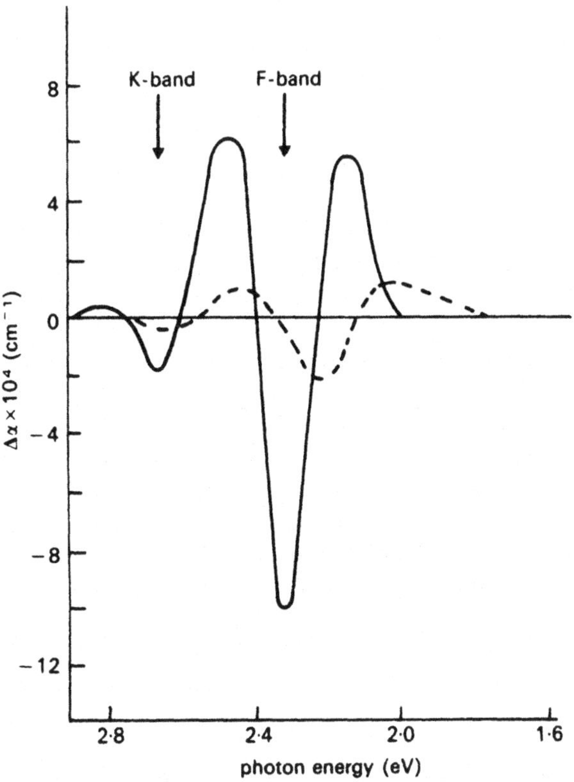

FIG. 7.9. Illustrating the electric-field-induced dichroism in the region of the F- and K-absorption bands in potassium chloride at 77 K (——) and 250 K (---). (After Chiarotti *et al.* 1969.)

the admixture of $|2s\rangle$ and $|2p\rangle$ states separated in energy by about 0.1 eV. This result is of particular significance for the measurements of the effect of electric field on the radiative lifetime. Recall that to account for the excessively long lifetime Bogan and Fitchen (1970) had proposed that the $|2s\rangle$ and $|2p\rangle$ levels are mixed by long-wavelength phonons. For phonons propagating along the z-direction the mixed wavefunctions are to first order in perturbation theory

$$|2s'\rangle = (1+\alpha^2)^{-1}(|2s\rangle + \alpha|2p_z\rangle) \quad (7.9a)$$

$$|2p'\rangle = (1+\alpha^2)^{-1}(|2p_z\rangle - \alpha|2s\rangle) \quad (7.9b)$$

where α depends on the matrix element of the Stark interaction between the $|2s\rangle$ and the $|2p_z\rangle$ states and is linear in the electric field. Since the $|2s'\rangle$ state is the lower in energy it is the emissive state. Following eqn (4.91) we write the radiative lifetime as

$$\frac{1}{\tau_R} = \text{const.} \times |\langle 1s|z|2s'\rangle|^2$$

$$= \text{const.} \times \frac{\alpha^2}{1+\alpha^2} |\langle 1s|z|2p_z\rangle|^2. \qquad (7.10)$$

Only the $|2s'\rangle$ level is populated at low temperature, giving rise to the long lifetime. The temperature dependence of the lifetime below ~80 K then arises from the increased population of the higher-lying $|2p'\rangle$ states. This model also predicts a quadratic decrease of τ_R with the electric field, as is observed experimentally. In general the magnitude of α^2 decreases through a halide series MF to MI, where M is Na, K, or Rb, since it reflects the way that spin–orbit coupling varies with atomic mass. This is observed experimentally. Note that for potassium fluoride, a 60 per cent fraction of the emitting state is derived from the $|2s\rangle$ state, whereas in potassium iodide this fraction is of order 90 per cent. That the emitting state is predominantly $|2s\rangle$ in character is also consistent with Stark effect measurements on the emission band. There is an induced linear dichroism, the magnitude of which varies quadratically with applied field, but which is constant at low temperature, typically below 40 K in potassium chloride. The dichroism arises because as a consequence of the additional field-induced s–p_z mixing, the probability for transitions parallel to the applied field is increased relative to that for transitions perpendicular to the field.

Finally we note the results of the numerous magnetic circular dichroism experiments. The philosophy has been to measure the spin–orbit structure of the excited states of the F-centre which is normally masked by the very large width of the F-band. As shown in Section 6.5.2 the circular dichroism, Σ, given by $\Sigma = -(I_+ - I_-)/I_{DC}$, is a function of temperature and magnetic field, from which the various moments of the bandshape may be computed. There is essentially no change in the zeroth moment, as indicated by the moments calculation in the absence of coupling to states lying higher than $|2s\rangle$ and $2p\rangle$. However, the difference in the first moments between σ_+ and σ_- light is given by

$$\Delta M_1^+ - \Delta M_1^- = -2|\langle y|\boldsymbol{L}_z|x\rangle| \, (\mu_B B + \zeta\langle \boldsymbol{S}_z\rangle) \qquad (7.11)$$

where $\langle \boldsymbol{S}_z\rangle$ is the spin polarization of the ground state, ζ is the spin–orbit coupling constant of the excited state, and $\boldsymbol{L}_z$ is the orbital angular momentum operator. It can then be shown that

$$\Delta M_1^+ - \Delta M_1^- = 2g_l\mu_B B - \zeta S \tanh(g\mu_B B/2kT) \qquad (7.12)$$

where g_l is the orbital g-value for the diamagnetic Zeeman effect on the $|2p\rangle$ state (Henry *et al.* (1965)). The first term in eqn (7.12) is temperature independent, whereas the second term is the temperature-dependent paramagnetic term.

Thus g_l and ζ have been determined for most of the alkali and caesium halides. The results are given in Table 7.5. Two features are clear: the spin–orbit constant is in all cases large and negative. From eqn (2.43) it will be recalled that for hydrogen-like atoms the spin–orbit constant scales as Z^4/n^3. In the case of the F-centre there is no central nuclear charge and it becomes clear that ζ in this case arises from the shells of ions with which the F-centre wavefunction has a significant overlap. To take this into account it is necessary to orthogonalize the F-centre wavefunction with respect to all shells of neighbours involved in the hyperfine structure. The predominant effect will be from the nearest-neighbour cations, e.g. in lithium fluoride there are the six Li^+ ions. For F-centres in lithium fluoride $\zeta \approx 27\,cm^{-1}$ which is actually larger than the $^2P_{1/2}$–$^2P_{3/2}$ splitting on an isolated Li^+ ion. Since the interaction is summed over six nearest-neighbour cations this is, perhaps, not too surprising. However, the values of ζ increase markedly not only with the atomic number of the cation but also of the anion. This indicates that any sensible theoretical model of the F-centre must as a very minimum include effects due to the electronic structure of nearest (cation) and next-nearest (anion) neighbour ions. The spin–orbit structure of the F-centre is a natural

Table 7.5

Values of the orbital g-value, g_l, and the spin–orbit splitting, ζ, of the unrelaxed excited states of F-centres in alkali halides and F^+-centres in oxides.

	$\zeta(cm^{-1})$	g_l
LiF	−27.4	0.59
MgO	−12	
NaCl	-62 ± 8	0.38
NaBr	-228 ± 48	—
CaO	−24	
KF	25.8	1
KCl	81.5 ± 8	0.62
KBr	240 ± 32	0.47
KI	460 ± 64	0.83
SrO	−185	
RbCl	−122	—
RbBr	−214.5	0.47
RbI	−400	—
BaO	−265	
CsCl	−300	
CsBr	−330	

consequence of the inside-out nature of the defect. Note that the orthogonalization procedure leads to electronic currents on the ions which are in opposite sense to those in the vacancy. This is the basis of the negative sign of the spin–orbit coupling constant.

In summary, the experimental evidence from a wide range of spectroscopic investigations confirms the 'single electron in an anion vacancy' model of the F-centre. Both the ground-state ESR/ENDOR and MCD emphasize the role played by extensive wavefunction overlap onto several shells of neighbouring ions. Analysis of the moments of optical bandshapes show that broadening of the optical transitions predominantly reflects coupling to breathing mode vibrations. Nonetheless, coupling to both tetragonal and trigonal modes is important. Stress, electric field, and magnetic field measurements on the absorption band show that the $^2P_{1/2}$ level lies below the $^2P_{3/2}$ level and both are lower than the $^2S_{1/2}$ level by about 0.1 eV. The unrelaxed excited state is a fairly compact ^2P state. After relaxation by phonon creation the lowest state is predominantly a ^2S state from which the emission takes place. This state may then be either compact or diffuse. The diffuse nature of the relaxed excited-state configuration is decisively confirmed by excited-state ESR/ENDOR and theoretical work.

7.2.7 F^-*-centres*

We have already seen that the charge distribution of the F-electron is not confined to the anion vacancy: rather, it spreads out over many shells of nuclei centred on the vacancy. Thus the effective positive charge of the vacancy is imperfectly screened by the trapped electron. This suggests that the vacancy may trap a second electron and so form an F^--centre. Indeed, the consequences of bleaching F-centres at low temperature with F-band light (Fig. 7.3) are the formation of the α-band and a broad band which overlaps the F-band on both its long and short wavelength sides. This band is ascribed to F^--centres which are formed from F-centres already present in the crystal. The evidence for this is that coincident with the formation of F^--centres, the F-band decreases in intensity and the α- (or F^+)-band grows in intensity. This photochemical process may be summarised by the reactions

$$F \xrightarrow{h\nu} F^+ + e \quad \text{and} \quad F + e \rightarrow F^-. \tag{7.13}$$

It is apparent from Fig. 7.3(b) that the F^--band is extremely broad and asymmetric in shape. Indeed, the bandshape is not thought to be associated with transitions to a single bound state, but to a continuum of states resonant with the conduction band. Hence the F^--centre is unstable optically even below room temperature. Obviously the binding energy of the second electron in the vacancy is quite small, so the F^--centre is also thermally unstable. Since the precursor of the F^--centre, F-centre ionization, is thermally assisted it follows that the F^--centre is formed only over a very limited temperature range. Illumination in the F-band, at such low temperatures that photoconductivity does not occur, does not lead to the formation of the F^--centres. However, optical bleaching anywhere in the F^--band yields a photocurrent even at liquid helium temperature. Because the F- and F^--bands overlap, irradiation with light in the F-band always leads to F^--centre ionization. In consequence, optical conversion of F-centres into F^--centres may never proceed to completion. However, it is always possible to completely eliminate F^--centres by bleaching with light at longer wavelengths than the F-band. In view of the formation mechanism it is evident that two F-centres are lost in producing a single F^--centre. In other words, the conversion efficiency can be as high as 2, the actual value being a strong function of temperature below about 170 K in potassium chloride.

7.3 Single vacancy centres in other crystals

7.3.1 One-electron centres

The alkali halides assume a position of particular importance in defect studies because of the state of completion of both experimental and theoretical work in these crystals. Various phenomena are greatly simplified both by the high symmetry and essentially ionic bonding. However, in recent years systematic studies have been made of defects in many other solids. In the caesium halides studies of F-centres have paralleled those in the other alkali halides, despite their different crystal structures. One important difference to emerge in these crystals was that in the excited state spin–orbit coupling is so large that resolved structure is observable on the F-centre absorption band. The F-centre in the alkaline earth fluorides also differs from that in the alkali halides in respect of the symmetry around the anion site. Such symmetry changes are easily recognized in the ESR and ENDOR spectra. The absorption/luminescence bands have been identified in the fluorides of calcium, strontium, and barium, and the effects of electric field and magnetic fields have been studied (Hayes and Stoneham 1974). In consequence much of the same understanding is apparent for F-centres in alkaline earth halides as in the alkali halides. Magnesium fluoride is rutile-structured; F-centres have been identified in this material by ESR/ENDOR, optical absorption, and polarized luminescence. Similarly detailed studies have been made of F-centres in potassium magnesium fluoride and related materials.

The alkaline earth oxides, beryllium oxide apart, are the divalent structural analogues of the alkali halides. In none of these crystals has there been evidence that bare anion vacancies may be produced in any significant concentrations. However, in each crystal F^+-centres may be produced by additive colouration, electron irradiation, or neutron irradiation. F^+-

centres are the oxide isomorphs of the F-centre in the alkali halides—a single electron is trapped in an oxygen vacancy. However, a complication arises from the divalent nature of the materials since the negative-ion vacancy may bind two electrons very strongly. Just which defect, F^+- or F-centre, is observed predominantly depends upon competing electron traps already present in the crystal. In neutron-irradiated crystals F^+-centres always predominate, whereas in other crystals F^+- and F-centres coexist.

The ESR spectra of F^+-centres in alkaline earth oxides are beautifully simple and unambiguous, requiring only a table of natural abundances of magnetic nuclides for their interpretation. The only naturally-occurring oxygen isotope with $I>0$ is ^{17}O $(I=\frac{5}{2})$ which has a natural abundance of only 0.037 per cent. Thus the ESR spectra show resolved hyperfine components of first-shell cations only. The natural abundances of magnetic nuclides for the cations is also quite low varying from 0.13 per cent for ^{43}Ca to 11.32 per cent for ^{137}Ba. The intensities of the components in the hyperfine spectrum are then determined by the probability, P_m, of finding magnetic nuclides in a particular near-neighbour shell. P_m is given by

$$P_m=\binom{n}{m}x^m(1-x)^{n-m} \tag{7.14}$$

where x is the fractional abundance of magnetic nuclide, n is the number of near-neighbour ions, and m is the number of ions in this shell which is a magnetic nuclide. For example, in magnesium oxide, the sole magnetic nuclide is ^{25}Mg $(I=\frac{5}{2})$ which is 10.11 per cent abundant. Thus a fraction $(0.8989)^6$ or 52 per cent of F^+-centres have no magnetic nuclides in nearest cation sites, in which case the most intense line in the ESR spectrum is due to that fraction of F^+-centres with no magnetic nuclides. The next most intense lines are from those F^+-centres, 35.6 per cent in magnesium oxide, which have one nearest-neighbour cation site occupied by a magnetic nuclide. Such first-shell interactions are entirely responsible for the hyperfine structure observed for F^+-centres in magnesium oxide. There are 10 per cent of F-centres in magnesium oxide which have two ^{25}Mg nuclei in the nearest-neighbour cation shell. A similar situation obtains in strontium oxide. In barium oxide there is first-shell hyperfine structure from both ^{135}Ba and ^{137}Ba nuclei. In general, one does not expect to observe hyperfine structure in the F^+-centre ESR spectrum of calcium oxide since ^{43}Ca is only 0.13 per cent abundant. However, in some calcium oxide crystals the F^+-centre ESR linewidth is as little as 15 mG: resulting from the enhanced sensitivity associated with such narrow lines, hyperfine structure due to both ^{43}Ca and ^{17}O is observed (Henderson and Tomlinson 1969). F^+-centres have been reported in beryllium oxide and zinc oxide, both crystals with the wurtzite structure. The ESR spectra show clearly the effect of F^+-centres being formed in two non-equivalent anion sites. ESR and ENDOR spectra have also been reported for F^+-centres in aluminium oxide and Naβ–Al_2O_3. In the latter case the F^+-centre is in the conduction plane, so that an axially symmetric hyperfine structure due to two equivalent ^{27}Al nuclei is an indelible fingerprint.

ENDOR studies have been reported for the oxides of beryllium, magnesium, strontium, and barium. In magnesium oxide ENDOR lines from cation nuclei in shells I, III, and V have been observed, and in ^{17}O-enriched samples ENDOR lines from anion nuclei in shell II. Analysis of the hyperfine splitting makes an interesting comparison with the alkali halides. An immediate and obvious point of dissimilarity is the positive charge on the defect which results in large distortions of the lattice near the F^+-centre. This enhances the electric field gradients at the neighbouring nuclei, which are readily measured in ENDOR studies. Nuclear quadrupole effects yield estimates of a 5–7 per cent outward relaxation of the nearest cation shell. The other consequence of the positive charge on the F^+-centres is that the F^+-centre wavefunction is much more localized inside the vacancy than for F-centres in the alkali halides (Halliburton *et al.* 1975).

Well-characterized optical spectra have been reported only for the cubic alkaline earth oxides. However, the results of such studies have proved controversial. The problems were twofold: high impurity levels which contributed a variety of photochromic reactions, and the photoconversion reactions between F- and F^+-centres. Hence the philosophy of theoretical work and magneto-optic studies was rather different from that in alkali halide F-centres. In both cases the emphasis was on the identification of the F^+-centres; the spin–orbit structure was something of a bonus (Hughes and Henderson 1972). In magnesium oxide none of the absorption bands in the visible region correlated with the intensity of the F^+-centre ESR spectrum, although a clear correlation was achieved between the ESR spectrum and a broad band with peak at $h\nu = 4.95$ eV in neutron-irradiated crystals. However, in additively coloured or electron-irradiated crystals the 4.95 eV absorption band was not accompanied by the F^+-centre ESR spectrum. This apparent conundrum arises because F^+-centres and F-centres have both strongly allowed optical transitions close to $h\nu = 4.95$ eV. Furthermore, neutron irradiation produces predominantly F^+-centres, whereas additive colouration and electron irradiation produce both F^+- and F-centres. Faraday rotation spectra of both additively coloured and irradiated magnesium oxide show that the defects associated with the 250 nm band are indisputably paramagnetic, but are they F^+-centres? This was proved by making use of the fact that the rotation pattern is sensitive to ESR in the F^+-centre ground state (see Section 12.6). The ground state of the F^+-centre is a Kramers doublet and spin–lattice relaxation times are very long at low temperature. Hence it is relatively easy to saturate the spin system with resonant microwaves; in this condition $\langle S_z \rangle = 0$. In other words, the paramagnetic component of the Faraday rotation pattern has been eliminated. That saturation of the F^+-centre ESR signal eliminated the paramagnetic component of the Faraday rotation pattern of the 4.95 eV band means that identification of the F^+-band is assured. Similar absorption/Faraday rotation studies have been reported for the oxides of calcium, strontium, and barium. The trends are similar to those in the alkali halides.

Emission bands have been reported for F^+-centres in magnesium oxide, calcium oxide and strontium oxide. These bands are strongly Stokes shifted to lower energies from the absorption band positions. Further evidence of the compact nature of the F^+-centre wavefunction, in this case in the excited state, comes from the measurements of the radiative lifetime. In calcium oxide the radiative lifetime at 4.2 K is 5.2 ns, i.e. three decades shorter than in halide F-centres. Indeed this value is very similar to that for 2p–1s transitions on atomic hydrogen. Analogous measurements for magnesium oxide and strontium oxide yield values of 10 ns and 35 ns, respectively. Clearly, strongly allowed transitions are involved so that the situation is very different from that in the alkali halides where the emissive state is predominantly $|2s\rangle$ in character. Thus for F^+-centres in the alkaline earth oxides, emission is from a compact $|2p\rangle$-like excited state.

The major surprise in the optical spectroscopy of the F^+-centre in oxides arises from the bandshapes in absorption. They are without exception broader than their counterparts in the alkali halides and also somewhat asymmetric in shape. This suggests a major difference in the nature of the vibronic coupling. In the alkali halides the coupling is strongest to the breathing mode (A_{1g}) vibrations, although coupling to tetragonal (E_g) and trigonal (T_{2g}) modes of vibration is significant. Strong coupling is represented by Huang–Rhys parameters in the range $S \simeq 18$–50 for the alkali halides. In the oxides non-symmetric lattice modes appear to be most strongly coupled as evidenced by the asymmetric absorption bands. The situation is highlighted by the absorption and emission spectra of F^+-centres in calcium oxide. As we show in Fig. 7.10, both absorption and emission bands show a clearly resolved zero-phonon line at a wavelength $\lambda = 355.7$ nm, with associated vibronic structure. There are several anomalous features associated with these spectra, in particular, the conflicting values of the Huang–Rhys parameters and the spin–orbit constant determined from the broad band and zero-phonon lines (see Table 7.6). To recapitulate, values of S can be determined from the strength of the zero-phonon line relative to the broad band ($P_{00} = \exp(-S)$), from the separation of band centroid from zero-phonon line ($\Delta = S\hbar\langle\omega\rangle$), and from the second moment of the broad band ($M_2 = S\hbar^2\langle\omega^2\rangle$) at $T = 0$ K. The value of S derived from the absorption band half width (Table 7.6) is clearly out of line with all other values. Similarly the values of ζ determined from circular dichroism of the zero-phonon line and broad band are surprisingly different. These data for some time led to

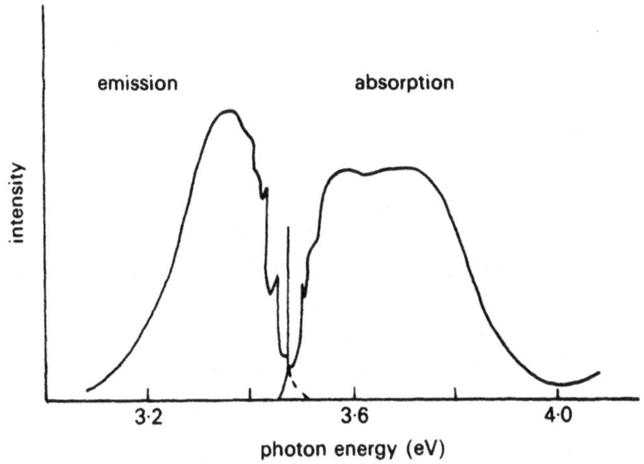

FIG. 7.10. Absorption and emission bands of F^+-centres in calcium oxide, measured at $T=4$ K. (After Henderson *et al.* 1972.)

Table 7.6

Measured values of the Huang–Rhys parameter and spin–orbit splittings for F^+-centres in calcium oxide. (After Hughes and Henderson 1972.)

Experimental measurement used to determine S and ζ	Absorption	Emission
Zero-phonon line as a fraction of total transition	4–6	4.9
Stokes shift	5–6	3.5
Band half width	13–16	4
ζ (broadband)	-24 cm^{-1}	—
ζ (zero-phonon line)	-0.5 cm^{-1}	—

disagreement as to whether or not the zero-phonon line was actually associated with the F^+-centre. However, uniaxial stress experiments on the zero-phonon line confirm that an electric dipole transition between states transforming as the irreducible representations A_{1g} and T_{1u} at a centre with O_h symmetry is involved, which is exactly what one expects for the F^+-centre. The situation is resolved by recognizing that in the presence of weak spin–orbit coupling and weak electron–phonon coupling there is no Jahn–Teller splitting of the band. Note that the width of the zero-phonon line is about 4.0 cm^{-1} which is greatly in excess of the value of ζ measured from the zero-phonon line. In consequence there is no measurable structure even within the zero-phonon line. However, of particular importance is the fact that a dynamic Jahn–Teller effect is involved which couples the T_{1u} state equally strongly to modes with E_g and T_{2g} symmetry. In this case there can be a large quenching of the spin–orbit splitting in the zero-phonon line without producing a static distortion. Transitions in the zero-phonon line arrive directly in the ground vibronic level of the relaxed excited electronic state which has been influenced by the Jahn–Teller effect. On the other hand, broadband transitions take the F^+-centre to the unrelaxed excited state, i.e. the undistorted configuration parametrized by the unquenched coupling constants. Hence the Ham orbital reduction factor is given by the ratio of the spin–orbit coupling constant measured for the zero-phonon line divided by that measured for the broad band, i.e. $k(T_1)=0.02$. Within this model the strain coupling constants B' and C' are quenched only by factors of $\frac{2}{5}$. These constants have been determined from the uniaxial stress data for the zero-phonon line. Coupling to the breathing mode, given by the coefficient $A'=1.54\times10^{-4}\ \mathrm{cm}^{-1}$, is unquenched. The quenched coupling coefficients for E_g and T_{2g} modes, respectively, are $B'=0.48\times10^{-4}\ \mathrm{cm}^{-1}$ and $C'=0.77\times10^{-4}\ \mathrm{cm}^{-1}$. This model has been applied to the F^+-centre by Hughes (1970) who shows it to be consistent with all of the experimental results including the apparent discrepancy in the Huang–Rhys parameters. The important result is that vibronic coupling to non-symmetric modes is dominant as can be seen from the various Huang–Rhys parameters for the various phonon modes (viz. $S_A\simeq1.8$, $S_E\simeq S_{T_2}\simeq4.0$). Although there is far less data available for the other oxides it appears that coupling to non-symmetric modes is more important than symmetric modes in each case.

7.3.2 Two-electron centres in oxides

The F-centre in oxides consists of two electrons trapped in an oxygen ion vacancy. This defect is the oxide analogue of the F^--centre in the alkali halides. However, in view of the additional trapping potential of the oxygen vacancy the excited states of the F-centre are strongly bound (except perhaps for the case of barium oxide.) In the ground state the two electrons have their spins aligned antiparallel forming a spin singlet $^1A_{1g}$ state in O_h symmetry derived from the helium-like configuration ($1s^2$). Possible excited states are obtained from two electrons in the (1s)(2s) or (1s)(2p) configurations. The order of the resulting excited states, $^1T_{1u}$, $^3A_{1u}$ and $^3T_{1u}$, cannot be determined simply by inspection. Since the F-centre in its ground state is diamagnetic there is no ESR spectrum; its identification was initially via photoconversion processes much like those of the F^--centres in the alkali halides. So far the most complete data are for calcium oxide and magnesium oxide, in which $F\leftrightarrow F^+$ photoconversion reactions gave the first evidence for the existence of F-centres. In some ways there remain serious questions related to the interpretation of some experimental results, because impurities compete with vacancies as traps for free carriers. Indeed, the current understanding owes much to comparative studies of defects produced by different techniques. It is not our intention to review the mechanisms of defect production. Additive coloration in oxides is a technique fraught with difficulties, because of the very high temperatures and pressures required. Just which of the two defects predominates is determined by the relative concentrations of anion vacancies and impurities whose valence might be changed by trapping an electron. Roughly, both F- and F^+-centres are produced when the total anion vacancy concentration is less than the total concentration of impurity traps. When the vacancy concentration exceeds the impurity concentration then the dominant defect is the F-centre. It is also the case that F- and F^+-centres are produced during single-crystal growth of calcium oxide, presumably as a consequence of the reducing action of the graphite electrodes in the submerged arc furnace. We will discuss briefly the optical properties of F-centres in calcium oxide, since they are best understood.

In calcium oxide the F- and F^+-bands absorb at 400 nm and 370 nm, respectively. The F^+-band was identified by MCD studies, as we discussed earlier. That the 400 nm band is due to F-centres came initially from photoconversion and luminescence data. The interconversion process was first observed in additively coloured calcium oxide crystals. Optical bleaching at 77 K with optical photons having energy $h\nu > 2.5$ eV produced a metastable enhancement of the F^+-band at the expense of the F-band. Bleaching with light of energy $h\nu < 2.5$ eV reverses the effect. The interpretation of these data is that blue light ionizes the F-centres, the excited electron being released into the conduction band after relaxation. This electron is then trapped at some defect or impurity trap. Long-wavelength radiation releases the electron from its trap, reforming the F-centre at the expense of the F^+-centre. Although several impurity traps are known to be involved (Henderson and Wertz 1977; Henderson and Tomlinson 1969), Chen *et al.* (1988) have shown that the most important electron trap is the OH^- ion. Photoconversion is not the sole effect of excitation in the F-band, since one also observes a luminescence band with peak at 601 nm. ODMR experiments (Section 12.2) have shown that the absorption band is due to an allowed electric dipole transition, ${}^1A_{1g} \rightarrow {}^1T_{1u}$, whereas the emission is due to the spin-forbidden ${}^3T_{1u} \rightarrow {}^1A_{1g}$ transition. At low temperatures, $T < 100$ K, this luminescence band shows a resolved zero-phonon line at 574 nm and weak vibronic structure atop the broad band. The emission lifetime at 4.2 K is 3.4 ms; at higher temperatures the decay pattern of this band includes a non-exponential phosphorescence with a persistence of several minutes. The emission lifetime at low temperature is consistent with an oscillator strength of 5×10^{-7}, which is reasonable for a spin-forbidden transition. The long-lived phosphorescence involves the release of electrons from (mainly) OH^- traps assisted by thermal excitation into the conduction band (Chen *et al.*

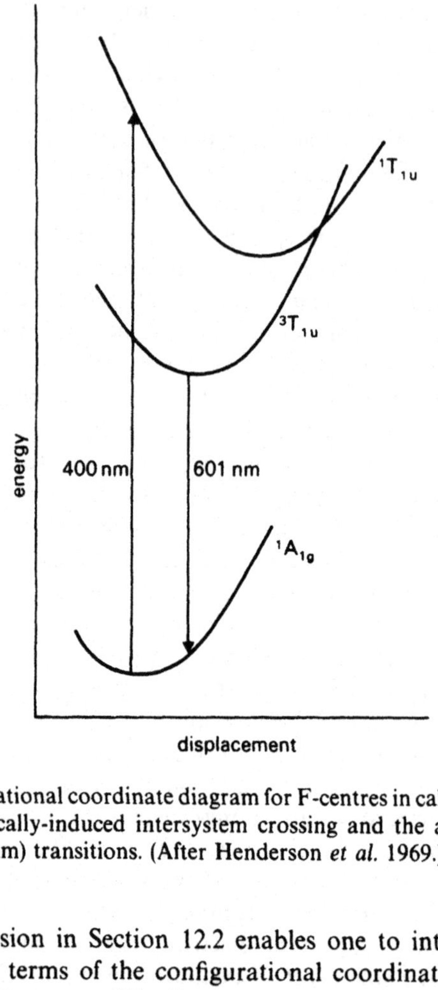

FIG. 7.11. Configurational coordinate diagram for F-centres in calcium oxide showing $^1T_{1u} \rightarrow {}^3T_{1g}$ vibronically-induced intersystem crossing and the absorption (400 nm) and emission (601 nm) transitions. (After Henderson *et al.* 1969.)

1988). The discussion in Section 12.2 enables one to interpret the optical pumping cycle in terms of the configurational coordinate diagram for the $^1A_{1g}$, $^1T_{1u}$ and $^3T_{1u}$ states (Fig. 7.11). At low temperature (4.2 K) only the $^3T_{1u} \rightarrow {}^1A_{1g}$ luminescence is observed. Hence optical pumping in the 400 nm singlet–singlet transition results either in photoionization of the F-centre (if the temperature is >40 K) or in population of the $^3T_{1u}$ state by radiationless decay from the $^1T_{1u}$ state. At higher temperatures ($T > 200$ K), however, emission is observed from both $^3T_{1u}$ and $^1T_{1u}$ states (Bates and Wood 1975).

Similar photoconversion processes occur in magnesium oxide, although they are much more difficult to follow because F- and F^+-absorption bands are almost coincident. This superposition led to some early controversy concerning the identification of the F- and F^+-bands. As we have already commented, both F- and F^+-centres exhibit broad optical absorption bands with peak wavelengths of *c.* 250 nm. In additively coloured or electron-irradiated crystals both defects exist, so that there is no correlation between the 250 nm band and the F^+-centre ESR spectrum. In fact the ESR spectrum is very weak although the 250 nm band is strong. Since in additively coloured crystals F^+-centres can be generated by X-irradiation or by ultraviolet irradiation at 77 K, whereas F-centres are favoured by a similar bleach at 300 K, it is quite clear that any lack of correlation is due to the presence of F-centres. Furthermore, a heat treatment of additively coloured crystals above 400 °C converts F^+-centres to F-centres. Using these interconversion processes Chen *et al.* (1969) have isolated two bands in the ultraviolet region. They concluded that the F-band has a half width of 35 nm and peak wavelength $\lambda = 247$ nm, with an oscillator strength of 1.25. The F^+-centre peaks at 250 nm with half width of 25 nm and oscillator strength of 0.8. The F-centre luminescence has been ascribed to a broad band with peak at 520 nm. This band has both a well-defined short component ($\tau_R \sim 7 \times 10^{-4}$ s) and a phosphorescence with multiple component lifetimes. This behaviour is very reminiscent of F-centre luminescence in calcium oxide (Chen *et al.* 1988; Williams *et al.* 1988).

7.4 Vacancy aggregate centres

The spectra shown in Fig. 7.3 were obtained using additively coloured crystals of potassium bromide. The primary defects involved, F- and F^--centres, both contain a single anion vacancy. Alkali halides may also be coloured by X-rays, γ-rays, reactor neutrons, and energetic charged particles (e^-, p^+). Irradiation of alkali halides also produces trapped hole centres, and if carried out at room temperature, a variety of bands associated with small aggregations of F-centres. These aggregate centres have optical absorption bands displaced to long wavelength from the F-band. Historically, work on these centres was begun in the 1940s by Molnar at MIT and Petroff in Göttingen. There are various recipes for the production of such centres. A crystal may be X-irradiated at low temperature to form largely F-centres, and subsequently illuminated with F-band light at room temperature. Similar bleaching of an additively coloured crystal will produce aggregate centres, as will X-irradiation at room temperature. Alternatively, an additively coloured crystal may be warmed to *c.* 400 K for a few hours. A typical absorption spectrum for potassium chloride irradiated at room temperature with X-rays is shown in Fig. 7.12. The principal features are the bands labelled F_2, F_3, and F_4, which are assigned to aggregates of two, three, and four F-centres in nearest neighbour sites. The evidence for these assignments are various, including optically-induced ESR, polarized absorption and luminescence, and perturbation spectroscopy.

Optically-induced ESR greatly enhances the identification of the structure of vacancy aggregate centres (Seidel and Wolf 1968); this technique involves

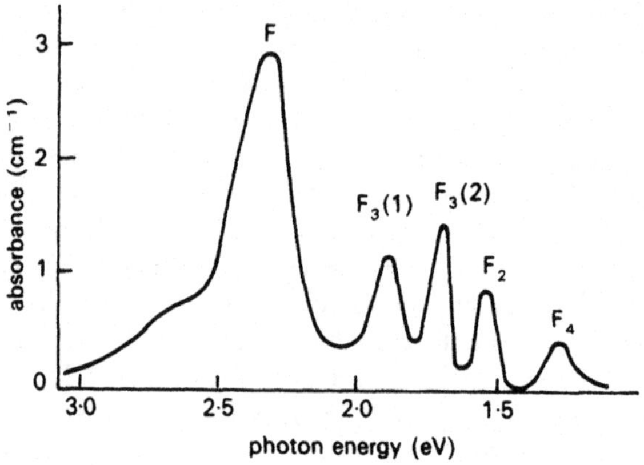

FIG. 7.12. Optical absorption spectrum of potassium chloride after X-irradiation at room temperature. Measurement temperature $T = 95$ K. (After Van Doorn 1962.)

ESR measurements in the presence of F-band light. In additively coloured and rapidly cooled potassium chloride cystals the ESR spectrum shows F-centres to be isolated. However, in the presence of light the shape and saturation behaviour of the F-centre ESR spectrum changes. The inhomogeneously broadened ESR line becomes more Lorentzian and the spectrum is less easily saturated by microwave power due to a shortening of the characteristic relaxation times. These changes are associated with exchange-narrowing of the ESR line because F-centres achieve higher local concentrations due to aggregation. The effects are not due to new centres. A minimum F-centre separation in these 'loose' aggregates is three to four nearest-neighbour distances.

When bands due to F_2-centres appear in the optical spectrum, no new ESR spectra appear. Although ESR studies on appropriate crystals are made difficult by the large F-centre concentration present, careful measurements made of spin concentration prior to and subsequent to aggregation reveal no evidence that the ground state of the F_2-centre is paramagnetic. This is to be expected since, as in the hydrogen molecule, the ground state of such a two-electron centre should be a spin singlet. However, there remains the possibility that optical excitation of F_2-centres may result in the population of excited triplet states. In the F_2-centre (Fig. 7.1) there are two F-centres in nearest-neighbour sites along a $\langle 110 \rangle$ direction so that the overall point symmetry in D_{2h}. Optical transitions may be thought of as between the (1s, 1s) ground state of two F-centres and excited states that are a product of a 1s state at one site and a 2p state at the other site. It is then usual to label the states using the terminology either of molecular chemistry or of group theory. The lowest energy states are shown in Fig. 7.13. In group-theoretical

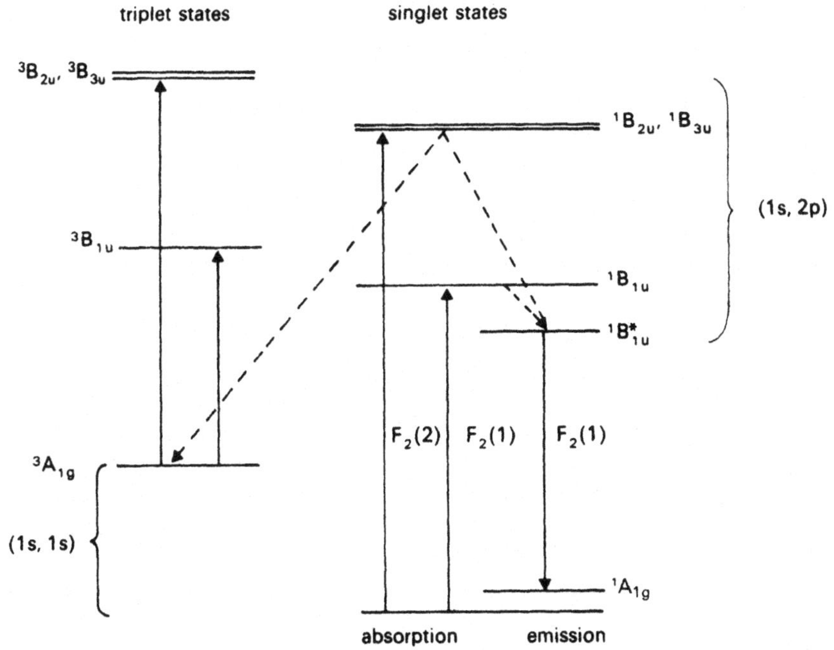

FIG. 7.13. Showing the lowest singlet and triplet levels of F_2-centres, and some transitions between them. The broken lines indicate non-radiative transitions.

language the $F_2(1)$-band corresponds to a ${}^1A_{1g} \rightarrow {}^1B_{1u}$ transition; it is excited by light polarized parallel to the F_2-centre axis. Other transitions, ${}^1A_{1g} \rightarrow {}^1B_{2u}$, ${}^1B_{3u}$, give rise to the $F_2(2)$-band, which is polarized perpendicular to the F_2-centre axis. The $F_2(2)$-band tends to overlap the F-band. The lowest triplet state, ${}^3A_{1g}$, is then populated by strong pumping in the $F_2(2)$-band. Because ${}^3A_{1g} \rightarrow {}^1A_{1g}$ is both spin- and parity-forbidden, the excited triplet state is very long-lived ($\tau_{NR} \simeq 50$ s in potassium chloride) so that a large fraction of F_2-centres may be prepared in the triplet state. Hence the $F_2(1)$-band absorption decreases and one observes additional, albeit transient, absorption bands from the excited triplet states. If such optical experiments are carried out in an ESR cavity then direct evidence of the F_2-nature of the centre is provided by an $S=1$ ESR spectrum. This spectrum has the same transient behaviour as the metastable absorption bands associated with the 50 s lifetime of the ${}^3A_{1g}$ state. The spectrum consists of three lines for rotations of B in the (100) plane and four lines for rotations in the (110) plane. These lines are due to zero-field splittings caused by the dipole–dipole interactions of the two electron spins. The spectrum fits the spin Hamiltonian

$$\mathscr{H} = g\mu_B \boldsymbol{B}.\boldsymbol{S} + DS_z^2 + E(S_x^2 - S_y^2)$$

in which $g=1.998$, $D/g\mu_B=-16.1$ mT and $E/g\mu_B=+5.4$ mT for F_2-centres in potassium chloride. One choice of principal tensor axes is $x=[110]$, $y=[001]$, $z=[1\bar{1}0]$; there are five other crystallographically equivalent sets. Clearly the spectrum is in agreement with the model of the F_2-centre. Indeed, the measured value of D is consistent with the dipolar splitting expected for electrons localized at neighbouring anion sites. Further confirmation of the correctness of the model comes from ENDOR measurements which show that the electronic wavefunction is the antisymmetrized linear combination of the corresponding F-centre wavefunctions (Seidel *et al.* 1963; Seidel and Wolf 1968).

Similar studies have been reported for the F_3-centres in potassium chloride. Again, in crystals containing large F_3-centre concentrations reversible suppression of the $F_3(1)$- and $F_3(2)$-bands is achieved by strong irradiation in the F-band, accompanied by new bands shifted from the known F_3-bands. On switching off the pumping light these bands decrease exponentially with a lifetime of 14.5 s. In this case the metastable excited state of the F_3-centres is a $^4A_2$ state ($S=\frac{3}{2}$) and the orientation dependence of the ESR spectrum fits the above spin Hamiltonian with $S=\frac{3}{2}$, $g=1.996$, $D/g\mu_B=16.85$ mT and $E/g\mu_B=0$. The principal axis is one of the four $\langle 111 \rangle$ axes of the crystal. The F_3 model is further substantiated by ENDOR data (Seidel and Wolf 1963, 1968). This work raises one apparent problem. Since three electrons are involved, why is there no ground-state ESR for the F_3-centre? As we shall see, the kinetics of aggregation and uniaxial stress effects on optical zero-phonon lines lend considerable support to the three F-centre model of the defect responsible for the F_3-absorption bands. And yet, despite the model predicting a paramagnetic ground state, ESR studies reveal no resonance spectrum. This problem was resolved by Krupka and Silsbee (1964) who showed that because the ground state is strongly coupled to the phonon field the ESR is normally broadened beyond observability, except in the presence of a large stress. The stress suppresses the relaxation processes and an ESR spectrum, with g-tensor axial about a $\langle 111 \rangle$ axis, may be observed for $T \simeq 2$ K. The g-tensor components are $g_{\parallel}=2.00$ and $g_{\perp}=2.06$, which are expected for an orbitally-degenerate ^2E state.

7.4.1 *Kinetics of aggregation*

The models of defects involving two, three, and four F-centres in aggregate were determined partially by studying the kinetics of their formation. Of the several possible recipes we consider only formation by radiation damage at *c.* 300 K. In general, ionizing radiation promotes photoconversion of F-centres into F^+- and F^--centres, and at temperatures in the range 250–300 K F^+-centres become mobile, diffusing through the crystal until they coalesce with F^--centres to yield F_2-centres. In addition there is also a finite probability that an F-centre will be produced in a near-neighbour site of an existing F-centre. A simple quantitative model of F_2-centre production from F-centres can be derived using the following argument. Suppose that at any instant in time the concentrations of F- and F_2-centres, [F] and [F_2] respectively, are changed by d[F] and d[F_2] during subsequent incremental intervals of irradiation. Since each F_2-centre has 12 anion neighbours along the different $\langle 110 \rangle$ directions, then out of the N halogen ion sites in unit volume of crystal only the $12 \times$[F] sites may be converted to an F-centre yielding an F_2-centre. Hence

$$\mathrm{d}[F_2] = 12[F]\,\mathrm{d}[F]/N. \tag{7.15}$$

Integrating eqn (7.15) using the boundary condition [F_2] = 0 when [F] = 0 yields

$$[F_2] = (6/N)[F]^2. \tag{7.16}$$

A quadratic dependence between [F] and [F_2] also results for additive colouration at elevated temperature. Extending this analysis to the formation of F_3-centres it is noted that the most straightforward mechanism is the creation of an F-centre in a halogen site neighbouring an F_2-centre. We can write the dynamic equilibrium as

$$F + F_2 \leftrightarrows F_3$$

which yields the following equilibrium constant:

$$K_3 = \text{const.}\,[F_3]/[F][F_2]. \tag{7.17}$$

These kinetic relationships have been tested for the formation of both F_2- and F_3-centres. Some data are shown in Fig. 7.14. Results for F_2 formation by additive coloration show that the quadratic dependence of [F_2] on [F] holds over two decades of coloration. Similarly good agreement has been obtained for X-ray formation of F_2-centres in potassium chloride and potassium bromide at both 77 K and 4 K. These data clearly support the model of the F_2-centre. They also give a means of measuring the oscillator strength for the F_2-centre, since from eqn (6.11)

$$[F_2] f_{F_2} = 0.87 \times 10^{17}\, \alpha_{F_2}(\nu_0)\Delta\nu_{F_2} \tag{7.18}$$

where f_2 is the oscillator strength of the F_2-band, $\alpha_{F_2}(\nu_0)$ in units of cm^{-1} is the absorption coefficient at the F_2-band at the peak, and $\Delta\nu_{F_2}$ in eV is the full width of the F_2-band at the half maximum. Substitution from eqn (7.16) gives

$$\alpha_{F_2}(\nu_0) = 0.87 \times 10^{17}\left(\frac{6}{N}\right)\left(\frac{\Delta\nu_F^2}{\Delta\nu_{F_2}}\right)\left(\frac{f_{F_2}}{f_F^2}\right)\alpha_F(\nu_0)^2. \tag{7.19}$$

Using values of $f_F = 0.48$ and 0.53 for F-centres in potassium bromide and potassium chloride respectively, oscillator strengths of 0.27 and 0.38 were obtained for the F_2-bands in these two crystals. Equation (7.17) has been

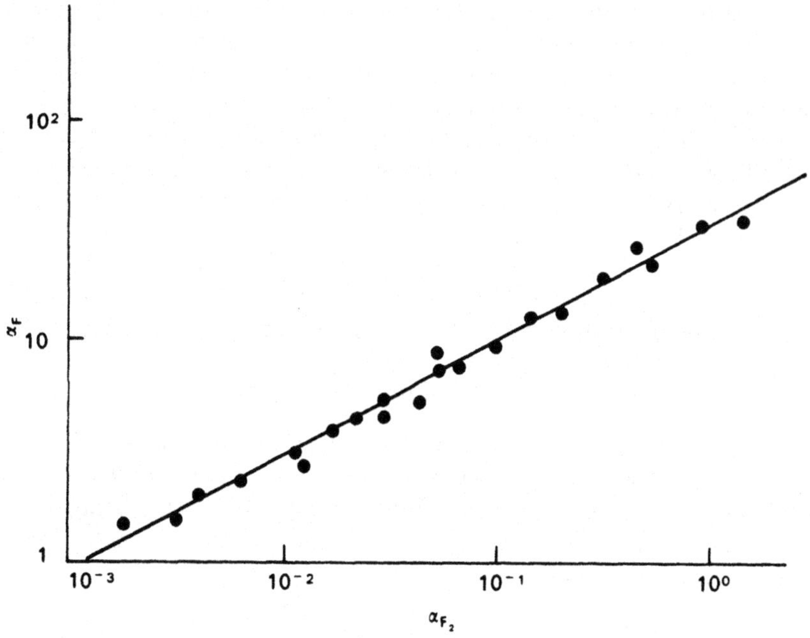

FIG. 7.14. The absorption coefficients α_F and α_{F_2} plotted logarithmically for equilibrium formation of F_2-centres from F-centres during additive coloration of potassium chloride at 700 °C. The slope of the straight line is 2. Absorption measurements at 4 K. (After Van Doorn 1962.)

tested for F_3-centre formation by X-irradiation near room temperature in several different crystals. The data in Fig. 7.15 are for potassium chloride; they show that the $F:F_2:F_3$ equilibrium is strongly dependent upon X-ray intensity. They strongly support the F_3-model for the associated defect.

Kinetic tests of the F_4-centre model are less compelling. Unlike the data for both F_3-centre bands, the two bands associated with the F_4-centres are somewhat contradictory in that the $F_4(1)$ band intensity is linear with F_2-centre intensity whereas the $F_4(2)$ band varies as the product $[F][F_3]$. This would appear to imply that the two bands are associated with different centres, one involving two F-centres and the other four centres.

7.4.2 *Positions and shapes of* F_2-, F_3-, *and* F_4-*bands*

As with F-centres the absorption bands of F-aggregate centres have peak wavelengths which vary in a regular manner with the unit cell size (***d***). The Mollwo–Ivey relationship takes the forms for F_2- and F_3-centres

$$F_2\text{-band: } \lambda_{max} = 1400\, d^{1.56}$$

$$F_3\text{-bands: } \lambda_{max} = k\, d^{1.84}$$

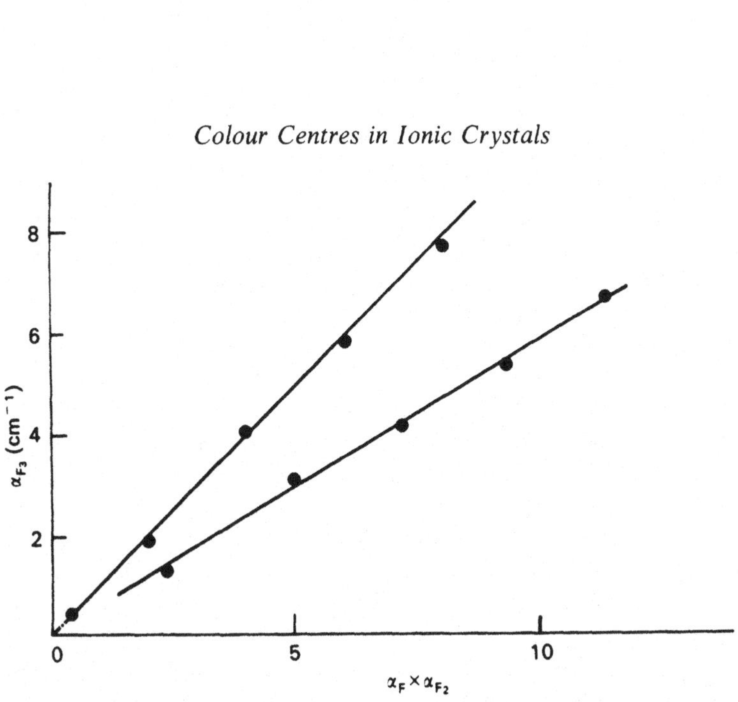

FIG. 7.15. Absorption coefficient for F_3-centres in potassium chloride plotted as a function of the product of the absorption coefficient for F- and F_2-centres, the centres having been produced at room temperature by X-irradiation at two different radiation intensities. Measurements made at 20K. (After King 1967, Ph.D. Thesis, University of Reading, unpublished.)

where for $F_3(1)$- and $F_3(2)$-bands, respectively, $k = 816$ and 884, d being measured in Å units. The deviations from the d^2 relation implied by the simple divacancy model of the centre again show the importance of terms in potential which vary as R^{-1}. Thus aggregate-centre absorption bands are shifted to long wavelengths relative to the F-band.

One interesting feature of F-aggregate centre spectroscopy is that resolvable structure may be observed on the absorption and emission spectra at low temperatures in most host crystals. This structure takes the form of a sharp zero-phonon line on the low-energy side of a broad absorption band which shows some resolved vibronic peaks. Charged versions of the F_2-centre (i.e. F_2^+- and F_2^--centres) as well as F_3^+-, F_3-, and F_3^--centres all have absorption and emission bands with well-resolved structure. The zero-phonon lines follow a Mollwo–Ivey relationship, similar to that for the appropriate broad band. The F_2 absorption band shown in Fig. 7.12 is due to transitions polarized parallel to the z-axis of the defect, i.e. it is the $F_2(1)$-band; it is well separated from other F_2-centre bands. A single emission band is observed, excited in either the $F_2(2)$ transitions or $F_2(1)$ transition. This emission is polarized parallel to the F_2-centre axis, and the Stokes shift relative to the $F_2(1)$ absorption band is relatively small (*c.* 0.3–0.4 eV). Thus the emission band is due to transitions between the same electronic states as the $F_2(1)$ absorption band. The polarization behaviour confirms that non-radiative decay processes from ${}^1B_{2u}$, ${}^1B_{3u}$ into ${}^1B_{1u}$ states are relatively efficient. Even at 300 K the luminescence efficiency remains high because the relaxed excited state is strongly localized in the vacancy. Hence the radiative lifetimes of F_2-centres are very short, being in the range 10–30 ns expected for strongly-allowed electric dipole transitions. The F_2-centre spectra do not show zero-phonon lines in absorption or in emission. Although the zero-phonon transition is allowed, the electron–phonon coupling is so strong that the transition probability is too small for experimental observations (i.e. $\exp(-S) \simeq 0$). Assuming coupling to a single vibrational mode then the Stokes shift between absorption and emission is $(2S-1)\hbar\omega$, whereas the zero-temperature halfwidth is given by $2.36S^{\frac{1}{2}}\hbar\omega$. Since the temperature dependence of the halfwidth follows a $\coth(\hbar\omega/2kT)$ relationship there still remain a number of ways of evaluating S and $\hbar\omega$. A comparison of S-values for aggregate centres in lithium fluoride gives $S \simeq 5.5$ for F_2^+- and F_3^+-centres, $S \simeq 18$ and 3.6 for F_2-centres and F_3-centres, $S \simeq 2$–3 for F_2^-- and F_3^--centres, and $S = 1$ for F_4-centres. Obviously the positively charged version is the more strongly coupled of the two charged versions of a particular defect species. There are numerous examples of near mirror symmetry between the vibronic structure of the absorption and emission bands. The mirror symmetry of the phonon sidebands in absorption and emission spectra of, for example, F_2^--centres in lithium fluoride (Fig. 7.16) indicates that in such cases the assumption of linear coupling theory is reasonable.

The shape of the optical spectrum is determined by the coupling strength, S, and the spectrum of coupled modes. The resolved structure can be analysed by a comparison with the phonon spectrum of the host crystal,

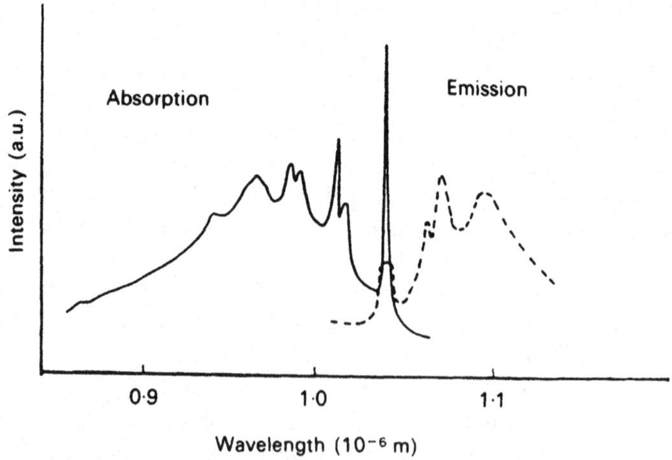

FIG. 7.16. The optical absorption and luminescence bandshapes of F_2^--centres in lithium fluoride at 4.2 K. (After Fitchen 1968.)

always bearing in mind that there are several possible contributants to the vibronic sideband. First of all the one-phonon contribution to the sideband is correlated with the high density of phonon states at van Hove singularities at zone boundaries. For $S>1$, the observed sideband structure is a superposition of one, two, and higher (n) phonon contributions (Chapter 5). Secondly there may be resonant modes with large amplitudes near to the defect with frequencies within the phonon continuum. Finally there may be localized modes with frequencies outside the lattice-mode spectrum. The total probability for all processes transferring energy $\hbar\omega$ to the phonon system is just a summation of the probabilities of the individual processes. In terms of the photon frequency, ν, the total transition probability, $W(\nu)$, for absorption of a photon of energy $h\nu = E_0 + \hbar\omega$ is given by

$$W(\nu) = \sum_{n=0}^{\infty} W_n(\nu) \tag{7.20}$$

the individual components $W_n(\nu)$ being given (Chapter 6) by

$$\int_0^{\infty} W_0(\nu)\,d\nu = \exp(-S)$$

and

$$\int_0^{\infty} W_n(\nu)\,d\nu = \frac{S^n}{n!}\exp(-S)$$

normalized so that $\int_0^{\infty} W(\nu)\,d\nu = 1$. In terms of the measured absorption coefficient $\alpha(\nu)$ these transition probabilities satisfy

$$W(\nu) = \frac{\alpha(\nu)/\nu}{\int_0^{\infty}[\alpha(\nu)/\nu]\,d\nu} \tag{7.21}$$

integrated over the full bandwidth. The application of these equations to the F_2^--absorption band is shown in Fig. 7.17. The agreement is very good, indicating the reasonableness of assuming that the phonons are excited independently with no restriction on how they combine in the n-phonon sideband. Such good agreement is not always observed. There are often cases of peaks in the vibronic sideband which do not correspond even approximately to peaks in the phonon density of states, especially when coupling to resonant modes is involved.

Absorption and emission spectra of charged and neutral F_3-type centres also show resolved structure. The trigonal (C_{3v}) symmetry of these centres allows orbitally degenerate electronic states and, consequently, incipient

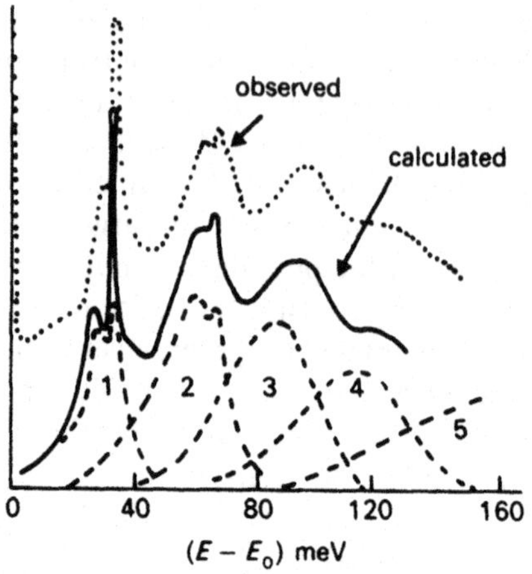

FIG. 7.17. A comparison of the optical absorption spectrum of F_2^--centres in lithium fluoride and the calculated one-, two-, etc. phonon probabilities normalized relative to the zero-phonon line. (After Fitchen 1968.)

instability against symmetry-lowering distortions which remove this degeneracy, i.e. the Jahn–Teller effect (Section 5.4.3). The lowest energy transitions of these F_3-centres are all of the type A ↔ E (Section 2.5.3). However, for the F_3^+-centre the lowest level is the ^1A state, whereas the F_3-centre and F_3^--centre the ground states are respectively ^2E and ^1A. In these cases the mirror symmetry of the F_2^--absorption/emission spectra in Fig. 7.16 is destroyed by Jahn–Teller coupling. A well-documented example is the F_3^--centre in lithium fluoride, where the A → E transition has a double-humped absorption band; the E → A emission shows no such structure (Fitchen 1968). Theoretical reconstruction of these bandshapes reflects the experimental trends quite nicely; in absorption the band is double-humped with unequal spacing between the vibronic peaks. However, the single-peaked emission band has equal spacing of the phonon-assisted structure.

Many optical zero-phonon lines have been reported in the different alkali halides; the symmetry properties of the associated defects were characterized using perturbation spectroscopy (Hughes 1966; Fitchen 1968; Von der Osten 1976). Such measurements involve the application of external perturbations, e.g. hydrostatic pressure, uniaxial stress, electric field, and magnetic field, which may remove certain kinds of degeneracy at the electronic centre. The experimental method (Section 6.4) then involves the measurement of the number, intensity, polarization properties, and energy shifts of the components induced by the perturbation. Many early studies were aimed at a determination of the symmetry properties of the defects and their associated electronic/vibronic states. In ionic crystals with the rocksalt structure centres were identified with monoclinic I, orthorhombic I and II, tetragonal, and trigonal symmetries. Despite such information detailed atomistic models in many cases were not forthcoming, for several reasons. First of all, a multiplicity of particular symmetry systems is observed in most crystals; in sodium fluoride alone there are four trigonal, five orthorhombic, and seven monoclinic centres. Secondly, although a particular perturbation experiment, e.g. uniaxial stress, may determine the symmetry of an optical centre, it may do so without revealing the point group that is involved. For example, uniaxial stress alone cannot distinguish between the C_{2v}, D_2, and D_{2h} point groups of the orthorhombic I symmetry class. Finally, the excessive width of inhomogeneously broadened zero-phonon lines may mask small splittings leading to incorrect identifications of defect symmetries. For a correct assignment of the point group symmetry it was necessary to apply high-resolution techniques such as optical holeburning (OHB) or fluorescence line-narrowing (FLN), in conjunction with externally applied perturbations. An example will be given in Section 7.4.3. Bearing in mind these cautionary comments, it is now appropriate to describe some experiments which led to detailed atomistic models of aggregate centres.

7.4.3 Uniaxial stress and Stark effects for orthorhombic centres

Several zero-phonon lines in each of the different alkali halides have been assigned to defects with orthorhombic symmetry on the basis of stress and electric field splitting patterns of these lines (Hughes 1966; Fitchen 1968; von der Osten 1976). In cubic crystals there are two orthorhombic symmetry classes (Kaplyanskii 1964); orthorhombic I having a $\langle 110 \rangle$ orientation and dipole moment along a $\langle 110 \rangle$ axis and orthorhombic II with a $\langle 100 \rangle$ orientation having dipole moments parallel with a $\langle 100 \rangle$ crystal axis. Defects with such symmetries can have orientational degeneracy only. In consequence, defects associated with these spectra seem most probably to be neutral or charged versions of the F_2- and F_4-aggregate centres (Figs. 7.1 and 7.2). Several features of piezospectroscopy are illustrated by the case of F_2^--centres. These centres consist of three electrons trapped within two anion vacancies in neighbouring sites along a $\langle 110 \rangle$ direction. In lithium fluoride such centres absorb light in a zero-phonon transition at 1040 nm. The electron–phonon coupling is relatively weak, as Fig. 7.16 makes clear. There are six equivalent orientations of the $\langle 110 \rangle$ axes (Fig. 7.18(a)) each equally populated with defects. With the uniaxial stress along the [001] direction dipoles 1 and 2, which are at an angle of 90° to the stress axis, respond differently to dipoles 3, 4, 5, and 6 (at 45° to [001]) causing a splitting of the zero-phonon line into two lines. With linearly polarized light propagating along the $[\bar{1}00]$ direction and E vector parallel to $[00\bar{1}]$, only dipoles 3, 4, 5, and 6 absorb light, so that only a single component is observed. However,

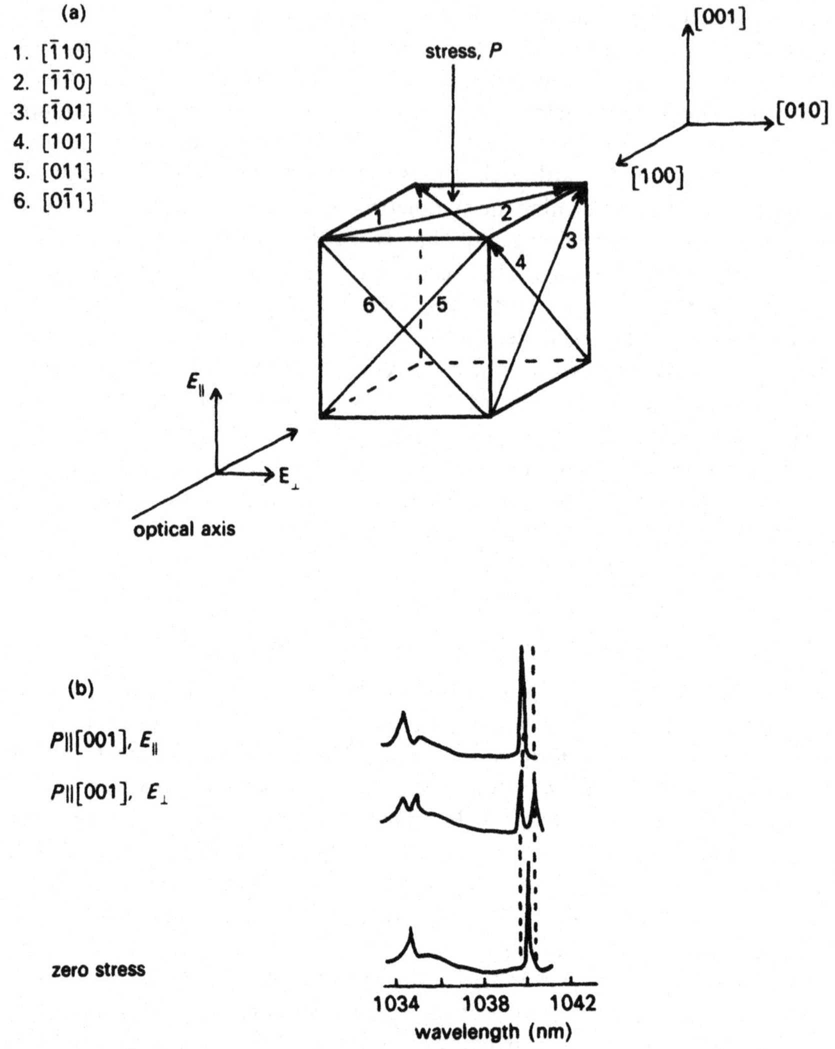

FIG. 7.18. (a) The six equivalent orientations of F_2^--centres in alkali halides are 1–[$\bar{1}$10], 2–[$\bar{1}\bar{1}$0], 3–[$\bar{1}$01], 4–[101], 5–[011] and 6–[0$\bar{1}$1]. (b) Polarization of zero-phonon and sideband absorption of F_2^--centres in lithium fluoride in the presence of a uniaxial stress parallel to the [001] direction. (After Fetterman 1968.)

using polarized light with E vector perpendicular to the stress direction. dipoles 1 and 2 and dipoles 5 and 6 each absorb radiation and both components are observed. The experimental results shown in Fig. 7.18(b) are entirely in accord with the predictions. Furthermore, the shifts in energy of the lines are linear with the applied stress. The spectra in Fig. 7.18(b) also show that the splitting pattern of the vibronic peak at 1034.3 nm is identical with that of the zero-phonon line. This is because the transition is between orbital singlet states, which can couple linearly only to symmetric (A_1) modes. Hence the vibronic levels have the same symmetry as the electronic states and the same selection rules as the zero-phonon lines.

Of the numerous zero-phonon lines assigned on the basis of piezospectroscopy to defects having orthorhombic I symmetry, only the F_2^+- and F_2^--centres are sure identifications. In other cases unambiguous identification of the associated point group would be helpful in determining atomistic models of the various centres. Such identification is possible either by combining high-resolution techniques and perturbation spectroscopy (Macfarlane *et al.* 1983) or by combining uniaxial stress and Stark effect measurements (Johannson *et al.* 1968). The determination of point group symmetry using such methods is discussed in terms of the zero-phonon lines observed at 695.5 nm in lithium fluoride and 607 nm in sodium fluoride. The 695.5 nm zero-phonon line is observed in lithium fluoride after X-ray or electron irradiation (Hughes 1966; Johannson *et al.* 1968). Figure 7.19 shows the π- and σ-polarization patterns of this line split by the application of a uniaxial stress or electric field along a $\langle 100 \rangle$ axis. The additional splitting in the Stark spectra arise from the odd parity of the electric field/electronic dipole Hamiltonian which enables electric field measurements to discriminate between centres which have, and those which do not have, an inversion centre. This can be seen by writing the perturbation Hamiltonian for an electric field $\boldsymbol{E}$ as

$$\mathscr{H} = -\boldsymbol{E}\cdot\boldsymbol{\mu}_e \tag{7.22}$$

where μ_e is the electronic dipole moment of the centre. Hence $\mathscr{H}$ is of odd parity. The first-order shifts in energy of the state ψ are then determined by evaluating the matrix element $\langle \psi | \mathscr{H} | \psi \rangle$. For centres having orthorhombic or lower symmetry, only singlet orbital states exist. If these centres have inversion symmetry there are no diagonal matrix elements of $\mathscr{H}$ and hence no linear Stark effect. Thus the defects associated with the 695.5 nm line are not from the D_{2h} point group. For centres lacking a centre of symmetry the symmetric direct product of the representation Γ_n of the state with itself is $\Gamma_n \times \Gamma_n = A_1$, so that all components of $\mathscr{H}$ have zero matrix elements, *except* those transforming like the representation A_1. For the D_2 group x, y, and z transform as the irreducible representations B_1, B_2, and B_3, respectively, and therefore there is no shift linear in electric field for D_2 either. Only in the case of C_{2v} symmetry, where the z-component transforms as the irreducible representation A_1, and x and y transform as B_1 and B_2, respectively is there an energy level shift, linear in electric field. A detailed analysis (Johannson *et al.* 1968) shows that the stress and electric field spectra can be accounted for only by an orthorhombic I centre (C_{2v}, point symmetry) with a $\langle 110 \rangle$ diad

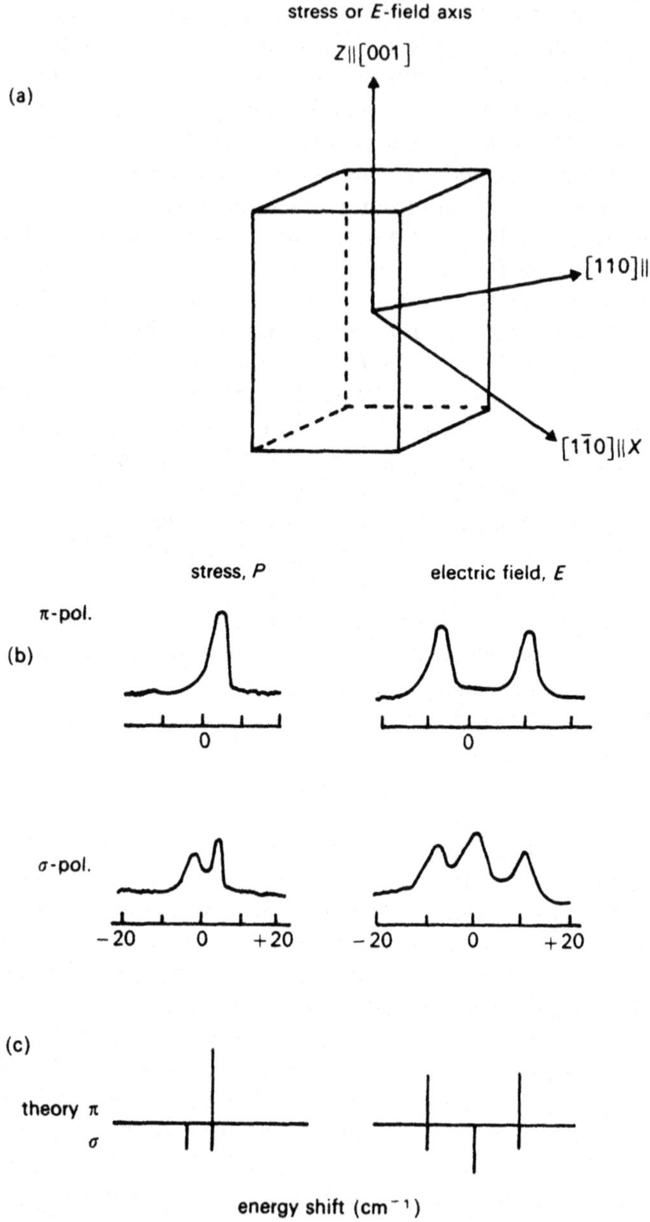

FIG. 7.19. (a) Illustrating the principal axes of an orthorhombic centre in a cubic crystal. Also shown are the σ- and π-polarization patterns of the 695.5 nm zero-phonon absorption line in lithium fluoride when (b) a uniaxial stress is applied and (c) an electric field is applied, in both cases parallel to an [001] direction. Measurement temperature 4 K. (After Johannson *et al.* 1968.)

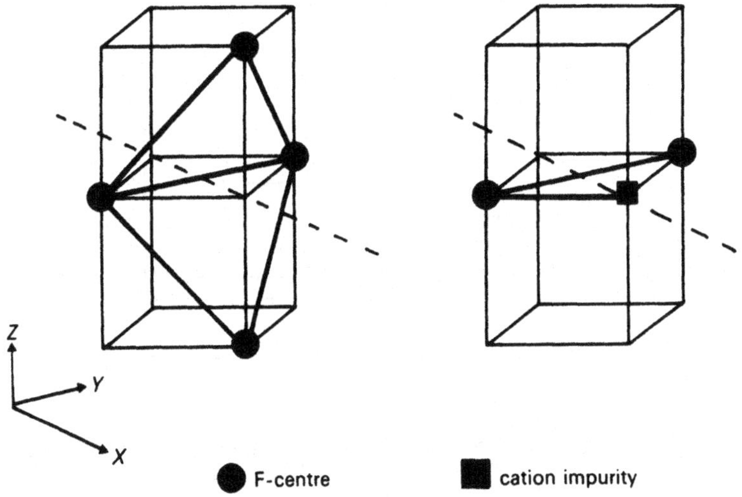

FIG. 7.20. Possible F-aggregate centres with $C_2(\langle 110 \rangle)$ symmetry which would give the stress and electric field splitting patterns shown in Fig. 7.19. (After Johannson *et al.* 1968.)

rotation. The two possible F-aggregate centres shown in Fig. 7.20 cannot be distinguished on the basis of these experiments.

Optical zero-phonon lines associated with defects are inhomogeneously broadened by the effects of random lattice strain. They are prime candidates for study using the high-resolution techniques of OHB and FLN (Section 6.7.3). Macfarlane *et al.* (1983) measured the Stark splitting of the hole burned in the 607 nm zero-phonon line in sodium fluoride (Fig. 7.21). The homogeneous width is only 21 MHz in comparison with the inhomogeneous width of 3 GHz. This line had been assigned by Baumann (1967) to F_4-centres, aggregates of four F-centres in nearest-neighbour sites on a (112) plane, on the basis of uniaxial stress effect measurements. The Stark splitting patterns in Fig. 7.21 are inconsistent with such a structure. Instead of the C_{2h} symmetry of the F_4-centre, they require the symmetry of the centre to be C_s with a (110) mirror plane. Obviously, perturbation spectroscopy alone is not sufficient to determine details of structural models of defects in many cases; it is usually necessary to have other evidence for confident assignments.

7.4.4 *Perturbation spectroscopy of trigonal centres*

The substantive evidence of the models of the F_3-type centres came not from ESR studies alone but from numerous optical studies. Thus for the charged and uncharged trigonal centres in most alkali halides very complete data exist for bandshapes, Stark, Zeeman, and stress effects—not just on the zero-

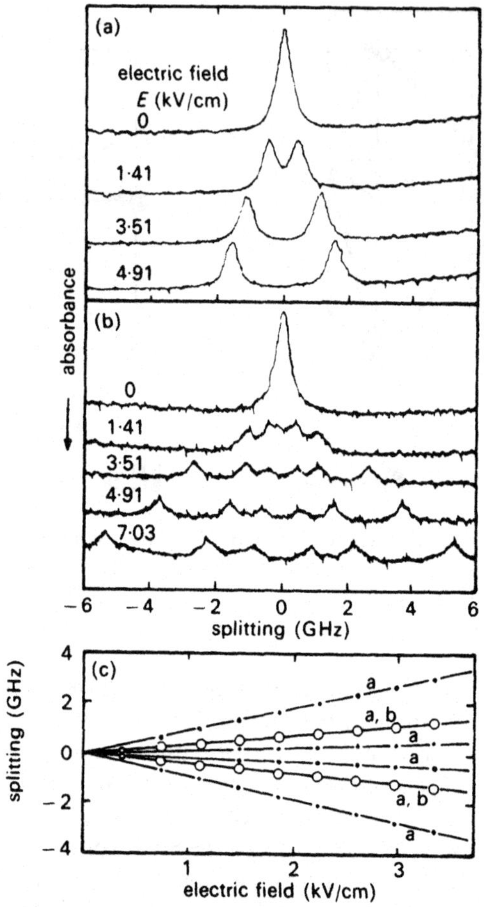

FIG. 7.21. Stark effect on the 607 nm line in sodium fluoride using OHB. The cases of E_S parallel to [111] is shown for two polarizations (a) E_L parallel to [111] and (b) E_L parallel to [110]. (After Macfarlane *et al.* 1983.)

phonon lines but also on the broad bands. Many experiments suggest that the structure of trigonal centres take the form of three nearest-neighbour anion vacancies located on the corners of an equilateral triangle in the (111) plane (Fig. 7.2). When one electron is associated with each vacancy the defect is the charge-neutral F_3-centre. This model has been determined on the basis of the kinetics of defect formation (Section 7.4.1) in additively coloured or irradiated crystals, polarized absorption/luminescence, perturbation, and ESR spectroscopy. In addition to the uncharged F_3-centre, charged centres may also be created. The F_3^+- and F_3^--centres, respectively, contain two and four electrons. As with F_2-centres, the coexistence of the various charge states of the F_3-aggregate centres in crystals leads to electron transfers between them under appropriate conditions of optical excitation. Such photochromism can be long-lived. Macfarlane *et al.* (1983) have made use of this in studies of OHB at F_3-type colour centres, especially in respect of their potential for frequency-domain optical storage (Moerner *et al.* 1986).

The F_3-centres have a large number of excited states which result in observable optical transitions. Figure 7.22 compares the optical absorption spectrum, linear dichroism of this spectrum under a uniaxial stress of 2.5 kg mm^{-2} in the [110] direction, luminescence excitation spectrum and the luminescence spectrum for a potassium chloride crystal containing aggregate centres including F_3-centres. Bands due to F_3-centres are evident in the same regions as the K-band of the F-centre, and the F_2- and F_4-centres,

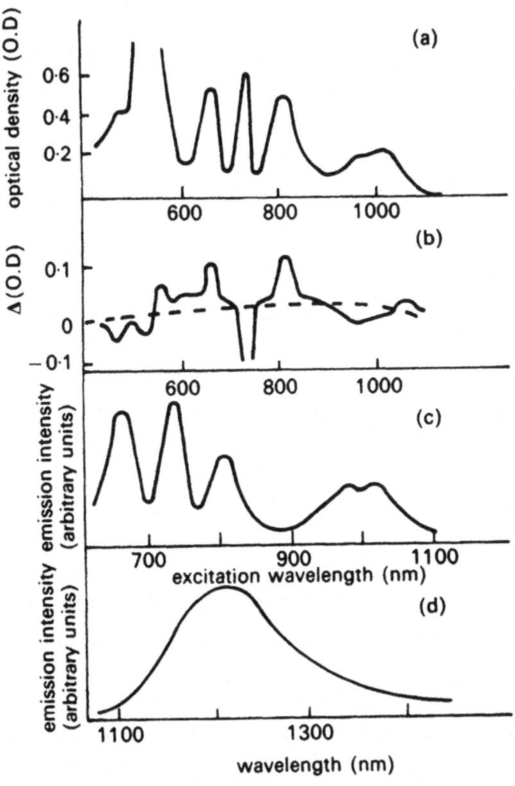

FIG. 7.22. Optical spectra of F_3-centres in potassium chloride. The optical density measured at 2 K is shown in (a), and (b) depicts the difference in optical density for light polarized parallel to the [110] and [1$\bar{1}$0] directions, i.e. $\Delta(OD) = OD[110] - OD[1\bar{1}0]$, when a stress $P = 2.5$ kg mm^{-2} is applied along [110]. In (c) the excitation spectrum of the luminescence spectrum plotted in (d) is shown. (After Silsbee 1965.)

(formerly referred to in the literature as M- and N-centres). Hence the designation of F_3(K)-, F_3(M)- and F_3(N)-bands. Clearly the stress-induced dichroism varies from band to band, and is most marked for the F_3(2)-band. The weak dichroism of the F_3(1)-band, long identified with the F_3-centre, indicates an $^2E \rightarrow ^2E$ transition. The absence of a clear splitting implies weak Jahn–Teller coupling in the excited ^2E state. Figure 7.22 shows that the F_3-centre has a significant absorption transition close to the F_2-centre absorption. Careful comparison shows that the F_3(M)-band and F_2(1)-band in potassium chloride are separated by only 13 nm; in potassium bromide the two bands are almost resolved. However, in view of the dichroic ratio $I[110]/I[1\bar{1}0]=3:1$ at very high stress this transition is identified as $^2E \rightarrow ^2A_1$. The weak dichroism in the spectral region of the F_4-centre bands is proportional to that in the regions of more obvious dichroism. This and detailed polarized bleaching experiments led to the conclusion that part of the absorption in this region is due to F_3-centres; hence the designation as F_3(N)-bands. The doublet structure of this band and the qualitative behaviour of the dichroism imply that the transition is between ground and excited states of E symmetry, the Jahn–Teller interaction in the excited E state being relatively strong. The dichroism on the high-energy side of the F-band is consistent with transitions to an even-parity state of A_2 symmetry (F_3(K) transition) and to an odd-parity states of E symmetry (F_3(F) transition). In view of the single emission band peaked at 1210 nm it appears that luminescence starts on the terminal electronic state of the F_3(4) absorption transition. This state may be directly excited in the F_3(4)-band or by radiationless transitions from the higher excited states of the centre.

Unlike the other optical transitions of the F_3-centre, the F_3(2)-band shows a well-resolved zero-phonon line at low temperature. The splitting of the zero-phonon line under stress indicates the removal of both orientational degeneracy and electronic degeneracy. Since the ^2E ground state is orbitally degenerate the splitting of the F_3-centre ground state by uniaxial stress results in striking stress and temperature-dependent intensities for transitions emanating from the two components of the stress-split ground state, the relative populations of which are determined by a Boltzmann factor. Obviously if there is no ground-state degeneracy, i.e. as for an orbital singlet state, there will be no population effect on the intensities. Hence uniaxial stress measurements of the absorption spectrum allow a determination of whether or not the ground state is orbitally degenerate. As was first shown by Silsbee (1965) the intensities of the components of the F_3(2) zero-phonon line are strongly dependent on the applied stress and the temperature. Hence the ground state is the orbital doublet and the transition is $^2E \rightarrow ^2A_2$, at a centre with trigonal symmetry and a [111] axis of symmetry. Other examples have been given by Hughes and Runciman (1965) for lithium fluoride and by Von der Osten (1976) for sodium fluoride.

As an example consider F_3-centres in lithium fluoride, for which the zero-phonon line occurs at 390.9 nm with half width of *c.* 15 cm^{-1} (0.36 nm) at 4.2 K. Figure 7.23 shows the splitting of the $F_3(2)$ zero-phonon line for stresses parallel to the [100], [110], and [111] directions. Experimentally there is a marked dichroism, i.e. the total π and σ intensities are very different at reasonable stresses. The theoretical stress spectra were calculated assuming the transition to be a $^2E \rightarrow {}^2A_1$ at a trigonal centre with a threefold axis along a $\langle 111 \rangle$ direction. The theoretical relative intensities of the π- and σ-

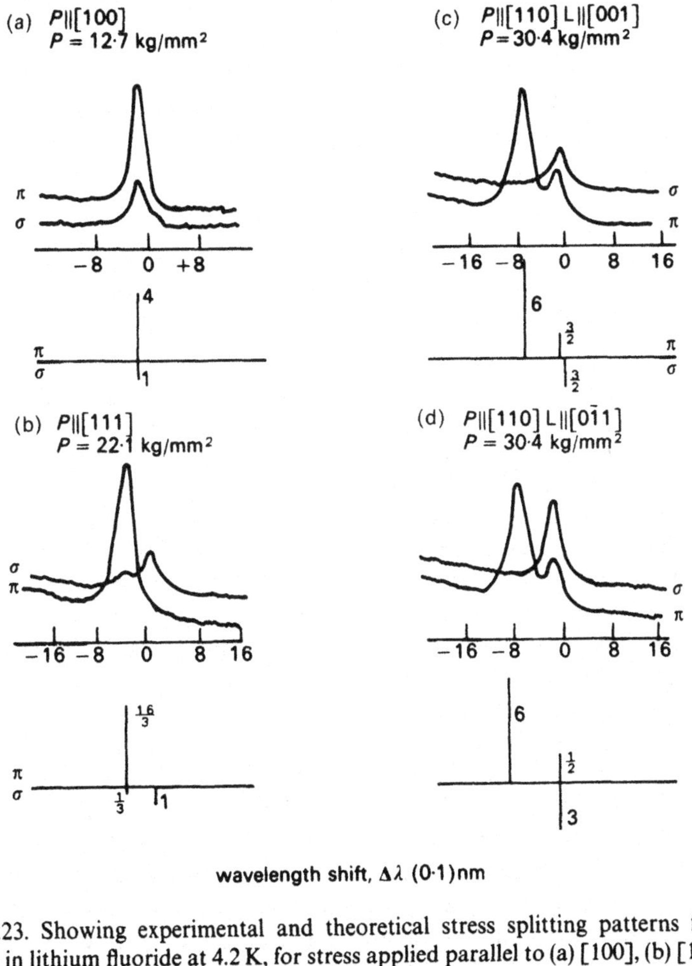

FIG. 7.23. Showing experimental and theoretical stress splitting patterns for F_3-centres in lithium fluoride at 4.2 K, for stress applied parallel to (a) [100], (b) [111], (c) and (d) [110]. For stress parallel to the [110] direction there are two viewing directions with light parallel to [001] or [011] directions i.e. $L \| [001]$ or $L \| [011]$. The theoretical intensity distribution assumes $^2E \rightarrow {}^2A_1$ absorption transitions, and includes allowance for the Boltzmann factor in the ground state levels. (Adapted from Hughes 1966.)

Table 7.7

Stress coupling coefficients (A_1, A_2, B and C) defined in terms of the shifts of π and σ components of stress spectra of $E \rightarrow A_1$ transitions at trigonal centres.

Stress direction	Polarization intensity π	σ	Theoretical shift (per unit stress)
[100]	4	1	$A_1 - 2B$
		3	$A_1 + 2B$
[111]		1	$A_1 + A_2$
	$\frac{8}{3}$	$\frac{1}{6}$	$A_1 - \frac{1}{3}A_2 - \frac{4}{3}C$
		$\frac{3}{2}$	$A_1 - \frac{1}{3}A_2 + \frac{4}{3}C$
[110]†		3	$A_1 + \frac{1}{2}A_2 - B + C$
	1		$A_1 + \frac{1}{2}A_2 + B - C$
		1	$A_1 - \frac{1}{2}A_2 + B - C$
	3		$A_1 - \frac{1}{2}A_2 - B - C$

† Measured with light propagating along [001]. Alternatively the spectra could be measured for light along [$1\bar{1}0$]: the shifts are the same and only the polarization intensities vary.

polarized components and the shifts of the lines under uniaxial stress are given in Table 7.7, which also serves to define the stress splitting parameters, eqn (6.20), in terms of the calculated shifts. If the observed splittings are due to the removal of orientational degeneracy only, the π and σ intensities would be identical since defects do not reorientate at low temperatures. The dichroism therefore arises from an orbitally-degenerate ground state, the degeneracy being removed by the applied stress, except for defects whose trigonal axis is parallel to the stress axis. The splitting is due to the energy difference between electric dipole moments parallel to $\langle 112 \rangle$ and $\langle 110 \rangle$ axes in the plane of the defect (Hughes 1966). The two different σ-spectra for stress parallel to a [110] direction are to be expected for electric rather than a magnetic dipole transition (see Section 6.3). If the stress is P then the splitting is given by $\delta_n P$, where the δ_n refers to shift per unit stress. The population ratio is then given by $\exp-(\delta_n P/kT)$, where T is the absolute temperature. Thus whichever of the dipoles lies lowest will be indicated by an enhanced intensity relative to the higher lying state. The observed dichroism increases with increasing stress and saturates at high stress, with the overall behaviour being that of a Langevin function (Hughes 1966).

Figure 7.23 shows that the number of lines and their intensities/polarizations are as expected for trigonal centres in the presence of thermal depopulation effects. The shifts are linear with respect to the applied stress, from which are determined the shifts per unit stress given in Table 7.8. Each parameter may be determined twice from the data; despite quite large discrepancies the results are self-consistent. The discrepancies presumably represent the effects of non-uniform stress in the sample. Unfortunately the stress results do not distinguish between $^2A_1$ and $^2A_2$ excited states. However, theoretical work shows that a compression of the equilateral triangle, which occurs under a [100] stress, lowers the E_x state relative to the E_y state. The measured polarization data confirms that the transition from the lower state has an optical dipole moment along the [112] direction. Such can be the case only if $^2A_2$ is the excited state.

For a transition between orbital singlet states the vibrational sidebands and zero-phonon line exhibit the same stress-induced dichroism. This is because only symmetric modes can couple linearly to these states so that the vibronic levels have the same symmetry as the electronic states, and the same selection rules. However, for an orbital doublet, coupling to non-symmetric modes can occur and the vibronic levels may have different symmetry and selection rules. This difference is observed in the sideband dichroism rather than the zero-phonon line because the lowest vibronic level retains the same symmetry character as that electronic state from which it originated. The effect has been used by Silsbee (1965) in his studies of the $F_3(2)$ transition in potassium chloride; his data for stress parallel to the [110] direction are shown in Fig. 7.24(a). The broadband dichroism is given approximately by $OD[110]/OD([\bar{1}10])=\frac{1}{3}$; in the sharp peak, which appears only in π-polarization, the dichroism is approximately zero. The explanation lies with the selection rules shown in Fig. 7.24(b). The stress raises the degeneracy of the vibronic pair E_x and E_y, so that at low temperature one of them is preferentially populated. The selection rules for linearly polarized light in the x- or y-direction of the centre show that only transition involving excitation of an even number (including zero) of E-mode vibrational quanta are dichroic. The

Table 7.8

Stress coupling coefficients for trigonal centres in lithium fluoride and magnesium oxide (in units of $cm^{-1}\,kg^{-1}\,mm^2$).

	A_1	A_2	B	C
F_3^+	0.35 ± 0.05	-1.40 ± 0.10	-0.26 ± 0.025	-0.25 ± 0.025
LiF F_3	0.51 ± 0.13	-0.65 ± 0.3	-0.17 ± 0.1	-0.35 ± 0.18
F_3^-	0.22 ± 0.04	-0.75 ± 0.06	-0.22 ± 0.03	-0.10 ± 0.03
F_3^+	0.077 ± 0.05	-0.190 ± 0.05	-0.13 ± 0.02	-0.19 ± 0.05
MgO F_3	0.18 ± 0.05	-0.18 ± 0.03	-0.08 ± 0.04	-0.13 ± 0.02
F_3^-	0.055 ± 0.05	-0.165 ± 0.01	-0.04 ± 0.02	-0.11 ± 0.01

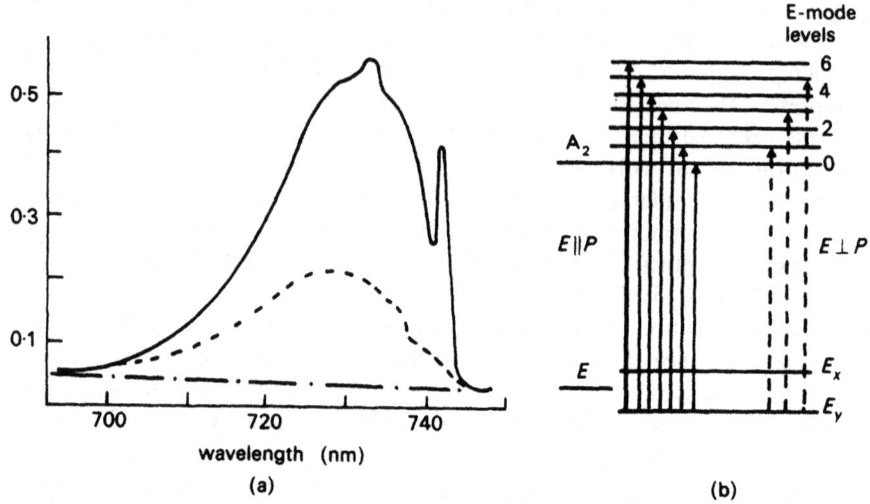

FIG. 7.24. (a) Stress-induced dichroism of the $F_3(2)$-band in potassium chloride for stress $P=0.4\ \mathrm{kg\,mm^{-2}}$ applied at $T=2\,\mathrm{K}$ in a [110] direction. (b) The selection rules for polarized absorption in the $F_3(2)$-band. (After Silsbee 1965.)

selection rules from the stress-split ^{2}E state to the vibronic levels of E symmetry in the excited electronic state (Fig. 7.24(b)) are appropriate to a $[1\bar{1}0]$ stress. Since transitions involving an odd number of vibronic quanta appear both for E∥[110] and E∥$[1\bar{1}0]$ there is no dichroism in this case. Note that transitions involving combinations of E- and A-modes are not any different since the selection rules depend only on the number of E-mode quanta.

In principle the ^{2}E ground state of the F_3-centre may be split even in the absence of an applied perturbation by spin–orbit coupling, by Jahn–Teller distortions, or random internal strains of E symmetry. Obviously such splittings must be less than the zero-phonon linewidth. At very low temperatures (1.2–4.2 K) the first moment of the lineshape will not change because of lattice expansion. Any change will be due to a thermal redistribution between the E_x and E_y states. Hence careful measurements of the first moment as a function of temperature should reveal such a splitting. These measurements in potassium chloride and lithium fluoride yield values of $0.4\pm0.2\ \mathrm{cm^{-1}}$ and $<2.0\ \mathrm{cm^{-1}}$, respectively. That both results were sample-dependent confirms the importance of random internal strains in the ^{2}E ground state of the F_3-centre.

Random internal strain is also important in the detection of the ^{2}E ground-state ESR spectrum. In the absence of an external uniaxial stress no resonance is observed. However, at large applied stresses, $P\geqslant 3\ \mathrm{kg/mm^2}$, a resonance is observed, the linewidth and g-values of which are strongly dependent on the magnitude of the applied stress. Interpretation of the data require detailed understanding of the Jahn–Teller effect *and* the influence of random internal strain. Krupka and Silsbee (1964) find that the Jahn–Teller coupling strength (Section 5.4.3) $k^2 = 2E_{JT}/\hbar\omega = 3.0 \pm 0.1$, together with a mean random internal stress of $P_0 \simeq 1\ \text{kg mm}^{-2}$, explain the data rather well. This result suggests that for the ^{2}E state the reduced Landé factor and spin–orbit coupling constant are given by the inequality $6.4\ \text{cm}^{-1} < (g_0\zeta)' < 4.7\ \text{cm}^{-1}$. Indeed, the internal stress is responsible for the failure to observe the line at zero stress, since a range of stresses between 0 and $1\ \text{kg mm}^{-2}$ would so broaden the line that at zero stress a linewidth of almost 30 mT would be observed, whereas at 2–3 kg mm^{-2} the linewidth is only $\simeq 5$ mT. At only $P = 0.3\ \text{kg mm}^{-2}$ the linewidth is 10 mT, and the signal-to-noise ratio is about 8:1. Since the signal height is proportional to $(\Delta H)^{-2}$, a spectrum with linewidth of 30 mT is undetectably weak.

Let us now return to the optical dichroism in the presence of uniaxial stress, since the strength of the Jahn–Teller coupling may be estimated from the dichroism in the zero-phonon line relative to the broad band. This is because, by the Franck–Condon principle, transitions at the peak of the broad band correspond to transitions at $Q = 0$ on the configuration coordinate diagram. Hence such distortions do not 'feel' the Jahn–Teller effect. However, the zero-phonon line involves a transition to one of the energy minima produced by the Jahn–Teller distortion. Hence for the zero-phonon transition electronic operators (e.g. spin–orbit coupling, Coulomb interaction) are reduced relative to the broad band. For weak coupling ($k^2 \leqslant 1$) the broadband dichroism is expected to follow that of the zero-phonon line. This is the case for the F_3-centre in lithium fluoride. However, for F_3-centres in potassium chloride the situation is very different, as Fig. 7.24 shows; in the case of the zero-phonon line $I_\sigma/I_\pi < 0.1$ whereas for the broad band this ratio is roughly $\frac{1}{3}$. This is the limiting value of a strong Jahn–Teller effect and is consistent with a value $7 > k^2 > 1$.

It is of some interest to compare the dichroism in the presence of a magnetic field with the linear dichroism induced by stress. Figure 7.25(a) shows the selection rules for absorption of circularly polarized light, and Fig. 7.25(b) shows the absorption band and circular dichroism. Note that the selection rules for the absorption of σ_+ and σ_- light are reversed when an odd number of E-mode phonons is created. The most obvious point in Fig. 7.25(b) is the difference between the sideband observed in the absorption band and in circular dichroism. The selection rules show that transitions involving the excitation of no E-mode quanta, e.g. the zero-phonon line at 741.7 nm and the symmetric-mode sidebands at 738 nm and 734 nm, have positive circular polarizations which are resolved in both instances, whereas odd E-mode vibronic bands have a negative dichroism which causes, for example, the dip

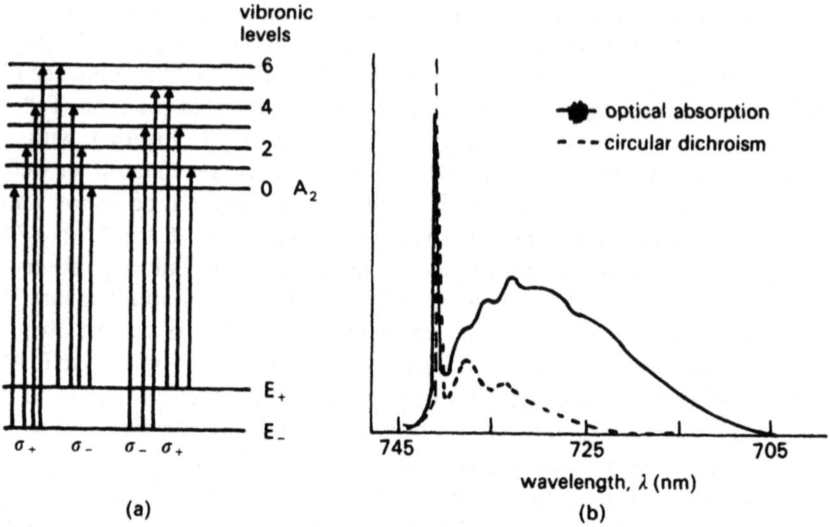

FIG. 7.25. (a) Selection rules for circular polarization of F_3-centres and (b) optical absorption and circular polarization of absorption for F_3-centres in potassium chloride measured at 1.6 K and high magnetic field. (After Merle d'Aubigné 1976.)

at 736 nm. At higher energies the opposite senses of polarization involving odd and even numbers of E-mode quanta tend to cancel. The temperature dependences of the changes in moments of the absorption band have been used to calculate for the zero-phonon line the Ham reduction factors associated with the Jahn–Teller effect in the ^2E level.

It is evident that the combination of ESR, uniaxial stress splitting, Stark, and Zeeman effect studies have resulted in a very detailed understanding of the electronic structure of the F_3-centre in potassium chloride (and to a lesser extent in lithium fluoride). There is no reason to believe that the situation is very different in any other of the alkali halides. Thus we conclude that the electronic ground state for this three-electron centre is ^2E; the centre has C_{3v} symmetry with linear electric dipole moments parallel to $\langle 110 \rangle$ and $\langle 112 \rangle$ directions in the plane of the F_3-centre for the $F_3(2)$-band. Armed with all the experimental data we can construct the energy level diagram for F_3-centres; Fig. 7.26 shows such a diagram for potassium chloride. The $F_3(1)$- and $F_3(2)$-bands are now well understood in most alkali halides, but especially in lithium fluoride, sodium fluoride, and potassium chloride, where there have been many studies of the effect of external perturbation on the optical bandshapes. The molecular orbital method (see Section 2.5) leads to the ^2E ground state. The lowest one-electron orbitals are an orbital singlet (a_1) with an orbital doublet (e) about 1–2 eV above it. Thus for the F_3-centre the ground state configuration (a_1^2) (e^1) leads to the ^2E state.

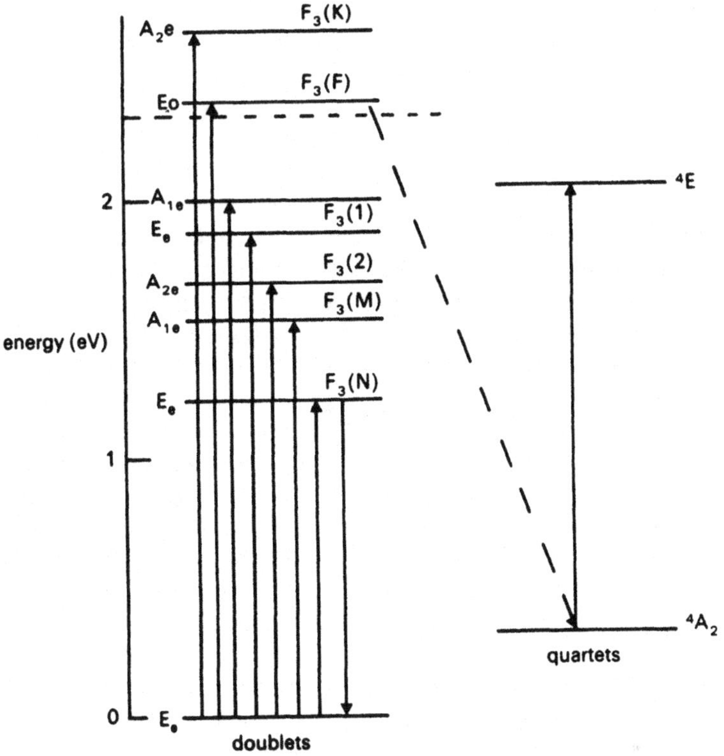

FIG. 7.26. Proposed energy-level diagram for F_3-centres in potassium chloride. The subscripts e and o on the various states indicate that they are of even and odd parity respectively for reflections in the plane of the triangle of vacancies.

Since charged versions of the F_3-centres, viz F_3^+- and F_3^--centres, are produced in alkali halides, it is pertinent to comment on their electronic structure. In the F_3^+-centre the ground configuration is obviously (a_1^2), leading to a $^1A_1$ ground state. Moreover, the configuration of the lowest excited state, (a_1) (e), leads to a ^1E state. Thus the fundamental optical absorption transition is $^1A_1 \rightarrow {}^1E$. In lithium fluoride Hughes and Runciman (1965) used piezospectroscopy to identify the zero-phonon line at 487.4 nm with this transition. In other crystals the spectral positions of the zero-phonon lines are close to those predicted by an empirical Mollwo–Ivey plot (Fitchen 1968). In lithium fluoride the sharp line has a halfwidth of 5 cm^{-1} at 4 K, and the stress spectra are consistent with the trigonal $^1A_1 \rightarrow {}^1E$ transition. Furthermore there is no stress-induced dichroism for the line in absorption, showing that contrary to the situation in the F_3-centre the ^1E state is the excited state rather than the ground state. Moreover, the ^1E state is the lowest excited state since a zero-phonon line at the same wavelength is observed in emission also. This is also the situation in sodium fluoride where the F_3^+-centre zero-phonon line occurs at 545.6 nm (Fig. 7.27(a)). The orbital degeneracy of the ^1E excited state of this centre is confirmed by a stress-induced dichroism in the emission line. These results can be interpreted as resulting from defects having D_{3h}, D_3, C_3, or C_{3v} point symmetry groups, which may be distinguished between on the basis of electric field measurements on the sharp line. The Stark splitting of the 545.6 nm line (Bauman 1967) confirms that the electric field removes the orbital (Stark effect) and orientational degeneracy (pseudo-Stark effect) of the E-state. That the splittings are linear with the applied electric field, for F_3^+-centres in lithium fluoride and sodium fluoride shows that the centre lacks inversion symmetry, in accord with the C_{3v} point symmetry for the F_3^+-centre.

The electronic states of the F_3^--centre, are derived from the (a_1^2) (e^2) configuration. Electron–electron interaction splits this configuration into three states, $^1A_1$, ^1E, and $^3A_2$. Hund's rules would then favour the $^3A_2$ state as the ground state. The excited states resulting from the (a^1) (e^3) configuration are ^1E or ^3E. Hence we might anticipate transitions of the type $^1A_1 \rightarrow {}^1E$ or $^3A_2 \rightarrow {}^3E$. No ESR transitions have been observed for crystals containing F_3^--centres, so that apparently the ground state is the spin singlet $^1A_1$. The zero-phonon lines occur with characteristic positions indicated by a Mollwo–Ivey plot. Piezospectroscopy confirms that A→E transitions at centres with trigonal symmetry are involved. The energy shifts of the components (Table 7.7) in the stress splitting pattern for a trigonal A↔E transition may be expressed in terms of four stress coupling coefficients A_1, A_2, B, and C, where A_1 and A_2 describe the removal of orientational degeneracy and B and C the removal of electronic degeneracy (Hughes and Runciman 1965). The stress coupling coefficients of the various charge states of the F_3-centres have the same sign and are of similar magnitudes, as can be seen by inspection of Table 7.8.

Since the zero-phonon lines of the F_3^+-, F_3-, and F_3^--centres are inhomogeneously broadened they are ideal candidates for OHB and FLN studies. The first OHB studies were reported for F_3^+-centres in sodium fluoride by Macfarlane and Shelby (1979). Absorption and emission spectra of the $^1A \rightarrow {}^1E$ transition are approximately mirror images of one another in the 545.6 nm zero-phonon line (Fig. 7.27(a), (b)). Holes burned in the zero-phonon line by bleaching for *c.* 2 s with 200 mW/cm^2 of single-frequency dye laser light give the spectrum shown in Fig. 7.27(c). The homogeneous linewidth of the $^1A \rightarrow {}^1E$ transition, 17 MHz corresponds to a dephasing time, $T_2 = 20$ ns. The hole recovery showed two components, one of several seconds associated with a low-lying spin triplet state of the centre and the other of order 70 minutes due to photoionization. Long-lived holes have

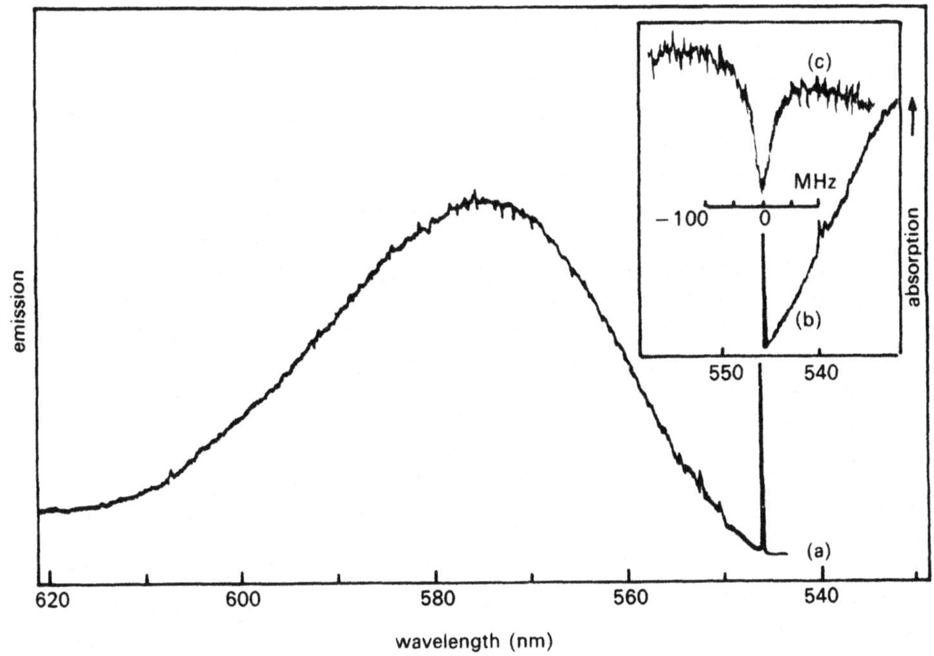

FIG. 7.27. The $^1A \leftrightarrow {^1E}$ transition of the F_3^+-centre in sodium fluoride. (a) the emission spectrum, (b) excitation spectrum near the zero-phonon line, (c) a hole burned in the zero-phonon line. (After Macfarlane and Shelby 1979.)

possible applications in information storage (Ortiz *et al.* 1981). OHB has also been studied in F_3^--centres in lithium fluoride (Moerner *et al.* 1982, 1986).

7.4.5 F-*centre aggregates in other crystals*

For over several decades the major attention in point defect studies has concentrated on the alkali halides. As the previous discussion has indicated a very thorough understanding has emerged of the nature of vacancy aggregate centres. To some extent guided by events in the alkali halides, there began to emerge during the 1960s a concerted effort to extend this understanding to defects in the alkaline earth oxides, the alkaline earth fluorides, and related materials such as magnesium fluoride and potassium magnesium fluoride.

In magnesium oxide neutron irradiation provides the most convenient means of producing F-aggregate centres. Zero-phonon lines and their attendant phonon sidebands have been subjected to detailed analysis, and their behaviour under uniaxial stress measured. Three classes of defects are indicated by the uniaxial stress results; centres with orthorhombic I symmetry, monoclinic I symmetry, and trigonal centres with electronic degeneracy in one of the participating states. Arguments have been put forward proposing that such centres have the F_2, F_3, and F_4 structures demonstrated for the alkali halides. Lacking comparable photochemical results to the alkali halides, the detailed models are not as securely based although recent polarized luminescence and optically detected magnetic resonance have given satisfying reassurance in respect of some of the proposed assignments. Excitation spectroscopy of the F_2-centre in magnesium oxide was discussed in Section 6.2.2. The zero-phonon line from this centre is at 361.8 nm in both absorption and emission. Uniaxial stress studies and polarized luminescence confirm that the centre has orthorhombic symmetry with linear electric dipole moments parallel to a $\langle 110 \rangle$ crystal direction, as demanded by the model. The trigonal centres have been recognized as having A$\leftrightarrow$E transitions with transition moment in a $\{111\}$ plane. For centres with zero-phonon lines at 642.0 nm and 649.0 nm no stress-induced dichroism is observed so that the orbital singlet state lies lowest. However, a zero-phonon line at 524.8 nm does have an E ground state. Removal of the twofold degeneracy associated with the E ground state results in transitions to the excited A state which are weighted by a Boltzmann factor at low temperatures. Hence a temperature-dependent dichroism is observed in the spectrum. This is very reminiscent of the trigonal centres in the alkali halides, Table 7.8 compares the stress coupling coefficients of the various F-type centres in magnesium oxide and lithium fluoride. The striking feature is that for each defect type in magnesium oxide and lithium fluoride the signs of the coefficients are identical in both materials. Furthermore, of the four coefficients required to describe the stress splitting results, A_2 is always very much the largest. In general for both materials the coefficients behave in a rather consistent fashion.

The symmetry of fluorite crystals (calcium fluoride, strontium fluoride, and barium fluoride) is quite different from the rock-salt structure of the alkali halides. The cube corners are occupied by anions, whereas only alternate cube centre positions are occupied by cations (Ca^{2+}, Sr^{2+}, etc.). The text by Hayes and his colleagues (Hayes and Stoneham 1974) summarizes much of the present understanding of the F-centre aggregates in these materials. There are two possible F_2-centre structures labelled $F_2(100)$- and $F_2(110)$-centres, the parentheses indicating the orientation of the centre in the lattice. In calcium fluoride the $F_2(110)$-centres predominate whereas in strontium fluoride the $F_2(100)$-centres appear exclusively. The spectroscopic data now accumulated includes polarized absorption and luminescence, radiative life-times, optical dichroism due to optical and thermal reorientation, stress splitting of zero-phonon lines, and optically-induced ESR in metastable triplet states. Similar optical studies have been carried out on F_3- and F_3^--centres in both calcium fluoride and strontium fluoride. In contrast to the triangular structure of F_3-centres in the alkali halides, in fluorite-structured crystals the F_3-complex is composed of a linear array of three nearest-neighbour F-centres parallel to a ⟨100⟩ axis. The expected $S=\frac{3}{2}$ metastable excited state of the F_3-centres has not so far been observed.

7.4.6 F_A-centres

In Fig. 7.1 one simple modification of the F-centre is shown as an F-centre with a single nearest-neighbour cation replaced by an impurity alkali ion. The local symmetry of this centre, the F_A-centre, is reduced to C_{4v} from the O_h symmetry appropriate to the F-centre. This model was deduced from optical and photochemical experiments and fully vindicated by ENDOR measurements. The formation of F_A-centres can be regarded as a model case of defect aggregation. They are observed in crystals containing a concentration of impurity alkali ions of smaller size than the host alkali halide. F-centres are introduced by additive coloration followed by quenching to room temperature. Then F-centres are photoconverted to F^--centres by irradiation at high intensity in the F-band at $\simeq 220$ K. No aggregation takes place at this stage. However, if the crystal is kept in the dark at this temperature for several hours F_A-centres are produced. Thus F_A-centres are produced by the prolonged presence of F^+, F^--centre pairs produced by light absorption rather than directly by the optical absorption. This process is thermally activated, the diffusing entity being the F^+-centre. The impurity ions are merely trapping sites for the mobile F^+-centres. There is an obvious similarity between F_A-centre formation and F_2-centre formation which was discussed in Section 7.4.1.

The consequences of the locally-reduced symmetry of the F_A-centre relative to the F-centre is shown in Fig. 7.28. The ground state, being spherically symmetric in the F-centre, is almost unaffected by the perturbing ion. In C_{4v} symmetry the ground state transforms as the irreducible representation $^2A_1$. For an impurity ion along the z-direction, the $2p_z$ orbital is lowered relative to the $2p_x$ and $2p_y$ orbitals which remain degenerate. Thus the excited state, which in the F-centre is $^2T_{1u}$, splits into a singlet $^2A_1$ and a doublet ^2E. There are two electric dipole transitions, which give rise to two absorption bands, viz.

$$F_{A1}: {}^2A_1 \rightarrow {}^2A_1, \ z\text{-polarized}$$

$$F_{A2}: {}^2A_1 \rightarrow {}^2E, \ x, y\text{-polarized.}$$

The latter absorption band should have twice the integrated intensity of the former band provided that there is an equal population of centres in each of the three defect orientations. In general the F_{A2}-band is almost coincident with the F-band, whereas the F_{A1}-band is displaced to lower photon energies (Lüty 1968).

We show in Fig. 7.29 the absorption spectrum of F_A(Na)-centres in potassium chloride. Note that the splitting in energy between the F_{A1}-band and the F_{A2}-band is roughly 0.25 eV. The excited-state splitting is determined

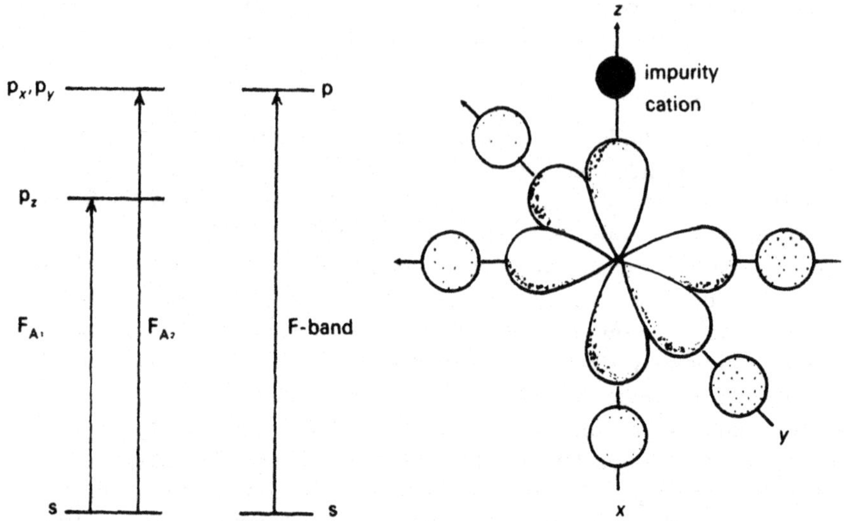

FIG. 7.28. Schematic representation of (a) the excited 2p state of the F_A(Na)-centre in potassium chloride and (b) the energy-level splitting in the F_A-centre 2p state relative to the F-centre. (After Henderson 1972.)

largely by the host anions, being somewhat larger for the chlorides than for the bromides. The more surprising feature is that the relative intensities of the two bands vary markedly. Such dichroism can result from either reorientation of the centres or ionization of one orientation in preference to another. In Fig. 7.29 (a)–(d) the different spectra were achieved by completely converting F-centres to F_A(Na)-centres by optical bleaching in the F-band at 220 K followed by storing and bleaching in the F-band at 295 K ((a) and (b)). The F_A-centres were then reoriented by bleaching in the F_{A1}-band with [100] polarized light at 220 K. Subsequently the absorption spectrum was measured at 90 K using light polarized in the [100] (spectrum (c)) or [010] (spectrum (d)) directions. The direction of propagation is [001]. The initial high degree of dichroism may be reduced by pulse annealing at $\simeq$285 K for

---

FIG. 7.29. Dichroic absorption of the F_A (Na)-centres in potassium chloride. (a) Shows the F-band produced by additive coloration of the crystal followed by quenching to room temperature to produce 10^{17} F-centres cm^{-3}, (b) absorption spectrum after completely converting F-centres to F_A(Na)-centres measured with unpolarized light propagating along the [001] direction. (c) Shows the absorption of [100] polarized light at 90 K following reorientation bleaching at 290 K with F-band light polarized along [100]. And (d) as in (c) but measured using [010] polarized light. (After Henderson and Gebler 1973, unpublished.)

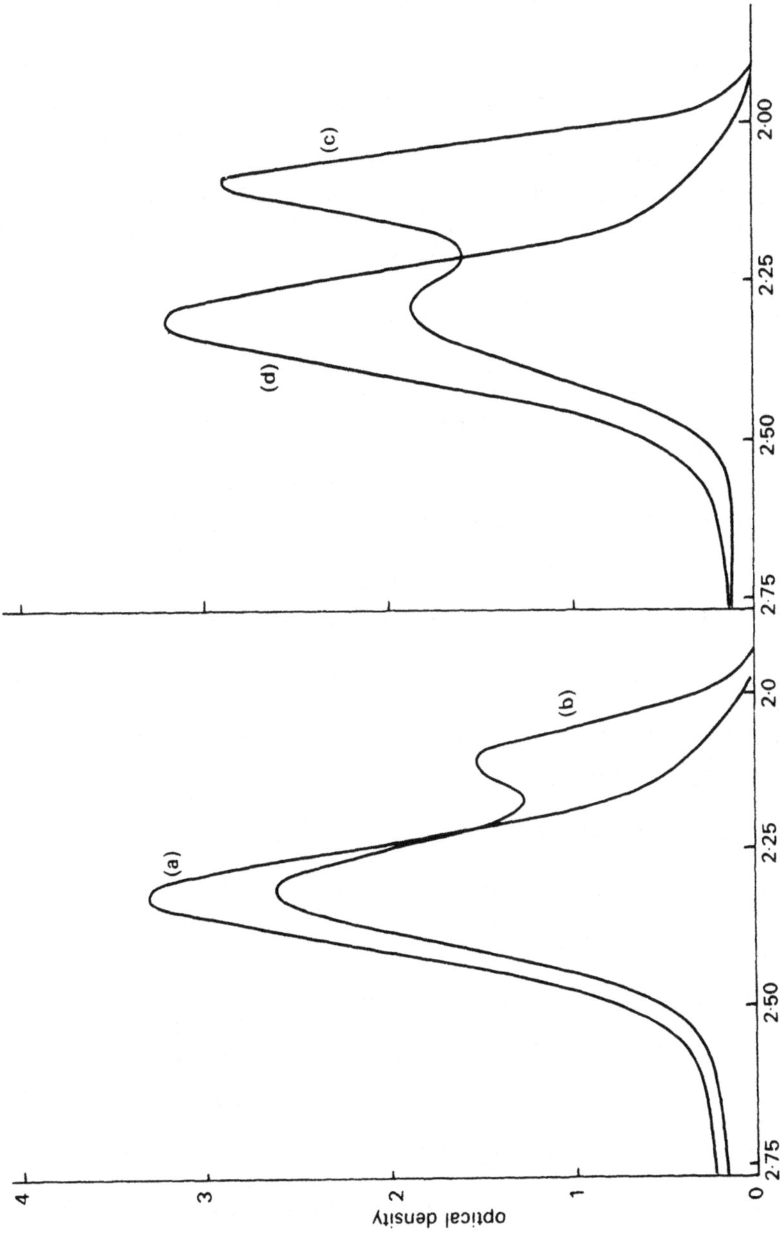
optical density
0
1
2
3
4
2·75
2·50
2·25
2·0
2·00
photon energy (eV)
(a)
(b)
(c)
(d)

varying times; the dichroism is reduced exponentially with time until the absorption spectrum corresponds to an equal distribution of F_A-centres along each ⟨100⟩ direction, i.e. spectrum (b). Such a condition is signalled by the intensity ratio F_{A2}/F_{A1} being approximately equal to 2.0. To a very good approximation no centres are destroyed during reorientation. It is convenient to use the selective reorientation of excited F-centres to analyse the strength of the partially-overlapping F_{A1}- and F_{A2}-bands. As might be anticipated, the integrated absorption of the F_A-bands is identical with the integrated absorption of the F-centres from which they were produced. And furthermore the oscillator strength of the F_{A1}-band is equal to that of either one of the components of the F_{A2}-band. A summary of optical properties of F_A-type centres is given in Table 7.9.

Perhaps the most surprising feature of the F_A-centre behaviour is that the emission properties depend very much on the size of the impurity ion. Irrespective of the impurity, one always observes a single emission band no matter whether absorption is in the F_{A1}- or F_{A2}-band. However, the emission characteristics fall into two quite distinct categories: emission is either essentially unpolarized and relatively long-lived ($\tau_R \sim 10^{-6}$ s), or polarized with a short lifetime ($\tau_R \sim 10^{-7}$ s). Thus the emission behaviour distinguishes between two categories of F_A-centre, F_A(I)- and F_A(II)-centres having the luminescence properties noted above. Which type of behaviour is observed depends upon the size of the impurity cation. We can understand these

Table 7.9

Absorption and emission properties of F_A-centres in alkali halides. (Adapted from Lüty 1968.)

System	F_A-type	Absorption peak (eV) F_{A1}	F_{A2}	Emission peak eV	τ_R (10^{-8} s)
KCl					
F		2.31		1.24	58
F_A(Na)	I	2.35	2.12	1.12	53
F_A(Li)	II	2.25	1.98	0.46	8.5
KBr					
F		2.06		0.92	111
F_A(Na)	I	2.07	1.90	0.84	100
F_A(Li)	II	2.00	1.82	0.75	10
RbCl					
F		2.05		1.09	60
F_A(Na)	I	2.09	1.85	0.93	60
F_A(Li)	II	1.95	1.72	0.45	9

different behaviours in terms of the configurational coordinate model (Sections 2.5, 5.3). For each category of F_A-centre the two absorption transitions giving rise to the F_{A1}- and F_{A2}-bands terminate on different excited-state parabolae. The difference between the two categories, F_A(I) and F_A(II), arises from the nature of the relaxed excited state. For the F_A(I)-centre the relaxed excited state is rather like that of the F-centre, as is indicated by the relatively long emission lifetime. As a consequence of the large orbital radius the electronic excited state does not feel the perturbation due to the impurity ion, so eliminating the excited state splitting and giving rise to a single (essentially) unpolarized emission band. The F_A(Na)-centre in potassium chloride is an example of an F_A(I)-centre. The Stokes shift for F_A(Na)-centres in the different crystals is about 1.0 eV.

The F_A(II)-centres, of which F_A(Li)- in potassium chloride is an example, have very different properties. There is an enormous Stokes shift and the lifetime is much shorter, by a factor of about 10. These differences have their origins in the relaxation process. Unlike the situation in the F_A(I)-centres, where the reorientation in the optically-excited state is thermally activated and can be frozen in at low temperatures, reorientation of the F_A(II)-centres is temperature independent even down to very low temperatures ($T < 4$ K). Reorientation is independent of pumping in the F_{A2}- and F_{A1}-bands. The polarization is constant, the ratio of emission intensity parallel to the defect axis relative to that perpendicular to the axis being 1.8. The explanation for this is associated with the emitting state being the saddlepoint configuration between the two orientations before and after reorientation. The situation is illustrated in Fig. 7.30; in the ground state the single vacancy configuration is stable. After optical excitation and relaxation an anion neighbouring the F_A-centre has moved between the Li^+ impurity and one of its nearest-neighbour ions in the cation shell immediately surrounding the anion vacancy, so producing a double-well configuration. After emission this anion may return to its former site or into the original anion vacancy with equal probability. In the latter case the F_A-centre has rotated through 90°.

Thus the two different F_A-centre behaviours are determined by which of two ionic configurations applies in the excited state. For the F_A(I)-centre the vacancy configuration produces a single-well potential which binds the electron in a spatially compact s-like ground state and a spatially diffuse excited state. The diffuse excited state leads to a reduced oscillator strength in emission and a relatively long radiative lifetime. One might reasonably conjecture that the emissive state is also of mixed s–p character. In the saddlepoint configuration there is a double-well potential. The lowest electronic states are then symmetric, Ψ_S, and antisymmetric, Ψ_A, combinations of the ground states of two separated potential wells. (This situation is analogous to the case of the F_2^+-centre emission discussed in Chapter 2.) The electron is able to follow adiabatically the ionic reorientation from single

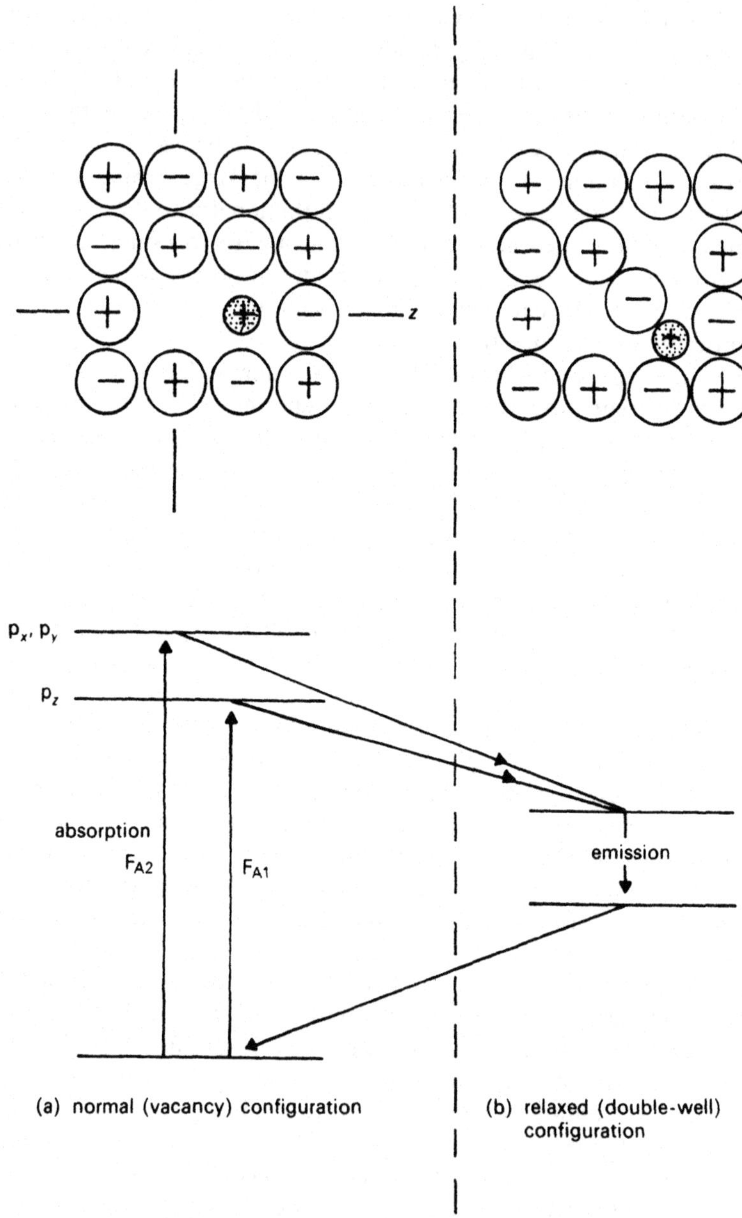

FIG. 7.30. (a) The configuration of F_A(II)-centres in the ground and relaxed excited states. (b) Absorption and emission transitions associated with these states. (From Mollenauer and Olson 1974.)

vacancy to saddlepoint configuration. A transition from single- to double-well potential transforms the s-ground state into the lowest symmetric state and the excited p-state into the lowest antisymmetric state of the double-well potential. As the Madelung energy of the double-well potential is considerably less than the single-well potential the Ψ_S state lies considerably higher in energy than the ground state of the vacancy. The situation is reversed in the electronic excited state. The anion vacancy is a very shallow trap for the p-state, hence the spatial diffuseness, whereas the double-well potential binds the electron in a low-lying compact state. Thus the F_A(II) emission near $h\nu \sim 0.5$ eV is an electric dipole transition between the Ψ_A and Ψ_S states of the F_A-centre in the saddlepoint configuration. The states are concentrated in the double well; they have a large overlap integral giving rise to the observed short radiative lifetime. The saddlepoint configuration has $\langle 110 \rangle$ symmetry and will emit characteristically polarized radiation parallel to this axis. The superposition of four equally-probable $\langle 110 \rangle$ dipole emitters should produce an emission polarization $I_{\parallel}/I_{\perp} = 2.0$ relative to the $\langle 100 \rangle$ F_A-centre axis. The experimental result, $I_{\parallel}/I_{\perp} = 1.8 \mp 0.3$, is in excellent agreement with theoretical expectations.

7.5 Optical properties of trapped hole centres

The products of irradiation of alkali halides include the F-like centres. Concomitant with the production of trapped electron centres are numerous hole centres, the structures of which depend critically upon irradiation type and temperature as well as crystal purity. These hole centres are also found in alkali halides heated in halogen gas to produce a stoichiometric excess of halogen. The major optical absorption bands of these centres lie on the short wavelength side of the F-band in the region *c.* 340–390 nm. For many years they were collectively referred to as V-centres. The principle intrinsic hole centres are now referred to as $[X_2^-]$-centres, H-centres, and V-centres. This latter centre is the antimorph of the F-centre, being a positive hole trapped in the neighbourhood of a cation vacancy. The $[X_2^-]$ ion is a molecular ion, $[F_2^-]$, $[Cl_2^-]$, etc., depending upon whether the ion is in a fluoride, chloride, or other halide lattice. The H-centre is an interstitial $[X_2^-]$-centre stabilized in a split-interstitial configuration centred on a single halide ion vacancy. Each of these defects may in its own way be associated with an impurity. (The $[X_2^-]$- and H-centres are shown in Fig. 7.1).

7.5.1 Self-trapped hole or $[X_2^-]$-centre

Molecular ions such as F_2^-, Cl_2^-, Br_2^- are stable in free space. In an alkali halide crystal such an ion is referred to as a self-trapped hole since no other defects or impurities are involved. The self-trapping is a spontaneous consequence of the hole being localized in a covalent bond between two adjacent halide ions. In consequence the molecular axis is a crystal ⟨110⟩ direction. The electronic wavefunction of the trapped hole is strongly localized on the two anions, and we may treat the centre as a slightly perturbed X_2^- ion frozen into a specific crystallographic orientation. In the Sonder–Sibley labelling scheme the square brackets imply that the molecular ion is present in a crystal. The $[X_2^-]$-centre has been observed in several other materials including cubic perovskites (e.g. $KMgF_3$), rutile-structured alkaline earth fluorides (magnesium fluoride) the ammonium halides and PH_4I. In such other crystals the X_2^- ions occupy sites determined by the crystal structure e.g. in the caesium chloride and calcium fluoride lattices the $[X_2^-]$ ions lie along a ⟨100⟩ direction. In all cases the centres may be produced by ionizing radiation at a sufficiently low temperature that hole migration is quenched. Furthermore the crystal must contain a sufficient density of electron traps to short-circuit hole annihilation by recombination. These traps may be either impurity ions or other colour centres (e.g. F^+-centres). A positive hole in the valence band may become self-trapped in an otherwise perfect crystal by a combination of three interactions: localization of the charge at some lattice point, polarization of the surrounding lattice, and finally the bonding energy due to the formation of the molecular ion. There is a net energy reduction of order 0 to -1.5 eV favouring the self-trapping mechanism. Since the model involves a positive hole in a p-shell, the centre has a single unpaired electron spin. It is, therefore, paramagnetic and is identified by the ESR spectrum. This was first reported by Castner and Kanzig (1957) for potassium chloride; subsequent studies have been reported for the alkali fluorides, bromides, and iodides (see e.g. Schoemaker 1976).

The ESR spectra of $[X_2^-]$-centres are highly anisotropic in terms of the Zeeman and of the nuclear hyperfine interactions. This anisotropy does show the ⟨110⟩ orientation of the centres and the involvement of two nuclides in the hyperfine structure. In the crystal the orbital angular momentum is strongly quenched so that we anticipate $g \simeq 2.00$. However, g-shifts may arise from coupling of the ground state to excited states of finite orbital momentum by spin–orbit interaction. The g-shift is approximately axial about ⟨110⟩ with $g_{\parallel} \simeq 2.00$; $g_{\perp}$ experiences a significant positive shift. The hyperfine structure is also approximately axial about the molecular axis.

The hyperfine constants have the following significance. The Fermi contact constant a_i scales as the electronic charge density at the nucleus, i.e. $|\Psi_i(0)|^2$, and in consequence measures the involvement of s-like orbitals in the electronic structure. The anisotropic hyperfine interaction b_i enters through the electron–nuclear dipolar coupling at separation r_i averaged over the p-like orbital only. Since a_i is finite in all the $[X_2^-]$-centres we seek a quantitative explanation of the ESR results by constructing molecular orbitals of the form

$$\Psi = \sum_{i=1}^{2} [\alpha_i |S\rangle_i + \beta_i |P\rangle_i]$$

where the α_i, β_i are admixture coefficients which should satisfy the usual normalization condition. Only very small differences are observed between the spin Hamiltonian parameters of isoelectronic centres in the different host lattices, indicating that it is the molecular entity which is of paramount importance.

To illustrate the importance of electronic structure we may consider that the molecular ion is constructed from a halide ion (X^-) and a halide atom (X). These entities have terms $^1S_0$ and $^2P_{3/2,\,1/2}$ respectively. Since the spin–orbit constant, ζ, for a more than half-filled shell is negative the $^2P_{3/2}$ level lies below the $^2P_{1/2}$ level. At large separations the states of $[X_2^-]$ are those of a free halide ion and a halogen atom in its lowest $^2P_{3/2}$ or $^2P_{1/2}$ state. As Fig. 7.31(a) shows, this gives two states at large separations, which split into four molecular orbitals, two of which are bonding orbitals and two antibonding orbitals. The states with twofold orbital degeneracy have this degeneracy raised by the combined effects of spin–orbit coupling and crystal field interaction. In atoms the $^2P_{1/2}$–$^2P_{3/2}$ splitting is $\frac{3}{2}\zeta$, mixing of p_x, p_y and p_z states contributing $\zeta/2$ each. The splitting is only $\sim\zeta$ in molecules because only the p_x and p_y orbitals contribute. Figure 7.31 indicates by broken lines the states admixed by spin–orbit coupling and which contribute to the g-shift, together with these optical transitions (full lines) in which the hole is transferred to states of opposite parity. The point symmetry of the $[X_2^-]$-centres (Fig. 7.1) is D_{2h}; such symmetry raises all orbital degeneracy and the electronic states are orbital singlets which transform as irreducible representations of the D_{2h} point group. Thus the highest energy transition (Fig. 7.31) takes place between the $^2B_{1u}$ ground state and the $^2B_{1g}$ excited state. The two other transitions $^2B_{1u}$ to $^2B_{2g}$ and $^2B_{3g}$ occur in the infrared: because of their large widths they are not always resolved, especially in the lighter halide crystals.

The polarization properties of the optical absorption transitions of the self-trapped hole centres are revealed by measurements of the dichroism of the absorption spectrum in certain circumstances. Such dichroic absorption arises from the reorientation of the centres during illumination with light in the ultraviolet band. The absorption strength of a particular transition varies as $\cos^2\theta$, where θ is the angle between the polarization vector, E, of the light and the molecular axis of the centre. Figure 7.31(b) shows the dichroic absorption spectrum of $[Br_2^-]$ centres in potassium bromide. In this case the self-trapped hole centres were created by X-irradiation at 77 K; optical dichroism was introduced by bleaching at 77 K with a 400 W mercury lamp and interference filter at $\lambda = 365$ nm. A suitable bleaching/measuring geo-

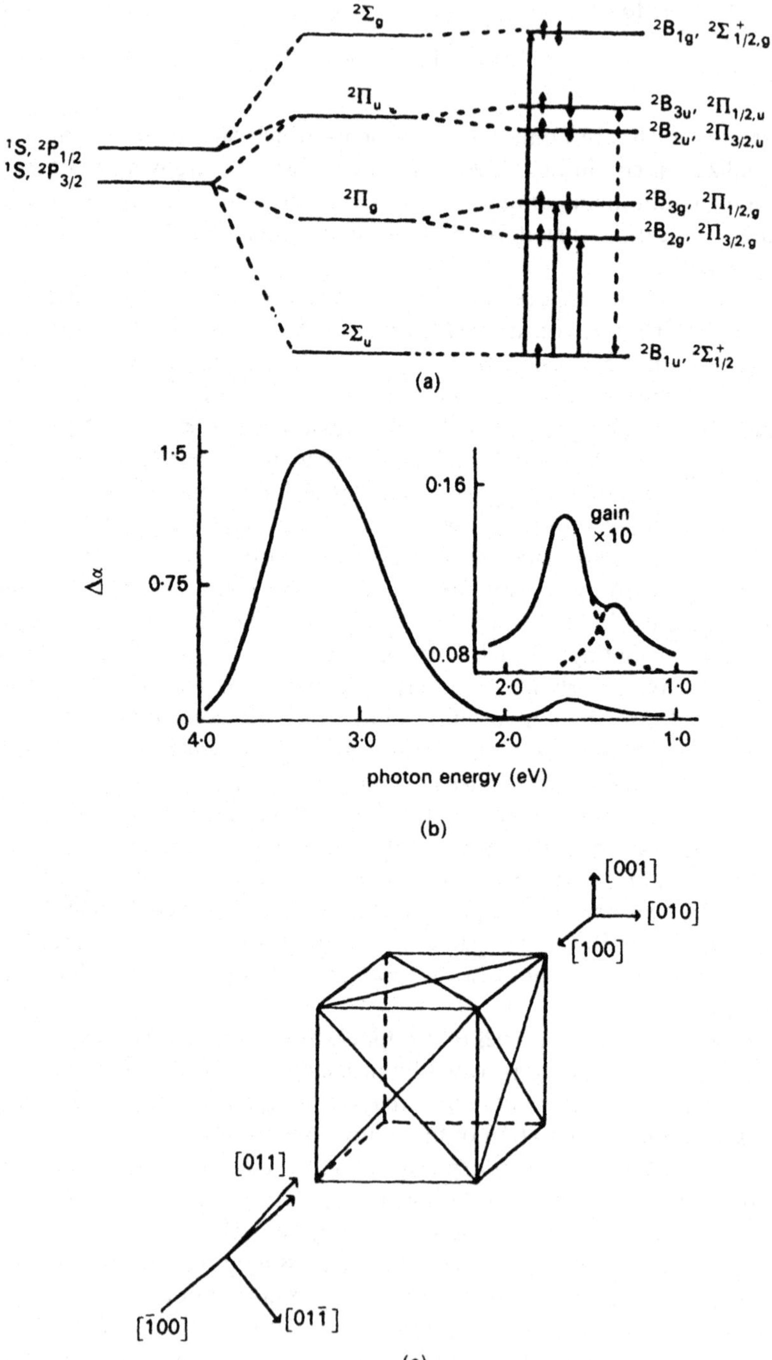
$^2\Sigma_g$
$^2\Pi_u$
$^1S, ^2P_{1/2}$
$^1S, ^2P_{3/2}$
$^2\Pi_g$
$^2\Sigma_u$
$^2B_{1g}, ^2\Sigma^+_{1/2,g}$
$^2B_{3u}, ^2\Pi_{1/2,u}$
$^2B_{2u}, ^2\Pi_{3/2,u}$
$^2B_{3g}, ^2\Pi_{1/2,g}$
$^2B_{2g}, ^2\Pi_{3/2,g}$
$^2B_{1u}, ^2\Sigma^+_{1/2}$
(a)
$\Delta\alpha$
1·5
0·75
0
4·0
3·0
2·0
1·0
photon energy (eV)
0·16
0.08
gain
×10
2·0
1·0
(b)
[001]
[010]
[100]
[011]
[$\bar{1}$00]
[01$\bar{1}$]
(c)

metry (Fig. 7.31(c)) uses light propagating along a crystal [100] direction with polarization vector, E, parallel to the [011] axis of the crystal. Those centres with molecular axis perpendicular to E, i.e. $[0\bar{1}1]$ centres, do not absorb [011]-polarized light in the energy band and so do not reorientate. All other centres absorb such radiation and undergo 60° or 90° reorientations of their molecular axis. In consequence the $[0\bar{1}1]$ direction is preferentially populated with defects whose molecular z-axis is parallel to this direction. By measuring the difference in the optical density of the crystal for light polarized in the [011] and $[0\bar{1}1]$ directions we obtain the dichroic absorption due only to the polarized optical transitions of $[X_2^-]$-centres, since transitions of centres which do not reorientate or are optically isotropic are then subtracted out. The dichroic absorption spectrum (Fig. 7.31(b)) identifies three absorption bands with very different strengths and half widths. The highest energy band is due to the transition

$$^2B_{1u} \rightarrow {}^2B_{1g};\ z\text{-polarized } [110];\ 380\ \text{nm} \equiv 3.25\ \text{eV},$$

an allowed electric dipole transition in which strong charge density oscillations occur along the molecular axis. Such behaviour is confirmed experimentally for $[F_2^-]$, $[Cl_2^-]$, $[Br_2^-]$, and $[I_2^-]$ molecular ions. The infrared transitions are identified as

$$^2B_{1u} \rightarrow {}^2B_{2g};\ y\text{-polarized } [1\bar{1}0];\ 930\,\text{nm} = 1.33\ \text{eV}$$

$$^2B_{1u} \rightarrow {}^2B_{3g};\ x\text{-polarized } [001];\ 770\ \text{nm} = 1.00\ \text{eV}:$$

Unfortunately, the predicted senses of polarization of these infrared transitions are not entirely consistent with the experimental results for $[Br_2^-]$ centres (Fig. 7.31(c)) or more generally (Kabler 1972). Indeed the data in Fig. 7.31(c) imply a net σ-polarization ($[110] \| x$, $[1\bar{1}0] \| y$) pattern. Note that, as Fig. 7.31(a) shows, the $^2B_{1u}$ and $^2B_{3g}$ states are derived mainly from the p atomic orbitals, whereas the $^2B_{2g}$ hole state come from the p_x and p_y orbitals. Thus the $^2B_{1u} \rightarrow {}^2B_{2g}{}^2B_{3g}$ transitions are forbidden in the electric dipole approximation except that admixtures of other orbitals into the $^2B_{2g}$ orbital may occur through either configuration mixing or spin–orbit coupling. The former admixes higher-lying S orbitals which once again gives a π-polarized transition moment. On the other hand spin–orbit interaction mixes $^2B_{1u}$ into

FIG. 7.31. (a) The energy levels of $[X_2^-]$-centres in alkali halide crystals. Both molecular orbital and group theoretical descriptions are given on the right-hand side. (b) Optical dichroism in the optical absorption spectrum of $[Br^{2-}]$-centres in potassium bromide measured at 20 K. (Adapted from Delbecq *et al.* 1961). (c) Geometry used for bleaching/dichroism measurements.

$^2B_{3u}$, so contributing σ-polarization. The oscillator strength of this transition is reduced by $(\zeta/\Delta E)^2$ relative to the $^2B_{1u} \rightarrow {}^2B_{1g}$ transition. For halides with low atomic number the π and σ contributions to the infrared bands are approximately equal so that they are approximately unpolarized. In the other extreme, $[I_2^-]$ centres, where $\zeta/\Delta E \sim 1$, the σ-polarized infrared band is almost as strong as the ultraviolet band.

The splitting between the infrared bands in Fig. 7.31(c) is just the spin–orbit coupling energy (~ -0.3 eV). Similar behaviour is observed for the $[I_2^-]$ centres in potassium iodide, where the spin–orbit energy is of order -0.5 eV, the splitting between the infrared bands is of the expected size. Only a single infrared absorption band has been observed for $[F_2^-]$ and $[Cl_2^-]$ centres, because the spin–orbit energy is rather small. However, in potassium chloride, measurement of the difference in the optical density for light polarized along [100] and [011] axes produces a splitting of the infrared band which is much larger than the spin–orbit interaction. This splitting between the infrared bands of *c.* 0.14 eV is the crystal field splitting.

Optically-induced reorientation by 60° or 90° rotation of the defect axes leads to an enhancement of the occupation of one orientation of centres over other possible orientations. Parallel studies of the changes in the optical and ESR spectra following polarized bleaching have been used to specifically tie the ESR spectrum to a particular optical band. They are an established feature of hole centre folklore and constitute an unmatched aid to identification of centres. Reorientation of centres may also be thermally activated. By measuring ESR and optical spectra following thermal reorientation of a pre-aligned spectrum it has been shown that 90° jumps do not occur with significant probability in either potassium chloride or potassium iodide. The measured activation energy for reorientation by 60° jumps is 0.54 eV in potassium chloride and 0.27 eV in potassium iodide (Keller *et al.* 1967).

7.5.2 The V*-centre—The antimorph of the* F*-centre*

In alkali halide crystals the V-centre is a positive hole trapped in the neighbourhood of a cation vacancy; a lower energy configuration is achieved by localization of the hole in a covalent bond between two anions adjacent to the vacancy. The resulting structure is a molecular ion $[X_2^-]$ adjacent to a cation vacancy. V-centres have been reported only in lithium fluoride, sodium chloride, potassium chloride, and rubidium chloride. They are produced directly by X-irradiation at temperatures at which $[X_2^-]$- and H-centres are unstable, or by thermal bleaching of self-trapped holes. These centres anneal irreversibly above ~ 230 K. Optical studies of V-centres are not well documented, probably for the reason that optical bands of these and other related centres strongly overlap because of their very great similarity. However, ESR results confirm that, although the g-tensor is essentially axial about a ⟨110⟩ direction, the principal axes of the hyperfine tensors are not the ⟨110⟩ directions. Instead, the molecular bond is bent through a small angle of about 4°. Hence the [X_2^-]-molecule ion has the symmetry of an isosceles triangle in an (001) plane with the base being a [110] direction.

The analogue of this defect in the alkaline earth oxides is the V^--centre, a hole trapped on an oxygen ion adjacent to a cation vacancy. In this case, however, the hole is associated with a single anion. There is, therefore, a marked dissimilarity between the V^--centre in the oxides and the V-centre in the alkali halides. The V^--centre has tetragonal symmetry whereas the alkali halide V-centre has orthorhombic symmetry due to the location of the hole in the molecular bond between two halogens. One consequence of the lower symmetry of the alkali halide V-centre is that the hole remains in the vicinity of the vacancy up to the dissociation temperature of *c.* 230 K, and there is no evidence of motional broadening up to at least 130 K in lithium fluoride. However, in oxides the ESR spectra show very clear evidence of motional averaging to the extent that a single isotropic line is observed at 300 K whereas at 77 K the spectrum shows a marked axial symmetry about a crystal ⟨100⟩ axis. Nonetheless, the hole remains in the immediate vicinity of the cation vacancy well above 300 K (Chen and Abraham 1975). The ground configuration for a free O^- ion is $2p^5$, corresponding to a single ^2p term. In the octahedral symmetry appropriate to the anion site in a cubic oxide, the ^2p state is an orbital triplet. However, in tetragonal symmetry, as is implied by the ESR spectra, the orbital triplet splits into an orbital singlet ground state, $^2A_1$, and an orbital doublet, ^2E, excited state. In principle optical transitions, $^2A_1 \rightarrow {}^2E$, are strictly forbidden since they involve a redistribution of electrons within the same type of orbital in a single quantum shell. Nevertheless they are observed with high probability, $f \sim 0.1$. Furthermore, these transitions should be linearly polarized perpendicular to the z-axis of the defect. Such polarization is not observed even at the lowest temperatures; this is attributed to hole migration between the six equivalent sites at rates which are rapid on the timescale of optical measurements. However, polarization may be induced by application of external perturbations, e.g. electric field and uniaxial stress. Such studies draw attention to the importance of internal stresses and electric fields (Rose and Cowan 1974; Rius *et al.* 1976; Henderson 1976a).

A family of related defects in the alkaline earth oxides involve a hole trapped on an oxygen ion neighbouring a unipositive alkali impurity (e.g. Li^+, Na^+, K^+): such defects are referred to as [Li]°, [Na]°, etc., centres. They have much in common with the V^--centres (Henderson and Garrison 1973; Chen and Abraham 1975). It should be noted that the ESR results are well understood; the optical properties have been subjected to several theoretical analyses, but cannot be regarded as satisfactorily understood.

7.5.3 Interstitial trapped hole centres

Although the [X_2^-]-centre is the simplest trapped hole centre in halide lattices, its formation in irradiated crystals is seldom sufficiently efficient that it is the predominant defect in pure crystals. Since it is charged relative to the host lattice, electron trapping impurities are required to prevent electron–hole recombination short-circuiting the build-up of X_2^--centres. Rather than these centres, interstitial halogen atoms or H-centres are the predominant species to be produced by irradiation at low temperature. As Fig. 7.1 shows, the H-centre is stable in a split interstitial configuration, which consists in essence of an X_2^- ion centred on a single anion site. Obviously such a centre produces a certain strain in the host lattice and on this account one might anticipate that the H-centre would have the greater space available to it in the ⟨111⟩ direction. At least in potassium chloride and potassium bromide the energy compensation from weak covalent bonding to the two halide ions at either end of the molecular ion must offset the increased strain energy. In these crystals the H-centre might be loosely described as an [X_4^{3-}] molecule ion. This structure is demanded by the observed ESR spectra. Nuclear hyperfine structure indicates that the unpaired spin interacts only weakly with the outermost two equivalent Cl^- nuclei. In contradistinction, the H-centre in lithium fluoride has been shown by ESR and ENDOR to be oriented along ⟨111⟩ directions. Thus in lithium fluoride the release of strain energy due to orientation along ⟨111⟩ directions offsets the reduction in energy due to the additional weak bonding effects (Kabler 1972; Schoemaker 1976). The analysis of optical absorption transitions for the H-centre proceeds much as in the case of the [X_2^-]-centres, except that the H-centre transitions occur at higher photon energies due to the closer spacing of the molecular ion in the H-centre relative to the self-trapped hole centre.

As discussed in Section 7.5.1, when [X_2^-] centres reorientate only 60° reorientations take place with significant probability. Since successive thermal reorientations of [X_2^-]-centres involve translations of the centres, reorientation leads to diffusion of [X_2^-]-centres. In contrast, when H-centres reorient they do so about the molecular centre, which coincides with a normal anion site. Since such reorientation does not involve translation of the centre, the H-centre does not diffuse through the lattice. If the H-centre in potassium chloride crystals is excited with [011] polarized light in the σ-polarized 336 nm band, most of the centres are rotated into the [$0\bar{1}1$] orientation. The resulting dichroic absorption (Fig. 7.32) is a consequence of a spatial anisotropy in the distribution of H-centres which results in the large difference ($\alpha_{011} - \alpha_{001}$) shown. This anisotropy remains indefinitely at 4 K, showing that the H-centre does not move or rotate thermally at this temperature. However, if the temperature is raised to 10.9 K the dichroism disappears. This corresponds to an activation energy for reorientation of

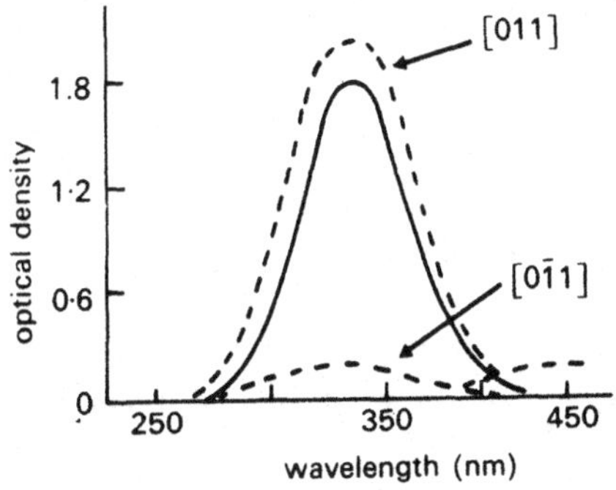

FIG. 7.32. Dichroic absorption of H-centres in potassium chloride at 4 K. (Adapted from Schoemaker 1976.)

0.031 eV. Dichroism of the optical absorption in the presence of a uniaxial stress confirms that reorientation proceeds predominantly by 60° jumps, as was the case for X_2^--centres. Despite the rapid reorientation, the H-centre does not decay until the temperature reaches 40–45 K. The H-centre can diffuse parallel to its molecular axis, this jump mechanism no doubt reflecting the weak molecular bonds with the two Cl^- anions along ⟨110⟩. The fate of such mobile H-centres is that they annihilate at F-centres and so restore a perfect lattice in this region. Alternatively they may be trapped at impurities. Indeed, the V_1-centre, whose structure was for more than two decades shrouded in mystery, is an H-centre trapped at a univalent alkali impurity.

7.5.4 Impurity-related hole centres

In mixed alkali halide crystals a wide variety of impurity-related spectra may be observed. For example, in mixed crystals containing two different halide ions [XY^-]-centres can be produced by suitable irradiation and thermal treatment. Consider the case of potassium chloride doped with (say) 0.1 per cent potassium bromide; [Cl_2^-] centres are created by irradiation at 77 K. Annealing the crystal at 200 K results in the holes migrating, some of which are trapped on substitutional Br^- ions. The [$BrCl^-$] centres are aligned in ⟨110⟩ directions. In potassium chloride these centres have broad absorption bands at 382 nm and 760 nm, relative to the [Cl^{2-}] centre bands at 365 nm and 750 nm. Similar ⟨110⟩-orientated defects are found in KBr:I and KCl:I. However, in KCl:F, KBr:F and KI:F these ⟨110⟩-orientated [XF^-]-centres are not observed. Instead, interstitial [XF^-]-centres occupying a single anion site with their internuclear axes along ⟨111⟩ are observed. That it is an interstitial centre is made certain by optically exciting F-centres; this releases electrons into the crystal which eliminates X_2^--centres by recombination. Hence the [XF^-]-centres are not destroyed because they are neutral relative to the lattice. Irrespective of their particular configuration [XY$^-$]-centres may optically or thermally reorientate, and so give rise to dichroism. They are also stable at rather higher temperatures than the conventional [X_2^-]- and H-centres. The intrinsic centres may be stabilized in another way, by replacing one of the alkali ions neighbouring the [X_2^-]- or H-centres by an impurity alkali ion. Such centres are usually called [X_2^-]$_A$-centres or H_A-centres. The [X_2^-]$_A$-centre differs only slightly from the self-trapped hole centre. There are numerous H_A-centres, their behaviour and properties being dependent on both the particular alkali halide and the impurity alkali halide. In potassium chloride the $H_A(Na^+)$-centre was originally called the V_1-centre, its distinct near ultraviolet absorption band having been known for many years. Originally the V_1-band was thought to be due to an intrinsic centre. This was incorrect. The difficulty in unravelling the structure of this centre lay with the elusive character of its ESR spectrum.

The $H_A(Na^+)$-centre in potassium chloride, as with most H_A-centres, is produced by X- or γ-irradiation of the suitably-doped alkali halide at 77 K. At this temperature the interstitial halide atoms are mobile, and as they move through the crystal are rapidly trapped at the alkali ion impurity. The electronic and ionic structure of this centre became clear only after its ESR spectrum had been discovered. This spectrum is only observable over a narrow temperature range near 35 K; below 25 K the ESR spectrum is very strongly saturated, whereas above 45 K motional averaging smears out the characteristic hyperfine structure. The structure deduced from the ESR spectrum differs from that of the unperturbed H-centre in that, although the molecular species lies in a (100) plane, the molecular axis makes an angle of 5.7° with a $\langle 110 \rangle$ direction. However, the hyperfine interaction on the four contributing nuclei are unequal. Emphasizing the similarity with the H-centre a strong $^2\Sigma_u + \rightarrow {}^2\Pi_g$ transition at 560 nm has been observed. Reorientation of the $H_A(Na^+)$-centre leads to optical anisotropy which decays at a temperature of $T = 16.8$ K. The $H_A(Na^+)$-centre in potassium chloride decays above 113 K.

A variety of other H_A- and (H_{AA})-centres have been reported, e.g. $H_A(Li^+)$ in potassium chloride and sodium fluoride, $H_A(Na^+)$ in lithium fluoride, $H_{AA}(Na^+)$ in potassium chloride. These different centres have certain common features to the $H_A(Na^+)$-centre in potassium chloride. However, the detailed geometrical arrangements, including, for instance bent bonds, leads to a very different reorientational motion. The basic principles are little different from those involved in H-centre motion (Schoemaker 1976; Kabler 1972).

8

Spectroscopy of lanthanide (rare-earth) and actinide ions in solids

IN the fifth period of the periodic table of elements after the element lanthanum (atomic number 57) the filling of the 4f shell takes place, from cesium with the outer configuration $5s^2 5p^6 4f^1 5d^1 6s^2$ to ytterbium with configuration $5s^2 5p^6 4f^{13} 5d^1 6s^2$. These are the lanthanide or rare-earth elements, which occur as doubly or triply charged ions in ionic solids. In the triply charged rare-earth ions all 5d and 6s electrons are removed and the 4f shell is only partially occupied; Table 8.1 gives the number (n) of the electrons in the unfilled 4f shell. The many energy levels of the unfilled $4f^n$ shell between which radiative transitions may occur cover a spread of energies of $\simeq 40\,000\ \mathrm{cm}^{-1}$. The next highest configurations are $4f^{(n-1)}5d$, $4f^{(n-1)}6s$ and $4f^{(n-1)}6p$, which for the triply charged ions are usually well above the low-lying $4f^n$ energy levels. The divalent rare-earth ions contain one more electron (e.g. Sm^{2+} has the same configuration as Eu^{3+}) and in these divalent ions the higher-lying energy configurations overlap some of the low-lying $4f^n$ energy levels (Dieke 1968). In a similar way the filling of the 5f shell occurs after the element actinium (atomic number 89). These actinide ions, which have an incomplete 5f shell, have many low-lying energy levels in solids. In consequence they also have radiative transitions in the optical region.

Table 8.1

Number of electrons (n) in the 4f shell of triply charged rare-earth ions

Ce^{3+}	1
Pr^{3+}	2
Nd^{3+}	3
Pm^{3+}	4
Sm^{3+}	5
Eu^{3+}	6
Gd^{3+}	7
Tb^{3+}	8
Dy^{3+}	9
Ho^{3+}	10
Er^{3+}	11
Tm^{3+}	12
Yb^{3+}	13

The optically active 4f electrons of the rare earth ions are shielded by the outer, though less energetic, 5s and 5p shells of electrons. As a result the optically active 4f electrons of rare-earth ions in solids are not strongly affected by neighbouring ligands. If the neighbouring ligands are neglected the energy levels of the 4f electrons are just the free ion levels, characterized by L, S, J values, with allowance made for some term mixing (Chapter 2). The energy levels of free ions are very sharp. Figure 8.1 shows the energy levels of trivalent rare-earth ions in lanthanum trichloride which were determined by Dieke and his co-workers; we refer to this as the *Dieke diagram*. In solids the effect of the neighbouring ligands is to split these free ion levels; the extent of this crystal field splitting for the rare-earth ions in lanthanum trichloride is indicated by the widths of the levels in the Dieke diagram. The centre of gravity of each crystal field multiplet approximately locates the free ion levels. In general the splitting of the free ion levels is less than the separation between free ion levels. Hence for rare-earth ions in different host materials the gross features of the energy level diagram are unchanged. The crystal field splitting will vary from host to host, reflecting the different symmetries and strengths of the different crystal fields.

8.1 Energy levels and eigenstates of the $4f^n$ electrons

In the central field approximation the eigenstates of 4f electrons are calculated by neglecting the interaction between 4f electrons and by assuming that each 4f electron moves in a spherically symmetric electrostatic field due to the nucleus and the other filled shells of electrons. Each 4f electron state is then characterized by the four quantum numbers $n(=4)$, $l(=3)$, m_l, and m_s. The orbital wavefunction is $R'_{nl}(r)\, Y_l^{m_l}(\theta, \phi)$, where $R'_{nl}(r)$ can be found by direct Hartree–Fock calculation and the prime indicates that these radial functions are not simple hydrogen-like functions. The electrostatic interaction between the 4f electrons is next taken into account. This interaction

$$\mathscr{H}' = \sum_{i>j} \frac{e^2}{4\pi\varepsilon_0 r_{ij}} \tag{8.1}$$

splits the energy level of the $4f^n$ configuration into different LS terms. The resultant wavefunctions are characterized by the quantum numbers L, S, M_L, M_S, as described in Chapter 2. The values of this electrostatic interaction in the case of each LS level can be expressed as a sum of Slater integrals, $F^{(k)}$,

---

FIG. 8.1. The energy level diagram for trivalent rare-earth ions in lanthanum chloride in which the width of a level is a measure of its crystal field splitting. Pendant semicircles indicate emitting levels. This figure represents work carried out over many years by the members of Professor Dieke's research group at Johns Hopkins University, and is reproduced here with the permission of Professor H. M. Crosswhite.

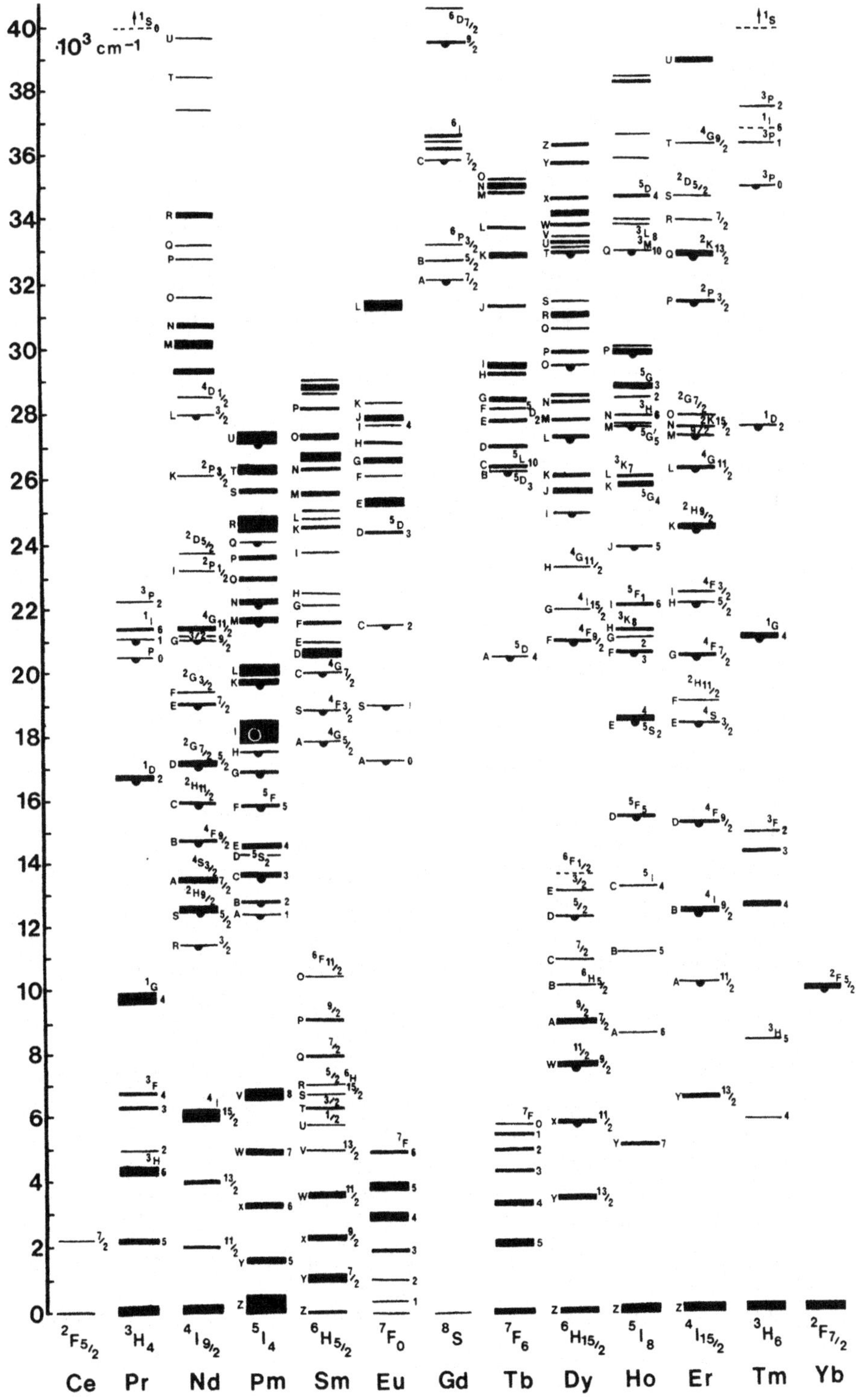

$\cdot 10^3\ cm^{-1}$
40
38
36
34
32
30
28
26
24
22
20
18
16
14
12
10
8
6
4
2
0
$^2F_{5/2}$
$^3H_4$
$^4I_{9/2}$
$^5I_4$
$^6H_{5/2}$
$^7F_0$
^8S
$^7F_6$
$^6H_{15/2}$
$^5I_8$
$^4I_{15/2}$
$^3H_6$
$^2F_{7/2}$
Ce
Pr
Nd
Pm
Sm
Eu
Gd
Tb
Dy
Ho
Er
Tm
Yb

where

$$F^{(k)} = \frac{e^2}{4\pi\varepsilon_0} \int_0^\infty \int_0^\infty \frac{r_<^k}{r_>^{k+1}} [R'_{4f}(r_i) R'_{4f}(r_j)]^2 r_i^2 r_j^2 \, dr_i \, dr_j \tag{8.2}$$

and where $r_>$ is the greater and $r_<$ is the lesser of r_i and r_j. Details of these calculations are found in Judd (1963), Wybourne (1965), Dieke (1968), Carnall *et al.* (1977), Reisfeld and Jørgenson (1977), and Pappalardo (1978). For the $4f^n$ configuration only $F^{(0)}$, $F^{(2)}$, $F^{(4)}$, and $F^{(6)}$ appear in the calculations. The $F^{(0)}$ term contributes equally to all LS states of the same $4f^n$ configuration and so can be ignored. Sometimes the energy values are expressed in terms of modified Slater integrals, F_k, which for 4f electrons are given by

$$F_2 = \frac{F^{(2)}}{225}, \quad F_4 = \frac{F^{(4)}}{1089}, \quad F_6 = \frac{25F^{(6)}}{184041}. \tag{8.3}$$

If the radial wavefunctions were hydrogen-like the following ratios would apply:

$$F_4/F_2 = 0.138 \qquad \text{and} \qquad F_6/F_2 = 0.0151 \tag{8.4}$$

so that only one parameter, F_2 say, would be needed. Values of these Slater integrals calculated using Hartree–Fock wavefunctions have been tabulated in Carnall *et al.* (1977). Generally one regards the modified Slater integrals, F_2, F_4, and F_6, as adjustable parameters whose values are found by comparison with spectroscopically-measured energy levels. Values of the modified Slater integrals obtained by fitting to energy levels of rare-earth ions in lanthanum chloride and lanthanum fluoride are also given in Carnall *et al.* (1977). For rare-earth ions in crystals the values of the various F_k parameters are reduced by about 2 per cent from those obtained from free-ion spectra. This reduction is attributed to a spreading outwards of the 4f-electron charge distribution, called by Jørgensen (1960) the *nephelauxetic effect*. The values of the F_k parameters also vary slightly with the nature of the surrounding ions.

The spin–orbit coupling operator $\mathscr{H}_{so} = \sum_i \zeta_i(r) \boldsymbol{l}_i \cdot \boldsymbol{s}_i$ commutes with $\boldsymbol{J}$, hence when $\mathscr{H}_{so}$ is taken into account the wavefunctions can be expressed as eigenfunctions of $\boldsymbol{J}^2$, J_z. If the energy separation between different LS terms is large compared with the spin–orbit coupling energy there is little mixing of LS terms by the spin–orbit coupling operator. If this mixing is neglected we are in the Russell–Saunders approximation and the spin–orbit coupling operator can be written as $\zeta \boldsymbol{L} \cdot \boldsymbol{S}$. The wavefunctions then can be characterized by the quantum numbers L, S, J, M_J. In this approximation spin–orbit coupling causes a splitting of each LS term into a number of states $|LSJM_J\rangle$ each with different values of J and M_J. These are the *J-multiplets*, and the additional energy of each is

$$\frac{\zeta}{2}[J(J+1)-L(L+1)-S(S+1)] \tag{8.5}$$

where the value of the spin–orbit coupling parameter, ζ, is normally determined by comparison with measurement. The separation between two adjacent J levels is ζJ, where J now refers to the larger of the two J values. This is the *Landé interval rule* for the energy interval between adjacent J multiplet levels. Figure 8.2 illustrates this rule for the ^{7}F ($L=3$, $S=3$) LS term split by spin–orbit coupling in which the Russell–Saunders approximation is assumed to hold and ζ is assumed positive. J ranges from $J=6$ to $J=0$, and the Landé interval rule gives the ratio of the splittings as 6:5:4:3:2:1. Each J level is $(2J+1)$-fold degenerate, these states being characterized by the value of M_J.

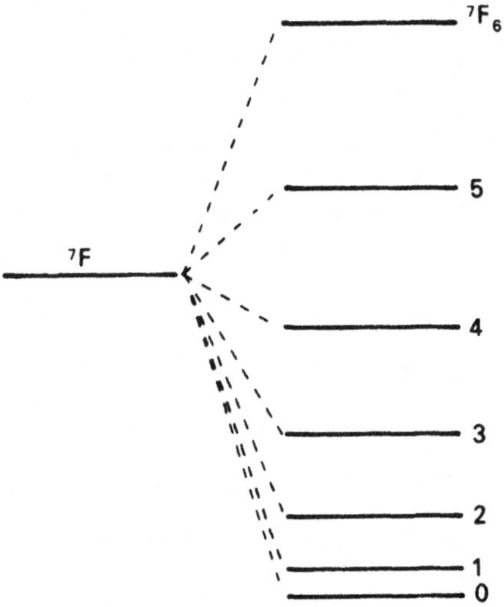

FIG. 8.2. Splitting of the ^{7}F ($L=3, S=3$) term by spin–orbit coupling in the Russell–Saunders approximation. The spin–orbit coupling parameter is assumed positive, as in the ground state of Eu^{3+} (see Fig. 8.1).

If the term separation is not large then the spin–orbit coupling operator mixes states with the same JM_J value but with different LS values. The resultant *intermediate coupling* eigenstates are given by

$$\psi(\gamma LSJM_J)=\sum_{\gamma' L' S'} c(\gamma' L' S')|(4f^n)\gamma' L' S' J M_J\rangle \tag{8.6}$$

where γ' describes any other quantum number required to distinguish the states, and $|(4f^n)\gamma' L' S' J M_J\rangle$ are Russell–Saunders states (i.e. no term mixing). We note that the intermediate coupling eigenstate is labelled by a specific set of LS values, as is indicated by the left-hand side of eqn (8.6), whereas the form of the eigenstate involves a summation over different sets of LS values. It is usual to employ, as the labelling LS pair, the set belonging to the $|LSJM_J\rangle$ Russell–Saunders state to which ψ reduces when spin–orbit coupling is made vanishingly small. This usually corresponds to the set of LS values for which $c(\gamma LS)$ is largest. In general the coefficients $c(\gamma LS)$ are determined by computer diagonalization of the matrix for the electrostatic interaction $\mathscr{H}'$ and spin–orbit coupling $\mathscr{H}_{so}$, and then fitting the calculated energy levels to the experimentally measured values. In addition to $\mathscr{H}'$ and $\mathscr{H}_{so}$ there are a number of smaller effects which must be taken into account to obtain good agreement between observed and calculated energy levels (Wybourne 1965).

As long as one is dealing with pure f states the $(14-n)$ electron system can be regarded as n *positively* charged 4f electrons in a filled 4f shell. Hence, except for a constant energy term which we can ignore, the Coulomb energy of mutual interaction between the $(14-n)$ negative electrons is the same as that between n positively charged electrons—which is the same as the energy of interaction between n negatively charged electrons. As a result, the same LS terms in the same energy order occur for the $4f^{(14-n)}$ and $4f^n$ configurations. For this reason in discussing the spectroscopic properties of individual rare-earth ions (Section 8.3) a particular $4f^n$ ion is grouped with the analogous $4f^{14-n}$ ion. The values of the Slater integrals tend to be larger for ions in the upper half of the rare-earth series, consequently the energy separation between LS terms will be larger for these ions.

Since spin–orbit coupling is the energy of interaction of a spinning electron with the magnetic field of the orbiting nucleus, the coupling energy changes sign when the sign of the electronic charge is reversed. As a result, the splitting of the LS levels into J multiplets is reversed in going from $4f^n$ to $4f^{(14-n)}$. ζ can be taken as negative for the second half of the rare-earth series. Further, the magnitude of ζ increases by a factor of 3–4 across the rare-earth series. As an example of this reversal we observe in the Dieke diagram (Fig. 8.1) the reversal in the order of the J levels in the ground ^7F_J states of $Eu^{3+}(4f^6)$ and $Tb^{3+}(4f^8)$. On the same diagram a comparison of the energy levels of $Ce^{3+}(4f)$ and $Yb^{3+}(4f^{13})$ shows clearly the much larger magnitude of ζ for Yb^{3+} as well as the reversal in the sign of ζ.

Very extensive spectroscopic analyses have been carried out for all trivalent rare-earth ions in lanthanum chloride and lanthanum fluoride, and the values of the various parameters are tabulated in Carnall *et al.* (1977). Values of the parameters for specific rare-earth ions in other host materials are available in the literature. The magnitude of ζ determined experimentally varies significantly through the rare-earth series, from $\zeta \simeq 744\ \mathrm{cm}^{-1}$ for $Pr^{3+}(4f^2)$ to $\zeta \simeq 3000\ \mathrm{cm}^{-1}$ for $Yb^{3+}(4f^{13})$ in the case of the rare-earth ions in lanthanum chloride. There is a much smaller variation in the experimentally determined values of the Slater integrals through the series.

Next we consider the interaction of the 4f electrons with the crystal field. As in Chapter 2 (eqn 2.57) we write the one-electron crystal field energy as

$$\mathscr{H}_c(r)=\frac{1}{4\pi\varepsilon_0}\sum_l\sum_{k=0}^{\infty}\sum_{t=-k}^{k}\frac{Z_l e^2}{a_l^{k+1}}r^k Y_k^t(\theta_l,\phi_l)C_t^{(k)}(\theta,\phi) \qquad (8.7)$$

where the summation l is over the neighbouring point charges. To calculate the energy levels in the presence of the crystal field the matrix elements of $\mathscr{H}_c(r)$ between various 4f-electron wavefunctions must be evaluated. The value of $\langle r^k\rangle$ will be a common factor, and if we confine our attention to the angular properties of the crystal field we have

$$\mathscr{H}_c(\theta,\phi)=\frac{1}{4\pi\varepsilon_0}\sum_l\sum_{k,t}^{\infty}\langle r^k\rangle\frac{Z_l e^2}{a_l^{k+1}}Y_k^t(\theta_l,\phi_l)C_t^{(k)}(\theta,\phi). \qquad (8.8)$$

When the sum over l is made we can write

$$\mathscr{H}_c=\sum_{k,t}B_t^k C_t^{(k)}(\theta,\phi) \qquad (8.9)$$

and this is the form of the crystal field energy that is commonly employed for rare-earth ions. As in the case of the transition metal ions one regards the B_t^k as parameters to be determined by fitting calculated levels to the observed spectra. In the case of 4f-electron wavefunctions we need only consider values with $k\leqslant 6$. Further, odd-k values (due to odd-parity components of the crystal field) will not contribute to matrix elements between 4f-electron wavefunctions and so are often omitted from the $\mathscr{H}_c$ formula. Not all even-k values are required; the number which is required is determined by the symmetry of the crystal field. We give below the forms of $\mathscr{H}_c$ for a number of the more common symmetries:

$$\mathscr{H}_c^{O_h}=B_0^4\left[C_0^{(4)}+\sqrt{\frac{5}{14}}(C_{-4}^{(4)}+C_4^{(4)})\right]+B_0^6\left[C_0^{(6)}-\sqrt{\frac{7}{2}}(C_{-4}^{(6)}+C_4^{(6)})\right]. \qquad (8.10)$$

(A comparison with the octahedral crystal field term for transition metal ions (eqn 2.93) is interesting.)

$$\mathscr{H}_c^{D_{3h}}(\text{even terms only})=B_0^2C_0^{(2)}+B_0^4C_0^{(4)}+B_0^6C_0^{(6)}+B_6^6(C_{-6}^{(6)}+C_6^{(6)}) \qquad (8.11)$$

$$\mathscr{H}_c^{C_{3v}}=B_0^2C_0^{(2)}+B_0^4C_0^{(4)}+B_3^4(C_{-3}^{(4)}-C_3^{(4)})+B_0^6C_0^{(6)}$$
$$+B_3^6(C_{-3}^{(6)}-C_3^{(6)})+B_6^6(C_{-6}^{(6)}+C_6^{(6)}). \qquad (8.12)$$

Although the La^{3+} site symmetry of lanthanum trichloride is C_{3h} it can be reasonably well approximated as D_{3h}. In lanthanum trifluoride the La^{3+} site symmetry is still a matter of debate, but to good approximation it can be taken as D_{3h}. Values of the crystal field parameters for all rare-earth ions in lanthanum trichloride and for some rare-earth ions in lanthanum trifluoride are given in Carnall *et al.* (1977).

If the crystal field energy is small compared with the spin–orbit splitting we expect that the crystal field will cause a splitting of the J levels but will not mix states from different J multiplets. Some mixing of states from different J levels (*J-mixing*) may, however, occur which can affect the J-selection rules. For example, the $^7F_0 \leftrightarrow {}^5D_0$ transition of Eu^{3+} ions is not allowed by the selection rule which says that $J=0 \leftrightarrow J=0$ transitions are forbidden. Yet this transition is seen weakly, and occurs because of J-mixing.

Divalent rare-earth ions can also be incorporated in a number of hosts (e.g. substitutionally for Ca^{2+} in calcium fluoride). The $4f^n$ levels are similar to those of the relevant trivalent rare-earth ion. The $4f^{(n-1)}5d$ levels, however, appear at lower energy and may overlap the higher $4f^n$ levels (Dieke 1968).

8.2 Selection rules for radiative transitions on rare-earth ions

Since the interaction of the 4f electrons with the crystal field is small, electron–phonon coupling effects are weak and optical transitions between $4f^n$ states are characterized by sharp lines. Furthermore, since these transitions occur between states of the same parity we expect them to occur by a magnetic dipole process. However, if the rare-earth ion is in a site which lacks inversion symmetry a mixing of opposite-parity states of the $4f^{(n-1)}5d$ configuration into the $4f^n$ states can occur. Even though the amount of this mixing is small and has a negligible effect on the energy values it may significantly change the intensity of transitions between the levels. This arises because even a small admixture of even- and odd-parity states allows a small electric dipole component to the transition, and since the electric dipole process is some five orders of magnitude stronger than the magnetic dipole process it may be the dominant component in the transition. Even when the rare-earth ion appears to occupy a site with inversion symmetry the observed transitions are sometimes found to be electric dipole. This may indicate that the substituting rare-earth ion tends to occupy a slightly distorted site where inversion symmetry no longer holds.

Without referring to a specific crystal field we will analyse the selection rules for transitions between the free ion J levels, where the states are described by intermediate coupling wavefunctions $\psi(\gamma, L, S, J)$. The *strength*, S, of a radiative transition between states a and b is defined as the square of the relevant dipole matrix element (Chapter 4). For a magnetic dipole transition this is

$$S_{ab}(\mathrm{MD}) = \sum_{M_{J_a}, M_{J_b}} |\langle \psi_b(\gamma_b S_b L_b J_b M_{J_b})| \mu_{\mathrm{m}} |\psi_a(\gamma_a S_a L_a M_{J_a})\rangle|^2. \quad (8.13)$$

From the Wigner–Eckart theorem we find that this matrix element is non-zero only for $\Delta J = 0, \pm 1$, with $0 \leftrightarrow 0$ forbidden. J mixing through crystal field interactions may relax this selection rule to some extent.

Electric dipole transitions between $4f^n$ states require an admixture of opposite-parity $4f^{(n-1)}5d$ states by an odd-parity crystal field component, $\mathscr{H}_c^{(u)}$. The effect of this mixing on the selection rules was investigated, independently, by Judd (1962) and Ofelt (1962). The matrix element of the electric dipole transition between states a and b is

$$\sum_\beta \frac{\langle b|\mathscr{H}_c^{(u)}|\beta\rangle\langle\beta|\mu_e|a\rangle}{E_b - E_\beta} + \text{converse} \tag{8.14}$$

where β signifies the opposite-parity $4f^{(n-1)}5d$ states. We do not have sufficient knowledge about the nature and position of these opposite-parity states, consequently we adopt the very helpful simplification of assuming that all have the same energy. The constant energy denominator can now be taken outside the summation and closure used. In this way one can regard $\sum_\beta \mathscr{H}_c^{(u)}|\beta\rangle\langle\beta|\mu_e$ as an even-parity electric dipole operator which operates between the free ion J states. The strength of the electric dipole transition between states a and b can be written as

$$S_{ab}(\mathrm{ED}) = e^2 \sum_{t=2,4,6} \Omega_t |\langle\psi_b(\gamma_b S_b L_b J_b)\| U^{(t)} \|\psi_a(\gamma_a S_a L_a J_a)\rangle|^2. \tag{8.15}$$

Ω_t are the Judd–Ofelt intensity parameters and they characterize the strength and nature of the odd-parity crystal field, and $U^{(t)}$ is a tensor operator of rank t. The reduced matrix elements $\langle(4f^n)\gamma SLJ \| U^{(t)} \| (4f^n)\gamma' S' L' J'\rangle$ between pure Russell–Saunders states can be calculated from published tables. Then the reduced matrix elements in the intermediate coupling scheme can be calculated once the numerical values of the $c(\gamma SL)$ coefficients are known. These reduced matrix elements have been calculated for transitions between various states of rare-earth ions in different host crystals and the sources are listed (Riseberg and Weber 1975). Carnall *et al.* (1977) give a complete listing of values for transitions on all trivalent rare-earth ions in lanthanum trichloride. By measuring the absorption strengths of a number of transitions from the ground state one can determine values for the Judd–Ofelt parameters, Ω_t. Once these have been determined for a given rare-earth–host crystal combination they can be used to calculate electric dipole transition strengths (absorption or emission) between any two levels of the system.

Some general selection rules for electric dipole transitions between rare-earth ion $4f^n$ states have been obtained from the Judd–Ofelt theory. These are:

(a) $\Delta J \leqslant 6$; $\Delta S = 0$, $\Delta L \leqslant 6$ (Russell–Saunders approximation)

(b) For a rare-earth ion with an even number of electrons:
1. $J=0 \leftrightarrow J'=0$ is forbidden
2. $J=0 \leftrightarrow$ odd J' values are weak
3. $J=0 \leftrightarrow J'=2, 4, 6$ should be strong.

8.3 Specific trivalent rare-earth ions

Ce^{3+} ($4f^1$) In its lowest energy state this ion has a single 4f electron outside closed shells. Spin–orbit coupling causes a splitting into a $^2F_{5/2}$ ground state and a $^2F_{7/2}$ state approximately 2000 cm^{-1} higher in energy, as Fig. 8.1 shows. Not shown in Fig. 8.1 are the next higher configurations in which, in the case of Ce^{3+}, the outer electron is in the 5d orbital. The lowest lying 5d level is approximately 30000 cm^{-1} above the ground state. The 5d–4f transitions, being parity-allowed, have large transition probabilities. Since the 5d electron orbital is much more strongly affected by the neighbouring ions, the 4f–5d absorption transition is a strong broad band. The emission is similarly broad with two peaks corresponding to the $^7F_{5/2}$ and $^7F_{7/2}$ terminal states. Excited Ce^{3+} ions can also act as *sensitizers* and transfer their excitation to other suitable rare-earth ions, e.g. Tb^{3+}, which can be introduced as co-dopants in the host crystal (Blasse 1988).

Yb^{3+} ($4f^{13}$) Because of the larger value of the spin–orbit coupling parameter the $^4F_{5/2}$ state of Yb^{3+} is about 10000 cm^{-1} above the ground $^4F_{7/2}$ state. When substituted for Ca^{2+} in the cubic site in calcium fluoride we expect (Table 3.10) the upper and lower states to be split into two and three levels, respectively. Figure 8.3 shows these levels as well as the low temperature absorption and luminescence transitions on the cubic site ion in a low concentration sample (Voron'ko *et al.* 1969). In addition to the transitions on cubic site ions one finds other lines (shown in dark outline in Fig. 8.3) some of which originate on Yb^{3+} ions in sites of lower symmetry while others originate on clusters of Yb^{3+} ions. A cooperative luminescence process in which two adjacent excited Yb^{3+} ions simultaneously deexcite, emitting a single photon of green light, was first demonstrated in ytterbium phosphate by Nakazawa and Shionoya (1970).

Pr^{3+} ($4f^2$) The full set of *LS* terms in order of ascending energy levels (with *J* values associated with the *LS* values given as subscripts) are:

$$^3H_{6,5,4} \quad ^3F_{4,3,2} \quad ^1G_4 \quad ^1D_2 \quad ^1I_6 \quad ^3P_{2,1,0} \quad ^1S_0$$

The positions of the *LS* levels, calculated in the absence of spin–orbit coupling, are shown on the left-hand side of Fig. 8.4. $\mathscr{H}_{so}$ splits each *LS* level into its *J* sublevels, and may also mix states of the same *J* from different *LS* terms. In the case of Pr^{3+} the spin–orbit coupling causes a mixing of $^3H_4$ and $^3F_4$. A similar mixing occurs between $^1I_6$ and $^3H_6$, and between $^1D_2$, $^3P_2$, and $^3F_2$. $^1S_0$ and $^3P_0$ are sufficiently far apart that mixing of these states is quite weak.

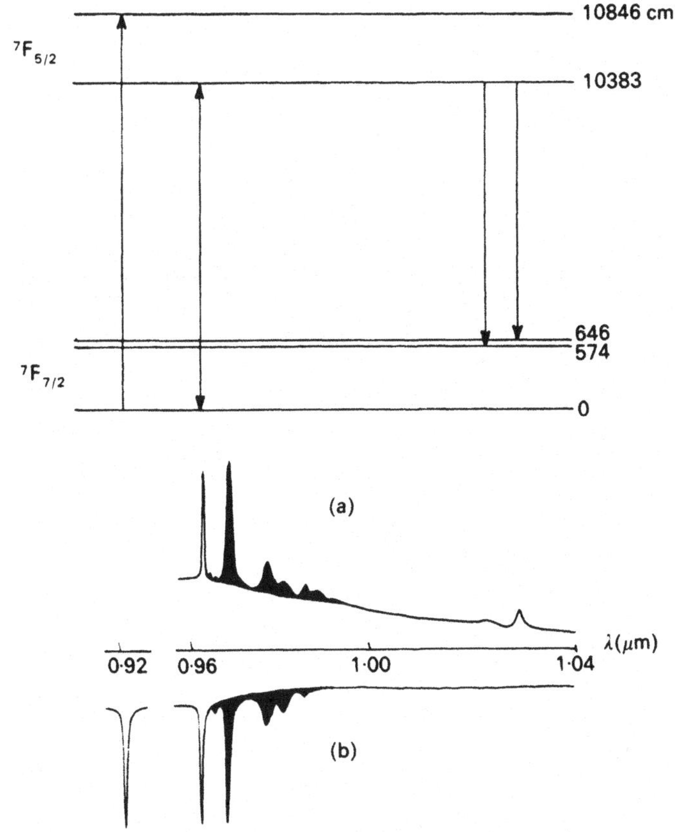

FIG. 8.3. Energy levels of Yb^{3+} ($4f^{13}$) in cubic sites in calcium fluoride. Low-temperature luminescence and absorption spectra are seen in (a) and (b). The shaded transitions are due to Yb^{3+} ions in perturbed sites (After Voron'ko *et al.* 1969.)

The splittings of the *LS* terms of Pr^{3+} in lanthanum fluoride are shown schematically in Fig. 8.4. Because of term mixing there are clear departures from the Landé interval rule among the *J* levels. Finally, these *J* levels are split by the crystal field.

At low temperature and with appropriate optical pumping visible luminescence is observed from the $^3P_0$ and $^1D_2$ levels of Pr^{3+} in lanthanum trichloride (Fig. 8.1). Similar luminescence is also observed from Pr^{3+} ions in other hosts. The energy separation between the $^3P_0$ level and the next lower $^1D_2$ level is sufficiently small, *c.* 4000 cm^{-1}, that non-radiative decay across this small gap may compete with the radiative decay processes from the $^3P_0$ level. In lanthanum trichloride and lanthanum trifluoride, as Fig. 5.32 shows, the phonons which are effective in coupling to these levels are of energy

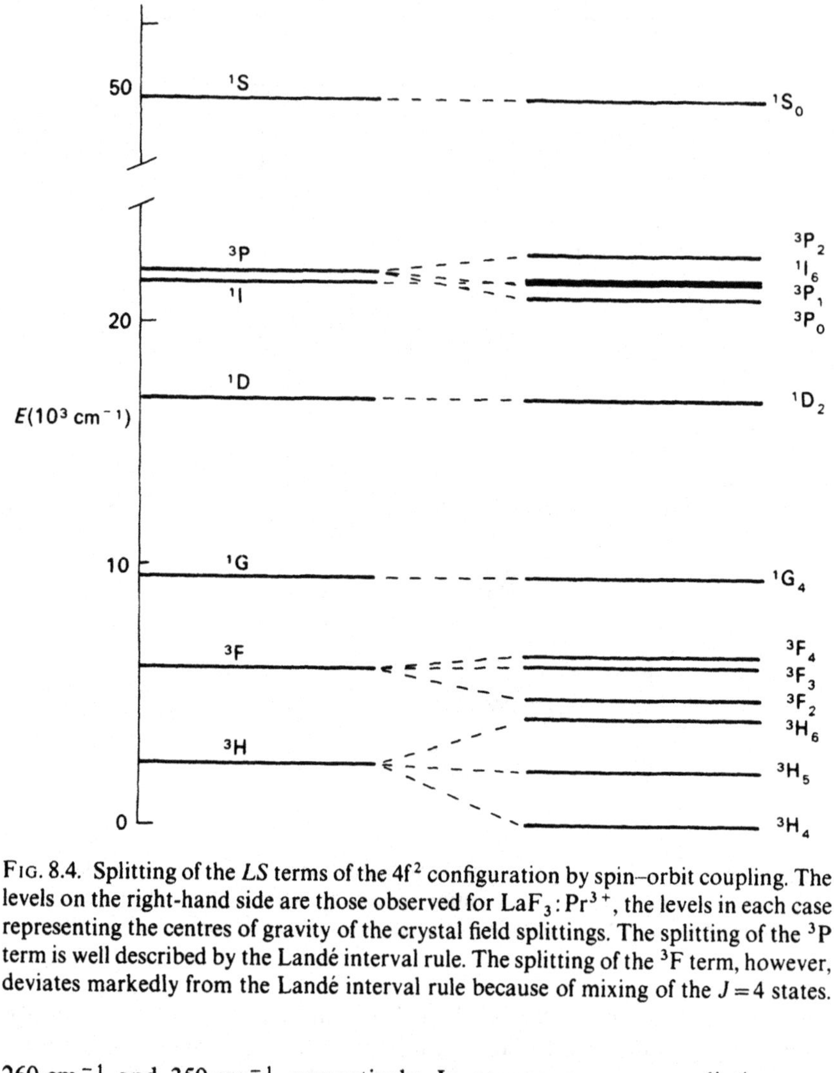

FIG. 8.4. Splitting of the *LS* terms of the $4f^2$ configuration by spin–orbit coupling. The levels on the right-hand side are those observed for $LaF_3:Pr^{3+}$, the levels in each case representing the centres of gravity of the crystal field splittings. The splitting of the ^3P term is well described by the Landé interval rule. The splitting of the ^3F term, however, deviates markedly from the Landé interval rule because of mixing of the $J=4$ states.

260 cm^{-1} and 350 cm^{-1}, respectively. In consequence, non-radiative processes tend to be rather slow, since they would involve some 15 and 11 phonons in the chloride and the fluoride, respectively. The $^3P_0 \rightarrow {}^3H_6$ *radiative* transition at 598.5 nm in $LaF_3:Pr^{3+}$ is sufficiently efficient for laser action to occur in this transition. The separation between $^3P_0$ and $^1D_2$ in $LaAlO_3:Pr^{3+}$ is around 3750 cm^{-1} and, assuming that 600 cm^{-1} phonons are effective in this material one might expect a lower-order phonon-induced non-radiative process giving a much faster non-radiative decay rate in $LaAlO_3:Pr^{3+}$ than is found in $LaF_3:Pr^{3+}$. This non-radiative decay in $LaAlO_3:Pr^{3+}$ has been measured by Delsart and Pelletier-Allard (1973) and

its temperature dependence is that expected for a six-phonon process. The transfer of $^3P_0$ excitation among Pr^{3+} ions in $LaF_3:Pr^{3+}$ and in praseodymium trifluoride has been the subject of detailed experimental and theoretical analysis (Chapter 10). Laser techniques have been employed to measure hyperfine structure in the $^1D_2$ excitated state of $LaF_3:Pr^{3+}$ (Erickson 1977a, b), where the hyperfine splitting of around 4 MHz are normally hidden by the inhomogeneous broadening of the $^1D_2$ level ($\sim$4 GHz).

Tm^{3+} ($4f^{12}$) The emitting $^3P_0$ level of Tm^{3+} is much higher in energy than the $^3P_0$ level of Pr^{3+} (see Fig. 8.1). The mixing of $J=4$ levels by spin–orbit coupling upsets the normal ordering of the low-lying ^3H_J levels. The infrared transition between $^3H_4$ and $^3H_6$ of Tm^{3+} in some host materials is sufficiently efficient for use as a laser transition.

Nd^{3+} ($4f^3$) As Fig. 8.1 shows, the energy level structure of Nd^{3+} shows a multitude of levels, many of which are emissive. The spectroscopic properties of the Nd^{3+} ion in many host materials have been intensively studied, particularly Nd^{3+} in YAG ($Y_3Al_5O_{12}$) and in various glasses where the 1.06 μm emission line is an efficient laser transition. Extensive investigations of Nd^{3+} ions in a wide variety of crystalline host materials have been undertaken by Kaminskii and his co-workers (Kaminskii 1981). The low-temperature luminescence transitions from the $^4F_{3/2}$ level to all the $4I_J$ levels of Nd^{3+} in YAG are shown in Fig. 8.5. Each of the ^4I_J free ion levels is split by the crystal field. This spectrum with its abundance of sharp lines is typical of the rich spectra found in many rare-earth systems.

Er^{3+} ($4f^{11}$) The many luminescent levels of the Er^{3+} ion are also shown in Fig. 8.1. Because of this rich level structure Er^{3+} exhibits a very detailed sharp line luminescence spectrum, and its spectroscopy in several different host crystals (e.g. $LaCl_3:Er^{3+}$ (Dieke 1968), $LaF_3:Er^{3+}$ (Carnall *et al.* 1977), $CaF_2:Er^{3+}$ (Pollack 1964)) has been studied extensively. When manganese fluoride is doped with trace amounts of Er^{3+}, efficient transfer of excitation occurs from the $^4T_1$ level of Mn^{2+} ions into the $^4F_{9/2}$ level of Er^{3+} and an intense sharp line Er^{3+} luminescence is observed (Flaherty and Di Bartolo 1973; Wilson *et al.* 1979). The transitions are shown in Fig. 8.6 along with the $^4F_{9/2}\rightarrow{}^4I_{15/2}$ and $^4I_{11/2}\rightarrow{}^4I_{15/2}$ visible spectra. No luminescence occurs from the $^4I_{9/2}$ level because of the small gap to the $^4I_{11/2}$ level. The $^4I_{13/2}\rightarrow{}^4I_{15/2}$ transition occurs in the infrared at 1.5 μm.

Pm^{3+} ($4f^4$) This highly radioactive element is of limited spectroscopic interest. However, Carnall *et al.* (1976) have published absorption and luminescence spectra of $LaCl_3:Pm^{3+}$.

Ho^{3+} ($4f^{10}$) Detailed analysis of the absorption and emission of $LaCl_3:Ho^{3+}$ have been published by Dieke and Pandey (1964), and a similar investigation of $LaF_3:Ho^{3+}$ has been carried out by Caspers *et al.* (1970). Laser action has been achieved in the $^5I_7\rightarrow{}^5I_8$ transition of Ho^{3+} in a number of lattices.

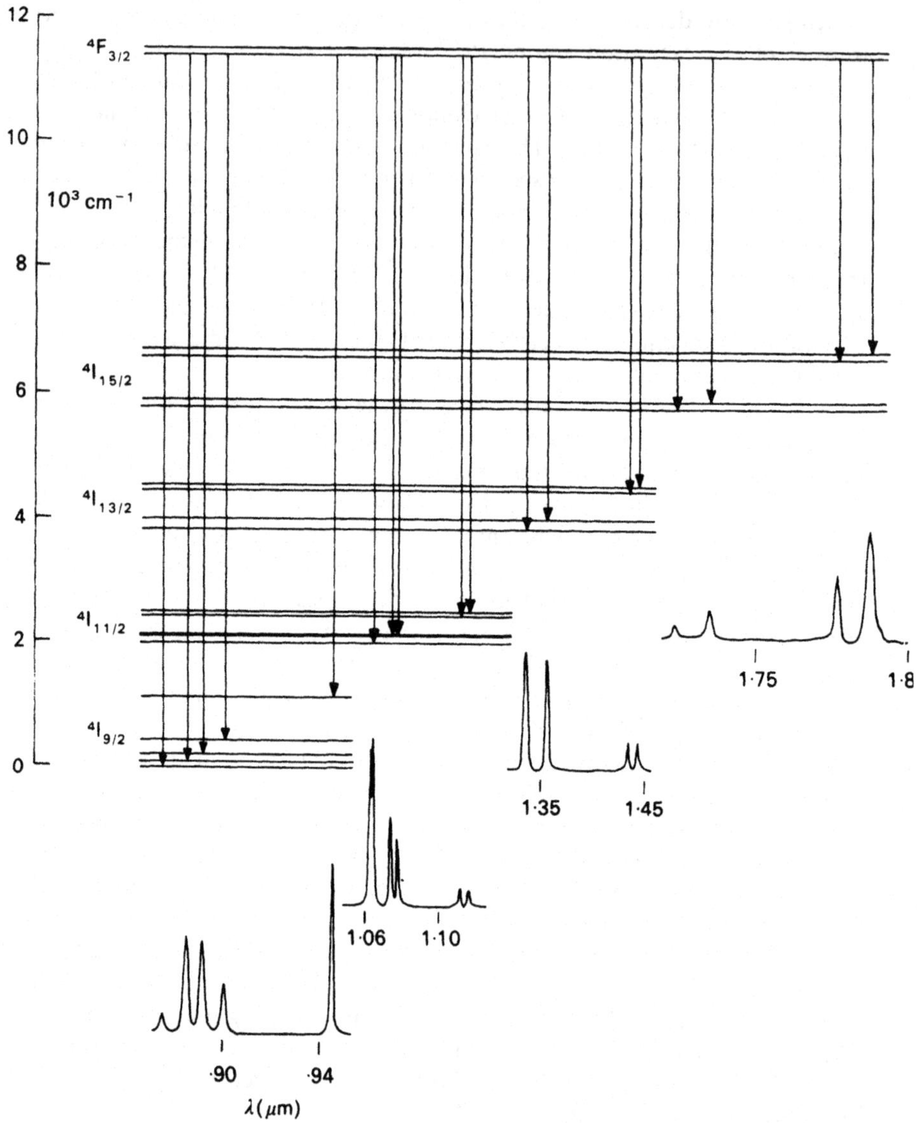

FIG. 8.5. The low-temperature luminescence spectrum from the $4F_{3/2}$ level of Nd^{3+} in YAG.

Sm^{3+} ($4f^5$) The emission and absorption spectra of Sm^{3+} in lanthanum trichloride have been measured by Dieke and his co-workers (Dieke 1968). The spectroscopy of $LaF_3:Sm^{3+}$ has been reported by Rast *et al.* (1967).

Dy^{3+} ($4f^9$) The spectrum of Dy^{3+} in lanthanum trichloride and in dysprosium trichloride has been described by Crosswhite and Dieke (1961).

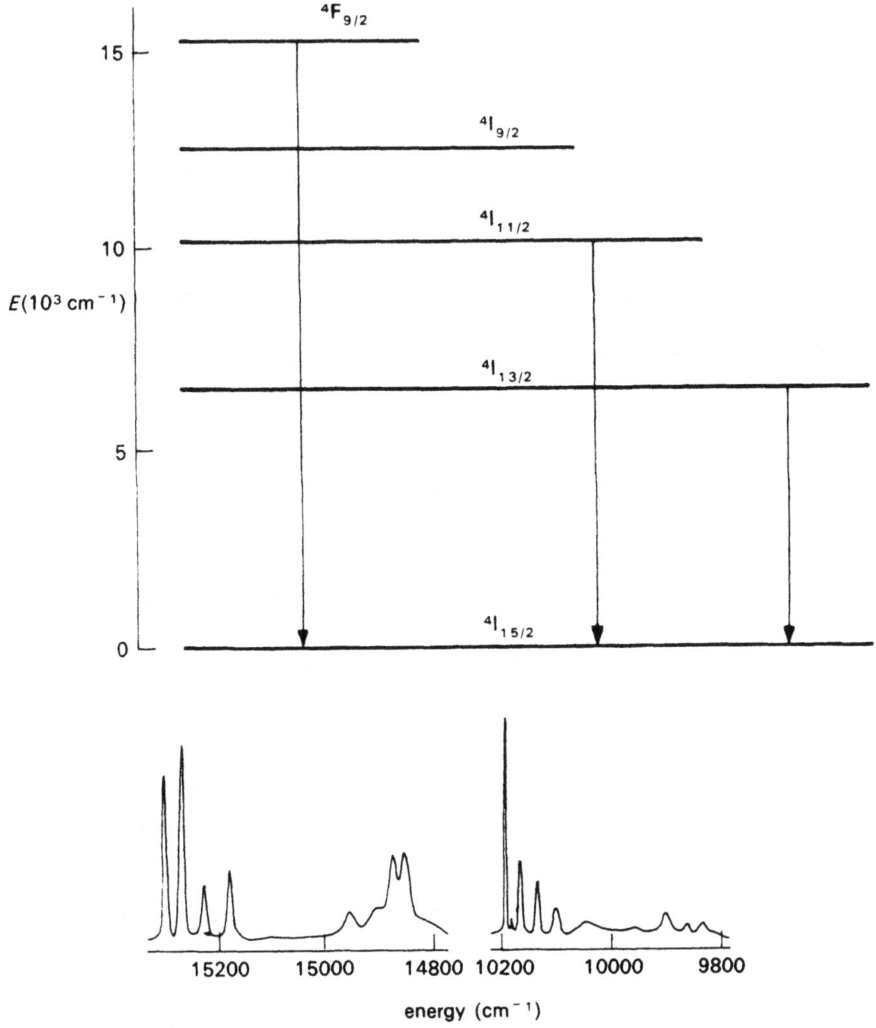

FIG. 8.6. The $^4F_{9/2} \to 4I_{15/2}$ and $^4I_{11/2} \to {}^4I_{15/2}$ emission spectra from two low-lying levels of Er^{3+} in manganese fluoride. Luminescence is not observed from the $^4I_{9/2}$ level. The emission from $^4I_{13/2}$ is observed in the infrared at around 1.5 μm.

Details of the levels of Dy^{3+}in lanthanum trifluoride are given by Carnall *et al.* (1977).

Eu^{3+}(4f^6) This is a commercially valuable luminescent ion for red-emitting phosphors in which the red emission is due to the transition from the $^5D_0$ level. The spectroscopic properties of several europium-doped phosphors have been discussed by Blasse (1978, 1984, 1988). Figure 8.7 shows the luminescence spectrum for Eu^{3+} ions in $EuZrF_6$ (Poulain *et al.* 1977) due to

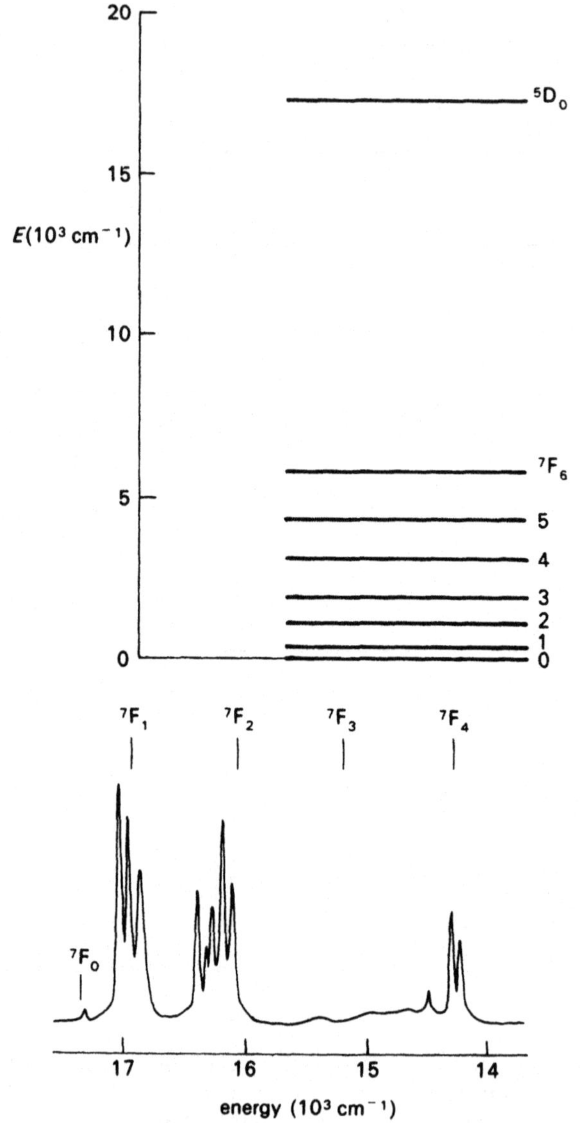

FIG. 8.7. $^5D_0 \rightarrow {}^7F_J$ transitions of Eu^{3+} ions in $EuZrF_7$ (after Poulain *et al.* 1977).

transitions from the $^5D_0$ level into some of the levels of the ^7F_J manifold. The $^5D_0 \rightarrow {}^7F_0$ transition is very weak, as is expected from the selection rules. It appears in the spectrum only because of J-mixing. The $^5D_0 \rightarrow {}^7F_1$ group of transitions is allowed by the magnetic dipole process. This is the only magnetic dipole transition, in agreement with the selection rules. All the other transitions are electric dipole in character being induced by odd-parity components of crystal field. Both $^5D_0 \rightarrow {}^7F_2$ and $^5D_0 \rightarrow {}^7F_4$ transitions are strong, whereas the $^5D_0 \rightarrow {}^7F_3$ transitions are weak, in general agreement with the Judd–Ofelt selection rules for electric dipole transitions.

$Tb^{3+}(4f^8)$ The *LS* terms are higher in energy for Tb^{3+} than for Eu^{3+} and consequently luminescence from the $^5D_4$ level to the ^7F_J ground manifold is in the green. The material $MgAl_{11}O_{19}$:Ce, Tb is an efficient green-emitting phosphor in which the Tb^{3+} ions are excited by energy transfer from the Ce^{3+} ions (Blasse 1984).

$Gd^{3+}(4f^7)$ The Dieke diagram (Fig. 8.1) shows that this ion has no absorption or luminescence spectra in the visible or infrared. However, absorptions and emission transitions do occur in the near ultraviolet. A detailed spectroscopic analysis of the optical properties of $GdCl_3.6H_2O$ has been published by Dieke and Leopold (1957). Strong emission at around 300 nm is observed from the $^6P_{7/2}$–$^8S_{7/2}$ transition. Gd^{3+} is weakly coupled to the lattice and the spectra consist of quite sharp lines even at room temperature. Gd^{3+} ions play an important supportive role in some efficient phosphor materials. For example, GdF_3: Ce, Tb is an efficient green phosphor. Excitation of the Ce^{3+} ion is rapidly transferred nonradiatively to the Gd^{3+} ions of the host crystal, the 5d–4f transition on the Ce^{3+} ion overlapping absorption transitions on the Gd^{3+} ion. The excitation migrates rapidly through the Gd^{3+} ions until it is transferred to Tb^{3+} ions from which the green luminescence is emitted (Blasse 1988). (The phenomenon of excitation transfer is discussed in Chapter 10.)

8.4 Emission from divalent rare-earth ions

The $4f^{(n-1)}5d$ states of the divalent rare-earth ions interact strongly with the lattice and, since they occur at relatively low energy, they overlap many of the $4f^n$ levels. Transitions between the $4f^n$ and $4f^{(n-1)}5d$ levels are intense and broad (see e.g. Anderson 1974), and generally do not display the same rich spectrum of sharp lines as is observed from the trivalent rare-earth ions. Of particular interest because of their luminescence properties are Sm^{2+}, Dy^{2+}, Eu^{2+}, and Tm^{2+}. $Sm^{2+}(4f^6)$ emits strongly at wavelengths near 700 nm, the emission intensity being due to the strong broadband absorption transitions. The emission transitions originate on either a low-lying $4f^55d$ level or on the $^5D_0(4f^6)$ level—depending on the host material—and terminates on the $^7F_1$ level. Laser action has been achieved in this transition. Sharp line emission at *c.* 2.4 μm arising from the $^5I_7 \rightarrow {}^5I_8$ transition of Dy^{2+} in calcium fluoride has been used as the basis for a single-frequency laser.

The Eu^{2+} ion has the $^8S_{7/2}$ ground level of the $4f^7$ configuration. Emissive transitions occur from the next highest levels derived from the $4f^65d$ states. The positions of these energy levels and, consequently, the wavelength of the emission vary with the host crystal. Some examples of the emission colours are $Sr_2P_2O_7:Eu^{2+}$ (violet), $BaAl_{12}O_{19}:Eu^{2+}$ (blue), $SrAl_2O_4:Eu^{2+}$ (green), $Ba_2SiO_5:Eu^{2+}$ (yellow). Blue-emitting phosphors based on Eu^{2+} ions are commercially important (Blasse 1984).

Tm^{2+}, like Yb^{3+}, has but two $4f^{13}$ energy levels, $^2F_{7/2}$ and $^2F_{3/2}$ which are separated by approximately 10 000 cm^{-1}. The $^2F_{5/2}$ level can be easily populated by pumping in the low-lying strong $4f^{13} \rightarrow 4f^{12}5d$ broadband absorption transitions, and strong luminescence occurs in the $^2F_{5/2} \rightarrow {}^2F_{7/2}$ transition. Laser action has been achieved on this transition in $CaF_2:Tm^{2+}$.

8.5 Charge transfer absorption bands

Our discussion so far has considered only those transitions which take place between levels on isolated ions. Many rare-earth-doped materials exhibit an intense broad absorption band with its peak in the ultraviolet, assigned to transitions in which electrons are excited from a neighbouring anion orbital onto the dopant ion. Such absorption bands are known as *charge transfer bands*. (Similar absorption bands occur for transition-metal-doped crystals.) These transitions have an important influence on the luminescence properties of some rare-earth-doped solids (Blasse 1978). The intensity of the charge transfer band arises from the movement of electronic charge across a typical interatomic distance so producing a large transition dipole moment and a concomitant large oscillator strength for the absorption process. A detailed discussion of transfer spectra has been given by both McClure (1975) and Jørgensen (1970).

8.6 Defect site spectroscopy

Not all impurity dopants in host crystals occupy substitutional lattice sites. Indeed, charge compensation required when a trivalent dopant ion substitutes for a divalent cation on a regular crystal site may take the form of local or remote lattice vacancies, interstitial ions or the incorporation of other impurity cations and anions, in order to maintain overall charge neutrality. Furthermore, unintentional lattice defects are always present to some extent even in undoped crystals. If such defects occur in the immediate environs of dopant ions they perturb the local site symmetry and introduce new and distinct features in the spectroscopic properties of the dopant ion. In addition, dopant ions may form clusters in the host crystal, which may have their own spectral characteristics. For example, Er^{3+} ions in calcium fluoride occupy up to 20 distinct lattice sites (Tallant and Wright 1975). By combining narrow-band tuned laser excitation and high-resolution spectroscopy, Wright (1977) demonstrated that such distinct features may be used to probe the thermodynamics of the defect–defect interactions in doped solids.

8.7 Spectroscopy of actinide ions

With the exception of uranium the ubiquitous radioactivity of actinides has restricted study of their spectroscopy to a small number of specially equipped laboratories (Pappalardo 1978). Analysis of the spectra is complicated by the stronger interaction of the 5f electrons with the host environment, compared to the 4f electrons of the rare-earth ions. The actinide ions are also likely to form molecular complexes, which may also give rise to greater spectral complexity. Figure 8.8 shows the energy level assignments for trivalent

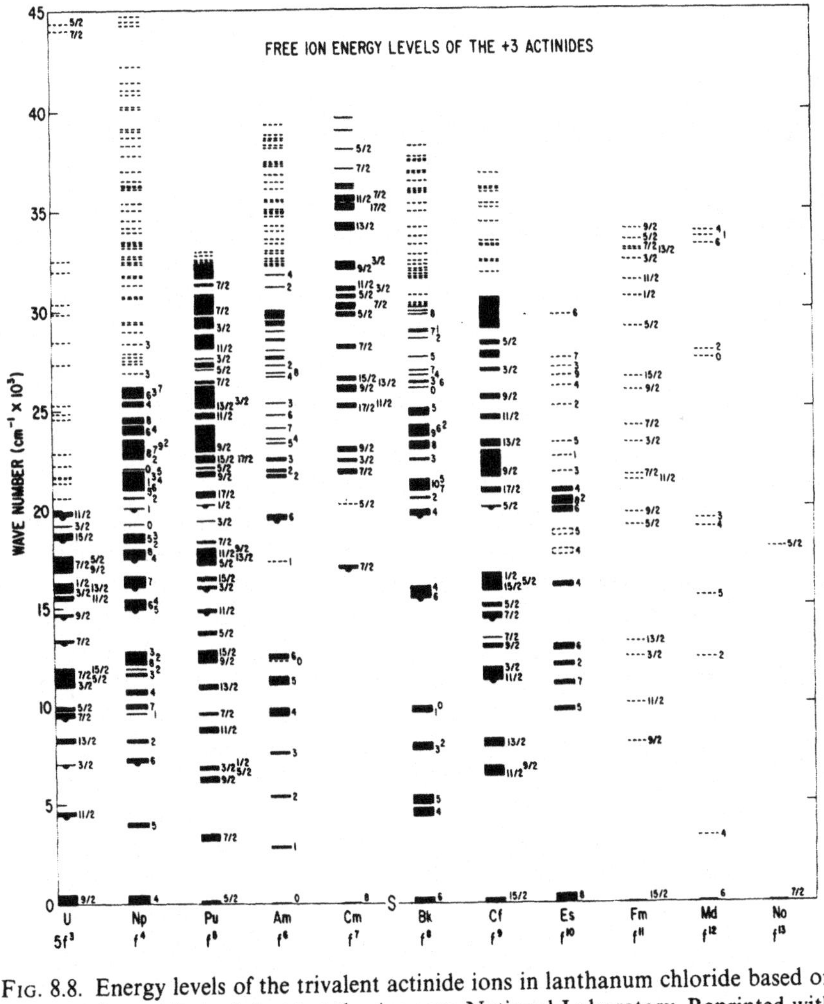

FIG. 8.8. Energy levels of the trivalent actinide ions in lanthanum chloride based on the extensive work carried out at the Argonne National Laboratory. Reprinted with the permission of Professor H. M. Crosswhite.

actinide ions in lanthanum trichloride, based on the extensive studies of Carnall, Crosswhite, and co-workers. In contrast to the other actinide ions, uranium luminescence has been studied for over a hundred years. Indeed, Becquerel discovered radioactivity while investigating the luminescence properties of uranium salts. The optically active uranium centre can be any of the ions U^{3+}, U^{4+}, U^{5+}, or U^{6+} in the form of molecular ion groups such as UO_2^{2+}, UO_4^{2-}, UO_6^{6-}, $UO_2Cl_4^{2-}$, etc. The spectroscopy of uranyl compounds has been described by Rabinowitch and Belford (1964). More recent work is detailed by Paszek 1978 (U^{3+}); Genet *et al.* 1977 (U^{4+}); Lupei *et al.* 1976 (U^{5+}); and Blasse *et al.* 1979, Jørgenson 1979 (U^{6+}).

8.8 Rare-earth ions in glass

Glass may be regarded as a random network of polyhedra, such as silicate tetrahedra in silicate glasses. The more useful glasses contain additional compounds, e.g. soda (Na_2O) or lime (CaO), which disrupt the network of polyhedra and create sites at which larger ions, e.g. rare-earth ions, may be incorporated into the structure. The arrangements of anions around the dopant rare-earth ions will, in general, have no particular order; this is to be contrasted with the situation in crystals where the sites available for dopants are, in principal at least, identical. Thus, instead of the sharp line transitions ($\Gamma \simeq 1\ cm^{-1}$) typical of rare-earth ions in crystals, spectral lines of rare-earth ions in glasses are inhomogeneously broadened ($\Gamma \simeq 100\ cm^{-1}$), the resulting broad lines being composites of the sharp line emissions at different frequencies from dopant ions in distinct sites.

Because glass melts at a lower temperature than the corresponding crystalline material it can be cast and worked into large pieces with far greater ease than is the case for the corresponding crystalline material. In addition, many rare-earth-doped glasses are efficient emitters of luminescence: Nd^{3+}-doped glass is an excellent laser medium, the success of which has been one factor in recent years to have stimulated more intensive research into the spectroscopic properties of doped glasses. Although the large inhomogeneous broadening masks the spectroscopic fine structure when conventional techniques (Sections 6.1–6.5) are used, laser excitation spectroscopy can be employed to obtain quite detailed information on the microscopic arrangement of ions around the dopant sites (Brecher and Riseberg 1980). The principal technique for this study is FLN (Section 6.7), in which a narrowband laser beam selectively excites a small subset of the dopant ions, those whose narrow absorption transitions are in resonance with the laser, resulting in a sharp luminescence signal from the excited ions. The application of FLN to the study of rare-earth ions in glasses is reviewed by Weber (1981) and Yen (1986).

FLN, absorptive hole burning (Macfarlane and Shelby 1983), and cumulative photon echoes (Hesselring and Wiersma 1979) permit accurate measurements to be made of the *homogeneous* broadening of transitions in glass. A surprising discovery in such studies (Selzer *et al.* 1976, Hegarty and Yen 1979) was the observation at low temperatures of a broadening process at low temperatures, varying approximately linearly with temperature, which apparently has no counterpart in crystalline materials. This particular broadening mechanism is attributed to the interaction of the dopant ion with low frequency 'tunnelling modes' that are a feature of amorphous materials (Weber 1987). These tunnelling modes, a local arrangement of ions that can switch between two quasi-equilibrium configurations which are adjacent in energy, have been predicted both by Anderson *et al.* (1972) and Phillips (1972) to explain the anomalously large values of specific heat capacity and thermal conductivity at low temperatures in amorphous materials.

9

Optical spectroscopy of transition metal ions in solids

THE transition metals (iron group) occur after the element calcium (atomic number 40) in the fourth period of the periodic table when the 3d shell is being filled. In ionic solids such elements lose the outer 4s electrons and possibly some 3d electrons in forming ionic bonds, and their resulting electronic configurations are $1s^2 2s^2 2p^6 3s^2 3p^6 3d^n$ where $n < 10$. Some common valence states of these ions are given in Table 9.1. Ions with incomplete 3d shells have a number of low-lying energy levels between which optical transitions may occur. Since the optically-active 3d electrons are outside the ion core they interact strongly with electric fields of nearby ions and in consequence the associated crystal field effects are stronger than those experienced by rare-earth ions. Similar series of elements occur in the fifth period, where the 4d shell is being filled (the palladium group), and in the sixth period, where the 5d shell is being filled (the platinum group). Compared with the 3d electrons the 4d and 5d electrons are less tightly bound to the parent ion, and *charge transfer transitions* (in which an electron is transferred between a cation and neighbouring ligand) occur rather easily. Such transitions give rise to strong absorption bands. For iron-group ions the charge transfer bands occur in the ultraviolet region. The ions of the palladium and platinum groups have lower-lying charge transfer states and the resulting intense visible region absorption bands may overlap spectra due to low-lying

Table 9.1
Number of electrons (n) in the 3d shell of commonly found transition metal ions

		Ti^{3+}	V^{4+}	1
		V^{3+}	Cr^{4+}	2
	V^{2+}	Cr^{3+}	Mn^{4+}	3
	Cr^{2+}	Mn^{3+}		4
	Mn^{2+}	Fe^{3+}		5
	Fe^{2+}	Co^{3+}		6
Fe^{+}	Co^{2+}	Ni^{3+}		7
Co^{+}	Ni^{2+}			8
Ni^{+}	Cu^{2+}			9

crystal field levels. This chapter discusses only crystal field spectra of iron-group ($3d^n$) ions.

9.1 Energy levels and radiative transitions on transition metal ions

The energy levels of transition metal ions in octahedral and tetrahedral crystal fields were examined in Chapters 2 and 3, and the crystal field energies plotted for each $3d^n$ configuration. The effects of weaker lower-symmetry crystal fields and of spin–orbit coupling can be treated afterwards in each specific case by perturbation theory. If the arrangement of neighbouring ions around the transition metal ion has inversion symmetry the crystal field potential energy term is of even parity, and so also are the low-lying eigenstates of the transition metal ion. Optical transitions between such states can occur only by magnetic dipole processes. However, if there is an odd-parity component of the crystal field, e.g. as in a tetrahedral crystal field, this can mix some higher-lying odd-parity states from an odd-parity configuration (e.g. $3d^{n-1}4p$) into the even-parity $3d^n$ states. In consequence, electric dipole transitions can occur between the low-energy crystal field states. As a result, even though the amount of the odd-parity admixture may be small the electric dipole process dominates the spectroscopic behaviour of the ion, since it has much the greater transition probability. The effect of such admixture on the energy levels is very small and is usually neglected. Even when the arrangement of neighbouring ions has inversion symmetry lattice vibrations can introduce odd-parity crystal field terms and so induce electric dipole vibronic processes (Section 5.6). The selection rules for radiative transitions can be evaluated using the Wigner–Eckart theorem once the symmetry and crystal field states have been established (Section 4.6).

9.2 Non-radiative transitions on transition metal ions in solids

Because they are strongly coupled to the neighbouring ligands, transition metal ions are more strongly affected by static lattice distortions and by lattice vibrations than are the rare-earth ions. Consequently non-radiative transitions are usually stronger than for rare-earth ions. An analysis of non-radiative decay processes on transition metal ions was presented in Chapter 5. As a crude generalization we can say that when the energy gap between two adjacent levels is less than about $10\,000\ cm^{-1}$ non-radiative effects are likely to be effective. However, we must note that the rate of non-radiative decay across this gap depends not only on the gap size but also on the lattice distortion (characterized by the Huang–Rhys parameter, S) which occurs when the ion changes its electronic state, as well as on the phonon energies of the host lattice. Thus we find that $Ti^{2+}(3d^2)$, $Co^{2+}(3d^7)$, and $Ni^{2+}(3d^8)$ in some host materials can emit luminescence at wavelengths greater than 1 μm.

9.3 Spectroscopy of Cr^{3+} ($3d^3$) ions

This ion, particularly in the material ruby ($Al_2O_3:Cr^{3+}$), has been exhaustively studied by spectroscopists for over a hundred years. In a treatise on light published in 1867 Edmond Becquerel devoted almost a chapter to the optical spectroscopy of ruby. (A short account of the early work on ruby was given by Mollenauer (1965).) Deutschbein (1932) made an extensive survey of Cr^{3+} luminescence in many materials. The Cr^{3+} ion has strong visible absorption bands, and the luminescence transitions can be so sharp and of such high quantum efficiency that trace amounts of Cr^{3+} are easily detected and identified in many solids.

9.3.1 Energy levels of Cr^{3+} ions

The Cr^{3+} free-ion energy levels are known from experiment (Moore 1950); only limited agreement is found with theoretical calculations (Wood 1969). Calculations usually start by neglecting both crystal field and spin–orbit effects and taking into account only the Coulomb interaction between the 3d electrons. The resulting energies of the free ion *LS* terms, expressed in terms of the Racah parameters, *A*, *B*, *C* (Griffith 1961) are listed in Table 9.2. Since we are interested only in energy differences the common term $3A$ in these expressions is ignored, and only two Racah parameters, *B* and *C*, are needed to categorize the free-ion levels. For pure d electrons we expect the ratio C/B to be about 4.0, and independent of both atomic number and of *n*. In many crystals Cr^{3+} ions occupy octahedral or near octahedral cation sites. The octahedral crystal field, characterized by the parameter Dq, must now be taken into account. The splittings of each free ion level by an octahedral crystal field are also indicated in Table 9.2. The new crystal field levels are

Table 9.2

Free-ion *LS* term energies for d^3 configuration (Griffith 1961) and splitting of free-ion levels in an octahedral crystal field.

LS term energy	Splitting into octahedral field terms
$^4F = 3A - 15B$	$^4A_2 + {}^4T_1 + {}^4T_2$
$^4P = 3A$	$^4T_1$
$^2H = 3A - 6B + 3C$	$^2E + 2\,{}^2T_1 + {}^2T_2$
$^2P = 3A - 6B + 3C$	$^2T_1$
$^2G = 3A - 11B + 3C$	$^2A_1 + {}^2E + {}^2T_1 + {}^2T_2$
$^2F = 3A + 9B + 3C$	$^2A_2 + {}^2T_1 + {}^2T_2$
$a^2D = 3A + 5B + 5C + \alpha^{\dagger}$	$^2E + {}^2T_2$
$b^2D = 3A + 5B + 5C - \alpha^{\dagger}$	$^2E + {}^2T_2$

$^{\dagger}\alpha = (193B^2 + 8BC + 4C^2)^{\frac{1}{2}}$

classified by irreducible representations of the O_h group. The crystal field splitting is shown in the energy level diagram in Fig. 3.20. The ^4F ground free ion state splits into $^4A_2$, $^4T_2$, and $^4T_1$ states. The free ion ^4P state is unsplit by the octahedral field and transforms as the $^4T_1$ irreducible representation of the O_h group. Hence there are two $^4T_1$ states, one from ^4F and one from ^4P, and there are matrix elements of the octahedral crystal field which connect the two $^4T_1$ states. The lowest-lying spin doublet term, ^2G, yields $^2A_1$, ^2E, $^2T_1$, and $^2T_2$ terms; of these the ^2E level lies lowest (Fig. 3.20).

Instead of starting with free-ion states we can calculate the effect of the octahedral field on the *individual* d electrons and then add the Coulomb interaction between the 3d electrons as a perturbation. In this strong field approach there are four different crystal field states; t_2^3, t_2^2e, t_2e^2, and e^3, at crystal field energies, 0, $10Dq$, $20Dq$, $30Dq$, each with a large energy degeneracy. (We measure energies from the lowest level.) When account is taken of the Coulomb interaction between electrons this degeneracy is partially removed. There are diagonal and off-diagonal matrix elements between these states. The matrix elements of the Coulomb interaction in the strong field scheme for all d^n configurations were calculated by Tanabe and Sugano (1954), and these are tabulated in the text by Sugano *et al.* (1970). Some of the matrix elements of the octahedral crystal field and of the Coulomb interaction for the $3d^3$ system are given in Table 9.3. There is a single $^4A_2$ state, which has only a diagonal matrix element with energy $-15B$. There are, similarly, only single $^4T_2$, $^2A_1$, and $^2A_2$ states whose energies are given in Table 9.3. The energies of the two $^4T_1$ states derived from ^4P and ^4F free-ion states are obtained by diagonalization of the 2×2 matrix shown in Table 9.3. Similarly, there are four ^2E states, and the 4×4 matrix shown in Table 9.3 must be diagonalized to obtain the ^2E energy values. The lowest energy ^2E state is important since strong sharp luminescence originates on this state for many Cr^{3+} systems. Matrices for the $^2T_1$ and $^2T_2$ states are given in the text by Sugano *et al.* (1970). The energies of the various octahedral crystal field levels of all $3d^3$ electronic systems depend on the parameters Dq, B, C. Values of these parameters are obtained by comparing the experimental energy levels with the calculated energies. The measured value of Dq varies with the environment of the Cr^{3+} ion. There is a pronounced variation in Dq with the neighbouring ligand type (Jørgensen 1960), the order of increasing Dq being $I^- < Br^- < Cl^- < F^- < H_2O < O^{2-} < S^{2-}$. The measured value of B for Cr^{3+} is found to be reduced from its free-ion value (*the nephelauxetic effect*, Jørgensen 1960) but the trends in the variation with crystalline environment are not so clear (Wood 1969).

The Tanabe–Sugano diagrams for $3d^n$ ions were reproduced in Figs. 3.18–3.24. In constructing such diagrams for specific ions an approximate value of C/B is assumed and since there are only two variable parameters, Dq and B, E/B may be plotted as a function of Dq/B. Such energy-level diagrams

Table 9.3

Matrix elements of the crystal field and Coulomb interaction for some low-lying d^3 states

	$^2E(a^2D, b^2D, {}^2G, {}^2H)$		
t_2^3	$t_2^2(^1A_1)e$	$t_2^2(^1E)e$	e^3
$-6B+3C$	$-6\sqrt{2}B$	$-3\sqrt{2}B$	0
$-6\sqrt{2}B$	$10Dq+8B+6C$	$10B$	$\sqrt{3}(2B+C)$
$-3\sqrt{2}B$	$10B$	$10Dq-B+3C$	$2\sqrt{3}B$
0	$\sqrt{3}(2B+C)$	$2\sqrt{3}B$	$30Dq-8B+4C$

$^4T_1(^4P, {}^4F)$	
$t_2^2(^3T_1)e$	$t_2e^2(^3A_2)$
$10Dq-3B$	$6B$
$6B$	$20Dq-12B$

$t_2^3\,{}^4A_2(^4F)$	$-15B$
$t_2^2(^3T_1)e\,{}^4T_2(^4F)$	$10Dq-15B$
$t_2^2(^1E)e\,{}^2A_1(^2G)$	$10Dq-11B+3C$
$t_2^2(^1E)e\,{}^2A_2(^2F)$	$10Dq+9B+3C$

are very useful in the initial comparison with experimental spectra. Once the various spectra have been identified a more refined comparison can then be made between theory and experiment, and more accurate values of *Dq*, *B*, and *C* deduced. The Tanabe–Sugano diagram constructed in this manner for $3d^3$ ions assuming $\gamma = 4.8$ (Macfarlane 1967) is shown in Fig. 9.1; the vertical broken line drawn at $Dq/B = 2.8$ is appropriate for Cr^{3+} in ruby. Values of Dq/B for this ion fall in the range of 1.5–3.0. When the value of Dq/B is less than about 2.3 the $^4T_2$ state is the lowest excited state, whereas for larger values of Dq/B the lowest excited state is ^2E. This forms a convenient means of distinguishing between Cr^{3+} ions in weak crystal fields, for which the $^4T_2$ level is below ^2E, and Cr^{3+} ions in strong crystal fields which have the ^2E level below $^4T_2$. The value of Dq/B at the crossover point between ^2E and $^4T_2$ depends slightly on the value of *C*.

The σ-polarized absorption spectrum of ruby at wavelengths in the visible region is compared with the Tanabe–Sugano diagram in Fig. 9.1. The expected values of the excited levels are seen to coincide with the absorption transitions. The absorption spectrum is dominated by the two strong spin-allowed transitions $^4A_2(t_2^3) \rightarrow {}^4T_2(t_2^2e)$, $^4T_1(t_2^2e)$. Since the initial and final states of these transitions are derived from different crystal field orbitals (t_2^3 and t_2^2e, respectively) the energy separations are very sensitive to the value of *Dq* as Fig. 9.1 shows. In consequence the transitions are broad and charac-

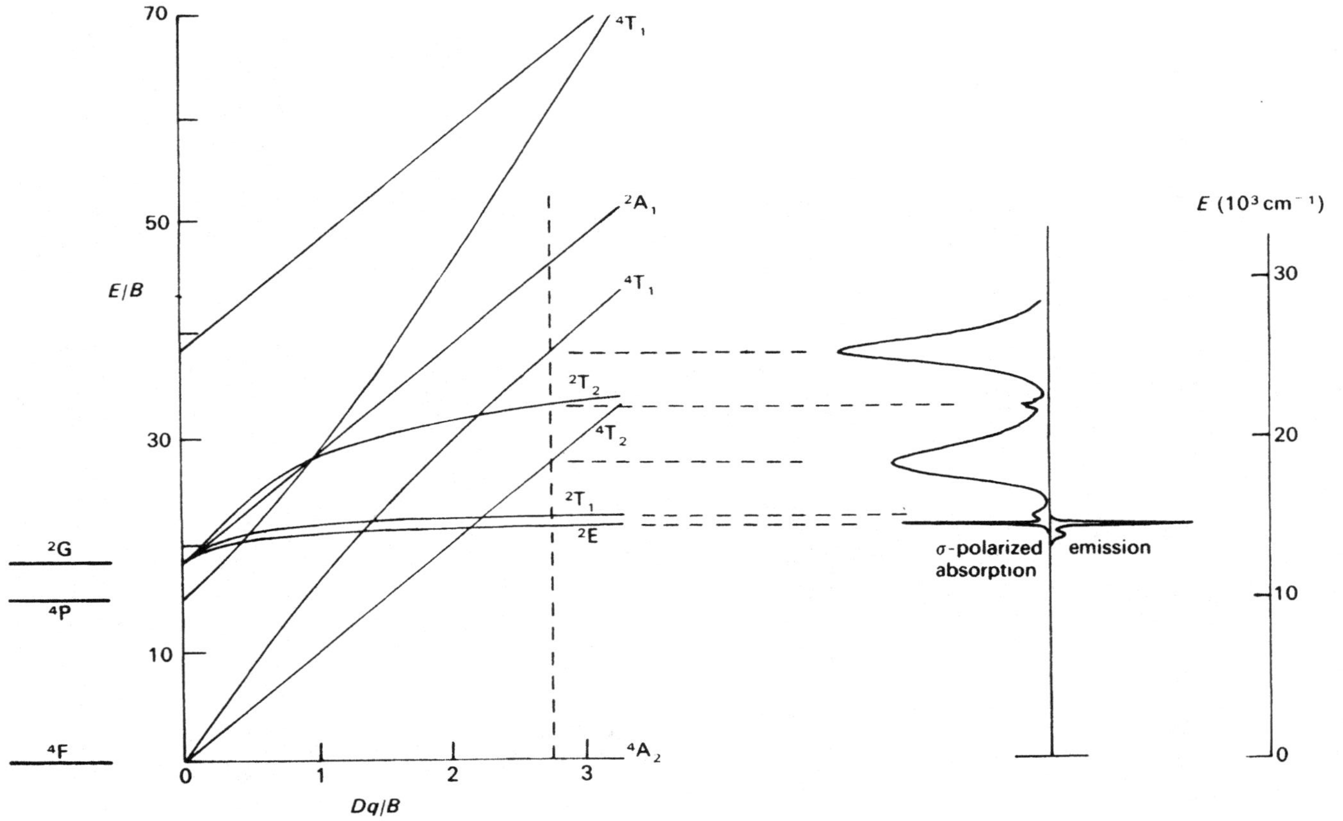

FIG. 9.1. Plot of E/B against Dq/B for $\gamma = 4.8$, a value appropriate for Cr^{3+} in aluminium oxide. The vertical broken line is drawn for $Dq/B = 2.8$, the ruby value. The observed levels of ruby (ignoring splittings due to trigonal crystal field) are shown on the right, and are seen to be in good agreement with the predictions of the crystal field analysis. At low temperatures luminescence is observed only from the ^2E level.

terized by large values of the Huang–Rhys parameter. In ruby these strong absorptions occur in the yellow-green and blue regions, hence the red colour of the material. For most chromium-doped oxides the value of Dq/B is smaller than for ruby, the absorption bands are more usually in the orange-red and blue, and the colour of the material is green. Examples are emerald ($Be_3Al_2(SiO_6)_3:Cr^{3+}$), alexandrite ($BeAl_2O_4:Cr^{3+}$), and all chromium-doped oxide glasses. Absorption transitions from $^4A_2$ to the low-lying doublet states, ^2E, $^2T_1$, $^2T_2$ are spin-forbidden and are relatively weak. These doublet states as well as the $^4A_2$ ground state are formed from the same t_2^3 crystal field orbitals, and for Dq/B values greater than around 1.5, the spin doublet energy separations from the $^4A_2$ ground state do not vary greatly with Dq. In consequence these transitions have small Huang–Rhys parameters and their spectra are dominated by sharp zero-phonon lines. Spin–orbit coupling and, if they are present, lower-symmetry crystal fields split some of the levels shown in Fig. 9.1. The centre of gravity of the split transition is then taken to represent the octahedral field levels and these are used to determine the parameters Dq, B, C.

Accurate values of these parameters are determined using the absorption data. The magnitude of Dq is obtained directly from the energy corresponding to the peak of the $^4A_2 \rightarrow {}^4T_2$ absorption band which, as Table 9.3 shows, is equal to 10 Dq. The second strong absorption band is due to a transition to the lower $^4T_1$ level, the energy of which is calculated by diagonalization of the appropriate 2×2 matrix (Table 9.3) and depends on both Dq and B. If we write ΔE as the energy difference between $^4T_2$ and the lowest $^4T_1$ state, i.e. the energy separation between the two strong visible absorption bands in Fig. 9.1, then the diagonalization procedure gives the following relationship between B, Dq and ΔE:

$$\frac{B}{Dq} = \frac{(\Delta E/Dq)^2 - 10(\Delta E/Dq)}{15(\Delta E/Dq - 8)}. \tag{9.1}$$

By substituting the measured value of ΔE and the calculated Dq value into eqn (9.1) we obtain the value of B. Determination of C requires a measurement of the position of one of the orbital doublets, all of which have a dependence on C. The lowest doublet (i.e. ^2E) is the easiest to measure experimentally and is generally used to determine C. To calculate the energy, $E(^2E)$, of this state, we diagonalize the 4×4 ^2E matrix in Table 9.3 and select the lowest-energy eigenvalue. The $E(^2E)$ value depends on the values of Dq, B, C. For the range of values $1.5 < Dq/B < 3.5$ and $3 < C/B < 5$, $E(^2E)$ is determined to vary *approximately* as

$$E(^2E)/B \simeq 3.05(C/B) + 7.90 - 1.80(B/Dq) \tag{9.2}$$

which gives a value of C to within 0.5 per cent of the value obtained by exact diagonalization. Having found accurate values of Dq, B, and C the positions of all the other levels can be calculated by diagonalizing the appropriate matrices. A full theoretical energy-level diagram can then be drawn for all the states of these ions.

As Fig. 9.1 shows, the energy separations of adjacent excited states of Cr^{3+} are not larger than about 5000 cm^{-1}. Consequently the higher excited states of Cr^{3+} ions decay non-radiatively to successively lower levels until the lowest excited state is reached. For Cr^{3+} ions in oxide crystals the gap from the lowest excited state to the ground state is larger than 12 000 cm^{-1}, which is too large for effective non-radiative decay, and radiative decay from the lowest excited state occurs generally with high quantum efficiency. When the Cr^{3+} ions are in high-field sites the lowest excited state is ^2E and the luminescence is characterized by sharp zero-phonon lines accompanied by one-phonon vibrational sidebands. However, if the Cr^{3+} ions are in low-field sites the $^4T_2$ state is lowest, the $^4T_2 \rightarrow {}^4A_2$ transition is characterized by a large Huang–Rhys parameter and is observed as a broad luminescence band.

9.3.2 Spectroscopy of Cr^{3+} ions in magnesium and aluminium oxides

When Cr^{3+} ions substitute for Mg^{2+} in magnesium oxide they may occupy high-field sites which have perfect octahedral symmetry. The ^2E level of such ions, which lies below $^4T_2$, is the level on which the luminescence originates. Since neither the ^2E level nor the $^4A_2$ level is split by spin–orbit coupling the $^2E \rightarrow {}^4A_2$ luminescence is characterized by one zero-phonon line, the *R-line*. Furthermore, the Mg^{2+} sites in the magnesium oxide lattice have inversion symmetry and consequently the pure electronic $^2E \rightarrow {}^4A_2$ transition is induced by a magnetic dipole process. This can be demonstrated by comparing the Zeeman patterns of the R-line luminescence under π-, σ-, and α-polarizations (see Section 6.3, Fig. 6.11). The absolute magnitude of the magnetic dipole absorption strength for the R-line was calculated accurately by Macfarlane (1970) and is in good agreement with the experimental value (Larkin *et al.* 1973).

The nature of the $^2E \rightarrow {}^4A_2$ transition for octahedral Cr^{3+} has been analysed using a simple perturbation approach (Sugano *et al.* 1960). In the absence of spin–orbit coupling this transition is spin-forbidden. Spin–orbit coupling, however, mixes the ^2E and $^4T_2$ states, while the magnetic dipole operator mixes $^4T_2$ and $^4A_2$ states. The matrix element of the $|^2E\phi M_s\rangle \rightarrow |^4A_2, M_s'\rangle$ transition is

$$\sum_{\psi, M_s''} \frac{\langle {}^4A_2 M_s' | \boldsymbol{\mu}_m \cdot \hat{\boldsymbol{B}} | {}^4T_2 \psi M_s'' \rangle \langle {}^4T_2 \psi M_s'' | \mathcal{H}_{so} | {}^2E \phi M_s \rangle}{E(^2E) - E(^4T_2)} \tag{9.3}$$

where ϕ and ψ are the orbital functions of the ^2E and $^4T_2$ states, respectively, and $^4A_2$ is an orbital singlet. The spin functions are denoted by M_s, M_s'', M_s', $\boldsymbol{\mu}_m$ is the magnetic dipole operator, $\Sigma(e/2m)(\boldsymbol{l}_i + 2\boldsymbol{s}_i)$, and $\hat{\boldsymbol{B}}$ is a unit vector denoting the direction of the magnetic field of the radiation. Only the $\boldsymbol{l}_i$ part

Table 9.4

Classification of some common functions as basis functions belonging to irreducible representations of the octahedral group. The functions are unnormalized. l_x, l_y, l_z are the components of the orbital angular momentum operator.

(i) *Tetragonal and cubic axes*	
A_{1g}	$x^2+y^2+z^2$
A_{2u}	xyz
$E_g u$, $E_g v$	$(3z^2-r^2)$, (x^2-y^2)
$T_{1u}\alpha$, $T_{1u}\beta$, $T_{1u}\gamma$	x, y, z
$T_{1g}\alpha$, $T_{1g}\beta$, $T_{1g}\gamma$	l_x, l_y, l_z
$T_{2g}\xi$, $T_{2g}\eta$, $T_{2g}\zeta$	yz, zx, xy
(ii) *Trigonal axes*	
$T_{1u}a_+$, $T_{1u}a_-$, $T_{1u}a_0$	Y_1^1, Y_1^{-1}, Y_1^0
$T_{1u}a_+$, $T_{1u}a_-$, $T_{1u}a_0$	$-\dfrac{x+iy}{\sqrt{2}}, \dfrac{x-iy}{\sqrt{2}}, z$
$(T_{1u}a_- - T_{1u}a_+)/\sqrt{2}$	x
$i(T_{1u}a_+ + T_{1u}a_-)/\sqrt{2}$	y

of the magnetic dipole operator, an operator of type T_{1g} (Table 9.4), can couple the $^4A_2$ state to the $^4T_2$ state. Using the Wigner–Eckart theorem we can evaluate the relative values of the matrix elements of the magnetic dipole operator connecting $|^4A_2M_s'\rangle$ with $|^4T_2\psi M_s''\rangle$. Since spin is not involved there is no change in M_s, hence $M_s''=M_s'$. The ^2E state has two degenerate orbital states, and these can be taken as u, v (cubic bases) or u_+, u_- (trigonal bases) orbitals, while there are three $^4T_2$ orbital states, of type x, y, z (cubic bases) or x_+, x_-, x_0 (trigonal bases) (Chapter 3). The spin states are $M_s=\pm\frac{1}{2}$ and $M_s'=\pm\frac{3}{2}$, $\pm\frac{1}{2}$ referred to some Zeeman z-axis. The matrix elements of the spin–orbit operator connecting $|^2E\phi M_s\rangle$ and $|^4T_2\psi M_s'\rangle$ using cubic bases and for arbitrary direction of the magnetic field z-axis are tabulated by Sugano *et al.* (1960). These are reproduced in Table 9.5(b). A constant magnetic field removes the $M_s=\pm\frac{1}{2}$ and $M_s'=\pm\frac{3}{2}$, $\pm\frac{1}{2}$ spin degeneracy of ^2E and $^4A_2$, respectively (Fig. 9.2). Assuming both states to have g values close to 2.0 leads to a simple Zeeman spectrum of three lines. The relative intensities of individual Zeeman transitions have been calculated by Sugano *et al.* (1960), and the predicted intensity patterns in the π- and σ-polarizations for the case where the Zeeman field is along the crystallographic [001] direction are shown in Fig. 9.2. These are in very good agreement with the experimentally measured Zeeman spectra (see Fig. 6.11).

The one-phonon vibrational sideband which accompanies the sharp zero-phonon R-line has already been illustrated (Fig. 5.20(a)). According to the

Table 9.5

(a) Spin–orbit matrix elements $\langle {}^2E\phi' M_s' | \mathscr{H}_{so} | {}^4T_2 \phi M_s \rangle$ using trigonal basis functions. The spins are quantized along the trigonal z-axis. (All values are to be multiplied by $-2\sqrt{2}i\zeta$.)

	ϕ	x_+				x_-				x_0			
ϕ'	M_s' | M_s	$\frac{3}{2}$	$\frac{1}{2}$	$-\frac{1}{2}$	$-\frac{3}{2}$	$\frac{3}{2}$	$\frac{1}{2}$	$-\frac{1}{2}$	$-\frac{3}{2}$	$\frac{3}{2}$	$\frac{1}{2}$	$-\frac{1}{2}$	$-\frac{3}{2}$
u_+	$\frac{1}{2}$	0	$\frac{i}{3\sqrt{2}}$	0	0	0	0	$-\frac{i}{6}$	0	$\frac{i}{2\sqrt{3}}$	0	0	0
	$-\frac{1}{2}$	0	0	$\frac{i}{3\sqrt{2}}$	0	0	0	0	$-\frac{i}{2\sqrt{3}}$	0	$\frac{i}{6}$	0	0
u_-	$\frac{1}{2}$	$\frac{-i}{2\sqrt{3}}$	0	0	0	0	$\frac{-i}{3\sqrt{2}}$	0	0	0	0	$\frac{-i}{6}$	0
	$-\frac{1}{2}$	0	$\frac{-i}{6}$	0	0	0	0	$\frac{-i}{3\sqrt{2}}$	0	0	0	0	$\frac{-i}{2\sqrt{3}}$

Table 9.5 (*continued*)

(b) Spin–orbit matrix elements $\langle {}^2E\phi' M_s' | \mathscr{H}_{so} | {}^4T_2 \phi M_s \rangle$ using octahedral basis functions. The spins M_s and M_s' are quantized along the direction θ, ϕ with respect to the octahedral z-axis. $a = \cos\theta/2$, $b = \sin\theta/2$. (Taken from Sugano *et al.* 1960.) The spin–orbit coupling parameter is labelled ζ' in this table. Take the lower sign for $i\eta$. All values tabulated for ξ (or $i\eta$) are to be multiplied by $(-i\zeta'/\sqrt{6})$. All values tabulated for ζ are to be multiplied by $(2\sqrt{2}i\zeta'/3)$.

		ϕ	ξ(or $i\eta$)				ζ			
ϕ'	M_s'	M_s	$\frac{3}{2}$	$\frac{1}{2}$	$-\frac{1}{2}$	$-\frac{3}{2}$	$\frac{3}{2}$	$\frac{1}{2}$	$-\frac{1}{2}$	$-\frac{3}{2}$
u	$\frac{1}{2}$		$\sqrt{3}(\pm b^2e^{i\phi} - a^2e^{-i\phi})$	$2ab(\pm e^{i\phi} + e^{-i\phi})$	$\pm a^2e^{i\phi} - b^2e^{-i\phi}$	0	0	0	0	0
	$-\frac{1}{2}$		0	$\pm b^2e^{i\phi} - a^2e^{-i\phi}$	$2ab(\pm e^{i\phi} + e^{-i\phi})$	$\sqrt{3}(\pm a^2e^{i\phi} - b^2e^{-i\phi})$	0	0	0	0
v	$\frac{1}{2}$		$(b^2e^{i\phi} \mp a^2e^{-i\phi})$	$(2ab/\sqrt{3})(e^{i\phi} \pm e^{-i\phi})$	$(1/\sqrt{3})(a^2e^{i\phi} \mp b^2e^{-i\phi})$	0	$\sqrt{3}ab$	$a^2 - b^2$	$-ab$	0
	$-\frac{1}{2}$		0	$(1/\sqrt{3})(b^2e^{i\phi} \mp a^2e^{-i\phi})$	$(2ab/\sqrt{3})(e^{i\phi} \pm e^{-i\phi})$	$(a^2e^{i\phi} \mp b^2e^{-i\phi})$	0	ab	$a^2 - b^2$	$-\sqrt{3}ab$

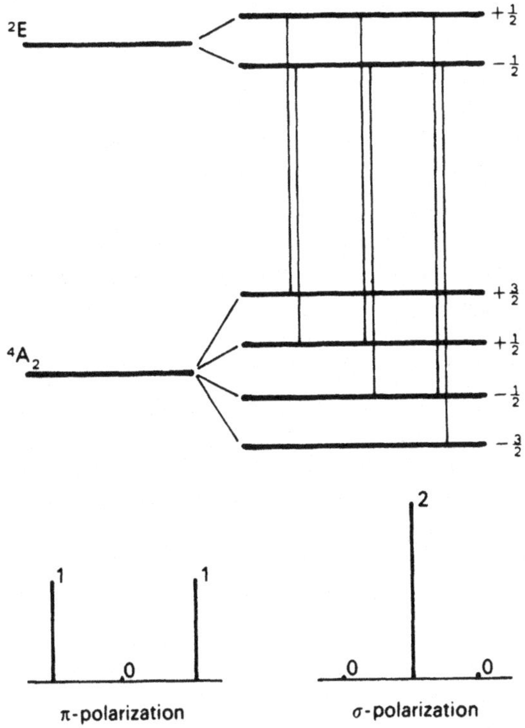

FIG. 9.2. Zeeman splitting of the R-line of $MgO:Cr^{3+}$ calculated on the basis of a magnetic dipole transition. The magnetic field is directed along the [001] crystallographic direction. The g-values of the two levels are taken to be approximately equal. as is found experimentally.

analysis of Chapter 5 the one-phonon sideband gives a weighted density of phonon states, which may be compared with the density of phonon states measured by neutron spectroscopy (Fig. 5.20(b)) (Peckham 1967). There are measurable differences in position between the peaks in the two spectra. Since the sideband measures the vibrational spectrum in the vicinity of the heavy impurity ion, Cr^{3+}, it displays the perturbed vibrational spectrum of magnesium oxide. The perturbations of the magnesium oxide phonon spectrum due to the different mass and coupling constants of the impurity ion were analysed by Sangster (1972) and related to observed differences between the phonon spectrum of pure magnesium oxide and the frequencies in the sidebands accompanying sharp lines of transition metal ions in magnesium oxide. Since the Huang–Rhys parameter for the $^2E \rightarrow {}^4A_2$ transition is small the one-phonon sideband is expected to be weak. Nevertheless at low temperature the integrated intensity of the one-phonon sideband of the R-line is about four times that of the R-line. As we discussed in Chapter 5, the sideband appears to be an electric dipole vibronic transition caused by lattice vibrations of odd parity at the site of the Cr^{3+} ion. The relative weakness of the R-line is attributable to its magnetic dipole nature.

The charge imbalance which occurs when Cr^{3+} ions substitute for Mg^{2+} ions in magnesium oxide is compensated for by the occurrence of cation vacancies—three Mg^{2+} ions are removed for every two Cr^{3+} ions incorporated into the crystal. A fraction of the Cr^{3+} ions then find themselves in the immediate neighbourhood of a cation vacancy, and the symmetry of the crystal field at the sites of these Cr^{3+} ions is lower than octahedral. These ions exhibit their own distinct ESR spectra (Wertz and Auzins 1957; Griffiths and Orton 1959) and optical spectra (Henry *et al.* 1976). In one such distorted site the next-nearest neighbour site to the Cr^{3+} ion along the crystal [110] direction is vacant. The value of Dq at this orthorhombic site is reduced relative to that in octahedral sites and as a result Dq/B is lower than the crossover point between the ^2E and $^4T_2$ states. The broad luminescence from Cr^{3+} ions in these sites is due to $^4T_2 \rightarrow {}^4A_2$ transitions. Figure 9.3 compares the luminescence spectra of Cr^{3+} ions in both octahedral and orthorhombic sites in magnesium oxide.

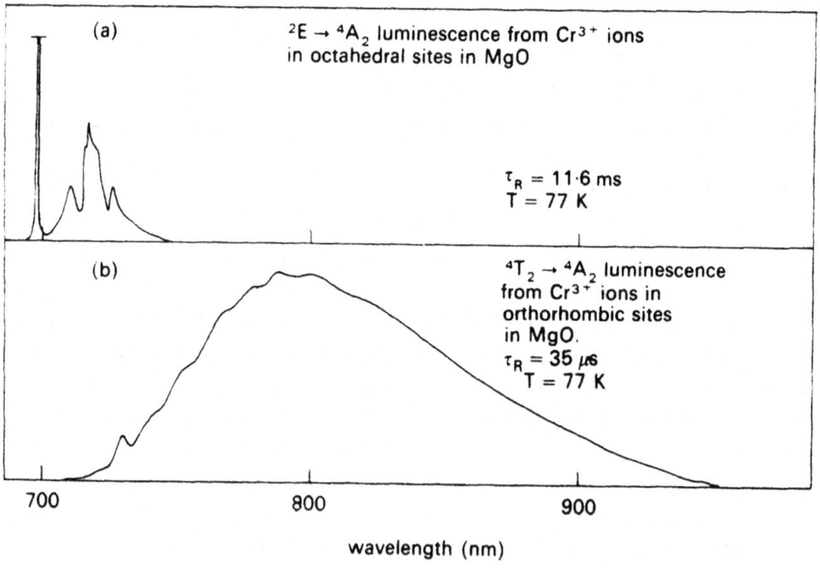

FIG. 9.3. A comparison of the luminescence from Cr^{3+} ions in high-field and low-field sites in magnesium oxide. These emission spectra overlap but have quite distinct radiative lifetimes. This difference in lifetime allows the two spectra to be separated from each other using phase-sensitive techniques, as discussed in Section 6.3. (Henry *et al.* 1976.)

In $Al_2O_3:Cr^{3+}$ the Cr^{3+} ion substitutes directly for Al^{3+}, entering a site of trigonally-distorted octahedral symmetry. Part of the alumina crystal structure is shown in Fig. 9.4. The arrangement of oxygen ions around the Cr^{3+} ion is shown in Fig. 9.5, viewed along the trigonal crystal axis. There are three oxygen ions in a triangle in the plane above the Cr^{3+} ion, and three ions in the triangle in the plane below the Cr^{3+} ion. The two triangles are of slightly different size and rotated out of exact symmetry through the angle ϕ. The

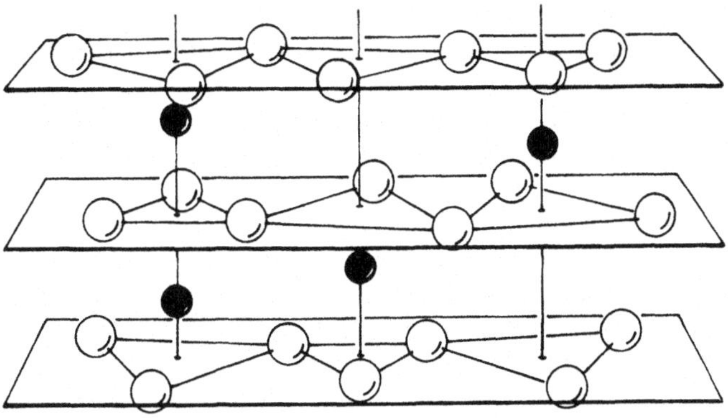

FIG. 9.4. A portion of the alumina lattice in which the Al^{3+} ions (dark circles) are shown between layers of O^{2-} ions (open circles). The optic axis of the crystal (trigonal axis at the site of the Al^{3+} ions) is perpendicular to the planes of oxygen ions. (After Geschwind and Remeika 1961.)

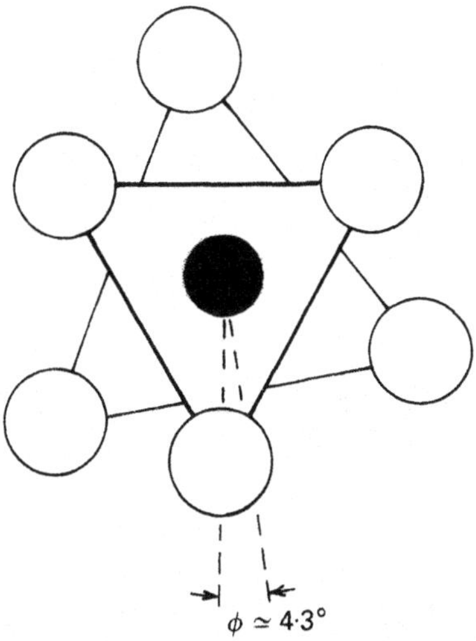

FIG. 9.5. The Cr^{3+} ion (dark circle) surrounded by its six neighbouring oxygen ions in ruby, as viewed along the trigonal crystal axis (optic axis of the sapphire or ruby crystal). (After Shinada *et al.* 1966.)

distortions from perfect octahedral symmetry give rise to the following additional electrostatic energy terms (Shinada *et al.* 1966; Klauminzer 1970):

1. An even-parity energy term of trigonal symmetry caused by stretching of the triangles along the trigonal axis (see Section 3.1.5). This energy term is of type $T_{2g}x_0$, and is written $V_{trig}(T_{2g}x_0)$.
2. An odd-parity term of type $T_{1u}a_0$ caused by a displacement of the Al^{3+} (and Cr^{3+}) ions along the trigonal axis from a position midway between the two planes of oxygen ions. This is written $V(T_{1u}a_0)$.
3. An odd-parity term of type A_{2u} caused by the reduction in size of one triangle relative to the other.
4. An odd-parity term of type $T_{2u}x_0$ caused by the rotation of one triangle relative to the other triangle through the angle $\phi = 4.3°$.

The even-parity trigonal term plus spin–orbit coupling causes a splitting of the octahedral energy levels of the Cr^{3+} ion. The odd-parity terms are unimportant in respect of the splitting of levels but are important in permitting electric dipole radiative transitions between the Cr^{3+} levels. Term (2) appears to be the dominant odd-parity term, and the normal absorption and luminescence transitions have been analysed with reasonable success on the basis of this odd-parity term only (Sugano and Tanabe 1958). To discuss absorptions originating on the excited ^2E level the other odd-parity terms must be considered (Shinada *et al.* 1966; Fairbanks *et al.* 1975). The point symmetry about the Cr^{3+} ion in ruby is C_3. If term (4) is ignored the symmetry is C_{3v}, and this is often adapted as the site symmetry in ruby. If all odd-parity terms are ignored the symmetry is $D_{3h} = D_3 \times i$.

We consider first the effect on the octahedral crystal field levels of the even-parity trigonal term $V_{trig}(T_{2g}x_0)$ which reduces the site symmetry to trigonal (we take the reduced symmetry to be C_{3v}) and causes a first-order splitting of the $^4T_1(^4F)$ and $^4T_2$ levels (Sugano and Tanabe 1958; Macfarlane 1963). The trigonal field parameters defined in terms of single d electron functions are

$$v = \tfrac{3}{2}\langle t_2x_0 | V_{trig} | t_2x_0 \rangle = -3\langle t_2x_+ | V_{trig} | t_2x_+ \rangle = -3K$$

$$v' = \langle t_2x_+ | V_{trig} | eu_+ \rangle \qquad (9.4)$$

where K is the parameter used by Sugano and Tanabe. When the symmetry is reduced to trigonal C_{3v}, the octahedral $^4T_2$ state splits into trigonal $^4A_1$ and ^4E states, as is shown by a comparison of the character tables of the O and C_{3v} groups (Tables 3.1 and 3.3, respectively). For purposes of analysing the decomposition of the T_2 irreducible representation of O_h into E and A_2 irreducible representations of C_{3v} we use the fact that the σ_v reflection operator in the C_{3v} group corresponds in the O_h group to a C_2' rotation followed by inversion in the O_h group. The trigonal $^4A_1$ and ^4E states are of type $^4T_2x_0$ and $^4T_2x_+$, $^4T_2x_-$, respectively. The octahedral $^4T_1(t_2^2e)$ state similarly splits into trigonal $^4A_2(^4T_1a_0)$ and $^4E(^4T_1a_+, {}^4T_1a_-)$ states. The splittings of the $^4T_2$ and $^4T_1$ levels are, to good approximation, given by $v/2$ and $v/2+v'$, respectively.

Similar first-order splittings of the $^2E(t_2^3)$, $^2T_1(t_2^3)$, and $^2T_2(t_2^3)$ states cancel because they belong to the half-filled t_2^3 configuration. However, these states split in second order through a combination of trigonal crystal field distortion and spin–orbit coupling. The character tables of the O_h and C_{3v} double groups (Tables 3.5 and 3.7 respectively) show that $^4A_2$ splits into two states, of type $2\bar{A}$ and $\bar{E}$; likewise ^2E splits into two states of type $2\bar{A}$ and $\bar{E}$, whereas $^2T_1$ and $^2T_2$ each split into three states. Macfarlane (1967) took into account many perturbation loops in determining the splittings of the octahedral states belonging to the t_2^3 configuration in obtaining the following approximate analytical expressions for the splitting of the $^4A_2$ and ^2E levels:

$$E[^4A_2(2\bar{A})]-E[^4A_2(\bar{E})] = -1.44\times10^{-8}\zeta^2 v' + 0.09\times10^{-8}\zeta^2 v$$
$$E[^2E(2\bar{A})]-E[^2E(\bar{E})] = 20.9\times10^{-5}\zeta v - 1.1\times10^{-5}\zeta v' \qquad (9.5)$$

in which ζ is the spin–orbit parameter. The splitting in the ground $^4A_2$ state is determined principally by v', while the splitting in the ^2E state is determined principally by v. There is, therefore, no correlation between the observed splittings of these two states. Similar analytical expression cannot be found for the splittings of the $^2T_1(t_2^3)$ and $^2T_2(t_2^3)$ states. The full set of energy level parameters for ruby are $Dq = 1810\ \text{cm}^{-1}$, $B = 650\ \text{cm}^{-1}$, $C = 3120\ \text{cm}^{-1}$, $\zeta = 170\ \text{cm}^{-1}$, $v = 800\ \text{cm}^{-1}$ and $v' = 680\ \text{cm}^{-1}$ (Macfarlane 1967).

The transition strength of the radiative processes on Cr^{3+} ions in ruby may be calculated, assuming the transitions to be electric dipole processes allowed by the static odd-parity crystal field terms. The dominant odd-parity term is of type $T_{1u}a_0$. The matrix element for the electric dipole transition between states $|^4A_2M_s\rangle$ and $|^4T_2\phi M_s\rangle$ induced by interaction with radiation polarized with the electric vector along the E direction is

$$\sum_{\Gamma_u,\phi'} \frac{\langle ^4T_2\phi M_s|\boldsymbol{\mu}_e.\hat{\boldsymbol{E}}|^4\Gamma_u\phi' M_s\rangle\langle ^4\Gamma_u\phi' M_s|V(T_{1u}a_0)|^4A_2M_s\rangle}{E(^4A_2)-E(^4\Gamma_u)} + \text{converse} \qquad (9.6)$$

where in the converse term the operator $\boldsymbol{\mu}_e.\hat{\boldsymbol{E}}$ and $V(T_{1u}a_0)$ are interchanged and the energy denominator is $E(^4T_2)-E(^4\Gamma_u)$. There are no spin components in the operators so that the M_s values of all states are the same. A similar matrix element occurs for the transition between $|^4A_2M_s\rangle$ and $|^4T_1\phi M_s\rangle$.

If all odd-parity states, Γ_u, are sufficiently distant in energy that all energy denominators can be assigned the same constant value ΔE then closure can be invoked and the electric dipole matrix element written as $\langle ^4T_2\phi M_s|\tilde{\boldsymbol{\mu}}_e.\hat{\boldsymbol{E}}|^4A_2M_s\rangle$. $\tilde{\boldsymbol{\mu}}_e.\hat{\boldsymbol{E}}$ is the effective dipole operator being defined as

$$\tilde{\boldsymbol{\mu}}_e \cdot \hat{\boldsymbol{E}} = \sum_{\Gamma_u, \phi'} \frac{\boldsymbol{\mu}_e \cdot \hat{\boldsymbol{E}} | ^4\Gamma_u \phi' \rangle \langle ^4\Gamma_u \phi' | V(\mathrm{T}_{1u} \mathrm{a}_0)}{\Delta E} = \frac{\boldsymbol{\mu}_e \cdot \hat{\boldsymbol{E}} \times V(\mathrm{T}_{1u} \mathrm{a}_0)}{\Delta E}. \tag{9.7}$$

When the electric vector of the radiation is parallel to the trigonal axis (π-polarization), $\boldsymbol{\mu}_e \cdot \hat{\boldsymbol{E}}$ is a vector of type $\mathrm{T}_{1u}\mathrm{a}_0$ (Table 9.4) and the odd-parity crystal field term is of type $\mathrm{T}_{1u}\mathrm{a}_0$, so that the operator $\tilde{\boldsymbol{\mu}}_e \cdot \boldsymbol{E}$ is of type $\mathrm{T}_{1u}\mathrm{a}_0 \times \mathrm{T}_{1u}\mathrm{a}_0$. From the table of Clebsch–Gordan coefficients (see Appendix 3B, Table 3B.2) we find that this contains even-parity components of type A_{1g} and $\mathrm{T}_{2g}x_0$. The Clebsch–Gordan coefficients (Table 3B.2) give the fractional amounts ($-1/\sqrt{3}$ and $-\sqrt{2}/\sqrt{3}$, respectively) of these two basis functions contained in the effective dipole operator, $\tilde{\boldsymbol{\mu}}_e \cdot \hat{\boldsymbol{E}}$. Since $\mathrm{A}_2 \times \mathrm{A}_1 = \mathrm{A}_2$ and $\mathrm{A}_2 \times \mathrm{T}_2 x_0 = \mathrm{T}_1\mathrm{a}_0$ (Table 3B.2) we see that there are no matrix elements of this π-polarized dipole operator between $^4\mathrm{A}_2$ and $^4\mathrm{T}_2$. However, there is a matrix element of $\tilde{\boldsymbol{\mu}}_e \cdot \hat{\boldsymbol{E}}$ between $^4\mathrm{A}_2$ and $^4\mathrm{T}_1\mathrm{a}_0$, the relevant Clebsch–Gordan coefficient being $-\sqrt{2}/\sqrt{3}$. For radiation polarized linearly in the x-direction perpendicular to the trigonal axis (σ-polarization) the operator $\boldsymbol{\mu}_e \cdot \hat{\boldsymbol{E}}$ is of type $T_{1u}(-\mathrm{a}_+ + \mathrm{a}_-)/\sqrt{2}$ (Table 9.4) and $\tilde{\boldsymbol{\mu}}_e \cdot \hat{\boldsymbol{E}}$ is of type $T_{1u}(-\mathrm{a}_+ \;\; + \mathrm{a}_-)/\sqrt{2} \times T_{1u}\mathrm{a}_0 = E_g\mathrm{u}_+(1/\sqrt{6})$, $E_g\mathrm{u}_-(-1/\sqrt{6})$, $T_{1g}\mathrm{a}_+(-\mathrm{i}/2)$, $T_{1g}\mathrm{a}_-(-\mathrm{i}/2)$, $T_{2g}\mathrm{x}_+(-1/2\sqrt{3})$, $T_{2g}\mathrm{x}_-(1/2\sqrt{3})$, where the Clebsch–Gordan coefficients derived from Table 3B.2, give the fractional amounts of the above functions contained in the effective dipole operator for σ-polarized transitions. Because of the $T_{1g}\mathrm{a}_+$ and $T_{1g}\mathrm{a}_-$ components there are non-zero matrix elements of $\tilde{\boldsymbol{\mu}}_e \cdot \hat{\boldsymbol{E}}$ between $^4\mathrm{A}_2$ and $^4\mathrm{T}_2\mathrm{x}_+$ and between $^4\mathrm{A}_2$ and $^4\mathrm{T}_2\mathrm{x}_-$. Similarly and because of the $\mathrm{T}_{2g}\mathrm{x}_+$ and $\mathrm{T}_{2g}\mathrm{x}_-$ components there are non-zero matrix elements of $\tilde{\boldsymbol{\mu}}_e \cdot \hat{\boldsymbol{E}}$ in σ-polarization between $^4\mathrm{A}_2$ and $^4\mathrm{T}_1\mathrm{a}_+$ and between $^4\mathrm{A}_2$ and $^4\mathrm{T}_1\mathrm{a}_-$. The predicted intensities are in the ratio of the squares of the relevant Clebsch–Gordan coefficients. Figure 9.6 shows the predicted intensities in π- and σ-polarization, and it is indicated that these transitions are allowed by the $\mathrm{T}_{1u}\mathrm{a}_0$ odd-parity static crystal field.

Next we carry out the analogous calculation of the electric dipole transition intensities allowed by the weaker $T_{2u}\mathrm{x}_0$ odd-parity static crystal field. For π-polarized transitions the effective dipole operator is of type $T_{1u}\mathrm{a}_0 \times T_{2u}\mathrm{x}_0 = \mathrm{A}_{2g}(-1/\sqrt{3})$, $T_{1g}\mathrm{a}_0(\sqrt{2}/\sqrt{3})$. This allows a transition between $^4\mathrm{A}_2$ and $^4\mathrm{T}_2\mathrm{x}_0$ but no transition between $^4\mathrm{A}_2$ and $^4\mathrm{T}_1$ states. For σ-polarized transitions (where we assume that the electric vector vibrates in the x-direction) the effective dipole operator is of type $T_{1u}(-\mathrm{a}_+ + \mathrm{a}_-)/\sqrt{2} \times T_{2u}\mathrm{x}_0 = E_g\mathrm{u}_+(-\mathrm{i}/\sqrt{6})$, $E_g\mathrm{u}_-(-\mathrm{i}/\sqrt{6})$, $T_{1g}\mathrm{a}_+(1/2\sqrt{3})$, $T_{1g}\mathrm{a}_-(-1/2\sqrt{3})$, $T_{2g}\mathrm{x}_+(-\mathrm{i}/2)$, $T_{2g}\mathrm{x}_-(-\mathrm{i}/2)$. Because of the $T_{1g}\mathrm{a}_+$, $T_{1g}\mathrm{a}_-$ components there are σ-polarized absorption transitions between $^4\mathrm{A}_2$ and $^4\mathrm{T}_2\mathrm{x}_+$, $^4\mathrm{T}_2\mathrm{x}_-$, and because of the $T_{2g}\mathrm{x}_+$, $T_{2g}\mathrm{x}_-$ components there are σ-polarized absorption transitions between $^4\mathrm{A}_2$ and $^4\mathrm{T}_1\mathrm{a}_+$, $^4\mathrm{T}_1\mathrm{a}_-$. The predicted transition intensities allowed by the $T_{2u}\mathrm{x}_0$ odd-parity static crystal field, also shown in Fig. 9.6, are in the ratio of the squares of the Clebsch–

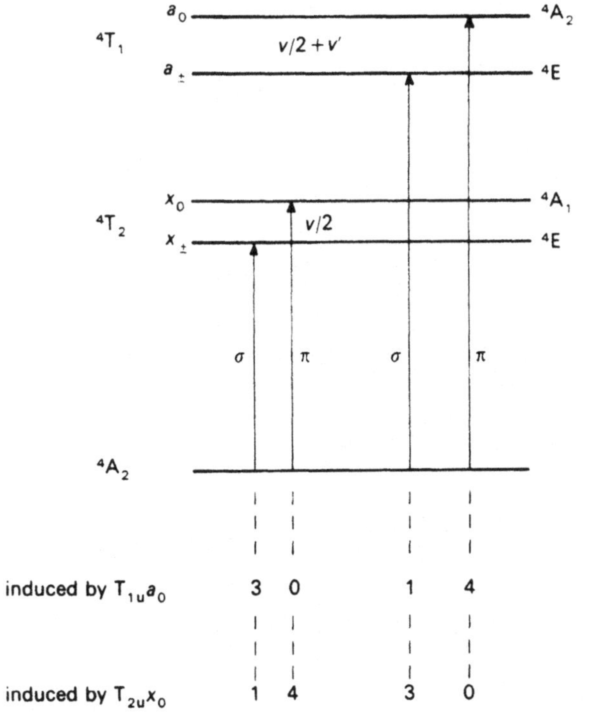

FIG. 9.6. Predicted polarizations for the strong spin-allowed transitions of $Al_2O_3:Cr^{3+}$ in the visible. The octahedral $^4T_1$ and $^4T_2$ levels are split by the trigonal crystal field term (of type $T_{2g}x_0$), the corresponding splitting of the $^4A_2$ level is negligible by comparison with the splittings of the $^4T_1$ and $^4T_2$ levels and is omitted in this figure. The transitions are assumed to be electric dipole induced by the static odd-parity crystal field of either type $T_{1u}a_0$ or type $T_{2u}x_0$. The labels on the right of the split levels are the C_{3v} symmetry labels.

Gordan coefficients. The observed π and σ absorption spectra of ruby are shown in Fig. 9.7, and there is general agreement with the calculated intensities. The weakness of the $^4A_2 \rightarrow {}^4T_2$ absorption in π-polarization is as predicted by the calculations; it occurs only because of the weaker $^4T_{2u}x_0$ odd-parity crystal field term. There are, undoubtedly, vibronic processes contributing also to the broad absorption bands. The splittings of the $^4T_2$ and $^4T_1$ levels show up clearly in the polarized spectra since the π and σ absorption transitions terminate on the separate split components.

In comparison with the broad spin-allowed absorption transitions the spin-forbidden transitions are sharp, and finer details of the level structure can be studied. We analyse the $^2E \leftrightarrow {}^4A_2$ transitions in ruby (the R-transitions) which have been studied in both absorption and in luminescence. The

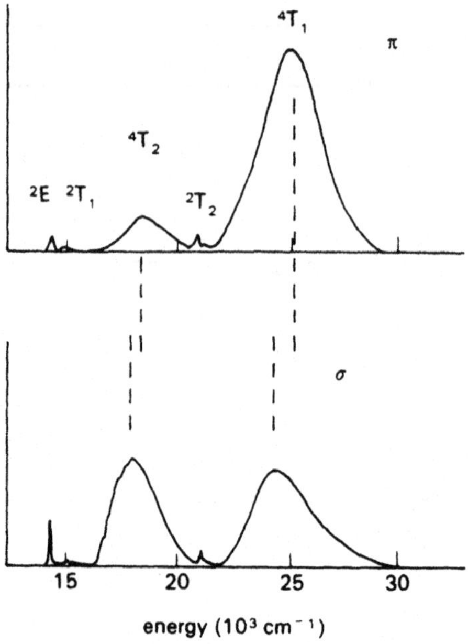

FIG. 9.7. Observed π- and σ-polarized absorption spectra of ruby at 77 K (after McClure 1959). The difference in the peak positions of the absorption bands measured in the two senses of polarization give the trigonal field splittings of the $^4T_2$ and $^4T_1$ levels.

appropriate energy levels are shown in Fig. 9.8. In the absence of a magnetic field only the twofold Kramers degeneracy remains. The ^2E and $^4A_2$ levels are split by 29 cm^{-1} and 0.38 cm^{-1}, respectively, by a combination of even-parity trigonal crystal field and spin–orbit coupling. The Zeeman splitting for a field of 1.3T parallel to the trigonal axis is also shown in the figure. A very complete ESR analysis of the ground-state splitting pattern of Cr^{3+} in ruby has been made by Schulz du Bois (1959).

FIG. 9.8. Splittings of the ^2E and $^4A_2$ states of Cr^{3+} in aluminium oxide in a magnetic field of 1.3 T along the C_3 axis. The Zeeman splittings and the ground state crystal field splitting (δ) are drawn to scale. The linearly polarized intensities of the transitions are calculated on the basis of odd-parity crystal field terms of type $T_{1u}a_0$ and $T_{2u}x_0$ of which the former should be the dominant term. In σ-polarization both odd-parity terms predict the ratio of components shown, where σ represents some normalizing intensity factor. In π-polarization only the $T_{2u}x_0$ is effective, the intensities should therefore be weaker, and the allowed transitions are shown by the broken vertical lines, where π represents some normalizing intensity factor. We expect $\sigma \gg \pi$. The observed Zeeman pattern for the R_1 luminescence line ($\bar{E}(^2E) \rightarrow {}^4A_2$) taken at 4.2 K is shown in the figure. Allowing for the population ratio in the two $\bar{E}$ levels there is good agreement with the predicted intensity ratios.

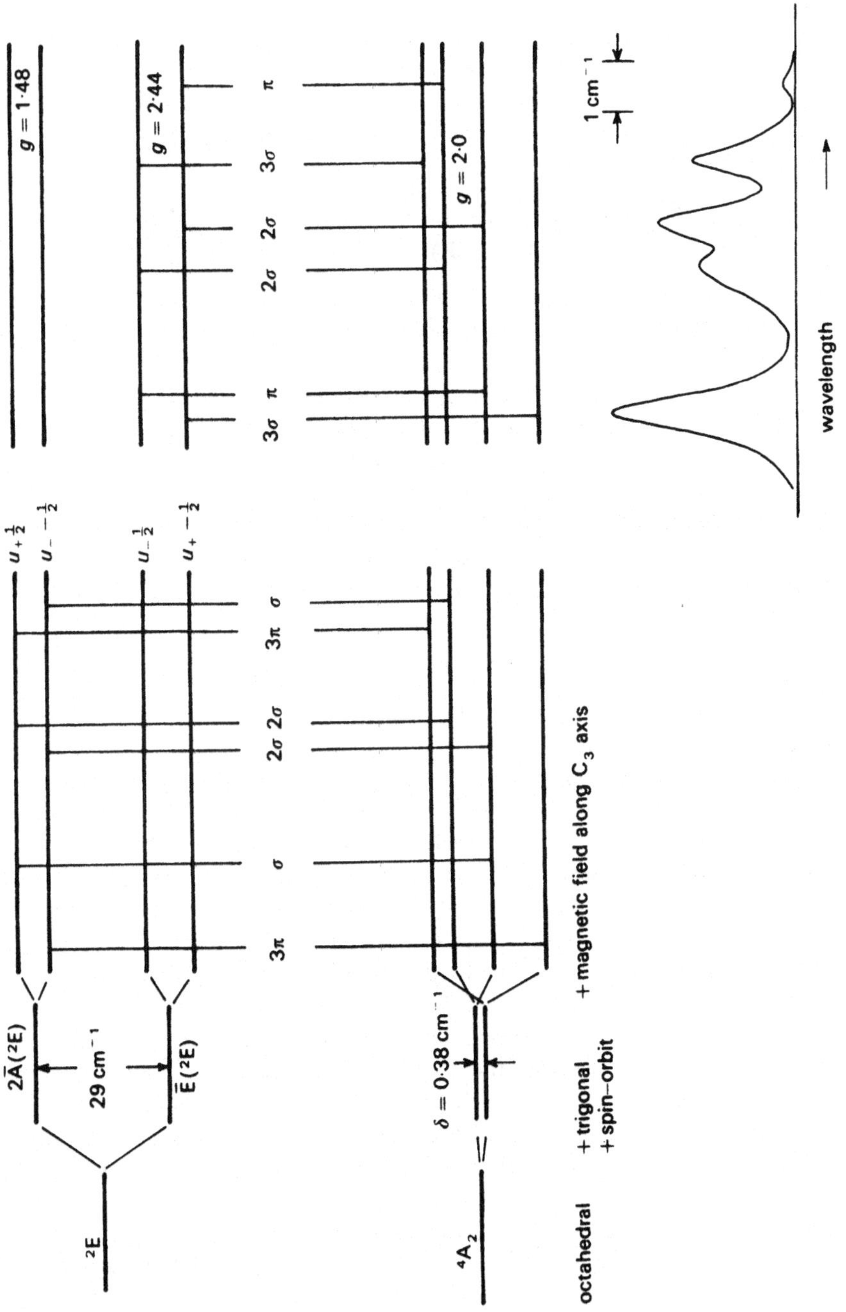

²E
2Ā(²E)
29 cm⁻¹
Ē(²E)
⁴A₂
δ = 0·38 cm⁻¹
octahedral
+ trigonal
+ spin–orbit
+ magnetic field along C₃ axis
g = 1·48
g = 2·44
g = 2·0
1 cm⁻¹
wavelength

Radiative transitions between ^2E and $^4A_2$ are spin-forbidden but become weakly allowed when spin–orbit coupling is invoked. The spin–orbit coupling operator $\mathscr{H}_{so}$ connects ^2E and $^4T_2$, while $^4T_2$ and $^4A_2$ are connected by the electric dipole operator $\tilde{\mu}_e . E$. Thus the intensity of the $^4A_2 \leftrightarrow {}^2E$ transitions are derived from the intensity of the $^4A_2 \leftrightarrow {}^4T_2$ transitions. The matrix element for an electric dipole transition between $|{}^2E\phi' M_s'\rangle$ and $|{}^4A_2 M_s\rangle$ for E-polarized radiation is

$$\frac{\langle {}^2E\phi' M_s' | \mathscr{H}_{so} | {}^4T_2 \phi M_s\rangle \langle {}^4T_2 \phi M_s | \tilde{\mu}_e . \hat{E} | {}^4A_2 M_s\rangle}{E({}^2E) - E({}^4T_2)} \tag{9.8}$$

The matrix elements of $\tilde{\mu}_e . \hat{E}$ have been discussed in connection with the $^4A_2 \rightarrow {}^4T_2$ broadband absorption transitions. The spin–orbit matrix elements can be evaluated from the tables in the text by Sugano *et al.* (1970). Values of $\langle {}^2E\phi' M_s' | \mathscr{H}_{so} | {}^4T_2 \phi M_s\rangle$ for trigonal and octahedral bases are given in Table 9.5. We thus have all the factors required to calculate the relative intensities of the Zeeman transitions. Figure 9.8 shows that such a calculation predicts that the σ components should be much stronger than the π components. The observed Zeeman pattern for the R_1-luminescence line ($\bar{E}({}^2E) \rightarrow {}^4A_2$) measured at 4.2 K is shown in Fig. 9.8. If allowance is made for the equilibrium populations in the two $\bar{E}$ levels there is good agreement with the calculated intensity ratios.

9.3.3 *Luminescence from Cr^{3+} ions in other crystalline hosts*

Figure 9.9 is a simplified diagram of the low-lying energy levels of Cr^{3+} in octahedral crystal fields. The vertical lines correspond approximately to the cases of (a) ruby, (b) alexandrite ($BeAl_2O_4 : Cr^{3+}$), (c) $KZnF_3 : Cr^{3+}$, and the shaded area represents the range of Dq/B values for a typical chromium-doped borate glass. In ruby the energy difference Δ between the $^4T_2$ and ^2E levels is 2300 cm^{-1} while for alexandrite it is 800 cm^{-1} and for emerald 400 cm^{-1}. Relaxation between the $^4T_2$ and ^2E levels is exceedingly rapid in all chromium systems so the excited ion populations in these levels thermalize in a time much shorter than the decay time to the ground state. For ions in high-field sites ($\Delta > 0$) only the ^2E level can maintain a sizeable equilibrium population at low temperatures and luminescence occurs only from this level. As the temperature is raised the $^4T_2$ level becomes increasingly populated and luminescence occurs from the $^4T_2$ level also. The luminescence from Cr^{3+} ions in alexandrite at liquid nitrogen temperatures shows sharp R-lines and a one-phonon sideband (Fig. 9.10(a), with a decay time of *c.* 1.5 ms. At room temperature, however, some 6 per cent of the excited ions are in the $^4T_2$ state, and since the $^4T_2 \rightarrow {}^4A_2$ transition probability is about two orders of magnitude larger than that of the $^2E \rightarrow {}^4A_2$ transition, it is the broadband $^4T_2$ emission which dominates the observed luminescence spectrum. The much shorter decay time of the luminescence at room temperature (*c.* 220 μs) is

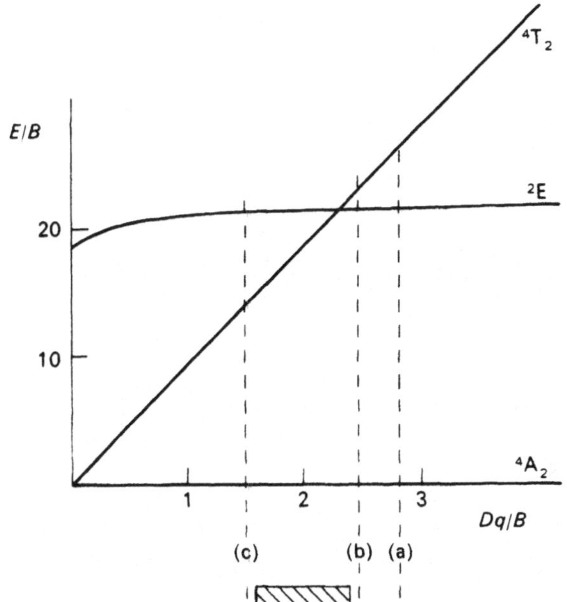

FIG. 9.9. Simplified energy level diagram of some of the low-lying energy levels of $3d^3$ ions in octahedral crystal fields. At $Dq/B \simeq 2.3$ the ^2E and $^4T_2$ levels cross. Above and below this value ions are said to be in high-field and low-field sites, respectively. The vertical lines correspond approximately to the cases of (a) ruby, (b) alexandrite ($BeAl_2O_4:Cr^{3+}$), (c) $KZnF_3:Cr^{3+}$ and $MgF_2:V^{2+}$. The shaded area represents the range of values of Dq/B found for Cr^{3+} ions in a typical borate glass; the ions can occupy both high-field and low-field sites in this glass.

consistent with the increased population of the $^4T_2$ level. Fig. 9.10(b) shows the room temperature luminescence spectrum, and the sharper ^2E emission is seen to be superimposed on the broader $^4T_2$ emission. The quantum efficiency is close to unity. Cr^{3+} can substitute for Zn^{2+} in $KZnF_3$ where the ion is surrounded by an octahedron of F^- ions. The values of Dq for Cr^{3+} ions in fluoride hosts are lower than in oxide hosts, and the ions are in low-field sites (Fig. 9.9). The luminescence of $KZnF_3:Cr^{3+}$ consists of broadband $^4T_2 \rightarrow {}^4A_2$ emission at all temperatures.

The decay time of the lower luminescent levels of Cr^{3+} ions vary from many milliseconds in some high-field cases to tens of microseconds in some low-field cases. While they are in this metastable state the ions can undergo further absorption transitions raising them to higher excited levels. The phenomenon of excited-state absorption (ESA) can have a significant influence on the efficiency of luminescence from the metastable state. When the ion is in the excited ^2E level the strongest absorption transitions are to other

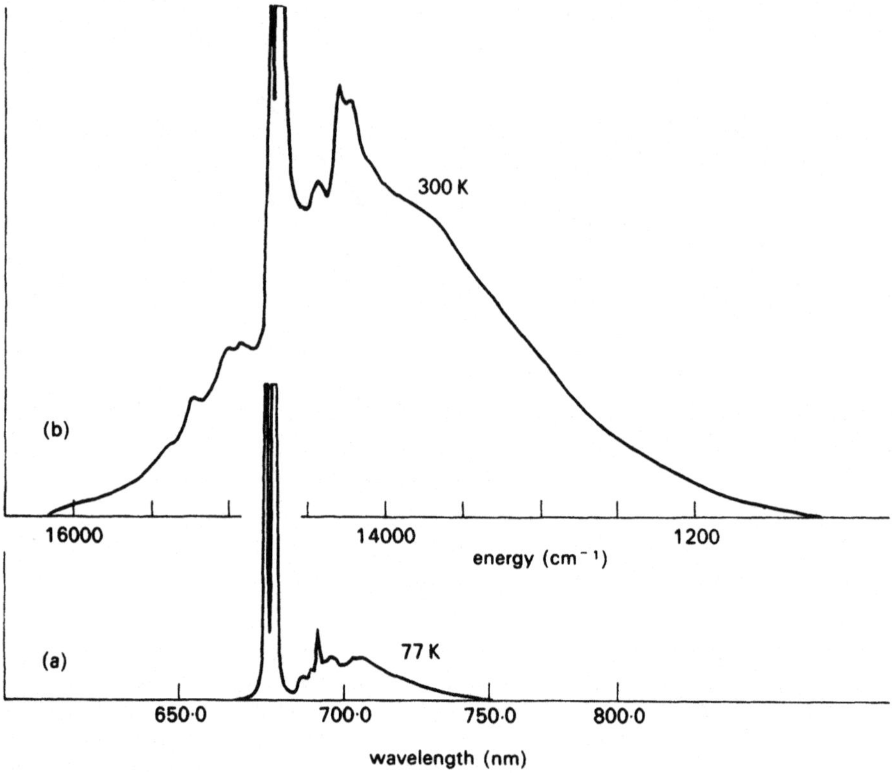

FIG. 9.10. Luminescence from alexandrite ($BeAl_2O_4{:}Cr^{3+}$) at 77 K (a) and at room temperature (b).

doublet levels, consequently, the ESA spectrum bears little resemblance to the ground state absorption. ESA in chromium systems has been studied experimentally by Gires and Meyer (1961), Klauminzer *et al.* (1966), Kushida (1966), Shinada *et al.* (1966), Fairbank *et al.* (1975), and Andrews and Hitelman (1986).

9.3.4 Spectroscopy of other 3d^3 *dopant ions*

V^{2+} is isoelectronic with Cr^{3+} and substitutes directly for Mg^{2+} in magnesium oxide without charge compensation. The spectroscopy of V^{2+} in magnesium oxide is very similar to that of the Cr^{3+} ion in high-field octahedral sites (Sturge 1965). In this material a relatively weak zero-phonon transition accompanies the $^4A_2 \rightarrow {}^4T_2$ broadband absorption transition, and this consists of an easily-resolved doublet of separation around 50 cm^{-1}. Group theory predicts that the $^4T_2$ state will split into four levels through spin–orbit coupling. The Jahn–Teller effect in the $^4T_2$ state of this and other V^{2+} systems has been analysed by Sturge (1970, and references therein).

In $MgF_2:V^{2+}$ the V^{2+} ion is in a low-field site of distorted octahedral symmetry, $Dq/B \simeq 1.5$ (Fig. 9.9). The luminescence and absorption spectra at 77 K are shown in Fig. 9.11 along with the low-lying *zero-vibrational* energy levels. Since the luminescence is across an energy gap of around 10 000 cm^{-1} the possibility of nonradiative relaxation out of the $^4T_2$ level must be considered. The decay time at low temperatures is 2.4 ms which remains constant up to approximately 200 K, but above this temperature the decay time begins to decrease significantly due to non-radiative relaxation. In addition, as Fig. 9.11 indicates, ESA out of the $^4T_2$ to higher levels can occur, and this process is sufficiently strong to seriously impair the laser performance of this material.

Mn^{4+} also has the same $3d^3$ electronic configuration as Cr^{3+}. The optical and electron paramagnetic resonance spectroscopy of $Al_2O_3:Mn^{4+}$ has been reported by Geschwind *et al.* (1962); it is very similar to that reported for Cr^{3+} in alumina.

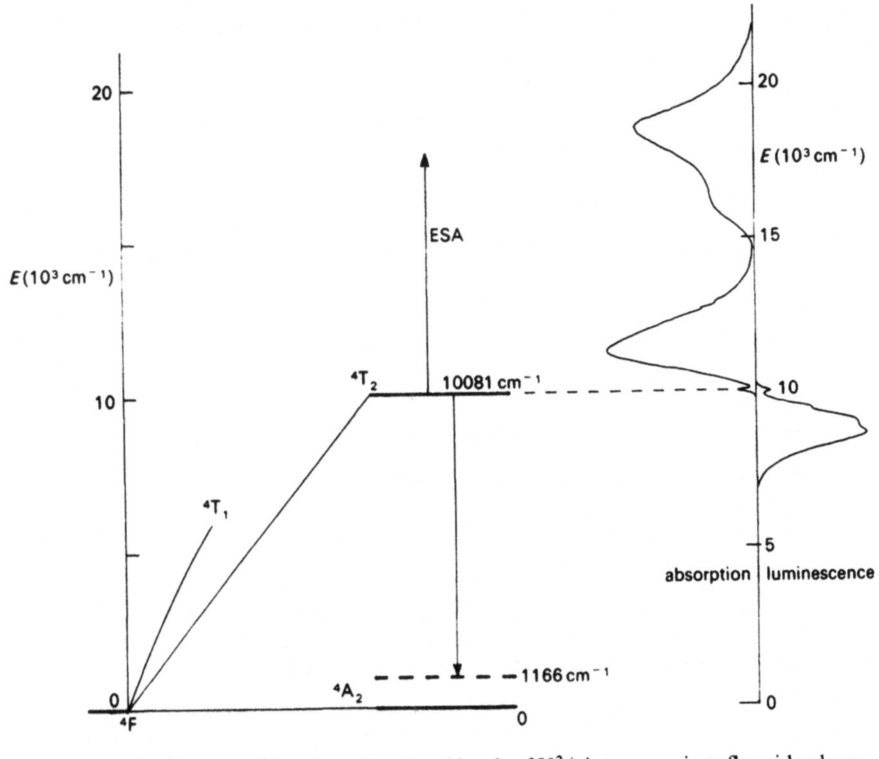

FIG. 9.11. The $^4A_2$ and $^4T_2$ *zero-vibrational levels* of V^{2+} in magnesium fluoride along with the absorption and luminescence spectra at 77 K.

9.4 A survey of other transition metal ions

9.4.1 The 3d^1 *configuration*

This is the simplest electronic system, containing a single 3d electron outside the closed shells. The Ti^{3+} ion tends to form octahedral complexes in which the single d level is split into ^2E and $^2T_2$ states, with an energy separation of $10\,Dq$. The spectroscopy of Al_2O_3:Ti^{3+} has been analysed in some detail. The $^2T_2$ ground state is split into $^2A_1$ and ^2E states by the trigonal crystal field (C_{3v} group symmetry labels); the energy separation is written as v, where v is the trigonal crystal field parameter defined in eqn (9.4). For transition metal ions in alumina one expects $v \simeq 700$–$1000\,cm^{-1}$. Taking spin–orbit coupling into account one finds that the ^2E state additionally splits into $\bar{E}$ and $2\bar{A}$ states (C_{3v} double group symmetry labels) separated by around $100\,cm^{-1}$. The $^2A_1$ state is not split by spin–orbit coupling; in the C_{3v} double group symmetry scheme it is labelled $\bar{E}$. The splitting in the ground state of Al_2O_3:Ti^{3+} has been measured by Nelson *et al.* (1967) and the two $\bar{E}$ states ($\bar{E}(^2E)$ and $\bar{E}(^2A_1)$) are at $37.8\,cm^{-1}$ and $108\,cm^{-1}$ above the $2\bar{A}$ ground state, in contradiction of the predictions to crystal field theory. This discrepancy has been shown by Macfarlane *et al.* (1968) to be caused by a dynamic Jahn–Teller effect which strongly quenches both spin–orbit and trigonal field splittings (Ham 1965). The excited ^2E state undergoes a static Jahn–Teller distortion of the Ti^{3+} ion and its surroundings to a new configuration of lower minimum energy. As a result the splitting in the excited state is much larger than predicted for an undistorted Al^{3+} site. These Jahn–Teller effects and a configurational coordinate model of the absorption and emission transitions are shown on a schematic configurational co-ordinate diagram on the right of Fig. 9.12. Optical absorption and luminescence spectra of Al_2O_3:Ti^{3+} measured at room temperature are shown in Fig. 9.13. The $^2T_2 \rightarrow {}^2E$ absorption near $20\,000\,cm^{-1}$ consists of two broad bands separated by around $2500\,cm^{-1}$. Luminescence occurs only from the lower excited state, the peak of the luminescence band occurs near $14\,000\,cm^{-1}$ (Gächter and Koningstein 1974). At low temperatures weak zero-phonon lines accompany the broad emission band; these correspond to transitions to the three ground electronic states.

9.4.2 The 3d^2 *configuration*

The variation in energy of the 3d^2 levels in an octahedral crystal field were shown in Fig. 3.18. V^{3+} can substitute for Al^{3+} in aluminium oxide and its spectrum has been investigated by a number of workers. In this material $B = 610\,cm^{-1}$, $C = 2500\,cm^{-1}$, and $Dq = 1800\,cm^{-1}$, giving $Dq/B = 2.95$. Broad spin-allowed absorption bands from the ground $^3T_1(t_2^2)$ state to the excited triplet states are observed (McClure 1962). The $^3T_1(t_2^2) \rightarrow {}^3T_2(t_2e)$ transition has been investigated by Scott and Sturge (1966) who explained the

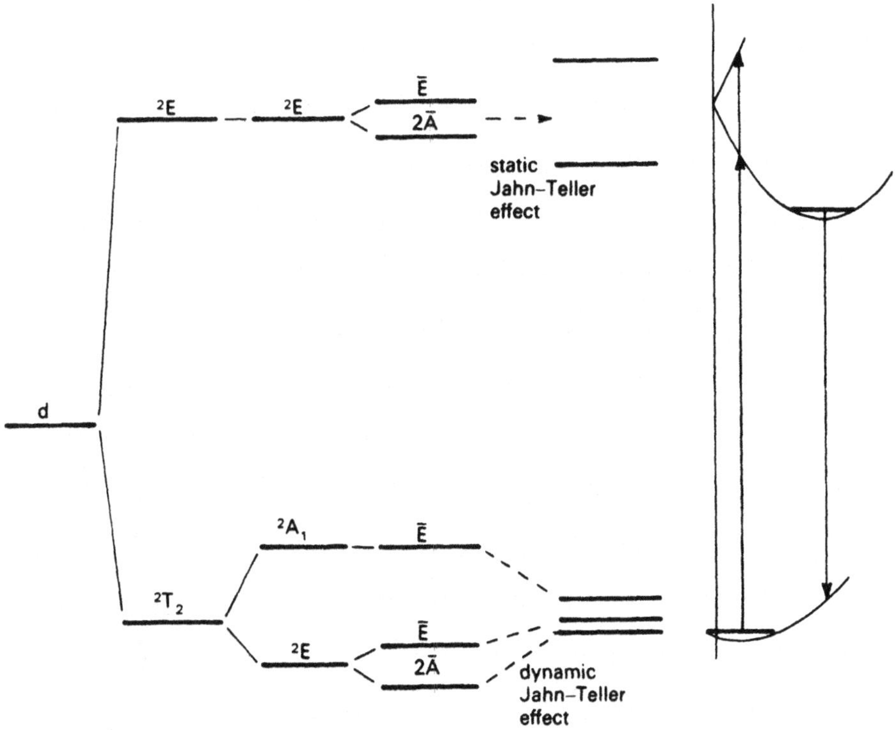

FIG. 9.12. Splitting of the 3d level in a static crystal field of trigonally distorted octahedral symmetry appropriate to Ti^{3+} ions in the Al^{3+} site in alumina. Jahn–Teller effects occur in the ground and excited states. On the right the levels are shown on a configurational coordinate diagram. For simplicity the ground-state splitting has been ignored. The arrows show the transitions corresponding to the two broad absorption bands and the single emission band. (After Gächter and Koningstein 1974.)

unusually-small splitting of the zero-phonon transition as a manifestation of the dynamic Jahn–Teller effect. Although the energy gap between the lowest excited states and the ground state of the $3d^2$ ion is small (Fig. 3.18), luminescence across this gap has been studied for Ti^{2+} in $MgCl_2$ by Jacobsen *et al.* (1986). We note that this transition, since it occurs between states of the same t_2^2 strong field configuration, is characterized by a small value of the Huang–Rhys parameter, S, and in addition the phonons in the $MgCl_2$ lattice are of low energy.

9.4.3 The $3d^4$ *configuration*

The best-known $3d^4$ ion is Mn^{3+}. This ion occupies octahedral sites in crystals. The change in energy of the various electronic levels with octahedral

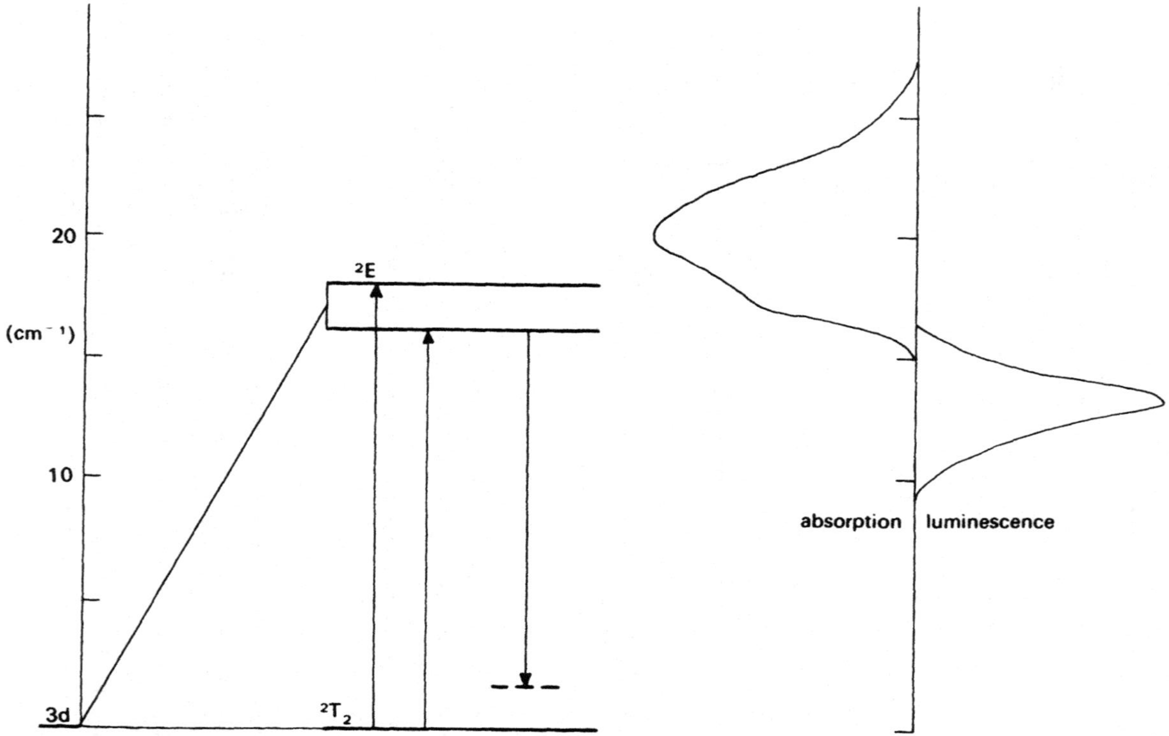

FIG. 9.13. Room temperature absorption and luminescence spectra of Ti^{3+} in aluminium oxide. The energy-level diagram represents the *zero-vibrational* levels. The small splitting in the ground state is omitted. The peak of the luminescence ends on a higher vibrational level of the ground state and so this system is suitable for use as a tunable four-level laser. The absence of ESA and the strong broad absorption in the visible are obvious advantages from the viewpoint of laser action.

crystal field strength was shown in Fig. 3.22. The free ion high-spin state, ^5D, splits into ^5E and $^5T_2$ states, the ^5E state being the lower. The higher energy spin triplet states also split in octahedral fields. For weak crystal fields the ground state is the high-spin ^5E state and a strong broad absorption transition is expected between the ^5E and $^5T_2$ levels which are separated in energy by $10\,Dq$. For strong octahedral crystal fields the $^3T_1$ level is lowest and the absorption spectrum should be quite different. In most materials the energy separations between adjacent levels are too small to allow strong luminescence from this ion.

9.4.4 The $3d^5$ *configuration*

Mn^{2+} and Fe^{3+} are the best-known examples of this configuration. The free ion ground state, ^6S, is the only spin sextet state, and it is not split by an octahedral crystal field (Fig. 3.24). Hence in weak fields there are no spin-allowed absorption transitions of the $3d^5$ configuration. The splittings in a tetrahedral crystal field are identical to those in an octahedral crystal field (Section 3.2.3). Because the absorption transitions are spin-forbidden they are weak, and consequently crystals doped with small concentrations of Mn^{2+} ions in octahedral sites are not strongly coloured.

Mn^{2+} ions in concentrated systems, such as the rutile-structured manganese fluoride, have been extensively investigated (Stout 1959). In this crystal each Mn^{2+} ion is surrounded by six equidistant F^- ions but, as Fig. 9.14

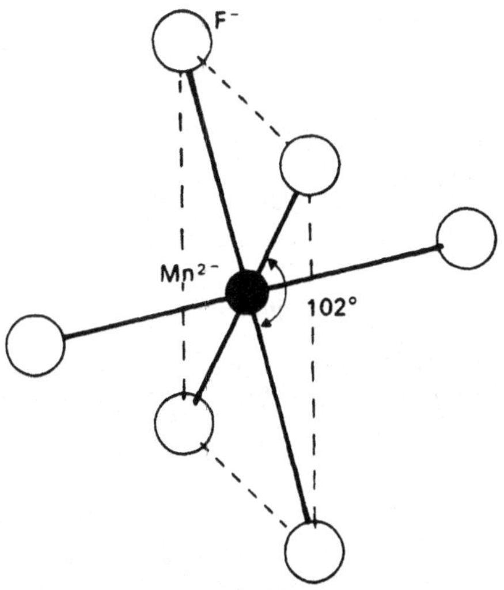

FIG. 9.14. Arrangement of fluorine ions around the manganese ion in manganese fluoride. The symmetry is reduced from octahedral because of the deviation of bond angles from 90° in the plane indicated in the figure.

shows, the angles between some of the adjacent Mn–F bonds are not right angles, and the resultant symmetry at the Mn^{2+} site is D_{2h}. This crystal field has a major component of octahedral symmetry and a smaller component of orthorhombic symmetry. The way in which the energy levels of a $3d^5$ electronic system are split by the octahedral crystal field is shown in Fig. 3.24. For Mn^{2+} in manganese fluoride a value of $Dq/B = 1.2$ is appropriate. At room temperature the optical absorption spectrum of manganese fluoride (Fig. 9.15) shows absorption band peaks which correspond very closely to those predicted from octahedral crystal field theory. The octahedral levels are split to a lesser extent by spin–orbit coupling, orthorhombic crystal field, and exchange interaction among the Mn^{2+} ions. These splittings have little effect on the positions of the absorption peaks, but they can be observed at low temperature as structure on the ${}^6A_1 \rightarrow {}^4T_1({}^4G)$ zero-phonon transition. The low temperature luminescence spectrum of manganese fluoride is discussed in some detail in Chapter 10.

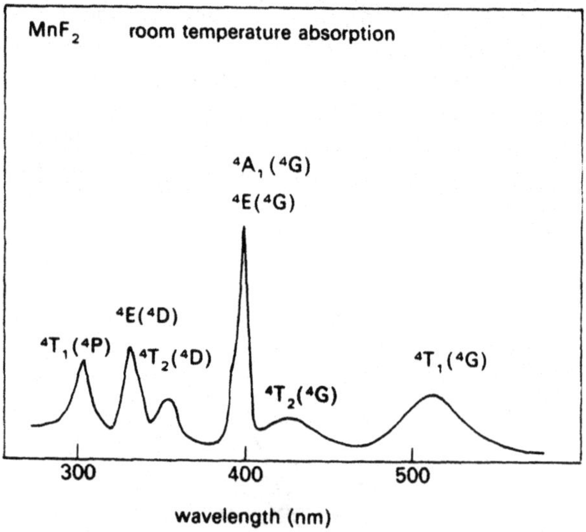

FIG. 9.15. Absorption spectrum of manganese fluoride at room temperature. The octahedral crystal field labels of the upper levels of the absorption transitions are indicated. (After Stout 1959.)

The ferric ion, $Fe^{3+}(3d^5)$, is found in octahedral and tetrahedral coordination. The identification of the Fe^{3+} absorption transitions is not as simple as are those of the Mn^{2+} ion. Luminescence spectra from Fe^{3+} reported in the literature (Pott and McNicol 1972; Melamed *et al.* 1972) are assigned to the ${}^4T_1({}^4G) \rightarrow {}^6A_1({}^6S)$ transition. When Fe^{3+} is incorporated into $LiGa_5O_8$ it substitutes for Ga^{3+} mainly in octahedral sites but with a small fraction of the Fe^{3+} ions substituting for Ga^{3+} in tetrahedral sites. The level splitting in the tetrahedral field is similar to that in the octahedral field (Section 3.2.3), the value of $|Dq|$ is smaller in the tetrahedral case. The octahedral Fe^{3+} ions dominate the absorption spectrum while the luminescence from this material observed near 15 000 cm^{-1} originates on tetrahedral Fe^{3+} ions. In contrast to the $^4T_1(^4G) \rightarrow {}^6A_1(^6S)$ luminescence transition from octahedral Mn^{2+} in manganese fluoride, which has weak magnetic dipole zero-phonon lines accompanying the broadband emission, luminescence from Fe^{3+} in tetrahedral sites in $LiGa_5O_8$ has a strong zero-phonon transition as is to be expected for an electric dipole process from tetrahedrally-coordinated Fe^3 (McShera *et al.* 1982).

9.4.5 *The* 3d^6 *configuration*

The free ion ground state is ^5D, the only spin quintet state, higher energy states being spin triplets and singlets. The splittings of free ion levels by an octahedral field are shown in Fig. 3.23. The ^5D state splits into $^5T_2$ and ^5E states of which the $^5T_2$ state lies lowest, and in weak crystal fields the $^5T_2$ state is the ground state. This is the situation generally found for Fe^{2+} in octahedral sites. Strong $^5T_2 \rightarrow {}^5E$ absorption occurs in the near infrared. Absorption transitions in the visible spectrum are spin-forbidden and weak. Because the adjacent levels are close together luminescence is not expected from Fe^{2+} ions.

Co^{3+} has the same 3d^6 configuration as Fe^{2+}. Co^{3+} substitutes for Al^{3+} in alumina in a strong crystal field configuration so that the $^1A_1$ level is lowest (McClure 1962). Two strong absorption bands, one to the $^1T_1$ level at 15 560 cm^{-1} and the other to the $^1T_2$ level at 22 980 cm^{-1}, are observed along with a number of weaker transitions to the triplet states. Co^{3+} enters tetrahedral sites in YAG (Wood and Remeika 1967). The energy level diagram for a 3d^6 ion in a tetrahedral crystal field is the same as that of a 3d^4 ion in an octahedral field (Fig. 3.22). A strong absorption band at around 8000 cm^{-1} is found in YAG:Co^{3+} and is attributed to the $^5E \rightarrow {}^5T_2$ transition on the Co^{3+} ion in tetrahedral sites.

9.4.6 *The* 3d^7 *configuration*

The more commonly encountered valence state of cobalt is Co^{2+} which is observed in sites of octahedral and tetrahedral symmetry. The energy levels of Co^{2+} in octahedral symmetry were shown in Fig. 3.21. In weak crystal field sites, which Co^{2+} ions normally occupy, the free ion term, ^4F, splits into $^4T_1$, $^4T_2$, and $^4A_2$ levels with $^4T_1$ level as the ground state (Fig. 9.16). The principal absorption bands of Co^{2+} in magnesium fluoride are due to the $^4T_1 \rightarrow {}^4T_2$ transition at around 8000 cm^{-1} and the $^4T_1 \rightarrow {}^4A_2$ transitions at around 18 000 cm^{-1} (Pappalardo *et al.* 1961a). Ions in the lowest excited $^4T_2$ state decay with the emission of near-infrared luminescence; at low tempe-

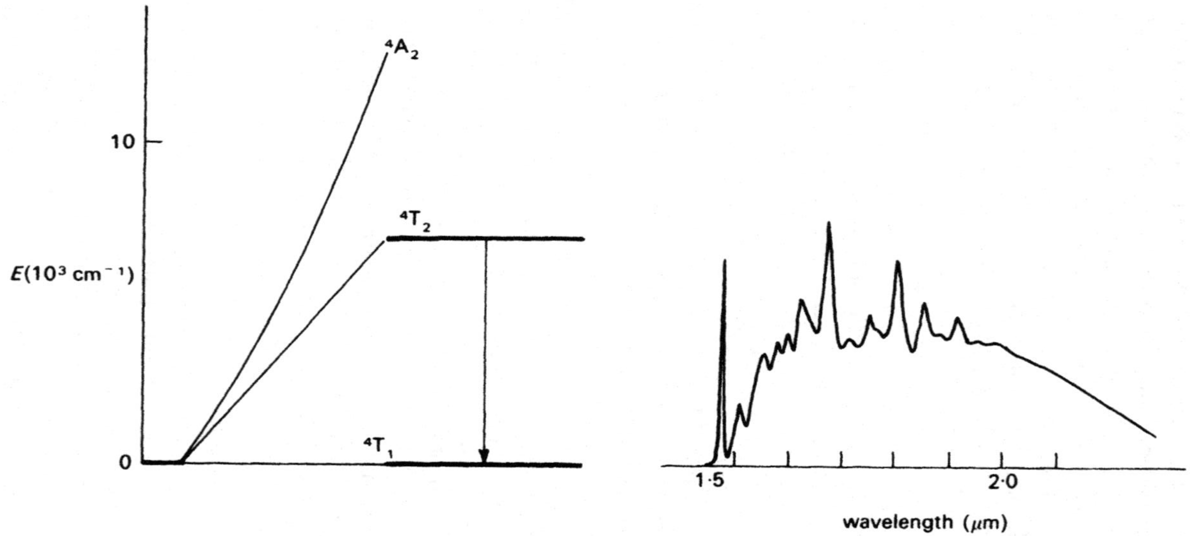

FIG. 9.16. The low-lying energy levels of Co^{2+} in an octahedral crystal field and the luminescence spectrum of $MgF_2{:}Co^{2+}$ at 77 K.

ratures this luminescence consists of sharp zero-phonon lines and a broad sideband, all being predominantly magnetic dipole in nature. The luminescence spectrum of Co^{2+} in magnesium fluoride at 77 K is shown in Fig. 9.16. The broadband luminescence in MgF_2:Co and in some other fluoride crystals at low temperatures is sufficiently efficient to be the basis for a continuously tunable laser over the wavelength range 1.6 μm to 2.3 μm.

This $^4T_2 \rightarrow {}^4T_1$ transition on Co^{2+} ions in octahedral sites is interesting because of its long wavelength. In view of the small gap between the $^4T_2$ and $^4T_1$ levels and the weak magnetic dipole nature of the transition one might expect that non-radiative processes would have dominated the decay out of the $^4T_2$ level. Sturge (1973) has made a detailed analysis of the non-radiative relaxation from the $^4T_2$ level of Co^{2+} ions in octahedral sites in $KMgF_3$. The observed decay time varies from 3.3 ms at 1.5 K to 2.5 μs at 550 K. At the lowest temperatures the non-radiative relaxation rate is about half the radiative decay rate while at high temperatures non-radiative decay dominates the relaxation process. Sturge argued that the non-radiative process is unusually weak because of anharmonicity in the coupling of the Co^{2+} ion to the lattice.

The spectroscopy of the Co^{2+} ion in tetrahedral sites in zinc oxide has been investigated by Pappalardo *et al.* (1961*a*). The splitting of the $Co^{2+}(3d^7)$ levels in tetrahedral sites is the same as the splitting of the $Cr^{3+}(3d^3)$ levels in octahedral sites (Fig. 3.20). However, Dq is much weaker in the tetrahedral sites, Dq/B being around 0.5 rather than 2–3 as is found for Cr^{3+} in octahedral sites. Pappalardo *et al.* (1961*a*) found the oscillator strengths of the Co^{2+} transitions for tetrahedral sites to be approximately three orders of magnitude greater than those found in octahedral sites, as is consistent with the absence of inversion symmetry in the tetrahedral site. Strong spin-allowed absorption transitions occur from the $^4A_2$ ground state to the $^4T_2$, $^4T_1(^4F)$, $^4T_1(^4P)$ excited levels. This last transition is in the yellow-green region of the visible spectrum resulting in the blue coloration of so many cobalt systems. At the value of $Dq/B \simeq 0.5$ we note that the $^2E(^2G)$ and $^2T_1(^2P)$ levels are very close. For Co^{2+} ions in tetrahedral sites in $ZnAl_2O_4$ the $^2E(^2G)$ level is the lower and the luminescence occurs from this level (Ferguson *et al.* 1961). In the case of Co^{2+} ions in tetrahedral sites in $LiGa_5O_8$ the $^4T_1(^4P)$ level is the lower and luminescence occurs from this level. Figure 9.17 shows the $^4A_2 \leftrightarrow {}^4T_1(^4P)$ absorption and luminescence transitions of $LiGa_5O_8:Co^{2+}$ at 77 K (Donegan *et al.* 1984). The radiative decay time is 200 ns, consistent with a spin-allowed electric dipole process on a tetrahedral ($3d^3$) ion.

9.4.7 *The* 3d^8 *configuration*

Divalent nickel, $Ni^{2+}(3d^8)$, occurs in octahedral and tetrahedral sites. The expected energy level diagrams are given in Figs. 3.19 and 3.18, respectively. A study by Pappalardo *et al.* (1961*b*) on Ni^{2+} ions in octahedral sites in

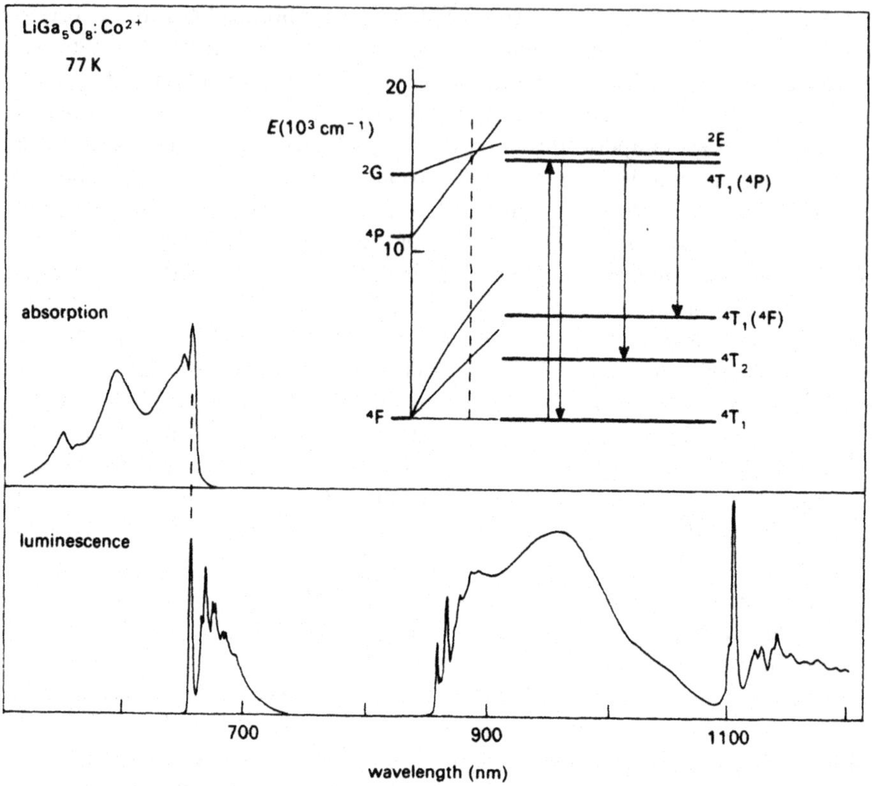

FIG. 9.17. Energy level diagram for Co^{2+} in a tetrahedral crystal field appropriate to $LiGa_5O_8:Co^{2+}$. The $^4A_2 \rightarrow {}^4T_1$ (^4P) absorption spectrum is shown along with the luminescence spectrum from the $^4T_1$ (^4P) state. The spectra are taken at 77 K. (Donegan *et al.* 1984.)

magnesium oxide showed that the positions of the levels are in good agreement with the predictions of crystal field theory, with $Dq/B = 0.9$. Ni^{2+} in octahedral sites in potassium zinc fluoride and magnesium fluoride (Iverson and Sibley 1979) have slightly smaller values of Dq. Figure 9.18 shows the low temperature absorption and luminescence transitions in $KMgF_3:Ni^{2+}$ which are characteristic of Ni^{2+} in octahedral fluoride crystals. Luminescence occurs in the near infrared ($^3T_2 \rightarrow {}^4A_2$), the red ($^1T_2 \rightarrow {}^3T_2$), and green ($^1T_2 \rightarrow {}^3A_2$). Because of the small gap ($7000\ cm^{-1}$) between $^3T_2$ and $^3A_2$ one might expect non-radiative processes to be effective. In $MgF_2:Ni^{2+}$ the decay time of the $^3T_2$ level decreases with increasing temperature which indicates the presence of non-radiative decay at high temperatures. ESA from the $^3T_2$ level appears to occur in these Ni^{2+}-doped

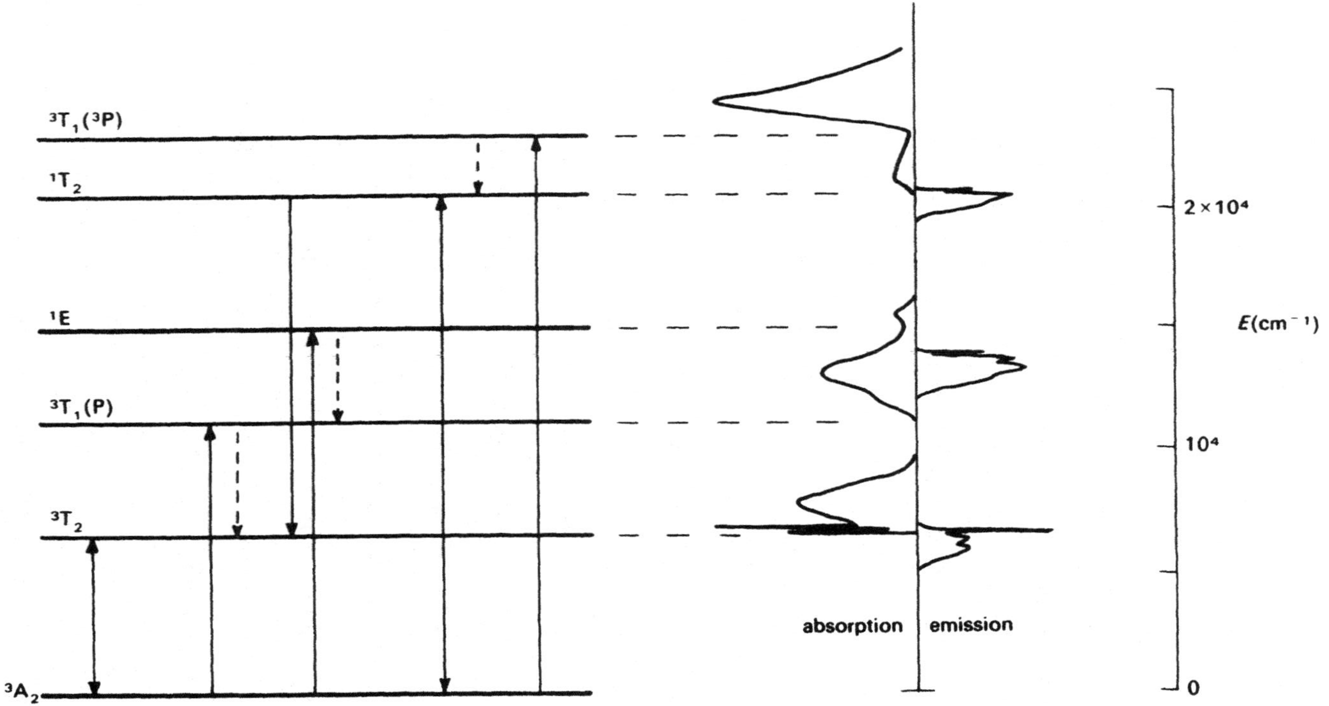

FIG. 9.18. Zero-vibrational levels as well as radiative and nonradiative transitions of Ni^{2+} in an octahedral crystal field appropriate to $KMgF_3{:}Ni^{2+}$. Low-temperature absorption and luminescence spectra of $KMgF_3{:}Ni^{2+}$ are shown on the right.

materials. The absorption spectrum of Ni^{2+} in a tetrahedral cation site in zinc oxide was analysed by Pappalardo *et al.* (1961*b*) who showed the observed energy levels to be in good agreement with the predictions of crystal field theory. As expected the oscillator strengths of transitions on the tetrahedral Ni^{2+} ions are much stronger than are found in the case of the octahedral Ni^{2+} ion.

The optical and ESR spectra of the isoelectronic ion Cu^{3+} in aluminium oxide were reported by Blumberg *et al.* (1963).

9.4.8 The $3d^9$ *configuration*

This configuration is complementary to the $3d^1$ configuration. The free ion ^2D state splits into $^2T_2$ and ^2E in an octahedral crystal field, the ^2E state now being lowest. The energy separation in $10\,Dq$. The $^2E \rightarrow {}^2T_2$ absorption bands due to transitions of $Cu^{2+}(3d^9)$ cause the strong colour of many copper salts.

9.5 Transition metal ions in glass

Glass is a continuous *random* network of ionic polyhedra (the glass *formers*) which may be joined together to form simple glass systems or which may be bound into the glass network by additional compounds, called network *modifiers* (Elliot 1984). Examples of glass formers are silicate tetrahedra, phosphate tetrahedra, and borate triangles. The polyhedra themselves have a definite geometry but they are connected in no particular order. Amorphous silica (SiO_2) is an example of a very simple oxide glass based on silicate (SiO_4) tetrahedra connected at the corners (Zallen 1983). More complicated glasses employ network modifiers which may be oxides of lithium, sodium, magnesium, etc. Most glasses in common use are based on oxides, but glasses based on halides have their own special properties and interests.

When doped with rare-earth or transition metal ions the dopant ions are generally expected to enter substitutionally for network modifier cations. In contrast to the case of dopant ions in crystals, where the environments of all ions are essentially identical and these ions experience the same crystal field and have the same transition frequencies, dopant ions in glass experience a wide range of crystal fields and display a spread of transition frequencies. The spectroscopy of rare-earth ions and of transition metal ions in glass have been reviewed by a number of workers (Reisfeld and Jørgensen 1975; Weber 1981). The radiative transitions in glass are broader than in crystals, reflecting the inhomogeneous broadening which results from the wide range of sites with different crystal fields. Dopant transition metal ions, being more sensitive to electrostatic crystal fields than are rare-earth ions, exhibit the broader transitions.

Cr^{3+} is the transition metal ion whose spectroscopy in glass has been most widely investigated. The luminescent properties of a wide range of chromium-doped glasses were investigated by Andrews *et al.* (1981). The shaded region in Fig. 9.9 shows the range of values of Dq/B found in a typical

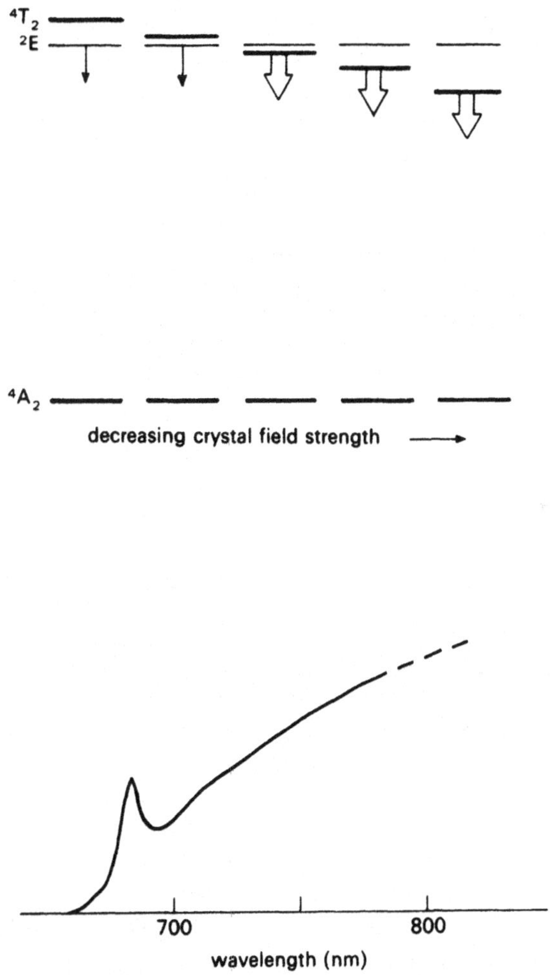

FIG. 9.19. Schematic representation of some of the energy levels of Cr^{3+} ions in sites of different crystal field strength in an oxide glass. The luminescence spectrum is that of chromium-doped ED-2 glass at 77 K. The feature at around 690 nm is the composite ^2E luminescence from Cr^{3+} ions in high-field sites. The broad luminescence from the $^4T_2$ levels of ions in low-field sites stretches to 1.5 μm.

borate or phosphate glass. Figure 9.19 shows in a schematic way how the ^2E and $^4T_2$ energy levels of Cr^{3+} might vary in a glass host with a continuous range of Dq values. The position of the $^4T_2$ level varies strongly with Dq while that of the ^2E level is much less affected. In this example Cr^{3+} ions occupy both high-field and low-field sites, and the overall luminescence spectrum (Fig. 9.19) consists of a very broad band, comprising the collective $^4T_2$ emissions, as well as a sharper band comprising the collective ^2E emissions.

The spectroscopy of individual Cr^{3+} ions in glass is not necessarily the same as that of individual Cr^{3+} ions in crystals, and there are many aspects of the spectroscopy of doped glasses which need investigation. In the glass network the dopant Cr^{3+} ions may act as nucleating agents resulting in clustering and the appearance of chromium-rich phases. Further, the oscillating modes of the glass network differ from crystals in that a high density of low-frequency 'tunnelling modes' exist in the glass but do not occur in crystals (Anderson *et al.* 1972; Phillips 1981). It is found that the temperature-dependent broadening of zero-phonon lines of dopant rare-earth ions in glass is anomalously large at low temperatures in comparison with that observed in crystals (Selzer *et al.* 1976; Hegarty and Yen 1979). A similar effect is found for the $^2E \rightarrow {}^4A_2$ zero-phonon lines of Cr^{3+} in glass (Bergin *et al.* 1986), and in comparison with most chromium-doped crystals, whose luminescence quantum efficiency is close to unity, the luminescence quantum efficiency of all chromium-doped glasses so far investigated is disappointingly low.

An interesting host material intermediate between a crystal and a glass is a transparent glass ceramic. These can be fabricated by thermally inducing devitrification of appropriately formulated glasses to produce submicron-size crystallites embedded in a matrix of residual glass. It is found (Andrews *et al.* 1984) that a transparent glass ceramic based on the precipitation of the aluminosilicate mineral mullite ($2Al_2O_3.SiO_2$) transfers Cr^{3+} ions from the glassy phase to the crystal phase during ceramic formation. The quantum efficiency of the Cr^{3+}-doped mullite in this ceramic is considerably higher than in the glasses. Such ceramic materials can be fabricated at temperatures well below that of the corresponding crystals. The maximum dimensions of the mullite crystallites in the ceramic are around 40 nm, and despite the mismatch in refractive index between the mullite and the host glass scattering is low because of the small size of the crystallites. The spectra of the Cr^{3+}-doped mullite ceramics are difficult to interpret bcause of the width of the transitions. There seems to be a range of high-field and low-field sites, and the luminescence decay patterns are non-exponential.

10
Spectroscopy at high dopant concentrations

So far this treatment of the spectroscopy of optically active centres in solids has assumed these centres to be isolated from each other. When the concentration of centres is increased, or when a second type of optically active centre is also introduced, allowance must be made for the possibility that two centres are sufficiently close to each other to interact. This interaction may be too weak to have a detectable effect on the energy levels so that the centres retain their individual identities but may yet be sufficiently strong to enable excitation to be transferred from one to another. The occurrence of *energy transfer* between centres can significantly affect the luminescence properties of a material. In some cases two or more centres may be near enough to interact strongly, so forming a new spectroscopic centre. In fully concentrated crystals such as manganese fluoride, chromic oxide, or praseodymium trifluoride the inter-ion coupling is so strong that the excitation must be interpreted as shared by all the ions acting collectively. Such states of collective excitation are *exciton* states, which propagate through the crystal with a characteristic wavevector k.

In this chapter we investigate the interactions between optically active ions under three headings: the transfer of optical excitation between ions, strongly interacting ion pairs which constitute new spectroscopic centres, and 100% concentration of optically active ions.

10.1 Energy transfer between optically active ions

To analyse the phenomenon of energy transfer as it is experimentally observed one must look at two distinct aspects, one microscopic, the other macroscopic. First we seek to understand the *microscopic* process in which energy on one ion is transferred to a nearby ion, then estimate the *macroscopic* behaviour of a very large number of randomly distributed ions in the material, since this is what is observed experimentally.

10.1.1 Theoretical analysis of energy transfer

The microscopic energy transfer process in inorganic materials was first considered theoretically by Förster (1948) and Dexter (1953). The excitation is transferred from a donor ion, D, to an acceptor ion, A, separated by a distance R. (The terms *sensitizer ion* and *activator ion* are used as alternatives

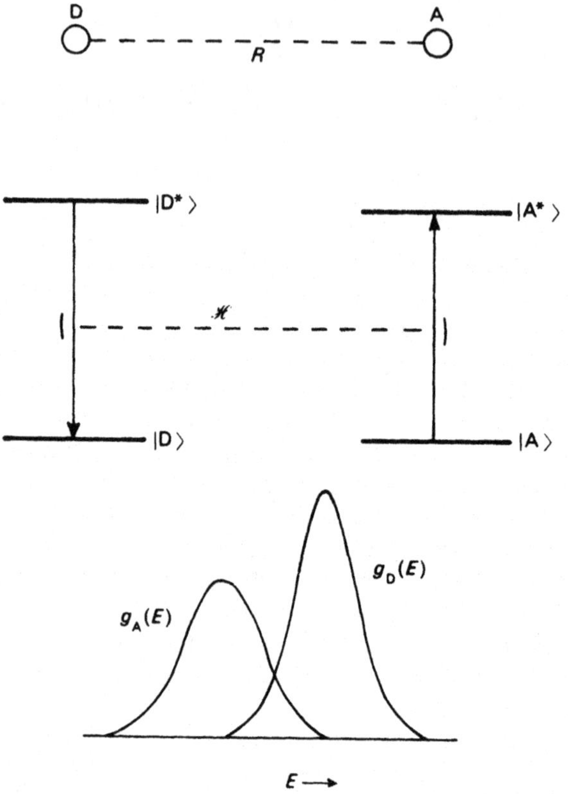

FIG. 10.1. Energy may be transferred from donor ion, D, to acceptor ion, A, by a non-radiative process which is analogous to a simultaneous emission process on D and an absorption process on A. Overlap of the corresponding emission and absorption bandwidths is necessary for conservation of energy.

for D and A.) Schematic energy levels and transitions are shown in Fig. 10.1, in which an asterisk indicates an excited state. We assume that the radiative emission transition $D^* \to D$ and the radiative absorption transition $A \to A^*$ have the normalized lineshape functions $g_D(E)$ and $g_A(E)$ appropriate to homogeneously broadened transitions. Initially the two ions are in the state $|D^*, A\rangle$. An interaction $\mathscr{H}'$ between the ions causes a transition from $|D^*, A\rangle$ to $|D, A^*\rangle$. The transition probability is

$$W_{DA} = \frac{2\pi}{\hbar} |\langle D, A^* | \mathscr{H}' | D^*, A\rangle|^2 \int g_D(E) g_A(E) \mathrm{d}E \qquad (10.1)$$

and the overlap integral reflects the requirement of energy conservation. This process is non-radiative; it does *not* involve the emission of light by D and the subsequent absorption by A. Rather is it a simultaneous deactivation of D and activation of A. The interactions which cause energy transfer are electrostatic coupling, magnetic coupling, and/or exchange coupling between ions.

The electrostatic interaction between the electrons on D and on A may be written (Fig. 10.2)

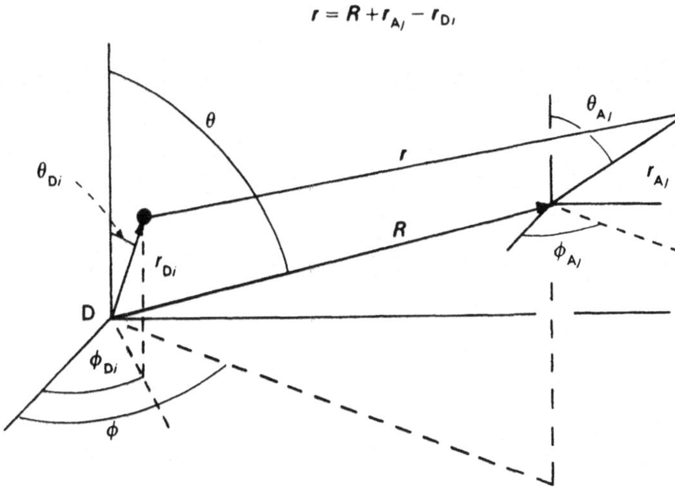

FIG. 10.2. Relationship between the coordinates of the electrons on the D and A ions.

$$\mathscr{H}'_{\mathrm{ES}} = \frac{1}{4\pi\varepsilon_0}\frac{1}{\kappa}\sum_{i,j}\frac{e^2}{|\boldsymbol{R}+\boldsymbol{r}_{\mathrm{A}j}-\boldsymbol{r}_{\mathrm{D}i}|} \tag{10.2}$$

where the sum is over the electrons on the two ions and a dielectric constant term has been included. When $\boldsymbol{R}$ is larger than $\boldsymbol{r}_{\mathrm{D}i}$ or $\boldsymbol{r}_{\mathrm{A}j}$ this potential term can be expanded (Carlson and Rushbrooke 1950) as follows:

$$\mathscr{H}'_{\mathrm{ES}} = \frac{1}{4\pi\varepsilon_0\kappa}\sum_{i,j}\sum_{\substack{k_1,k_2\\q_1,q_2}}\frac{G(k_1,k_2,q_1,q_2)}{\boldsymbol{R}^{1+k_1+k_2}} \times [C^{(k_1+k_2)}_{q_1+q_2}(\theta,\phi)]^*\mu^{(k_1)}_{q_1}(\boldsymbol{r}_{\mathrm{D}})\,\mu^{(k_2)}_{q_2}(\boldsymbol{r}_{\mathrm{A}}) \tag{10.3}$$

where

$$G(k_1,k_2,q_1,q_2) = (-1)^{k_1}\sqrt{\frac{(2k_1+2k_2+1)!}{(2k_1)!\,(2k_2)!}}\begin{pmatrix} k_1 & k_2 & k_1+k_2 \\ q_1 & q_2 & -q_1-q_2 \end{pmatrix} \tag{10.4}$$

$$C^{(k)}_q(\theta,\phi) = \sqrt{\frac{4\pi}{2k+1}}\,Y^q_k(\theta,\phi) \tag{10.5}$$

$$\mu^{(k)}_q(\boldsymbol{r}_{\mathrm{D}}) = \sum_i e r^k_{\mathrm{D}i}\,C^{(k)}_q(\theta_{\mathrm{D}i},\phi_{\mathrm{D}i}). \tag{10.6}$$

There is a corresponding formula for $\mu_q^{(k)}(\boldsymbol{r}_A)$. The 3j symbol is defined in terms of Clebsch–Gordan coefficients as follows:

$$\begin{pmatrix} k_1 & k_2 & k_3 \\ q_1 & q_2 & q_3 \end{pmatrix} = (-1)^{k_1-k_2-q_3}(2k_3+1)^{-1/2}\langle k_1 k_2 q_1 q_2 | k_1 k_2 k_3 - q_3\rangle. \quad (10.7)$$

In the summation k takes on integer values between 0 and ∞, while q has integer values from $-k$ to $+k$. The $k_1=0$ and $k_2=0$ terms denote constant monopolar terms. The first term in the expansion contributing to energy transfer comes from the values $k_1=1$, $k_2=1$; this term contains the electric dipole–dipole interaction $er_{Di}er_{Aj}/R^3$. Next is the electric dipole–electric quadrupole term which varies as R^{-4}, and subsequently the electric quadrupole–electric quadrupole term varies as R^{-5}.

The probability of energy transfer from D to A due to $\mathscr{H}'_{ES}$ varies as the square of the matrix element of $\mathscr{H}'_{ES}$. If this matrix element squared is averaged over angles θ, ϕ, neglecting cross-terms, we obtain a simplified general formula (Kushida 1973):

$$|\langle D, A^*|\mathscr{H}'_{ES}|D^*, A\rangle|^2 = \left(\frac{1}{4\pi\varepsilon_0\kappa}\right)^2 \sum_{k_1,k_2}\left(\frac{1}{R^{1+k_1+k_2}}\right)^2 \times \frac{(2k_1+2k_2)!}{(2k_1+1)!\,(2k_2+1)!}|\mu_D^{(k_1)}|^2\,|\mu_A^{(k_2)}|^2 \quad (10.8)$$

where

$$|\mu_D^{(k_1)}|^2 = \sum_{q_1}|\langle D|\mu_{q_1}^{(k_1)}(\boldsymbol{r}_D)|D^*\rangle|^2 \quad (10.9)$$

with a similar formula for $|\mu_A^{(k_2)}|^2$.

The transition probability for energy transfer by electric dipole–dipole interaction is obtained from the $k_1=1$, $k_2=1$ term in eqn (10.8). We find

$$W_{DA}^{dd} = \frac{4\pi}{3\hbar}\left(\frac{1}{4\pi\varepsilon_0\kappa}\right)^2 \frac{1}{R^6}|\langle\mu_D^{(1)}\rangle|^2\,|\langle\mu_A^{(1)}\rangle|^2 \int g_D(E)g_A(E)\,dE. \quad (10.10)$$

The factor $|\langle\mu^{(1)}\rangle|^2$ in eqn (10.10) is identified with the factor $|\langle a|\boldsymbol{\mu}_e|b\rangle|^2$ which is the square of the matrix element for the electric dipole radiative transition between states b and a, defined in Chapter 4. This is proportional to the oscillator strength of the radiative transition,

$$f_D(ED) = \frac{2m\omega}{3\hbar e^2}|\langle\mu_D^{(1)}\rangle|^2 \quad (10.11)$$

where ω is the central frequency of the transition. We can express W_{DA}^{dd} in terms of the oscillator strengths of the $D^* \rightarrow D$ and $A \rightarrow A^*$ transitions:

$$W_{DA}^{dd} = \left(\frac{1}{4\pi\varepsilon_0}\right)^2 \frac{3\pi\hbar e^4}{n^4 m^2\omega^2}\frac{1}{R^6} f_D(ED) f_A(ED)\int g_D(E)g_A(E)\,dE \quad (10.12)$$

using $\kappa = n^2$, where n is the refractive index, and ω is taken as the average frequency of the transitions involved in the transfer process. f_D and f_A can be determined from the radiative decay rate of the $D^* \rightarrow D$ radiative transition and from the absorption strength of the $A \rightarrow A^*$ transition, respectively, as described in Chapter 4.

To estimate the transfer rate expected by this process for typical rare-earth ions in solids we use $f_D(\mathrm{ED}) = f_A(\mathrm{ED}) = 10^{-6}$, and assume that the absorption transition on A and the emission transition on D are fully overlapping each other, each with an homogeneous width of $10\,\mathrm{cm}^{-1}$. This gives an overlap integral of $\simeq 10^{22}\,\mathrm{J}^{-1}$. Taking $\omega \simeq 5 \times 10^{15}\,\mathrm{rad\,s}^{-1}$, $R = 10^{-9}\,\mathrm{m}$, and $n = 1.7$ we find that $W_{\mathrm{DA}}^{\mathrm{dd}} \simeq 10^3\,\mathrm{s}^{-1}$, which is approximately the radiative decay rate for the $D^* \rightarrow D$ transition. Hence an excited ion with another unexcited ion nearby is as likely to lose its energy by transfer as by radiative emission.

Up to this point the analysis has been valid for the case in which there is a resonance between reasonably sharp electronic transitions on two nearby ions. Does the same formalism hold in the case where the transitions occur by phonon-assisted processes, described in Section 5.4, and where $g_D(E)$ and $g_A(E)$ describe broad bands? Since the phonon-assisted process involves modulation of the electronic energy levels by lattice vibrations and since the *same* lattice vibration modulates the processes on nearby donor and acceptor ions, there can be interference terms which reduce the transfer rate below that predicted by eqn (10.12). This point has been discussed by Orbach (1975).

If the electric dipole oscillator strength on one or both ions is small then higher-order terms, e.g. dipole–quadrupole and quadrupole–quadrupole, must be taken into consideration in eqn (10.8). For the electric dipole–quadrupole transfer mechanism we put $k_1 = 1$ and $k_2 = 2$ obtaining

$$W_{\mathrm{DA}}^{\mathrm{dq}} = \frac{2\pi}{\hbar}\left(\frac{1}{4\pi\varepsilon_0 \kappa}\right)^2 \frac{1}{R^8} |\langle \mu_{\mathrm{D}}^{(1)} \rangle|^2 \sum_{q_2} |\langle \mathrm{A}^* | \mu_{q_1}^{(2)} | \mathrm{A} \rangle|^2 \int g_{\mathrm{D}}(E) g_{\mathrm{A}}(E)\,\mathrm{d}E. \tag{10.13}$$

A comparison of $W_{\mathrm{DA}}^{\mathrm{dq}}$ with $W_{\mathrm{DA}}^{\mathrm{dd}}$ is interesting. We find

$$\frac{W_{\mathrm{DA}}^{\mathrm{dq}}}{W_{\mathrm{DA}}^{\mathrm{dd}}} = \frac{3}{2}\frac{1}{R^2} \frac{\sum_{q_2} |\langle \mathrm{A}^* | \mu_{q_2}^{(2)} | \mathrm{A} \rangle|^2}{\sum_{q_2} |\langle \mathrm{A}^* | \mu_{q_2}^{(1)} | \mathrm{A} \rangle|^2} \simeq \left(\frac{a_0}{R}\right)^2 \tag{10.14}$$

assuming allowed electric dipole and electric quadrupole matrix elements with values ea_0 and ea_0^2, respectively, where a_0 is the Bohr radius. Taking $R = 10a_0$ this ratio is $\simeq 10^{-2}$, whereas the ratio of the allowed electric quadrupole *radiative* transition to the allowed electric dipole *radiative* transition is $\simeq (a_0/\lambda)^2$, λ being the radiation of the wavelength involved. That this ratio is very small, $\simeq 10^{-7}$, means that quadrupolar radiative processes are unimportant in solids whereas quadrupolar energy transfer processes can be important particularly if the electric dipole strengths are weak. As a general shorthand notation we write the energy transfer rates as

$$W_{DA}^{dd} = \alpha_{DA}^{(6)}/R^6,\ W_{DA}^{dq} = \alpha_{DA}^{(8)}/R^8,\ W_{DA}^{qq} = \alpha_{DA}^{(10)}/R^{10}. \qquad (10.15)$$

The transition rate for energy transfer by magnetic dipole–dipole processes may be calculated from eqn (10.10) by replacing the electric dipole matrix elements by magnetic dipole matrix elements, by replacing n^4 by unity and by replacing $(4\pi\varepsilon_0)^{-1}$ by $(\mu_0/4\pi)$. This rate also varies as R^{-6}. Higher-order magnetic interaction terms are too small to be of consequence.

If the ions are close enough for direct overlap of the electron clouds to occur there can be a direct exchange interaction between the ions. Even if direct overlap does not occur, weaker superexchange interactions via intervening ions may still be present. Exchange interactions, $\mathscr{H}'_{EX}$, are very short range, and the transfer matrix element $\langle D, A^*|\mathscr{H}'_{EX}|D^*, A\rangle$ may reasonably be written as $J_0 \exp(-R/L)$, where J_0 is often approximated by the diagonal exchange term between D and A ions in nearest-neighbour positions, and L is a typical nearest-neighbour distance ($\simeq 10^{-10}$ m).

Finally we consider energy transfer when there is an energy mismatch between the transitions on the donor and acceptor ions (Fig. 10.3). Such an energy mismatch may be bridged by lattice vibrations, in which case the electron–phonon coupling must be invoked together with the multipole interaction between D and A ions.

Energy transfer finds application in the common phosphor $Ca_5(PO_4)_3(F, Cl){:}Sb^{3+}, Mn^{2+}$ which has long been used as a coating

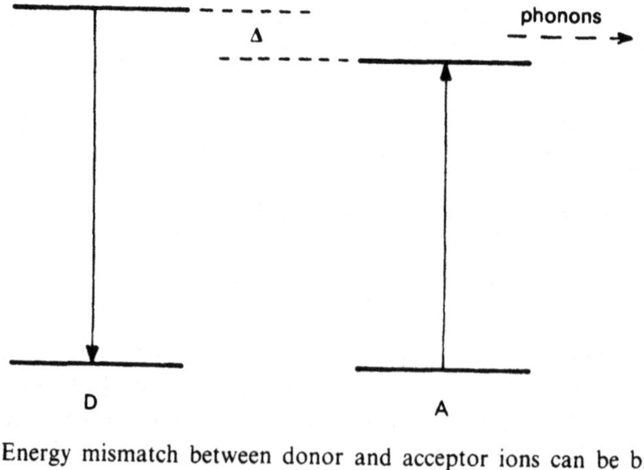

FIG. 10.3. Energy mismatch between donor and acceptor ions can be bridged by lattice vibrations.

around the mercury discharge in standard fluorescence lamps. The radiation from the mercury discharge excites the Sb^{3+} ions which in turn can emit in the blue. The Mn^{2+} dopant ions are poor absorbers but some excited Sb^{3+} ions transfer excitation to the Mn^{2+} ions which then emit a characteristic yellow-orange luminescence. By varying the manganese content of the phosphor one can change the balance between broadband emission in the blue and orange regions, thereby modifying the nature of the 'white' light emitted by the fluorescent lamp. Energy transfer is also involved in the red-emitting television phosphor $YVO_4:Eu^{3+}$, in which transfer occurs from the broad-band-emitting vanadate groups (VO_4^{3-}) to the Eu^{3+} ions which emit in the red. Transfer of energy also occurs among vanadate groups and this can facilitate the excitation of the Eu^{3+} ions. In this case the transfer among the vanadate groups is termed donor–donor transfer, the transfer from VO_4^{3-} to Eu^{3+} is termed donor–acceptor transfer. In manganese fluoride the Mn^{2+} excitation migrates very rapidly among the Mn^{2+} and tends to end up being transferred to unintentional dopant ions which occur in trace amounts in the manganese fluoride crystal. So efficient is this transfer that even when the unintentional dopants are present in concentrations of only parts per million, the regular manganese luminescence is strongly 'quenched'. To study energy transfer experimentally one might selectively excite a particular ion species in a material and observe the resulting luminescence. In such an experiment one studies the excitation of and luminescence from a *macroscopic* volume of the material which contains a large number of randomly distributed donor and acceptor ions. Relating the macroscopic experimental measurements to the *microscopic* formulae we have developed requires statistical averages to be taken over the macroscopic volume of the material.

10.1.2 Statistical aspects of energy transfer

In a statistical treatment of energy transfer the starting point is the set of coupled equations (Huber 1979):

$$\frac{dP_n(t)}{dt} = -\left(\frac{1}{\tau_R} + X_n + \sum_{n'} W_{nn'}\right) P_n(t) + \sum_{n'} W_{n'n} P_{n'}(t) \tag{10.16}$$

in which $P_n(t)$ is the probability that the nth donor ion is excited at time t; if the nth ion is excited P_n takes on the value unity. τ_R is the intrinsic decay time of the excited donor ion, $W_{nn'}$ is the donor–donor (D–D) transfer rate from the nth donor to the n'th donor, X_{nl} is the donor–acceptor (D–A) transfer rate from the nth donor to the lth acceptor, and $X_n = \Sigma X_{nl}$ where the summation is over all acceptors. The quantities $W_{nn'}$ and X_{nl} are related to our previous transfer rates as follows:

$$W_{nn'} = W_{Dn,\,Dn'}; \quad X_{nl} = W_{Dn,\,Al} \tag{10.17}$$

and the value of X_n depends on the arrangement of acceptor ions in the vicinity of the nth donor ion. Back transfer from acceptors to donors is omitted from eqn (10.15) as it can generally be neglected. The number of excited donor ions is found by multiplying the total number of donor ions (N_D) by the configurational average of $P_n(t)$ over all arrangements of donors and acceptors. This configuration average is

$$\langle P(t)\rangle_c = \frac{1}{N_D}\sum_n P_n(t) \tag{10.18}$$

The intensity of donor luminescence is proportional to $N_D\langle P(t)\rangle_c$.

We first consider the 'static' case where no D–D transfer occurs, i.e. $W_{nn'} = W_{n'n} = 0$ in eqn (10.16). In this case one obtains an exact solution for $\langle P(t)\rangle_c$ (Huber 1979):

$$\langle P(t)\rangle_c = \exp(-t/\tau_R)\prod_l(1 - C_A + C_A\exp(-X_{0l}t)) \tag{10.19}$$

where C_A is the probability that a site is occupied by an acceptor and l refers to all the sites in the lattice. If a number of donors is excited by a sharp excitation pulse the donor luminescence decays as $\langle P(t)\rangle_c$. Thus the decay is in general non-exponential, the initial part varying as

$$\text{rate}\,(t = 0) = (\tau_R^{-1} + C_A\sum_l X_{0l}). \tag{10.20}$$

Using a continuum approximation Inokuti and Hirayama (1965) derive the following equation for $\langle P(t)\rangle_c$:

$$\langle P(t)\rangle_c = \exp(-t/\tau_R)\left[\frac{1}{V}\int \exp[-W_{DA}(R)t]\,4\pi R^2\,dR\right]^{N_A} \tag{10.21}$$

where N_A is the number of acceptors in the volume V of the material, and $W_{DA}(R)$ is the transfer rate from a donor to an acceptor a distance R away: For energy transfer by a multipolar interaction we can write

$$W_{DA}(R) = \frac{\alpha_{DA}^{(n)}}{R^n} \tag{10.15$'$}$$

where $n = 6$, 8, 10 for dipole–dipole, dipole–quadrupole, and quadrupole–quadrupole processes, respectively. Using *reasonable* approximations one finds (Watts 1975) a non-exponential decay of $\langle P(t)\rangle_c$ with time:

$$\langle P(t)\rangle_c = \exp(-t/\tau_R)\exp\left[-\frac{n_A}{n_0}\Gamma\left(1 - \frac{3}{n}\right)(t/\tau_R)^{3/n}\right]. \tag{10.22}$$

Here n_A is the concentration of acceptor ions, i.e. the number per m^3. n_0 is a critical concentration defined in terms of a range parameter R_0, where

$$1/\tau_R = \frac{\alpha_{DA}^{(n)}}{R_0^n} \tag{10.23}$$

and

$$n_0 = \frac{1}{\frac{4}{3}\pi R_0^3}. \tag{10.24}$$

For D–A transfer by dipole–dipole interaction eqn (10.22) becomes

$$\begin{aligned} \langle P(t)\rangle_c &= \exp\left[-\frac{t}{\tau_R} - \frac{n_A}{n_0}\sqrt{\pi}\left(\frac{t}{\tau_R}\right)^{\frac{1}{2}}\right] \\ &= \exp\left[-\frac{t}{\tau_R} - \frac{4}{3}\pi^{\frac{3}{2}} n_A (\alpha_{DA}^{(6)} t)^{\frac{1}{2}}\right]. \end{aligned} \tag{10.25}$$

Figure 10.4 shows the decay of $\langle P(t)\rangle_c$ after excitation by a short pulse in the absence of D–D transfer for two limiting cases, (a) in which no D–A transfer occurs, and (b) in which D–A transfer occurs by multipolar interactions at the critical concentration, $n_A = n_0$. This latter condition means that the transfer rate between a donor and an acceptor with the *average* D–A separation equals the radiative decay rate. Because of the random distribution of the acceptors some donors will find acceptors in very close proximity. These donors will decay rapidly by energy transfer causing the initial fast decay seen in Fig. 10.4. As time progresses only the more distantly separated D–A pairs remain excited and the decay rate decreases with increasing time, without ever reaching the 'isolated' value of $1/\tau_R$.

The decay patterns in the presence of acceptors are different for the three mechanisms, dipole–dipole, dipole–quadrupole, and quadrupole–quadrupole, as Fig. 10.4 shows. In principle the nature of the transfer mechanism may be determined, assuming it to be multipolar, if very accurate decay measurements are available. An expression for the decay of excited donors in the presence of D–A transfer mediated by exchange interaction is given by Watts (1975). In this case the decay is also non-exponential.

The use of a continuum approximation overstates the decay rate at very early times. This occurs because the continuum approximation admits vanishingly small D–A separations (with correspondingly large transfer rates) whereas in fact the smallest separation has a finite value determined by the lattice structure. These large transfer rates contribute to the initial decay. The correct formula for the initial decay rate is given by eqn (10.20).

If D–D transfer occurs in addition to D–A transfer this migration of the donor excitation can greatly add to the complexity of the problems. The starting point is again the set of coupled equations (10.16). An exact solution for $\langle P(t)\rangle_c$ can be found (Huber 1979) if the D–D transfer rate is exceedingly

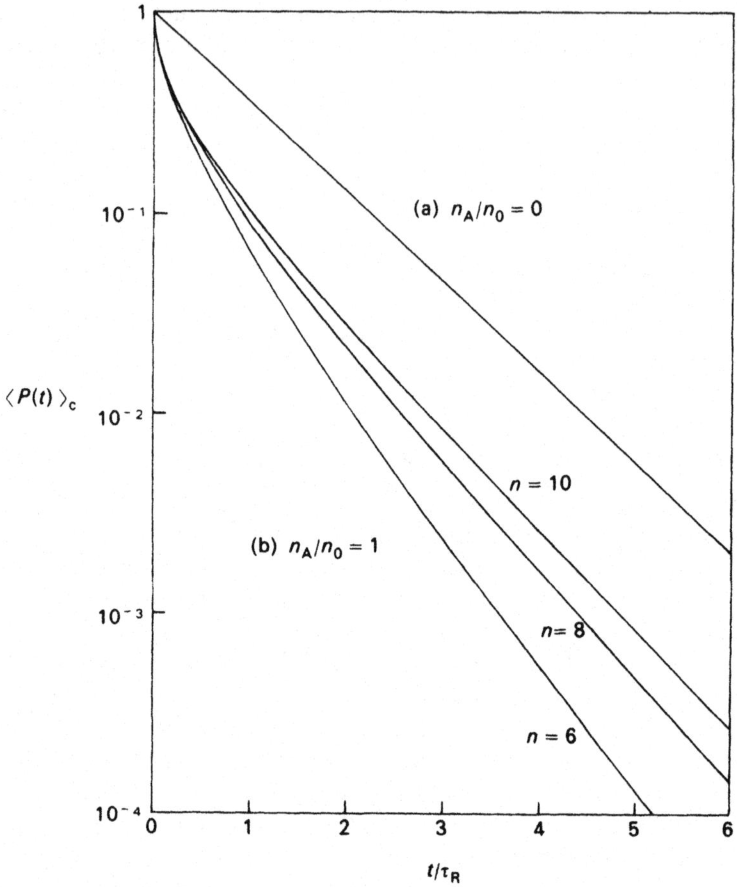

FIG. 10.4. Variation in $\langle P(t) \rangle_c$ after excitation by a short pulse in the absence of D–D transfer: (a) when no D–A transfer occurs, (b) D–A transfer occurs by dipole–dipole ($n=6$), dipole–quadrupole ($n=8$), and quadrupole–quadrupole ($n=10$) interactions at the critical concentration $n_A = n_0$. The decay patterns are seen to be different for the three interactions. The patterns are also noticeably different at very short times after the pulse (not apparent in this figure).

fast relative to the D–A transfer rate. In this case the excitation migrates between so many donor ions that it senses the average environment of acceptor ions. The average D–A transfer rate is now $\Sigma C_A X_{0l}$ where the summation is over *all the acceptor sites* in the solid, and the decay of the donors is exponential with a rate $1/\tau_R + C_A \Sigma X_{0l}$. This rate is exactly the same as the initial decay rate of the excited donors in the absence of D–D transfer (eqn 10.20). Physically the situation is that immediately after the exciting pulse all donors have an equal probability of being occupied and, as a result, the donor excitation senses the average acceptor environment.

The regime intermediate between the static case of no D–D transfer and the case of very rapid D–D migration is difficult to treat. Two approximate approaches are employed, the first of which treats the migration of excitation among donors as a diffusion process (Yokota and Tanimoto 1967). In this theory $P_n(t)$ is replaced as the physical quantity of interest by $\rho(\boldsymbol{R}, t)$, the density of excited donors. $\langle P(t)\rangle_c$ is replaced by the analogous quantity $\phi(t) = \int \rho(\boldsymbol{R}, t) d^3\boldsymbol{R}$, and eqn (10.16) is replaced by

$$\frac{\partial \rho(\boldsymbol{R}, t)}{\partial t} = -\frac{1}{\tau_R}\rho(\boldsymbol{R}, t) + D\nabla^2\rho(\boldsymbol{R}, t) - \sum_{l=1}^{N_A} W_{DA}(\boldsymbol{R} - \boldsymbol{R}_l)\,\rho(\boldsymbol{R}, t). \tag{10.26}$$

D is the *diffusion constant* which characterizes the migration among donor ions. An approximate solution to eqn (10.26) for the case of dipole–dipole D–A transfer has been given by Yokota and Tanimoto; they find

$$\phi(t) = \exp\left[-\frac{t}{\tau_R} - \frac{4}{3}\pi^{\frac{3}{2}} n_A (\alpha_{DA}^{(6)} t)^{\frac{1}{2}} \left(\frac{1 + 10.87y + 15.5y^2}{1 + 8.743y}\right)^{\frac{3}{4}}\right] \tag{10.27}$$

where

$$y = D\,(\alpha_{DA}^{(6)})^{-\frac{1}{3}} t^{\frac{2}{3}}. \tag{10.28}$$

When there is no diffusion $\phi(t)$ reduces, as expected, to the form given by eqn (10.25).

The diffusion constant, D, depends on both D–D transfer rate and on the density of donors. When the donor concentrations is low, Huber (1979) gives for a multipolar process

$$D = \frac{1}{2}(n-5)^{-1}\left(\frac{4}{3}\pi n_D\right)^{\frac{n-2}{3}} \alpha_{DD}^{(n)} \tag{10.29}$$

so that for dipole–dipole transfer ($n = 6$) one has

$$D = \frac{1}{2}\left(\frac{4}{3}\pi n_D\right)^{\frac{4}{3}} \alpha_{DD}^{(6)}. \tag{10.30}$$

At sufficiently long times, t, the decay becomes exponential, and if the D–A transfer is a dipole–dipole process this asymptotic decay rate is

$$\frac{1}{\tau} = \frac{1}{\tau_R} + W_{DA}^{(\mathrm{diff})} = \frac{1}{\tau_R} + 4\pi D n_A R_s \tag{10.31}$$

where R_s is a scattering length given by Huber (1979):

$$R_s = 0.91\left(\frac{\alpha_{DA}^{(6)}}{D}\right)^{\frac{1}{4}}. \tag{10.32}$$

The asymptotic decay rate can then be written as

$$\frac{1}{\tau}=\frac{1}{\tau_R}+W_{DA}^{(\mathrm{diff})}=\frac{1}{\tau_R}+21n_Dn_A(\alpha_{DD}^{(6)})^{\frac{3}{4}}(\alpha_{DA}^{(6)})^{\frac{1}{4}}$$

$$=\frac{1}{\tau_R}+8.5n_A(\alpha_{DA}^{(6)})^{\frac{1}{4}}D^{\frac{3}{4}}. \qquad (10.33)$$

The second approach to the problem treats the migration of excitation among donor ions as a random walk process (Burshtein 1972). The approach is discussed by Watts (1975) and Di Bartolo (1984). The decay function of the donors, $\phi(t)$, is now given by a renewal equation

$$\phi(t)=\langle P(t)\rangle_c\exp(-t/\tau_0)+\frac{1}{\tau_0}\int_0^t\phi(t-t')\langle P(t')\rangle_c\exp(-t'/\tau_0)\mathrm{d}t' \quad (10.34)$$

where $\langle P(t)\rangle_c$ is the decay function of the donors in the absence of D–D transfer, i.e. eqn (10.19). τ_0 is the average time that the excitation resides on a donor ion before hopping to another donor ion. This is related to the density of donors n_D and to α_{DD}. For dipole–dipole transfer (Watts 1975) $1/\tau_0$ is given by

$$\frac{1}{\tau_0}=\left(\frac{2\pi}{3}\right)^3 n_D^2\alpha_{DD}^{(6)} \qquad (10.35)$$

We assume that τ_0 is much smaller than τ_R. The solution to eqn (10.32) can be obtained by numerical methods. It is found that $\phi(t)$ is initially non-exponential but for large t values it becomes exponential; Huber (1979) gives the asymptotic decay rate as

$$\frac{1}{\tau}=\frac{1}{\tau_R}+W_{DA}^{(\mathrm{hopping})}=\frac{1}{\tau_R}+C_A\sum_l X_{0l}(1+\tau_0X_{0l})^{-1} \qquad (10.36)$$

where the summation is over all sites. At very short times after the excitation pulse, i.e. before migration becomes effective, $\phi(t)$ shows the same decay rate as $\langle P(t)\rangle_c$, i.e. $1/\tau_R+C_A\Sigma X_{0l}$. When the hopping rate is very rapid, $\tau_0\to 0$, the decay is exponential at all times with this same decay rate.

Both the diffusion model and the hopping model are approximate, but not equivalent, approaches to the general problem of D–A energy transfer in the presence of D–D migration. Huber (1979) has investigated the regions of validity of both models, and the general problem has been extensively reviewed by Burshtein (1985). As a general rule we can say that the diffusion model is valid if the D–D transfer mechanism is much weaker than the D–A transfer mechanism whereas the hopping model is valid if the D–D transfer mechanism is greater than the D–A transfer mechanism. If both processes proceed by dipole–dipole mechanisms we find

$$\text{diffusion model valid: } \alpha_{DD}^{(6)} \ll \alpha_{DA}^{(6)},$$
$$\text{hopping model valid: } \alpha_{DD}^{(6)} \geqslant \alpha_{DA}^{(6)}. \tag{10.37}$$

Figure 10.5 shows the time evolution of donor luminescence under various conditions and illustrates the theoretical results derived in the preceding paragraphs. Curve (a) gives the luminescence decay in the absence of any energy transfer; this shows a single exponential with decay rate $1/\tau_R$. Curve (b) is derived from eqn (10.25) and shows the luminescence decay when D–A transfer occurs by dipole–dipole process under the conditions $n_A/n_0=2$. D–D transfer does not occur. The decay is non-exponential at all times. Curve (c) describes the situation where a weak D–D transfer occurs in addition to D–A transfer and radiative decay. This situation is described by eqn (10.27). For illustration we choose a value of the diffusion constant D such that $6D\tau_R=R_0^2$ or $D=(1/6)(\alpha_{DA}/R_0^4)$. This means that in a time τ_R the excitation can migrate a distance R_0 through the donor system. This value corresponds to $y=1/6(t/\tau_R)^{2/3}$, which is used in eqn (10.27) to calculate the decay pattern of curve (c). We note that it reaches its asymptotic decay rate within a few τ_R periods. Next we consider a more rapid migration. For this we use the hopping model and choose $\tau_0=\tau_R/10$. The prediction of the hopping model is shown in curve (d). We notice that as migration increases the decay becomes exponential at earlier times except for a small region around $t=0$. In the limit of exceedingly fast migration the decay becomes a single exponential with a decay rate given by eqn (10.20). This is shown *schematically* in curve (e). We note from the curves that if D–A transfer occurs, whether or not D–D transfer occurs, all decays have the same initial decay rate which is given by eqn (10.20).

If diffusion is weak the decay will be non-exponential for a detectable period. This is the *diffusion-limited* case. When the migration is very fast the decay rapidly establishes an exponential behaviour and we are in the *fast migration* regime. In the exponential regime the donors appear to behave collectively with an effective transfer rate of W_{DA}, where expressions for W_{DA} are given in eqns (10.31), (10.33), and (10.36).

All the formulae so far developed describe the decay patterns of the excited donors, $N_D^{(b)}$. Theoretical formulae for the decay of excited acceptors, $N_A^{(b)}$, can similarly be developed and can be very complicated. A great simplification occurs if the donor decay is exponential since then the transfer from donors can be described by the collective rate W_{DA}. One can then use rate equations to describe the behaviour of the donors and accpetors. If we neglect back transfer the decay of excited donors and excited acceptors, after some initial excitation, is given by (Fig. 10.6)

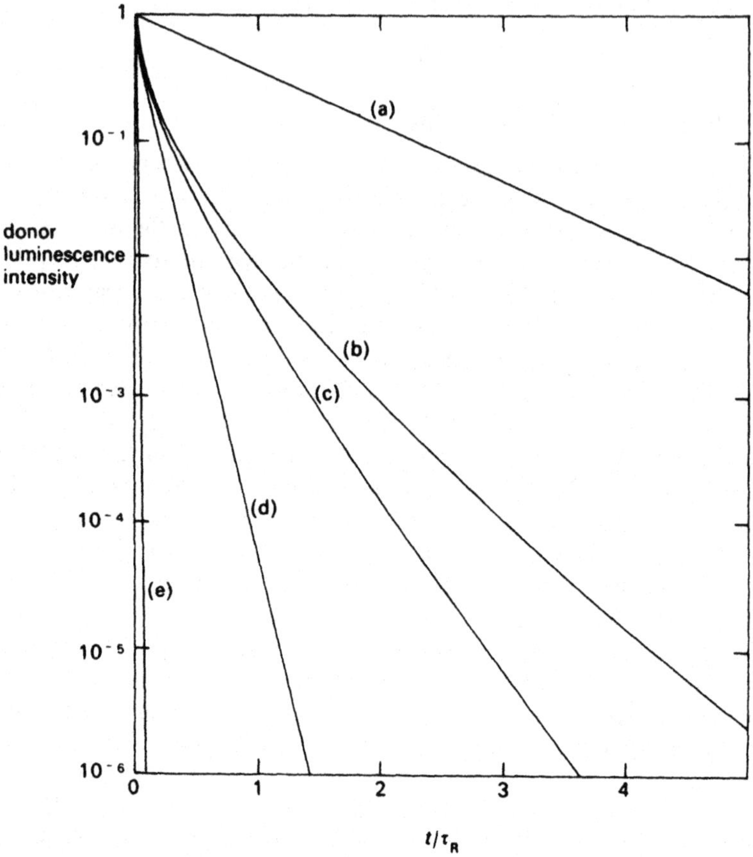

FIG. 10.5. This illustrates schematically the time evolution of donor decay under various conditions:
(a) No energy transfer, $\alpha_{DD}=\alpha_{DA}=0$. The decay is exponential with decay time τ_R. (b) D–A transfer in the absence of donor–donor transfer, with $n_A/n_0=2$. The decay is non-exponential at all times. This is the *static* case. (c) D–A transfer at a rate similar to (b) in the presence of weak D–D transfer (see text). This is the *diffusion-limited* case. The decay ultimately becomes exponential and the decay rate is then given by eqn (10.31). (d) D–A transfer at the same rate as for cases (b) and (c), and with fast migration through the donors characterized by $\tau_0=\tau_R/10$ on the hopping model. After a short initial non-exponential part the decay becomes single exponential. (e) Similar D–A transfer with exceedingly rapid migration through the donors. In this case the excited donor ions always sense an average acceptor environment and the decay is single exponential given by $\tau_R^{-1}+C_A\sum_l X_{0l}$. Conditions (b)–(e) differ only in the migration rate through the donors, all have the same D–A rate, and, consequently, all have the same initial decay rate.

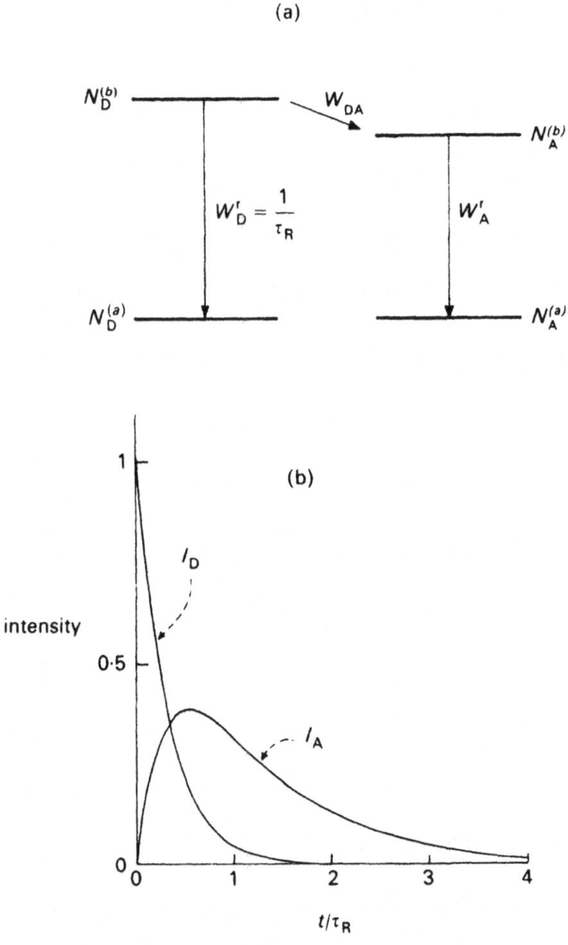

FIG. 10.6. (a) Rates of radiative decay and energy transfer in the fast migration regime when the D→A transfer is described by a collective rate, W_{DA}. (b) Decay of donor and acceptor luminescence after pulsed excitation of the donor system at $t=0$ for the case where $W_D^r = W_A^r$, $W_{DA} = 2W_D^r$.

$$\frac{dN_D^{(b)}}{dt} = -W_D^r N_D^{(b)} - W_{DA}\,dN_D^{(b)}$$
$$\frac{dN_A^{(b)}}{dt} = W_{DA} N_D^{(b)} - W_A^r N_A^{(b)} \qquad (10.38)$$

where W_D^r and W_A^r are the radiative decay rates of the excited donors and acceptors, respectively, and W_D^r is $1/\tau_R$ of the previous discussion. The solutions in the case where $N_D^{(b)}(0)$ excited donors are created at $t=0$ and where no acceptors are excited, are

$$N_D^{(b)}(t) = N_D^{(b)}(0)\exp[-(W_D^r + W_{DA})t]$$

$$N_A^{(b)}(t) = N_D^{(b)}(0)\frac{W_{DA}}{(W_D^r + W_{DA}) - W_A^r}[\exp(-W_A^r t) - \exp[-(W_D^r + W_{DA})t]] \tag{10.39}$$

If the acceptors decay radiatively the acceptor luminescence varies as $N_A^{(b)}(t)$; it rises from zero at $t=0$, reaches a maximum, and then decays exponentially at either its own decay rate W_A^r or the decay rate of the donors $(W_D^r + W_{DA})$, whichever is the smaller (Fig. 10.6).

10.1.3 Experimental studies of energy transfer

We now review some experimental studies of energy transfer which illustrate the theoretical analyses outlined in the previous sections. In an early and significant study Weber (1971) investigated $Eu^{3+} \to Cr^{3+}$ transfer in a Cr^{3+}-doped europium phosphate glass. Eu^{3+} ions excited to the $^5D_0$ metastable level emit luminescence in the $^5D_0 \to {}^7F_J$ transitions in the 0.6–0.7 μm wavelength region. These transitions overlap the broad $^4A_2 \to {}^4T_2$ absorption band of the Cr^{3+} ions. Even though only the Eu^{3+} ions are initially excited, both Eu^{3+} and Cr^{3+} luminescence is observed, indicating $Eu^{3+} \to Cr^{3+}$ transfer. In the absence of Cr^{3+} ions the Eu^{3+} fluorescence decay is exponential with a decay time of 2.1 ms at 77 K, and this is the intrinsic decay time of the europium $^5D_0$ state. As the concentration of Cr^{3+} ions in the material is increased the Eu^{3+} decay becomes non-exponential, as shown in Fig. 10.7. That the fluorescence ultimately decays exponentially means that D–D migration as well as D–A transfer is occurring in this material. At 77 K, however, the Eu^{3+}–Eu^{3+} D–D migration is slow compared with the Eu^{3+}–Cr^{3+} D–A transfer, so that the decay behaviour in Fig. 10.7 represents diffusion-limited transfer.

To understand the weakness of the donor migration at low temperatures we note that at low temperatures only the $^7F_0$ ground state is occupied and D–D transfer involves $^5D_0 \leftrightarrow {}^7F_0$ transitions which, being a $J=0 \to J=0$ transition, has a very small transition probability (see insert in Fig. 10.7). Since at higher temperatures the higher J states of the ground ^7F_J manifold are occupied the Eu^{3+}–Eu^{3+} migration involves the stronger $^5D_0 \leftrightarrow {}^7F_{1,2}$ transitions and becomes more effective. Weber (1971) studied the Eu^{3+} decay patterns as a function of temperature, and in so doing observed the effects of varying the donor migration rate. The data were analysed using the diffusion model. From the measured oscillator strengths of the $^5D_0 \to {}^7F_J$ transitions and the equilibrium populations in the ^7F_J levels he estimated the temperature dependence of the diffusion constant. According to the diffusion

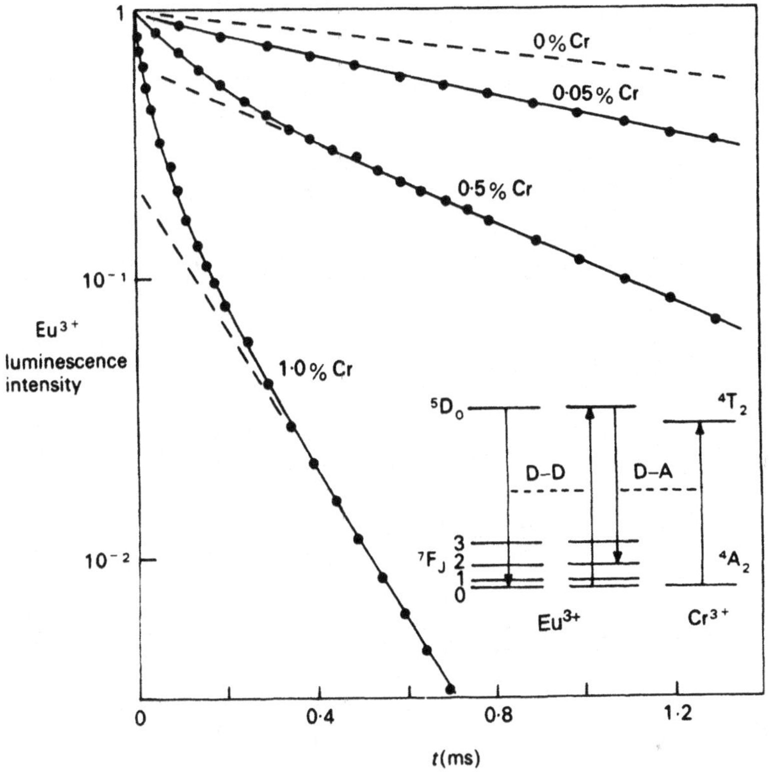

FIG. 10.7. Variation in the time dependence of Eu^{3+} luminescence intensity from a chromium-doped europium phosphate glass $Eu(PO_3)_2$ with increasing chromium content. (After Weber 1971.)

model the decay pattern ultimately becomes exponential and this asymptotic decay rate is (eqn 10.33) $1/\tau = 1/\tau_R + W_{DA}^{(diff)}$ where

$$W_{DA}^{(diff)} = 8.5 n_A (\alpha_{DA}^{(6)})^{\frac{1}{4}} D^{\frac{3}{4}} \tag{10.40}$$

assuming that the Eu^{3+}–Cr^{3+} transfer proceeds by a dipole–dipole process. Weber (1971) showed experimentally that the measured values of W_{DA} were accurately proportional to the $\frac{3}{4}$ power of the estimated values of D, in agreement with the diffusion model.

A later study is that of Hegarty, Huber, and Yen on $LaF_3{:}Pr^{3+}$ (Hegarty *et al.* 1982). Figure 10.8 shows the measured values of the luminescence intensity from the $^3P_0$ state after pulsed excitation at different temperatures. The doping of Pr^{3+} ions is 20 at. %. The broken line shows the experimental decay pattern in the case of dilute doping; this represents the intrinsic decay

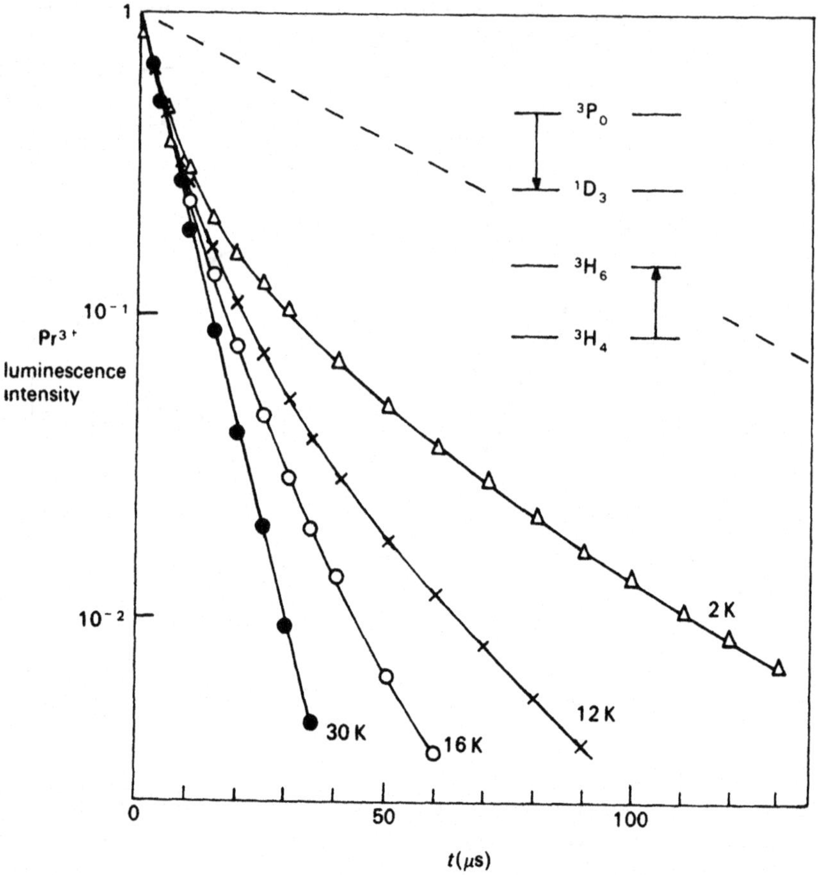

FIG. 10.8. Variation in the time dependence of the luminescence intensity from the $^3P_0$ state of Pr^{3+} in $LaF_3:Pr^{3+}$ (20 per cent) at different temperatures after pulsed excitation. At this doping level cross-relaxation, as illustrated in the figure, occurs and acts as a D–A process in that it removes $^3P_0$ excitation from the Pr^{3+} ions. The broken line shows the exponential decay rate in the case of very dilute doping and gives the intrinsic decay rate. In this material the D–A transfer is constant while the D–D transfer increases with increasing temperature. (After Hegarty *et al.* 1982.)

of the $^3P_0$ state ($1/\tau_R$). With a doping of 20 per cent cross-relaxation between nearby Pr^{3+} ions manifests itself. This process is illustrated in Fig. 10.8; on one ion the excited $^3P_0$ state decays to $^1D_3$ while an adjacent unexcited ion is raised to $^3H_6$. The second ion quickly decays to the ground state. This cross-relaxation is regarded as a D–A transfer in that it removes $^3P_0$ excitation from the Pr^{3+} system. The rate is expected to be a constant independent of temperature up to 30 K, the temperature region of interest. In addition to the

D–A process a D–D process also operates in which an ion in an excited $^3P_0$ state decays to the *ground state* while an adjacent ion is raised from the ground state to the $^3P_0$ state. Thus D–D transfer and D–A transfer both take place within the Pr^{3+} ion system.

The D–D transfer is strongly dependent on temperature, varying as T^3 up to 30 K. (We discuss D–D transfer in this material in the next section.) At 2 K this D–D transfer is negligible, only D–A transfer is effective, and the decay has the non-exponential pattern predicted for such a situation. By increasing temperature one can increase the donor migration rate from the static case, i.e. no D–D transfer, at 2 K to extremely rapid migration and subsequent exponential decay at temperatures of $\simeq 30$ K. The similarity between the experimental measurements (Fig. 10.8) and the theoretical decay curves (Fig. 10.5) confirm this interpretation of the behaviour of the $LaF_3:Pr^{3+}$ system.

A classic series of experiments were performed by Hegarty *et al.* (1981) on $PrF_3:Nd^{3+}$ in which D–D transfer occurs between the Pr^{3+} ions and D–A transfer occurs from Pr^{3+} to Nd^{3+} ions. We first consider the pure praseodymium trifluoride system. Pr^{3+} ions in the excited $^3P_0$ state decay by a combination of radiative emission and cross-relaxation and the observed decay time (750 ns) stays constant up to around 40 K. D–D transfer can occur in this material. This was studied by fluorescence line narrowing (FLN) techniques, to be described in the next section, and the rate varies with temperature at $T^{4.3}$ between 5 K and 40 K. The D–D transfer rates measured for pure praseodymium trifluoride were also found to apply for $PrF_3:Nd^{3+}$, where the Nd^{3+} concentration was 5 at. %. In contrast to the temperature-dependent D–D process, the D–A process in $PrF_3:Nd^{3+}$ is expected to be independent of temperature.

At 5 K, D–D transfer in $PrF_3:Nd^{3+}$ is negligible; only D–A transfer is effective and the intensity of Pr^{3+} luminescence after pulsed excitation, $I(t)$, should vary with time as described by eqn (10.19). If this transfer operates through a dipole–dipole mechanism the intensity is given by eqn (10.25):

$$\ln I(t) + \frac{t}{\tau_R} = -\tfrac{4}{3}\pi^{\frac{3}{2}} n_A (\alpha_{DA}^{(6)} t)^{\frac{1}{2}}$$

$$= -\tfrac{4}{3}\pi^{\frac{3}{2}} n_A (R_{0l})^3 (X_{0l} t)^{\frac{1}{2}} \qquad (10.41)$$

where R_{0l} is the distance between nearest-neighbour Pr^{3+} ions, and X_{0l} is the nearest-neighbour D–A transfer rate. The measured values of $\ln I(t) + t/\tau_R$ plotted against $t^{1/2}$ are shown in Fig. 10.9. This shows an excellent straight-line fit, confirming that the D–A transfer occurs by a dipole–dipole mechanism. From these measurements one determines $X_{0l} = 11.3 \times 10^6\ s^{-1}$ and a value for $\alpha_{DA}^{(6)} = X_{0l}/(R_{0l})^{(6)}$ can be found. These rates are expected to be independent of temperature.

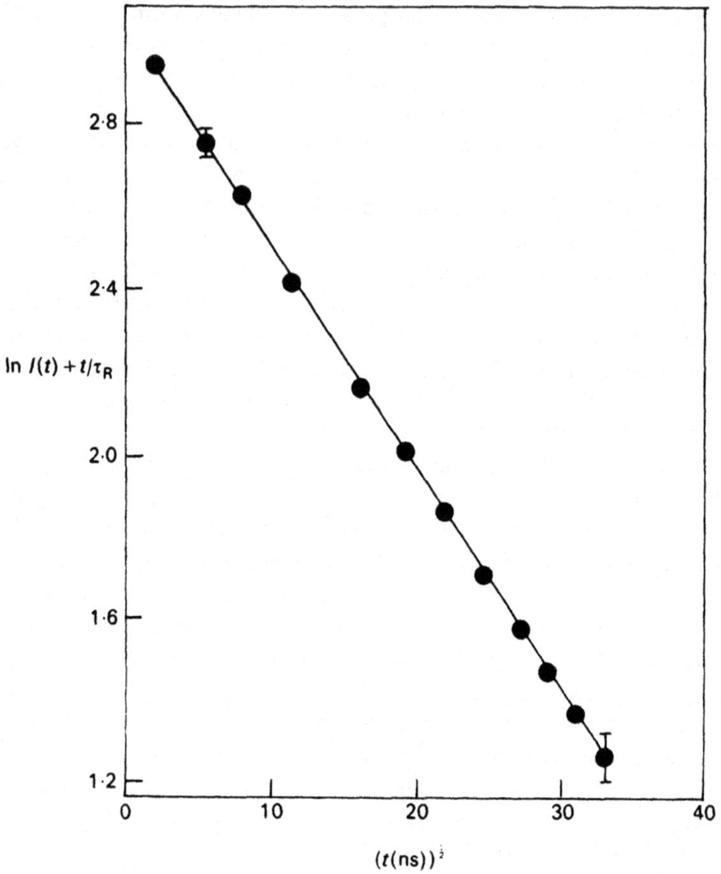

FIG. 10.9. Variation with time of the luminescence intensity of Pr^{3+} ions in PrF_3:Nd^{3+} (5 at.%) at 5 K. The data are in agreement with the model in which $Pr^{3+} \rightarrow Nd^{3+}$ transfer occurs by an electric dipole–dipole process. (After Hegarty *et al.* 1981.)

By raising the temperature the D–D transfer rate increases until at 40 K this D–D transfer rate is much greater than the D–A transfer rate. We are now in the exceedingly fast migration regime where the decay of Pr^{3+} luminescence should be single exponential with a decay rate given by eqn (10.36) with $\tau_0 = 0$:

$$\frac{1}{\tau} = \frac{1}{\tau_R} + C_A \sum_l X_{0l} = \frac{1}{\tau_R} + C_A \alpha_{DA}^{(6)} \sum_l (R_{0l})^{-6}. \tag{10.42}$$

The observed decay at 40 K was found to be single exponential, and the value of X_{0l} obtained from this measurement agreed with the value obtained at 5 K, thus confirming the temperature-independence of the D–A transfer. By varying the Nd^{3+} concentration the decay rate was found to be a linear function of C_A as expected from eqn (10.14).

With this knowledge of the D–D and D–A mechanisms Hegarty *et al.* undertook to study the Pr^{3+} luminescence decay in the intermediate temperature regime. Here the Pr^{3+} luminescence decay should be non-exponential initially, becoming single exponential at later times, and we write this asymptotic decay rate as $1/\tau = 1/\tau_R + W_{DA}$. At 15 K the measured values of $\alpha_{DD}^{(6)}$ and $\alpha_{DA}^{(6)}$ are approximately equal. Consequently the diffusion model should be valid to within a few degrees of 15 K while the hopping model should be valid above 15 K.

At low temperatures when the diffusion model is expected to be valid W_{DA} is given by eqn (10.33):

$$W_{DA} = 21\, n_A n_D (\alpha_{DD}^{(6)})^{\frac{3}{4}} (\alpha_{DA}^{(6)})^{\frac{1}{4}}. \qquad (10.33')$$

The only temperature-dependent factor in this formula is $(\alpha_{DD}^{(6)})^{3/4}$. Figure 10.10 shows the experimental values of W_{DA} plotted against $\alpha_{DD}^{(6)}/\alpha_{DA}^{(6)}$ on a log–log scale. Between 10 K and 13 K the dependence is linear with a slope which varies as $(\alpha_{DD}^{(6)})^{3/4}$ to within experimental accuracy, in agreement with the diffusion model, eqn (10.33′).

Above 15 K we expect the hopping model to be valid and W_{DA} to be given by

$$W_{DA} = C_A \sum_l \frac{X_{0l}}{1 + \tau_0 X_{0l}} \qquad (10.36')$$

where τ_0 is proportional to $(\alpha_{DD}^{(6)})^{-1}$ (eqn 10.35). From the previously determined values of $\alpha_{DD}^{(6)}$ a value for τ_0 was calculated for different temperatures, the values of W_{DA} were estimated using eqn (10.36′) above and the calculated values are given by the solid curve in the high temperature region of Fig. 10.10. The agreement is quite good. There appears to be a region where $\alpha_{DD}^{(6)} \simeq \alpha_{DA}^{(6)}$ for which neither the diffusion model nor the hopping model appears to be adequate.

As a final example of D–A energy transfer we describe the study by Kallendonk and Blasse (1981) on $TbAl_3B_4O_{12}$. This material contains most of its Tb^{3+} ions in regular sites but some Tb^{3+} ions occupy irregular sites with distinct absorption and emission transitions. If these regular Tb^{3+} ions are selectively excited by a pulsed excitation they can decay both radiatively and by transferring excitation to irregular Tb^{3+}. The luminescence patterns of the regular Tb^{3+} ions (donors) and of the irregular Tb^{3+} ions (acceptors) after such pulsed excitation are seen in Fig. 10.11. The behaviour of the acceptor luminescence is as predicted in eqn (10.37), rising from zero to a maximum and then decaying at the slower intrinsic acceptor rate, W_A^r.

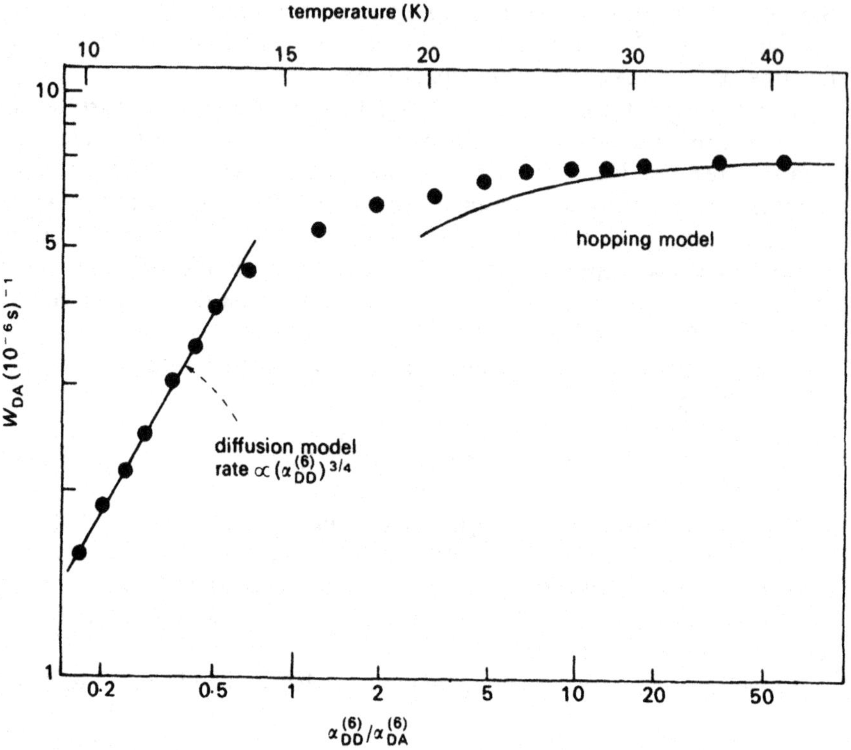

FIG. 10.10. By varying temperature the transfer of Pr^{3+} excitation in $PrF_3{:}Nd^{3+}$ (5 at. %) can be studied in the region where the diffusion model is expected to be valid, in the region where the hopping model is expected to be valid, and in intermediate regions. In the asymptotic limit the decay becomes exponential with a rate given by $\tau_R^{-1} + W_{DA}$. The measured values of W_{DA} are shown with the predictions of the diffusion model (straight line) and predictions of the hopping model (solid curve on right). (After Hegarty *et al.* 1981.)

10.1.4 Intraline (D–D) *transfer*

That D–D transfer can play an important role in causing an efficient D–A transfer is clear from the preceding section. This D–D transfer, however, is difficult to measure directly; its presence can often only be inferred from measurements of D–A transfer. There is a number of interesting questions attached to the D–D transfer. Is it, for example, a resonant or a non-resonant process? What are the effects of random strains and lattice defects, since these can shift the electronic levels of adjacent ions out of resonance with each other? The resultant inhomogeneous broadening of a transition can be much larger than the homogeneous broadening, particularly at low temperatures.

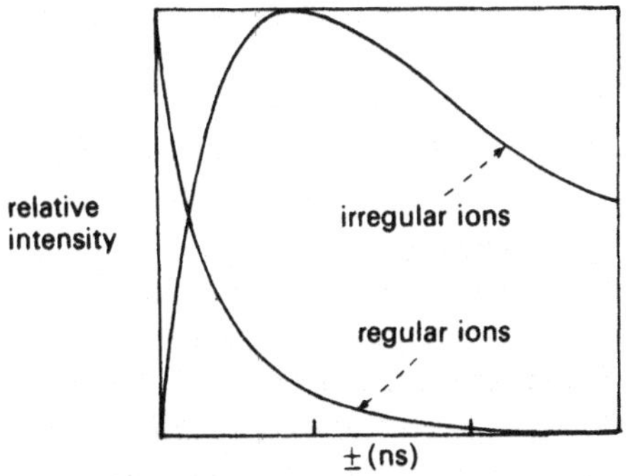

FIG. 10.11. Selective excitation of regular Tb^{3+} ions in $TbAl_3B_4O_{12}$ by a pulse of excitation results in the decay of the excited regular Tb^{3+} ions with an exponential decay rate. This decay occurs partly by excitation transfer to irregular Tb^{3+} ions. The subsequent luminescence from the irregular Tb^{3+} ions is predicted by eqn (10.39). (After Kallendonk and Blasse 1981.)

This inhomogeneously broadened lineshape is a composite of many transitions from ions in different strain environments, each with its own much narrower homogeneous width. The situation is illustrated schematically in Fig. 10.12. If we excite into such an inhomogeneously broadened absorption transition with a narrowband laser of bandwidth ΔE_{L} we selectively excite ions in a particular strain environment. After the laser is switched off one expects to observe only the luminescence emitted by those ions excited by the laser; the width of this luminescence line can be very small, and is given by (Selzer 1981).

$$\Delta E = \Delta E_{\mathrm{L}} + 2\Delta E_{\mathrm{homo}} \tag{10.43}$$

where ΔE_{homo} is the homogeneous width. This is the phenomenon of FLN and it can be used to investigate whether energy transfer occurs from ions in a

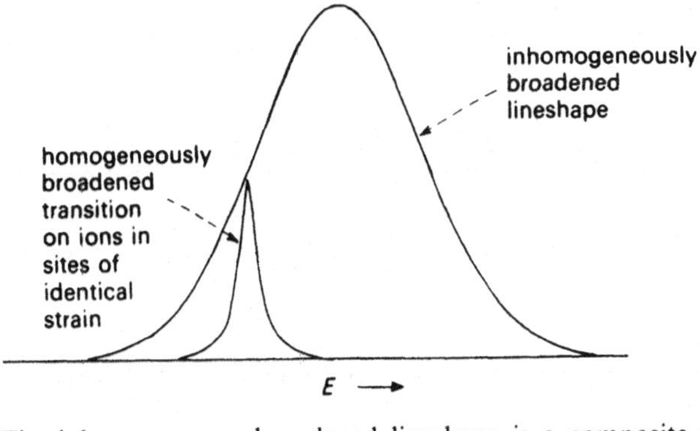

FIG. 10.12. The inhomogeneous broadened lineshape is a composite of narrower transitions from ions in different strain environments.

particular strain environment (excited by the laser) to other unexcited ions in different strain environments.

Figure 10.13 shows the FLN signal from Pr^{3+} ions in lanthanum trifluoride investigated by Huber *et al.* (1977). The narrowband laser pulse is tuned to excite those Pr^{3+} ions which absorb at the edge of the inhomogeneous line (Fig. 10.13). Observation of the fluorescence immediately after the laser pulse reveals the sharp FLN signal. As the time interval after the laser pulse progresses, however, luminescence is observed from other ions whose emissions occur at different wavelengths within the inhomogeneous profile. Thus one observes a gradual rise of the *entire* inhomogeneous line, indicating that energy is being transferred from the laser-excited Pr^{3+} ions to Pr^{3+} ions with different energies in other strain environments. This is a direct observation of D–D transfer. In analysing this intraline D–D transfer one must take back-transfer properly into account. The points in Fig. 10.13 are the experimental measurements, the solid curves are the theoretical predictions of Huber *et al.* (1977). Since the entire inhomogeneous line grows the transfer is independent of the size of the energy mismatch. A similar behaviour is found in transition metal doped crystals and in rare-earth doped glasses where the inhomogeneous width can be $\simeq 100\,\text{cm}^{-1}$ (Weber 1981).

The microscopic energy transfer process responsible for this intraline spectral transfer is phonon-assisted. There are two types of phonon-assisted mechanisms which can operate here; a one-phonon mechanism involving the absorption or creation of a low energy phonon to make up the energy mismatch and a two-phonon mechanism involving two higher energy phonons which differ by the energy mismatch between the Pr^{3+} ions. These various mechanisms have been examined in detail by Holstein *et al.* (1981). All these processes become much more efficient with increasing temperature.

To examine the intraline transfer from a theoretical viewpoint we return to the general set of coupled equations (10.16) and we consider the case of D–D transfer only; we put $X_{0l}=0$. If back-transfer is omitted the solution for $\langle P(t)\rangle_c$ is (Huber *et al.* 1977):

$$\langle P(t)\rangle_c = \exp(-t/\tau_R)\prod_n (1-C_D+C_D\exp(-W_{0n}t)) \qquad (10.44)$$

where C_D is the probability that a site is occupied by a donor, and the summation is over all sites. It is convenient for this discussion to write $\langle P(t)\rangle_c=\exp(-t/\tau_R)R(t)$, where $R(t)$ is the decay pattern of the initially excited Pr^{3+} ions due only to D–D transfer. Hence we have

$$R(t) = \prod_n (1-C_D+C_D\exp(-W_{0n}t)). \qquad (10.45)$$

For the case where $C_D=1$, viz. praseodymium trifluoride, this becomes

$$R(t)=\exp\left(-\sum_n W_{0n}t\right) \qquad (10.46)$$

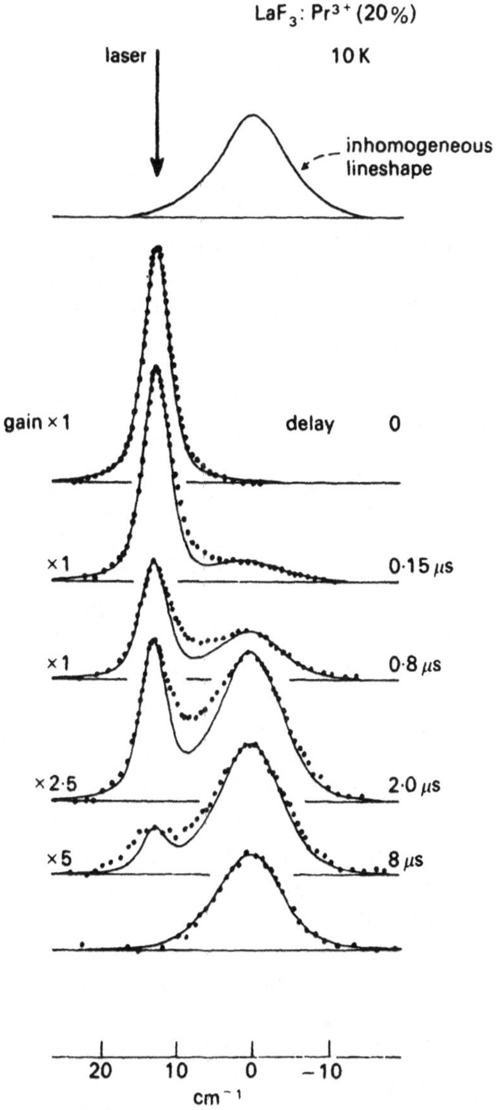

FIG. 10.13. FLN and intra-line energy transfer in $LaF_3:Pr^{3+}$ (20 per cent) at 10 K. A narrowband laser excites a subset of ions to the $^3P_0$ state. After the laser pulse the FLN signal appears. As time progresses luminescence from the main body of Pr^{3+} ions begins to grow, indicating that energy is being transferred from the initially excited subset of ions to the general body of ions. (After Huber *et al.* 1977.)

When back-transfer is taken into account the formula for $R(t)$ is modified by introducing a multiplicative factor (Huber *et al.* 1977).

If the lattice sums are replaced by integrals in a continuum approximation one obtains (Huber *et al.* 1977), for D–D transfer by a multipolar interaction mechanism,

$$R(t) = \exp\left[-\frac{4\pi}{3} n_D \Gamma\left(1-\frac{3}{n}\right) R_{min}^3 (W_{0l}t)^{\frac{3}{n}} f_{BT}\right] \qquad (10.47)$$

where n_D is the density of donors, $n=6, 8, 10$ for dipole–dipole, dipole–quadrupole, and quadrupole–quadrupole mechanisms, R_{min} is the closest distance between nearest-neighbour cations, and W_{0l} is the transfer rate between Pr^{3+} ions in nearest-neighbour sites. f_{BT} is a factor to take back-transfer into account; its form depends on the model used to describe the back-transfer process.

The reduction with time in the sharp FLN signal due to radiative decay and D–D transfer is described by $\langle P(t)\rangle_c$, (eqn (10.44)). If there are no other non-radiative decay channels for these Pr^{3+} ions, which is assumed in this mathematical treatment, the narrowband intensity, I_N, decreases with time as $\langle P(n)\rangle_c = \exp(-t/\tau_R)R(t)$ and the total Pr^{3+} luminescence, I_N plus the broad inhomogeneous background, I_B, decreases with time as $\exp(-t/\tau_R)$. If other decay channels exist for the Pr^{3+} ions, e.g. cross-relaxation, these will reduce the intensity of I_N and I_B at the same rate and will not affect the ratio $I_N/(I_N+I_B)$. Therefore the FLN technique allows one to investigate D–D transfer even when D–A processes also occur.

The D–D transfer among Pr^{3+} ions in $LaF_3:Pr^{3+}$ (Huber *et al.* 1977), and in praseodymium trifluoride (Hamilton *et al.* 1977), were investigated by this FLN technique, and from these experiments the variation with temperature of the D–D transfer rate was accurately established. The ability to vary the migration rate over orders of magnitude by simply changing the temperature was used to good effect in the analyses of D–A energy transfer in $PrF_3:Nd$ described in the previous section.

The D–D transfer we have been describing in this section can be termed non-resonant D–D transfer. It is felt that it is this non-resonant D–D transfer which is responsible for the migration of donor excitation to the vicinity of acceptor ions. Hence it is the non-resonant transfer rates determined by these FLN experiments which are used in the analyses of the D–A transfer studies described in the previous section.

The question remains of whether true resonant transfer, i.e. without the assistance of phonons, takes place among the donors. For such transfer to occur the relevant excited electronic levels of the two ions must be close enough for overlap for occur between the *homogeneously* broadened levels. The homogeneous widths may be very small, particularly at low temperatures. Further, the two resonant ions must be spatially close enough for an effective interaction to occur between them. Whether these two conditions can be satisfied depends on the nature of the strains in the material, since in solids it is the strains which shift the energy levels of individual ions. If the strains are random and microscopic the energy levels of two adjacent ions can be shifted in energy by amounts of up to the inhomogeneous linewidth. Indeed it can be argued that the presence of dopant ions in a host material introduces local strains of microscopic dimensions due simply to the size difference between the impurity and host ions which they replace, so that the introduction of microscopic strains should be regarded as an inherent part of the doping process. On the other hand, if the strains are macroscopic the regions of strain are large and the difference in the magnitude of the strain between two nearby sites will be small.

We have been describing the situation from the single-ion point of view. If, however, two similar dopant ions are sufficiently close for a near-resonant interaction between them they are more correctly regarded as constituting a single quantum system. When several similar ions are close enough to interact near-resonantly they may constitute a single *extended* quantum system; the excited state of this extended quantum system is an *extended state*. The D–A transfer is then visualized as a transfer of the excitation of this extended state to an acceptor ion within the region of the extended state.

In this context the paper by Anderson (1958) concerning excitation transport in a lattice of random centres is of some importance. Anderson considered a three-dimensional lattice in which the active centres are randomly distributed and the energy of a centre varies randomly from site to site. This randomness in energy gives an inhomogeneous width to the observed energy level, ΔE. The interaction between centres j and k is written $V_{jk}(R)$ where R is the separation between the centres. Anderson concludes that if $V_{jk}(R)$ falls off faster than $1/R^3$ then there exists a critical concentration of centres below which transfer cannot occur and the excitation stays localized, i.e., does not belong to an extended state. The criterion for localization is

$$\frac{\Delta E}{V} \geqslant \simeq 2 \tag{10.48}$$

where V is the average interaction between adjacent centres. V increases as the average separation between centres decreases, that is, as the concentration increases. As the concentration of centres increases through this critical concentration there should be an abrupt transition from a localized to a delocalized state. In order for this Anderson transition to occur the interaction must be of shorter range than dipole–dipole. Experiments have been carried out to detect such a sharp transition between localized and delocalized states in optically active systems but unambiguous evidence of the effect has not been demonstrated.

10.1.5 Transient grating experiments

A technique employing degenerate four-wave mixing to form a transient grating can be employed to obtain a direct experimental measurement of energy migration (Eichler *et al.* 1985; Eichler 1986). In these experiments two laser beams, derived from the same laser, are focused onto the same portion of the material where they interfere to produce a periodic spatial variation in the density of excited centres. This creates a spatial modulation of the refractive index of the material of the same periodicity which can act as a diffraction grating for a third laser beam. When the two laser beams which create the grating are switched off the grating decays with a decay time which depends on the intrinsic decay time of the excited centres and on the diffusion of excitation among the excited centres. The situation is illustrated in Fig. 10.14. k_1 and k_2, derived from the same laser, interfere to form a grating which we represent by the periodic shaded areas. If θ is the angle between k_1 and k_2 the grating spacing is

$$d = \frac{\lambda}{2 \sin \theta/2} \tag{10.49}$$

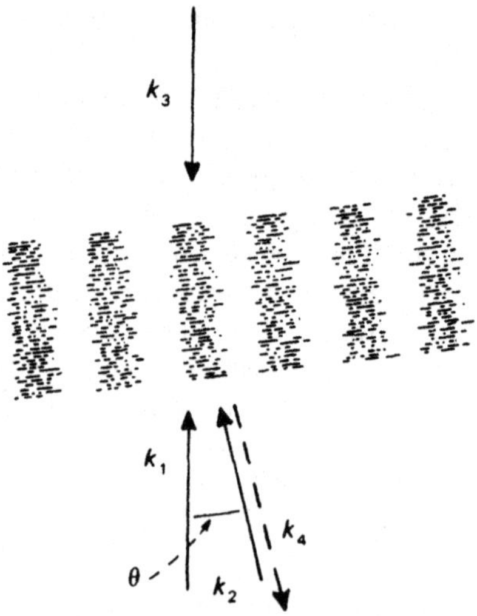

FIG. 10.14. A representation of the formation of a grating by the interference of coherent beams k_1 and k_2, showing the relationship between the probe beam k_3 and the diffracted beam k_4.

where λ is the laser wavelength. If beams k_1 and k_2 are switched off and a laser beam k_3 focused onto the grating this laser beam will be diffracted by the grating. If k_3 is derived from the same laser as k_1 and k_2 and if k_3 is along the negative k_1 direction (a typical geometry) the resultant diffracted beam, k_4, will be along the negative k_2 direction (Fig. 10.14). The gradual fading of the grating means that the backward diffracted beam k_4 decreases in intensity as

$$I_4(t) = I_4(0)\exp(-t/\tau) \tag{10.50}$$

where

$$\frac{1}{\tau} = \frac{2}{\tau_R} + 2D\left(\frac{4\pi}{\lambda}\sin\frac{\theta}{2}\right)^2. \tag{10.51}$$

τ_R is the intrinsic decay time of the excited centres and D is the diffusion constant characterizing the transfer of excitation among the centres. An example of the use of this technique is the study of Lawson *et al.* (1981) of D–D migration in $La_{1-x}Nd_xP_5O_{14}$.

Another laser technique employed to study intraline energy transfer is *holeburning*, in which an intense narrowband laser depletes the ground-state population of those ions whose absorption transitions are in resonance with the laser, thereby burning a narrow 'hole' in the overall absorption spectrum (Szabo 1975). The adjustment of the ions is studied by a second probe beam (Chapter 6).

10.1.6 Radiative transfer

In addition to the non-radiative transfer mechanisms described in the preceding sections, optical energy may also be transferred between similar ions by a radiative process in which one ion emits a photon which can be reabsorbed by another similar unexcited ion before the photon can leave the material. Such *radiative trapping* in a solid material was first observed to occur in ruby by Varsanyi *et al.* (1959). This trapping can significantly increase the observed decay time in ruby samples of a few millimetres linear dimension. Radiative energy transfer in ruby has been studied by FLN techniques (Selzer and Yen 1977).

In this section on energy transfer we have confined our attention to a small number of examples which illustrate basic aspects of the phenomenon. Very full accounts exist in the literature of other energy transfer studies in insulating crystalline crystals (Blasse 1984, Auzel 1984a, b, Boulon 1984), in glasses (Reisfeld 1984), and in semiconductors (Klingshirn 1984).

10.2 Cooperative behaviour, upconversion schemes

In their study of $LaCl_3:Pr^{3+}$ Varsanyi and Dieke (1961) showed that a single photon could be absorbed by a pair of Pr^{3+} ions, each ion being raised to an excited state. This occurs when a weakly-interacting pair of Pr^{3+} ions acts as a single system. This *cooperative absorption* process is illustrated in Fig. 10.15(a). Cooperative absorption processes involving Cr^{3+} and Eu^{3+} ions in $EuAlO_3:Cr^{3+}$ were also demonstrated by van der Ziel and Van Uitert (1969).

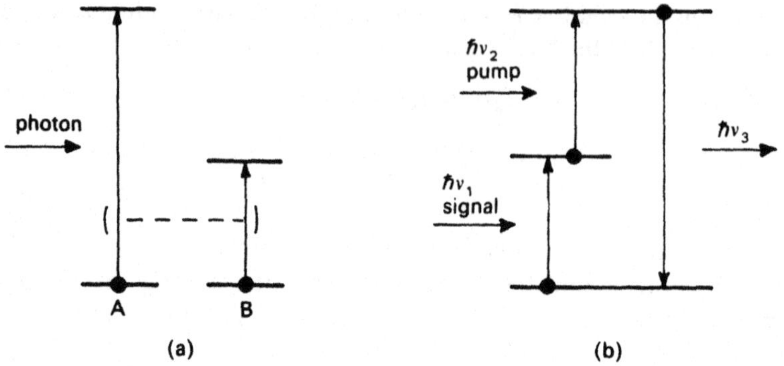

FIG. 10.15. (a) Cooperative absorption process. A single photon is absorbed raising two ions to excited states. (b) Sequential absorption of two photons by a single ion raising the ion to a higher excited state.

A converse process in $LaCl_3:Pr^{3+}$ was found by Porter (1961): two photons can be absorbed sequentially by a single ion raising it to an upper excited state, as illustrated in Fig. 10.15(b). Such a process can provide a mechanism for the operation of an *infrared quantum counter* (Bloembergen 1959) in which an ion first absorbs an infrared 'signal' photon of energy $h\nu_1$. The ion is now irradiated with photons of energy $h\nu_2$ from the 'pump' source. Absorption of the pump photon by the already excited ion raises this ion to a higher excited state from which it decays with the emission of a visible photon. Ideally the scheme results in the emission of a visible photon for every infrared signal photon absorbed. Such a process is termed *upconversion.* In practice this upconversion process is not efficient.

Cooperative luminescence, in which a pair of excited ions simultaneously lose excitation and emit a single photon, was first demonstrated by Nakazawa and Shionoya in ytterbium phosphate (1970). This process is shown schematically in Fig. 10.16(a). Yb^{3+} has a simple energy level structure with a $^2F_{7/2}$ ground state and a $^2F_{5/2}$ excited state at around 10000 cm^{-1}. If

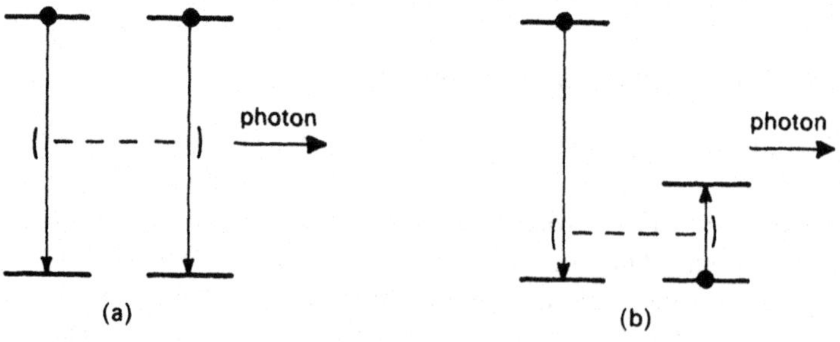

FIG. 10.16. (a) Cooperative luminescence process. A pair of excited ions simultaneously lose excitation emitting a single photon. (b) Cooperative Raman process.

ytterbium phosphate is illuminated with intense infrared radiation corresponding to the $^2F_{7/2} \rightarrow {}^2F_{5/2}$ transition, two adjacent excited Yb^{3+} ions may simultaneously decay emitting a single photon at around 500 nm in accordance with the process:

$$Yb^{3+}(^2F_{\frac{5}{2}}) + Yb^{3+}(^2F_{\frac{5}{2}}) \rightarrow Yb^{3+}(^2F_{\frac{7}{2}}) + Yb^{3+}(^2F_{\frac{7}{2}}) + h\nu(\text{blue}). \qquad (10.52)$$

This process is very weak.

A *cooperative Raman luminescence* was observed by Feofilov and Trofimov (1969) and is illustrated in Fig. 10.16(b). Such a process was also reported by van der Ziel and Van Uitert (1973) in $EuAlO_3$:Cr^{3+}.

Cooperative energy transfer is a process in which two excited ions simultaneously transfer their excitations to another ion. This was first observed by Livanova *et al.* (1969) in calcium fluoride and strontium fluoride doped with Tb^{3+} and Yb^{3+}. Energy was transferred from the $^2F_{5/2}$ levels of two Yb^{3+} ions to the $^5D_4$ level of Tb^{3+} resulting in luminescence from this Tb^{3+} level; the process is illustrated in Fig. 10.17(a). A cooperative transfer process involving three excited Pr^{3+} ions in $LaF_3:Pr^{3+}$ (illustrated in Fig. 10.17(b)) was observed by Lee *et al.* (1984).

The cooperative energy transfer process is very inefficient. A much more efficient upconversion process involving *sequential energy* transfer was first demonstrated by Auzel (1966) and is shown in Fig. 10.17(c): two Yb^{3+} excitations are transferred sequentially to an Er^{3+} ion. In this scheme it is possible to use a high concentration of donor ions to absorb the 'signal' photons and transfer the energy to a smaller concentration of acceptor ions. Upconversion involving the sequential transfer of up to five Er^{3+} $^4I_{13/2}$ excitations to a single Er^{3+} ion to raise it to a higher excited state has been reported by Auzel (1984b).

10.3 Multicentre complexes

If two or more optically-active centres are close enough to interact strongly with each other the resultant complex may act as a single optical centre with distinct spectroscopic properties while the individual centres lose their separate identities. Examples of such complexes are ion pairs and triads coupled through a strong exchange interaction, dopant ions in the vicinity of colour centres, and various other aggregate defects. In this section we will confine our attention to exchange-coupled pairs; aggregate defects involving anion and cation vacancies are discussed in Chapter 7.

10.3.1 Exchange-coupled ion pairs

When the concentration of optically-active ions is increased there is an increasing probability that two ions may occupy sites sufficiently close to each other that a strong exchange interaction occurs between them. The

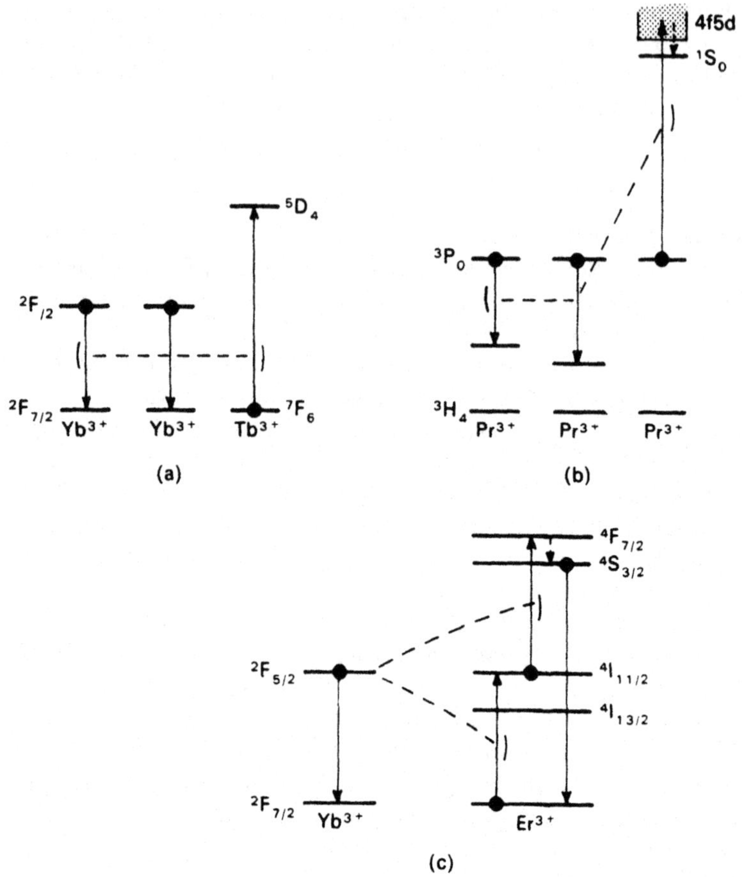

FIG. 10.17. (a) Cooperative energy transfer. Two excited Yb^{3+} ions simultaneously transfer their excitations to an unexcited Tb^{3+} ion. (b) Cooperative energy transfer among three Pr^{3+} ions in $LaF_3:Pr^{3+}$ each in the excited $^3P_0$ state resulting in the excitation of one ion to the $^1S_0$ state. It is likely that this Pr^{3+} ion becomes excited into the 4f 5d band and then quickly decays to the $^1S_0$ level. (c) The Auzel process. Two Yb^{3+} excitations are transferred sequentially to an Er^{3+} ion.

spectroscopic properties of such pairs are distinct from those of the isolated ions (Parrot 1978). Optical transitions on exchange-coupled pairs were first recognized by Schawlow *et al.* (1959) in heavily concentrated ruby ($Al_2O_3:Cr^{3+}$). In Fig. 10.18(a) the luminescence spectrum of a dilute ruby at 77 K shows both the R-lines at 691.9 nm and 693.4 nm and their vibrational sidebands. Figure 10.18(b) shows that in the luminescence spectrum of heavily-doped ruby at 77 K there are additional sharp lines. These lines are attributed to emission by exchange-coupled pairs of Cr^{3+} ions, comprising one ion in the $^4A_2$ ground state and one ion in the excited ^2E metastable

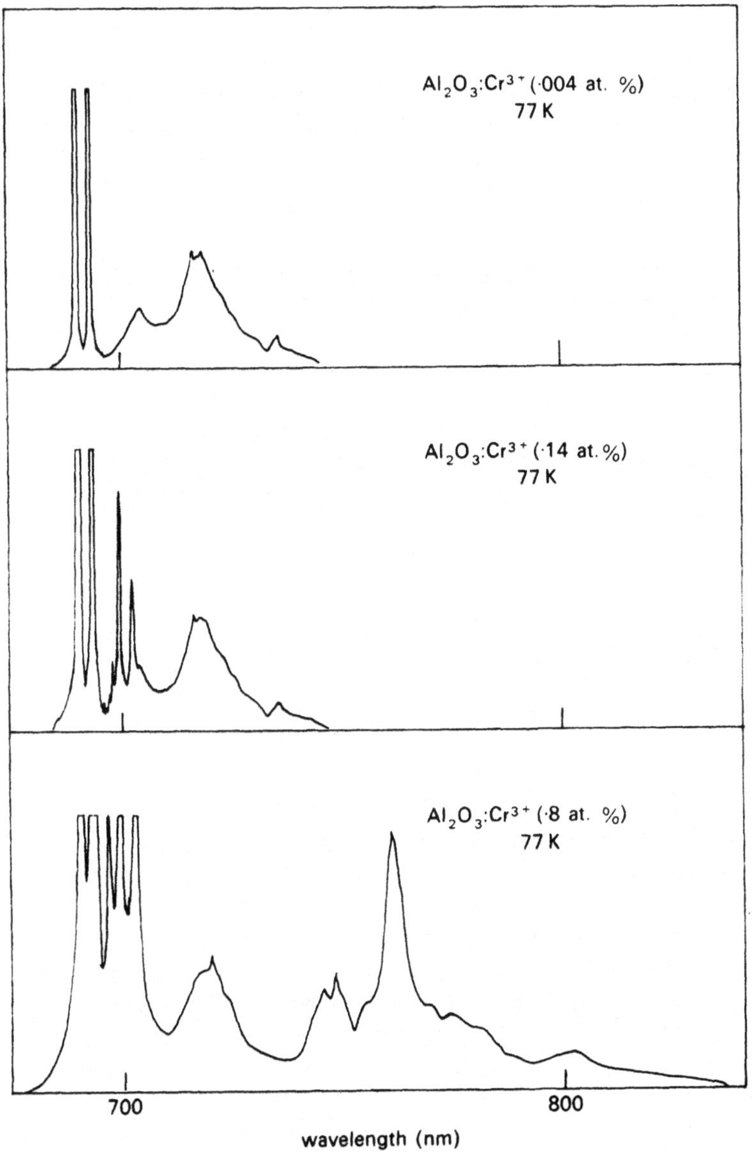

FIG. 10.18. The luminescence spectrum of ruby at 77 K with different concentrations of chromium.

state. In a very heavily doped ruby (Fig. 10.18(c)) additional broad bands observed at longer wavelength are believed to originate on triads and larger clusters of chromium ions.

An exchange interaction occurs between two ions when their electronic charge clouds overlap, either directly or indirectly through the medium of intervening ions. In the latter process, so-called *superexchange*, the interaction strength is greatly reduced. Neglecting constant terms the exchange interaction between ions A and B can be written as (Section 2.3.5)

$$-\sum_{i,j} J_{Ai,Bj} s_{Ai} \cdot s_{Bj} \tag{10.53}$$

where the summation i, j is over all the electrons on ions A and B, respectively (Kisliuk *et al.* 1969; Birgenau 1969). $J_{Ai,Bj}$ is a parameter related to the strength of the Coulomb interaction between two electrons, one on each ion (the exchange integral). Under certain conditions (Orbach 1975; McClure 1975) this interaction can be written in the Heisenberg form, $-J\boldsymbol{S}_A . \boldsymbol{S}_B$, where $\boldsymbol{S}_A$ and $\boldsymbol{S}_B$ are the total spins of ions A and B, respectively, and J is the effective exchange parameter. Even when the Heisenberg form is not fully justified it may be a reasonable description for the calculation of energy levels. But in describing transitions between various coupled ion states the original form, eqn (10.53), should be used.

The forms of the exchange interaction described so far are isotropic, depending only on the relative orientation of the spins, and are valid for free ions. Where the ions are in a solid there may be preferred spin orientations, and anisotropic coupling terms should be added. And in a solid there is a magnetostriction term which can be written $j(\boldsymbol{S}_A . \boldsymbol{S}_B)^2$. If anisotropic coupling terms are small we can omit them and take as our exchange Hamiltonian

$$\mathcal{H}_{EX} = -J\boldsymbol{S}_A . \boldsymbol{S}_B + j(\boldsymbol{S}_A . \boldsymbol{S}_B)^2. \tag{10.54}$$

We refer to these two terms as the quadratic and biquadratic terms, respectively[†]. The full Hamiltonian for the two exchange-coupled ions is $\mathcal{H} = \mathcal{H}_A + \mathcal{H}_B + \mathcal{H}_{EX}$ where $\mathcal{H}_A$ and $\mathcal{H}_B$ are the Hamiltonians for the isolated ions, A and B, respectively. $\mathcal{H}_{EX}$ commutes with the total spin operator of the pair, $\boldsymbol{S} = \boldsymbol{S}_A + \boldsymbol{S}_B$. If spin–orbit coupling terms are neglected in $\mathcal{H}_A$ and $\mathcal{H}_B$ the full Hamiltonian $\mathcal{H}$ commutes with $\boldsymbol{S}$ and the energy levels of the pair can be classified by the allowed values of $\boldsymbol{S}$. Ferromagnetism and antiferromagnetism are manifestations of the collective excitations of all atoms or ions in crystals coupled together by means of the exchange interaction between nearby ions. The spectra of exchange-coupled ion pairs enables one to study the fundamental magnetic interaction at the most basic level, involving only two ions rather than the entire assembly acting collectively in the crystal.

[†] The quadratic formula is sometimes written $-2J\boldsymbol{S}_A \cdot \boldsymbol{S}_B$

We take as an example an exchange-coupled pair of Cr^{3+} ions. The ground state of each Cr^{3+} ion is an orbital singlet and orbital effects are small; its spin is $\frac{3}{2}$. Equation (10.54) is an adequate Hamiltonian to describe the exchange interaction between the ions when both are in the ground state. The total spin takes on values $S=0$, 1, 2, 3 with energies

$$\begin{aligned} S &= 0, \quad E = \frac{15}{4}J + \frac{225}{16}j \\ S &= 1, \quad E = \frac{11}{4}J + \frac{121}{16}j \\ S &= 2, \quad E = \frac{3}{4}J + \frac{9}{16}j \\ S &= 3, \quad E = -\frac{9}{4}J + \frac{81}{16}j. \end{aligned} \tag{10.55}$$

If $j=0$ the ground state splitting follows the Landé interval rule, the separations of the $S=3$, 2, 1, 0 levels being in the ratio 3:2:1. Biquadratic terms cause a small deviation from this rule. In the emitting state of the pair one ion has $S=\frac{1}{2}(^2E)$ and the other $S=\frac{3}{2}(^4A_2)$, thus there are two possible emitting states characterized by the values of $S=2$, 1 of the total spin quantum number. However, in the excited pair state the effects of lower symmetry crystal fields and of spin–orbit coupling are equally as strong as the exchange interaction, and the resultant energy-level structure is very complicated. (Kisliuk *et al.* 1969; Powell *et al.* 1967.) The optical studies indicate that the excited pair levels can still be classified according to total spin. In this case the spins of $\frac{1}{2}$ and $\frac{3}{2}$ give total spin states of $S=1$, 2.

Exchange-coupled pairs of Cr^{3+} ions in alumina are particularly interesting since in this crystal first, second, third, and fourth nearest-neighbour pairs are characterized by small separations. Consequently exchange interactions on each of these pairs are large enough that heavily doped ruby exhibits a rich spectrum of pair transitions. These are the transitions shown in Fig. 10.18(b) in the vicinity of the R_1- and R_2-lines. Figure 10.19 shows the luminescence spectrum of a ruby of medium concentration at 2 K when luminescence occurs only from the lowest excited levels. The line at $\lambda=693.4$ nm is the single ion R_1-line, whereas the four lines near 7000 A originate on the fourth-nearest-neighbour pairs showing clearly the ground state splitting into $S=0$, 1, 2, 3 states. The lowest excited level of the fourth-nearest-neighbour pairs is classified as $S=1$. Since the $S=3$ level of the ground state is lowest the exchange parameter J for fourth-nearest-neighbours is positive. For third nearest neighbours J is negative (Mollenauer and Schawlow 1968; Kaplyanskii and Przhevuskii 1967).

Ferguson *et al.* (1966) carried out detailed studies of the spectrum of pairs of Mn^{2+} ions in $KZnF_3$. They showed that, despite the individual Mn^{2+} ions

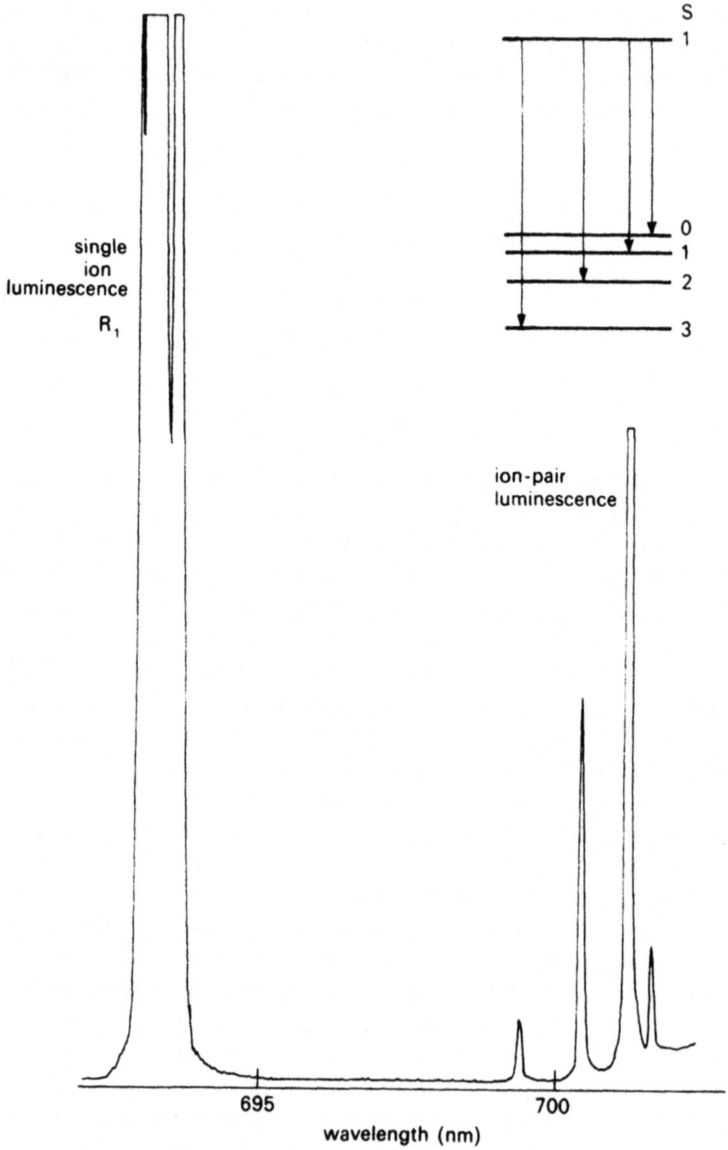

FIG. 10.19. Part of the sharp line luminescence spectrum of a medium concentration ruby at 2 K. The luminescence transitions at very low temperatures on a pair of exchange-coupled Cr^{3+} ions are shown in the case where the ions are ferromagnetically coupled in the ground state and antiferromagnetically coupled in the excited state.

occupying sites of octahedral symmetry, the strong optical transitions of the Mn^{2+} pairs were electric dipole in nature. This is an exchange-induced electric dipole process. In their analysis of the spectra Ferguson *et al.* employed an interaction between the ion pair and the electric field of the radiation, (E), of the form

$$\sum_{i,j} \pi_{Ai,Bj} \cdot E s_{Ai} \cdot s_{Bj}. \tag{10.56}$$

The form of the coupling parameter, π, is discussed by both Tanabe *et al.* (1965) and Gondaira and Tanabe (1966). The nature of the above interaction suggests that the total spin of the pair should be conserved during these electric dipole transitions. That is, one anticipates the selection rule $\Delta S = 0$. We note that in Fig. 10.19 the strongest luminescence line in the fourth-nearest-neighbour spectrum is the $\Delta S = 0$ transition.

The excited pair state under consideration here consists of ion A in the excited state exchange-coupled to ion B in the ground state. We may represent such a pair state by the product state $\psi_1 = |e_A, g_B\rangle$, where e and g denote excited and ground single-ion states. ψ_1 is degenerate in energy with $\psi_2 = |g_A, e_B\rangle$. Any interaction connecting ψ_1 and ψ_2 removes this degeneracy. If $\mathscr{H}'$ is such an interaction, and if we define

$$\mathscr{H}'_{AB} = \langle g_A e_B | \mathscr{H}' | e_A g_B \rangle \tag{10.57}$$

then the eigenstates for the excited pair states are

$$\psi_{\pm} = \frac{1}{\sqrt{2}}(\psi_1 \pm \psi_2) \tag{10.58}$$

and the splitting between them $2\mathscr{H}'_{AB}$. These eigenstates correspond to states in which the excitation is shared between the ions, and the rate of energy transfer between the ions is given by (Bohm 1951) $4\mathscr{H}'_{AB}/h$. The energy separation between ψ_+ and ψ_- is proportional to the rate of energy transfer between the ions of the pairs.

Exchange interactions between rare-earth ions are considerably weaker than between transition metal ions, and the transitions on exchange-coupled pairs of rare-earth ions are very close to the single-ion transitions. Nevertheless by the skilful use of tunable laser excitation techniques the spectroscopy of these pairs can be studied, as demonstrated, for example, by Vial, Buisson, and co-workers in their studies of exchange-coupled pairs in lanthanum fluoride (Buisson and Vial 1981).

10.4 Fully concentrated materials

In fully concentrated materials the inter-ion separation is small and ions may interact strongly with each other. As a result such materials may exhibit some form of cooperative magnetism at low enough temperatures. This is particularly the case for salts of the transition metals, e.g. manganese oxide, manganese fluoride, etc. where strong exchange interaction between the ions result in ferromagnetism, ferrimagnetism, or antiferromagnetism. For example, chromic oxide is antiferromagnetic with a Néel temperature (T_N) of 308 K, below which the Cr^{3+} magnetic moments are antiferromagnetically coupled so that the spins are aligned antiparallel to one another. Similarly, manganese fluoride is antiferromagnetic with $T_N = 67.3$ K. A high value of a Curie temperature or a Néel temperature is indicative of a strong exchange coupling between the ions, and this strong exchange interaction affects the spectroscopic properties of the ions in these materials. In materials such as $FeCl_2.4H_2O$ which contain transition metal ions in a fully concentrated state, the separation between the transition metal ions is sufficiently large that the exchange interaction small and cooperative magnetic effects only occur at very low temperatures. For this material the Néel temperature is near 1 K. Rare-earth ions, with their shielded 4f electron shells, also exhibit weak exchange interactions in the fully concentrated state. For example in $DyAlO_3$ the onset of antiferromagnetism occurs below the Néel temperature of 3.4 K. Since the exchange effects are strongest in the concentrated transition metal ion materials this discussion concentrates almost exclusively on them.

10.4.1 Electronic states in fully concentrated materials: exciton states

In addition to the static magnetic phenomena exhibited by these materials, the strong coupling between ions means that they respond collectively in their interaction with the radiation field. In this case we envisage the ensemble of ions as a single quantum-mechanical entity in its interaction with the radiation. As a simple theoretical example of a fully concentrated system we consider a linear lattice of N cells, each containing a single active ion (Fig. 10.20). The Hamiltonian, $\mathscr{H}_0$, is the sum of the energies of the individual ions in some molecular field approximation. In the simplest case each ion has a single ground state, g, and a single excited state, e. The ground state of the linear chain representing the fully concentrated crystal is then

$$|G\rangle = \prod_i |g_i\rangle \tag{10.59}$$

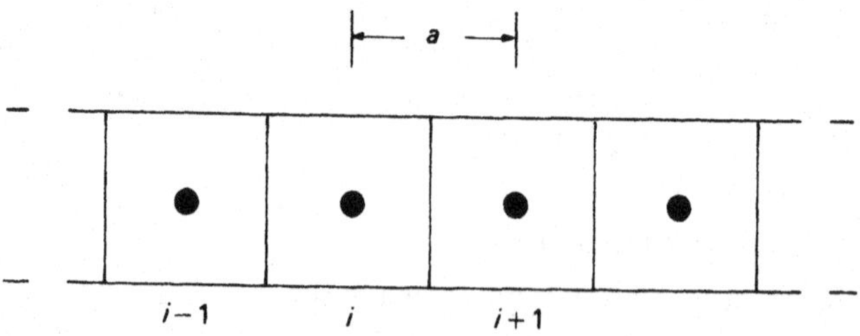

FIG. 10.20. Linear lattice of N cells each containing one active ion.

where the summation is over all N ions in the chain. The energy of this state defines the zero of energy. In the first excited state of this system only the ith ion is in the excited state e_i. Hence the excited state can be written as

$$|E_i\rangle = |e_i\rangle \prod_{j \neq i} |g_j\rangle \tag{10.60}$$

which has energy E_0. The N different states with this energy E_0 are given by eqn (10.60), in which i can have any one of N values.

We now consider the effect of an interaction, $\mathscr{H}'$, which couples these excited states. The matrix element which couples states $|E_i\rangle$ and $|E_j\rangle$ is

$$\mathscr{H}'_{ij} = \langle E_i|\mathscr{H}'|E_j\rangle = \langle e_i g_j|\mathscr{H}'|g_i e_j\rangle. \tag{10.61}$$

For simplicity we will assume that all diagonal matrix elements, $\mathscr{H}'_{ii}$, are zero.

The eigenstates of $\mathscr{H}_0 + \mathscr{H}'$ are linear combinations of the $|E_i\rangle$ states:

$$|E\rangle = \sum_i c_i|E_i\rangle. \tag{10.62}$$

Inserting $|E\rangle$ from eqn (10.62) into the Schrödinger equation and adopting periodic boundary conditions gives

$$c_i = \frac{1}{\sqrt{N}} \exp(ikR_i) \tag{10.63}$$

where R_i denotes the position of the ith ion. The allowed values of k are

$$k = \frac{2\pi}{Na} n \tag{10.64}$$

where a is the separation between adjacent ions and n is an integer. We need only consider N consecutive values of k chosen so that k spans the first Brillouin zone between $-\pi/a$ and $+\pi/a$. The eigenstates classified by their k values are written as

$$|E(k)\rangle = \frac{1}{\sqrt{N}} \sum_i \exp(ikR_i)|E_i\rangle. \tag{10.65}$$

These *exciton states* are states of collective excitation of the crystal. Any antisymmetrization requirements are understood to be included in the summation in eqn (10.65).

The energy of the state $|E(k)\rangle$ is

$$\langle E(k)|\mathscr{H}_0 + \mathscr{H}'|E(k)\rangle$$

$$= E_0 + \frac{1}{N} \sum_{i,j} \mathscr{H}'_{ij} \exp(ik(R_j - R_i)). \tag{10.66}$$

For simplicity we assume once more that nearest neighbour interactions only are non-zero. Hence $\mathscr{H}'_{ij}$ takes on the values $\mathscr{H}'_{i(i+1)}$ and $\mathscr{H}'_{i(i-1)}$ for a given i.

Assigning to these matrix elements the value V' we find that the energy of the state $|E(k)\rangle$ is

$$E_0 + \frac{1}{N}\sum_i V'(\exp(ika) + \exp(-ika))$$

$$= E_0 + 2V'\cos ka. \tag{10.67}$$

This result is plotted in Fig. 10.21 for k values in the first Brillouin zone, assuming V' to be positive.

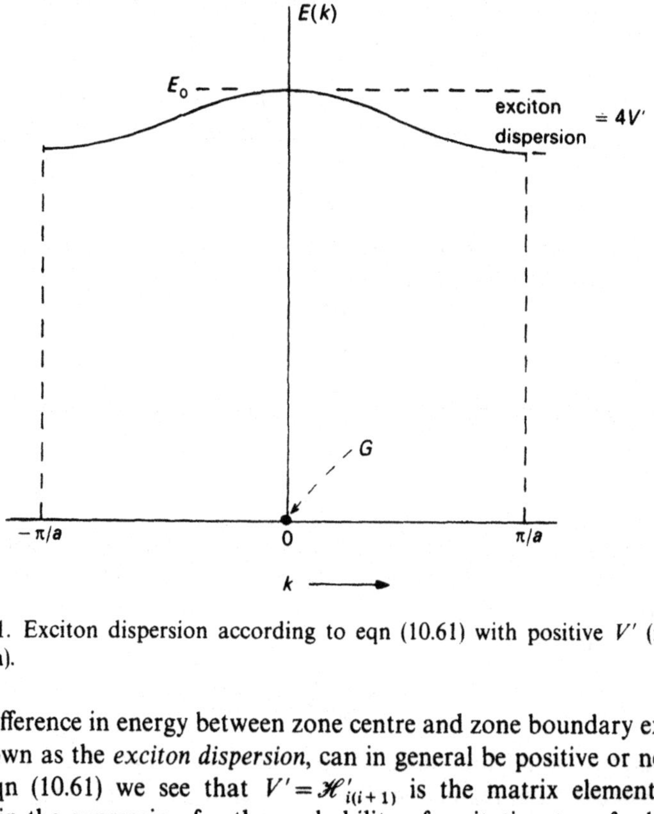

FIG. 10.21. Exciton dispersion according to eqn (10.61) with positive V' (negative dispersion).

The difference in energy between zone centre and zone boundary excitons, $4V'$, known as the *exciton dispersion*, can in general be positive or negative. From eqn (10.61) we see that $V' = \mathscr{H}'_{i(i+1)}$ is the matrix element which appears in the expression for the probability of excitation transfer between adjacent ions. Hence the magnitude of the dispersion is related to the rate of transfer of excitation among ions. A dispersion of around 1 cm^{-1} means a transfer rate between resonant adjacent ions of around 10^{11} s^{-1}.

Now consider the case that there are two identical ions A and B in each lattice cell, the line joining A and B is orthogonal to the line of cells. This system can be regarded as consisting of two sublattices, the A and B sublattices, respectively, each sublattice containing N ions (Fig. 10.22). We can similarly develop exciton theory for this two-sublattice system.

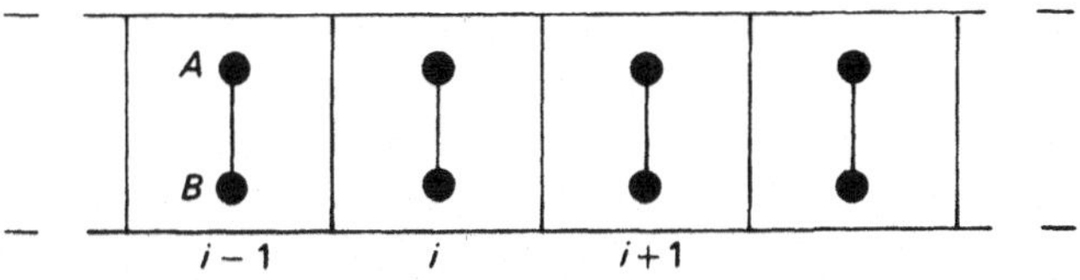

FIG. 10.22. Linear lattice of N cells each containing two active ions A and B. The line joining A and B is perpendicular to the line joining the similar ions.

Accordingly the energy of the kth exciton state is

$$E(k) = E_0 + 2V' \cos ka \pm 2V'' \tag{10.68}$$

where the intrasublattice coupling matrix element is given by

$$V' = \langle e_i^{\mathrm{A}} g_{i+1}^{\mathrm{A}} | \mathscr{H}' | g_i^{\mathrm{A}} e_{i+1}^{\mathrm{A}} \rangle \tag{10.69}$$

and the intersublattice coupling is

$$V'' = \langle e_i^{\mathrm{A}} g_i^{\mathrm{B}} | \mathscr{H}' | g_i^{\mathrm{A}} e_i^{\mathrm{B}} \rangle. \tag{10.70}$$

The intrasublattice coupling, V', leads again to exciton dispersion while the intersublattice coupling, V'', leads to the exciton band being split into two bands. The energy separation between the bands, the *Davydov splitting*, is related to the transfer of excitation between sublattices A and B. If this intersublattice coupling term is zero the excitons are confined to one sublattice or the other.

Such a one-dimensional model can be extended to three dimensions to include interactions other than nearest-neighbour, and to include more complicated arrangement of the two ions within the cell. McClure (1975) shows that the general form for the energy of the kth exciton state is

$$E(k) = E_0 + V_1(\boldsymbol{k}) \pm V_2(\boldsymbol{k}) \tag{10.71}$$

where $V_1(\boldsymbol{k})$ and $V_2(\boldsymbol{k})$ are the intrasublattice coupling energy and the intersublattice coupling energy, respectively, which are responsible for the exciton dispersion and Davydov splitting.

The exciton states described above are collective states of electronic excitation. Phonons are collective states of vibrational excitation. Collective states of magnetic excitation also occur in which a disturbance of the magnetic system from the fully aligned state is spread coherently through the lattice of magnetic ions. These disturbances are known as *magnons* (quantized spin waves). Both phonon and magnon states are classified by k values, they can properly be regarded as excitons, but for our purposes it is more useful to reserve the term exciton for the collective states of electronic excitation with energy in the optical region.

We consider now the creation of an exciton by the absorption of a photon which occurs through the interaction of the radiation with the electrons. The analysis of optical transitions in fully concentrated materials differs from that of isolated dopant ions in that the excited state is described by an additional parameter, its wavevector k, and a k-selection rule appears. To see this we again use our one-dimensional model and look at the matrix element of the electric dipole transition between the ground state, G, and state $E(k)$ due to the interaction, $\mathscr{H}_{\text{int}}$, of the line of ions with the radiation field

$$\mathscr{H}_{\text{int}} = -\sum_l p_l E_0 \exp \pm \mathrm{i}(KR_l - \omega t) \tag{10.72}$$

where E_0 is the maximum strength of the electric field of the radiation, K is the wavevector of the radiation, and we choose the $+\mathrm{i}$ form for absorption (Section 4.1). For an electric dipole process the relevant matrix element is

$$\begin{aligned} &\langle E(k)|\mathscr{H}_{\text{int}}|G\rangle \\ &= -\frac{1}{\sqrt{N}} \sum_i \langle \mathrm{E}_i \exp(\mathrm{i}kR_i)| \sum_l p_l E_0 \exp(\mathrm{i}KR_l)|G\rangle \\ &= \frac{1}{\sqrt{N}} \sum_i \langle \mathrm{E}_i | \sum_l p_l E_0 \exp(\mathrm{i}KR_l)|G\rangle \exp-(\mathrm{i}kR_i) \end{aligned} \tag{10.73}$$

According to eqns (10.59) and (10.60) E_i differs from G only in that the ith ion is in the excited state, whereas it is in the ground state in G. Hence the ith matrix element in eqn (10.73) is

$$\langle e_i \exp(\mathrm{i}kR_i)| p_i E_0 \exp(\mathrm{i}KR_i)|g_i\rangle = M_i \exp \mathrm{i}(K-k)R_i \tag{10.74}$$

and M_i is independent of i. The full matrix element of the transition, eqn (10.73), is

$$-M \sum_i \frac{1}{\sqrt{N}} \exp \mathrm{i}(K-k)R_i = -M\sqrt{N}\delta_{K-k,G} \tag{10.75}$$

where the subscript G in the Kronecker delta refers to the general reciprocal lattice vector. We will only be concerned with normal processes for which $G=0$. Hence the k-selection rule is $k=K$, which in three dimensions becomes $\boldsymbol{k}=\boldsymbol{K}$.

This selection rule holds for electric dipole transition. If a transition occurs by a magnetic dipole process then in eqn (10.72) μB replaces pE, B being the magnetic field strength of the radiation. However, the $\boldsymbol{k}$-selection rule is unchanged, applying irrespective of the nature of the transition. Since the wavelength for optical radiation, $\lambda = 2\pi/K$, is much larger than the separation between atoms, a, we can assume $\boldsymbol{K} \simeq 0$ for an optical photon. This means that if the absorption of an optical photon results in the creation of an exciton only a $\boldsymbol{k}=0$ exciton is created. Even though the exciton dispersion may be large one still observes only the sharp line corresponding to the creation of a $\boldsymbol{k}=0$ exciton.

An absorption transition in which an exciton and a phonon or an exciton and a magnon are created must also obey the $\boldsymbol{k}$-selection rule. In this transition the exciton of wavevector $\boldsymbol{k}$ can be created if a phonon (or magnon) of wavevector $-\boldsymbol{k}$ is simultaneously created. Such transitions are observed as sidebands accompanying the sharp, pure exciton transitions. The pure exciton absorption transition and the exciton-plus-magnon absorption transition from the ground state, G, with $\boldsymbol{k}=0$, are shown schematically in Fig. 10.23. In this figure the exciton states are shown as having a small negative dispersion. The magnon states have a larger dispersion, with energy rising from a very small value ($\simeq 0$) at $\boldsymbol{k}=0$ to a maximum at the zone boundary. All transitions originate at $\boldsymbol{k}=0$ in the ground state, G. In Fig. 10.23 vertical arrow 1 indicates the pure exciton absorption transition in which only the $\boldsymbol{k}=0$ exciton is created. This transition is observed as the sharp absorption line in the hypothetical absorption spectrum shown in the figure. Arrows 2 and 3 indicate the creation of a magnon and an exciton of wavevectors $\boldsymbol{k}$ and $-\boldsymbol{k}$, respectively. The simultaneous creation of such exciton–magnon pairs at all $\boldsymbol{k}$ values gives us the broad magnon sideband which accompanies the sharp pure exciton line; this is illustrated in the hypothetical absorption

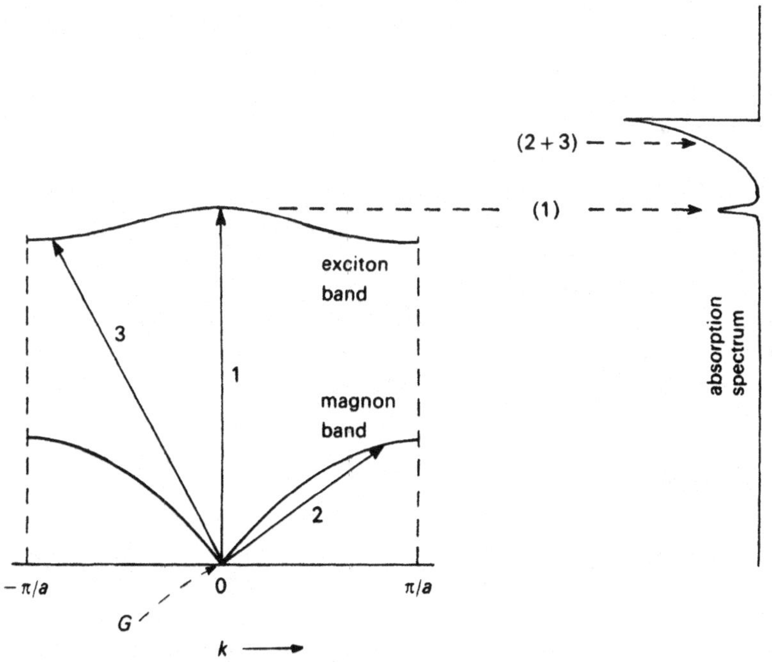

FIG. 10.23. Exciton creation (arrow 1) and exciton-plus-magnon creation (arrows 2 and 3) in a hypothetical material. The hypothetical absorption spectrum due to these transitions is shown on the right.

spectrum in the figure. The sideband shape reflects the density of magnon states.

A similar diagram can be drawn for exciton–phonon transitions which also obey the $\Delta \boldsymbol{k}=0$ selection rule. The mechanisms responsible for phonon sidebands were considered in detail in Chapter 5 where, however, no mention was made of a $\boldsymbol{k}$-selection rule. In that case we were dealing with radiative transitions on an *isolated* dopant ion in a host crystal such as the isolated Mn^{2+} ion in zinc fluoride. This electronic system does not have translational symmetry, the states are not characterized by specific $\boldsymbol{k}$ values, and no $\boldsymbol{k}$-selection rule applies. On the other hand, in manganese fluoride the electronic state is a coupled state characterized by a $\boldsymbol{k}$ value and $\boldsymbol{k}$-selection rules apply.

10.4.2 *Absorption spectra of fully concentrated material*

As an illustrative example we now discuss the optical absorption spectrum of the fully concentrated solid, manganese fluoride, which illustrates many of the ideas developed above. This antiferromagnetic solid has the rutile structure in which the Mn^{2+} ions occupy sites with inversion symmetry on two interpenetrating simple tetragonal sublattices which we label the A and B sublattices (Fig. 10.24). All the Mn^{2+} ions have identical arrangements of F^- ions about them, but the anion sublattice about the A cation is rotated by 90° about the optic axis relative to the anion sublattice around the B ions. Hence the A sublattice is crystallographically distinct from the B sublattice. The crystal field at the site of the Mn^{2+} ion is mainly octahedral but has a small orthorhombic component.

The energy-level diagram of a Mn^{2+} ion in an octahedral crystal field (Fig. 10.25) shows the ground state to be an orbital singlet with sixfold spin degeneracy, i.e. the $^6A_1$ state. The exchange field in manganese fluoride removes this spin degeneracy leaving as the lowest level the spin state with $|M_s|=\frac{5}{2}$, the axis of quantization being the optic axis of the crystal. The lowest excited electronic level, $^4T_1$, is situated about 18 000 cm^{-1} above the ground state. This state has orbital and spin degeneracy which is removed by the combined effects of spin–orbit coupling, orthorhombic crystal field, and the exchange field. The two lowest energy excited states labelled in order of increasing energy, as E_1 and E_2, may to good approximation be assigned $|M_s|=\frac{3}{2}$ (Fig. 10.25).

Below the Néel temperature (67.3 K) the ions are ordered magnetically. In the ground magnetic state the spins on the A sublattice are aligned parallel to each other and point along the optic axis ($M_s=\frac{5}{2}$), whereas the spins on the B sublattice are parallel to each other but antiparallel to the spins on the A sublattice (i.e. $M_s=-\frac{5}{2}$); thus the crystal is in an antiferromagnetic state, the A and B sublattices being crystallographically and magnetically distinct.

We first consider the exciton states of manganese fluoride in the ordered antiferromagnetic state. These are formed from the single-ion ground and

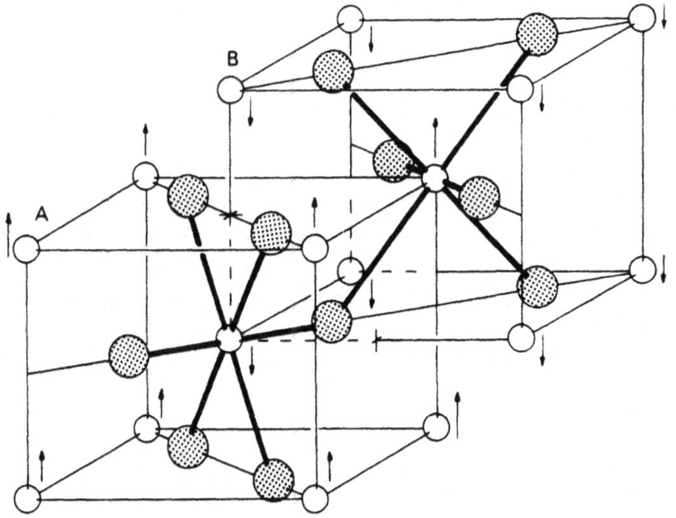

FIG. 10.24. Showing the crystal structure of manganese fluoride. The Mn^{2+} ions (open circles) form two interpenetrating simple tetragonal sublattices (labelled A, B). The spins of the Mn^{2+} ions in sublattice A point in the same direction; those in sublattice B point in the opposite direction. The spins are indicated in the figure. The arrangement of six F^- ions (shaded circles) about two Mn^{2+} ions, one in each sublattice, are shown. These arrangements are identical but are rotated through 90° about the optical axis (vertical direction in the figure). The crystal field at each Mn^{2+} ion is mainly octahedral with a small orthorhombic component.

excited states in accordance with formulae (10.59), (10.60), and (10.65). The lowest electronic state has all the Mn^{2+} ions in the ${}^6A_1(|M_s|=\frac{5}{2})$ state, while in each of the two lowest energy exciton states, ${}^4T_1(E_1, |M_s|=\frac{3}{2})$ and ${}^4T_1(E_2, |M_s|=\frac{3}{2})$, one single ion excitation is created. It is convenient to label these exciton states by the single-ion excitation. Thus we refer to the ${}^4T_1(E_1, |\frac{3}{2}|)$ exciton state and to the ground ${}^6A_1(|\frac{5}{2}|)$ state. The exciton states are characterized by k values in the first Brillouin zone and exhibit an exciton dispersion across the zone (shown schematically in Fig. 10.23). The exciton dispersion measures the ability of the excitation to travel from one Mn^{2+} ion to an adjacent Mn^{2+} ion on the same sublattice. The transfer of excitation from an excited A ion to a nearby unexcited A ion involves the simultaneous transitions, ${}^4T_1(\frac{3}{2})\rightarrow{}^6A_1(\frac{5}{2})$ on one ion and ${}^6A_1(\frac{5}{2})\rightarrow{}^4T_1(\frac{3}{2})$ on the other ion. There is no change in total spin in this process. On the other hand, the transfer of excitation from an excited A ion to a nearby unexcited B ion involves the transitions, ${}^4T_1(\frac{3}{2})\rightarrow{}^6A_1(\frac{5}{2})$ and ${}^6A_1(-\frac{5}{2})\rightarrow{}^4T_1(-\frac{3}{2})$ which results in a total spin change, $\Delta M_s=2$. Such a transition is strongly spin-forbidden. Consequently it is a good approximation to assume that a ${}^4T_1$ exciton, once created, is confined to either the A or the B sublattice. This considerably simplifies the spectroscopic analyses.

Next we consider the magnon states of manganese fluoride. In the ground magnetic state at $T=0$ K the spins are essentially all aligned in the ordered antiferromagnetic state. The lowest excited magnetic state—a magnon state—is one of coherent spin deviation from the fully aligned arrangement; it contains one $^6A_1(|\frac{3}{2}|)$ excitation. The magnon states are characterized by k values in the first Brillouin zone, and in each magnon state, irrespective of its k value, the total spin deviation equals that of a single spin flip, $|\Delta M_s|=1$. The magnon dispersion and the density of magnon states in manganese fluoride are shown in Fig. 10.26. Low k magnons involve spin deviations spread over both A and B sublattices, but at the zone boundary the magnons are confined to one or other sublattice.

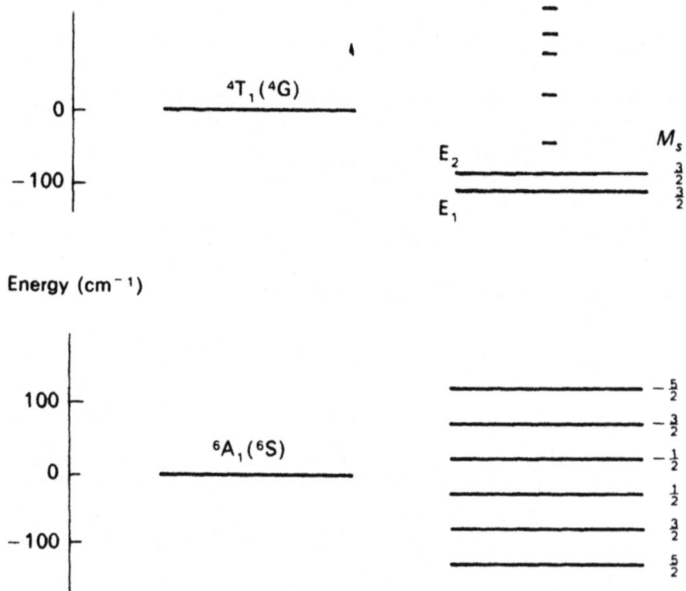

FIG. 10.25. Energy levels of the ground and lowest excited states of Mn^{2+} (a) in an octahedral crystal field, (b) with additional terms: spin–orbit coupling, orthorhombic crystal field, and exchange interaction appropriate to Mn^{2+} in manganese fluoride. The exchange interaction is the major perturbation in the ground state where it removes the M_s degeneracy. The splitting in the $^4T_1$ state is very complicated. The two lowest levels, E_1 and E_2, are assigned the value $M_s=\frac{3}{2}$.

The experimentally observed absorption and luminescence spectra of manganese fluoride at $T = 2$ K in the 500 nm–700 nm region are shown in Fig. 10.27. These are the $^6A_1 \leftrightarrow {}^4T_1$ transitions, they are characterized by a large Stokes shift, and are predominantly phonon-assisted transitions. We

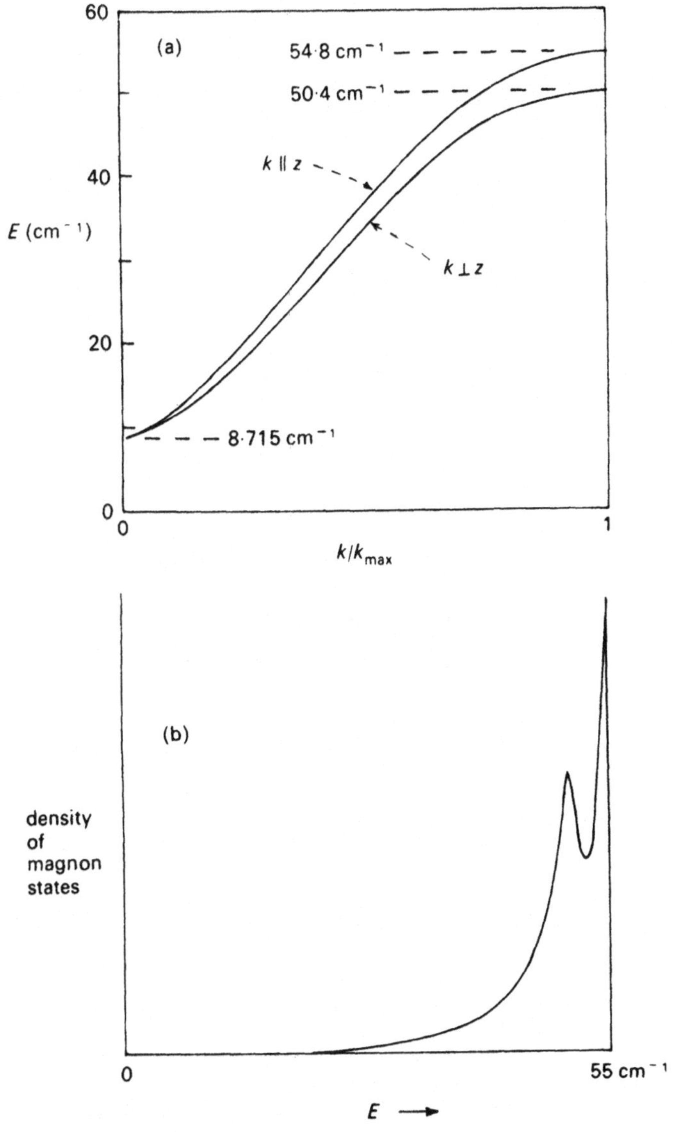

FIG. 10.26. (a) Magnon dispersion in manganese fluoride. The optic axis is along the $\hat{z}$-direction. (b) Density of magnon states in manganese fluoride (After Loudon 1966.)

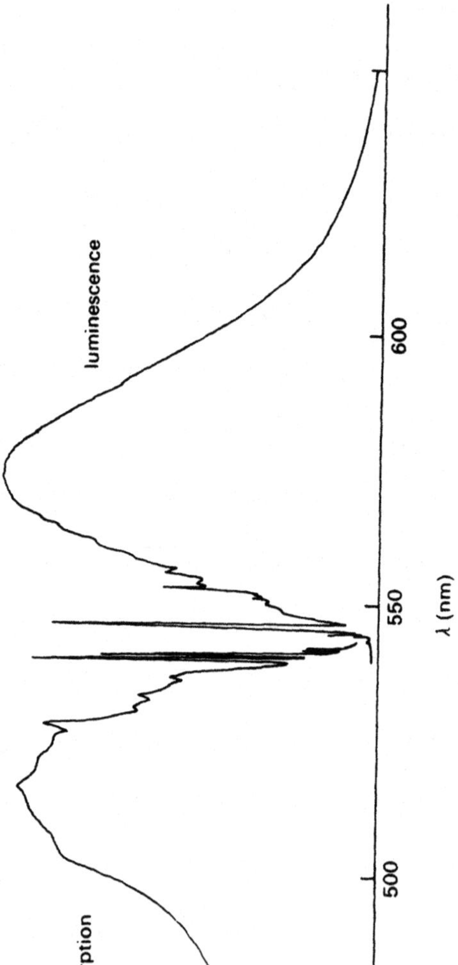

FIG. 10.27. The $^6A_1 \rightarrow {}^4T_1$ absorption spectrum and the $^4T_1 \rightarrow {}^6A_1$ luminescence spectrum of manganese fluoride at 2 K.

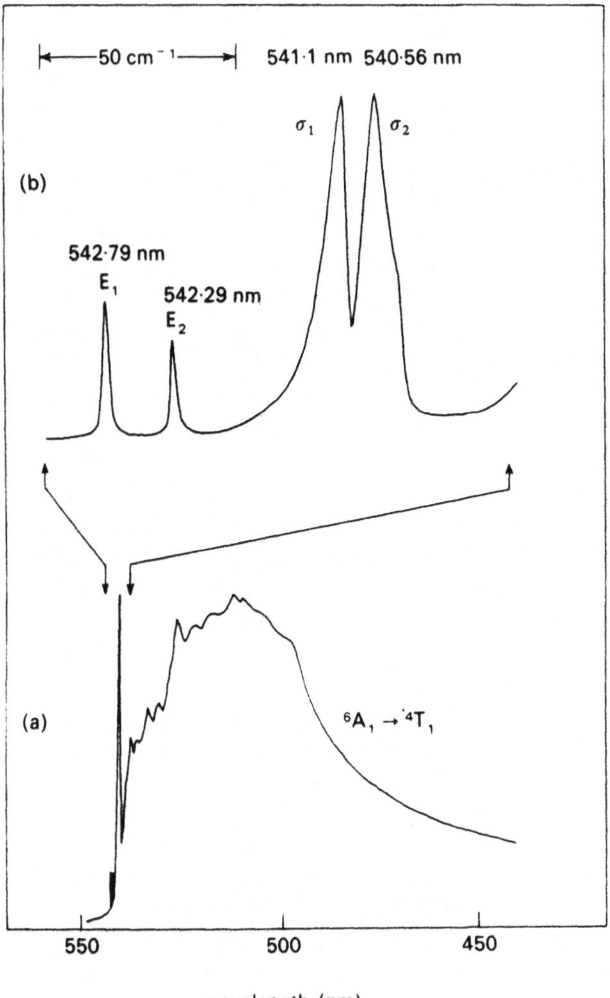

FIG. 10.28. (a) $^6A_1 \rightarrow {}^4T_1$ absorption spectrum of manganese fluoride at 2 K in σ-polarization. (b) Details of the sharp absorption transitions on an expanded wavelength scale.

first analyse the absorption transition. Figure 10.28(a) shows the $^6A_1 \rightarrow {}^4T_1$ transition in σ-polarization at 2 K. The sharp features are seen in greater detail in Fig. 10.28(b).

The two sharp lines, E_1 and E_2, are the pure exciton transitions $^6A_1(\frac{5}{2}) \rightarrow {}^4T_1(E_1, \frac{3}{2})$, $^4T_1(E_2, \frac{3}{2})$, on sublattice A and the corresponding transitions with opposite M_s values on sublattice B. Since these involve spin changes $\Delta M_s = -1$ and $+1$ on sublattices A and B, respectively, they are spin-forbidden. In addition, since the Mn^{2+} ions occupy sites of inversion symmetry these transitions are parity-forbidden. Hence the E_1 and E_2 transitions are very weak. The broader and stronger absorptions, σ_1 and σ_2, are one-magnon sidebands accompanying lines E_1 and E_2, respectively. These transitions involve the creation of a magnon and an exciton; their shapes approximately resemble the density of magnon states for manganese fluoride (Fig. 10.26(b)). These magnon sidebands are much stronger than the pure exciton transitions and are found to be electric dipole transitions.

The existence of an exchange-induced electric dipole radiative mechanism was inferred by Ferguson *et al.* (1965, 1966) from their studies of the strengths of the radiative processes on exchange-coupled Mn^{2+} ion pairs. The interaction operator is given by eqn (10.56). One can expect an analogous process in magnetically-ordered materials (Tanabe *et al.* 1965). We recall from the discussion on the exchange-induced electric dipole transition process (Section 10.3.1) that the total spin must be conserved, and for magnetically-ordered materials the analogous selection rule is $\Delta M_s = 0$. Such a spin-conserving transition in magnesium fluoride is one involving the creation of an exciton and a magnon. If an exciton is created on sublattice A ($\Delta M_s = -1$) and a magnon is created with $\Delta M_s = +1$ then total spin is conserved, and such a transition may proceed through the exchange-induced electric dipole process. This is just the magnon sideband process, and the exchange-induced electric dipole process is responsible for the strong electric dipole nature of these sidebands. To satisfy the $\boldsymbol{k}$-selection rule, the magnon of wavevector $\boldsymbol{k}$ must be created along with an exciton of wavevector $-\boldsymbol{k}$. For example, the creation of a zone-boundary magnon on the A sublattice ($\Delta M_s = +1$) will require the simultaneous creation of an exciton on the B sublattice ($\Delta M_s = -1$), so that in the case of the zone boundary magnon sideband absorption transition the exciton and magnon are created *on opposite sublattices.* This spin-conserving process is illustrated in Fig. 10.29(a) which shows the *ion-pair analogue* of the exciton–magnon absorption process. This involves a ${}^6A_1(-\frac{5}{2}) \rightarrow {}^4T_1(E_1, -\frac{3}{2})$ transition on a B ion and a magnetic spin-flip (e.g. ${}^6A_1\,(\frac{5}{2}) \rightarrow {}^6A_1\,(\frac{3}{2})$ transition on an adjacent A ion. There is no change in total spin in this absorption process.

The sideband shape should reflect the density of magnon states, modified to take exciton dispersion into account. In addition, since the magnon is being created in the vicinity of the exciton, and since the exciton creates a spin defect (a ${}^4T_1(\frac{3}{2})$ state instead of a ${}^6A_1(\frac{5}{2})$ state), the magnon is perturbed by this spin defect and this also affects the sideband shape (Elliott *et al.* 1968).

Similar absorption spectra are found in other fully concentrated manganese compounds, such as $RbMnF_3$, Rb_2MnF_4. Interesting absorption spectra which reflect the strong exchange interaction between the ions are also found in other transition metal compounds: ferrous fluoride, cobalt fluoride, and chromic oxide. A detailed analysis of the absorption transitions

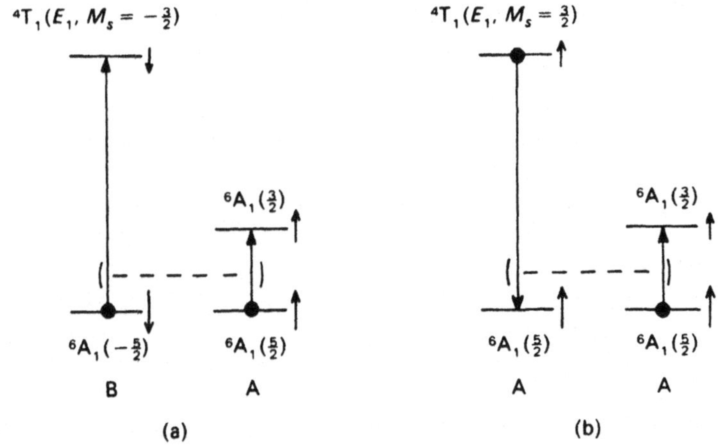

FIG. 10.29. Single-ion and ion-pair analogues of the pure exciton and exciton–magnon transitions. The $^6A_1(M_s = -\frac{5}{2}) \rightarrow {}^4T_1(M_s = -\frac{3}{2})$ transitions on an ion in sublattice B (or the equivalent transition on an ion in sublattice (A) is a single-ion analogue of the E_1 pure exciton absorption transition. The absorption process involving the E_1 absorption transition on a B ion and a magnetic spin–flip (i.e. $M_s = \frac{5}{2} \rightarrow M_s = \frac{3}{2}$) on an A ion is the ion-pair analogue of the exciton–magnon absorption process in manganese fluoride. This is illustrated in (a). There is no change in the value of total spin in this process. The corresponding emission processes are illustrated in (b). In order to conserve total spin in the exciton–magnon emission process the annihilation of the E_1 exciton and spin–flip transition must occur on ions belonging to the same sublattice.

in the rutile structured antiferromagnets (manganese fluoride, ferrous fluoride, and cobalt fluoride) has been given by Loudon (1966) and by Eremenko and Petrov (1977).

10.4.3 Luminescence from fully concentrated materials

When present at low concentration in inorganic insulators Cr^{3+} and Mn^{2+} ions are efficient luminescence centres, as is evident from the efficiency of ruby, Al_2O_3: Cr^{3+}, the first laser material, and by the many technically useful manganese phosphors. Chromic oxide, however, exhibits no intrinsic luminescence, and manganese fluoride at temperatures above that of liquid nitrogen exhibits no manganese luminescence. A similar absence of luminescence is observed in some rare-earth systems. For example, although $LaCl_3$: Nd^{3+} is a strong emitter no luminescence is observed from neodymium chloride. On the other hand, the efficiency of Pr^{3+} luminescence in praseodymium chloride is sufficiently high that laser action can be generated in that material.

Frequently as the concentration of luminescent centres is increased the luminescence efficiency of the material decreases; the luminescence is said to be *quenched*. Such luminescence quenching at higher concentrations occurs because interactions between centres may allow efficient energy transfer away from the excited ion to other centres which do not exhibit luminescence ('traps'). Cross-relaxation between the nearby ions in fully concentrated materials may also cause quenching. Since the energy transfer can be very efficient in concentrated materials the quenching traps need only be present in very dilute amounts for a total quenching of the luminescence. These traps may be unintentional dopants which are present at concentrations of a few parts per million.

Again we take manganese fluoride as a model system in which to study these quenching effects. Under optical pumping the Mn^{2+} ions are excited to the lowest excited $^4T_1$ level. At low temperatures only the E_1 level is occupied, and the radiative decay rate from this level is around $30\ s^{-1}$. In pure manganese fluoride the excited state is an exciton state and the low temperature emission process should be one in which an E_1 exciton is annihilated. In this pure exciton process the ground state G has $\boldsymbol{k}=0$ so the process involves the annihilation of a $\boldsymbol{k}=0$ exciton. Hence at low temperatures one expects to find an E_1 exciton emission transition coincident in frequency with the E_1 absorption transition. This pure exciton emission transition may be accompanied by magnon or phonon sidebands. Since the pure exciton transition involves a spin change $\Delta M_s = \pm 1$ one expects it to be accompanied by a strong magnon sideband transition in which the creation of a magnon accompanies the exciton annihilation in such a way that there is no total spin change. The $\boldsymbol{k}$-selection rule must also be obeyed.

Consider, for example, the annihilation of a zone-boundary $(k=\pi/a)$ exciton on the A sublattice involving a spin change $\Delta M_s = +1$. This is accompanied by the creation of a zone-boundary magnon with opposite $\boldsymbol{k}$ value. This satisfies the $\Delta \boldsymbol{k}=0$ selection rule. To conserve spin the creation of the magnon must involve a spin change $\Delta M_s = -1$. We recall that zone-boundary magnons are confined to one of the sublattices, so the creation of a zone boundary magnon on the A sublattice involves a spin change $\Delta M_s = -1$. Consequently the spin-conserving exchange-induced electric dipole transition can proceed by the annihilation of a zone boundary exciton and the creation of a zone boundary magnon *on the same sublattice*. The ion pair analogue of this exciton–magnon emission process is illustrated in Fig. 10.29(b). Spin-conserving processes can involve other than zone-boundary excitons and magnons, and the full sideband spectrum can cover the range of magnon energies.

The absorption and luminescence spectra from a nominally pure manganese fluoride crystal at 2 K are seen in Fig. 10.27. The sharp features in these spectra are shown in greater detail in Fig. 10.30, which also shows the relevant energy levels. The positions of the E_1 and E_2 levels are determined by the E_1 and E_2 absorption lines. At such low temperatures one would expect

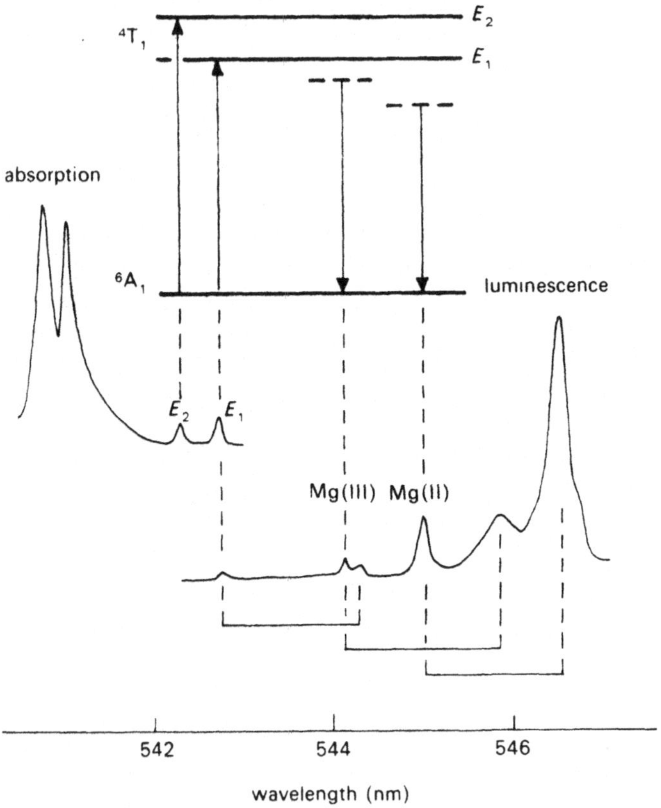

FIG. 10.30. Low-temperature sharp line absorption and luminescence transitions in nominally pure manganese fluoride. The luminescence is predominantly from Mn^{2+} ions perturbed by adjacent Mg^{2+} ions. Each sharp luminescence line is accompanied by a magnon sideband at longer wavelength, as indicated by the horizontal lines.

the luminescence to consist of a sharp line at the E_1 frequency accompanied by magnon and phonon sidebands. This is not what is observed. Only a very weak luminescence line is seen at the E_1 frequency and it is accompanied by a correspondingly weak magnon sideband. Further, such intrinsic emission is observed only at very low temperatures and then only from very pure manganese fluoride. In general the sharp luminescence lines occur at longer wavelength. Two such lines, labelled Mg(111) and Mg(11), are shown in Fig. 10.30. These lines originate on perturbed Mn^{2+} ions whose lowest excited electronic levels are below E_1, and these levels are represented by the broken horizontal lines in Fig. 10.30. The perturbations are due to the presence of nearby Mg^{2+} ions which are unintentional dopants at concentrations of a few parts per million; the Mg(111) line originating on a Mn^{2+}

ion which has a Mg^{2+} ion in a third-nearest-neighbour manganese site, the Mg(11) line originating on a Mn^{2+} ion which has a Mg^{2+} ion as a second nearest neighbour. Since the lowest excited levels of these perturbed Mn^{2+} ions are below the E_1 exciton level they can act as traps and draw excitation from the regular Mn^{2+} ions, which is then released as luminescence characteristic of these perturbed Mn^{2+} ions. The Mg(111) and Mg(11) lines are accompanied by magnon sidebands (Fig. 10.30); but most of the luminescene is contained in the phonon sidebands of these sharp transitions (Fig. 10.27).

In other nominally pure manganese fluoride crystals Zn^{2+} or Ca^{2+} ions may be the predominant trace impurities. The low temperature luminescence from these materials originates on Mn^{2+} ions perturbed by these impurity ions (Greene *et al.* 1968). And the sharp transitions of these Zn^{2+} or Ca^{2+}-perturbed Mn^{2+} ions are distinct from the sharp transitions of the Mg^{2+}-perturbed Mn^{2+} ions. Figure 10.31 shows the pure electronic transition and the magnon sideband from Mn^{2+} ions perturbed by Ca^{2+} ions. By comparing this spectrum with Fig. 10.21(b), the shape of the magnon sideband is seen to be a very good replica of the magnon density of states.

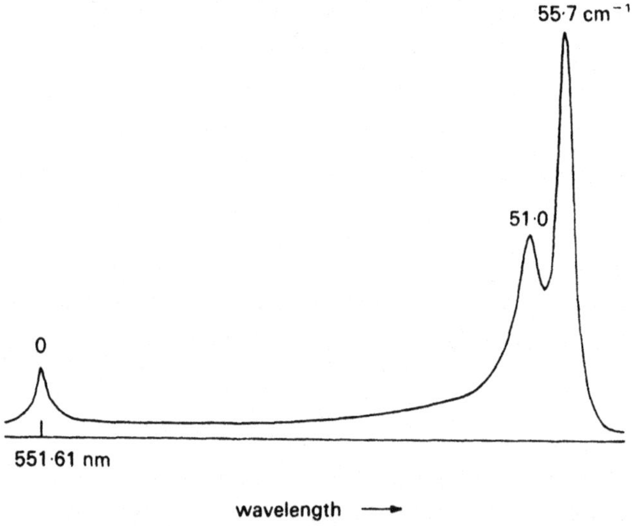

FIG. 10.31. Pure electronic transition and magnon sideband transition in the luminescence spectrum of Mn^{2+} ions perturbed by nearby Ca^{2+} ions in a nominally pure crystal of manganese fluoride.

Nominally pure manganese fluoride crystals may also contain trace amounts of iron and nickel. These impurities also act as traps. However, their trap levels are well below the E_1 level and they de-excite either by radiative transitions in the infrared or by non-radiative emission. These are the so-called *quenching traps*. As the temperature is raised the luminescence from nominally pure manganese fluoride crystals decreases and it is essentially quenched at 77 K.

In manganese fluoride crystals doped intentionally with small amounts of rare-earth ions, excitation may be transferred from the Mn^{2+} ions to the rare-earth ions which then emit their characteristic luminescence. The low-temperature luminescence spectrum of manganese fluoride doped with trace amounts of europium is seen in Fig. 10.32. The relatively sharp line at 547 nm and the broad band at longer wavelengths are characteristic of perturbed Mn^{2+} ions, the other sharp lines are due to ${}^5D_0 \rightarrow {}^7F_J$ transitions on Eu^{3+} ions. The ${}^5D_0$ level of Eu^{3+} is about 1150 cm^{-1} below E_1, and this level is efficiently populated by energy transfer from the E_1 level of the regular Mn^{2+} system.

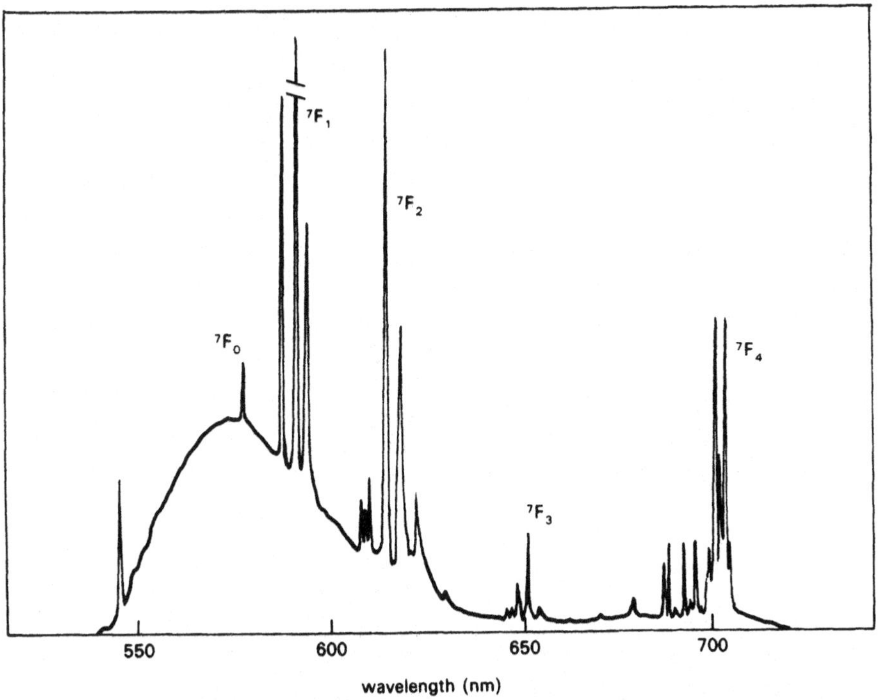

FIG. 10.32. Luminescence of $MnF_2{:}Eu^{3+}$ at 4.2 K showing both Mn^{2+} and Eu^{3+} transitions. The sharp transitions at 547 nm and the broad band at longer wavelength are from Mn^{2+} ions. The other sharp lines are ${}^5D_0 \rightarrow {}^7F_J$ transitions of Eu^{3+}.

At 77 K the Mn^{2+} luminescence is no longer observed but the Eu^{3+} luminescence continues to be observed. As the temperature rises further the Eu^{3+} luminescence decreases and can no longer be observed at room temperature. The temperature variation of the luminescence from MnF_2: Eu^{3+} can be explained with the aid of the simple energy-level diagram in Fig. 10.33. The Mn^{2+} trap levels, i.e. levels of Mn^{2+} ions perturbed by nearby Mg^{2+}, Zn^{2+}, Ca^{2+} ions, are around 100 cm^{-1} below E_1, the Eu^{3+} level is around 1500 cm^{-1} below E_1, and the quenching trap levels are 8000–12 000 cm^{-1} below E_1. The E_1 exciton level is populated by optical pumping, and this excitation can be transferred to the traps at rates f_1, f_2, f_3, as shown in the figure. At low temperatures all traps are populated, back transfer to the E_1 level is negligible, and both Mn^{2+} trap and Eu^{3+} emission is observed. As the temperature rises, boil-back can occur. The shallow Mn^{2+} traps are first to become depopulated, the excitation is transferred to Eu^{3+} ions and to quenching traps. Only Eu^{3+} luminescence is observed. As the temperature is further raised the Eu^{3+} ions transfer their energy back to the E_1 exciton level, and it is transferred to the quenching traps.

The populations of the excited traps can be calculated by rate equations (Greene *et al.* 1968, Hegarty 1976) from which the intensity and decay rate of the various traps can be estimated. These calculations give the following expression for the intensity, I, of the Eu^{3+} luminescence as a function of temperature:

$$I(T) = \frac{1}{C_1 \exp(-\Delta/kT) + C_2} \tag{10.76}$$

where Δ is the energy difference between the E_1 and $^5D_0$ levels. Figure 10.34 shows the variation with temperature of the Eu^{3+} luminescence. It is seen to exhibit the activation-type behaviour of eqn (10.76). The observed value of Δ (1365 cm^{-1}) is larger than the separation between $^5D_0$ and E_1 but this probably reflects the fact that the Eu^{3+} ions can boil back their excitation not only to the E_1 level but also to the higher energy $^4T_1$ levels. $\Delta = 1350$ cm^{-1} is the effective activation energy. The decay time of the Eu^{3+} luminescence, τ, is similarly found to obey the activation formula

$$\frac{1}{\tau} = \frac{1}{\tau_R} + \lambda \exp\left(-\frac{\Delta}{kT}\right) \tag{10.77}$$

predicted by the rate equations. τ_R is the low-temperature decay time of the Eu^{3+} ions.

The behaviour found for manganese fluoride is typical of many concentrated systems. For example, in Ba_2CaUO_6 the uranate group (UO_6^{6-}) is the optically-active centre. At low temperatures this material emits efficient luminescence from uranate groups near defects. The perturbed uranate groups have energies $\simeq 100$ cm^{-1} below the intrinsic uranate group. If the temperature is raised these traps are emptied and the excitation is transferred to quenching traps. A number of other interesting examples of concentration quenching effects in rare-earth systems are described by Blasse (1984).

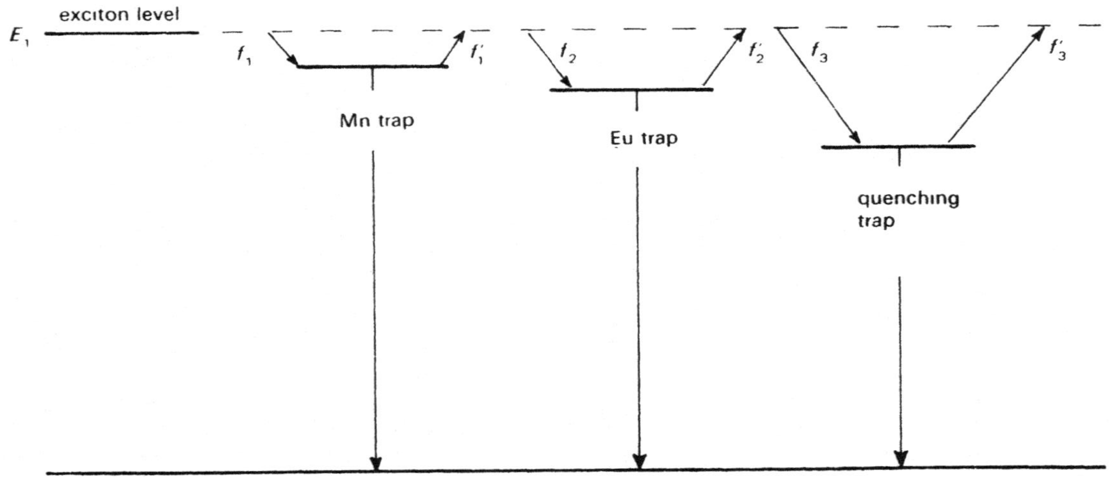

FIG. 10.33. Simple energy-level scheme to explain the luminescence behaviour of a crystal of manganese fluoride containing radiating and quenching traps. The boil-back rate from the ith trap, f'_i, is related to f_i through $f'_i = f_i \exp(-\Delta_i/kT)$ where Δ_i is the depth of the ith trap below the E_1 level.

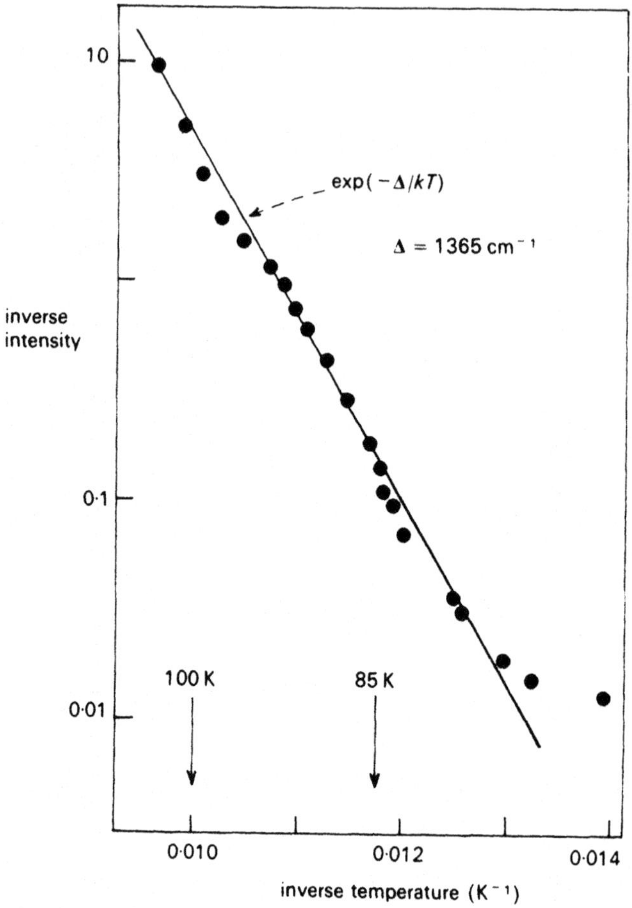

FIG. 10.34. Plot of the inverse of the Eu^{3+} luminescence intensity against inverse temperature in MnF_2:Eu.

The pure exciton E_1 and E_2 lines of manganese fluoride are transitions involving the $\boldsymbol{k}=0$ ground state, hence we are dealing with $\boldsymbol{k}=0$ excitons. In terbium hydroxide ($Tb(OH)_3$), which is a ferromagnetic insulator, Cone and Meltzer (1975) have studied the pure exciton luminescence from the excited $^5D_4, \mu'=2$ state to the $^7F_5, \mu=1$ state, which is situated about 20000 cm^{-1} above the ground state. Both of these states show exciton dispersion (Fig. 10.35). Under steady state optical pumping one expects a distribution of excitons across the $^5D_4, \mu'=2$ zone, and the luminescence transition displays the double-peaked lineshapes shown in the figure. This is a direct measurement of the (mainly $^7F_5, \mu=1$) dispersion. By exciting from the ground state directly into the $^5D_4, \mu'=2$ state one generates $\boldsymbol{k}=0$ excitons, and by

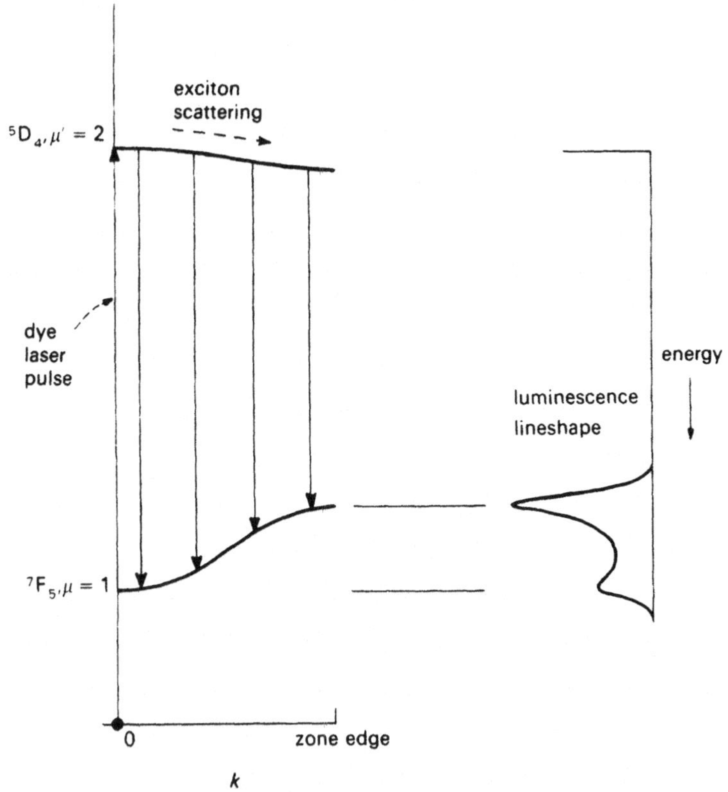

FIG. 10.35. The $^5D_4$, $\mu'=2 \rightarrow {}^7F_5$, $\mu=1$ luminescence transition of terbium hydroxide ($Tb(OH)_3$) at 1.3 K has a double-peaked shape due to exciton dispersion in the lower state. A magnetic field of 2 T is applied. Absorption directly into the $^5D_4$, $\mu'=2$ state creates $\boldsymbol{k}=0$ excitons. The scattering of excitons across the zone could be measured by monitoring the time evolution of the luminescence lineshape. (After Meltzer and Cone 1976.)

studying the time evolution of the luminescence lineshape Meltzer and Cone (1975) were able to determine the sign of the $^7F_5$, $\mu=1$ dispersion and to study the dynamics of the excitons in the $^5D_4$, $\mu'=2$ state.

In very pure manganese fluoride crystals the E_1 pure exciton luminescence can be detected with sufficient signal strength to carry out some interesting experiments on these excitons. By pumping directly into E_1 with an intense laser pulse the initial density of excitons can be sufficiently large for an exciton–exciton interaction to occur. This interaction can lead to a form of cross-relaxation in which two excitons combine to produce a single Mn^{2+} ion in a highly excited state which eventually relaxes to the E_1 state (Wilson *et al.* 1978). The probability that a single exciton will be removed by

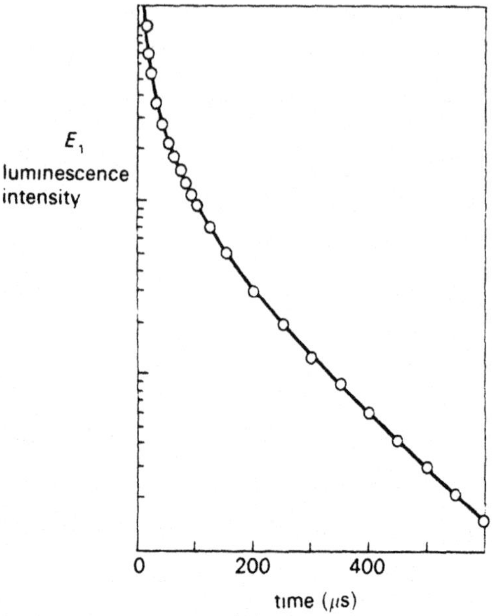

FIG. 10.36. Decay of E_1 excitons in pure manganese fluoride at 1.7 K after intense optical pumping by a pulse of resonant laser light. The curve is a fit to a theoretical model which includes biexciton decay. (After Wilson *et al.* 1976.)

this *biexciton decay* depends on the density of excitons, so we have $dN/dt = -k_2N^2$ for this process. The rate of decay of excitons due to all mechanisms can be written

$$\frac{dN}{dt} = -k_1N - k_2N^2 \tag{10.78}$$

where k_1 represents the sum of all linear decay rates. The solution is

$$N(t) = \frac{k_1}{k_2} \frac{N(0)}{\left(\dfrac{k_1}{k_2} + N(0)\right)\exp(k_1t) - N(0)}. \tag{10.79}$$

Since the E_1 luminescence intensity varies as $N(t)$, the luminescence decay will be non-exponential. Figure 10.36 shows the observed E_1 intensity after pulsed excitation into a very pure manganese fluoride sample at 1.7 K (Wilson *et al.* 1978). The curve is a plot of $N(t)$, eqn (10.79), and the agreement is excellent.

11
Solid state lasers

THE word *laser* is an acronym meaning Light Amplification by Stimulated Emission of Radiation. The importance of the spontaneous and stimulated emission processes first introduced by Einstein have already been stressed in Chapter 4. Amplification of electromagnetic radiation in the microwave region was proposed independently by workers in the USA and the USSR during 1953–55. Subsequently Townes and Schawlow (1958) discussed the very precise practical conditions required to obtain amplification of the stimulated emission process in the optical region. They also discussed a number of schemes for operational lasers in both gaseous and solid media. Indeed these workers first suggested the use of the fluorescent lines emitted by Cr^{3+} in ruby. Schawlow believed that because stimulated emission in the R_1-line terminated on the ground state, laser action at this wavelength would be difficult to achieve, and so Schawlow and Devlin (1961) demonstrated that laser action can be obtained from N_1- and N_2-lines in ruby, which are associated with exchange-coupled pairs of Cr^{3+} ions. However, the prize of producing the first operational laser went to Maiman (1960), who demonstrated R_1-line laser action in ruby by using the very high excitation intensities from a xenon flash lamp. Maiman's laser, therefore, operated in the pulsed mode. Since the ruby laser involves the sharp R_1-line it is essentially a single-frequency device. Lasers based on emission from rare-earth ions e.g. Nd^{3+}, Ho^{3+} etc., are also restricted to single-frequency output. In contrast the outputs from dye lasers are tunable over wide wavelength ranges. Early in the history of the laser Johnson *et al.* (1963, 1964, 1966) observed laser action in the broad phonon-assisted emission bands from transition metal ions in ionic crystals, indicating the potential of such materials for tunable solid state lasers. Commercial development of tunable solid state lasers, however, awaited exploitation of the colour centre laser by Burleigh Instruments and the alexandrite laser ($BeAl_2O_4:Cr^{3+}$) by Allied Chemicals.

11.1 Basic laser phenomena

Our discussion of the principles of laser operation has as its starting points the earlier accounts of spontaneous and stimulated emission (Section 4.4) and the optical absorption coefficient (Section 4.4.3). The acronym for laser tells us that it is the stimulated emission process which is of primary importance. Such stimulated transitions take place only in the presence of a radiation field. We first describe how gain arises for free atoms in a laser cavity. However, the principles are perfectly general; they apply as well to solid state systems as to any other medium.

11.1.1 Population inversion and laser gain

In Chapter 4 it was shown that the intensity of a beam of light changes on passing through a medium containing absorbing centres according to Lambert's law:

$$I(l)=I_0\exp(-\alpha(\nu)l) \tag{4.83'}$$

where $\alpha(\nu)$ is the absorption coefficient. Since optical transitions have a finite spectral width the absorption coefficient is frequency dependent and is given by

$$\alpha(\nu)=\frac{\lambda^2}{8\pi n^2\tau_R}[(g_b/g_a)N_a-N_b]g(\nu) \tag{4.89}$$

where λ is the vacuum wavelength at the peak of the band, N_a and N_b are the densities of atoms in the states a and b, respectively, n is the refractive index of the medium at the wavelength λ, $g(\nu)$ is the shape factor and $A_{ba}=1/\tau_R$. For a medium in thermal equilibrium $N_a \gg N_b$ and $\alpha(\nu)$ is positive. In this situation the beam is always attenuated due to absorption of the incident radiation by the sample.

To obtain amplification of the beam requires that $\alpha(\nu)$ be negative. For this to be so requires that $N_b>(g_b/g_a)N_a$, i.e. the excited-state population must exceed the ground-state population. This condition, referred to as *population inversion*, is not appropriate to a system in thermodynamic equilibrium. Population inversion is achieved by a suitable pumping mechanism which disturbs the system from thermal equilibrium such that the excess population in the excited state is given by

$$\Delta N=N_b-(g_b/g_a)N_a>0. \tag{11.1}$$

Under these conditions the beam intensity increases by a factor $\exp(\gamma(\nu)l)$, i.e.

$$I(l)=I_0\exp(\gamma(\nu)l) \tag{11.2}$$

where the *small-gain coefficient* $\gamma(\nu)$ is given by

$$\gamma(\nu)=\frac{\lambda^2\Delta N}{8\pi n^2\tau_R}g(\nu). \tag{11.3}$$

The small-gain coefficient results entirely from the atomic processes involved in absorption and stimulated emission. Although the process of spontaneous emission also adds to the total emitted intensity, it does not contribute significantly to the beam in the specific direction of the incident radiation because these photons are emitted isotropically rather than in a single direction. The non-equilibrium situation described by eqn (11.1) is often referred to as a state of *negative* temperature for the two levels. Assuming that the populations in levels a and b are related by a Boltzmann factor then

$$N_b = (g_b/g_a)N_a \exp(-h\nu/kT) \quad (11.4)$$

which yields an effective temperature

$$T_{\text{eff}} = -\frac{h\nu}{k\ln(g_a N_b/g_b N_a)} \quad (11.5)$$

which in the case of $N_b > N_a g_b/g_a$ is less than zero.

11.1.2 Threshold gain in a laser cavity

In its simplest form the laser consists of an active medium brought to the condition of population inversion when contained between a pair of parallel mirrors which constitute the laser cavity. The mirrors in solid state lasers may be highly reflecting metal coatings plated onto the end faces of the laser rod as shown schematically in Fig. 11.1. Alternatively they may be separated from the active medium. Obviously the cavity system introduces losses and it is necessary that the amplification of the medium exceeds the losses in the system. Under what conditions will this laser produce an amplified beam? Population inversion is usually achieved by an intense beam in the pump

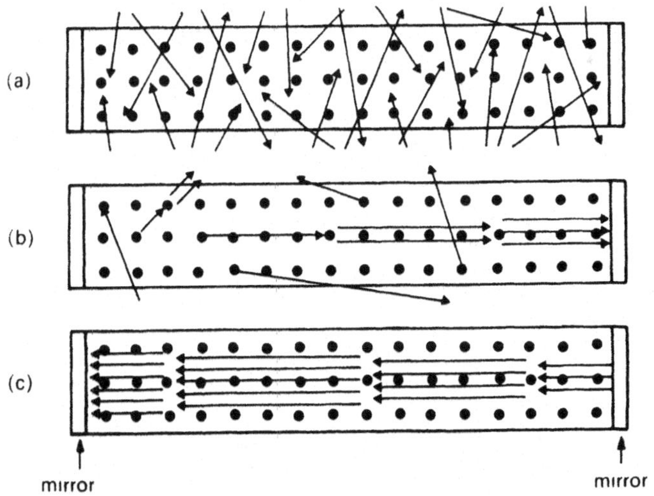

FIG. 11.1. The end mirrors in this simple laser cavity may, or may not, be attached to the laser rod. (a) Randomly emitted (spontaneous) photons are lost through the external walls. (b) Except for those travelling parallel to the cavity axis which may stimulate emission by other excited atoms. In this way they may contribute to the beam of stimulated photons (c) oscillating backwards and forwards parallel to the axis, between the two mirrors. (Adapted from Schawlow 1970.)

band. Spontaneous emission is an isotropic process, so that some photons will be travelling in a direction parallel to the axis of the laser. Such photons may be reflected backwards and forwards between the mirrors stimulating emission from other excited atoms as they pass through the active medium. Since those photons created by the stimulated emission process travel in the direction of the stimulating photons then for each passage of the beam through the active medium there is an average gain of intensity given by

$$G_M = I(L)/I_0 = \exp(\gamma(\nu)L), \tag{11.6}$$

L being the length of the active medium. If the single-pass gain G_M is high then only a small number of passes is necessary to achieve laser operation. Indeed in some cases G_M is sufficiently high that laser amplification can be obtained in a single transit and end mirrors are therefore not required. Factors which militate against gain include absorption by the medium at the emission wavelength, transmission at the end mirrors, scattering by the medium, and cavity diffraction effects. For lasers requiring multiple reflections it is the transmission by the mirrors which constitutes the most serious source of loss. When good-quality multilayer dielectric mirrors are used the reflectivity may be very close to unity. Hence one mirror is almost 100 per cent reflecting at the laser wavelength. However, since the useful output of the laser is that light which is transmitted through an end reflector, the output mirror is by design partially transmitting. (In the absence of absorption by the mirrors, the reflectivity (R) and transmittivity (T) are related by $R+T=1$.) For most laser configurations the transmittance of the partially-transmitting mirror is in the range $0.03<T<0.30$. The loss per round trip (i.e. two passes through the medium) due to losses at the mirrors may be written in the form

$$\frac{I(2L)}{I_0} = R_1 R_2 \simeq \exp[-(1-R_1 R_2)] \tag{11.7}$$

and the loss per pass due to other miscellaneous losses (e.g. diffraction, absorption, etc.) as

$$\frac{I(L)}{I_0} = \exp(-\beta). \tag{11.8}$$

The full loss in the cavity per pass can be written as

$$G_c = e^{-\Delta} \tag{11.9}$$

where the loss coefficient Δ is given by

$$\Delta = [\tfrac{1}{2}(1-R_1 R_2)+\beta]. \tag{11.10}$$

We can now express the cavity loss in terms of the time taken for the beam to decrease to $1/e$ of its original intensity. This is the *cavity decay time*, τ_c. In a time, t, the beam travels a distance $vt = ct/n$ in the medium, making ct/nL passes. The loss after this time can be written as a function of time as

$$\frac{I(t)}{I_0} = [\exp(-\Delta)]^{ct/nL} = \exp\left(\frac{-t}{\tau_c}\right) \tag{11.11}$$

where τ_c is given by

$$\tau_c = \frac{nL}{c\Delta} = \frac{nL}{c[\frac{1}{2}(1 - R_1 R_2) + \beta]}. \tag{11.12}$$

For example, in a ruby rod of length 0.18 m, reflection coefficients $R_1 = 0.99$, $R_2 = 0.90$, $n = 1.7$ and miscellaneous losses of 5 per cent per round trip we estimate $\tau_c \simeq 7 \times 10^{-10}$ s. By analogy with the definition of Q-factor for a resonant microwave cavity we similarly define $Q = 2\pi \times$ energy stored/energy loss per cycle. This gives

$$Q = 2\pi I(t) \Bigg/ \left|\frac{dI(t)}{dt}\right| \frac{1}{\nu} = 2\pi\nu\tau_c \tag{11.13}$$

using eqn (11.11). In the optical region $\nu \simeq 10^{15}$ Hz and taking $\tau_c \sim 7 \times 10^{-10}$ s we find $Q \sim 4 \times 10^6$, showing that the optical resonator is by some orders of magnitude a better store of electromagnetic radiation than the resonant microwave cavity, for which $Q \sim 10^4$.

Taking both the gain in the medium (G_M) and the cavity losses (G_c) into account we can write the condition for laser action in terms of the overall *single-pass gain*, G, as

$$G = G_M G_c = \exp(\gamma(\nu)L - \Delta) > 1 \tag{11.14}$$

which implies a minimum condition ($G \geqslant 1$) given by

$$\frac{\Delta}{L} \leqslant \gamma(\nu) = \frac{\lambda^2 \Delta N}{8\pi n^2 \tau_R} g(\nu). \tag{11.15}$$

Since from eqn (11.12) $\Delta = nL/c\tau_c$ we can define the threshold condition as

$$\frac{\lambda^2 c \Delta N_t \tau_c}{8\pi n^3 \tau_R} g(\nu) \geqslant 1. \tag{11.16}$$

This is the necessary condition which must apply before a beam can be produced. Should eqn (11.16) not hold, intensity amplification cannot occur at any frequency and laser action does not occur.

11.1.3 The laser output

The laser cavity consists of two reflectors facing one another as in the Fabry–Perot interferometer. In consequence, for the cavity losses to be minimized the beam must be reflected back and forth between the mirrors in such a way that the multiply-reflected beams interfere constructively. The condition for this to occur is

$$2nL = m\lambda \quad \text{or} \quad \nu_m = m\frac{c}{2nL}. \tag{11.17}$$

When this condition is satisfied the loss (G_c) is given by $\exp(-\Delta)$; Δ is defined by eqn (11.10). For other frequencies the value of G_c is much smaller because of destructive interference. Laser action, therefore, is expected to occur only at the cavity resonance frequencies, given by ν_m, eqn (11.17), for which the condition $G > 1$ applies. We note that the frequency spacing, $c/2nL$, is just the reciprocal of the cavity round-trip time. Furthermore the mode number (m) is normally large; in a ruby laser with $L = 0.1$ m, $n = 1.7$, $\lambda = 700$ nm substitution in eqn (11.17) gives $m \simeq 5 \times 10^5$.

Returning to the condition for laser operation $G = G_M G_c > 1$ we find that in general $G_M > 1$ and is frequency dependent through the lineshape function, $g(\nu)$. $G_c \sim 1$ and is frequency independent over the narrow frequency range around each allowed mode. The two contributions, G_M and G_c, as well as their product are plotted schematically in Fig. 11.2 as a function of mode number, m. From the very large set of values of m only three mode frequencies fall within the gain envelope, G_M, of the spectral line. Below threshold, eqn (11.15), all modes within the gain envelope are populated. However, as population inversion is gradually increased one mode with a large value of τ_c will satisfy the threshold condition before any others. For an homogeneously broadened transition all the excited atoms are then forced to emit into this mode by the stimulated emission process. It is this single mode which is amplified, and the energy contained in the mode reaches a level much higher than all the other modes. Hence the output beam can be intense, is highly monochromatic and is unidirectional.

11.1.4 Multilevel laser systems

In Fig. 4.2 we showed the spontaneous and stimulated transitions between two levels a and b. Since $W_{ba}/W_{ab} = (g_a/g_b)(1 + n_\omega)/n_\omega$, where n_ω is the photon occupancy at resonance, it is clear that population inversion can never be achieved by resonantly pumping such a two-level system. To obtain population inversion and achieve laser action one needs to involve transitions to other levels. Most practical lasers are three- or four-level systems. Figure 11.3 shows two different three-level and two different four-level pumping schemes. In Fig. 11.3(a), a four-level system, the fast non-radiative decay between levels 2 and 1 must be faster than the laser transition $3 \rightarrow 2$ so that there can be a population inversion $N_3 > N_2$. This very convenient arrangement can operate at relatively low pump powers since both excited levels (2 and 3) may have much smaller populations than the ground state. The opposite is true of the three-level pump scheme shown in Fig. 11.3(b) where population

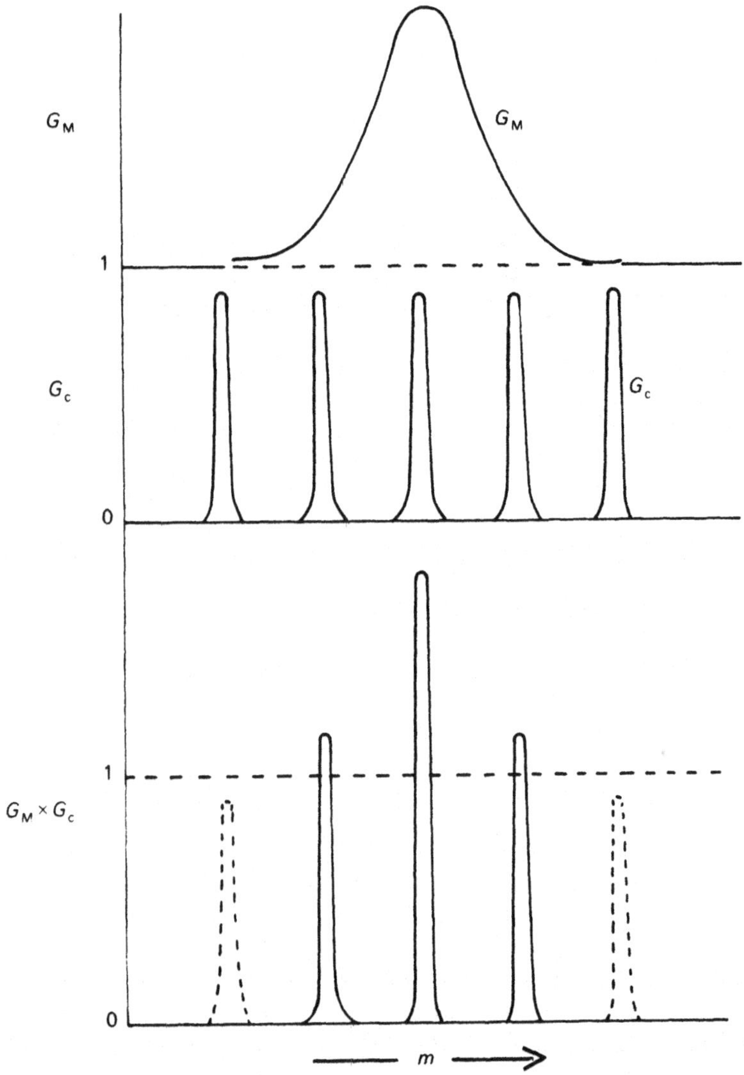

FIG. 11.2. The allowed modes of oscillation of a laser cavity. G_M and G_c are the gain per pass of the material and the Fabry–Perot, respectively. Since a population inversion exists only when $G_M \geqslant 1$ the laser oscillates only when $G_M \times G_c \geqslant 1.0$.

inversion must be effected between level 2 and ground state, level 1. In this case non-radiative decay from level 3 to level 2 must be fast and level 2 must be long-lived. Nevertheless very effective pumping is required since less than half of the atoms ions must be left in the ground state so that $N_1 < N_2$. The

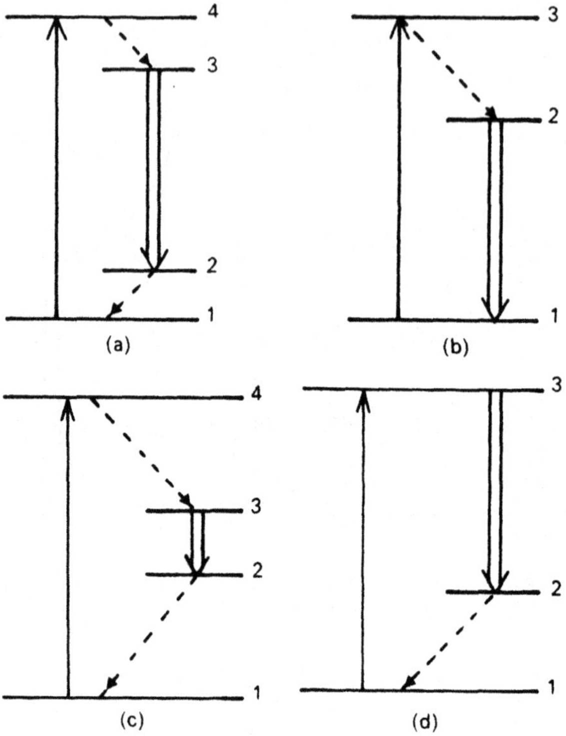

FIG. 11.3. Schematic optical pumping cycles for three- and four-level laser systems.

ruby laser, an example of this latter scheme, is the archetypal three-level system.

Of the two four-level systems shown in Fig. 11.3, scheme (a) is obviously more efficient than scheme (c) since much less of the pump energy is lost through non-radiative transitions 4→3 and 2→1. Provided that these two transitions are fast relative to the lifetime of the upper laser level then population inversion $N_3 > N_2$ may be achieved at quite modest pump levels. The diamond H_3-centre laser is a good example of scheme (a) whereas scheme (d) is typical of F_A-centre lasers in the alkali halides.

11.2 Optically pumped three-level lasers

Let us examine the conditions for obtaining population inversion with the three-level system involving the different transition possibilities shown in Fig. 11.4. Given the very short lifetime of level 3 against non-radiative 3→2 transitions (W_{32}^{nr}) we can assume $N_3 = 0$, implying that ions are immediately

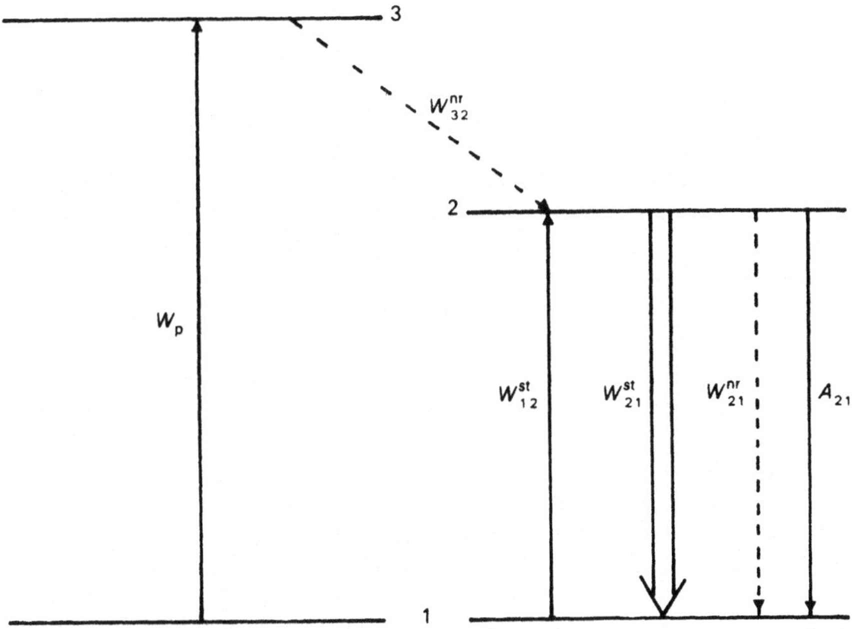

FIG. 11.4. Transitions involved in the optical pumping cycle of a three-level laser such as ruby. The superscript 'st' denotes a stimulated transition.

pumped into level 2 by the pump radiation. Since laser materials of necessity must possess a high radiative efficiency we assume $W^{nr}_{21} \simeq 0$, then only the pump rate, W_p, into level 2 via level 3, and the rates of transitions between levels 1 and 2 (Fig. 11.4) need be considered. The rate of change of population of level 2 is then given by

$$\frac{dN_2}{dt} = -(W^{st}_{21} + A_{21})N_2 + (W_p + W^{st}_{12})N_1. \qquad (11.18)$$

For the steady state condition $dN_2/dt = 0$, hence we find $(W_p + W^{st}_{12})N_1 = (W^{st}_{12} + A_{21})N_2$. Assuming that the statistical weights of levels 2 and 1 are equal we have $W^{st}_{12} = W^{st}_{21}$ (Chapter 4). If we write $N_1 + N_2 = N_0$, the total population density, then the population inversion is

$$\Delta N = N_2 - N_1 = \frac{W_p - A_{21}}{W_p + A_{21} + 2W^{st}_{12}} N_0. \qquad (11.19)$$

We require that $\Delta N > 0$ for net amplification. This condition obviously requires that $W_p > A_{21}$. In other words the pump rate to level 2 must exceed the decay rate by spontaneous emission.

11.2.1 The ruby laser

The optical absorption and luminescence spectra of ruby (Al_2O_3: Cr^{3+}) have been described in Chapter 9. For laser operation the concentration of Cr^{3+} ions is usually around 0.05 weight %. The pump transitions are the broad $^4A_2 \rightarrow {}^4T_1$, $^4T_2$ absorption bands with peaks at 410 nm and 560 nm, respectively. Once raised to the excited $^4T_1$ and $^4T_2$ levels, ions decay rapidly ($< 10^{-13}$ s) by non-radiative transitions to the metastable ^2E level. Despite its low oscillator strength ($f \sim 10^{-6}$), the $^2E \rightarrow {}^4A_2$ transition has a high quantum efficiency; this is the laser transition.

Maiman's (1960) ruby laser consisted of a laser rod containing *c.* 10^{19} Cr^{3+} ions cm^{-3} placed at the centre of a helical xenon flashlamp. The end faces of the laser rod were polished flat and parallel to a very high degree of accuracy. One end face was coated to give 100 per cent reflectivity, while the other end face (the output coupler) was partially transmitting. Because of the difficulty of maintaining over half of the Cr^{3+} ions in the excited ^2E state the laser was operated in a pulsed mode pumped by a short, very intense pulse of light from a xenon flashlamp activated by discharging a capacitor bank through the gas. The energy absorbed in the $^4A_2 \rightarrow {}^4T_1$, $^4T_2$ bands is channelled into the narrow R-line emission at a wavelength $\lambda = 694.3$ nm, which has a spectral bandwidth at room temperature of about 1.2 nm. Laser action has been achieved not only in the R_1-line at 694.3 nm, but also in the 692.9 nm (R_2-line), and in the 700.9 nm and 704.1 nm lines of exchange-coupled chromium pairs. The reasons then for the successful operation of the ruby laser are the long-lived emitting state, the large absorption cross-section for level 3, and the availability of an intense radiation source for optical pumping.

Xenon lamps are sources of white light, emitting radiation at many wavelengths not absorbed by the ruby. In consequence, the total power from this source required to produce an output power of about 1 kw from the laser is many times that absorbed by the ruby. This can be illustrated by an order of magnitude estimate; it is assumed that the ruby rod is uniformly illuminated throughout its volume. The pump power per unit area, P, can then be written as the product of photon flux, $\phi(m^{-2}\,s^{-1})$, and photon energy ($h\nu_p$), i.e. $P = \phi\, h\nu_p$, and the pump rate $W_p = \phi\, \sigma_p$ where $\sigma_p = \alpha(\nu_p)/N_1$ is the absorption cross section at the pump frequency, ν_p. Hence $P = W_p h\nu_p/\sigma_p$. Since for population inversion $W_p \geqslant A_{21}$, we determine that the minimum power required to produce an excess population in level 2 is

$$P_{min} = \frac{h\nu_p}{\sigma_p \tau_r}. \tag{11.20}$$

For ruby, $h\nu_p$ at the peak of the $^4A_2 \rightarrow {}^4T_1$ transition is about 3 eV, $\sigma_p = 2.5 \times 10^{-24}$ m^2 and $\tau_R \sim 4$ ms, so that the minimum pump power is of order 50 MW m^{-2}. Thus a laser rod of length 0.1 m and diameter 10^{-2} m requires at least 150 kW of pump power to achieve population inversion.

Since under laser output conditions there is very little spontaneous emission i.e. $W^{st}_{21} \gg A_{21}$, we further assume that one absorbed photon produces one output photon. Therefore the laser output power, P_0, is just $\nu_{21}/\nu_p \times$ absorbed pump power. The absorbed pump power is given by $P = N_1 V W_p h\nu_p$. Since the minimum requirement for inversion is that N_1 is just less than $N_0/2$ we can write the maximum output power as

$$P_0 = \frac{P\nu_{21}}{\nu_p} = \frac{(N_0 V)}{2} \frac{h\nu_{21}}{\tau_r}. \tag{11.21}$$

For a ruby laser rod of volume $V \sim 8\ \text{cm}^3$ containing 0.05 weight % of Cr^{3+} ions, i.e. $N_0 = 5 \times 10^{18}\ \text{cm}^{-3}$, we estimate the maximum output power as $P_0 = 0.7$ kW, since $h\nu_{21} \simeq 1.8$ eV. In fact once above threshold the stimulated emission rate increases so rapidly that soon it exceeds the rate at which the lamp provides pump radiation. Thus the population inversion is depleted until it falls below threshold. In consequence, the output from the ruby laser in pulsed or continuous-wave (CW) mode, consists of many sharp pulses (spikes). This shows that the actual population dynamics during laser action are more complicated than envisaged in the steady state model assumed in order to simplify the analysis of the power output of the three-level laser.

11.3 Optically pumped four-level lasers

Most solid state lasers based on rare-earth ions, colour centres, or transition metal ions are four-level systems. The different transitions involved in the pumping scheme appropriate to four-level lasers are shown in Fig. 11.5. The pump light which drives the absorption transition $1 \to 4$ results in population N_3 of level 3 by virtue of very efficient non-radiative de-excitation of level 4 into level 3. Furthermore, non-radiative transitions from level 2 to level 1 make for an efficient return to the ground state. Hence in analysing the dynamics of the pumping cycle we may make the simplifying assumption that $N_4 = 0$. We also make the practical assumption that W^{nr}_{32} is negligible. Then the rate equations are

$$\frac{dN_1}{dt} = -W_p N_1 + N_2 W^{nr}_{21} \tag{11.22}$$

$$\frac{dN_3}{dt} = W_p N_1 + N_2 W^{st}_{23} - N_3(W^{st}_{32} + A_{32}) \tag{11.23}$$

where the pump rate, $W_p N_1$, is the number of ground-state atoms undergoing absorption transition $1 \to 4$ and thence to level 3 per second.

If steady state conditions are assumed and the system is being pumped at a constant rate then the time derivatives of the state populations are equal to zero. Equations (11.22) and (11.23) then yield

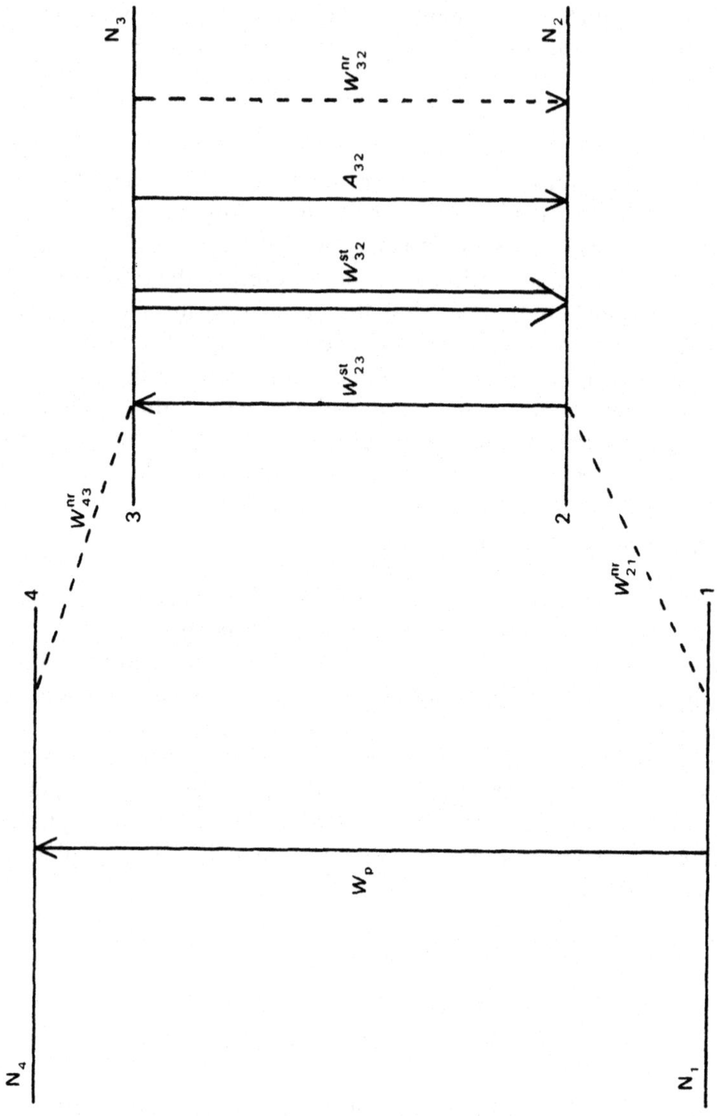

FIG. 11.5 Showing the radiative and non-radiative transitions in a typical four-level laser.

$$\frac{N_3}{N_2} = \frac{W_{23}^{\text{st}} + W_{21}^{\text{nr}}}{W_{32}^{\text{st}} + A_{32}}. \tag{11.24}$$

Using eqns (11.22) and (11.24) the population inversion can be written

$$\Delta N = N_3 - N_2 = \frac{N_1 W_\text{p}}{W_{21}^{\text{nr}}} \frac{(W_{21}^{\text{nr}} - A_{32})}{(W_{32}^{\text{st}} + A_{32})} = N_1 W_\text{p} \frac{(1 - A_{32}/W_{21}^{\text{nr}})}{(W_{32}^{\text{st}} + A_{32})} \tag{11.25}$$

where we assume that $W_{23}^{\text{st}} = W_{32}^{\text{st}}$. Equation (11.25) then implies that unless $A_{32} < W_{21}^{\text{nr}}$ no population inversion can take place. Since in most laser systems transition $2 \rightarrow 1$ occurs by very efficient phonon-assisted non-radiative decay, we can write $W_{21}^{\text{nr}} \gg A_{32}$, W_{32}^{st} and $(1 - A_{32}/W_{21}^{\text{nr}}) \simeq 1$. This being the case we can assume that $N_2 \simeq 0$ so that at low pump rates, i.e. below threshold, where the rate of simulated emission, W_{32}^{st}, is very small, the population inversion is given by

$$\Delta N \simeq N_3 \simeq \frac{W_\text{p} N_1}{A_{32}}. \tag{11.26}$$

ΔN increases linearly with pump rate until the threshold condition (eqn 11.16) is reached, at which we may write the population inversion ΔN_t as

$$\Delta N_\text{t} = \frac{W_\text{p}^\text{t} N_1}{A_{32}}. \tag{11.27}$$

At threshold the stimulated emission rate is still small. Irrespective of the pump rate above W_p^t the population inversion never rises above ΔN_t because W_{32}^{st} now starts to increase rapidly. Equation (11.24) then becomes

$$\Delta N_\text{t} = \frac{W_\text{p} N_1}{(W_{32}^{\text{st}} + A_{32})}.$$

For a directed beam the intensity density is given by $I(\nu) = u(\nu)v$, where $u(\nu)$ is the energy density in the beam and v is the velocity of the radiation.

Recalling (eqn 4.63) that $W_{32}^{\text{st}} = B_{32} u(\omega) = B_{32} u(\nu)/2\pi = B_{32} I(\nu)/2\pi v$ we find, on solving for $I(\nu)$,

$$I(\nu) = \frac{2\pi v A_{32}}{B_{32}} \left(\frac{W_\text{p} N_1}{A_{32} \Delta N_\text{t}} - 1 \right) \tag{11.28}$$

Showing that the stimulated emission intensity increases linearly with pump power once $W_\text{p} > W_\text{p}^\text{t}$. Below threshold the gain associated with stimulated emission never exceeds the losses of the laser system so that the output intensity is zero. The behaviour of ΔN and I are shown in Fig. 11.6. We note that the additional power above threshold is not spread over all modes of oscillations of the cavity but instead is channelled into a few modes with high Q-factor.

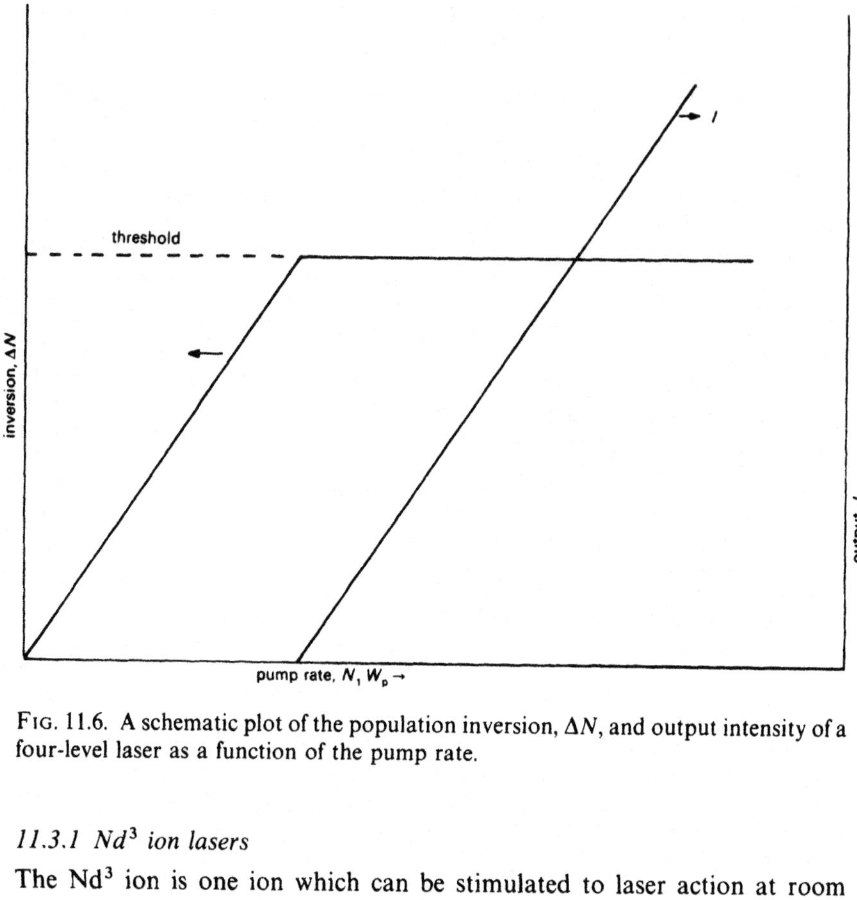

FIG. 11.6. A schematic plot of the population inversion, ΔN, and output intensity of a four-level laser as a function of the pump rate.

11.3.1 Nd^3 ion lasers

The Nd^3 ion is one ion which can be stimulated to laser action at room temperature in almost any ionic solid state host material. Also, given appropriate cooling, Nd^{3+} ions can be operated in continuous or quasi-continuous fashion. The spectroscopic properties of Nd^{3+} ions in solid host materials were discussed in Chapter 8. The lowest energy levels of Nd^{3+} involved in laser transitions are shown in Fig. 11.7. Because the energy levels derived from the $4f^3$ configuration are well shielded from the crystal field there are only slight variations in the laser wavelengths from one host crystal to another. The most commonly used laser transition involves the ${}^4F_{3/2} \rightarrow {}^4I_{11/2}$ transition, which in all hosts occurs at an energy-level separation $\sim 9400\ cm^{-1}$. The situation in Nd^{3+}-glass lasers is in one respect different from single crystal lasers. In a glass the Nd^{3+} ions occupy sites of slightly different crystal field, and hence ${}^4F_{3/2} \rightarrow {}^4I_{11/2}$ transitions have slightly different wavelengths. The result is that the linewidth is somewhat larger than for single crystals. Because of the large energy gap between ${}^4F$ and ${}^4I$ levels the quantum efficiency of the ${}^4F_{3/2} \rightarrow {}^4F_{11/2}$ transitions is high. However, the

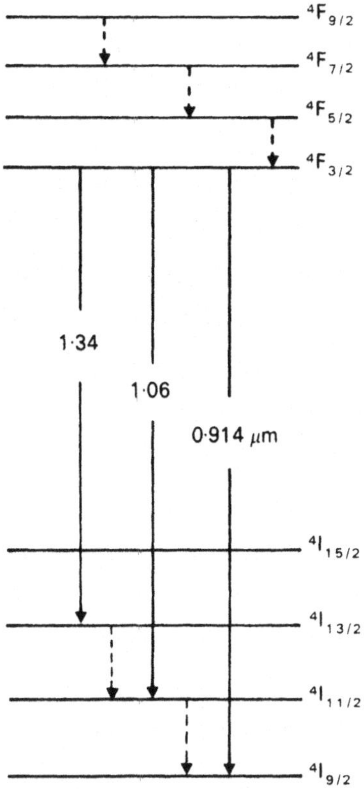

FIG. 11.7. Showing the energy levels, radiative and non-radiative transitions of Nd^{3+} ions in typical ionic crystals and glasses.

$^4I_{11/2}$ terminal level in the laser transition is only 2000 cm^{-1} above the $^4I_{9/2}$ ground state, and this gap can be bridged fairly efficiently by multiphonon non-radiative decay.

The two most commonly used Nd^{3+} ion lasers available on the commercial market are YAG : Nd^{3+} and glass : Nd^{3+}. YAG : Nd^{3+} is mainly used in a quasi-continuous mode, with peak output power in the kilowatt range. These lasers are often *mode-locked* (Section 11.5), emitting a train of pulses each of ~100 ps duration. A single pulse may then be isolated and amplified to produce peak powers of about 10^6 MW! Nd^{3+} : glass lasers, however, are usually operated as single pulses, as is the ruby laser. Because of the large spectral bandwidth of the Nd^{3+} : glass laser, they are never operated as single-mode lasers. They are, however, often *Q*-switched or mode-locked to produce short pulses (10 ps) and very high power densities.

11.3.2 Other rare-earth ion lasers

The strong fluorescence exhibited by other rare-earth ions in ionic crystals has made these natural candidates in searches for efficient laser systems. Indeed the second successful solid state laser material was $CaF_2:Sm^{2+}$. Soon afterwards $CaF_2:Tm^{2+}$ and $CaF_2:Dy^{2+}$ were added to the growing list once it was realized how to produce stable divalent rare-earth ions in ionic lattices. Many rare-earth ions have served as optical centres in ionic materials used for lasers including Sm^{3+}, Eu^{3+}, Gd^{3+}, Dy^{2+}, Ho^{3+}, Er^{3+}, Tm^{3+}, Tm^{2+}, Yb^{3+}, and U^{3+}. All are examples of four-level laser systems. For successful operation even in the pulsed mode most rare-earth ions need cooling usually to the temperature of liquid nitrogen or lower. The number of host crystals is almost as diverse as the number of ions. Included on an ever-expanding list are calcium tungstate, lanthanum fluoride, strontium fluoride, barium fluoride, calcium fluoride, yttrium lithium fluoride, calcium molybdate, strontium molybdate, yttrium oxide and magnesium fluoride (see e.g. Weber 1986). In all cases laser action occurs in a sharp line spectrum so that wavelength tuning is not possible.

11.4 Tunable vibronic lasers

The ruby and rare-earth ion solid state lasers are essentially single-wavelength lasers. This places them at a serious disadvantage relative to liquid dye lasers, which may be tuned over very wide ranges of wavelength in the visible and infrared regions of the spectrum. In recent years there has been a successful search for tunable solid state lasers directed at colour centres and transition metal ions in ionic crystals. These lasers may be termed *vibronic lasers*, which in this context means that the laser transitions occur in broad vibrational sidebands (Chapter 5). This is depicted in Fig. 11.8, which demonstrates how in this four-level laser system the transition energy is partitioned between the emitted photon and lattice phonons. Because of the large width of the vibronically-broadened absorption transition such a system can be very efficiently pumped with a broadband source. The pump band is broad because the electron–lattice coupling in the excited state (4) is strong. Vibrational relaxation then takes the electronic system into the emitting state (3). Population inversion is achieved between level 3 and level 2, the latter being one of the excited vibrational levels of the electronic ground state. The laser wavelength is determined by whichever of the vibrational levels is the terminus of the transition. Essentially this is a four-level laser with small gain distributed over the range of frequencies associated with the broad emission band. Levels 1 and 2 in Fig. 11.8 belong to the same electronic state, similarly for levels 3 and 4, which is the case, for example, of the diamond H_3-centre laser. Alternatively all four levels may belong to different electronic

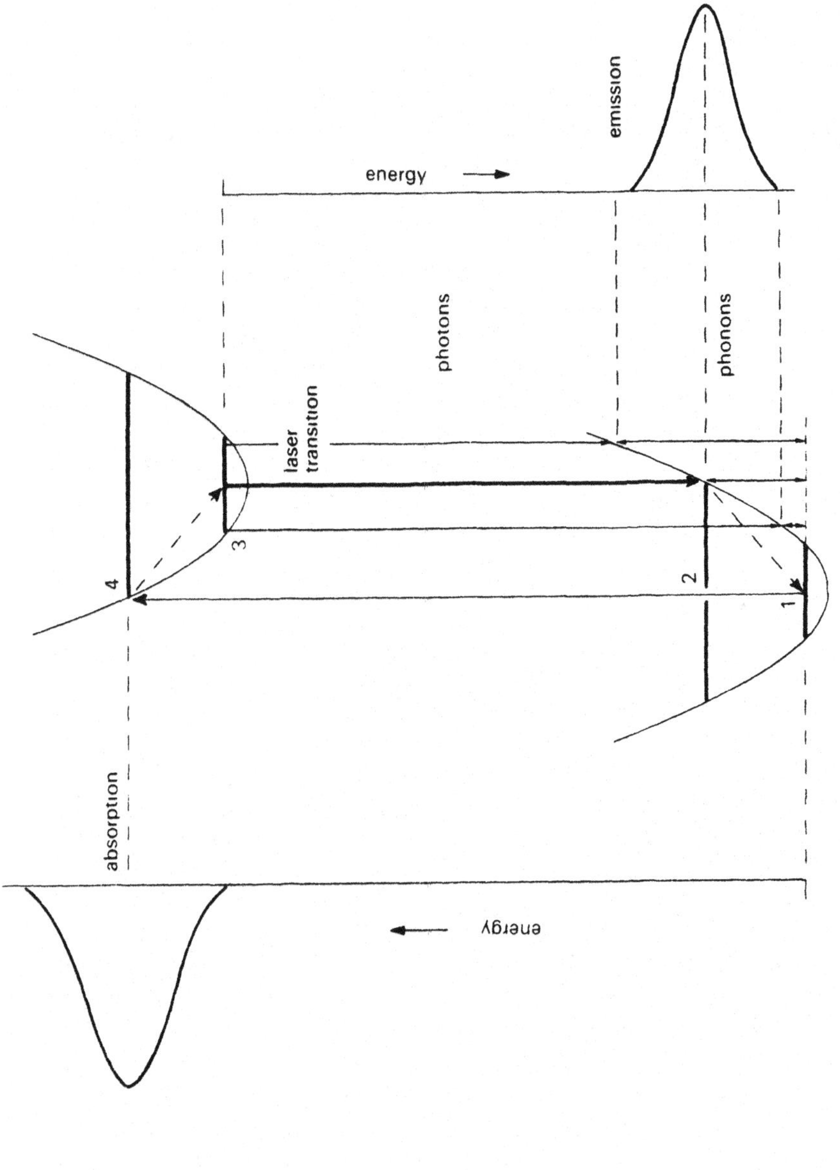

FIG. 11.8. Configurational coordinate diagram indicating the four-level nature of a vibronically tuned laser. The partitioning of the de-excitation energy between emitted radiation (photons) and crystal heating (phonons) is indicated.

states, as in the F_A-centre laser discussed below. The major merits of these vibronic lasers are the low threshold pump power needed for laser action and the wide tuning range. Infrared transitions between 3d levels of V^{2+}, Co^{2+}, and Ni^{2+} ions in such host crystals as magnesium fluoride, manganese fluoride or magnesium oxide provided the first examples of vibronic lasers. In general these lasers can be operated in the pulsed or CW mode, although they are often operated below room temperature. This is also the case for the variety of colour-centre lasers now available. However, the recently developed tunable Cr^{3+} lasers, of which the alexandrite laser is the prototype, are continuously operable well above room temperature.

11.4.1 Colour-centre lasers

We have already discussed the optical pumping cycle of the F-centre (Chapter 7), an apparently ideal four-level laser system. The non-radiative relaxation times, τ_{nr}, are very short, of order 10^{-13} s whereas the emission lifetime is relatively long ($\tau_r \sim 10^{-6}$ s). Hence population inversion of the relaxed excited electronic state is obtained at essentially any finite pump rate. The single pass gain, G, is given by eqn (11.14) in which $\Delta < \gamma(\nu)L$ for net gain. The gain variation with frequency is reflected by the luminescence bandshape. For a Gaussian band the small gain coefficient $\gamma(\nu_0)$ at the band peak is obtained from eqn (11.3) as

$$\gamma(\nu_0) = \frac{\Delta N}{8\pi n^2} \times \frac{\eta}{\tau} \times \frac{\lambda_0^2}{1.07\delta\nu} \tag{11.29}$$

where ν_0 and λ_0 are, respectively, the frequency and the wavelength at the band peak, n is the refractive index, τ is the observed decay time and $\delta\nu$ is the full width at the half-power points. Note that in eqn (11.29) we recognize the probable competition between radiative and non-radiative decay processes and substitute for the radiative lifetime $\tau_R = \tau/\eta$, in which η is the luminescence quantum efficiency. Substituting values appropriate to F-centres in potassium chloride, i.e. $\lambda_0 = 1000$ nm, $\tau/\eta = 600 \times 10^{-9}$ s, $\delta\nu = 6.3 \times 10^{13}$ Hz gives $\gamma(\nu_0) = 0.044$ for $\Delta N = 10^{16}$ cm^{-3}. ΔN is effectively the excited state population because in this system $W_{21}^{nr} \gg A_{32}$ in eqn (11.25). The calculated value for $\gamma(\nu_0)$ is small, hence the gain through the medium, G_M, is also small. For this reason the F-centre is inoperable as a tunable laser. However, as we shall see, other colour centres are almost ideal examples of tunable solid state lasers. As an example Fig. 11.9 shows that the power tuning curve for the $F_2^+:O^{2-}$ laser in sodium chloride is smooth and broad, the threshold power is only *c.* 600 mW at the band peak and the output varies linearly with pump power in accord with eqn (11.28) (Pollock *et al.* 1986).

The first colour-centre laser was developed by Fritz and Menke (1965) using flash-lamp pumped F_A(Li)-centres in potassium chloride. However, the real potential for colour-centre lasers was not realised until 1974 when

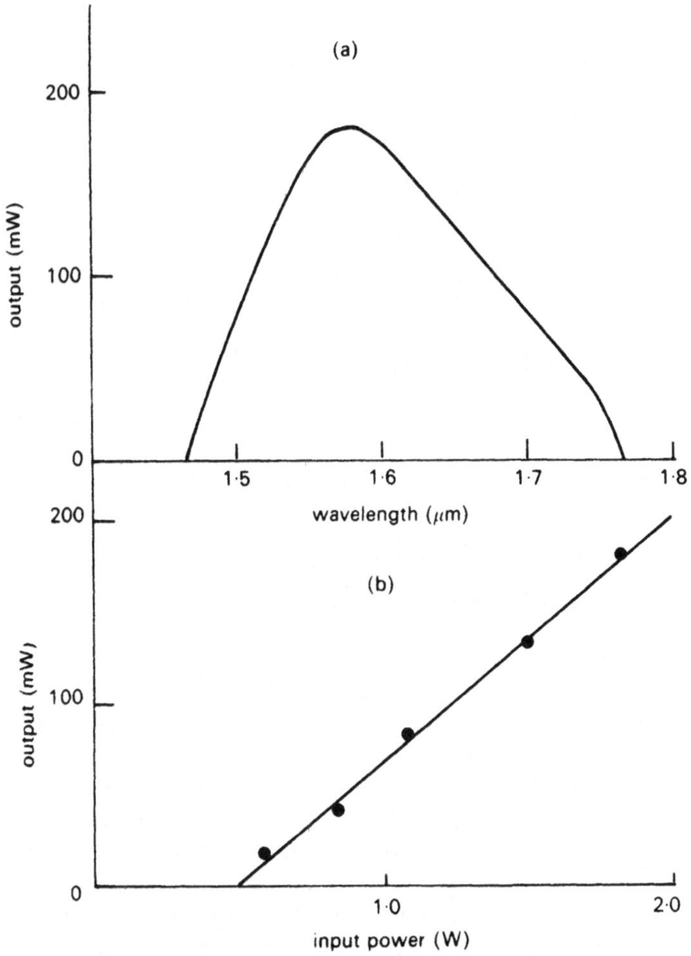

FIG. 11.9. Showing the CW operating characteristics of an $F_2^+:O^{2-}$ centre sodium chloride laser during pumping with 2 W of input power from a 1.06 μm Nd:YAG laser: (a) shows the power output versus wavelength using a 5 per cent output coupler, (b) output power versus input power at the band peak. (After Pinto *et al.* 1985.)

Mollenauer and Olson reported the first tunable laser operation with $F_A(II)$-centres in both KCl:Li and RbCl:Li. The optical design of the laser cavity used by Mollenauer is shown in Fig. 11.10; it owes much to the construction of dye lasers in that it utilizes a highly concentrated modal beam. The beam is focused to a diffraction-limited spot—the beam waist—using the mirrors M_0, M_1, and M_2 in a folded, astigmatically-compensated cavity layout. The crystal of thickness 1–2 mm, containing the laser-active defects, is held at the

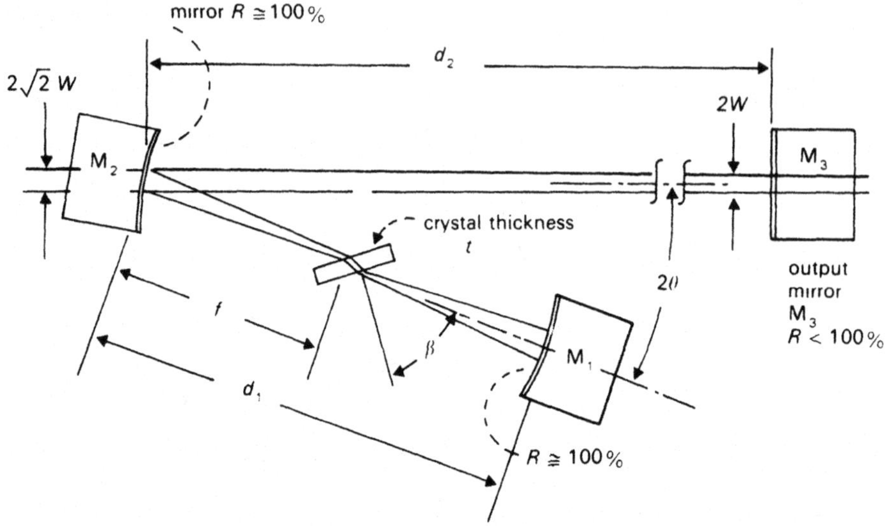

FIG. 11.10. Optical detail of an astigmatically-compensated laser cavity used in tunable colour centre lasers. (After Mollenauer and Olson 1974.)

beam waist on a cold finger at 77 K. The sample and most of the optics is contained in a vacuum enclosure. By mounting the crystal at the Brewster angle, β, to the pump beam one avoids reflection losses at the crystal surfaces without the use of antireflection coatings. Furthermore, by adjustment of the angle θ, the astigmatism due to M_2 can be exactly compensated by that due to placement of the laser crystal at the Brewster angle. The beam waist is at the radius of curvature of M_1 and just outside the focus of M_2. Using a flat output mirror guarantees that the beam has its largest diameter, w_1, at mirror M_2, slowly tapering to an output waist of diameter $w_1/\sqrt{2}$. Since in a practical device the mode length $L \gg w_1$ the mode is essentially a parallel beam: this greatly facilitates insertion of intracavity devices. The mode is stable as long as w_1 is of finite size. The layout of a commerical laser cavity marketed by Burleigh Instruments is shown in Fig. 11.11.

11.4.2 F_A- *and* F_B-*centre lasers*

In alkali halide crystals containing alkali impurity ions, F-centres may be produced with one (F_A) or two (F_B) impurities in the nearest-neighbour cation shell. When the impurity cation is smaller than the host cation these centres undergo a very unusual lattice relaxation following the 1s→2p-like excitation (Chapter 7). Because of the space available around the F_A- or F_B-centre, the defect may lower its total energy in the excited state by movement of one of the neighbouring anions into the saddlepoint configuration. The

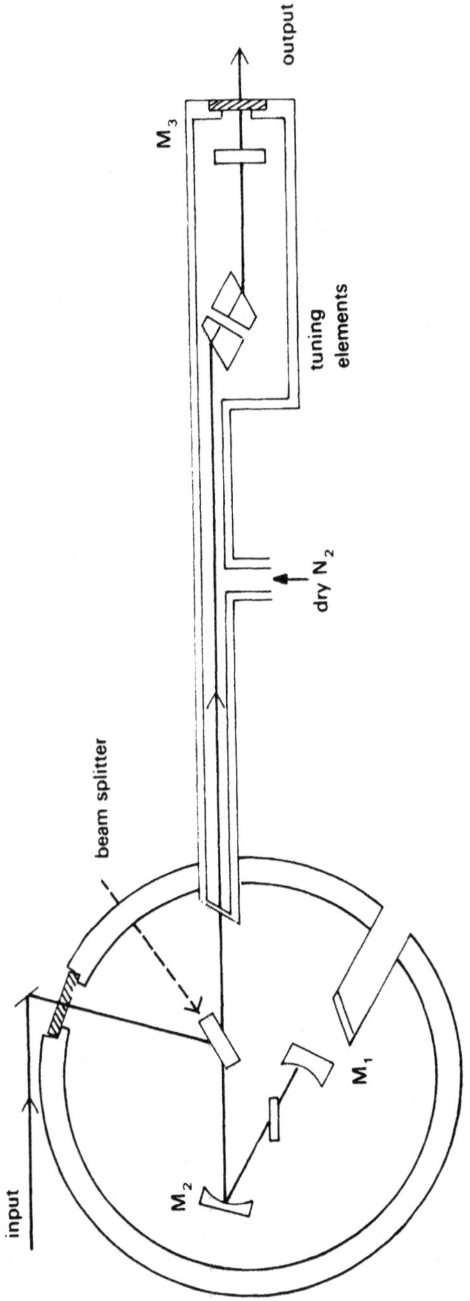

FIG. 11.11. Details of a commercial colour-centre laser (German 1984).

consequence of this ionic rearrangement is a double-well potential from which the emission takes place with a very large Stokes shift. This type of relaxation is characteristic of so-called type II centres. Lasers based on type II F_A- and F_B-centres are very reliable, optically stable, and have good stability at room temperature. The tuning range varies from 2.0–3.6 μm, depending upon the species of crystal used. In type I centres the excited state relaxation is reminiscent of that of the F-centre, and for much the same reasons as for the F-centre, (i.e. small emission cross-section and long radiative lifetime), type I F_A-centres are unsuitable as laser-active defects.

As an example of a type II system we consider the F_A(Li)-centre in potassium chloride: the absorption bands peak at 550 nm and 630 nm, whereas the emission band is shifted into the infrared region at 2700 nm (Fig. 11.12). Hence the laser may be pumped with Kr^+ or Ar^+ ion lasers. Mollenauer chose the 647 nm line from a Kr^+ ion laser. The luminescence bandwidth is about 1.5×10^{13} Hz and $\eta/\tau = 5 \times 10^{-6}$ s. Substitution in eqn (11.29) gives $\gamma(\nu_0) = 4.2\ \text{cm}^{-1}$ for an excited-state concentration of $10^{16}\ \text{cm}^{-3}$. Unfortunately the quantum efficiency is less than unity and decreases monotonically with increasing temperature; as a result the laser threshold (Fig. 11.6) is much lower at low temperature. Furthermore some 80–90 per cent of the input energy must be dissipated as heat, so that for an input power of 2.6 W a peak output power of 240 mW has been observed. A comparison of the absorption and luminescence properties of F_A(Li)-centres in potassium chloride and rubidium chloride is given in Fig. 11.12. In view of the dichroic absorption and emission transitions (Chapter 7), a judicious choice of defect axis relative to the pump and laser polarizations must be made for efficient operation of this laser.

11.4.3 F_2^+*-centre lasers*

The F_2^+-centre in the alkali halides is an anion vacancy pair containing a single trapped electron. Such centres are a by-product of F-centre production. The first systematic study of the formation kinetics and optical properties of these defects was reported by Aegerter and Lüty (1971). However, for application as a laser host the crystal must contain a high concentration of defects stable against thermal and optical degradation. Mollenauer (1979) stabilized the large F_2^+-centre concentration in alkali halide crystals by incorporating deep electron traps such as Mn^{2+}, Ni^{2+}, or Pd^{2+} in the crystal during growth from the melt. These crystals were irradiated with 1–2 MeV electrons at a few $\mu\text{A cm}^{-2}$ current density in the temperature range ~120–180 K, so producing F^+- and F-centres as well as monovalent impurities M^+. Warming to room temperature for 10–15 minutes enables F^+-centres to migrate to and aggregate with F-centres so forming F_2^+-centres. For laser operation the crystals are then cooled to 77 K, otherwise the F_2^+-

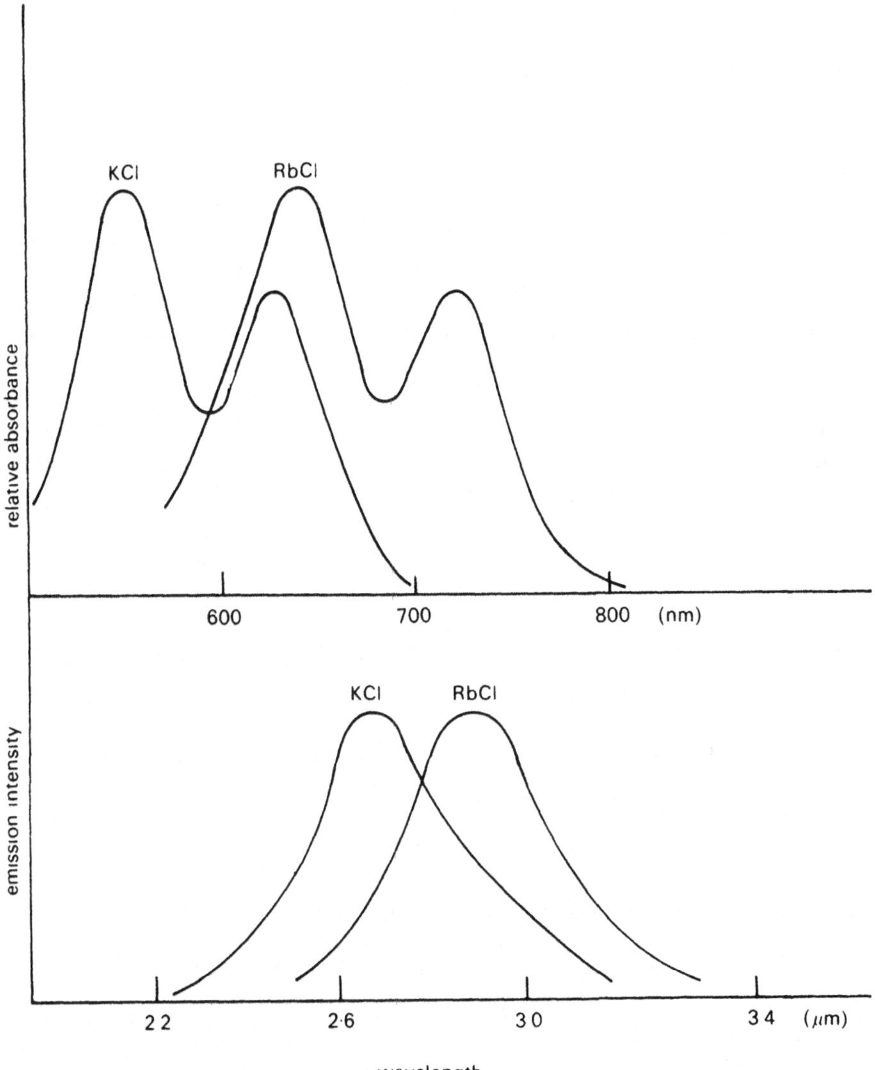

FIG. 11.12. A comparison of the optical absorption and luminescence properties of F_A(Li)-centres in potassium chloride and rubidium chloride.

centres dissociate. Gellerman *et al.* (1980a, b, 1981) added large and controlled amounts of OH^- or SH^- anions as alternative electron traps.

The energy levels of the F_2^+-centre resemble those of the H_2^+ molecule ion; they are accurately predicted using the dielectric continuum model

(Chapter 2). The scaling of properties from molecular ion to defect is derived from eqn (2.81) with $m/m^* = 1$. Hence

$$E(F_2^+) = E(H_2^+, R_{12})/\kappa^2 \tag{11.30}$$

and

$$R_{12} = r_{12}/\kappa \tag{11.31}$$

where κ is the dielectric constant, R_{12} is the vacancy–vacancy separation and r_{12} is the proton–proton separation in the molecular ion. These equations afford a very large tuning range since both R_{12} and κ vary markedly over all the alkali halides. Figure 11.13 shows the lowest energy levels of F_2^+-centres. In the nomenclature of molecular spectroscopy the two allowed absorption transitions are written as $\Sigma_g \rightarrow \Sigma_u$ and $\Sigma_g \rightarrow \Pi_u$. The higher energy emission $\Pi_u^* \rightarrow \Sigma_g^*$ is only observed at the lowest temperature ($T < 50$ K) because at higher temperature it is overwhelmed by non-radiative decay to the Σ_u^* level, so leading to $\Sigma_u^* \rightarrow \Sigma_g^*$ emission alone. However, the $\Sigma_g \rightarrow \Pi_u$ absorption transition is very important because by optically pumping in this band with linearly polarized light all F_2^+-centres can be aligned in a single $\langle 110 \rangle$ orientation, which in the laser allows all centres to contribute fully to one polarized laser mode. The alignment procedure is carried out prior to laser operation, in which the absorption and emission transitions involve the ground and first excited state ($\Sigma_g \leftrightarrow \Sigma_u$). The $\Sigma_g \rightarrow \Sigma_u$ transitions of the F_2^+-centres have almost ideal characteristics for a four-level laser. Both in absorption and emission the cross-sections are large, $\sim 10^{-16}$ cm^2, and this guarantees a large single-pass gain. For F_2^+-centres in potassium chloride crystals of 1–2 mm in thickness $\gamma(\lambda_0)$ is 3.5 cm^{-1}. The quantum efficiency is of order 100 per cent, and is independent of temperature. The laser emission is polarized parallel to the axis of the centre, and has an oscillator strength $f \sim 0.3$. Since the Stokes shift is comparatively small the pump energy conversion efficiency is high and may approach 80 per cent. Furthermore, there is no self-absorption of the laser emission due to the higher excited states. In Fig. 11.14 we summarize the absorption and emission bands for F_2^+-centres in nine of the alkali halides. The absorption bands range from 650 nm to 1650 nm whereas the laser emission spans the infra-red region from 920 nm to 1920 nm.

Despite their obvious promise as four-level laser systems, there are operational problems with F_2^+-centre lasers associated with optical bleaching and reorientation of these anisotropic defects under intense laser pumping. This latter process involves simultaneous absorption of two or three photons from the pump beam leading to a slow reorientation of centres into the five equivalent F_2^+-centre orientations which are optically less efficient than the principal laser orientation. Hence there is an eventual degradation of the centres, and stable laser operation from one region of the crystal is possible only for a limited period. In addition, once prepared, F_2^+-centres have

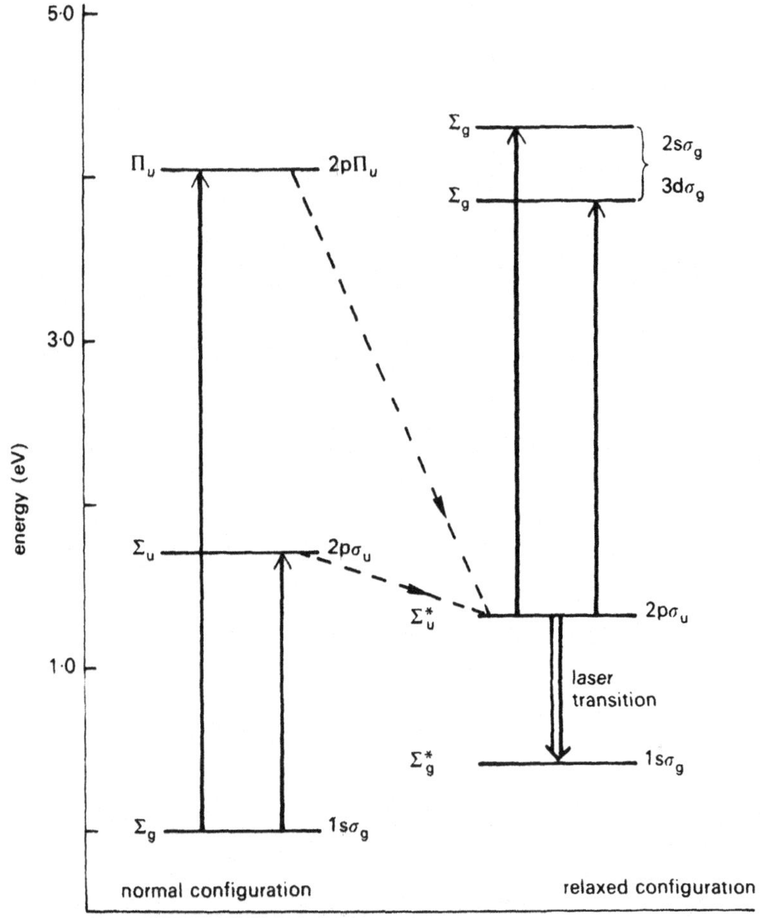

FIG. 11.13. The energy-level structure of F_2^+-centres in sodium fluoride determined from optical absorption measurements for transitions out of the ground and first excited states. (After Mollenauer 1979.)

relatively short shelf-life even when kept in the dark at room temperature, due partially to thermally-activated reorientation and dissociation.

Two ways of overcoming the problems have been proposed, both involving the association of the F_2^+-centres with other radiation damage products resulting in the formation of so-called $(F_2^+)^*$- and $(F_2^+)^{**}$-centres. Mollenauer (1979) showed that when sodium fluoride crystals doped with divalent electron traps such as Mn^{2+} are heavily irradiated at room temperature the F_2^+-band decays and is replaced by another band at longer wavelength. This band, labelled by Mollenauer the $(F_2^+)^*$-band, is due to (F_2^+)-centres perturbed by a cation vacancy (Hoffmann *et al.* 1985). The cation vacancies in

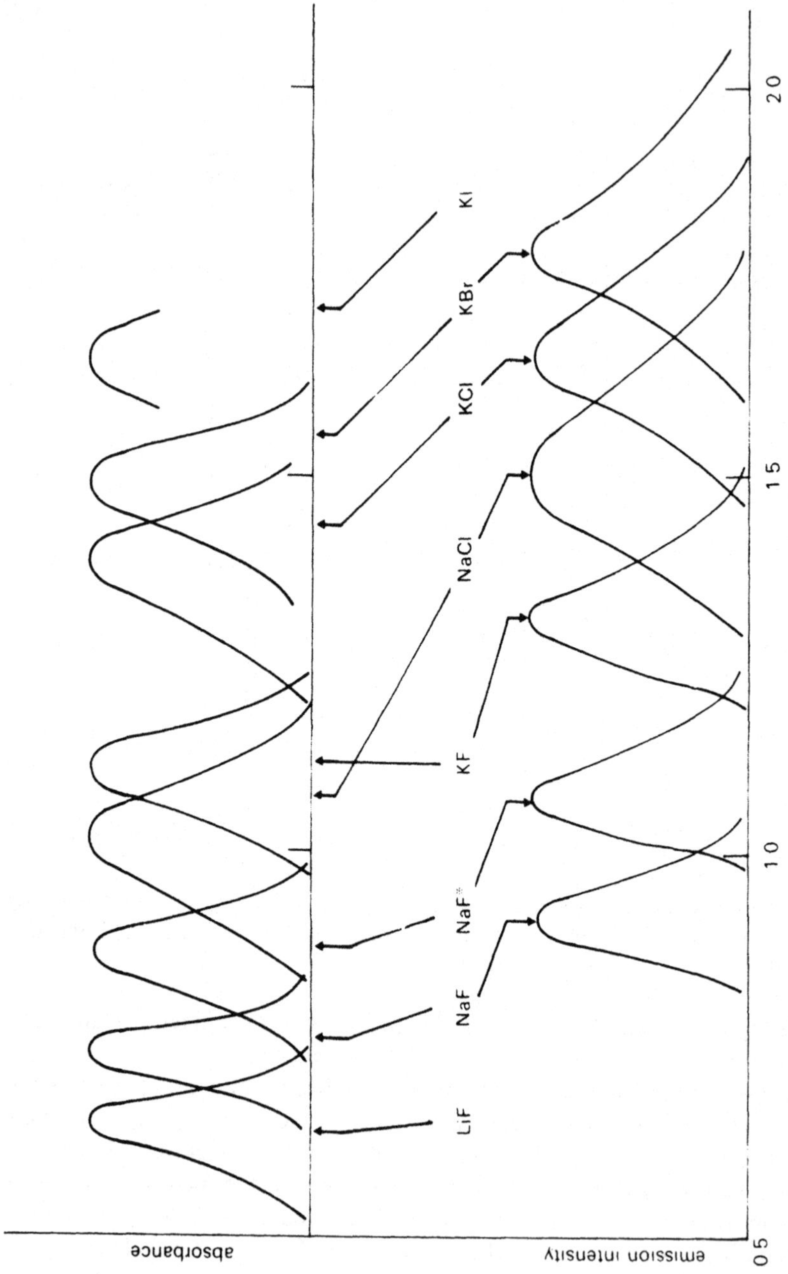

FIG. 11.14. Compilation of infrared absorption and emission spectra for F_2^+-centres in alkali halides (adapted from Mollenauer 1985).

these crystals are required to charge compensate for the divalent cation dopants. If NaF : OH^- crystals are used new bands (so-called $(F_2^+)^{**}$-bands) are observed at even longer wavelength. It is believed that in this case F_2^+-centres are associated with O^{2-} ions (Gellerman *et al.* 1986; Pollock *et al.* 1986), the basic optical nature of $F_2^+ : O^{2-}$ centres is F_2^+-like in every respect. The relative positions of their lowest energy absorption and emission bands in sodium chloride, potassium chloride, and potassium bromide are shown in Fig. 11.15. The $F_2^+ : O^{2-}$ centre laser in sodium chloride spans the wavelength range near 1.4–1.8 μm, which is of considerable interest both for optical communications and non-linear optics.

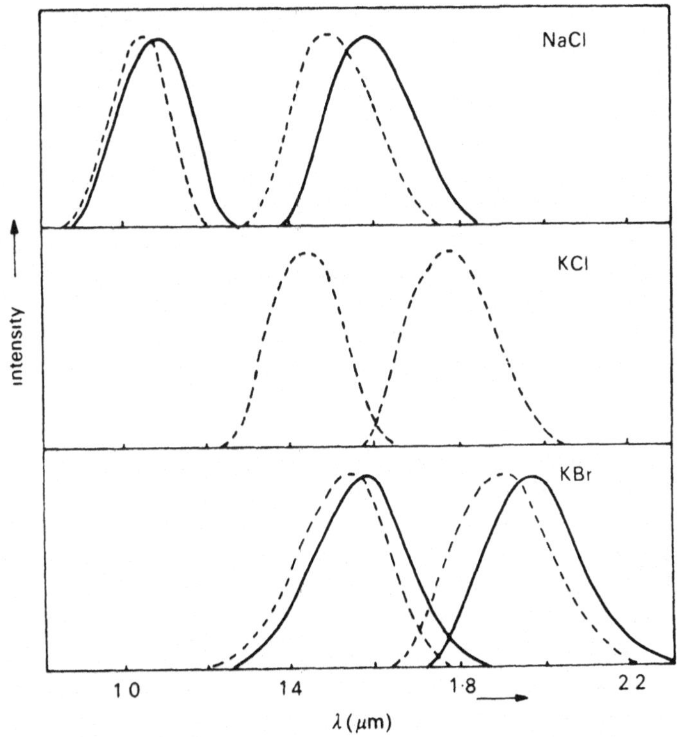

FIG. 11.15. A comparison of the absorption and emission spectra of F_2^+- (-----), F_2^+:O^{2-}-centres (———) in sodium chloride, potassium chloride, and potassium bromide. (After Gelterman *et al.* 1987.)

Schneider and his colleagues (1979, 1981, 1983) have studied F_{2A}^+-centres in KCl : Na^+, KCl : Li^+, as well as in potassium bromide, potassium iodide, and rubidium iodide doped with Li^+. The defects in this case are F_2^+-centres formed with a cation neighbour substituted by an alkali impurity Na^+ or Li^+. These defects extend the useful tuning range of the F_2^+-like centres out to about 3600 nm. Indeed by double-doping potassium chloride with Li^+ and Na^+ it has proved possible to tune continuously from 1670 to 2460 nm, due to emission from both F_{2A}^+(Li)- and F_{2A}^+(Na)-centres. A prime advantage of these lasers over the F_2^+-centre lasers is their relative stability. This arises because of the mode of formation. In their experiments Schneider *et al.* (1979, 1981, 1983) used single crystals grown with large alkali impurity concentrations which were additively coloured to give F-centre densities of about 3×10^{18} cm^{-3}. The F-centres were then converted to F_2- and F_{2A}-centres by irradiation with F-band light near room temperature (230–280 K), prior to irradiation below 200 K with light at 365 nm so converting F_2-centres to F_{2A}-centres. This conversion process involves diffusion of F_2-centres by multiple reorientations until the F_2-centres become trapped at alkali impurity ions. The laser crystal was then cooled to 77 K and re-illuminated with 365 nm light which converts F_{2A}-centres to F_{2A}^+-centres. Concomitant with F_{2A}^+-centre formation is the formation of F_A-centres which have narrow absorption bands that fortunately do not overlap the pump or emission bands. Clearly it is important that bleaching with 365 nm light at 77 K enables a dynamic equilibrium to be established between F_{2A}- and F_{2A}^+-centres in which the action of F_A-centres as electron traps is crucial. These laser materials have the particular merit that they may be stored at room temperature for several weeks and may deliver up to 400 mW of stable CW laser power. Essentially the F_{2A}^+-centres involving Li^+ or Na^+ impurities substituted for K^+ in potassium halides are shifted to longer wavelengths because of the extra space available to the single electron. If a larger impurity cation (K^+) is substituted for the host cation (Na^+) then the optical electron occupies a smaller potential well. The infrared emission is then shifted to shorter wavelengths relative to the pure F_2^+-centre.

So far we have discussed two basic types of colour centre laser in the alkali halides, i.e. lasers involving $F_A(11)$- and F_2^+-like centres. The principal virtues of $F_A(11)$-centres are simplicity of production, thermal stability, absence of degradation during high power laser operation, together with ease of handling even at room temperature. However, they usually operate only at high threshold power and low slope efficiency. In F_2^+-centre lasers the defect production mechanisms are complex and the centres bleach during laser operation. Nevertheless these lasers have advantageous operational properties of low threshold, high slope efficiency, and wide tuning range.

11.4.4 Tl^0-centre lasers

The operational advantages of $F_A(11)$- and F_2^+-centre lasers without any of the disadvantages are to be found in the Tl^0-centre lasers. Gellerman *et al.* (1979) originally referred to such centres as F_A(Tl)-centres. These infrared lasers have superior properties to both $F_A(11)$- and F_2^+-like systems. The overall defect system is charge neutral so that they do not require stabilizing electron traps. The centres are produced by extensive electron irradiation to produce large F-centre densities in crystals containing Tl^0 impurities held at temperatures below ~240 K. Bleaching with white light at this temperature results in a temporary photoionization of the F-centre. On warming to ~260 K the F^+-centres become mobile, diffusing randomly through the crystal until trapped at the Tl^+-ion, which has already trapped an electron released by the photoionization. The ensuing defect, a Tl^0 atom next to an F^+-centre, has axial symmetry about [100] direction. The basic optical properties of Tl^0-centres are discussed at length in Chapter 12. Suffice it to say here that irrespective of host crystal the emission bands are close to 1.5 μm. The lifetime of the emission is 1.6 μs and is independent of temperature below 150 K. The value for τ_R corresponds to an oscillator strength of $f = 7.5 \times 10^{-3}$, which when combined with the emission bandwidth (~650 cm^{-1}) yields a gain cross-section of about 10^{-17} cm^2. The optical properties of the $Tl_0(1)$-centre are strongly polarized parallel to the crystal [100] axes. Both pump and laser bands use the lowest energy ($\Phi \rightarrow \Psi$) transition, derived from the crystal field plus spin–orbit split states of the $6^2P_{3/2, 1/2}$ manifold. In potassium chloride the $Tl_0(1)$ absorption band peaks at a wavelength of 1.04 μm so that the laser is rather conveniently driven using the 1.06 μm line from a Nd-YAG laser.

11.4.5 Transition metal ion lasers

Several transition metal ions (Ti^{3+}, V^{2+}, Cr^{3+}, Co^{2+}, Ni^{2+}) have been used as the optical centres in vibronically tuned solid state lasers. The first demonstration of laser action in a broad phonon-assisted sideband was with $MgF_2:Ni^{2+}$ crystals (Johnson *et al.* 1963). Shortly afterwards Johnson *et al.* (1964, 1966) reported phonon-terminated laser action in other Ni^{2+}-doped ionic crystals. Such lasers could only be operated at 77 K. The theory of laser action in a transition ending on a higher vibrational level of the electronic ground state was developed by McCumber (1964). Interest in such lasers then waned until the late 1970s, when the announcement of tunable laser emission in alexandrite ($BeAl_2O_4:Cr^{3+}$) *at room temperature* stimulated renewed interest in transition metal ion systems generally (Moulton 1985). Much of the recent research has been directed at the Cr^{3+} ion in sites of weak crystal field where the broadband-emitting $^4T_2$ level is the lowest excited level. The basic energy level schemes and optical spectroscopy of these ions has already been developed in Chapters 3 and 9. We have also shown that the threshold condition for laser action required the single-pass gain, $\gamma(\nu)$, to exceed the cavity losses, i.e.

$$\frac{\Delta}{L} \leqslant \gamma(\nu) \quad \text{where} \quad \gamma(\nu) = \frac{\lambda^2 \Delta N}{8\pi n^2 \tau_R} g(\nu).$$

Hence, the gain spectrum is related to the shape function $g(\nu)$ of the emission spectrum, which is determined by the strength of the electron–phonon coupling. However, since $\tau_R \propto 1/\lambda^3$, $\gamma(\nu) \propto \lambda^5 g(\nu)$, we see that the spectral gain profile will peak at a different wavelength than the fluorescence spectrum. The difference in peak position of the fluorescence and gain curves is obviously more significant at long wavelengths than short wavelengths.

The simplest transition metal ion is Ti^{3+} which contains a single 3d electron outside closed shells. $Al_2O_3:Ti^{3+}$ would seem to be an excellent material for operation as a tunable laser: indeed, the laser behaviour of titanium sapphire is almost that to be expected from an ideal four-level system. Moulton (1982, 1984) first reported the performance of the $Al_2O_3:Ti^{3+}$ laser; he observed a tuning range of 660–990 nm. That $Ti^{3+}:Al_2O_3$ is an excellent gain medium is not surprising. The strong crystal field of sapphire provides for a large splitting between the ^2E and $^2T_2$ levels, and these are the only levels of importance for optical transitions. In consequence, there is a large Stokes shift between absorption and emission, so that there is no reabsorption of the luminescence by ground state ions. Nor is there any excited-state absorption of either pump beam or luminescence output. Furthermore, the two overlapping blue–green absorption bands at *c.* 485 nm and 550 nm provide for pumping over wide wavelength ranges from 400–600 nm (Section 9.4.1). For optimal performance crystals containing about 0.1–0.2 at. % Ti^{3+} in aluminium oxide are used in which situation the pump bands have absorption cross-sections of 6×10^{-20} cm^2 (485 nm) and 2×10^{-20} cm^2 (550 nm) for light polarized parallel to the optic axis of the sapphire crystal. The peak emission cross-section calculated from the room-temperature fluorescence spectrum and the radiative lifetime $\tau_R = 3.85\ \mu s$ is 4.5×10^{-19} cm^2. The laser characteristics are illustrated in Fig. 11.16, which shows the lifetime temperature dependence, threshold behaviour, wavelength tuning and input–output characteristics of the $Ti^{3+}:Al_2O_3$ laser. For these measurements pumping was effected by the all-lines output (7.6 W) from an Ar^+ laser; the long wavelength tuning was limited by the bandwidth of the mirrors used in the laser cavity. The low-temperature lifetime is constant up to 250 K, but temperature above this decreases to about 0.8 μs at 350 K (Fig. 11.16(a)). These data, and the optical absorption/emission characteristics (Fig. 9.13), are consistent with a Huang–Rhys parameter $S \simeq 12$, phonon energy $\hbar\omega = 385$ cm^{-1} and non-radiative decay rate of 2.5×10^{14} s^{-1} at 350 K. Obviously the laser threshold is a minimum (Fig. 11.16(b)) near to the peak in the emission cross-section versus wavelength curve. This is close to the peak in the output versus wavelength curve (Fig. 11.16(c)). Finally, the linear output power versus input power (Fig. 11.16(d)) is just as is predicted from the rate equation analysis for a four-level laser system (Section 11.3).

The emission bandshape for $3d^3$ ions is determined by whether the ^2E state or the $^4T_2$ state is lowest. If ^2E is the emitting state, as in ruby, $S < 1$ so that the zero-phonon lines are dominant and one is restricted to a single wave-

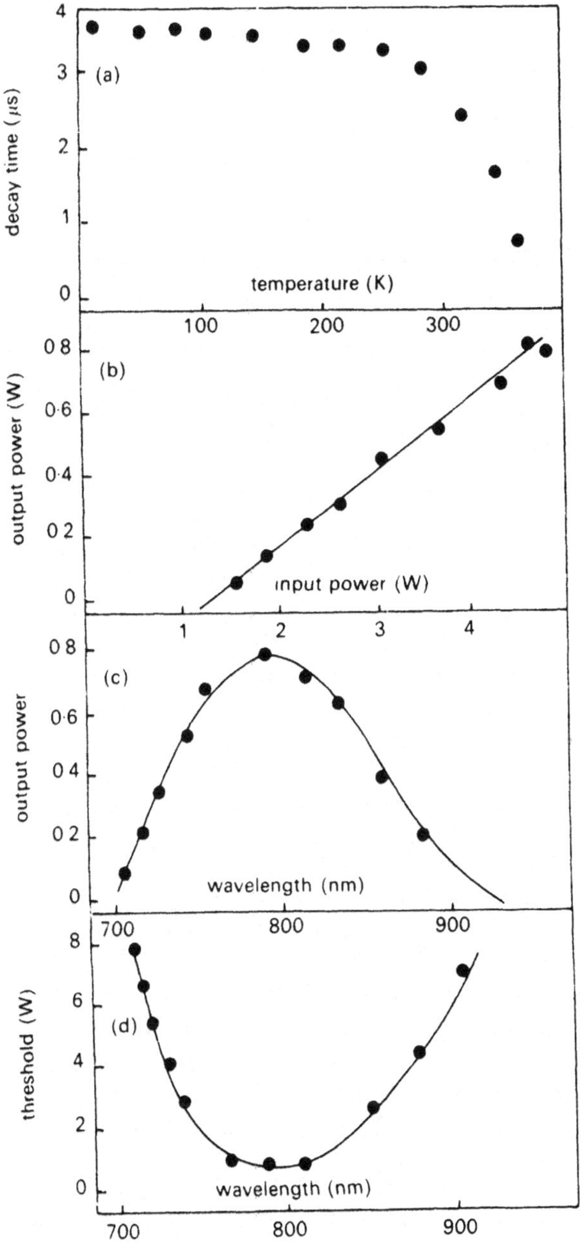

FIG. 11.16. Showing the (a) temperature dependence of the decay time, (b) input power–output power characteristics, (c) wavelength tuning, and (d) threshold behaviour for an 0.15 per cent Ti:Al_2O_3 laser in which the output coupler transmission was 0.8 per cent.

length laser. However, both of the $3d^3$ ions V^{2+} and Cr^{3+} have been used in broadband laser systems operating on the $^4T_2 \rightarrow {}^4A_2$ transition. Since the Huang–Rhys parameter for this transition is rather large ($S \sim 3$–7) a broad emission band is observed, which may be accompanied by weak zero-phonon lines. The decay times are typically tens of microseconds. In Fig. 9.9 we showed a simplified energy-level diagram for $3d^3$ ions in octahedral crystal fields indicating the values of Dq/B values for ruby, alexandrite, and $KZnF_3:Cr^{3+}$. When the $^4T_2$ level is much lower than ^2E then emission is exclusively due to the $^4T_2 \rightarrow {}^4A_2$ transition, and is broadband. This applies to $Cr^{3+}:KZnF_3$. However, when ^2E and $^4T_2$ levels are close together then both narrow-line, $^2E \rightarrow {}^4A_2$, and broadband, $^4T_2 \rightarrow {}^4A_2$, emissions occur as is the case for some garnets (Huber 1986). For alexandrite, $^4T_2$ and ^2E states are separated in energy by $\sim 800\ cm^{-1}$ and at room temperature the $^4T_2 \rightarrow {}^4A_2$ broadband emission becomes the dominant feature (Fig. 9.10). This material is a very successful vibronic laser at room temperature with a tuning range of 700–850 nm.

The very attractive laser properties of alexandrite (Walling *et al.* 1980; Walling 1982): room-temperature operation, broadband tunability, and high efficiency (50–60 per cent), stimulated considerable research activity into other potential laser materials containing $3d^3$ ions. To date at least sixteen Cr^{3+}- or V^{2+}-doped ionic crystals have been reported to sustain laser action, all operating on the broad vibronic tuning range of the $^4T_2 \rightarrow {}^4A_2$ emission (Caird 1986; Huber 1986). Obviously the materials search is directed at extended wavelength coverage, increased energy storage, better quantum yield, higher emission cross-section, and less excited-state absorption. Much research has been directed at rare-earth garnets, $R_3A_2B_3O_{12}$, such as gadolinium scandium gallium garnet (GSGG) ($Gd_3Sc_2Ga_3O_{12}$), in which Cr^{3+} substitutes on the octahedral (A) site (Huber and Peterman 1985). With these materials there are growth problems with compositional disorder (e.g. Ga^{3+} occupying Sc^{3+} sites in GSGG) and Cr^{3+} distribution in the grown boule. It is, therefore, sometimes difficult to avoid in-built strain and chemical inhomogeneity, both of which might have deleterious consequences for laser performance. Nonetheless these are very promising laser materials, and better laser properties will surely result from improved crystal quality.

There is also much interest in Cr^{3+}- and V^{2+}-doped perovskite crystals. An obvious candidate is $Cr^{3+}:YAlO_3$. Unfortunately the octahedral Al^{3+} position occupied by the Cr^{3+} ion is a strong crystal field site, so that luminescence is predominantly in the R-line and this material is not useful as the laser gain medium. However $Cr^{3+}:KZnF_3$ (Dürr *et al.* 1985, 1986) and $Cr^{3+}:ScBO_3$ (Lai *et al.* 1986) both sustain broadband laser action. In Fig. 11.17 we show the optical absorption and luminescence spectra measured at 300 K for scandium borate crystals containing 0.3 at. % Cr^{3+} substituting on the scandium site. In absorption the two spin-allowed bands

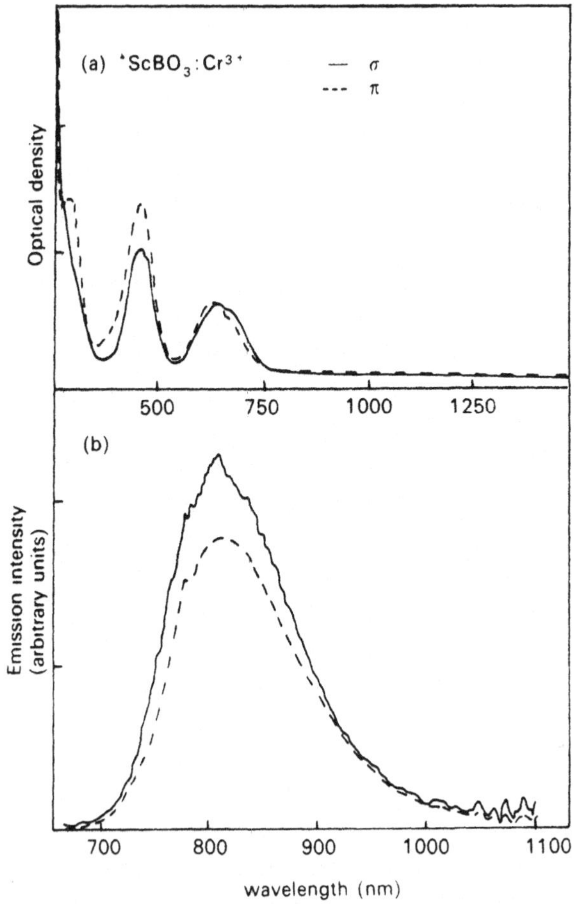

FIG. 11.17. Polarized absorption and luminescence of Cr^{3+} in scandium borate measured at 300 K. (After Lai *et al.* 1987.)

$^4A_2 \rightarrow {}^4T_2$, $^4T_1$ are clearly visible with peaks at *c.* 650 nm and 450 nm respectively. The $^4T_1 \rightarrow {}^4A_2$ emission spectrum is broadband at room temperature, without evidence of the R-line system. Clearly in this material Cr^{3+} occupies a weak crystal field site. The peak occurs at 815 nm, wheres the peak in the emission cross-section occurs at 840 nm with a maximum value of 1.2×10^{-20} cm^2. The fluorescence decay profiles show an exponential decay with characteristic time $\tau_R = 115$ μs at 300 K. The lifetime decay with increasing temperature (4–600 K) is shown in Fig. 11.18(a). The slow lifetime decay at temperatures up to 300 K is probably due to the effects of odd-parity vibrations. In the range 300–600 K the much faster decay is probably due to multiphonon non-radiative relaxation. Despite the lifetime decrease and non-

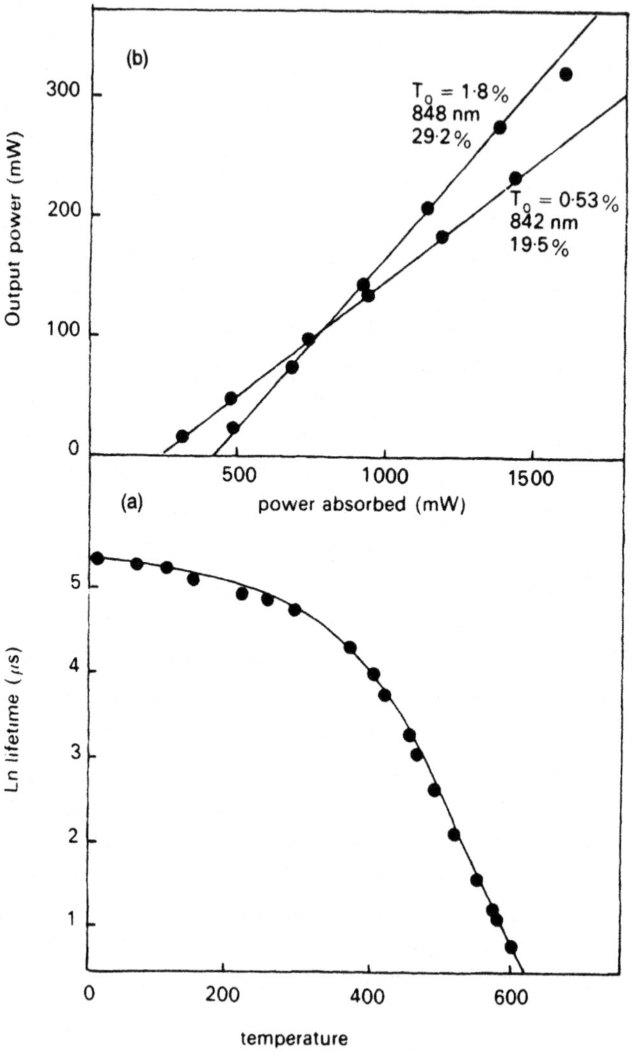

FIG. 11.18. (a) The lifetime versus temperature of Cr^{3+} emission in scandium borate and (b) the input–output characteristics of the $Cr^{3+}:ScBO_3$ laser. (After Lai *et al.* 1987.)

radiative relaxation the quantum yield is almost unity up to 300 K. The laser output characteristics are shown in Fig. 11.18(b). Obviously there is a linear relationship between output power and input power, as is expected for a four-level laser (eqn 11.28). For an output mirror transmitting 1.8 per cent of the light at the laser wavelength the slope efficiency was 29.2 per cent and the threshold occurs at 400 mW input power. However, with the output coupler reduced to 0.53 per cent transmission the slope efficiency was only 19.5 per cent and the threshold was at 250 mW. These data were measured at the band peak of 842 nm. The tuning range of the laser, obtained using an output coupler of 0.3 per cent transmission was from 787 nm to 892 nm (Lai *et al.* 1986). Other perovskite crystals which have been used as Cr^{3+} laser hosts include $KZnF_3$, $RbZnF_3$, $RbCdF_3$, and $CsCdF_3$ (Dürr *et al.* 1985). Indeed, $KZnF_3$ has become the second commercially available tunable solid state laser. The tuning range was from 785–865 nm with good slope efficiency under laser pumping.

The broadband luminescence of $Co^{2+}:MgF_2$ and other fluorite crystals is sufficiently efficient at low temperatures to be the basis for a continuously tunable laser at 77 K in the wavelength range 1.5–2.4 μm (Johnson *et al.* 1964; Moulton and Mooradian 1979; Moulton 1986). These lasers are pumped in the lowest ${}^4T_1 \rightarrow {}^4T_2({}^4F)$ band using the 1.32 μm line of Nd^{3+}-YAG or in the ${}^4T_1 \rightarrow {}^4T_2({}^4P)$ band using the 514 nm line form an Ar^+ ion laser. The pump threshold may be as low as 50 mW, and output power of order 100 mW has been observed for absorbed pump powers of about 1 W. The $Co^{2+}:MgF_2$ laser has been operated at temperatures up to 300 K, but with the penalty of slightly higher threshold and lower slope efficiency on account of the non-radiative decay processes (Moulton 1986; Welford and Moulton 1988).

The near-infrared transition ${}^3T_2 \rightarrow {}^3A_2$ has been used as the basis of several Ni^{2+}-doped solid state lasers. In $MgO:Ni^{2+}$ the observed decay time is constant with temperature up to 400 K, which is indicative of a high quantum efficiency. However, laser action does not occur at temperatures above 200 K. The decay time of Ni^{2+} in magnesium fluoride is about 10 ms at 77 K and decreases with increasing temperature due to the onset of non-radiative transitions. In this material laser action occurs only at low temperature. Laser action has also been demonstrated in $KMgF_3:Ni^{2+}$, $MgAl_2O_4:Ni^{2+}$, and $MgGa_2O_4:Ni^{2+}$. Taken collectively the tuning ranges of these materials span the near infrared from 1.15–1.8 μm. However, in all cases not only is non-radiative decay an inhibiting factor but so also is excited-state absorption. The very large bandwidth required for wide tunability implies a rather small emission cross-section. To some extent this can be compensated by relatively large concentrations of the dopant ion. The penalty is then that the emission band is at least partially broadened by inhomogeneous processes; the tuning band is not entirely continuous, in contrast to the behaviour of colour-centre lasers.

In selecting broadband emitting materials for tunable solid state lasers a compromise may be required for the value of the Huang–Rhys parameter, S. Large S values lead to broad emission bands and hence to broad tuning ranges, but they also lead to enhanced non-radiative transition probabilities. Nevertheless one anticipates a continued outgrowth of research into new laser host crystals. Perovskite crystals with various dopants have shown promise; the scheelite- and wolframite-structured materials also offer interesting possibilities. For example, tunable laser operation has been observed in $ZnWO_4:Cr^{3+}$ over the wavelength range 980–1090 nm. Extension to $CaWO_4$ and also to the metaphosphates, vanadates, niobates, and molyodates is desirable. More activity is promised in V^{2+}-doped fluorites because of the (potential) extended tuning ranges beyond 1 μm. An extension into the 4d transition metal ions and into charge transfer complexes is also predictable following recent measurements of tunable optical gain in $Rh^{2+}:RbCaF_3$ and in the blue-green charge transfer bands of $Ti^{4+}:Li_4Ge_5O_{12}$ (Powell 1986).

Considerable effort is being expended in improving the output of transition metal ion lasers. Single-mode ring laser operation has been reported for both Cr^{3+}:GSGG and $Cr^{3+}:KZnF_3$ in which stable single-frequency operation was achieved (Fuhrberg *et al.* 1986). Similar studies are expected of other solid state lasers in attempts to match the narrow linewidth performance ($\sim$100 kHz) of commerical dye lasers. Experiments involving injection-locking aimed at high-power ring laser operation in the pulsed mode have given very promising results for alexandrite (e.g. 500 mJ Q-switched operation and nanosecond pulse duration) including injection using a single-frequency diode laser (de Rougemont *et al.* 1986; Krasinski *et al.* 1986). Injection control is also a promising way of producing efficient narrow-linewidth operation in cases where the laser gain bandwidth is homogeneously broadened. Perhaps the greatest future impact will be in the area of miniaturization, an area where efficiency, reliability, and simplicity of design are critical requirements. In this context the progress in developing semiconductor diode lasers as pump sources for miniature tunable solid state lasers, especially those based on Cr^{3+} and Ti^{3+}, is important. If such miniature lasers make use of waveguiding techniques, either in the form of waveguides produced on bulk surfaces or on monomode crystal fibres, then the enhanced energy density will lead to milliwatts rather than watts of pump power being required to exceed laser threshold.

11.5 Production of ultra-short pulses

In recent years there has been increasing interest in the production of single laser pulses with very high peak power. For the shortest pulses the technique of *mode-locking* is used, in which some external perturbation forces the various modes to maintain fixed phases relative to one another. The output is then repetitive in time, consisting of regularly spaced pulses of very high peak power. To produce a short-duration waveform requires many frequencies. If at (say) $t=0$ all the phases of the modes are random then the instantaneous power is never large. However, when these modes, all with equal frequency spacing of $\Delta\nu=c/2nL$, where L is the cavity length, have the same phase at time $t=0$ there results a series of pulses of period $T=1/\Delta\nu$. This pulse period corresponds to the round-trip transit time of a pulse within the laser cavity. It is a general result that if all phases are constant, no matter what the mode then only a single pulse exists within the cavity round-trip time. The pulse duration τ is roughly given by the inverse of the frequency width of the laser output: for N modes oscillating this corresponds to $(N\times\Delta\nu)^{-1}$, i.e. $\tau=2nL/Nc$. Thus the number of modes is given approximately by the ratio of the pulse spacing to the pulse width, i.e.

$$\frac{T}{\tau}=N. \tag{11.32}$$

The condition that N must be large can be satisfied by using a broad laser transition and a long cavity, the latter to give a small intermode separation. Experimentally the modes can be forced into definite phase relationships by *passive mode-locking*, using a saturable dye, or by *active mode-locking* using an electro-optic shutter. Since tunable solid state lasers have usually been mode-locked by synchronously pumping with an actively mode-locked ion laser or Nd^{3+}-YAG laser we will briefly describe the technique of active mode-locking.

In principle all that is required is to modulate the losses of the laser cavity at a frequency equal to the reciprocal of the pulse round-trip time, i.e. $(c/2nL)$. An appropriate form of shutter is an electro-optic modulator between crossed polarizers placed before the output mirror. The shutter is open for only very short times every round-trip time. A pulse with a temporal width equal to the time that the shutter remains open and which arrives exactly as the shutter opens passes through the shutter unaffected. However, if the pulse arrives when the shutter is closed it bounces back and forth in the laser. Every time the pulse passes through the shutter and is reflected by the output mirror a portion of it is transmitted. In this case pulses of duration $2nL/Nc$ are produced as a periodic pulse train with time intervals of $2nL/c$ between pulses.

In mode-locking by synchronous pumping the length of the laser cavity is adjusted to be a multiple (or sub-multiple) of the length of pump laser. Then the violent gain oscillations result in many sidebands being excited with frequency separation equal to the mode spacing. These sidebands provide for the mutual phase-locking of pulses in both pump and pumped beams through the stimulated emission process. For synchronous pumping to be effective the gain cross-section must be large enough for the laser to be pumped above the threshold by a single pulse. Colour centres and $Tl^0(1)$-centres may exceed this necessary condition. Mode-locking of a colour-centre laser by synchronous pumping was first reported by Mollenauer and Bloom (1979); they used F_2^+-centres in lithium fluoride pumped by a mode-locking Kr^+ laser operating on the 647 nm line. Mode-locked colour-centre lasers using F_2^+-centres in potassium fluoride and sodium chloride and $Tl^0(1)$ lasers in potassium chloride were synchronously pumped using 80 ps wide pulses from a Nd-YAG laser at 1.06 μm. The pulse repetition rate was 100 MHz and time-averaged pump powers in excess of 5 W were used. The laser cavity used by Mollenauer *et al.* (1980) is shown in Fig. 11.19; the birefringent tuner plates effectively limit the pulse bandwidths to the minimum required by the uncertainty principle. For a $(\text{sech})^2$ pulse shape, the product $\Delta t \times \Gamma \sim 0.3$, where Δt and Γ are the FWHM of the pulse temporal width and frequency bandwidth respectively. Pulse widths of typically 4 ps could be maintained over most of the tuning band of the F_2^+-centres. Because of the lower gain in $Tl^0(1)$-centre lasers, good pulses can be maintained over a rather smaller fraction of the gain curve.

The NaCl:F_2^+ centre has a peak in the gain curve near 1.5 μm, making it an eminently suitable source for pulse propagation studies in optical fibres. This laser was used by Mollenauer *et al.* (1980) and Stolen *et al.* (1983) in experiments on extreme pulse compression and soliton effects in low-loss, single-mode, and polarization-preserving optical fibres. The various effects result from the interaction of a small non-linearity in the refractive index and the negative group velocity dispersion. Because of the non-linearity, pulses propagating in a fibre experience variable phase shifts across the pulse because of the range of intensities present in the pulse. A consequence of this variable phase shift is that the frequencies in the trailing half of the pulse are raised and those in the leading half are lowered. For wavelengths greater than 1.3 μm in optical fibres the group velocity dispersion is negative ($\partial v/\partial \lambda < 0$). Due to this negative dispersion the trailing half of the pulse, containing the raised frequencies, is advanced and the leading half is retarded. In other words, the pulse tends to collapse upon itself. To predict this pulse-narrowing one must solve a non-linear Schrödinger equation (Mollenauer *et al.* 1980); given the correct pulse input conditions, solution of this equation predicts a range of soliton effects. For propagation as a fundamental soliton the pulse amplitude must be such that the pulse-broadening effects of dispersion are exactly balanced by the pulse-narrowing effects of the non-linear refractive index. Thus the fundamental ($N = 1$) soliton is a pulse that in a zero-loss fibre never changes shape. At higher input powers the pulses experience an initial narrowing followed by periodic splittings of the pulse. The initial narrowing represents the promise of obtaining sub-picosecond pulses at wavelengths greater than 1.3 μm from pulses of width of order 5 ps. This first pulse narrowing is defined by the soliton order, N, and the soliton period, Z_0; for $N > 1$ the pulse splitting and narrowing is periodic with propagation along the optical fibre. In their work on soliton effects Mollenauer *et al.* (1983*b*) showed that a pulse, initially with a FWHM of 7 ps, can be compressed to 0.26 ps FWHM in a single-mode optical fibre of length 100 m, using some 300 W peak pulse at the input.

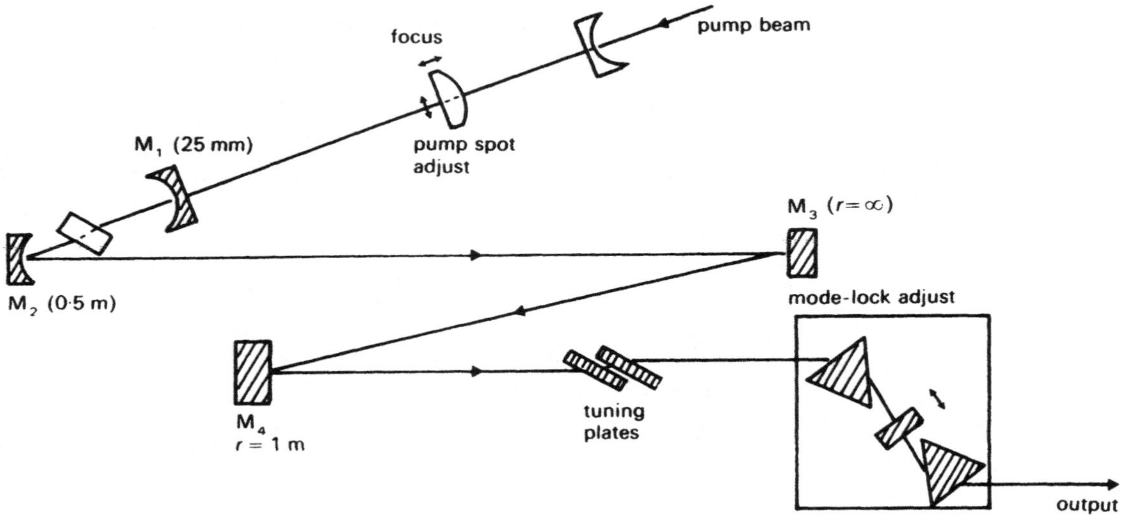

FIG. 11.19. Cavity used for mode-locking colour-centre lasers by synchronous pumping (After Mollenauer *et al.* 1984).

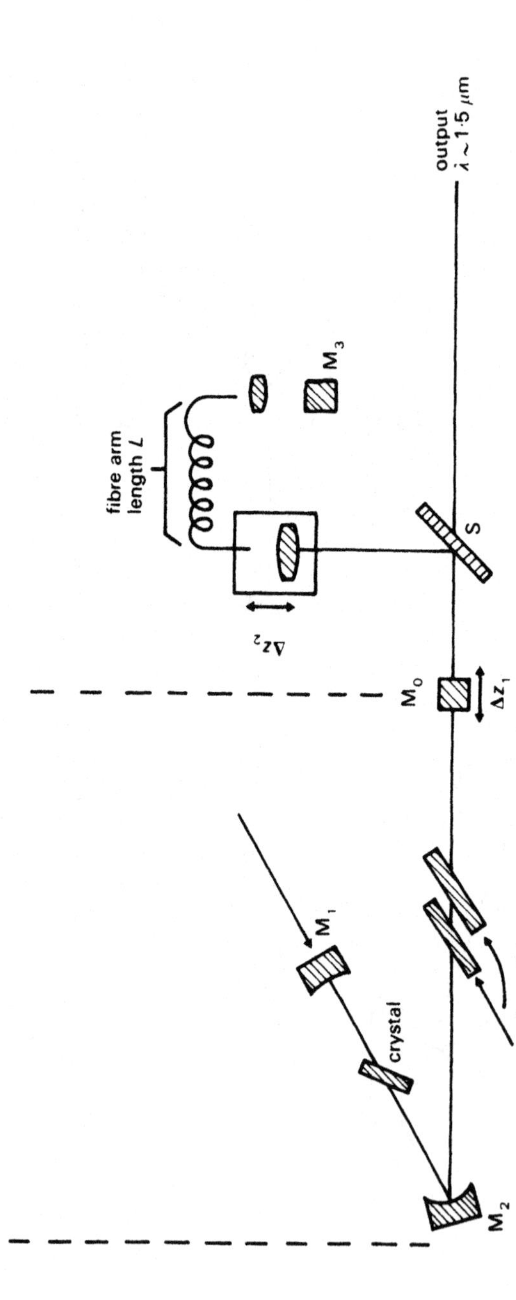

FIG. 11.20. A schematic of the soliton laser (after Mollenauer and Stolen, 1984). The basic laser is a three-mirror, astigmatically compensated cavity with the sample crystal positioned at the beam waist. Output pulses are injected into the fibre-arm of the cavity, where the retroreflector M_3 directs the pulses back into the mode-locked laser. Mirror M_0 and M_3 are adjusted to synchronize pulses in the fibre arm with the mode-locked laser. The output is then utilized for pulse diagnostics using an autocorrelator or for fast spectroscopy measurement.

In 1984 Mollenauer and Stolen used these ideas to develop the *soliton laser*. To the mode-locked $Tl^0(1)$-centre laser is added an optical fibre feedback loop (Fig. 11.20). Pulses from the $Tl^0(1)$-centre laser are used to launch $N = 2$ solitons in a length of optical fibre, and these are then injected back into the $Tl^0(1)$-centre laser, so forcing the laser itself to produce pulses of a definite shape and width. The narrowest pulses obtained by Mollenauer and Stolen using this technique were as short as 130 fs using a fibre arm of length 0.2 m. By compression in a second external fibre, pulses of only 50 fs FWHM are obtained. For very short pulses peak powers in the fibre appear large ($\sim$ 10 kW). However, the time-averaged powers are modest because of the long time between pulses.

12
Optical detection of magnetic resonance

ELECTRON spin resonance (ESR) is a technique normally used to measure the ground-state magnetization of an electronic spin system. In a typical experiment the sample is placed in a resonant microwave cavity and irradiated with microwave power. The spin degeneracy of the ground state is lifted by a magnetic field, the strength of which may be varied. When the energy separation between magnetic sublevels which differ in M_s value by ± 1 equals the energy of a microwave photon, $h\nu$, microwave power is absorbed by the spin system. Such an absorption of microwave power upsets a balanced microwave bridge circuit, and this permits the resonance condition to be detected. Most experiments are carried out at a microwave frequency near 9.5 GHz, 20 GHz, or 35 GHz at temperatures in the range 1.6–300 K. Simple aspects of such experiments were discussed in Chapter 6. More detailed accounts are given by Orton (1968) and Abragam and Bleaney (1970). ESR may also be observed in excited states of electronic centres if a sufficient population of centres can be maintained in the excited states. Clearly the thermal equilibrium population of a state (say) some 20 000 cm^{-1} above the ground state is very small and some technique for enhancing the excited-state population must be employed. The excess population in the excited state is achieved by irradiating the sample with optical photons of appropriate energy. Even so, the excited-state population achieved by optical techniques is often many times smaller than that detectable by conventional ESR methods. It is then necessary to observe the effects of microwave-induced ESR transitions on the optical absorption or luminescence spectrum of the centre. This technique which detects ESR transitions optically is usually referred to as Optical Detection of Magnetic Resonance or simply ODMR. The absorption of microwave power associated with ESR in the excited states upsets the equilibrium populations of the sublevels and so changes the pattern of optical transitions between excited and ground state. Such changes in luminescence indicate that the resonance condition is met. This is an example of so-called 'trigger-detection', since the absorption of one microwave photon triggers the emission of one optical photon. The enhanced sensitivity of ODMR relative to ground-state ESR, of the order $(\nu_{opt}/\nu_{micr}) \simeq 10^4$, can compensate for the relatively small excited-state population. Experimental details of ODMR were described in Chapter 6.

Historically, optical-microwave double resonance techniques were first carried out in the gas phase. The earliest studies of ODMR in solids were first reported by Geschwind *et al.* (1959) in the ^2E state of Cr^{3+} ions in ruby. These pioneering studies demonstrated the possible detection schemes that can be used, including high-resolution spectroscopy and measurements of circular polarization of luminescence. We first discuss the application of these methods to detect ESR in the ^2E state of $3d^3$ ions in aluminium oxide and magnesium oxide. We also show that ODMR may be used to study ground-state ESR under suitable circumstances. Subsequently ODMR studies of recombination processes in ionic crystals and semiconductors are described. Finally we discuss the measurements of transient effects that occur when spin systems interact.

12.1 ODMR of $3d^3$ ions

The first demonstration of solid state ODMR, in the luminescent ^2E state of Cr^{3+} in ruby, was soon followed by similar studies of V^{2+} and Mn^{4+} in aluminium oxide (Geschwind *et al.* 1965; Imbusch *et al.* 1967) and of V^{2+} and Cr^{3+} in magnesium oxide (Chase 1968*a*). Luminescence from the lowest-lying ^2E state of $3d^3$ ions in these hosts is in the form of strong zero-phonon lines and accompanying vibrational sidebands (Section 9.3.1). The steady ^2E state populations required for ODMR can be achieved by continuous optical pumping in the broad strongly-absorbing $^4A_2 \rightarrow {}^4T_1$ and $^4A_2 \rightarrow {}^4T_2$ bands. Ions raised to the $^4T_1$ and $^4T_2$ states decay rapidly by non-radiative transitions to the ^2E state. At low temperatures de-excitation of the ^2E state is purely radiative with a decay time of a few milliseconds. The $^2E \rightarrow {}^4A_2$ luminescence is monitored to detect the absorption of microwave power in the excited ^2E state. Some properties of the $3d^3$ ions relevant to the detection of excited-state ESR are noted in Table 12.1.

Table 12.1

Properties of $3d^3$ ions in magnesium oxide and alumina of relevance to ODMR measurements.

		^2E level cm^{-1}	R-line width at 0 K (cm^{-1})	Spin–orbit parameter (cm^{-1})	Radiative decay time (ms)
MgO	V^{2+}	11498	~0.25	168	100
	Cr^{3+}	14310	~0.25	270	11.4
	Mn^{4+}	15279	4.0	410	2.5
Al_2O_3	V^{2+}	11697 (R_1)	0.3	168	65
	Cr^{3+}	14418 (R_1)	0.1	270	3.5
	Mn^{4+}	14782 (R_1)	5.0	410	0.85

12.1.1 Energy levels and optical transitions of 3d^3 *ions*

The energy-level scheme of the ^{2}E and ^{4}A$_2$ states of 3d^3 ions in octahedral and nearly-octahedral sites are shown in Fig. 12.1. These states are not split in a purely octahedral crystal field. The ^{2}E state is an orbital doublet, described by octahedral basis functions u,v or u_+, u_-, these being suitable descriptions relative to tetragonal and trigonal axes, respectively (Chapter 2). Figure 12.1 indicates that Δ, the splitting of the orbital doublet when the crystal field is reduced from octahedral, O_h, to tetragonal, D_4, by an axial distortion along the cubic z-axis is much larger than that between the $M_s = \pm\frac{1}{2}$ and $M_s = \pm\frac{3}{2}$ spin states derived from the ^{4}A$_2$ ground state (90 cm^{-1} and 0.16 cm^{-1}, respectively, in the case of Cr^{3+} in tetragonal sites in magnesium oxide). Reduction of the symmetry of the crystal field from octahedral, O_h, to trigonal, C_{3v}, by an axial distortion along the trigonal z-axis also splits the ^{2}E and ^{4}A$_2$ states. However, in this case the splitting of the ^{2}E state is quite different from that which occurs in tetragonal symmetry, because in C_{3v} symmetry the ^{2}E state splitting is due to the combined effects of the axial field along the z-axis and spin–orbit coupling. The separation of the ^{2}E state into spin-orbitals which belong to the 2$\bar{\text{A}}$ and $\bar{\text{E}}$ irreducible representations of the trigonal group is shown on the right-hand side of Fig. 12.1. The remaining degeneracies in the ground and excited state are removed by the application of a magnetic field as shown in Fig. 12.1.

The energy separations between 2$\bar{\text{A}}$ and $\bar{\text{E}}$ states for V^{2+}, Cr^{3+}, and Mn^{4+} are 12 cm^{-1}, 29 cm^{-1}, and 80 cm^{-1}, respectively, and at low temperature a significant population can be detected only in the lower $\bar{\text{E}}$(^{2}E) state. In consequence ODMR measurements in the temperatures range 1.6–4.2 K detect spin-state population changes only in this state. Similarly for Cr^{3+} ions in tetragonal symmetry in magnesium oxide, the ^{2}Eu–^{2}Ev splitting is 90 cm^{-1} and magnetic resonance is studied only in the lower-lying ^{2}Ev state. In magnesium oxide the Cr^{3+} and V^{2+} ions also occupy octahedral sites so that ODMR studies are carried out in the orbitally degenerate ^{2}E state. This state is split by the random strains present in all magnesium oxide crystals, whereas the ^{2}Ev state of 3d^3 ions in tetragonal sites in magnesium oxide and the $\bar{\text{E}}$(^{2}E) state of 3d^3 ions in aluminium oxide cannot be further split by random strain fields. This sensitivity of the unsplit ^{2}E level to random strains has a significant effect on the shape and width of ODMR lines of V^{2+} and Cr^{3+} ions in octahedral sites in magnesium oxide (Chase 1968*a*; McDonagh *et al.* 1980*a*).

Figure 12.2 shows the relative intensities of the linearly (σ) and circularly polarized ($\sigma_\pm$) Zeeman components of the $\bar{\text{E}}$(^{2}E)$\rightarrow$^{4}A$_2$ transition of 3d^3 ions in aluminium oxide in a magnetic field along the trigonal axis. These electric dipole transitions are induced by an odd-parity crystal field of type $T_{1u}a_0$. The strengths of these electric dipole luminescence transitions between

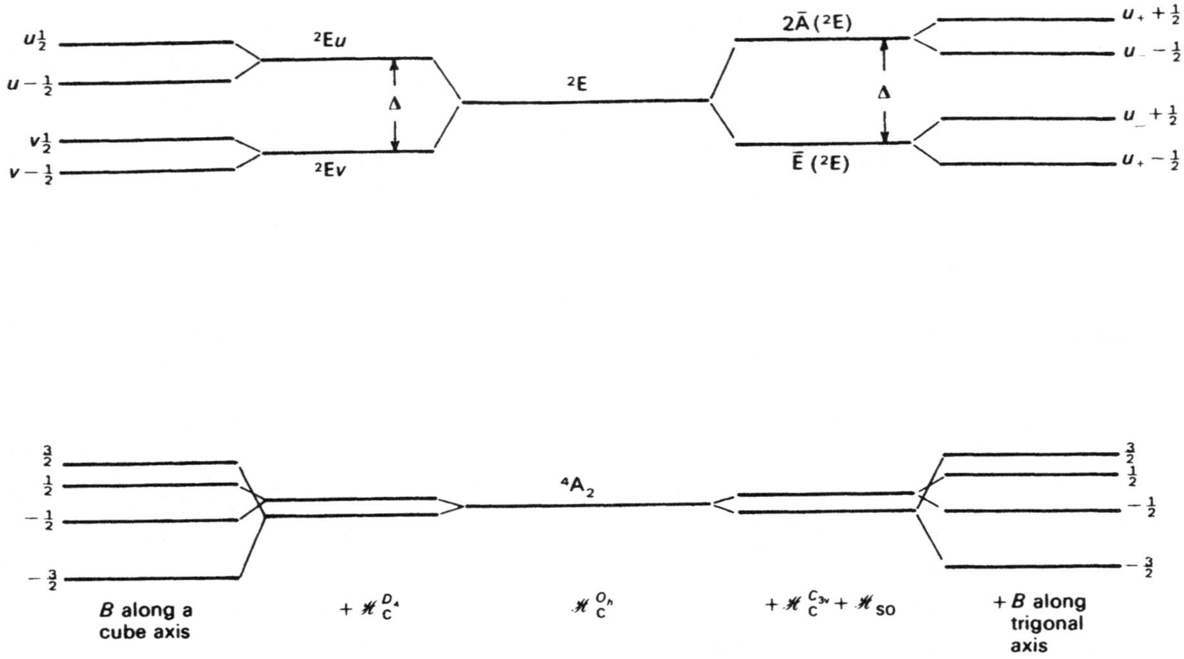

FIG. 12.1. Energy-level schemes of the ^2E and $^4A_2$ states of a $3d^3$ ion in octahedral, tetragonal, and trigonal crystal fields. In general the splitting of the ^2E level in the low symmetry cases is much larger than that of the $^4A_2$ level. The removal of spin degeneracy by a magnetic field applied along the symmetry axis is shown for the lower symmetry crystal field levels. In the octahedral case the $^2Eu\frac{1}{2}$ and $^2Ev\frac{1}{2}$ levels cannot be optically resolved from each other, nor can the $^2Eu-\frac{1}{2}$ and $^2Ev-\frac{1}{2}$ levels.

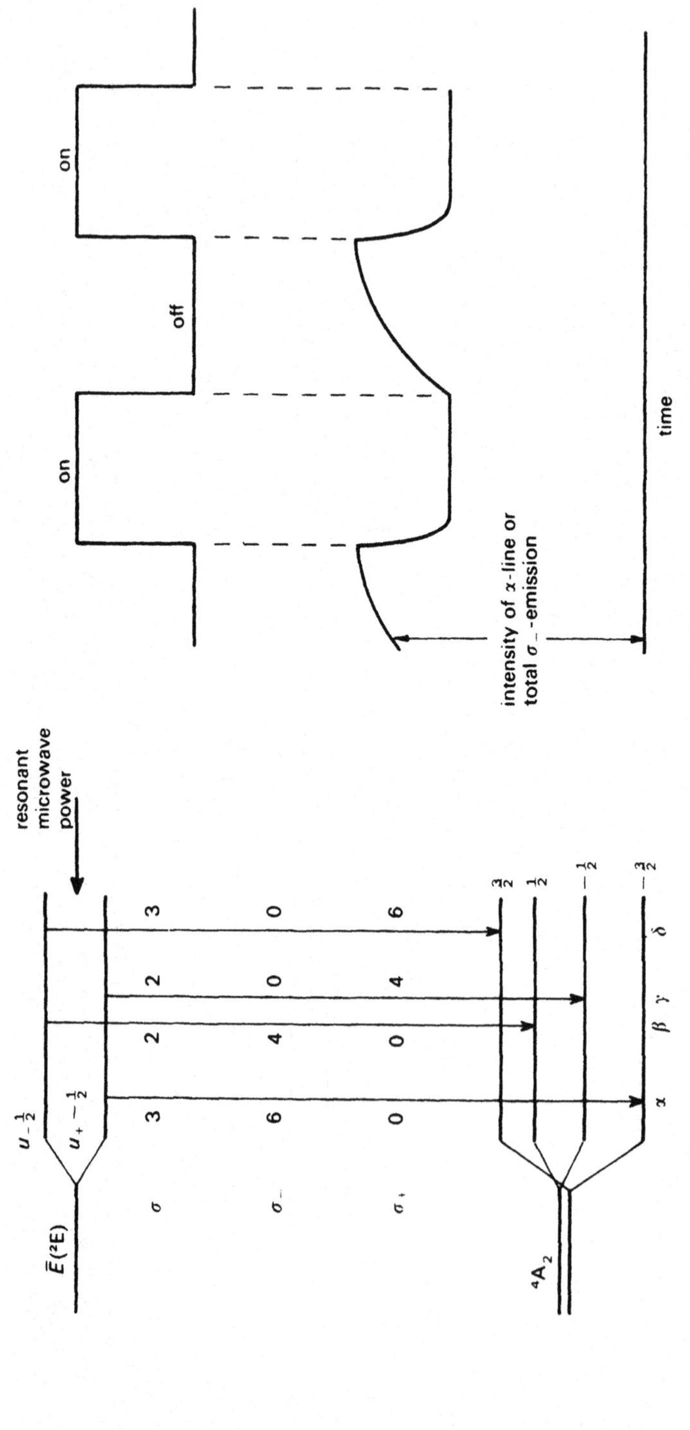

FIG. 12.2. The relative intensities of the Zeeman components of the R_1 emission line of Cr^{3+} ions in alumina in linear σ- and circular $\sigma_{\pm}$-polarizations. By modulating the resonance microwave power the population changes in the $\bar{E}(^2E)$ state are observed as a change in the intensity of an individual component or as a change in the total σ_+- or σ_--light intensity.

$|^2E\phi' M'_s\rangle$ and $|^4A_2 M_s\rangle$ levels are calculated by evaluation of the matrix element

$$\frac{\langle {}^4A_2 M_s|\tilde{\boldsymbol{\mu}}_e \cdot \boldsymbol{E}|{}^4T_2\phi M_s\rangle\langle {}^4T_2\phi M_s|\mathscr{H}_{so}|{}^2E\phi' M'_s\rangle}{E({}^2E) - E({}^4T_2)} \tag{12.1}$$

where $\tilde{\boldsymbol{\mu}}_e \cdot \boldsymbol{E}$ is the relevant effective even-parity electric dipole operator (9.3.2) and $\boldsymbol{E}$ is the electric field of the radiation for the particular polarization involved. The procedure for calculating such matrix elements for linearly polarized transitions was outlined in Chapter 9. For σ_+-polarization, where the direction of propagation is along the magnetic field direction, the odd-parity electric dipole operator in emission, $\sum_j e(x_j - iy_j)/\sqrt{2}$, is of type $T_{1u}a_-$, hence the effective even-parity electric dipole operator is of type $T_{1u}a_- \times T_{1u}a_0$. From Appendix 3B we see that this operator contains a component of type $T_{1u}a_-$ with Clebsch–Gordan coefficient $(i/\sqrt{2})$. With this effective dipole operator the only non-zero matrix elements of $\tilde{\boldsymbol{\mu}}_e \cdot \boldsymbol{E}$ in eqn (12.1) are between $|^4A_2 M_s\rangle$ and $|^4T_2 x_+ M_s\rangle$. The spin–orbit matrix elements between $|^2E\phi' M'_s\rangle$ and $|^4T_2 x_+ M_s\rangle$ states are tabulated by Tanabe and Kamimura (1958). Table 9.5 gives values of these matrix elements appropriate to trigonally-distorted octahedral sites, M_s being defined relative to the z-axis of the trigonal crystal field. The resultant intensities for σ_+-polarized emission are shown in Fig. 12.2, together with the intensities for σ_--emission, in which case the intermediate $|^4T_2 x_- M_s\rangle$ states are involved.

The detection techniques used in measuring the ODMR spectrum can now be described with the aid of Fig. 12.2. We will assume that spin–lattice interaction in the ^2E state is efficient relative to radiative decay to the ground state. In consequence the population of the $E(u_+ - \frac{1}{2})$ level exceeds that in the $\bar{E}(u_- + \frac{1}{2})$ level at low temperatures, resulting in the α-component of the linear polarized σ pattern being more intense than the δ-component. In addition, the *total* amount of circularly polarized σ_- emission exceeds that of the σ_+ radiation. When microwave power resonant with the Zeeman splitting of the $\bar{E}(^2E)$ state is absorbed by the sample, spin–flip transitions tend to equalize the populations in the two levels. In consequence the population in the $\bar{E}(u_+ - \frac{1}{2})$ level is reduced, so reducing the intensity of the σ-polarized α-component. There is a corresponding increase in the intensity of the δ-component. If the detection system is capable of resolving the individual components in the Zeeman pattern then the absorption of microwave power is measured via the change in intensity of α- or δ-components. This *high-resolution detection technique* was used in measuring the ODMR spectra of both Cr^{3+} and V^{2+} ions in aluminium oxide.

Even when the individual Zeeman components cannot be resolved the ODMR spectrum may still be recorded by monitoring changes in the total intensity of the circularly polarized emission from the $\bar{E}(^2E)$ state. Reduction in population of the lower-lying $\bar{E}(u_+ - \frac{1}{2})$ level by microwave absorption leads to a decrease in intensity of the σ_- component, with a concomitant increase in the amount of σ_+-polarized light. Observation of ODMR using the *circular polarization detection scheme* was particularly suited to detection of ODMR in the $\bar{E}(^2E)$ state of Mn^{4+} ions in aluminium oxide where the individual Zeeman components could not be directly resolved. The ESR spectrum of the $\bar{E}(^2E)$ state of Mn^{4+} ions in aluminium oxide measured with the static magnetic field, *B*, applied along the trigonal axis of the crystal is shown in Fig. 12.3. The spectrum shows the typical hyperfine structure of the 100 per cent abundant ^{55}Mn isotope which has a nuclear spin $I = \frac{5}{2}$. The magnetic fields at resonance were fitted to the spin Hamiltonian

$$\mathcal{H} = \mu_B \boldsymbol{B}.\hat{\boldsymbol{g}}.\boldsymbol{S}' + \boldsymbol{I}.\hat{\boldsymbol{A}}.\boldsymbol{S}' \tag{12.2}$$

in which $S' = \frac{1}{2}$ is the fictitious spin of the $\bar{E}(^2E)$ state. Since the selection rule

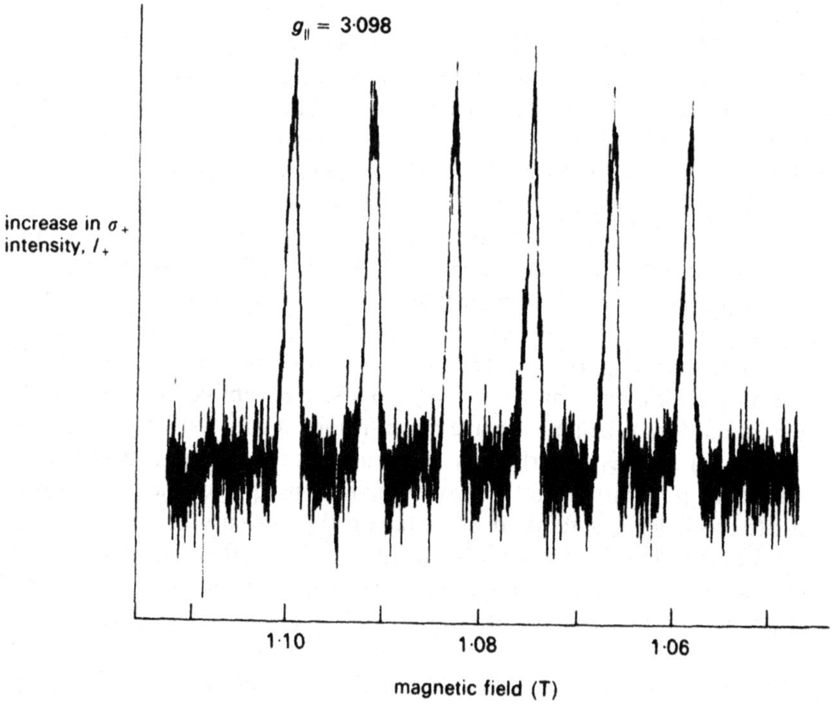

FIG. 12.3. The ODMR spectrum of Mn^{4+} ions in aluminium oxide measured with magnetic field, *B*, along the crystal *c*-axis at $T = 2.0$ K and microwave frequency $\nu = 46.758$ GHz. The signal is detected as an increase in total σ_+-intensity when resonant absorption of microwaves occurs. The six-line hyperfine structure is due to the 100% abundant ^{55}Mn nucleus which has $I = \frac{5}{2}$. (After Imbusch *et al.* 1967.)

for ESR transitions is $\Delta M_s = \pm 1$, $\Delta M_I = 0$, microwave absorption occurs at resonance fields, B, given by

$$B = h\nu / g_{\parallel} \mu_B + M_I A_{\parallel} \tag{12.3}$$

giving the pattern of $2I + 1 = 6$ hyperfine components. Similar hyperfine splittings are observed for V^{2+} and for isotopically enriched $^{53}Cr^{3+}$ in aluminium oxide.

In eqn (12.2) S', the fictitious spin of the $\bar{E}(^2E)$ state, taken as $-\frac{1}{2}$ for the *upper* level and $+\frac{1}{2}$ for the *lower* level, is opposite in sign to the actual spin. The reason for using this fictitious spin is that the upper and lower levels respectively transform as the Γ_5 and Γ_4 double-valued irreducible representations of the C_3 group, i.e. they transform as the $(+\frac{1}{2}, -\frac{1}{2})$ and $(+\frac{1}{2}, +\frac{1}{2})$ spin states of the free electron. Hence the use of the $-\frac{1}{2}$ and $+\frac{1}{2}$ effective spin labels (Imbusch *et al.* 1967). The spin Hamiltonian parameters of the $\bar{E}(^2E)$ states of V^{2+}, Cr^{3+} and Mn^{4+} in alumina are given in Table 12.2. We note the following significant results:

(1) $g_{\parallel}$ is noticeably greater than 2.0023,
(2) $g_{\perp}$ is very small, and
(3) the sign of $A_{\parallel}$ may be positive or negative.

Table 12.2
Spin Hamiltonian parameters for the $\bar{E}(^2E)$ state of $3d^3$ ions in aluminium oxide.

Ion	$g_{\parallel}$	$\Delta g_{\parallel}$	$g_{\perp}$	$A_{\parallel}$ (10^{-4} cm^{-1})
V^{2+}	(−) 2.2198	0.218	0.05	+ 46.3
Cr^{3+}	(−) 2.445	0.443	0.06	< 2
Mn^{4+}	(−) 3.0959	1.094	0.3	−123.3

The negative values for $g_{\parallel}$ follow from the use of the fictitious spin, S', and have no deeper significance. Since the hyperfine constant $A_{\parallel}$ is also referred to the fictitious spin, its sign is also opposite to that for real electron spins. The g-shifts, $\Delta g_{\parallel}$ and $\Delta g_{\perp}$, are related to the small amount of orbital angular momentum in the $\bar{E}(^2E)$ state. Since there are no matrix elements of L within the pure ^2E state the g-shift is due to the mixing of orbital angular momentum from $^2T_1$ and $^2T_2$ states into the ^2E state so giving a positive g-shift. In terms of the crystal field splittings between $\bar{E}(^2E)$ and the $2\bar{A}(^2E)$, $^2T_1$, and $^2T_2$ states (Δ, δ_1, and δ_2, respectively) the shift in $g_{\parallel}$ is given by

$$\Delta g_{\parallel} = 16 \frac{\Delta^2}{\zeta^2} \left(\frac{\delta_2}{\delta_1} \right) \tag{12.4}$$

where ζ is the one-electron spin–orbit coupling parameter, reduced by the crystal field. For V^{2+}, Cr^{3+}, and Mn^{4+} ions the experimental g-shifts require the values of ζ be roughly 100, 175, and 230 cm^{-1}, respectively, being reduced from the free-ion values by a factor of 2.5–3.0.

12.1.2 Optical detection of spin–lattice relaxation

The ODMR technique is easily modified to measure the spin–lattice relaxation time, T_1, between the Zeeman components of the ^2E-state (Geschwind *et al.* 1965; Imbusch *et al.* 1967). The principle of the technique may also be described by reference to the selection rules shown in Fig. 12.2. For illustrative purposes we consider the measurement of the intensity of the α-component, which is determined by the population in the lower $\bar{E}(u_+ - \frac{1}{2})$ level. With the magnetic field at resonance and the sample under continuous illumination the luminescence is monitored while the microwave power is switched on and then off. The absorption of microwave power will reduce the intensity of the α-component. After the microwave power is switched off the population of the $|u_+ - \frac{1}{2}\rangle$ level recovers with some characteristic relaxation time, which is measured from the time dependence of the recovery of the intensity of the α-component, as also shown schematically in Fig. 12.2. Monitoring the total σ_- light intensity will similarly permit the observation of the relaxation rate, as an examination of the selection rules shows. To obtain accurate data it is usually necessary to signal-average over many on/off cycles. Analysis of the appropriate rate equations (Geschwind *et al.* 1965; Imbusch *et al.* 1967) shows that the characteristic recovery time, τ, depends upon the spin–lattice relaxation time, T_1, and the radiative lifetime, τ_R, according to

$$\frac{1}{\tau} = \frac{1}{\tau_R} + \frac{1}{T_1}. \tag{12.5}$$

Thus if τ_R is measured independently T_1 is determined by a measurement of τ. We have already shown (Fig. 5.25) the observed results for Cr^{3+} in aluminium oxide and interpreted the T_1 versus temperature variation in terms of an Orbach process involving the higher-lying $2\bar{A}$ state. Thus $T_1 = \frac{1}{4}T_{2\bar{A}-\bar{E}}\exp(\Delta/kT)$, where $T_{2\bar{A}-\bar{E}}$ is the lifetime of the $2\bar{A}$ state against the spontaneous spin-flip transition from $2\bar{A}(^2E)$ to $\bar{E}(^2E)$ with emission of a phonon of energy Δ. The direct process relaxation is very weak in the $\bar{E}(^2E)$ level. In general the ^2E state does not interact strongly with lattice vibrations; indeed the *pure* ^2E state is unaffected by vibrations of either E or T_2 symmetry. The relaxation process between $2\bar{A}(^2E)$ and $\bar{E}(^2E)$ occurs only because the higher-lying $^2T_2$ state is mixed with the ^2E state by the combined effects of spin–orbit coupling and the trigonal crystal field.

In Table 12.3 we compare the experimental and theoretical values of the lifetime of the $2\bar{A}$ state for V^{2+}, Cr^{3+}, and Mn^{4+} in alumina. Since $T_{2\bar{A}-\bar{E}}$

Table 12.3

Properties of the ^{2}E-state of 3d^3 ions in ruby

	$\Delta(\text{cm}^{-1})$	$T^{(\text{theor})}_{2\bar{A}-\bar{E}}$ $(10^{-9}\,\text{s})$	$T^{(\text{exp})}_{2\bar{A}-\bar{E}}$ $(10^{-9}\,\text{s})$
V^{2+}	12.3	170	210
Cr^{3+}	29	30	15
Mn^{4+}	80	2	0.64

represents the relaxation from $2\bar{A}$ by spontaneous emission of a phonon of energy Δ this relaxation time should vary as Δ^{-3} because of the ω^2 dependence of the density of states at these values of Δ. The value of $T_{2\bar{A}-\bar{E}}$ will also depend on the sensitivity of the $\bar{E}(^2E)$ and $2\bar{A}(^2E)$ levels to strain. There is reasonably good agreement between estimated and measured values of $T_{2\bar{A}-\bar{E}}$ for V^{2+} and Cr^{3+} (Table 12.3). The larger discrepancy for Mn^{4+} may reflect the dispersion in the ω versus k curve at the large value of $\Delta(=80\,\text{cm}^{-1})$ appropriate to this ion.

12.1.3 ODMR of 3d^3 *ions in magnesium oxide*

Magnesium oxide has the cubic rock-salt structure and in the absence of local charge compensation Cr^{3+} ions substituted on the Mg^{2+} sublattice occupy sites of octahedral symmetry where the ^2Eu and ^2Ev components of the ^{2}E state are degenerate. The $^2E \rightarrow {}^4A_2$ fluorescence is then observed as a single R-line (Section 9.3), the width of which is strongly sample dependent due to random strain splittings of the ^{2}E state. In consequence the ODMR spectra in the ^{2}E state of 3d^3 ions in octahedral sites show an interesting interplay between magnetic and strain interactions. The most suitable procedure for observation of ODMR in the ^{2}E state of both Cr^{3+} and V^{2+} ions in octahedral sites in magnesium oxide is to measure the circular polarization of emission along the direction of the magnetic field (Chase 1968*a*). Figure 12.4 shows the results of such measurements for Cr^{3+} in magnesium oxide, recorded with the applied magnetic field along the [100] and [111] crystal directions. The spectra are broad and anisotropic. The line at $g = 1.98$ is due to ODMR in the $^4A_2$ *ground* state; it is observed only within a few degrees of the [100] orientation. The weak resonance at $g = 2.00$ is largely due to the ^2Eu state and that at 1.81 to the ^2Ev state. These resonances are symmetrically spaced about the single line observed in the ODMR spectrum when the magnetic field is directed along the crystal [111] direction. The three features which must be explained are the shape and orientation dependence of spectrum, the magnitude of the signal and the appearance of the ground-state signal in the ODMR spectrum.

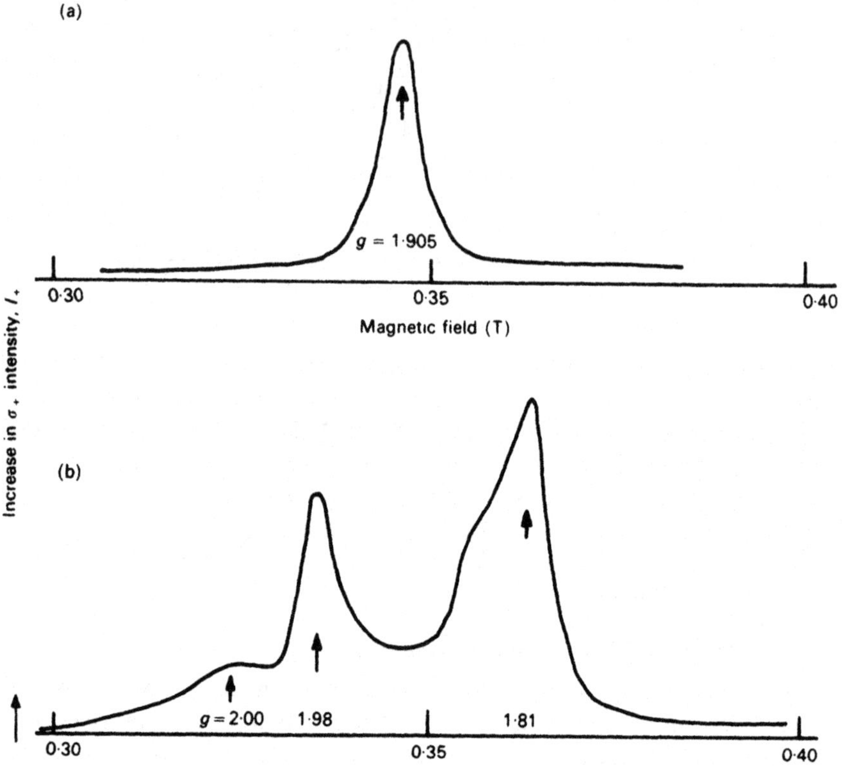

FIG. 12.4. The ODMR spectrum of Cr^{3+} ions in octahedral symmetry sites in magnesium oxide measured at $T=1.6$ K, microwave frequency $\nu \simeq 9.2$ GHz with magnetic field B along (a) a $\langle 111 \rangle$ axis and (b) a $\langle 100 \rangle$ axis.

Both the lineshape of the ODMR spectrum and its anisotropy arise out of the combined effects of admixtures of higher-lying states into ^{2}E and the random strain splitting of this state. In the absence of the usual perturbation terms which admix orbital momentum into the ^{2}E state all second-order terms of the form $\langle ^2\mathrm{E}|L|^2\Gamma\rangle\langle ^2\Gamma^2|\mathscr{H}_{\mathrm{so}}|^2\mathrm{E}\rangle$ are zero because none of the spin-doublet states, $^2\Gamma$, are coupled to ^{2}E by both L and $\mathscr{H}_{\mathrm{so}}$. However, Chase (1968*a*) showed that significant admixtures are produced by third-order interactions involving $\mathscr{H}_{\mathrm{c}}^{\mathrm{O_h}}$ and $\mathscr{H}_{\mathrm{so}}$, the most important of which are:

$$\frac{\langle ^2\mathrm{E}|\boldsymbol{L}|^2\mathrm{T}_1\rangle\langle ^2\mathrm{T}_1|\mathscr{H}_{\mathrm{c}}^{\mathrm{O_h}}|^2\mathrm{T}_1'\rangle\langle ^2\mathrm{T}_1'|\mathscr{H}_{\mathrm{so}}|^2\mathrm{E}\rangle}{(\mathrm{E}(^2\mathrm{T}_2)-\mathrm{E}(^2\mathrm{E}))(\mathrm{E}(^2\mathrm{T}_1')-\mathrm{E}(^2\mathrm{E}))}, \tag{12.6a}$$

$$\frac{\langle ^2\mathrm{E}|\boldsymbol{L}|^2\mathrm{T}_1\rangle\langle ^2\mathrm{T}_1|\mathscr{H}_{\mathrm{so}}|^2\mathrm{T}_2\rangle\langle ^2\mathrm{T}_2|\mathscr{H}_{\mathrm{so}}|^2\mathrm{E}\rangle}{(\mathrm{E}(^2\mathrm{T}_2)-\mathrm{E}(^2\mathrm{E}))(\mathrm{E}(^2\mathrm{T}_1')-\mathrm{E}(^2\mathrm{E}))}. \tag{12.6b}$$

These expressions, in which the $E(^2\Gamma)$ are the energies of the $^2\Gamma$ states, show which higher-lying states are admixed into the ^{2}E state by the combined effects of the orbital angular momentum operator, L, the spin–orbit interaction, $\mathscr{H}_{so}$, and the cubic crystal field, $\mathscr{H}_c^{O_h}$. The orientation dependence of the Zeeman splitting of the ^{2}E state and of the g-values for orientations of the magnetic field, B, having direction cosines (lnm) relative to the cubic axes were also calculated by Chase (1968*a*), who showed that the g-value of the ^{2}E state varies with orientation according to

$$g = 2 + 4\alpha[1 \pm (l^4 + m^4 + n^4 - l^2m^2 - l^2n^2 - m^2n^2)^{\frac{1}{2}}] \tag{12.7a}$$

where α is an empirical parameter calculated by summing over all interactions of the type given by eqns (12.6a) and (12.6b). The net result gives $\alpha = -0.025$, and the g-value for magnetic field orientations in a $\{110\}$ plane is given by

$$g = 2 + 4\alpha\,[1 \pm (\cos^2\theta - \tfrac{1}{2}\sin^2\theta)], \tag{12.7b}$$

θ being the angle between B and a $\langle 001\rangle$ axis. Thus one expects a single line when the magnetic field is parallel to a $\langle 111\rangle$ direction with $g = 1.90$; the experimental value is 1.905 (Fig. 12.4(a)). When $B \parallel \langle 001\rangle$ there are two resonances at magnetic fields corresponding to the theoretical values $g = 2.00$ for ^{2}Eu and $g = 1.81$ for ^{2}Ev in good agreement with experimental peaks shown in Fig. 12.4(b). The strain-broadened width of the Cr^{3+} R-line in magnesium oxide is strongly sample dependent. To first order the ^{2}E level is unaffected by lattice strains; the observed strain sensitivity comes from the configuration mixing and spin–orbit coupling. Strain is assumed to stabilize a particular linear combination of ^{2}Eu and ^{2}Ev orbitals, and the magnetic field splitting of such combinations is determined by the g-values

$$\begin{aligned} g_1 &= \cos^2\gamma\, g(^2\mathrm{E}u) + \sin^2\gamma\, g(^2\mathrm{E}v) \\ g_2 &= \sin^2\gamma\, g(^2\mathrm{E}u) + \cos^2\gamma\, g(^2\mathrm{E}v) \end{aligned} \tag{12.8}$$

where γ is related to the strain distribution at a given site (Chase 1968*a*). For all sensible strain amplitudes eqn (12.8) leads to a resonance shape with singularities at the g-values given by eqn (12.7b), in accord with the experimental data.

The magnitude of the microwave-induced changes in the R-line luminescence is calculated by evaluating the theoretical intensities of the various Zeeman components, as we have already done for $3d^3$ ions in aluminium oxide. For $3d^3$ ions in octahedral sites in magnesium oxide, luminescence transitions between $|^2\mathrm{E}\phi' M_s'\rangle$ states are magnetic dipole transitions with matrix element

$$\frac{\langle ^4\mathrm{A}_2 M_s | \boldsymbol{\mu}_\mathrm{m} \cdot \hat{\varepsilon}_\mathrm{B} | ^4\mathrm{T}_2 \phi M_s\rangle \langle ^4\mathrm{T}_2 \phi M_s | \mathscr{H}_{so} | ^2\mathrm{E}\Phi' M_s'\rangle}{E(^2\mathrm{E}) - E(^4\mathrm{T}_2)} \tag{12.9}$$

where $\boldsymbol{\mu}_m \cdot \hat{\varepsilon}_B$ is the relevant magnetic dipole operator for the polarization in question. The matrix elements $\langle {}^2E\phi' M'_s | \mathscr{H}_{so} | {}^4T_2 \phi M_s \rangle$ listed in Table 9.5 were evaluated by Sugano *et al.* (1960) using spin states defined with respect to a magnetic field direction with angles θ, ϕ relative to the cubic z-axis. For circular polarization the light is assumed to propagate along the magnetic field direction. When the magnetic field is along the cubic [001] direction the magnetic dipole operator for the emission of σ_+ radiation has the form $(e/2m)(l_x - il_y)/\sqrt{2}$, using the convention adopted in Chapter 4. This is an operator of type $T_{1g}(\alpha - i\beta)/\sqrt{2}$, which involves only the $|({}^4T_2(\xi + i\eta)/\sqrt{2})M_s\rangle$ state in the matrix element above.

Before giving the result of such a calculation for the luminescence of Cr^{3+} ions in octahedral sites in magnesium oxide we will digress briefly to consider Cr^{3+} ions in lower symmetry sites. Substitutions of Cr^{3+} ions on the Mg^{2+} sublattice creates a charge imbalance which may be compensated by the incorporation of one cation vacancy for every two Cr^{3+} ions. When such a vacancy occupies a nearest neighbour site along a cubic axis the symmetry at the Cr^{3+} ion site is reduced from octahedral to tetragonal. The resultant axial crystal field removes the orbital degeneracy of the ^2E state yielding ^2Eu and ^2Ev states separated in energy by about 90 cm^{-1}. ODMR experiments on the lower ^2Ev state of Cr^{3+} ions in tetragonal sites in magnesium oxide have been reported by McDonagh *et al.* (1980*b*).

In calculating the selection rules for magnetic dipole transitions on Cr^{3+} ions in these tetragonal symmetry sites it is necessary to take into account that the cation vacancies can occupy sites along any of the cubic axes. We choose the z-axis as the direction in which the vacancy occurs; the three possible orientations of such tetragonal centres, labelled A, B, and C, are shown in Fig. 12.5. The magnetic dipole operators for σ_+-polarized light, assuming the static magnetic field to be directed along a cube edge, are also indicated in this figure. The intensities of the individual Zeeman components in emission are calculated by evaluating the relevant matrix elements in eqn (12.9) using the appropriate magnetic dipole operator and from Table 9.4 the spin–orbit matrix elements. Figure 12.6 shows the relative intensities of the individual $^2EuM'_s \rightarrow {}^4A_2M_s$ and $^2EvM'_s \rightarrow {}^4A_2M_s$ transitions averaged over Cr^{3+} ions in A, B, and C sites. The g-values in these states are sufficiently similar as to be indistinguishable by purely optical techniques, as Fig. 12.6 implies. The small crystal field splitting in the $^4A_2$ state of Cr^{3+} ions in tetragonal symmetry has also been neglected in this figure. ODMR in the ^2Ev state of Cr^{3+} ions in tetragonal sites in magnesium oxide are interpreted in terms of these selection rules as discussed below.

For Cr^{3+} ions in octahedral sites in magnesium oxide the same selection rules are obeyed. However, the ^2Eu and ^2Ev states are degenerate so that microwave absorption can be induced in both states, albeit at different magnetic fields determined by the different g-values. The ODMR spectrum of

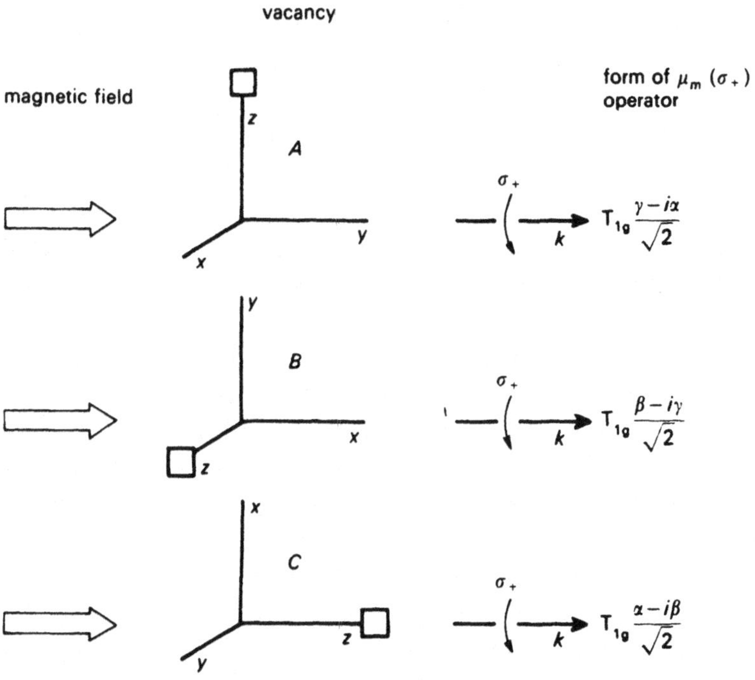

FIG. 12.5. Showing the three equivalent orientations A, B, and C of tetragonal centres in magnesium oxide and the sense of circular polarization of light when the magnetic field, *B*, is applied along a cubic axis. The magnetic dipole operator for the emission of σ_+-radiation in each centre is given.

Cr^{3+} ions in octahedral sites shown in Fig. 12.4 can therefore also be interpreted using these selection rules. Figure 12.7 summarizes the emission intensities from the Zeeman-split levels of the ^2Eu and ^2Ev states for linearly and circularly polarized emission due to magnetic dipole transitions when the magnetic field is parallel to a cubic axis. We also give the relative σ_+ intensities for electric dipole transitions induced by $T_{1u}\gamma$ and $T_{2u}\zeta$ distortions. We draw attention to the contrast between the intensities emitted by the upper and lower Zeeman split levels which are different for circular polarization but identical in linear polarization. As a result, monitoring the total intensity of some linear polarization will not detect the absorption of microwave power; this microwave absorption is, however, detected as a change in either the σ_+ or σ_- intensity.

The ODMR spectrum from the ^2E state of Cr^{3+} in magnesium oxide shown in Fig. 12.4 was measured at 1.6 K and at a microwave frequency of *c.* 9.28 GHz. For this spectrum about 150 mW of Ar^+ ion light at 488 nm was used to excite R-line fluorescence from Cr^{3+} ions in octahedral sites in a

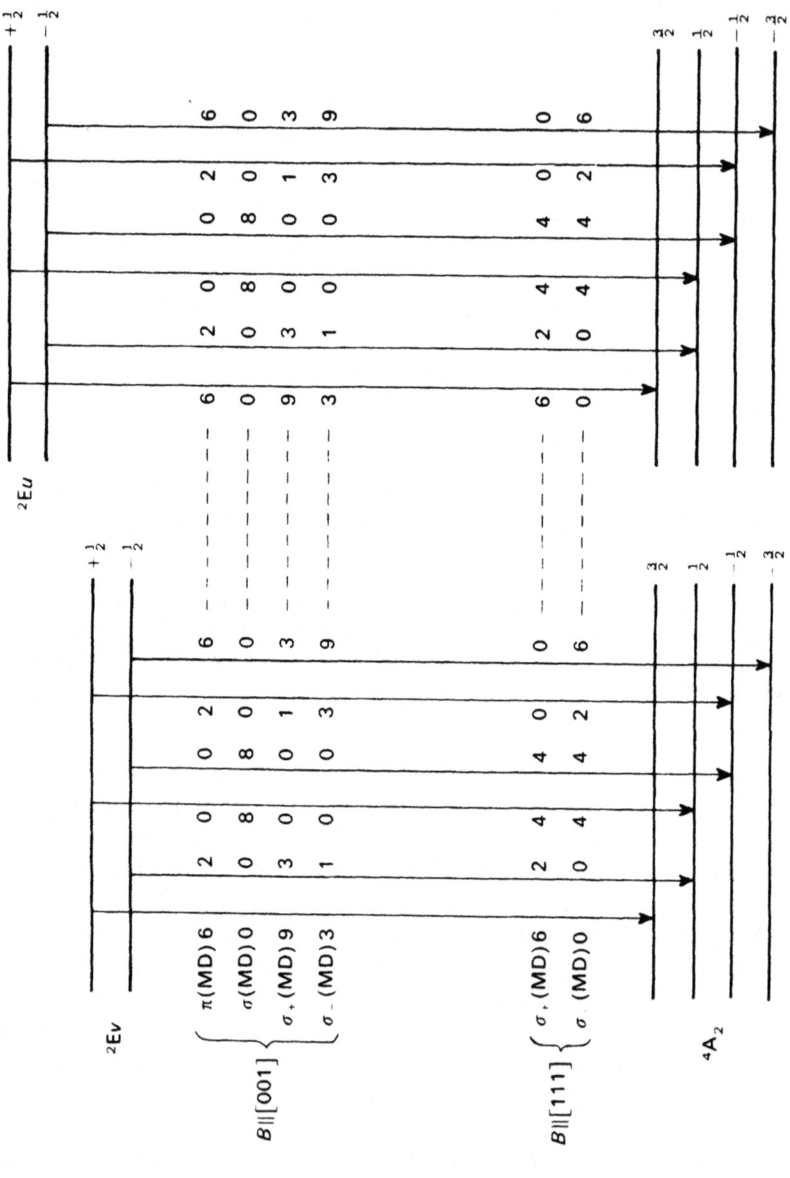

FIG. 12.6. Relative intensities of the Zeeman components for various senses of polarization from $3d^3$ ions in tetragonal sites in magnesium oxide summed over the three equivalent sites A, B, and C (shown in Fig. 12.6.).

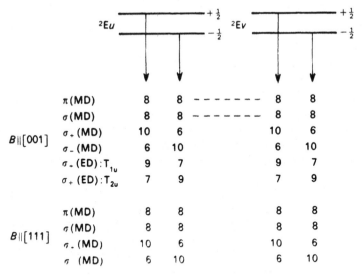

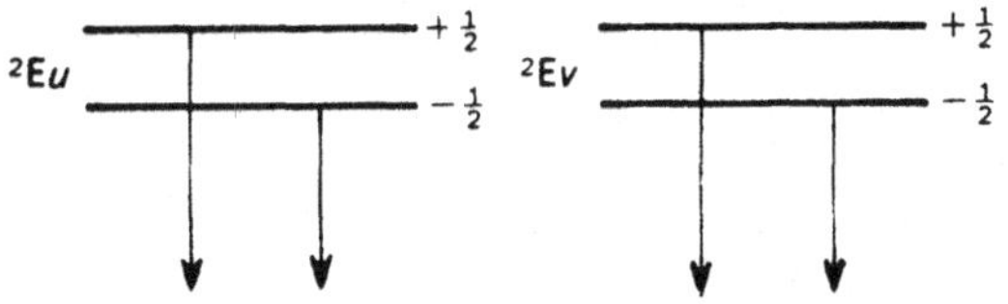

		^{2}Eu		^{2}Ev	
B‖[001]	π(MD)	8	8	8	8
	σ(MD)	8	8	8	8
	σ_+(MD)	10	6	10	6
	σ_-(MD)	6	10	6	10
	σ_+(ED):T_{1u}	9	7	9	7
	σ_+(ED):T_{2u}	7	9	7	9
B‖[111]	π(MD)	8	8	8	8
	σ(MD)	8	8	8	8
	σ_-(MD)	10	6	10	6
	σ (MD)	6	10	6	10

FIG. 12.7. Relative intensities of the total emitted light from the Zeeman split ^{2}Eu and ^{2}Ev levels of 3d^3 ions in tetragonal sites in magnesium oxide in various polarizations and summed over the three sites A, B, and C.

magnesium oxide crystal containing 0.13 at.% of Cr^{3+}. About 50% of the light was absorbed. Given that the radiative lifetime is about $\tau_R \simeq 11$ ms then the ^{2}E state population is estimated to be about 10^{14} cm^{-3}. An analysis of the strength of the ODMR signal leads to the conclusion that at resonance the saturation of the ^{2}E population is about 25% effective assuming spin–lattice relaxation to be fast relative to radiative decay. If $T_1 < \tau_R$ then the spin levels of the ^{2}E state are in thermal equilibrium and the population of the $|^2\mathrm{E}u, +\frac{1}{2}\rangle$ level is less than that of the $|^2\mathrm{E}u, -\frac{1}{2}\rangle$ level. Since $M_s = |-\frac{1}{2}\rangle \rightarrow |+\frac{1}{2}\rangle$ ESR transitions in the ^{2}Eu state tend to equalize these populations and since the $|^2\mathrm{E}u, +\frac{1}{2}\rangle$ level emits mainly σ_+-polarized light then the σ_+ intensity incrases, as Fig. 12.4 shows. Similarly, ESR transitions in the ^{2}Ev state equalize the $|^2\mathrm{E}v, -\frac{1}{2}\rangle$ and $|^2\mathrm{E}v, +\frac{1}{2}\rangle$ populations, so that the σ_+ emission is enhanced and the σ_- emission is decreased.

It is, perhaps, surprising that in Fig. 12.4(b) the $^4A_2$ ground-state ESR spectrum also appears as a change in circularly polarized luminescence intensity, and with only slightly less intensity than the ^{2}Ev excited-state resonance. This ground-state resonance signal varies markedly in intensity with Cr^{3+} ion concentration. In Chase's (1968a) original studies using samples with low Cr^{3+} ion concentrations the ground-state resonance, which was relatively weak and of opposite sense to the ^{2}E state resonances, was interpreted as due to selective reabsorption of the R-line by the Cr^{3+} ions. The resonance line at $g = 1.98$ in Fig. 12.4 is much too strong to be due to such an effect. Since the ^{2}E spin states are expected to be in thermal equilibrium ($T_1 = 600\ \mu s < \tau_R = 11$ ms), selective feeding of the emissive transitions cannot be responsible for the observed ground state resonance. It is more likely that the resonance occurs by mutual spin–flip transitions between pairs of Cr^{3+} ions, one in the ^{2}E state and one in the $^4A_2$ state. Such cross-relaxation is enhanced by a high Cr^{3+} ion concentration.

The R-line of Cr^{3+} ions in sites of octahedral symmetry ($\lambda = 698.1$ nm) is accompanied by a one-phonon vibronic sideband ranging from 710–740 nm (Fig. 12.8). Whereas the zero-phonon line is a purely magnetic dipole transition the one-phonon sideband is expected to be mainly electric dipole in nature, being induced by odd-parity vibrations. In the octahedral environment of the cation site in magnesium oxide electric dipole transitions may be induced on $3d^3$ ions by phonons of T_{1u} or T_{2u} symmetry. The selection rules in this case have been evaluated by Manson and Shah (1977) and McDonagh *et al.* (1980*c*) and were used by the latter authors to discuss the spectral variations in the ODMR spectrum. We note from Fig. 12.8 that the ratio of the intensities from the upper and lower Zeeman levels in σ_+-polarization varies with the nature of the odd-parity vibrations, whether they are of T_{1u} or T_{2u} symmetry. Further, the ratios are reversed in the σ_+- and σ_--polarizations. In consequence, the sign of the ODMR signal will depend on whether the sidebands are induced by T_{1u} or T_{2u} vibrations. The experimental data in Fig. 12.8 (McDonagh *et al.* 1980*c*) show that the T_{2u} modes are more effective than the T_{1u} modes in producing the vibronic sidebands, and that the acoustical sideband is absent from the spectral dependence of ODMR because the T_{1u} and T_{2u} modes make equal contributions to the vibronic structure but their ODMR signals are of opposite sense. Thus the ODMR results indicate that the optical modes of vibration which contribute to the sideband intensity are largely of T_{2u} symmetry. The earlier theoretical work of Sangster (1972) also concluded that the odd-parity vibrations active in inducing electric dipole character in the R-line sideband is mainly of the T_{2u} symmetry.

McDonagh *et al.* (1980*b*) also reported the ODMR spectra of Cr^{3+} ions in tetragonal symmetry sites, the experimental orientation dependence of which is consistent with an axial g-tensor, the principal components of which are $g_\parallel = 1.980 \pm 0.003$ and $g_\perp = 1.724 \pm 0.001$ relative to the impurity-vacancy axis, i.e. the crystal $\langle 100 \rangle$ axes. This spectrum is due to ESR transitions in the $|^2Ev\rangle$ state, which in axial symmetry lies some 90 cm^{-1} lower than $|^2Eu\rangle$ state. The observed ODMR line is narrower than that of the ^{2}E state of cubic centres because random strain effects on the linewidth are no longer a dominant factor. In tetragonal symmetry sites higher-lying orbital momentum states are mixed into the $|^2Ev\rangle$ orbital in second order, rather than in third order as was the case for Cr^{3+} ions in cubic symmetry, and the g-shift is more pronounced.

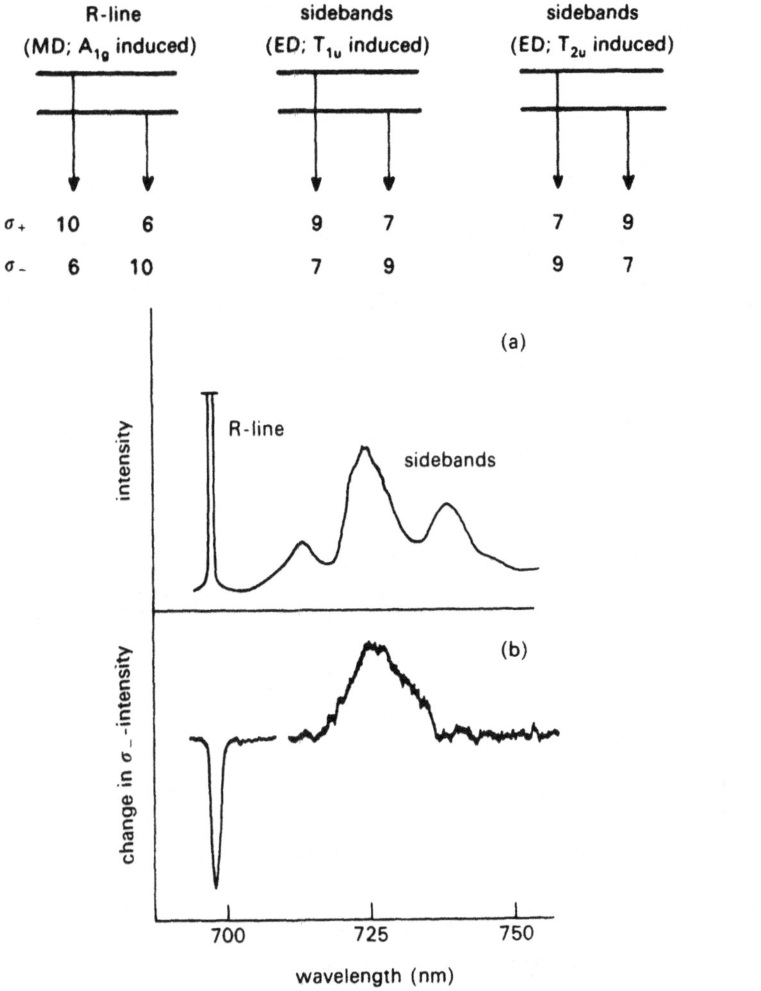

FIG. 12.8. The selection rules for σ_+- and σ_--emission in the R-line and its sideband. (a) shows the optically-pumped R-line and sideband intensities from Cr^{3+} ions in octahedral sites in magnesium oxide, (b) shows changes in σ_--intensities when resonant microwave power is also applied.

12.2 ODMR in the triplet state of defects in insulators

Anion vacancy centres in oxides may contain an even number of electrons (Fig. 12.9). Although the ground states of such even electron centres are diamagnetic (total spin $S = 0$), they also have fluorescent spin triplet states ($S = 1$) which are amenable to study by ODMR (Sections 6.6, 7.4). In calcium

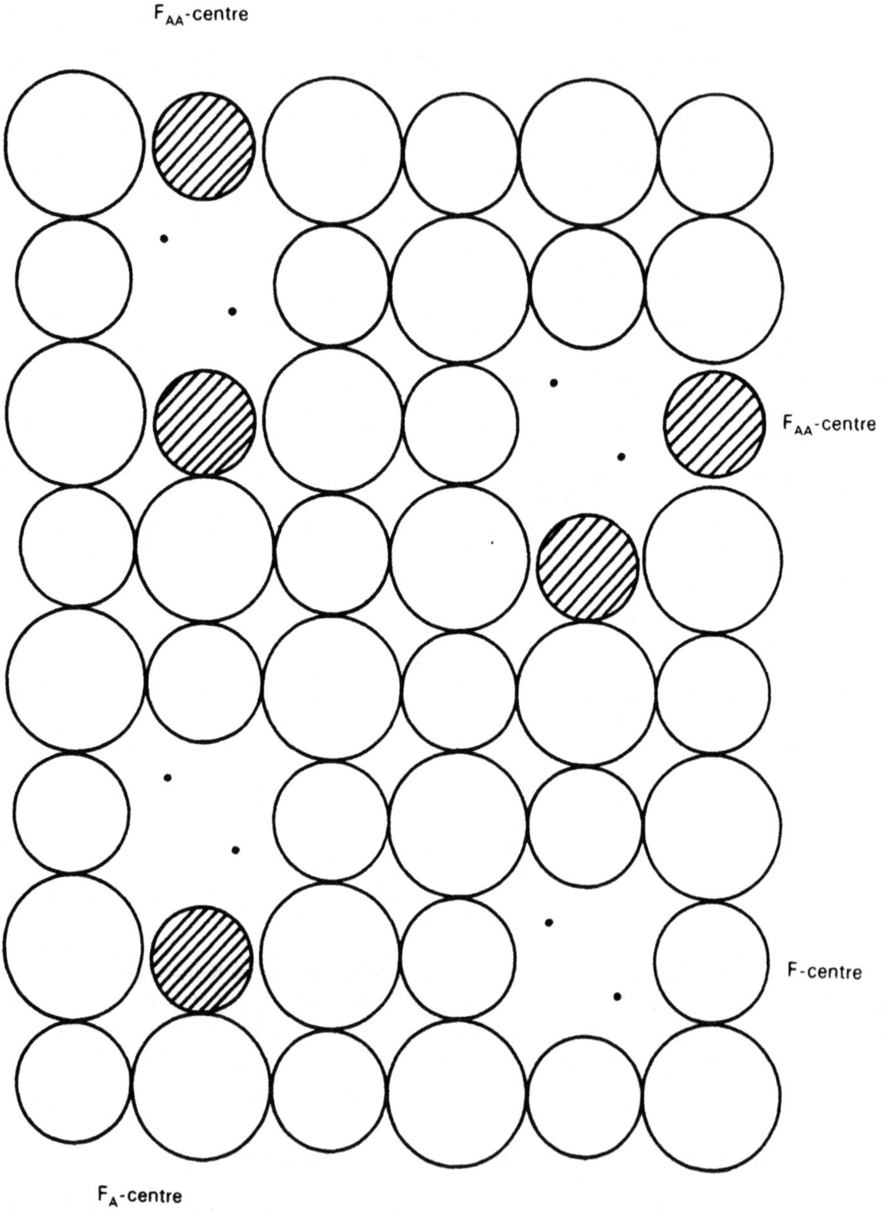

FIG. 12.9. The structures of some two electron anion vacancy defects in calcium oxide.

oxide the F-centre and vacancy-impurity aggregate centres, F_A(Mg)- and F_{AA}(Mg)-centres, have been identified (Fig. 12.9). The F-centre in magnesium oxide is enigmatic and there has been uncertainty in the interpretation of both luminescence and ODMR results. However, in both magnesium oxide and calcium oxide, triplet states due to F_2-type aggregate centres have been identified (Section 7.4.5). Exciton decay also proceeds via triplet → singlet luminescence in alkali halides, silver halides, and complex oxides.

12.2.1 The F*-centre in calcium oxide*

The F-centre in calcium oxide consists of two electrons trapped in the Coulomb field of a negative ion vacancy. The ground state is a spin singlet, $^1A_{1g}$, from which electric dipole absorption transitions are allowed into a $^1T_{1u}$ state derived from the (1s2p) configuration. Such $^1A_{1g} \rightarrow {}^1T_{1u}$ transitions are signified by a strong ($f \simeq 1$) optical absorption band centred at a wavelength $\lambda \simeq 400$ nm. Except at high temperature, de-excitation of this $^1T_{1u}$ state does not proceed via $^1T_{1u} \rightarrow {}^1A_{1g}$ luminescence. Instead, there is efficient non-radiative decay from $^1T_{1u}$ into the triplet $^3T_{1u}$ state also derived from the (1s2p) configuration (Henderson *et al.* 1969). The spin-forbidden $^3T_{1u} \rightarrow {}^1A_{1g}$ transition gives rise to a striking orange fluorescence, which occurs with a radiative lifetime $\tau_R = 3.4$ ms at 4.2 K. The ODMR spectrum of the F-centre and its absorption and emission spectral dependences are depicted in Fig. 6.15; other details are shown in Fig. 12.10(a). With the magnetic field along a $\langle 100 \rangle$ crystal axis there are just four lines. The outer pair of lines is due to centres which have the tetragonal axis parallel to [100] whereas the more intense inner pair of lines is due to two sets of centres which have their fourfold axes parallel to [010] and [001] directions. However, in more general orientations there are six lines. From the variation of the resonant fields with the orientation of the magnetic field in the crystal, Edel *et al.* (1972) identified the spectrum with the $S = 1$ state of a tetragonally-distorted F-centre. The orientation dependence of the spectrum was accounted for using the spin-Hamiltonian (Section 6.5)

$$\mathcal{H} = g\mu_B \boldsymbol{B}.\boldsymbol{S} + D[S_z^2 - \tfrac{1}{3}S(S+1)]. \tag{12.10}$$

Solving for the resonant fields to first order in perturbation theory gives

$$h\nu = g\mu_B B \pm \frac{D}{2}[3\cos^2\theta - 1] \tag{12.11}$$

where θ is the angle between the defect z-axis and the static magnetic field $\boldsymbol{B}$ (Abragam and Bleaney 1970). Hence the centre of gravity of the ODMR spectrum gives the g-value, and the line separation gives $2D$ for defects with $\theta = 0$ and D for $\theta = 90°$. The measured orientation dependence gives $g_\parallel \simeq g_\perp = 1.999$ and $D = 60.5$ mT.

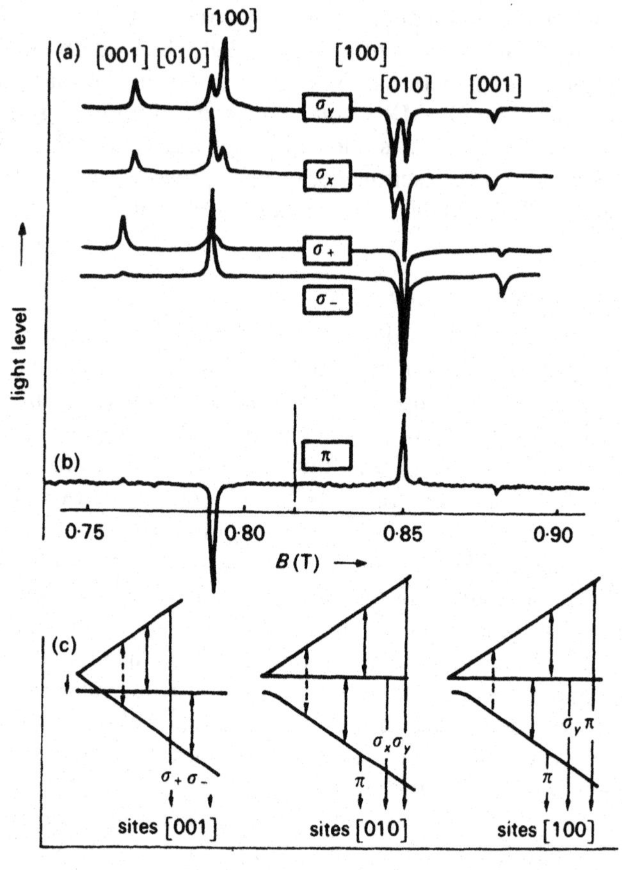

FIG. 12.10. Observation of the ODMR of F-centres in calcium oxide using different senses of circularly polarized emission. (a) shows the change in spectra for σ_x, σ_y, σ_+ and σ_--polarizations (schematic) whereas (b) shows the π-polarized spectrum (facsimile). In (c) are shown in selection rules for the ${}^3T_{1u} \rightarrow {}^1A_{1g}$ emission. The static magnetic field, B, is along the [001] axis of the crystal, except for the σ_x and σ_y spectra in which case there is a slight misalignment. (After Edel *et al.* 1972.)

Figure 12.10(c) shows the selection rules for emission of circularly polarized light by $S = 1$ states in axial crystal fields. The sign of the crystal field parameter, D, has been assumed positive i.e. $E(M_s = \pm 1) > E(M_s = 0)$ in zero field. We denote the populations of the $M_s = 0, \pm 1$ levels as N_0 and $N_{\pm 1}$. The low-field ESR line, corresponding to the $M_s = 0 \rightarrow M_s = +1$ transition, should be observed as an increase in σ_+ light because $N_0 > N_{+1}$ and ESR transitions enhance the $M_s = +1$ level. However, the high-field line is observed as a change in intensity of σ_- light. If spin–lattice relaxation is efficient, i.e. $T_1 < \tau_R$, then the spin states are in thermal equilibrium, $N_0 < N_{-1}$, and ESR transitions depopulate the $|M_s = -1\rangle$ level. Thus the high-field ODMR line is seen as a decrease in the intensity of σ_- light. Evidently the excited-state ESR spectrum confirms two further aspects of the F-centre in these crystals, viz. that for the lowest $^3T_1$ state D is positive and the spin states are in thermal equilibrium. It is worth noting that since the $|M_s = 0\rangle \rightarrow |M_s = \pm 1\rangle$ ESR transitions occur at different values of the magnetic field, ODMR may be detected simply as a change in the emission intensity at resonance; it is not necessary to measure specifically the sense of polarization of the emitted light.

The experimental data clearly establish the tetragonal symmetry of the F-centre in calcium oxide: the tetragonal distortion occurs in the excited $^3T_{1u}$ state due to vibronic coupling to modes of E_g symmetry resulting in a static Jahn–Teller effect (Edel *et al.* 1972). In principle the $^3T_{1u}$ orbitals are coupled to vibrational modes of A_{1g}, E_g, and T_{2g} symmetry. However, the coupling to A_{1g} and T_{2g} modes is weak; indeed the available experimental evidence is consistent with a Huang–Rhys parameter $S_E \geqslant 2.0$ and $S_A + S_T \leqslant 1.5$. The simple discussion of the Jahn–Teller effect given in Chapter 5 to illustrate application in defect spectroscopy is readily adapted to this case. In fact Ham (1965) specifically considered the problem of an orbital triplet coupled to E_g modes of vibration. It is convenient to label the electronic T_1 functions of the F-centre as linear combinations $|x\rangle$, $|y\rangle$, and $|z\rangle$, of the usual angular momentum states, $M_L = 0, \pm 1$. For any given electronic state, say $|x\rangle$, the energy is lowered by the stabilization energy, E_{JT}, due to the tetragonal distortion of the surrounding lattice in the x-direction. The $|y\rangle$ and $|z\rangle$ orbitals are similarly modified in energy by Y and Z distortions respectively. Thus the state energies are not simply dependent on electronic coordinates but also on the nuclear coordinates. In consequence T_1 states are described by three vibronic wavefunctions, which in the linear coupling regime (Chapter 5) are Born–Oppenheimer products of electronic and vibrational wavefunctions. For coupling to E_g modes the vibrational wavefunctions are eigenfunctions of a two-dimensional harmonic oscillator whose equilibrium position corresponds to a distortion along a cube axis (Ham 1965). The normal modes are written as Q_θ and Q_ε in the vibronic Hamiltonian

$$\mathscr{H} = E_0 + \left[\frac{1}{2\mu}(P_\theta^2 + P_\varepsilon^2) + \frac{m^*\omega^2}{2}(Q_\theta^2 + Q_\varepsilon^2)\right] + V_E(Q_\theta \varepsilon_\theta + Q_\varepsilon \varepsilon_\varepsilon) \tag{12.12}$$

where E_0 is the triplet state energy in the absence of distortion, P_θ and P_ε are the momenta conjugate of the normal modes, m^* is the effective mass of the oscillator with angular frequency, ω, V_E is the coupling coefficient between the E_g modes and the centre, and the ε's are electronic operators transforming as the representation E_g. The coordinates of the equilibrium position of the oscillator are then

$$Q_\theta = -V_E \varepsilon_\theta / m^* \omega^2 \quad \text{and} \quad Q_\varepsilon = -V_E \varepsilon_\varepsilon / m^* \omega^2 \tag{12.13}$$

at which point the oscillator has potential energy

$$E = E_0 - V_E^2/2m^*\omega^2 + (n+k+1)\hbar\omega, \tag{12.14}$$

n and k being vibrational quantum numbers with values 0, 1, 2, 3, etc. The quantity $E_{JT} = V_E^2/2m^*\omega^2$ is the Jahn–Teller energy and the vibronic wavefunctions are $\chi_{ikn} = |ikn\rangle$ where $i = x$, y, and z. The vibronic ground state, $|i00\rangle$ remains a triplet which possesses the same symmetry as the original triplet state. The distortion Q raises the orbital degeneracy, displacing the adiabatic potential energy surfaces by $3E_{JT}$ at the three minima, as is illustrated in Fig. 12.11 for one of the $|i00\rangle$ vibronic states.

We now evaluate the matrix elements of a general electronic operator, Γ, between vibronic levels $|xkn\rangle$ and $|ykn\rangle$. If such an operator does not act on the vibrational states ψ_n and ψ_k then

$$\begin{aligned}(xkn|\Gamma|ykn\rangle &= \langle (nk)_x|(nk)_y\rangle\langle x|\Gamma|y\rangle \\ &= K(\Gamma)\langle x|\Gamma|y\rangle.\end{aligned} \tag{12.15}$$

The vibrational overlap integrals $K(\Gamma)$, usually referred to as Ham reduction factors, are limited to the range $0 \leqslant K(\Gamma) \leqslant 1$. Electronic operators having diagonal matrix elements are unchanged by the Jahn–Teller effect, whereas off-diagonal matrix elements are reduced. Hence for electronic oscillators transforming as the irreducible representation A_1 or E_g the coupling coefficients $K(A_1)$ and $K(E_g)$ are unaffected by the Jahn–Teller effect, i.e.

$$K(A_1) = K(E_g) = 1. \tag{12.16}$$

On the other hand, off-diagonal matrix elements of the form

$$\begin{aligned}\langle x\,0\,0|\Gamma|y\,0\,0\rangle &= \langle x|\Gamma|y\rangle\langle 0_x|0_y\rangle_k\langle 0_x|0_y\rangle_n \\ &= \langle x|\Gamma|y\rangle \exp(-3S/2)\end{aligned} \tag{12.17}$$

involve the exponential reduction factors the arguments of which are proportional, through the Huang–Rhys parameter, S, to the square of the differences between the equilibrium positions of the oscillators. In this sense S is often referred to as the Jahn–Teller coupling strength, which has magnitude $S = (-3V_E^2/4\mu^2\hbar\omega^3) = 3E_{JT}/2\hbar\omega$. Since from eqn (12.17)

$$K(T_1) = K(T_2) = \exp(-3S/2) \tag{12.18}$$

we see that electronic operators which transform as the irreducible representations T_1 and T_2 are strongly quenched when S is large. Hence Jahn–Teller coupling to modes of E_g symmetry reduces the local symmetry to

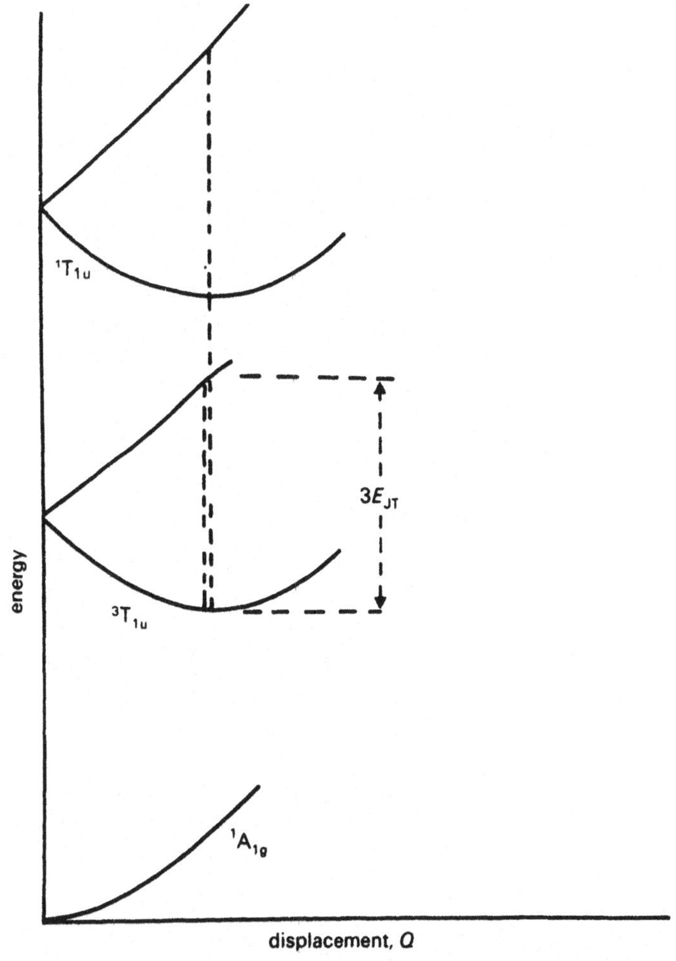

FIG. 12.11. Showing the splitting of the $^1T_{1u}$ and $^3T_{1u}$ energy levels of the F-centre due to Jahn–Teller coupling to vibrations of E symmetry.

tetragonal, the fourfold axis being a crystal ⟨100⟩ direction. Indeed each of the directions [100], [010], and [001] will be equally populated by tetragonally-distorted defects. This situation applies both to $^1T_{1u}$ and $^3T_{1u}$ states (Fig. 12.11). In the latter case, each vibronic state is a spin triplet, so that the lowest vibronic level of $^3T_{1u}$ is ninefold degenerate in the absence of other interactions. Coupling to internal strains of T_{2g} symmetry is strongly reduced by the Jahn–Teller effect. However, coupling to internal strains of E_g symmetry lifts the orbital degeneracy of the $^3T_{1u}$ ground vibronic level as Fig. 12.12 shows. This strain splitting is of the same order of magnitude as the

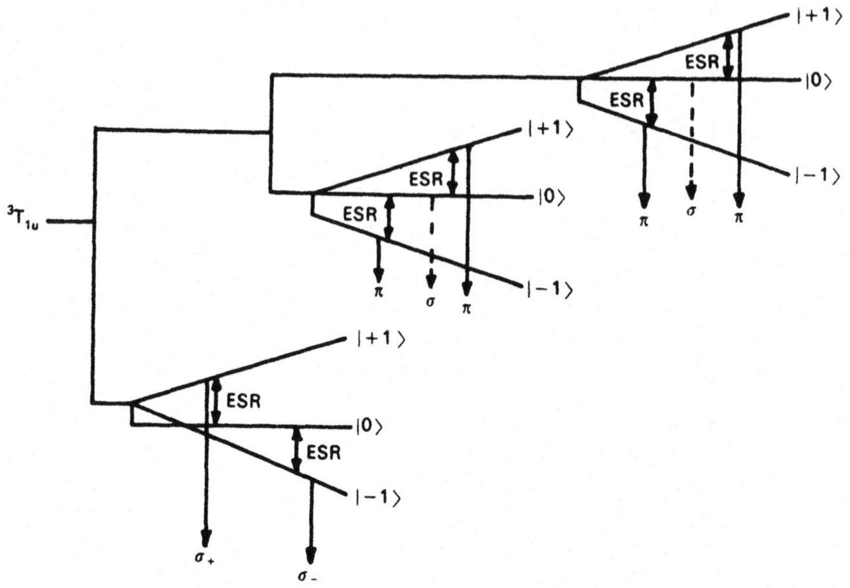

FIG. 12.12. The removal of vibronic degeneracy of the $^3T_{1u}$ state of the F-centre in calcium oxide by internal strains of E_g symmetry. The effects of a static magnetic field, B, parallel to a ⟨100⟩ axis are also shown, as are the selection rules for circularly polarized emission.

optical zero-phonon linewidth, i.e. 5–6 cm^{-1}. The raising of the spin degeneracy by an external magnetic field, and the selection rules for circularly polarized and linearly polarized light are indicated together with the relative transition rates.

Following Edel *et al.* (1972) we represent the coupling between the one-electron operators s_1 and s_2 by the spin-Hamiltonian

$$\mathscr{H} = \mu_B \boldsymbol{B}.\hat{\boldsymbol{g}}.(\boldsymbol{s}_1 + \boldsymbol{s}_2) + g_L \mu_B \boldsymbol{B}.(\boldsymbol{l}_1 + \boldsymbol{l}_2) + D_1 \boldsymbol{s}_1.\boldsymbol{s}_2 + 2\zeta(\boldsymbol{l}_1.\boldsymbol{s}_1 + \boldsymbol{l}_2.\boldsymbol{s}_2) \tag{12.19}$$

in which ζ is the spin–orbit coupling for an electron, g_L is the orbital g-value, and $D_1 \boldsymbol{s}_1.\boldsymbol{s}_2$ represents the spin–spin interaction. The spin–orbit coupling term may be written as

$$\mathscr{H}_{so} = 2\zeta(\boldsymbol{l}_1.\boldsymbol{s}_1 + \boldsymbol{l}_2.\boldsymbol{s}_2) = \zeta[\boldsymbol{L}.\boldsymbol{S} + (\boldsymbol{l}_1 - \boldsymbol{l}_2)\cdot(\boldsymbol{s}_1 - \boldsymbol{s}_2)] \tag{12.20}$$

in which $\boldsymbol{L} = \boldsymbol{l}_1 + \boldsymbol{l}_2$ is the total orbital angular momentum and $\boldsymbol{S} = \boldsymbol{s}_1 + \boldsymbol{s}_2$ is the total spin angular momentum. The first term in eqn (12.19) mixes states within the $^3T_{1u}$ level, whereas the second term mixes the $^1T_{1u}$ and $^3T_{1u}$ levels.

Ham (1965) shows that the contribution to D by spin–orbit coupling within the $^3T_{1u}$ state is given by

$$D_2 = \frac{\zeta^2}{3E_{JT}}[S_z^2 - \tfrac{1}{3}S(S+1)] \tag{12.21}$$

in the limit of strong coupling, whereas the coupling between $^3T_{1u}$ and $^1T_{1u}$ gives rise to a term

$$D_3 = \frac{\zeta'^2}{\Delta}[S_z^2 - \tfrac{1}{3}S(S+1)] \tag{12.22}$$

where ζ' is the effective spin–orbit coupling and Δ is the exchange splitting between the $^1T_{1u}$ and $^3T_{1u}$ levels. The total crystal field splitting, D, in eqn (12.9) is then the sum of three contributions $D = D_1 + D_2 + D_3$, where D_1 is the mutual dipolar coupling between the electron spins. The g-values also calculated within the strong coupling approximation are

$$g_{\parallel} = 2.0023, \quad g_{\perp} = 2.0023 - (2\zeta g_L/3E_{JT}). \tag{12.23}$$

The three parameters D, $g_{\parallel}$ and $g_{\perp}$ determined in the ODMR experiment, taken with the optical band peaks and radiative lifetime of the $^3T_{1u} \rightarrow {}^1A_{1g}$ transition may be used to determine D_3, ζ, ζ', E_{JT} and g_L. The radiative lifetime of the $^3T_{1u} \rightarrow {}^1A_{1g}$ transition, $\tau_R = 3.4$ ms, corresponds to an oscillator strength of $f = 5 \times 10^{-7}$. In fact, the magnitude of f is determined by spin–orbit mixing between $^1T_{1u}$ and $^3T_{1u}$ states and is given by $f \simeq (\zeta'/\Delta)^2$. Furthermore, the maximum value of Δ in eqn (12.21) is just the energy difference between absorption and emission band peaks (Fig. 6.15), i.e. $\Delta \simeq 8500$ cm^{-1}. Hence $\zeta' \simeq 6$ cm^{-1} and we estimate that $D_3 \simeq 4 \times 10^{-4}$ cm^{-1}. The magnitude of the Jahn–Teller energy, E_{JT}, may be estimated from the luminescence bandshape since $S = 3E_{JT}/2\hbar\omega$. Values of $S \simeq 3.5$ and $\hbar\omega \simeq 250$ cm^{-1} have been determined from the temperature dependence of the bandwidth (Henderson *et al.* 1969), so that $3E_{JT} \simeq 1500$ cm^{-1}. If we assume $\zeta \simeq \zeta'$ it appears that $D_2 > 10D_3$. Consequently D_3 may be neglected.

The dipolar coupling, D_1, may be represented by the spin–spin operator

$$\mathscr{H}_d = (4\mu_B^2/r_{12}^3)[\boldsymbol{s}_1 \cdot \boldsymbol{s}_2 - \{3(\boldsymbol{s}_1 \cdot \boldsymbol{r}_{12})(\boldsymbol{s}_2 \cdot \boldsymbol{r}_{12})/r_{12}^2\}] \tag{12.24}$$

which within the $|z, 0, 0\rangle\rangle$ level becomes

$$\mathscr{H}_d = D_1[S_z^2 - \tfrac{1}{3}S(S+1)] \tag{12.25}$$

in which $D_1 = \langle z|\Gamma(E_{g\theta})|z\rangle$ may be calculated if the wavefunctions are known. Wood and Wilson (1977) calculated D_1 using smoothly varying Slater-type orbitals normalized to the neighbouring ion core orbitals. They obtained a value $D_1 = -0.11$ cm^{-1}. Since D_2 is positive it is evident that the measured zero-field splitting D may be positive or negative. For F-centres in calcium oxide $D = +563 \times 10^{-4}$ cm^{-1} hence $D_2 \simeq 1600 \times 10^{-4}$ cm^{-1}. Such a value for D_2 is consistent with $\zeta \simeq 20\ \text{cm}^{-1}$, a not unreasonable value when compared with the value of $\zeta \simeq 30\ \text{cm}^{-1}$ for the F^+-centre. From $\Delta g_\perp = 4 \times 10^{-3} = 2g_L\zeta/3E_{JT}$ we determine $g_L \simeq 0.25$, whereas for spherical harmonics appropriate to atomic wavefunctions $g_L = 1.00$.

Other experiments have been aimed at elucidating the role that internal strains play in dynamic aspects of F-centre spectroscopy. These include measurements of zero-field (Krap *et al.* 1978) and high-field ODMR under uniaxial stress (Le Si Dang *et al.* 1978), spin–lattice relaxation times (Cibert *et al.* 1979) and optically detected spin dephasing and transient nutation (Gravesteijn and Glasbeek 1979). Such measurements make it clear that the energies and wavefunctions of the lowest vibronic state are sensitive to E_g and T_{2g} modes of vibration. Relaxation measurements show that the recovery of the signal after each microwave pulse contains both fast and slow components. Since the slow component decays with the radiative lifetime of the fluorescence, and is independent of the polarization of the monitoring light it is associated with the total population of the $^3T_{1u}$ state. The fast process decays with a time constant of the order of 4.5 μs due to relaxations which occur via jumps between different Jahn–Teller wells in which spin is either conserved or changed. Spin-conserving relaxation processes are induced by the trigonal T_{2g} component of the phonon-induced strain field, whereas phonons of E_g symmetry in the presence of spin–orbit coupling promote relaxations in which the spin state is changed. F-centres in parts of the crystal with large strains relax via the T_{2g} process; the E_g relaxation predominates in a region of small strain. Direct relaxations between spin levels within the same vibronic state occur only with exceedingly low probability. Rather does relaxation occur by Orbach process (Chapter 5) involving higher-lying, vibronic states. These relaxation mechanisms have repercussions on the ODMR intensities. The fast relaxation time $T_1 = 4.5\ \mu$s being very much shorter than $\tau_R = 3.4$ ms, leads to a Boltzmann distribution in the $M_s = 0, \pm 1$ spin levels, responsible for the intensity pattern in Fig. 12.10.

12.2.2 Centres related to the F*-centre in calcium oxide*

The ODMR spectra of the F_A(Mg) and F_{AA}(Mg) centres (Fig. 12.9) have been reported (Dawson *et al.* 1980; Ahlers *et al.* 1982). These centres, respectively, have one and two of the Ca^{2+} ions in the nearest-neighbour cation shell of the F-centre substituted by Mg^{2+} impurity ions. Note that the linear configuration of the F_{AA}-centre, i.e. the $-Mg^{2+}-F-Mg^{2+}$ structure, has only a 25 per cent probability of occurring relative to the right-angle configuration of the F_{AA}-centre. For simplicity we discuss only the linear F_{AA}-centre. The optical absorption spectrum of crystals containing F-, F_A-, and F_{AA}-centres showed the individual absorption bands to be masked by very broad absorption in the range 400–600 nm (Welch 1980). However, the luminescence spectrum clearly shows three luminescence bands on excitation in the region of the F-absorption band. ODMR experiments on such crystals can be carried out in several ways. Light emitted by the crystal can be detected through a broadband filter, allowing all three bands to pass. ODMR spectra then show triplet-state ESR lines from all three defects. Alternatively, appropriately simplified ODMR spectra may be recorded by using suitable interference filters which pass only one of these three luminescence bands. Finally, the luminescence may be detected through a monochromator. These latter techniques enable the ODMR spectra of F-, F_A-, or F_{AA}-type centres to be detected alone. Figure 12.13 shows the ODMR spectrum of F_A-centres detected through a monochromator as the change in the total light inten-

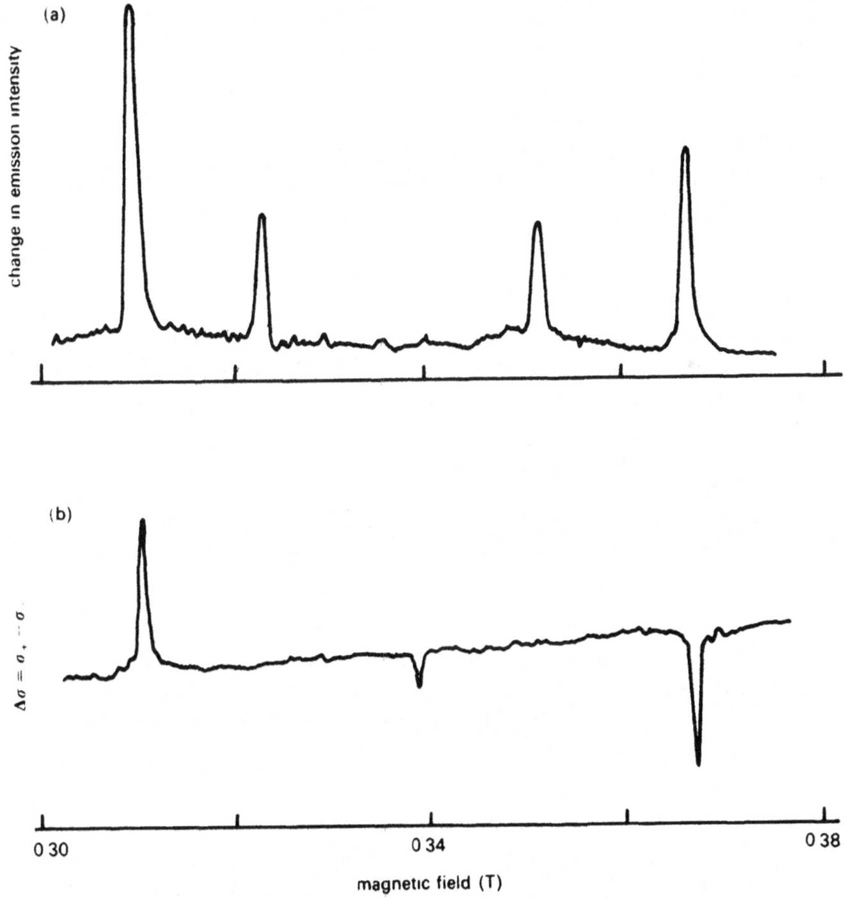

FIG. 12.13. The ODMR spectrum of F_A(Mg)-centres in calcium oxide measured at 1.6 K and $\nu = 9.3$ GHz. In (a) the resonances are detected as a change in total emission intensity whereas (b) shows changes in circular polarization, $\Delta\sigma = \sigma_+ - \sigma_-$.

sity (a) and as the change in the circular polarization of the emission, i.e. $\Delta\sigma = \sigma_+ - \sigma_-$. Note that in (a) an increase is observed in the light level at each value of the resonance field, showing that the spin states are not in thermal equilibrium. The measurement of $\Delta\sigma$ shows that the σ_+-polarized low-field line increases in intensity whereas the high-field line (σ_+-polarized) decreases, confirming that the zero-field splitting, D, is negative. A similar result was observed for F_{AA}-centres.

Table 12.4 compares the optical and ODMR data for F-type centres in calcium oxide. The 400 nm absorption band of the F-centre is due to transitions between the $^1A_{1g}$ ground state and the $^1T_{1u}$ excited state. There is a shift between the absorption peak of F- and F_A-centres arising at least partially from the elastic mismatch of the Mg^{2+} ion in the calcium oxide lattice. The photoluminescence evidence confirms that as a consequence of this elastic interaction a tetragonal distortion occurs at the F_A-centre, which raises the orbital degeneracy of the excited $^1T_{1u}$ state yielding ^1A and ^1E excited states. Polarized luminescence studies (Welch *et al.* 1981) show that the 440 nm band is due to a superposition of the $^1A \rightarrow {}^1A$ and $^1A \rightarrow {}^1E$ transitions, the latter occurring at very slightly higher energies. Presumably a similar situation obtains for the F_{AA}(Mg)-centres. The very small shift between the differently polarized absorption bands of the F_A-centre is related to spatially diffuse singlet excited states of F_A-centres. The shifts in zero-phonon line and band peaks are much more pronounced for the triplet state transitions because in the triplet state the wavefunction is more strongly localized in the vacancy state. The variation in lifetime represents a factor of five decrease in oscillator strength between F- and F_{AA}-centre.

We now consider the F_A-centre (Dawson *et al.* 1980) relative to the F-centre. The magnitude of D_2 can now be written as

$$D_2 = \frac{\zeta^2}{3E_{JT} + \delta} \tag{12.26}$$

where δ is due to the elastic distortion associated with the substitution of a Mg^{2+} for a Ca^{2+} ion in the first shell of cation neighbours of the F-centre. Combining the predicted value of $D_1 = -0.11\ \text{cm}^{-1}$ with the experimental values of $3E_{JT} = 1500\ \text{cm}^{-1}$ and $D = -440 \times 10^{-4}$ leads to $\delta = 3500\ \text{cm}^{-1}$. The same procedure applied to F_{AA}-centres gives $\delta = 6500\ \text{cm}^{-1}$, approximately twice that of a single Mg^{2+} ion. For the F_A- and F_{AA}-centres, fast inter-well jumps of the kind observed for F-centres are precluded since the depth of the potential well is large, in the range $\sim 3500 - 6500\ \text{cm}^{-1}$. Hence T_1 is expected to be very long. Indeed, that both high-field and low-field F_A-centre lines in Fig. 12.13 correspond to an increase in light level indicates that $T_1 > \tau_R$. In this situation population builds up in the $M_s = 0$ level so that both $M_s = 0 \rightarrow \pm 1$ transitions lead to increases in the total emitted light.

Table 12.4
Optical and EPR data of F-type centres in calcium oxide

Centre	Absorption peak (nm)	Emission ZPL (nm)	Emission peak (nm)	τ (ms)	$g_{\parallel}$	$g_{\perp}$	D (mT)	A (mT)
F[a]	400	574	601	3.4	1.9991	1.998	60.3	—
F_A(Mg)[b]	460	656	689	12.5	2.0038	2.0063	−48.3	1.96
F_A(Mo)[c]	460	657.1	689	—	1.998	2.006	50.2	1.99
F_{AA}(Mg)[d]	—	758	783	30	2.005	2.005	−72.5	1.90

[a] Edel *et al.* (1972)
[b] Dawson *et al.* (1980)
[c] Gravesteijn *et al.* (1977)
[d] Ahlers *et al.* (1982)

One additional factor to take into account is that Mg^{2+} has a greater electron affinity than Ca^{2+} so that the F centre electron will spend more of its time in the neighbourhood of Mg^{2+} ions than of Ca^{2+} ions. Since $\zeta(\text{Mg}) < \zeta(\text{Ca})$ and angular momentum generated in the excited states of F-centres is associated with the F-centre wavefunction overlap on the neighbouring cations we expect that $\zeta(F_{AA}) < \zeta(F_A) < \zeta(F)$ as is confirmed by the fluorescence lifetimes (Table 12.4). The change in D_2 due to this mechanism has not been calculated but seems unlikely to be more than *c.* 10–20 per cent. It is apparent that although not precisely determined the contributions to D and to $g_{\perp}$ as discussed above are internally consistent and physically reasonable.

12.2.3 Triplet states of divacancy centres

The basic structural unit for electron excess centres in the alkali halides and alkaline earth fluorides is the F-centre. The simplest aggregate centre in these crystals is the F_2-centre, formed by two nearest neighbour F-centres. Such centres have as many as six absorption bands in the optical region, whereas de-excitation occurs through a single luminescence band and/or non-radiative decay to a metastable triplet state. In the alkali halides and alkaline earth fluorides these triplet states are low lying relative to the ground state, and decay from them is also non-radiative. The triplet state is very long-lived; in KCl of the order 50 seconds and in CaF_2 approximately 1.6 seconds. Thus a steady state population may be built up in the spin triplet state well in excess of that required for conventional ESR detection. Indeed the original observation of these centres, apart from metastable bleaching of the F_2-centre absorption spectrum, came from the traditional ESR method.

Neutral F_2-centres in magnesium oxide and calcium oxide contain four electrons, two for each vacancy. The 361.8 nm zero-phonon line in magnesium oxide, observed in absorption and emission, was assigned to the F_2-centre on the basis of uniaxial stress and polarized luminescence spectroscopies (O'Connell *et al.* 1981) (see Chapters 6 and 7). In additively coloured crystals these centres may be ionized by optical pumping in the F-band producing F_2^+- and F_2^{2+}-centres. The latter is a two-electron centre, with high-lying triplet states which in magnesium oxide give rise to a characteristic blue luminescence. Excitation spectra show that the principal absorption maximum occurs at 320 nm; the band is broad with no resolved structure. The energy shift to the triplet-state emission band is large. However, the triplet emission band shows resolved structure, with a zero-phonon line at 424.7 nm and a Huang–Rhys parameter of $S = 4.0$. The large Stokes shift, weak vibronic coupling, and long-lived fluorescence ($\tau_R = 25$ ms) are characteristic of de-excitation processes involving singlet–triplet intersystem crossings.

For F_2-type centres in rock-salt-structured solids the local symmetry is D_{2h} and the electronic states transform as irreducible representations of this group. Thus the (1s, 1s) ground state of the two electron centre is the spin-singlet ^1A_g state, and the excited states derived from the spin singlet (1s, 2p) configuration transform as the irreducible representations $^1B_{1u}(z)$, $^1B_{2u}(y)$ and $^1B_{3u}(x)$. (By convention z is identified with the molecular axis for a two-centre system, and the x, y, and z-axes are defined to form a right-handed, mutually-perpendicular triad of basis vectors consistent with the local symmetry. In the present case a suitable basis set is $x \| [001]$, $y \| [110]$ and $z \| [\bar{1}10]$.) There is also a spin triplet state $^3B_{1u}$ derived from the (1s, 2p) configuration. Since the components of $S = 1$ transform as rotations about the x, y, and z-axes we see by reference to the character table for D_{2h} that

$$^3B_{1u} = B_{1u} \times (B_{1g} + B_{2g} + B_{3g}) = A_u + B_{3u} + B_{2u}. \qquad (12.27)$$

Selection rules for absorption transitions from the ^1A_g ground state may then be obtained by forming the triple product $^1A_g \times V \times (^1B_{1u} + {}^1B_{2u} + {}^1B_{3u})$ where V is the electric dipole operator. Since the components of the perturbation transform as x, y, and z we have transition moments transforming as, for example,

$$^1A_g \times B_{1u} \times (^1B_{1u} + {}^1B_{2u} + {}^1B_{3u}) = A_g + B_{3g} + B_{2g}. \qquad (12.28)$$

Thus we have an allowed electric dipole transition $^1A_g \rightarrow {}^1B_{1u}$ polarized parallel to the z-axis. Similarly $^1A_{1g} \rightarrow {}^1B_{2u}$, $^1B_{3u}$ are polarized parallel to y and x respectively. Transitions out of the $^3B_{3u}$ state are easily shown to be

$$A_u \rightarrow {}^1A_g \text{ (strictly forbidden) and}$$

$$B_{3u}, B_{2u} \rightarrow {}^1A_g \; x\text{- or } y\text{-polarized, respectively.} \qquad (12.29)$$

The effects of the D_{2h} symmetry on the $^3T_{1u}$ state is to remove the spin degeneracy of the A_u, B_{2u}, and B_{3u} states. A schematic energy-level diagram

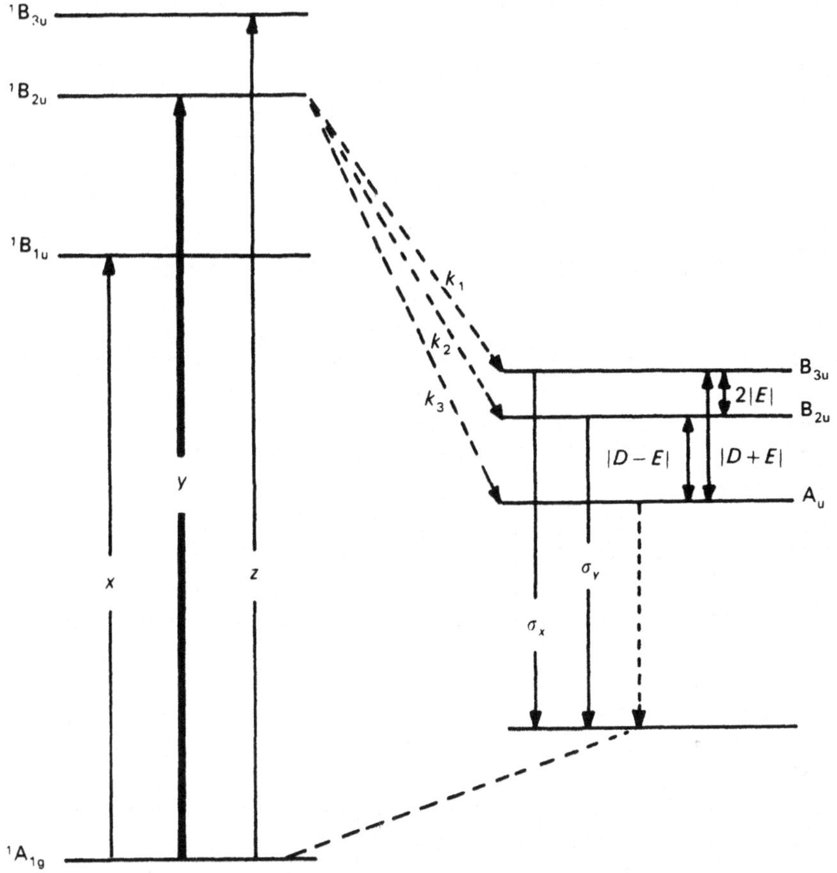

FIG. 12.14. Showing the polarized absorption and emission transitions of F^{2+}-type centres in crystals with the rocksalt crystal structure. (After Henderson *et al.* 1981.)

is shown in Fig. 12.14: also illustrated are the polarizations of the allowed optical transitions in zero magnetic field.

The energy-level splittings are quite small, and this enables optically detected magnetic resonance in the absence of a magnetic field. Zero-field ODMR has been reported for F_2^{2+}-centres in both calcium oxide and magnesium oxide (Gravesteijn and Glasbeek 1979; Krap *et al.* 1978; Glassbeek *et al.* 1980). As Fig. 12.14 shows, three zero-field transitions are anticipated, an expectation which is realized experimentally (Fig. 12.15), the lines in magnesium oxide being observed at 0.4722 GHz, 2.3186 GHz, and 2.7911 GHz. The spin Hamiltonian appropriate to zero-field and orthorhombic symmetry is

$$\mathcal{H} = D[S_z^2 - \tfrac{1}{3}S(S+1)] + E[S_x^2 - S_y^2] \qquad (12.30)$$

which gives the zero-field energy levels as $E(M_s=0)=-\frac{2}{3}D$, $E(M_s=\pm 1)=\frac{1}{3}D\mp E$. Hence the three microwave transitions observed at the frequencies given above lead to $|D|=2.5547\pm 0.0003$ GHz and $|E|=0.2361\pm 0.0002$ GHz. The selection rules in Fig. 12.14 suggest that the $|D-E|$ transitions are detected as a change in y-polarized light, and the $|D+E|$ transitions as a change in x-polarized light. These selection rules are strictly adhered to experimentally. The splittings of these lines in magnetic fields parallel to either [100] and [110] directions are consistent with $g=2.003\pm 0.001$ and principal axes of the D tensor along the crystal axes [001]-x, [$\bar{1}$10]-y, and [110]-z. Thus there are six magnetically inequivalent sites. Assuming dipolar coupling between electron spins at adjacent oxygen sites along a $\langle 110\rangle$ axis results in $D\simeq\frac{3}{2}(g\mu_B)^2/r^3$. For magnesium oxide, $r=0.297$ nm results in $D=-2.97$ GHz, some 16% larger than the experimental value. This discrepancy can be accounted for by a modest (5 per cent) outward relaxation of the lattice around the F_2^{2+}-centre.

These measurements are confirmed by ODMR studies in high magnetic field. The polarizations of the spectra then confirm that the sign of D is negative. Intensity changes at resonance amount to almost 300 per cent, although they show a marked dependence on orientation of the magnetic field relative to defect axes. Analysis of the rate equations in conjunction with the selection rules for ODMR shows that non-radiative de-excitation of the $^3B_{1u}$ state plays an important role in the decay of the triplet state. Such a large signal also guarantees that the spin levels are not in thermal equilibrium, and that their populations are determined by the selection rules for feeding into and emission out of these states. By judicious choice of crystal axes relative to the magnetic field and careful selection of polarizations we may select a particular defect orientation for detailed study. In Fig. 12.14 we indicate laser pumping of the B_{2u} level. Intersystem crossing is then facilitated by lattice vibrations and spin–orbit coupling. The spin–orbit coupling operator and breathing-mode phonons both transform as A_{1g}, so that they will lead to non-radiative decay into the B_{2u} level of the triplet. Phonons of B_{1g} and B_{2g} symmetry will lead to population of the A_u and B_{3u} levels respectively. Thus we identify feeding rates k_1, k_2, and k_3 to the triplet levels in Fig. 12.14. The rate equation analysis leads to the conclusion that $k_2>k_1+k_3$, implying that the intersystem crossing process in this orientation is assisted mainly by A_{1g} phonons (Henderson *et al.* 1981).

In calcium oxide, photoluminescence of an F_2^{2+}-centre consists of a sharp zero-phonon line at 683 nm with vibronic sidebands peaking at 700 nm. Again ODMR demonstrates the involvement of an $S=1$ emitting state. A striking difference between the centres in the two oxides, however, is in the orientation of the principal axes of the fine structure tensor. For the calcium oxide centre the directions of the x and z principal axes are interchanged with

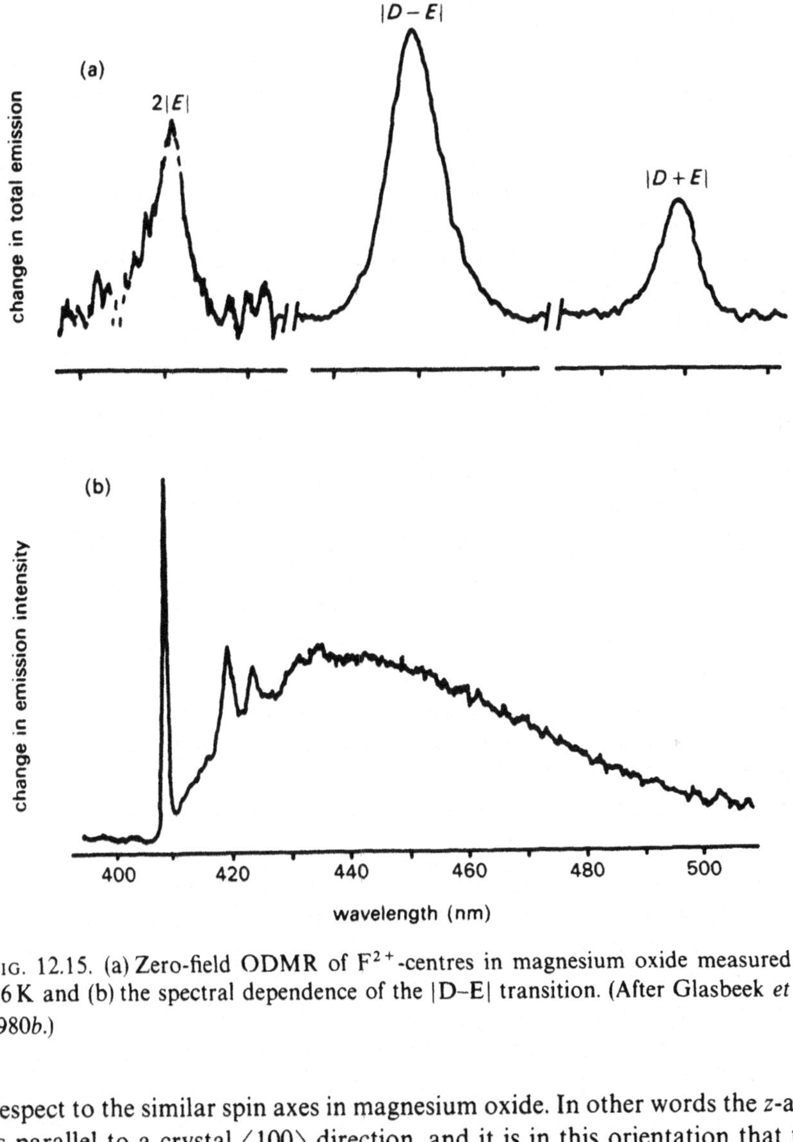

FIG. 12.15. (a) Zero-field ODMR of F^{2+}-centres in magnesium oxide measured at 1.6 K and (b) the spectral dependence of the |D–E| transition. (After Glasbeek *et al.* 1980*b*.)

respect to the similar spin axes in magnesium oxide. In other words the z-axis is parallel to a crystal ⟨100⟩ direction, and it is in this orientation that the maximum magnetic splitting of the ODMR spectrum arises. It was suggested that in calcium oxide the rotation of the D tensor axis is due to the presence of a Ca^{2+} vacancy along a ⟨100⟩ direction perpendicular to the molecular axis, i.e. this vacancy is equidistant from each anion vacancy along equivalent ⟨100⟩ directions. The principal distortion axis is then ⟨100⟩, which is the z-axis of the fine structure term.

12.3 Some phosphor ions in alkali halides

The heavy metal ions Tl^+, In^+, Ga^+, Sn^{2+}, and Pb^{2+} are of considerable technological interest because of their use in phosphors. In alkali halide crystals these ions may display several different charge states. For example, Tl^0, Tl^+ and Tl^{2+} ions have been reported in substitutional sites of O_h symmetry, Tl^0 and Tl^{2+} being produced photochemically from the monovalent ion (Gellerman *et al.* 1981; Mollenauer *et al.* 1983*a*). In this section we discuss only a small part of the vast literature on these ions. More extensive reviews are given by Boulon (1986) and Ranfagni *et al.* (1983).

12.3.1 Triplet states of impurity centres

Numerous even-electron ions in crystals (e.g. Tl^+, In^+, Ga^+, Sn^{2+} and Pb^{2+}) have paramagnetic excited states which are amenable to study by ODMR (Romestain *et al.* 1977). These ions have ground state configurations (ns^2). The excited configuration (ns) (np) in the limit of Russell–Saunders coupling yields $^3P_0$, $^3P_1$, $^3P_2$, and $^1P_1$ states. In the free ions spin–orbit coupling lifts the orbital degeneracy of the (ns) (np) configuration and mixes the $^1P_1$ and $^3P_1$ states together. (N.B. strictly speaking one must use *j–j* coupling for the very heavy ions.) Thus, although the $^1S_0 \rightarrow {}^1P_1$ transition is the only electric dipole transition, the spin allowed transition, $^1S_0 \rightarrow {}^3P_1$ is partially allowed by virtue of the singlet–triplet mixing. In crystalline solids the effects of the crystal field must be included, and the states labelled as irreducible representations of the appropriate symmetry group. For Tl^+ ions in alkali halide crystals, O_h group, the group theoretical labels are:

Free Ion:	$^1S_0$	$^3P_0$	$^3P_1$	$^3P_2$	$^1P_1$
Crystal Field:	$^1A_{1g}$	$^3A_{1u}$	$^3T_{1u}$	$^3E_u + {}^3T_{2u}$	$^1T_{1u}$

The situation is reminiscent of F-centres in CaO, except that spin–orbit coupling is very much stronger.

Obviously the strongest absorption band corresponds to the $^1A_{1g} \rightarrow {}^1T_{1u}$ transition. However, as Fig. 12.16 shows there are as many as four absorption bands in the ultraviolet spectrum of Tl^+-like ions in alkali halide crystals, usually labelled as the A, B, C, and D bands. These data refer to absorption by Tl^+ ions in order of increasing energy. The C-band is identified with the $^1A_{1g} \rightarrow {}^1T_{1u}$ transition. Next in order of observable intensity is the A-band which arises from $^1A_{1g} \rightarrow {}^3T_{1u}$ transitions with large oscillator strength 'borrowed' from the $^1T_{1u}$ state by virtue of spin–orbit coupling. This transition is much stronger than the corresponding transition for F-centres in alkaline earth oxides. The B-band transitions between the ground and ($^3E_u + {}^3T_{2u}$) states is induced by vibronic mixing of the $^3T_{1u}$ state and ^3E_u, $^3T_{2u}$ states.

A phenomenological model of the crystal field levels of Tl^+ ions treats all parameters in the free-ion Hamiltonian matrix as experimental variables (see

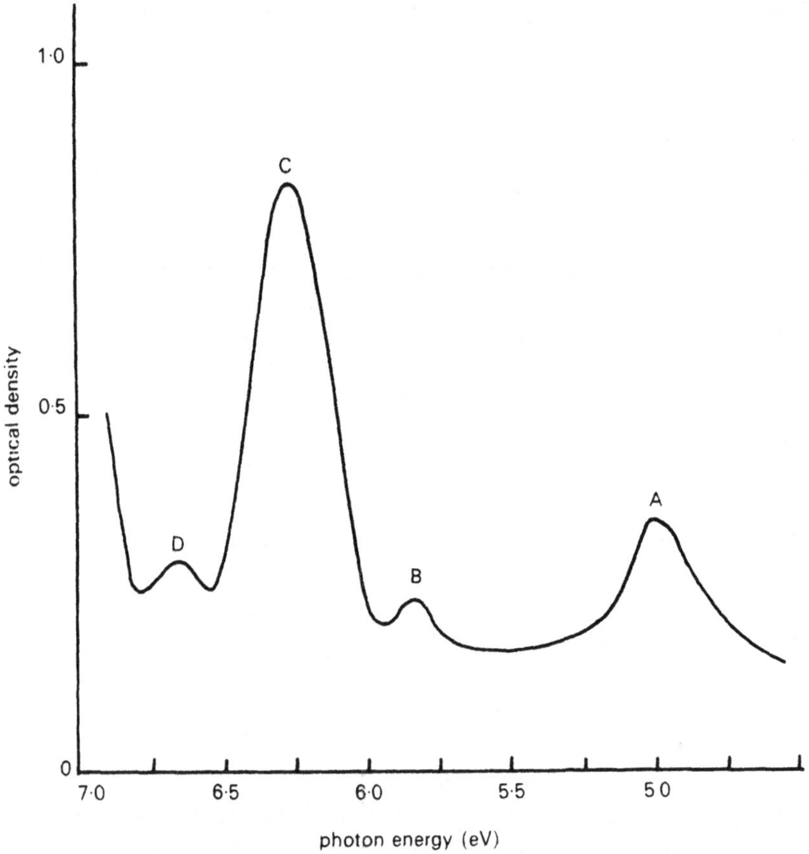

FIG. 12.16. The ultraviolet absorption spectrum of Tl^+ ions in potassium chloride measured at $T=77$ K. (After Delbecq *et al.* 1963).

e.g. Fowler 1968). An exact diagonalization of this Hamiltonian including spin–orbit and exchange coupling terms gives the following energies relative to the ground state:

$$\begin{aligned}
E(^3\mathrm{P}_0) &= E_0 - \zeta - E_1,\\
E(^3\mathrm{P}_1) &= E_0 - \zeta/4 - [(E_1 + \zeta/4)^2 + \zeta^2/2]^{\frac{1}{2}},\\
E(^3\mathrm{P}_2) &= E_0 + \zeta/2 - E_1,\\
E(^1\mathrm{P}_1) &= E_0 - \zeta/4 + [(E_1 + \zeta/4)^2 + \zeta^2/2]^{\frac{1}{2}}.
\end{aligned} \tag{12.31}$$

These equations apply only in the Russell–Saunders case; in general one cannot use either pure *LS* coupling or pure *j–j* coupling. Using this description of Tl^+ in the solid suggests that the one-electron energy difference E_0, exchange energy E_1, and spin–orbit coupling constant ζ can be measured from the positions of the A, B, and C-bands. Data for Tl^+ in potassium halides are given in Table 12.5. The reduction in E_0 suggests that the Madelung crystal field increases the ground state energy rather more than that of the excited state. The remarkable reduction in ζ probably results from several factors; a negative contribution from neighbouring anions as in F-centre problems, rather diffuse excited state wavefunctions for Tl^+ and a reduction of orbital operators by the dynamic Jahn–Teller effect. This latter interaction also plays an important role in the development of structure in both the A and C-bands, when measured at low temperature. As noted above, spin–orbit interaction admixes $^1T_{1u}$ and $^3T_{1u}$ states thus:

$$|^1T_{1u}\rangle' = \alpha|^1T_{1u}\rangle - \beta|^3T_{1u}\rangle$$

and

$$|^3T_{1u}\rangle' = \beta|^1T_{1u}\rangle + \alpha|^3T_{1u}\rangle$$

with $\alpha^2 + \beta^2 = 1$. The ratio of oscillator strengths for A- and C-bands was shown by Sugano (1962) to be

$$\frac{f_A}{f_C} = \frac{E_A}{E_C}\left(\frac{2 + 2x - [6 - 2(2x-1)^2]^{\frac{1}{2}}}{4 - 2x + [6 - 2(2x-1)^2]^{\frac{1}{2}}}\right) \tag{12.32}$$

where $x = (E_B - E_A)/(E_C - E_A)$. Hence we may obtain the relative dipole strength simply by measuring the photon energies at the band peak. A test of Sugano's model is shown in Fig. 12.17 for Ga^+, In^+, Tl^+, Sn^{2+}, and Pb^{2+} in potassium chloride, where we plot f_A/f_B, eqn (12.32), against x. The solid curve represents the theoretical values. Agreement is very good for the heavier ions but less good for ions of lower mass. The relatively broad absorption bands shown in Fig. 12.16 imply relatively strong electron–phonon coupling (Merle d'Aubigné *et al.* 1981). This is confirmed by the luminescence behaviour of these ions, since in general broad bands and rather large Stokes shifts are observed. Moreover the experimental

Table 12.5

Some properties of Tl^+ ions in potassium halide crystals. (All energies are given in eV.)

	E_0	E_1	ζ	A	B	C
Free ion	8.19	1.05	1.015	—	—	—
KCl	5.82	0.25	0.67	5.03	5.84	6.15
KBr	5.51	0.23	0.60	4.79	5.58	5.93
KI	4.93	0.16	0.61	4.40	5.08	5.31

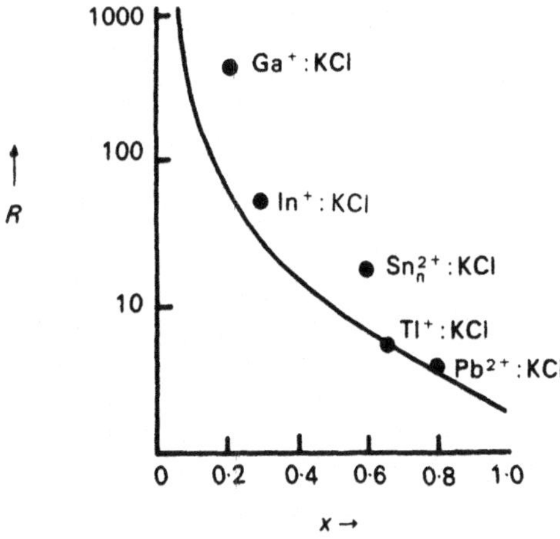

FIG. 12.17. A test of Sugano's crystal field model for isoelectronic $(ns)^2$ ions in alkali halides in which the ratio, R, of oscillator strengths of A and C absorption bands is plotted as a function of $x=(E_B-E_A)/(E_C-E_A)$. (Adapted from Fowler 1968.)

situation is complex, and several emission bands are observed depending upon the excitation process used (Fukuda *et al.* 1967, 1976). We will discuss only the spectra emitted by excitation in the A-band, i.e. the lowest spin-forbidden transition, ${}^1A_{1g} \to {}^3T_{1u}$.

In general such excitation results in two luminescence bands, A_X at low energy, and A_T at high energy (Fukuda 1970; Le Si Dang *et al.* 1977; Romestain *et al.* 1977). However for In^+ in sodium chloride or potassium chloride, or Tl^+ in potassium chloride, only the high-energy A_T-band is observed. In contrast Ga^+ ions in sodium chloride, potassium chloride, potassium bromide, and potassium iodide display both bands, since at low temperature the A_T-band appears whereas above *c.* 60 K only the A_X-band is observed. At intermediate temperatures both A_T and A_X are observed, although the total emitted intensity remains constant. For both In^+ and Tl^+ in potassium bromide both A_T and A_X appear, although at high temperature the behaviour is complex. The observed bandpeak energies are given in Table 12.6. In general the A_T emission is linearly polarized parallel to a cube edge upon excitation with linearly polarized light, whereas the A_X transition is either unpolarized or weakly polarized along the body diagonal of the cube (Wasiela *et al.* 1980). Recalling our discussion of the Jahn–Teller effect in the calcium oxide F^+- and F-centres, a strong Jahn–Teller coupling to E_g or T_{2g} modes will lead to either tetragonal or trigonal distortions respectively, but not both. Since excitation into the lowest ${}^3T_{1u}$ state leads to both A_T and A_X emissions, and since emission takes place from the minimum in the excited-state potential energy surface, there must be two non-degenerate minima in

Table 12.6

The peak energies (in eV) of A_T- and A_X-triplet state transitions for Ga^+, In^+ and Tl^+ ions in some alkali halide crystals.

		Crystal			
Ion	Transition	NaCl	KCl	KBr	KI
Ga^+	A_T	3.10	2.85	2.74	2.47
In^+	A_T	3.05	2.95	2.94	2.81
Tl^+	A_T		4.17	4.02	3.70
Ga^+	A_X	2.45	2.35	2.24	2.04
In^+	A_X			2.46	2.20
Tl^+	A_X			3.50	2.89

the $^3T_{1u}$ potential energy surface. However, a linear Jahn–Teller effect produces only one kind of minimum, so that spin–orbit mixing of $^1T_{1u}$ and $^3T_{1u}$ and quadratic Jahn–Teller coupling must be added to the usual linear effect, to provide co-existing and non-degenerate minima on the lowest potential energy surface of the relaxed excited state (Ranfagni 1972). The situation is depicted in Fig. 12.18, which nicely explains why the A_X luminescence requires an activation energy, B, and thus competes at high temperature with the A_T luminescence.

The most compelling evidence on the nature of the relaxed excited $^3T_{1u}$ level comes from the ODMR work reported by the Grenoble group (Le Si Dang *et al.* 1977; Romestain *et al.* 1977). They measured the ODMR spectra of Ga^+ in potassium bromide at 1.6 K at microwave frequencies of $\nu \simeq 35$ GHz and 70 GHz. The $^1A_{1g} \rightarrow {}^3T_{1u}$ absorption strength is about 10^{-3} relative to the $^1A_{1g} \rightarrow {}^1T_{1u}$ transition which peaks at around $\lambda = 270$ nm. Optical pumping in this band at *c.* 2 K gives a strong A_T-band centred at 455 nm and a much weaker A_X-band at 550 nm. The ODMR measurements were made on both the A_T- and A_X-bands; typical spectra are shown in Fig. 12.19, for magnetic field orientations parallel to [100] and [111] respectively. In both instances magnetic hyperfine structure due to ^{69}Ga and ^{71}Ga nuclides which have $I = \frac{3}{2}$ confirm that Ga^+ is involved. For more general field orientations the spectra are much more complex, there being 32 hyperfine lines for each site. The simplicity of the spectra along the particular directions shown suggest that the principal directions of the complexes are along these directions, i.e. [100] for A_T and [111] for A_X. However the A_X spectrum has only one 'perpendicular' spectrum for $B \parallel [110]$, whereas the A_T spectrum shows two 'perpendicular' components for $B \parallel [100]$. The significance of this observation is that A_X is a pure axial system about the $\langle 111 \rangle$ direction whereas A_T must be orthorhombic about a $\langle 100 \rangle$ direction. Furthermore only the σ_--polarized emission is sensitive to ODMR confirming

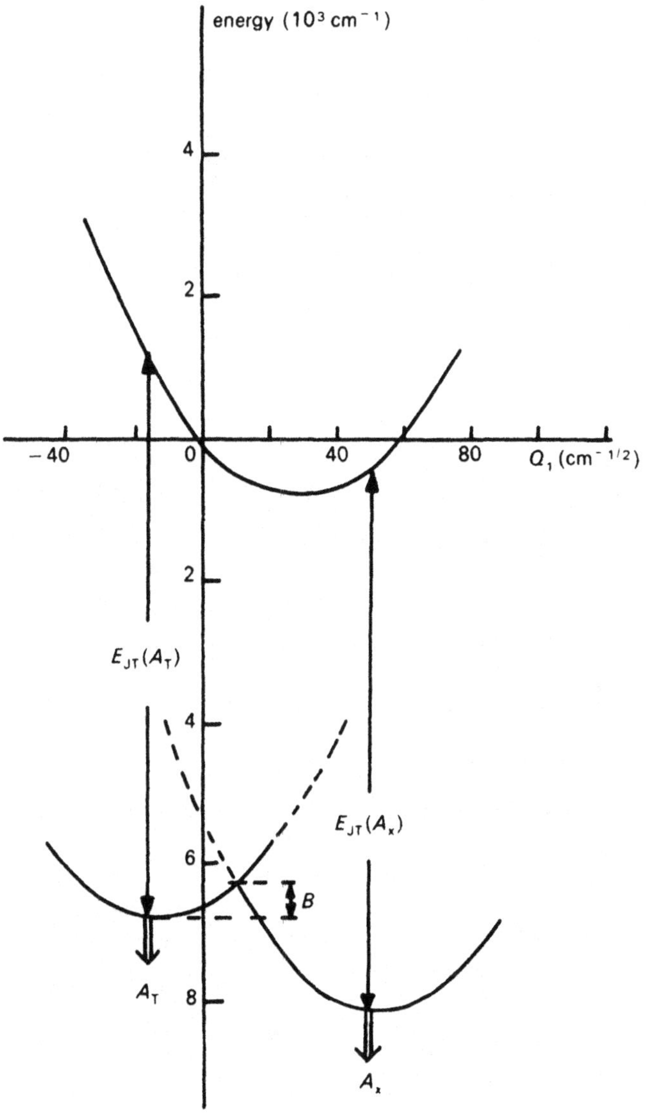

FIG. 12.18. Showing the development of two coexisting minima on the potential energy surface of the $^3T_{1u}$ state of $(ns)^2$-like ions. Emission occurs from either (or both) minima depending upon temperature, and upon the energy difference between these minima. (After Romestain *et al.* 1977.)

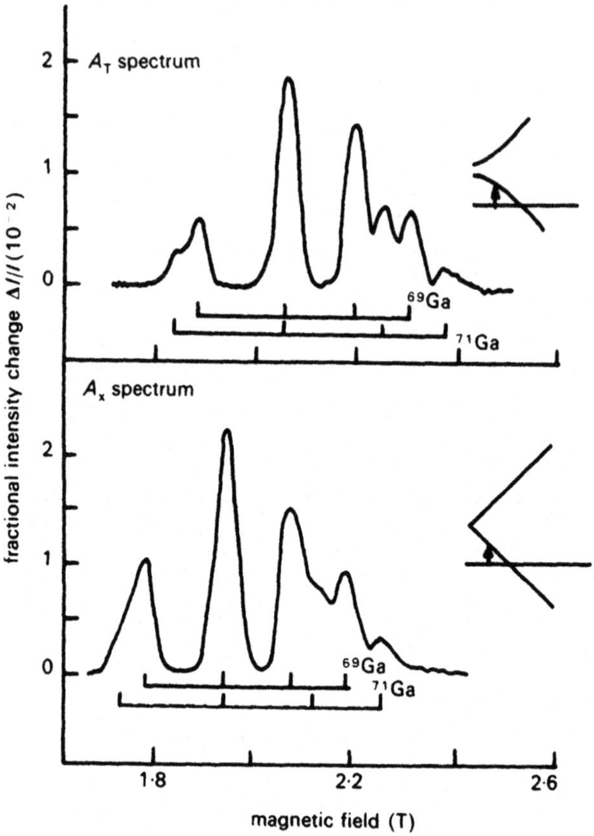

FIG. 12.19. The ODMR spectrum of Ga^+ centres in potassium bromide measured at $T = 2$ K, and $\nu = 35$ GHz with (a) $B \| [100]$ and (b) $B \| [111]$. (After Romestain *et al.* 1977.)

that the zero-field splittings are large enough to ensure that only the low-field $M_s = -1 \to 0$ transition could be reached at the fields available. The spectra were fitted to the spin Hamiltonian

$$\mathscr{H} = \mu_B \boldsymbol{B}.\boldsymbol{g}.\boldsymbol{S} + D(S_z^2 - \tfrac{2}{3}) + E(S_x^2 - S_y^2) + \boldsymbol{T}.\boldsymbol{A}.\boldsymbol{S} \qquad (12.33)$$

with the following spin Hamiltonian parameters (Le Si Dang *et al.* 1978):

	A_T	A_X
$g_x = g_y$	2.00 (5)	2.00 (5)
g_z	2.02 (1)	2.01 (1)
D(GHz)	93.6 (5)	90.0 (5)
E(GHz)	8.0 (5)	<0.2
$^{69}A_x = {}^{69}A_y$(GHz)	3.6 (1)	3.4 (2)
$^{69}A_z$(GHz)	4.04 (2)	3.80 (2)

Even though the symmetry of A_T is orthorhombic about ⟨100⟩ whereas A_X is trigonal about ⟨111⟩ the magnitude of these parameters is distinctly similar.

Finally we comment on the observation of both orthorhombic and trigonal distortions associated with the A_T and A_X emissions. The combination of spin–orbit mixing and linear coupling to E_g and T_{2g} modes leads to either trigonal or tetragonal distortions. To obtain an orthorhombic distortion Le Si Dang *et al.* (1977) assume bilinear coupling to E_g and T_{2g} modes *and* quadratic coupling to breathing-mode phonons (A_{1g}) which in the presence of spin–orbit mixing of the singlet and triplet levels enables the coexistence of two minima of different energies on the lowest energy surface of the relaxed excited state. In a highly simplified calculation these authors find that the experimental data are consistent with $E_{JT} = 0.95$ eV, $\Delta E = 0.17$ eV and $B = 0.07$ eV. Similarly detailed ODMR studies of the other Tl^+-like ions have not yet been reported. However they are expected, in view of the range of spin–orbit energies involved, to highlight the different effects of competition between Jahn–Teller and spin–orbit interactions.

12.3.2 Tl^0 centres in the alkali halides

Neutral Tl^0 has the ground configuration ($6p^1$), which yields ground ($6^2P_{1/2}$) and first excited ($6^2P_{3/2}$) states split by a spin–orbit interaction of almost 1 eV. Electric dipole transitions between these states are not allowed in the free atom or in cubic crystals. Tl^0 centres in octahedral symmetry sites in alkali halides were first reported by Delbecq *et al.* (1966, 1967). The $6^2P_{1/2} \rightarrow 6^2P_{3/2}$ transition was observed as a doublet absorption band whose strength increased rapidly with increasing temperature. Both the splitting and transition strength proceed as a result of vibronic coupling to odd-parity vibrations. Transitions from the odd-parity $6^2P_{1/2,\,3/2}$ to the even-parity $7^2S_{1/2}$ (7s) state are also observed, and with considerable oscillator strength.

Recent ground state ESR studies have shown that in Tl-doped crystals other Tl^0-centres are produced associated with one and two nearest-neighbour anion vacancies along ⟨100⟩ directions (Goovaerts *et al.* 1981). These centres, which have tetragonal symmetry, were dubbed the $Tl^0(1)$- and Tl^0 (2)-centres. In purely optical studies Gellerman *et al.* (1981) observed absorption bands near to F-centre bands which they attributed to F_A(Tl)-centres. They also demonstrated that these centres were laser-active, with the laser emission at *c.* 1.5 μm irrespective of the host crystal. Subsequent optical and MCD studies (Mollenauer *et al.* 1983*a*; Ahlers *et al.* 1983) showed that Tl^0 (1)- and F_A(Tl)-centres are one and the same centre.

Since in the $Tl^0(1)$-centre an anion vacancy occupies one of the nearest neighbour sites of the Tl^0 atom, there is an odd-parity crystal field acting on the atom. In consequence the theoretical model of this centre (Goovaerts *et al.* 1981; Mollenauer *et al.* 1983*a*) assumes crystal field terms of spherical, cubic, and odd symmetries. The odd-parity term splits the $^2P_{1/2,\,3/2}$ manifold into states between which electric dipole transitions of reasonable strength are allowed. There is no need to invoke odd-parity vibrations. In the crystal field expansion the anion vacancy is treated as an effective positive charge, $+q_e$, located at a distance r_0 from the Tl^0 nucleus. The potential energy expanded in spherical harmonics centred on the Tl nucleus has leading terms

$$\begin{aligned} V_1(r) &= q_e(r_</r_>^2)\cos\theta \\ V_2(r) &= q_e(r_<^2/r_>^3)(3\cos^2\theta - 1)/2 \end{aligned} \tag{12.34}$$

where $r_<$ and $r_>$, respectively, refer to the lesser and greater of the electron–nuclear separation, r, and the equilibrium separation, r_0. Since only $V_2(r)$ mixes states in the 6p manifold one needs only to diagonalize the one-electron Hamiltonian,

$$\mathscr{H} = \mathscr{H}_0 - eV_2(r). \tag{12.35}$$

Mollenauer *et al.* (1983*a*) determined the energy eigenvalues as a function of γ/Δ, in which the spin–orbit splitting, Δ is just the energy difference between the $6^2P_{1/2}$ and the $6^2P_{3/2}$ state ($\varepsilon_{3/2}$), and $\gamma = (-\frac{1}{5})bq_e$, where b is a radial integral for the 6P function and the factor $\frac{1}{5}$ arises out of the angular integrals. The eigenvalues of states derived from the 6P manifold (Goovaerts *et al.* 1981; Mollenauer *et al.* 1983a) yield the three lowest-lying Kramers doublets $\Phi_\pm$, $\Psi_\pm$, and $\Xi_\pm$ in order of increasing energy (Fig. 12.20). Also shown is a $\Sigma_\pm$ doublet derived from the 7s and 6d states admixed by the odd-parity term, V_2. However V_1 will admix the even-parity $\Sigma_\pm$ states with the odd-parity $\Phi_\pm$ and $\Xi_\pm$ states.

A typical optical absorption spectrum (Fig. 12.21(a)) shows bands at 1040, 830, 720, 635, 550, 405, and 340 nm. The intensities of the 830 nm and 635 nm bands are strongly sample dependent; they are clearly due to other centres. The principal optical transitions identified by Mollenauer *et al.* (1983*a*) with the $Tl^0(1)$-centres included a single luminescence band at 1520 nm and the absorption band peaks (Fig. 12.21(a)) at 1040, 720, 550, and 340 nm. Using only a single adjustable parameter $\gamma/\Delta = 0.5$, the luminescence peak and two lowest energy absorption bands were fitted to the simple theoretical model to within a few per cent. The model also accounts for the observed polarization properties and oscillator strengths of the absorption bands in terms of the wavefunction admixture coefficients. For example the 550 nm absorption band is strongly *z*-polarized since it involves $\Phi_\pm$ (p_z-like) to $\Sigma_\pm$ (s-like) transitions, whereas the 720 nm band is a transition from a small Σ component of $\Phi_\pm$ ground state into a pure $p_{xy}(\Xi_\pm)$ orbital, so that it is strongly *xy*-polarized.

Essentially the link between the ground-state ESR spectra and the optical data rests on their interpretation in terms of a common theoretical model. The application of spin-sensitive MCD provides in a single experiment direct experimental verification which does not rely in any way on the theoretical

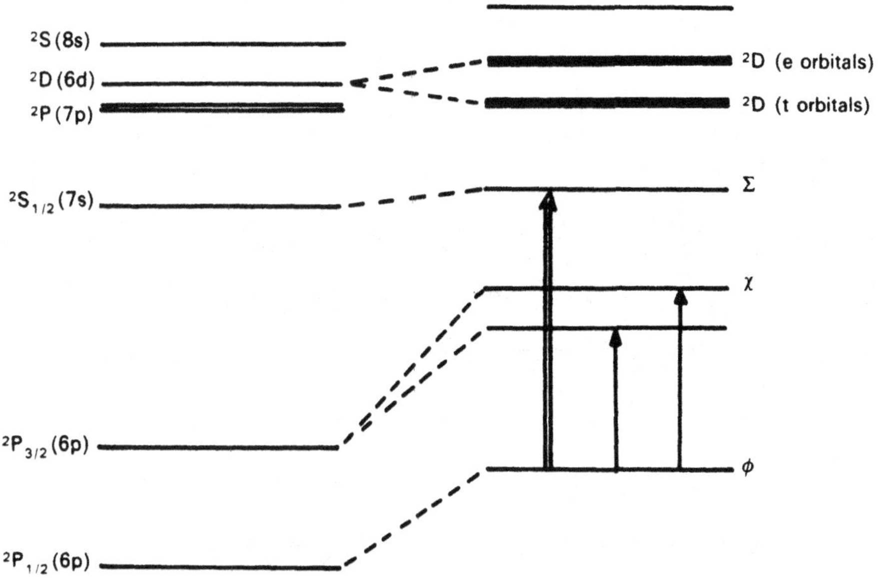

FIG. 12.20. The lowest-lying free atom states of neutral Tl (i.e. Tl^0) and their counterparts in the $Tl^0(1)$-centre. Strong absorption transitions from the ground state $^2P_{1/2}(\Phi_{\pm})$ are identified by arrows. (After Mollenauer *et al.* 1983*b*.)

model (Ahlers *et al.* 1983). If the monochromator is set at the peak of the 1040 nm band, the magnetic field applied along a ⟨100⟩ crystal axis is swept in the range 0.7–1.7 T, the sample being contained in a resonant cavity at microwave frequency $\nu \simeq 24$ GHz and $T = 1.5$ K, one observes the ODMR spectrum shown in Fig. 12.21(b). In this orientation there are two sets of $Tl^0(1)$-centres, those which have their z-axis parallel to B, and a twofold degenerate set perpendicular to B. The low-field set of lines corresponds to defects parallel to B and the high-field set to defects perpendicular to B. These transitions correspond to allowed $\Delta M_s = \pm 1, \Delta M_I = 0$ transitions of a centre with $S = \frac{1}{2}$, $I = \frac{1}{2}$. The very weak inner pairs of transitions which appear at high microwave power, are due to forbidden ($\Delta M_s = \pm 1$, $\Delta M_I = \pm 1$) transitions. The Hamiltonian parameters of this spectrum are identical to those of the $Tl^0(1)$-centre measured in ground-state ESR (Goovaerts *et al.* 1981; Ahlers *et al.* 1983). In other words the changes in the MCD signal arise from the microwave-induced decreases in the spin polarization of the ground-state Kramer's doublet.

Having established the involvement of the $Tl^0(1)$-centre ESR spectrum and the 1040 nm band it is relatively easy to observe which other absorption bands are attributable to these centres. The resonance field is set to one of the ODMR lines and the excitation wavelength scanned with the monochroma-

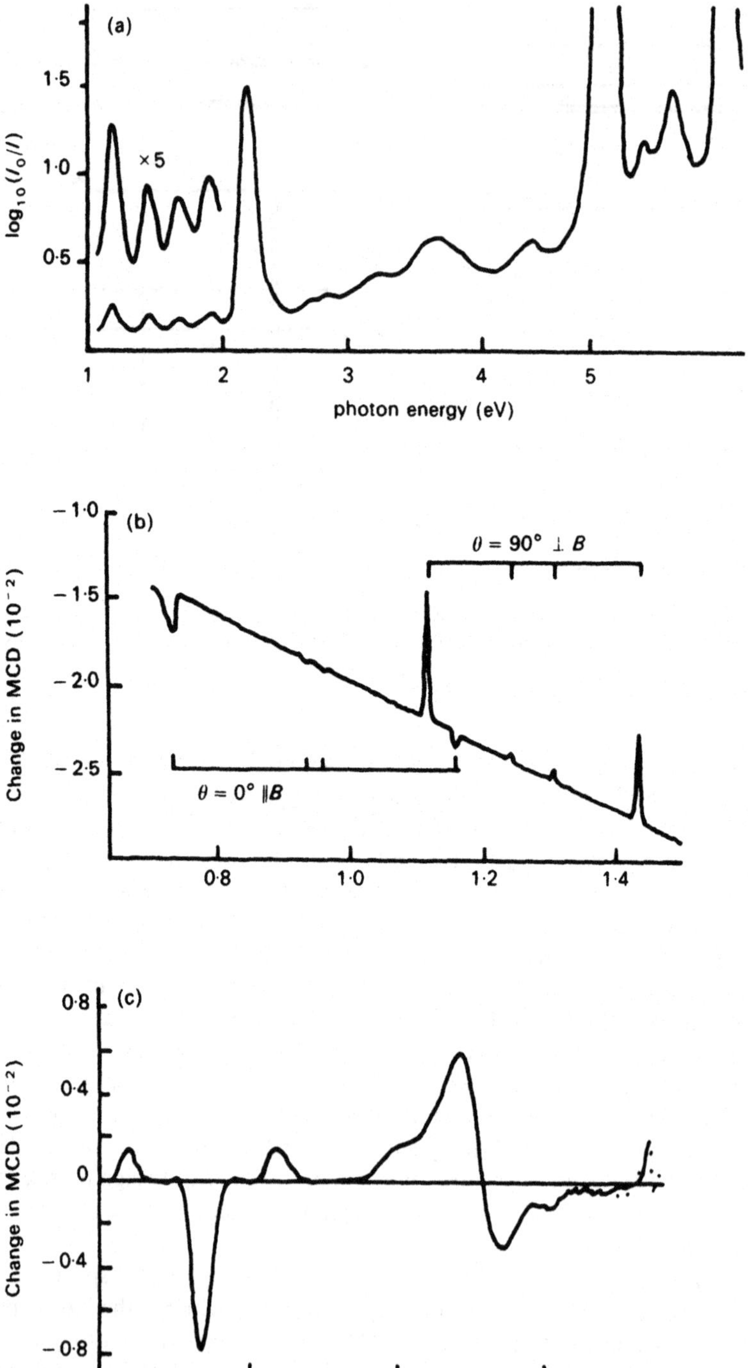
(a)
1·5
1·0
0·5
log₁₀(I₀/I)
×5
1
2
3
4
5
photon energy (eV)
(b)
−1·0
−1·5
−2·0
−2·5
Change in MCD (10⁻²)
θ = 90° ⊥ B
θ = 0° ‖B
0·8
1·0
1·2
1·4
(c)
0·8
0·4
0
−0·4
−0·8
Change in MCD (10⁻²)

tor. The resulting spectrum (Fig. 12.21(c)) shows very clearly optical absorption bands at 1040, 725, 550, 405, and 335 nm, and identifies each of them with the $Tl^0(1)$-centre. This is confirmed by a similar wavelength scan with the magnetic field set to each of the ODMR lines. Moreover, not only does the MCD experiment confirm the relationship between the $Tl^0(1)$-centre, the ESR spectrum, and several optical bands, it also discriminates against other troublesome overlapping bands. This technique of measuring the exciting or fluorescence wavelength of a centre while inducing ESR transitions at fixed magnetic field is of great power and generality; it has been used for a number of years by workers on organic triplet states, inorganic defect resonances, and semiconductor recombination luminescence.

12.4 Exciton recombination in crystals

Photoexcitation of electrons into the conduction band and holes into the valence band may lead to the formation of free or bound excitons in semiconductors and self-trapped excitons in ionic crystals (see Gibson and Elliot 1974). In such crystals the exciton energy of the electron–hole pair is simply the energy difference between the ground state of a filled valence band and empty conduction band and an excited state consisting of a hole in the valence band and an electron in the conduction band. To this extent electrons and holes are quasi-particles of opposite charge which interact mutually through Coulomb and exchange forces. It is usual to refer to an electron–hole pair with mutual interaction as an *exciton*. The mutual interaction between electron and hole leads to the exciton having bound energy levels just below the conduction band. Since the photoexcited carriers are charged they also interact with lattice ions through the Coulomb interaction and this interaction leads to a lattice distortion. There is a net decrease in energy since the electrostatic energy decreases linearly with displacement, whereas the elastic energy increases quadratically for small displacements. In semiconductors the ionic displacements are small, and many ions are involved. Thus the energy of the electron (hole) in the conduction (valence) band varies only slightly with ionic displacement, so that carriers may move in the presence of an applied field dragging their displacement field with them.

FIG. 12.21. (a) Optical absorption spectrum of a potassium chloride crystal containing $Tl^0(1)$-centres. (b) The ODMR spectrum from the same sample of circularly polarized light measured as the changes in the absorption coefficient at the peak of the 1040 nm band at 1.5 K induced by microwave transitions at frequency $\nu = 24$ GHz with $B \parallel [100]$. (c) The spectral dependence of the ODMR spectrum shown in (b). (After Ahlers 1983.)

The energy of the exciton, E_e, includes terms for an electron in the conduction band state n with wavevector k_e, a hole in the valence band state m with wavevector k_h together with the interaction terms; i.e.

$$E_e = E_n(\boldsymbol{k}_e) - E_m(\boldsymbol{k}_h) - \langle n\boldsymbol{k}_e, m\boldsymbol{k}_h | E | n\boldsymbol{k}_e, m\boldsymbol{k}_h \rangle$$
$$\pm 2\delta \langle n\boldsymbol{k}_e, m\boldsymbol{k}_h | E | n\boldsymbol{k}_h, m\boldsymbol{k}_e \rangle \quad (12.36)$$

where $E = e^2/4\pi\varepsilon_0 r_x$, $\boldsymbol{r}_x = \boldsymbol{r}_e - \boldsymbol{r}_h$, $\delta = 1$ for the singlet state (+), $\delta = 0$ for the triplet state (−) and $E(m, \boldsymbol{k}_h)$ is negative because the electron is excited out of the valence band. In the crystal ground state at $T = 0$ K the total momentum is zero for a full band and in the excited state the quasi-momentum of the pair is $\boldsymbol{k}_x = \boldsymbol{k}_e - \boldsymbol{k}_h$. Close to the upper and lower edges of the valence and conduction bands, respectively, the energies are

$$E_n(\boldsymbol{k}_e) = E_n(\boldsymbol{k}_e = 0) + \frac{\hbar^2 \boldsymbol{k}_e^2}{2m_e} \quad \text{and} \quad E_m(\boldsymbol{k}_h) = E_m(\boldsymbol{k}_h = 0) + \frac{\hbar^2 \boldsymbol{k}_h^2}{2m_h}$$

where m_e and m_h are the effective masses of electron and hole respectively, so that in the effective mass approximation E_e is written as

$$E_e = E_G + \frac{\hbar^2 \boldsymbol{k}_x^2}{2(m_e + m_h)} - \frac{\mu_x e^4}{8\varepsilon_0^2 \kappa^2 \hbar^2 n^2} \pm 2E_{ex} \quad (12.37)$$

in which $E_G = E_n(\boldsymbol{k}_e = 0) - E_m(\boldsymbol{k}_h = 0)$ is the energy gap between valence and conduction bands, E_{ex} is the exchange interaction and μ_x is the reduced mass of the exciton. Effective mass theory assumes spatially diffuse electron and hole wavefunctions, and represents the crystal as an homogeneous medium with static dielectric constant $\kappa(0)$. Consequently the interparticle Coulomb interaction is much reduced, so that the electron and hole are well separated. Such excitons as may be treated within the effective mass approximation with energies given by eqn (12.36) are known as *Wannier* excitons. It follows from eqn (12.36) that the creation of excitons should be signalled by a hydrogen-like absorption spectrum if the electron k-vector is conserved i.e. $\boldsymbol{k}_x = 0$. As discussed in Chapter 1 these transitions are *direct transitions* across the bandgap. For *indirect transitions*, where $\boldsymbol{k}_e \neq \boldsymbol{k}_h$ at the band extrema, corrections to the energies are necessary. For small $\boldsymbol{k}_x$ values the energy correction is the kinetic energy of the centre of mass of the exciton. Since $\boldsymbol{k}_e$ and $\boldsymbol{k}_h$ join points in the reciprocal lattice then $\boldsymbol{k}_x$ is a reciprocal lattice vector which characterizes exciton migration in phase space and, because of the relationship between a real crystal and its reciprocal lattice, in the crystal also. Excitons which migrate through the crystal with propagation vector $\boldsymbol{k}_x$ may be referred to as *free excitons*. As we have discussed in Chapter 10, excitons may be treated as elementary excitations of the crystal, and we may define ladder operators which create or destroy excitons with quasi-momentum vector $\boldsymbol{k}_x$ and pair separation $\boldsymbol{r}_x = \boldsymbol{r}_e - \boldsymbol{r}_h$. Such operators are analogous with those defined in earlier chapters for angular momentum, phonons, or magnons.

Note that free or Wannier excitons are observed in crystals which have relatively large values of $\kappa(0)$; elemental and compound semiconductors are just such materials. However, once created, excitons may be trapped at impurities and imperfections and so become *bound excitons*. The reduced exciton binding energy in materials with high $\kappa(0)$ is largely due to the polarization of the lattice ions with the consequent excitation of optical phonons. In solids with small $\kappa(0)$ the electron–hole separation is decreased so that the rotational frequency of one quasi-particle around the other increases. The *effective* dielectric constant is then a complex function of $\boldsymbol{k}_x$. When the rotational frequency of the exciton is higher than the frequency of longitudinal optical phonons then the lattice polarization can no longer follow the excitonic motion. This occurs when the electron–hole interaction is sufficiently strong that the exciton is localized on one or a small number of ions in the crystal. The exciton is then described in terms of excited atomic or molecular states rather than in terms of band states. Localized excitons are usually referred to as *Frenkel excitons*. The lattice polarization which results from the strong localization of the quasi-particles may then lead to self-trapping of the exciton. In general Frenkel excitons are observed in crystals with large lattice constant and small static dielectric constant, since then the pair cannot separate before recombination. These conditions are met by ionic and molecular crystals and solid rare gases.

12.4.1 Excitons in semiconductors

The hydrogenic term in eqn (12.36), shows that excitons are created in optical absorption transitions characterized by changes in the quantum number, n, which occur in the wavelength range close to the absorption edge of the crystal. Nevertheless the observation of an extended series of hydrogen-like absorption lines is unusual. In general exciton recombination in semiconductors is a complex process, and ODMR studies have given clear insights into the structure of many recombination centres. Usually use is made of spin-dependent selection rules for electron–hole recombination. In Fig. 12.22 electrons, holes, and excitons are assumed to have different g-values. For electrons and holes the spin states in a magnetic field, B, are labelled as $|s, m_s\rangle$ where $s = \frac{1}{2}$ and $m_s = \pm\frac{1}{2}$ whereas exciton states are labelled as $|S, M_s\rangle$ with $S = 0$ or 1. Assuming that electrons and holes recombine to form triplet excitons ($|1, M_s\rangle$ with $M_s = 0, \pm 1$) we expect that luminescence will be circularly polarized, $\sigma_{\pm}$, from the $|1, \pm 1\rangle$ levels. The populations of these levels are determined by the interplay between spin–lattice relaxation (T_1) and the decay rate (τ) of the triplet exciton. Except in the extreme cases ($T_1 \ll \tau$ when thermal equilibrium is maintained and $T_1 \gg \tau$ when the populations are rigorously determined by the selection rules) this competition

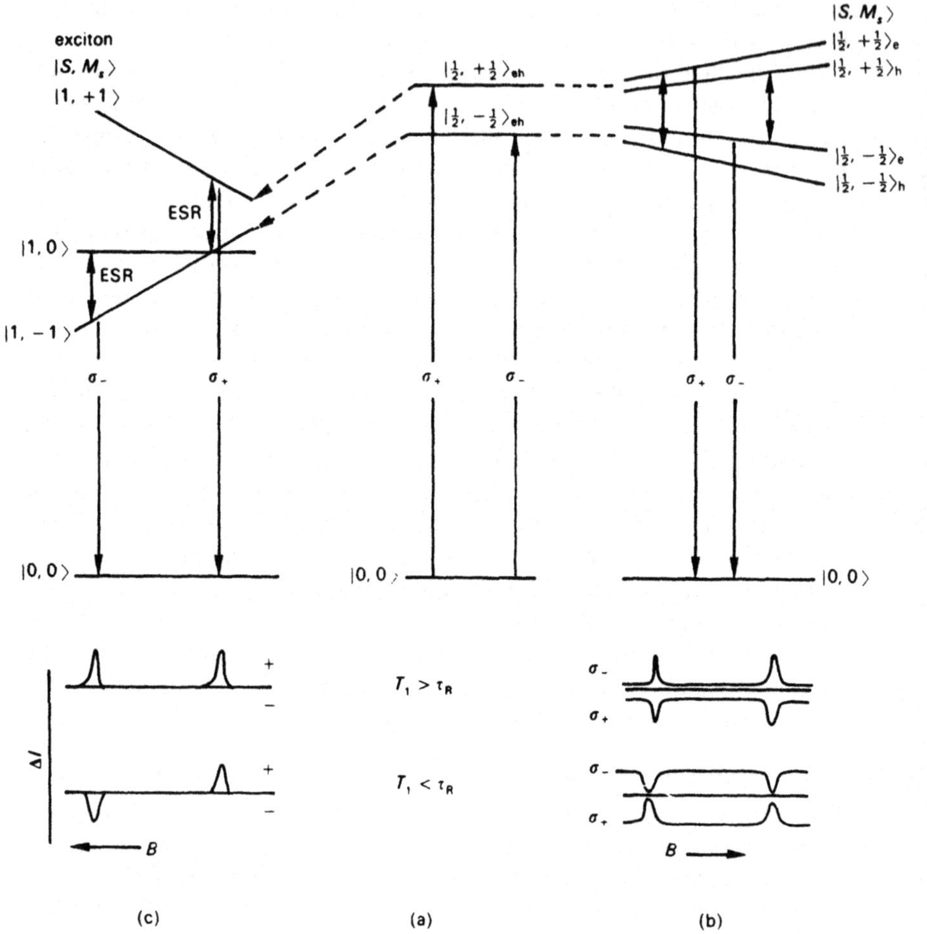

FIG. 12.22. (a) Production of spin-polarized electrons and holes by optical pumping using circularly polarized ($\sigma_{\pm}$) bandgap light. The selection rules for circularly polarized emission and ODMR of recombining electrons and holes and of triplet excitons are given in (b) and (c) respectively as are schematic ODMR spectra under the assumption of perfect spin memory ($T_1 > \tau_R$) and fast spin–lattice relaxation ($T_1 < \tau_R$).

results in different populations of the $|s, m_s\rangle$ levels of electrons and holes and of the $|S, M_s\rangle$ levels of the exciton. Irradiation with σ_+-polarized band gap light, therefore, enhances the populations $|\frac{1}{2}, +\frac{1}{2}\rangle_e$ and $|\frac{1}{2}, +\frac{1}{2}\rangle_h$ which exclusively feeds the population of the $|1, +1\rangle$ level of the exciton, if there is no loss of spin memory (Section 12.6). Assuming this to be the case, microwave-induced ESR transitions on the hole, $|\frac{1}{2}, +\frac{1}{2}\rangle \rightarrow |\frac{1}{2}, -\frac{1}{2}\rangle$, which increase the population of the $|\frac{1}{2}, -\frac{1}{2}\rangle$ state will be observed as a decrease in the intensity of the σ_+-polarized emission triplet and an increase in σ_- emission. Similar changes in the pattern of circularly polarized exciton emission are observed for ESR transitions on the electron.

As an example we consider free exciton formation in gallium selenide, a III–VI semiconductor with a layered crystal structure. The gallium selenide photoluminescence spectrum (Fig. 12.23(a)) is rich in structure. The ODMR spectrum obtained from the free-exciton line (Fig. 12.23(b)) at 588 nm shows ESR lines due to electrons and holes at resonance fields corresponding to g-values of $g_e = 1.13$ and $g_h = 1.72$. Both electron and hole resonances are observed as increases in the intensity of σ_+ emission and decreases in σ_- emission. This result is consistent with $T_1 < \tau_R$ so that the excited spin states are (approximately) in thermal equilibrium (Dawson *et al.* 1979; Cavenett 1981).

12.4.2 Triplet excitons in compound semiconductors

Excitons created in semiconductors by bandgap radiation may be trapped by imperfections and/or trace impurities. Indeed, much of the structure shown in Fig. 12.23 is due to triplet excitons bound to unknown traps (Cavenett 1981). As noted above, competition between spin–lattice relaxation, T_1, and radiative decay, τ_R, determines the populations $|s, m_s\rangle$ of electron and hole states and through them the populations of the $|S, M_s\rangle$ spin levels of triplet excitons. For $T_1 > \tau_R$, pumping with σ_+-polarized bandgap light preferentially populates the $|\frac{1}{2}, +\frac{1}{2}\rangle$ electron and hole states, leading to exciton formation in the $|1, +1\rangle$ level, and *vice versa* for the case of σ_- pumping. However, when $T_1 < \tau_R$ the exciton spin states are in thermal equilibrium and the population differences between $|1, +1\rangle_x$, $|1, 0\rangle_x$ and $|1, -1\rangle$ are much smaller. Since the $|1, +1\rangle$ and $|1, -1\rangle$ states emit σ_+ and σ_- light respectively, ESR transitions $|1, \pm 1\rangle \leftrightarrow |1, 0\rangle$ of the triplet exciton are observed at appropriate values of the magnetic field as changes in the intensity of circularly polarized emission from the decaying (triplet) exciton. The electron and hole ESR transitions may also be detected via the intensity of circularly polarized emission, is shown in the bound exciton ODMR from gallium selenide (Fig. 12.23(c)). The magnitudes of such changes are much larger in the presence of spin memory than when the spin states are in thermal equilibrium.

Numerous recombination processes have been identified in other compound semiconductors including excitons recombining at isoelectronic traps (N in gallium phosphide, O in zinc sulphide), excitons bound at neutral donors (X, D_0) or neutral acceptors (X, A_0). The optical spectroscopy was exhaustively discussed by Berg and Dean (1976), and the ODMR spectra reviewed by Davies (1976) and Cavenett (1981). The ODMR technique has been particularly useful in elucidating the nature of recombination centres in device materials, such as the II–VI and III–V semiconductors and their

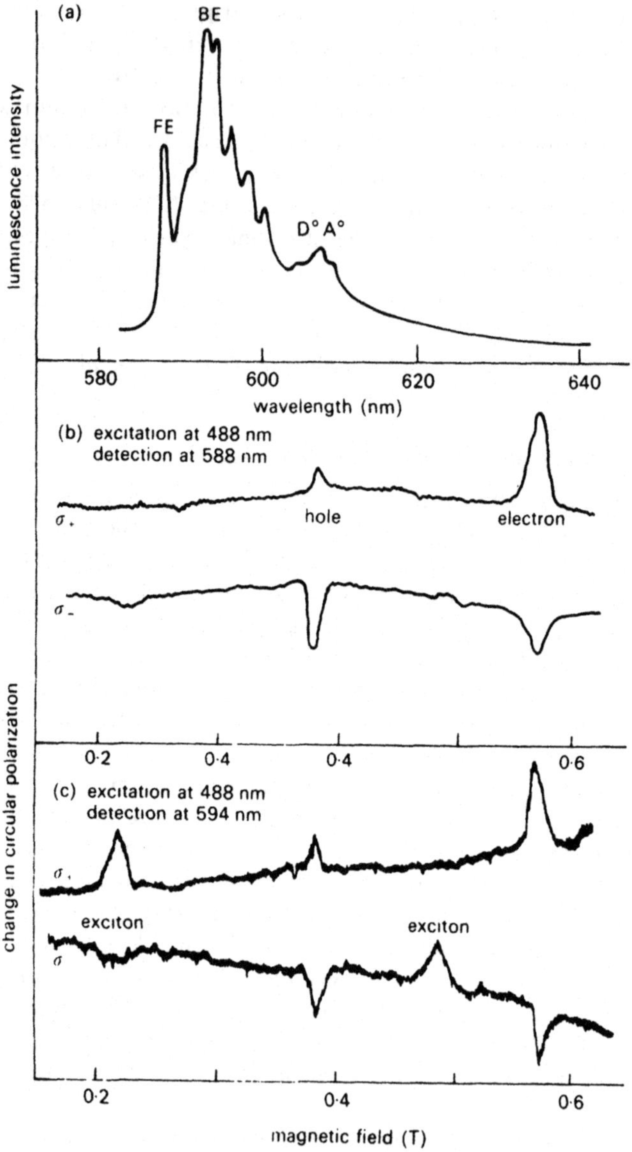

FIG. 12.23. (a) Recombination luminescence in gallium selenide shows structure due to free excitons (FE), bound excitons (BE) and donor–acceptor recombination (D^0A^0) measured at 1.7 K and excited using the 488 nm line from an Ar^+ laser. Optically detected magnetic resonance from the same sample measured at 9.5 GHz, $T = 1.6$ K and with $B \| c$-axis shows electron and hole resonances both on the free exciton line at 588 nm in (b) and on the bound exciton line at 594 nm in (c). In (c) lines due to triplet states of the bound excitons are also observed (Adapted from Cavenett, 1981).

respective alloys. Of special interest is the case of exciton recombination at deep N traps in gallium phosphide. The isoelectronic N, substituting on the P sublattice, acts as a very efficient recombination centre for bound excitons. In ODMR studies Cavenett *et al.* (1977a) observed microwave-induced intensity changes of 0.18% at magnetic fields corresponding to the conduction band g-value, which they explained in terms of spin-dependent exciton formation involving conduction band electrons (Fig. 12.22). We will restrict detailed discussion to ODMR studies of triplet excitons recombining at antisite defects in III–V compounds and at isoelectronic titanium centres in silicon carbide.

The antisite defect in a $A_{III}B_V$ compound involves either the Group V element being substituted on a site normally occupied by the a Group III element (the B_A-centre) or vice versa (the A_B-centre). Van Vechten (1975) first suggested that antisites in gallium arsenide, gallium phosphide, and their alloys may act as non-radiative 'killer traps'. This was confirmed by experiments which identified the ESR spectrum of the P_{Ga}^+ antisite centre in gallium phosphide. The spectrum was shown to be due to a centre with $S = \frac{1}{2}$; the resolved hyperfine structure associated with the central ionized P^+ atom was split by transferred hyperfine interaction with four equivalent P nuclides occupying normal tetrahedral phosphorus sites. A similar spectrum was observed from the antisite defect $P_{Ga}Y_p$ in which one of the nearest-neighbour P atoms of the antisite P_{Ga} is replaced by an unknown impurity atom, Y_p. As an example of the application of the ODMR technique to these problems we present in Fig. 12.24 optical data obtained from undoped gallium phosphide (Lee 1983). Relatively broad emission bands (Fig. 12.24(a)) in the near-infrared region extend from 0.8 μm to 1.5 μm. In zinc-doped gallium phosphide 1.20 μm emission is greatly enhanced and the ODMR spectrum then consists of a complex series overlapping ESR lines. However, the several contributing centres each have different spin–lattice relaxation times, and the component spectra may be differentiated between using the phase setting of the lock-in detector to 'null out' unwanted lines as described in Section 6.2. The ODMR spectrum from GaP:Zn shown in Fig. 12.24(b) was recorded in this way (Lee 1983). Orientation dependence studies confirmed that the spectrum is due to an $S = 1$ state of a defect occupying a site with C_{3v} symmetry. The resolved hyperfine structure comes both from the central P nucleus and three equivalent P nuclei each with nuclear spin, $I = \frac{1}{2}$. These lines are due to antisite $P_{Ga}Y_p$ defects in orientations parallel to the magnetic field, $\boldsymbol{B}$. Defects in other orientations give lines of rather low intensity. There are two fine structure transitions, $\Delta M_s = 0 \rightarrow \pm 1$, each split into two lines by the hyperfine interaction with the central P nucleus ($\Delta M_I = 0$). Transferred hyperfine interaction due to the three equivalent P ligands with $I = \frac{1}{2}$, splits each of these four transitions into $2(I_1 + I_2 + I_3) + 1 = 4$ lines with relative intensities 1:3:3:1. The four sets

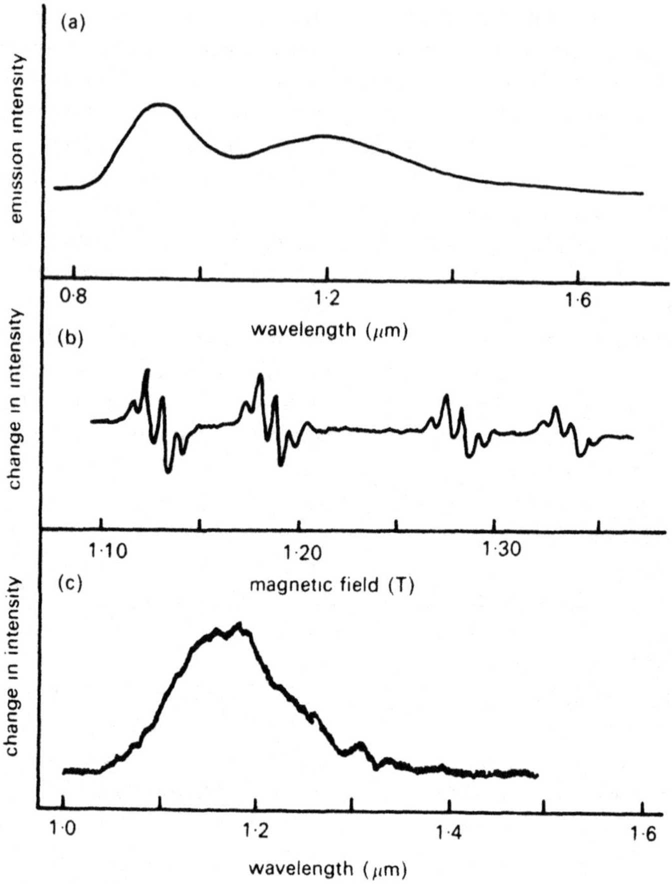

FIG. 12.24. (a) Luminescence spectrum of undoped gallium phosphide at $T=1.7$ K (b) ODMR spectrum of the $P_{Ga}Y_p$ centre $S=1$ state in GaP:Zn for $B\|[111]$, $\nu=35$ GHz at $T=1.7$ K and (c) spectral dependence of this ODMR spectrum. (Adapted from K. M. Lee 1983.)

of four superhyperfine lines, shown in Fig. 12.24(b) may be fitted to the spin Hamiltonian,

$$\mathscr{H} = g\mu_B \boldsymbol{B}.\boldsymbol{S} + D[S_z^2 - \tfrac{1}{3}S(S+1)] + A\boldsymbol{S}.\boldsymbol{I} + \sum_i \boldsymbol{S}.\boldsymbol{T}_i.\boldsymbol{I}_i \qquad (12.38)$$

with spin Hamiltonian parameters $g=2.007$, $D=+717\times10^{-4}\,\text{cm}^{-1}$, $A=+530\times10^{-4}\,\text{cm}^{-1}$ and $|T|=67\times10^{-4}\,\text{cm}^{-4}\,\text{cm}^{-1}$. The positive signs of D and A parameters were determined unambiguously from the polarization behaviour of the lines. From the spectral dependence of the ODMR, Fig. 12.24(c), it is clear that in GaP:Zn there is a luminescence spectrum from the $P_{Ga}Y_p$ centre which peaks at $\lambda \simeq 1.15$ μm. The most probable impurities substituting on the phosphorus site are carbon and silicon, although there is no means of differentiating between them. That bandgap radiation excites infrared luminescence suggests that the recombining exciton is trapped at a deep trap. Other recombination processes which contribute to the infrared emission from GaP : Zn, Fig. 12.24(a), involve the P_{Ga}^{+} antisite defect, whereas in 'pure' gallium phosphide a deep donor centre, probably O_p, is observed.

The properties of silicon carbide are intermediate between those of silicon and diamond. There are several crystallographic polytypes due to different Si–C stacking sequences, the commonest structures being cubic (3C), hexagonal (4H and 6H) and rhombohedral (15R). These materials display a wide range of energy gaps varying from *c.* 2.4–3.3 eV. Almost without exception in these polytypes, recombination luminescence occurs with high efficiency. The polytypism also leads to various electronic centres occupying different inequivalent sites in silicon carbide. The well-known green emission from silicon carbide (6H) is characterized at low temperature by three zero-phonon lines, associated with bound exciton decay at neutral isoelectronic centres. Selective isotopic doping experiments confirmed that the isoelectronic trap is titanium. Similar spectra from the other polytypes were observed in which the number of zero-phonon lines could be correlated with the number of site inequivalences. That the positions of the zero-phonon lines in the various polytypes are almost independent of the bandgap energies led Patrick and Choyke (1974) to suggest that the trap energy is tied to the valence band edge.

Figure 12.25 is a compilation of optical data for the green emission in SiC(6H):Ti. The intense luminescence band centred at $\lambda \simeq 500$ nm shows weak but well-resolved zero-phonon lines A_0, B_0, and C_0, and their phonon replicas. This emission is excited over a broad absorption band which peaks at 390 nm. There is a compelling correlation between these emission and excitation spectra and the spectral dependences shown in Fig. 12.26 of the ODMR spectrum. Indeed there are two distinct ODMR spectra, labelled Ti^A and Ti^C as shown in Fig. 12.26. Both are triplet-state resonances: their orientation dependences for magnetic field orientations in the (1120) plane show that each pair of lines cross at an orientation of $\theta = 54.7°$, where θ is the angle between the magnetic field and the *c*-axis. The very large magnetic splitting between the pairs of lines in Fig. 12.26 confirms that there is a rather large zero-field splitting between $M_s = 0$ and ± 1 levels, implying that recombination occurs at a single titanium site. When examined in detail each $\Delta M_s = \pm 1$ transition shows remarkable structure (Fig. 12.27), which Lee *et al.* (1983) attributed to different *D*-values of the *five* naturally occurring isotopes of titanium, three of which are non-magnetic with $I = 0$ (^{46}Ti, ^{48}Ti, and ^{50}Ti), one with $I = \frac{5}{2}$(^{47}Ti) and one with $I = \frac{7}{2}$(^{49}Ti). The structure associated with the different *D*-values of the even isotopes shows additional splittings due to

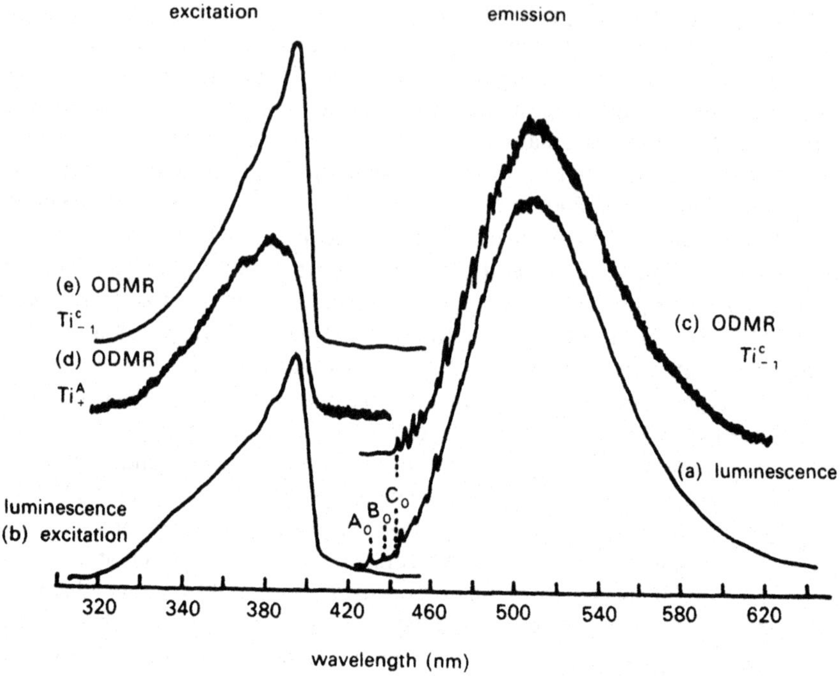

FIG. 12.25. The wavelength dependence (a) of luminescence, (b) excitation of luminescence and (c)–(c) of fine structure transitions of the triplet state ODMR, of Ti^{3+} in silicon carbide (polytype 6H) measured at 1.6 K. (After Lee *et al.* 1981.)

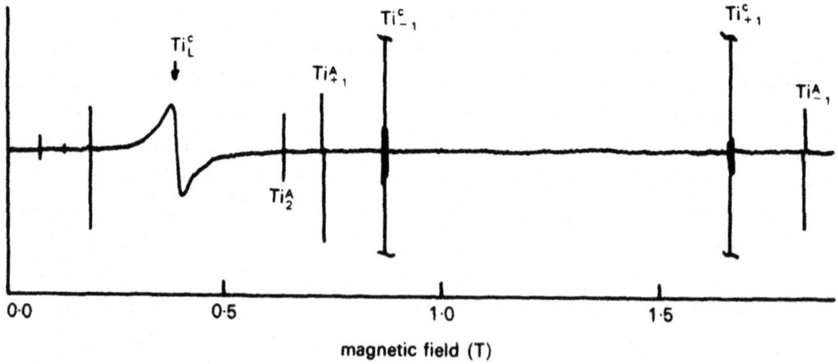

FIG. 12.26. Typical ODMR spectrum of 6H SiC:Ti showing the fine structure transitions of the A and C spectra, measured at $\nu \simeq 35$ GHz, $T = 1.6$ K with B parallel to the c-axis. The subscript $+1(-1)$ indicates that the ESR transition involves the $|1, +1\rangle$ $|1, -1\rangle$ state. Transitions labelled L are level-crossing transitions and subscript 2 represents a $\Delta M = 2$ transition. (After Lee *et al.* 1981.)

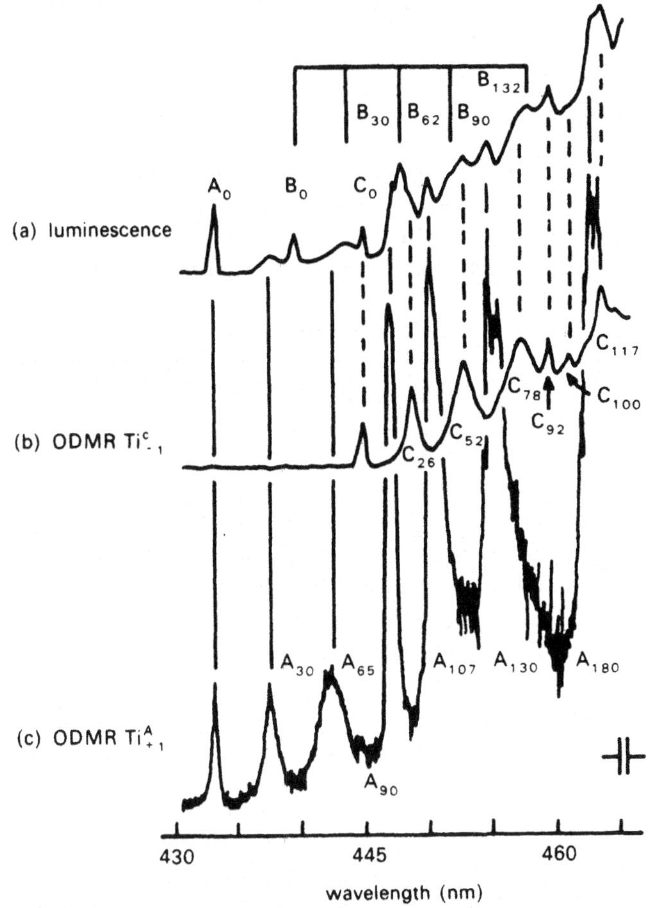

FIG. 12.27. High-resolution spectral dependences of the (a) luminescence and (b), (c) of ODMR spectra of the Ti^{3+} triplet resonance in 6H SiC:Ti. (After Lee *et al.* 1981.)

isotopic shifts in the *D*-values by the ^{28}Si, ^{29}Si, and ^{30}Si nuclides in neighbouring sites. No evidence of shifts or splittings due to carbon were observed. Indeed the entire structure is a convolution of two naturally-abundant isotopic species of titanium and silicon, the silicon nuclei being six of the twelve nearest neighbours to the central titanium atom. Note that the A-specturm, which lacks this structure, is only a smeared-out version of the C-spectrum.

The intensities of the ODMR lines are of interest. Both low- and high-field lines in the spectrum of Fig. 12.26 correspond to increases in intensity for the Ti^C centre, whereas for the Ti^A centre the low-field line increases and the high-field line decreases in intensity. The intensities are a maximum with *B* parallel to the crystallographic c-axis ($\theta=0°$) and decrease to zero at $\theta=54.7°$. The relative intensities of the Ti^A, $\Delta M_s = \pm 1$ lines reverse in sign after crossing at this orientation, whereas the Ti^C lines do not. These results show that the triplet states are in thermal equilibrium in high magnetic field for the Ti^A-centre, implying that $T_1 < \tau_R$. However in the Ti^C-centre the spin sub-levels are not in equilibrium, i.e. $T_1 > \tau_R$, and since radiative decay from $M_s = 0$ is forbidden, population builds up there at the expense of the emitting levels $M_s = \pm 1$. The signs of the intensity changes of the ODMR transitions show that for the A-spectrum D is positive, whereas for the C-spectrum D is negative.

In Fig. 12.27 we show a high-resolution spectral dependence of the Ti^A and Ti^C ODMR spectra for comparison with the luminescence. The emission shows A_0, B_0, and C_0 zero-phonon lines and a number of imperfectly resolved phonon sidebands. When the magnetic field is tuned to the $\Delta M_s = 0 \leftarrow 1$ ODMR line of the Ti^C spectrum and the monochromator scanned through the emission spectrum, one observes only the C_0 zero-phonon line and its phonon sideband. The strong 26 meV phonon replication occurs up to third order, whereas the 117 meV line is probably a harmonic mixture of the 26 meV and 92 meV phonons. This vibronic interaction appears to be largely due to a local mode of the titanium impurity. In contrast the spectral dependence of the Ti_A ODMR spectrum shows weak transverse acoustical phonons at 30 and 65 meV, strong longitudinal acoustical phonons at 90 meV as well as optical phonon replicas at 107 meV. The resolution afforded by this ODMR spectrum is a dramatic demonstration of the power of this technique.

In summary, three distinctly different triplet excitons are observed bound to titanium impurities in silicon carbide. The ODMR is characterized by a large crystal field splitting, which shows an unusually large isotopic shift, some 50 times larger than any values for isotopes with comparable masses. The results (Lee *et al.* 1981; Lee 1983) show that an isolated titanium atom occupying a site with six equivalent silicon neighbours is involved. Thus titanium cannot be substituted on the carbon sublattice, at which the carbon site has only four nearest-neighbour silicon sites. The silicon site has twelve silicon neighbours; to reduce this to six requires vibrational coupling to titanium modes of vibration along the c-axis having A_1 symmetry. This is consistent with the ligand field model of Patrick and Choyke (1974) which yields an empty localized d_e orbital in the bandgap for substitutional titanium; this level acts as an efficient exciton trap by first localizing an electron and then binding the hole through the Coulomb attraction. Thus the hole state is rather more diffuse than the electron state. For a two-particle system, in which one particle is localized and the other diffuse, the central hyperfine structure is expected to be less than that of a single particle system. For the exciton trapped at a titanium centre in silicon carbide the observed central hyperfine constant is $A \simeq 5 \times 10^{-4}$ cm^{-1}, roughly half that of the Ti^{2+} ion in tetrahedrally coordinated II–VI compounds. Furthermore, a delocalized hole state on the titanium atom should lead to stronger vibronic coupling for the hole than for the electron. Assuming a p_z-like orbital for the hole state, which transforms as the irreducible representation A_1 in hexagonal symmetry, only coupling to A_1 phonon modes is expected. Such coupling involves predominantly the six silicon atoms, three above and three below the plane containing the titanium atom. This would explain why the isotopic effect involve six rather than twelve silicon atoms.

12.4.3 *Ionic crystals*

Let us return briefly to the analogy between the energy levels of an exciton and the hydrogen atom. The 'orbital radius' of the exciton can be written in terms of the radius of the 1 s orbital of atomic hydrogen, a_0, as

$$a_n = \kappa \left(\frac{m}{\mu}\right) n^2 a_0 \tag{12.39}$$

where $n = 1, 2, 3$, etc. Since eqns (12.37) and (12.39) involve the macroscopic dielectric constant, the Wannier exciton concept implies an orbital radius, a_x, much larger than the diameter of the first Bohr-like orbit of hydrogen. In fact scale of orbit is a very convenient means of differentiating between excitons in various materials. For ionic crystals, where the electrostatic interactions are large, the distortion around a charge carrier involves fewer ions, the displacements of which are large. The lowering of free energy may then depend strongly on the position of the charge; this may lead to *self-trapping.* Although the possibility of self-trapping of electrons in ionic solids was studied theoretically over forty years ago (Gurney and Mott 1937) there is no experimental evidence for this process. However, in some ionic crystals positive holes may be self-trapped: in the alkali halides and alkaline earth fluorides the hole is bound to two adjacent halide ions resulting in the $[X_2^-]$ molecular ion (Kabler 1972; Hayes and Stoneham 1974). In silver halide crystals, however, the hole is self-trapped at Ag^+ ions so forming Ag^{2+} ions. A similar situation occurs in caesium fluoride (Hayes 1978).

In ionic crystals the self-trapped hole may trap an electron in an excited orbital resulting in the formation of a self-trapped exciton. Such excitons play an important role in defect creation in the alkali halides, where the primary products of radiolysis (F-centre $+X_2^-$-centre) result from non-radiative decay of the exciton. Generally, irradiation using photons with energy larger than the band gap is necessary for exciton formation in an excited state. If we use the molecular orbital approach, developed in Chapter 7 for the $[X_2^-]$-centre, then the (unstable) ground state of the self-trapped exciton is $(\sigma_g, np)^2$ $(\pi_u, np)^4(\pi_g, np)^4(\sigma_u, np)^2$, giving a $^1\Sigma_g$ state. Excited states are obtained by promoting an electron from the $(\sigma_u, np)^2$ orbital to give $(\sigma_u, np)\,(\sigma_g, (n+1)s)$.

This configuration yields $^1\Sigma_u$ and $^3\Sigma_u$ states, and it is evident that a ladder of triplet and singlet states of both Σ and π symmetry exist up to an ionization limit. (Alternatively we may use the group-theoretical description as used above in the discussion of divacancy centres.)

Recombination of electrons and holes created by X-irradiation of some alkali halides leads to strong luminescence. Generally decay to the ground state is a competitive process between luminescence and defect formation. The luminescence may be in two broad bands. We show in Fig. 12.28(a) as examples the X-ray-induced luminescence from potassium bromide and sodium chloride. For potassium bromide the emission bands at $\lambda = 230$ nm and 380 nm, respectively, have radiative lifetimes of 2.8 ns and 0.3 ns at liquid helium temperature. The high-energy band is σ-polarized with the electric vector parallel to the molecular axis of the centre (i.e. $\langle 110 \rangle$ directions), whereas the low-energy band is partially π-polarized due to $^1\Pi_u$ admixture into the triplet state by spin–orbit coupling. That the $^3\Sigma_u$ state is metastable enables one to measure transient absorption to higher-lying states. Such a measurement for sodium chloride is shown in Fig. 12.28(b). Absorption in the

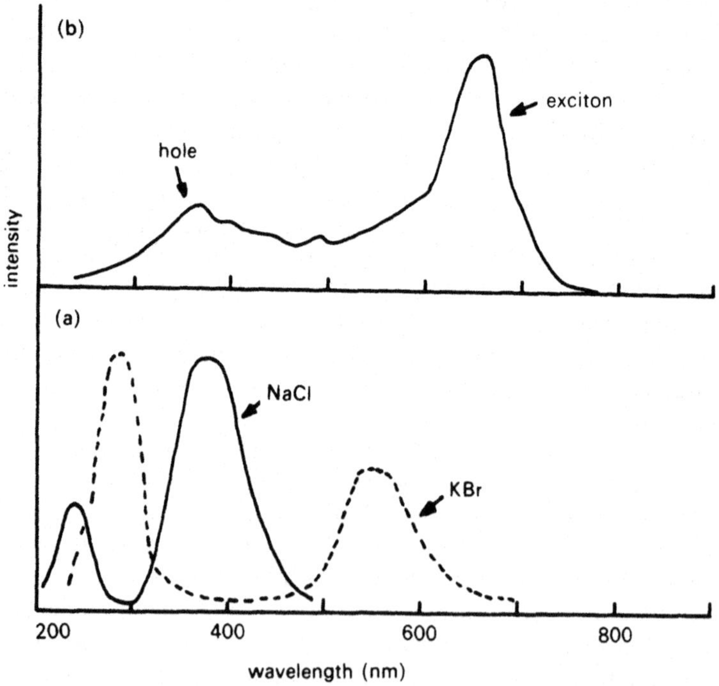

FIG. 12.28. (a) X-ray induced luminescence from excitons in sodium chloride and potassium bromide measured at 4.2 K (after Kabler 1964). (b) shows metastable absorption due to transitions out of the lowest lying triplet, $^3\Sigma_u$, of the trapped exciton in sodium chloride. (After Fuller *et al.* 1970.)

range 700–500 nm is interpreted as due to a Rydberg-like series of transitions of the exciton, whereas the higher energy absorption is interpreted as excitation of the hole. To a greater or lesser extent similar behaviour has been reported for self-trapped excitons in alkali, silver, and caesium halides, alkaline earth fluorides, and even large-bandgap oxides.

That the low-energy luminescence bands involve triplet→singlet transitions is suggested by the large Stokes shifts and long lifetimes. ODMR spectra have now been reported for most of the materials listed above. In silver chloride under X-irradiation the emission band from the triplet state occurs at $\lambda = 490$ nm at 4.2 K with a bandwidth of 50 nm. The ODMR spectrum in Fig. 12.29(a) shows two intense lines due to fine structure transitions ($M_s = 0 \rightarrow \pm 1$) of those excitons with the distortion axis parallel to B. Note that in Fig. 12.29(b) the signal represents the circular polarization $P = (I_- - I_+)/(I_- + I_+)$; since this quantity is negative for the low-field line. D is negative and $M_s = 0$ lies below $M_s = \pm 1$. From the orientation dependence of the spectrum Hayes *et al.* (1977) determine that the distortion axis is parallel to the crystal $\langle 100 \rangle$ direction confirming the excition to be self-trapped at a single Ag^+ ion. Excitons in $KMgF_3$ and $YLiF_4$ behave much like the alkali halides, whereas in alkaline earth fluorides electron recombination at a self-trapped hole leads to the centre occupying an interstitial site. Thus the self-trapped exciton is related to the H-centre rather than the self-trapped hole, in that the molecular axis is parallel to a crystal $\langle 111 \rangle$ direction. Triplet excitons have been observed in $YAlO_3$ and SiO_2 during X-irradiation but not in aluminium, magnesium, and calcium oxides.

12.5 Donor–acceptor recombination in semiconductors

Although bandgap excitation generates electrons and holes in the conduction and valence bands, respectively, and may result in radiative decay via triplet exciton states, other decay routes exist involving donor (D^+) and acceptor (A^-) centres. The electrons and holes released by bandgap excitation are trapped at donors (D^+) and acceptors (A^-), respectively, according to the reaction $D^+ + A^- + \text{photon} \rightarrow D^0 + A^0$. The recombination process, $D^0 + A^0 = D^+ + A^- + \text{photon}$, then involves energy changes associated with electron-hole annihilation of

$$E = E_G - (E_D + E_A) + e^2/4\pi\varepsilon_0 \kappa r \tag{12.40}$$

in which E_D and E_A are the donor and acceptor binding energies, κ is the dielectric constant, and r is the D–A separation. The Coulomb term is positive because the final pair state after recombination involves ionized donors (D^+) and acceptors (A^-). Recombination radiation then occurs according to the equation

$$(D^0, A^0) \rightarrow (D^+, A^-) + h\nu(r) \tag{12.41}$$

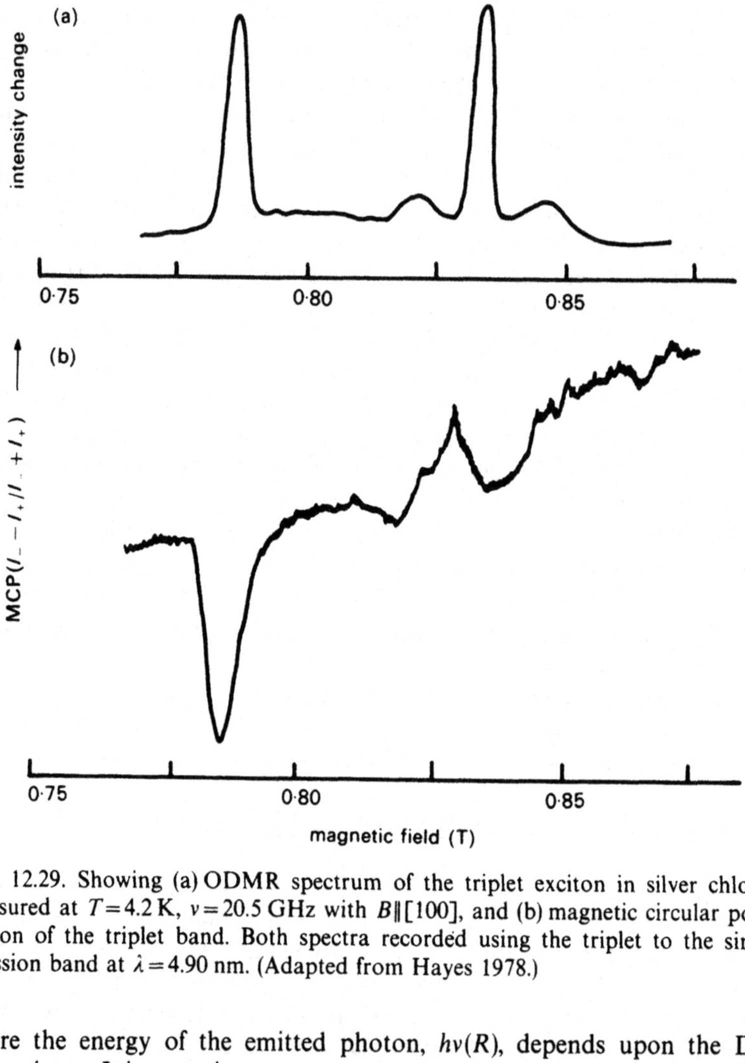

FIG. 12.29. Showing (a) ODMR spectrum of the triplet exciton in silver chloride measured at $T=4.2\,\text{K}$, $\nu=20.5\,\text{GHz}$ with $B\|[100]$, and (b) magnetic circular polarization of the triplet band. Both spectra recorded using the triplet to the singlet emission band at $\lambda=4.90\,\text{nm}$. (Adapted from Hayes 1978.)

where the energy of the emitted photon, $h\nu(R)$, depends upon the D–A separation, r. It is sometimes necessary to modify eqn (12.40) by including the energy of any photon(s) absorbed or emitted in the recombination process. Discrete emission lines are observed for close pairs, whereas recombining distant pairs emit in broad bands. The radiative transition probability, $P(r)$ also depends upon the D–A separation through

$$P(R)=P_0\exp(-2r/a_0) \tag{12.42}$$

where a_0 is the Bohr radius of the shallowest defect and P_0 is the radiative transition probability in the limit that $r\to 0$. The many sharp luminescence lines observed in gallium phosphide (Dean *et al.* 1970) were interpreted as

D–A recombination. However, in wide-gap II–VI compounds broad luminescence bands observed at longer wavelengths were also attributed to D–A recombination. Three types of observation confirmed this rather convincingly. Measurement of the radiative lifetime, τ_R, at different wavelengths across the broad band showed that τ_R increased at longer wavelengths. This followed from the fact that the longer wavelength emission involves the more distant pairs; since τ_R varies as the reciprocal probability, $P(r)^{-1}$, it should increase exponentially with r in view of eqn (12.42). Time-resolved luminescence measurements showed that the emission peak shifts to longer wavelengths as the time delay is increased. Secondly, the longer the radiative lifetime the easier is it to saturate the transition. Consequently the peak in the emission band should move to shorter wavelengths with increasing excitation power, because the most distant pairs saturate at the lowest power. Such shifts are indeed observed. Thus the D–A nature of the luminescence, even though broadband, may be determined. However, except where specific dopants were used, it was often not possible to recognize the chemical species involved in recombination simply from a luminescence spectrum. Once again the spin selectivity of ODMR was of great advantage. This will be illustrated with a few examples of ODMR studies of recombination processes in II–VI and III–V semiconductors.

In general, there are four categories: shallow donor–shallow acceptor pairs, deep donor–deep acceptor pairs, shallow donor–deep acceptor pairs and shallow acceptor–deep donor pairs. The principle of D–A ODMR is shown in Fig. 12.30(a). Both donor D^0, and acceptor, A^0, are assumed to have $S=\frac{1}{2}$ with different g-values so that the Zeeman splittings of these spin doublets are unequal. The four possible recombination paths, two $\Delta M_s=0$ transitions and two $\Delta M_s=\pm 1$ transitions are shown. Even though donors and acceptors are physically separated, a weak exchange interaction will influence the energies of the magnetic sub-levels. If spin–orbit coupling is small, and since the final state of the pair (D^+, A^-) has $S=0$, the selection rule for allowed electric dipole transitions is $\Delta S=0$; i.e. $\Delta M_s=\pm 1$ transitions are forbidden. However, spin–orbit coupling mixes the spin states and will provide different but non-zero transition probabilities in all four decay channels. Since the radiative transition probability, $P(r)$, is determined by the overlap integral of electron and hole wavefunctions, it is advantageous for either donor or acceptor to be spatially diffuse. This is most likely for effective mass-like states. The populations of the magnetic sub-levels are determined by the relative pump rates, radiative transition rates (τ_R^{-1}) and spin–lattice relaxation rates (T_1^{-1}) between the Zeeman levels. When $T_1<\tau_R$, we expect the populations of the four levels to be in thermal equilibrium. Application of microwaves resonant with either the donor or acceptor Zeeman splittings, will induce ESR transitions and disturb the thermal equilibrium populations. Since the transition probabilities are in general different, both donor and acceptor resonances increase

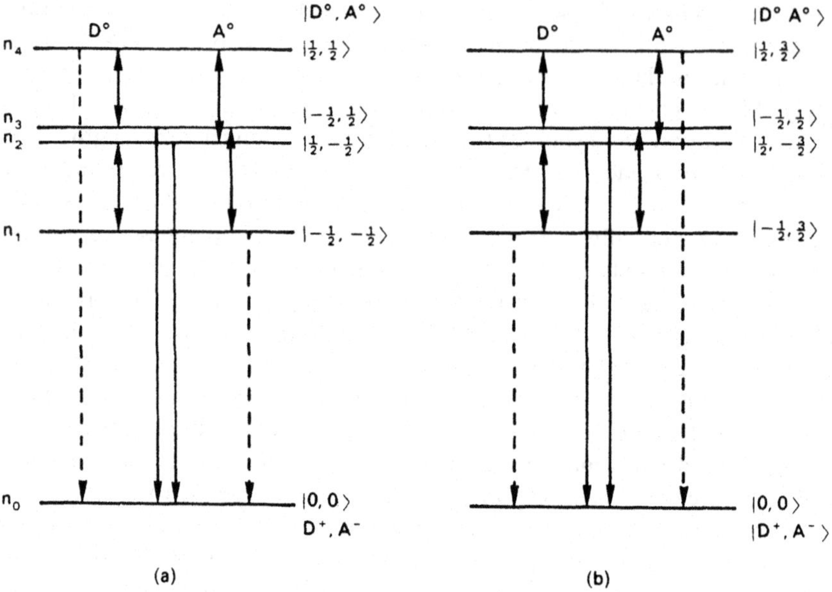

FIG. 12.30. The selection rules for D^0A^0 recombination ODMR for (a) $S=\frac{1}{2}$ and small spin–orbit coupling, and (b) $J=|\frac{3}{2}, \pm\frac{3}{2}\rangle$ acceptors.

the total luminescence intensity. However, the luminescence changes are usually quite small. Indeed, there may be no change in the total light intensity. It is then necessary to detect changes in the σ_+ and σ_- polarized intensities. For spin states with $T_1 > \tau_R$, we observe a population build up in the $|-\frac{1}{2}, -\frac{1}{2}\rangle$, $|\frac{1}{2}, \frac{1}{2}\rangle$ levels, i.e. populations n_1 and n_4, because optical transitions from these levels are relatively weak. Donor and acceptor ESR then decrease the population of these levels (i.e. n_1 and n_4), resulting in intensity decreases for σ_+- and σ_--polarized transitions at appropriate values of the magnetic field. There are concomitant increases in the intensity of the $\Delta S=0$ transitions. The effects may be quite large and 1–2 per cent increase in total intensity have been observed. In consequence, whether or not the spin states are in equilibrium both donor and acceptor resonances change the emission intensity, and so both may be detected.

12.5.1 Shallow donor–shallow acceptor pairs

Generally for shallow donor–shallow acceptor pairs the magnetic sub-levels are in thermal equilibrium. However, in cubic crystals the ESR spectra of shallow acceptors are not normally observed because of the degeneracy of the $J=\frac{1}{2}$ and $J=\frac{3}{2}$ valence bands and their sensitivity to crystal distortions. In hexagonal crystals this degeneracy is raised, and acceptor resonances have been observed. The first ODMR spectra were reported for shallow donor–acceptor pairs in cadmium sulphide, zinc selenide, and zinc telluride (Brunwin *et al.* 1976; Davies 1985; Killoran *et al.* 1981). The crystal structure of cadmium sulphide is hexagonal, C_{3v}; the $S=\frac{1}{2}$ conduction band is slightly anisotropic. The acceptor level is derived from the upper valence band ($J=\frac{3}{2}$; $\pm\frac{3}{2}$). The selection rule for circularly polarized light with B parallel to the c-axis is $\Delta M_s=\pm 1$; these are shown by the unbroken arrows in Fig. 12.30(b). The green edge emission from cadmium sulphide, measured at $T=2$ K with light propagating parallel to both the c-axis and static magnetic field, is shown in Fig. 12.31(a). The ODMR signal measured as the change in σ_--polarized light is shown in Fig. 12.31(b), and the spectral dependence of the resonance is shown in Fig. 12.31(c). The change in light intensity at resonance is about 0.05 per cent. The measured g-values of $g_\parallel=1.789$ and $g_\perp=1.769$ confirm that donor resonances are involved. The spectral dependence of the ODMR shows that only the D–A pair bands are involved.

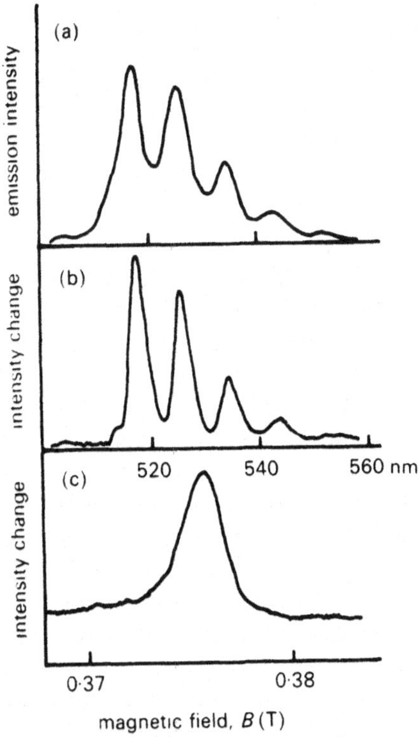

FIG. 12.31. Showing (a) the green edge emission in cadmium sulphide, (b) the spectral dependence of the ODMR spectrum, and (c) the ODMR spectrum measured at $T=1.6$ K, $\nu\cong 9.2$ GHz with B parallel to [001]. (After Brunwin *et al.* 1976.)

Subsequent measurements with B in directions away from the c-axis reveal the acceptor resonances (Patel *et al.* 1981). This followed the first ODMR observation of a shallow acceptor in a semiconductor, viz., the shallow aluminium acceptor in hexagonal silicon chloride (Lee *et al.* 1980).

12.5.2 Shallow acceptors in gallium phosphide

The difficulties in observing the ESR spectrum of shallow acceptors in cubic crystals by magnetic resonance techniques are associated with the degeneracy existing at the top of the valence band. The wavefunction of effective-mass-like acceptors is sensitive to strains, inevitable in materials grown at high temperature; this results in excessive line broadening of ESR transitions. The problem was first solved for shallow acceptors in silicon by Feher *et al.* (1960) using conventional ESR. They applied a uniaxial stress larger than the mean random internal stress in order to raise the degeneracy at the top of the valence band. Thus they were able to select a Kramers doublet in which to observe the ESR. Nowadays improved crystal growth techniques for silicon have eliminated this source of difficulty. For other materials, however, there has been no such improvement and one must resort to the uniaxial stress technique. This is the case also in ODMR.

The first detection of the ODMR of shallow acceptors in a cubic semiconductor under stress was observed in p-type GaP:Zn (Fig. 12.32). The luminescence spectrum of GaP:Zn (Fig. 12.33(a)) also shows the strong band at 1.15 μm observed in 'pure' gallium phosphide (Fig. 12.24), and which is associated with the decay of triplet excitons at $P_{Ga}Y_p$ centres (at least, in part). The ODMR spectra in Fig. 12.32 were detected by monitoring the total luminescence output in the wavelength range 0.8–1.8 μm with the magnetic field, B, in a crystal [100] direction (Lee 1983). At zero stress, Fig. 12.32(a), the strong P–P_4 spectrum is entirely due to donors. The interesting feature is the effect of an increasing applied stress (Fig. 12.32 (b)–(f)), manifest by the evolution of the resonance near $B = 1.1$ T attributed to A_{Zn}. The spectral dependences of the A_{Zn} resonances, although rather complex (Fig. 12.33 (b)–(e)), show several overlapping bands due to D–A processes including deep donors; they may be interpreted as the sum of three bands at 950 nm, 1.35 μm, and 1.47 μm. Some simplification is afforded by detecting the ODMR spectrum through a low-resolution monochromator, which acts as a variable bandpass filter centred on the wavelength shown (Fig. 12.34). The A_{Zn} resonance is common to each spectrum: the spectrum obtained using the 1.5 μm region (not shown here) is the same as that obtained by detecting the total infrared output. However, when the 950 nm band is monitored a new weak donor resonance with $g = 2.025$ is observed. In addition, observations using the visible emission confirm two bands centred at wavelengths of *c.* 570 nm and 620 nm to be involved. The longer wavelength visible band exhibits the D_s and weak A_{Zn} resonances, whereas the 2.0 eV band shows the

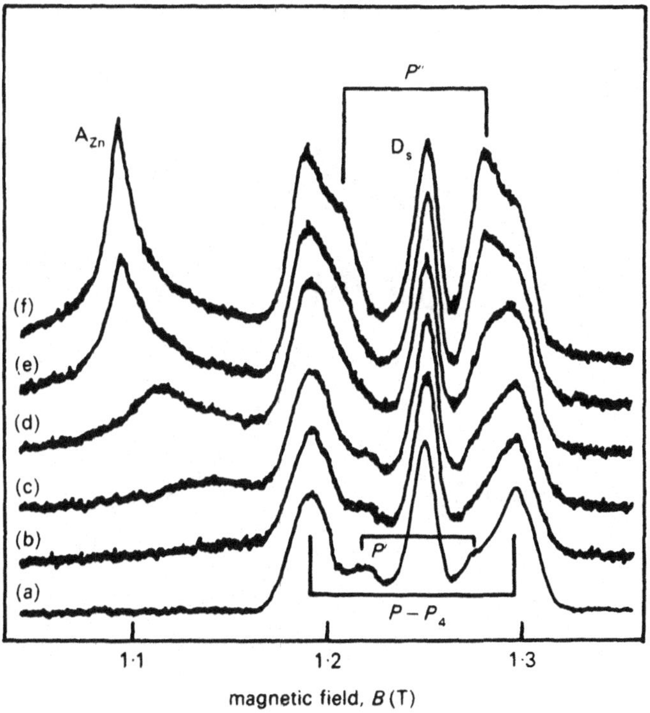

FIG. 12.32. The ODMR spectrum of GaP:Zn under uniaxial stress (a) $P=0$, (b) $P=50$ MPa, (c) $P=160$ MPa, (d) $P=270$ MPa, (e) $P=330$ MPa, and (f) $P=500$ MPa, recorded at $T=1.6$ K, $\nu \sim 35$ GHz with B along the [001] axis. (After Lee 1983.)

shallow acceptor (A_{Zn}) and a broader resonance D_Y at $g \simeq 2.01$. By comparing ESR g-values the 950 nm band was assigned to D–A pair recombination between shallow zinc acceptors and deep oxygen donors on phosphorus sites (O_p centres).

The spectral dependences of these resonances and of the bound exciton resonances (Section 12.4.2) show that recombination processes in gallium phosphide are complex. As discussed earlier, bound excitons recombine via triplet emission in the $\lambda = 1.15$ μm band associated with the $P_{Ga} \cdot Y_p$ pair. The D_s ODMR signal observed in both the 570 nm and 1.2 μm bands is associated with shallow ($S=\frac{1}{2}$) donors—probably sulphur—and unknown deep acceptors. The 1.35 μm band and its associated antisite resonance and stress-induced A_{Zn} signal is attributed to D–A pair recombination involving the double donor P_{Ga}^{+} and the shallow zinc acceptor. Fig. 12.34(c) shows that the A_{Zn} resonance is also associated with a band at 1.0 μm, which also involves the broad unknown deep donor resonance D_x. Two D–A processes are

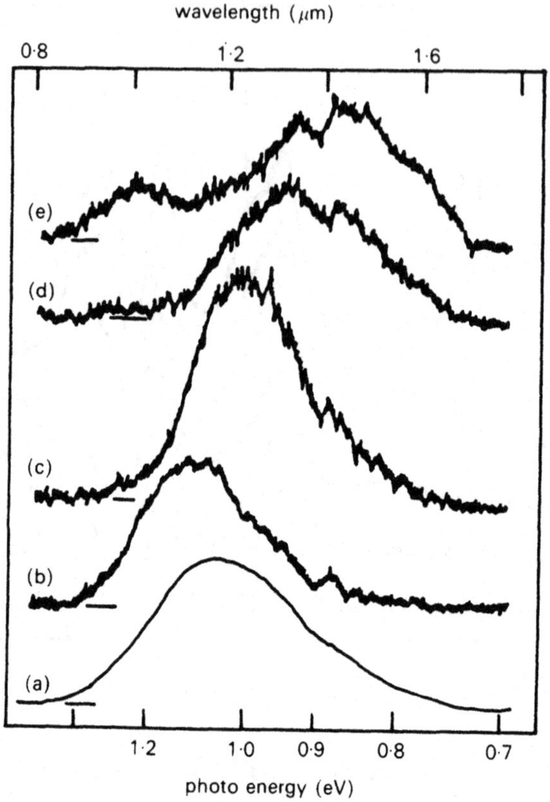

FIG. 12.33. (a) The total infrared emission signal in GaP:Zn. The spectral dependencies of the ODMR signals in this sample are shown in (b) $P_{Ga}Y_P$ spectrum, (c) isotropic D_s spectrum, (d) P-P4, and (e) stress-induced A_{zn} resonance. (After Lee 1983.)

observed in the visible spectrum, the bands having peak wavelengths of 620 nm and 570 nm (Fig. 12.34). The former involves the shallow zinc acceptor, A_{Zn}, and an unknown donor, D_y. In view of the very weak A_{Zn} resonance and the strong D_s resonance, it is not clear whether the shallow zinc acceptor is involved in the 570 nm D–A pair band. Although these D–A processes are very complex, the ODMR method once again affords a very detailed description of the D–A recombination process, irrespective of whether the donors and/or acceptors are deep or shallow.

12.6 Spin memory effects in ODMR

In Chapter 6 we discussed how spin memory may produce the excited-state population difference needed for the observation of ODMR. Spin memory is

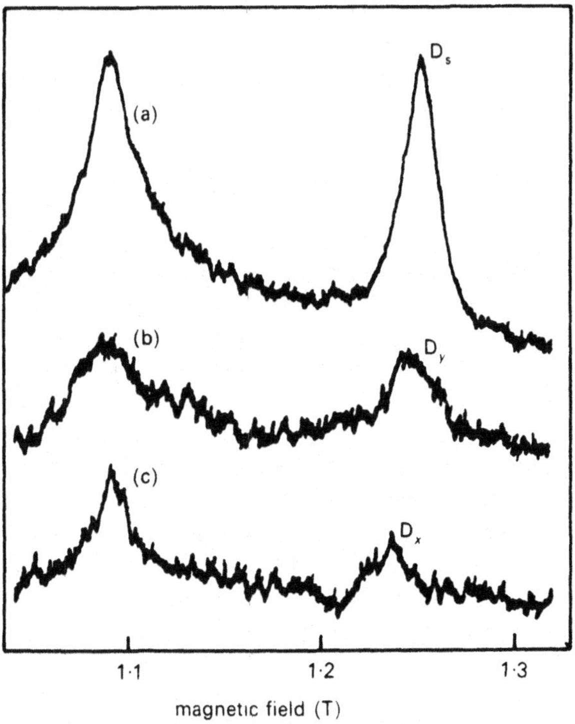

FIG. 12.34. ODMR spectra of GaP:Zn measured at $\nu \sim 35$ GHz, $T = 1.6$ K under a uniaxial stress of 300 MPa, obtained by monitoring light in wavelength regions: (a) visible range (550–800 nm) (b) 640 nm (c) $\lambda \sim 1$ μm. (After Lee 1983.)

a rather general process which selectively populates magnetic sublevels of the excited states as a consequence of there being spin selection rules for the absorption and emission transitions and also for the non-radiative decay process. Spin memory may be observed as a change in the intensity pattern of the Zeeman transitions from an excited state when the *ground state* undergoes saturated microwave pumping at the resonance frequency. The first direct experimental evidence of spin memory was in measurements on the $^2E \rightarrow {}^4A_2$ fluorescence from $3d^3$ ions in Al_2O_3. Imbusch and Geschwind (1966) used unpolarized light to pump the $^4T_2$ absorption bands. The $^4A_2 \rightarrow {}^4T_2$ electric dipole transition conserves electron spin, i.e. $\Delta M_s = 0$. In consequence such transitions transfer $|^4A_2 - \frac{3}{2}\rangle$ spin population to the $|^4T_2 - \frac{3}{2}\rangle$ level, $|^4A_2 - \frac{1}{2}\rangle$ population to $|^4T_2 - \frac{1}{2}\rangle$, etc, etc, so that the population distribution among the $|^4T_2 M_s\rangle$ sublevels is just that of the $|^4A_2 M_s\rangle$ ground state levels. However the non-radiative $^4T_2 \rightarrow {}^2E$ transitions occur by virtue of spin–orbit coupling, for which the selection rule is $\Delta M_s = 0, \pm 1$. Consequently the $|^4T_2 - \frac{3}{2}\rangle$ spin level can transfer population only to the

$|^2E-\frac{1}{2}\rangle$ level, whereas $|^4T_2-\frac{1}{2}\rangle$ can transfer population to both $|^2E\pm\frac{1}{2}\rangle$ levels. Similar arguments apply to $|^4T_2+\frac{1}{2}\rangle$ and $|^4T_2+\frac{3}{2}\rangle$ levels. The overall result is that the $|^2E\pm\frac{1}{2}\rangle$ population distribution reflects that of the ground state spin system in thermal equilibrium, when accessed using broad-band unpolarized pumping into the $^4T_2$ state.

Optical pumping with broad-band unpolarized light is not necessarily the most effective means of producing the population difference in the excited state. Since the same selection rules for linearly and circularly polarized light apply for the direct $^4A_2 \rightarrow {}^2E$ absorption transitions as apply to the $^2E \rightarrow {}^4A_2$ luminescence transitions we see from Fig. 12.2 that pumping directly into the R_1 line in ruby with σ_- light will selectively populate the $|u_+ -\frac{1}{2}\rangle$ state. This occurs because of the 6:4 ratio in the strengths of the two σ_- transitions and because of the greater population in the $|^4A_2-\frac{3}{2}\rangle$ state. Judicious use of the selection rules to populate the excited state may therefore be expected to produce a much larger population difference. Note that spin memory is observed only if the spin–lattice relaxation time (T_1) in the excited state is longer than the radiative decay time (τ_R). If τ_R is much longer than T_1 the populations in the excited levels have time to come to equilibrium before luminescence occurs, resulting in the loss of memory of the ground state spin system by the excited state spin system. For the ^2E level of Cr^{3+} in aluminium oxide Imbusch *et al.* (1967) found that at 3 K $T_1 \simeq \tau_R$, so that spin memory effects were observed only below 3 K. Similar effects were observed for V^{2+} and Mn^{4+} ions in aluminium oxide. In an analogous fashion, spin memory has been demonstrated in the optical pumping cycle of Tm^{2+} ions in calcium fluoride where the selection rules lead to excited-state population inversion and *nuclear* spin memory (Anderson and Sabisky, 1966).

That spin memory is a rather general phenomenon in solids is apparently surprising, since during the absorption–luminescence cycle the excited electron may transfer some 1–2 eV of energy to the lattice. Since phonon energies are much smaller than this, de-excitation proceeds by multiphonon processes which may involve crossovers from one electronic level to another (so-called intersystem crossings). Phonon transitions between different electronic states are induced by an interaction in which phonons modulate the orbital angular momentum of states of the electron. Obviously since this orbit–lattice interaction operates only on orbital coordinates it does not *directly* affect real spins. However, spin–orbit coupling in conjunction with orbit–lattice coupling can cause spin transitions during non-radiative transitions. For crystal field energies much stronger than spin–orbit coupling energies, so that M_l and M_s are good quantum numbers, the fractional mixing of one M_s state with another is of order ζ/Δ, where ζ and Δ are the spin–orbit coupling parameter and crystal field splitting respectively. Consequently the larger the crystal field splitting, the greater the selective feeding of M_s states and hence memory of the ground-state spin polarization. The ratio ζ/Δ is small for transition metal ions, colour centres, and triplet states of aromatic molecules. That spin selection rules operate during non-radiative transitions was first reported for exchange-coupled pairs of F^+-centres in calcium oxide by Tanimoto *et al.* (1965). Similar behaviour was subsequently reported for analogous defects in magnesium oxide (Henderson 1976) and for F_2^{2+}-centres in magnesium oxide (Henderson *et al.* 1981) and calcium oxide (Gravesteijn *et al.* 1977). Spin memory is an important effect in the measurement of excited state ESR for F-centres and Tm^{2+} ions in halide crystals and of conduction electron spin resonance in semiconductors.

12.6.1 Spin memory and magnetic circular dichroism of F*-centres*

In Section 6.6.3 the magnetic circular dichroism (MCD) signal for transitions between two Kramers doublets was shown to be

$$\frac{\Delta I}{I_{\mathrm{DC}}} = -\alpha_0(\nu)g(\Delta\nu l)(n_- - n_+) \tag{12.43}$$

This expression applies when spin–lattice relaxation in the excited Kramers doublet is very efficient relative to radiative decay, so that all spin memory is destroyed. The population difference $(n_- - n_+)$ is then the thermal equilibrium value, i.e. $\tanh(g_g\mu_B B/2kT)$. However, in the presence of spin memory the dynamics of the optical pumping cycle are rather different, and eqn (6.52) must be modified. A detailed analysis of the effects of spin memory on the MCD signal of F-centres in alkali halides (Mollenauer *et al.* 1969) was generalized by Geschwind (1972) for application to such systems as Cr^{3+} in alumina and Tm^{2+} in alkaline earth fluorides (Anderson and Sabisky 1966). In these systems spin memory in the optical pumping cycle makes it possible to detect both ground- and excited-state ESR spectra in measurements of the MCD signal. We discuss only the case of F-centres, and assume that the ground and relaxed excited states may be treated as Kramers doublets. The discussion given in Section 6.6.3 is then appropriate.

The optical pumping cycle of F-centres in alkali halides is shown in Fig. 12.35. Circularly polarized light is used to pump the 1s–2p-like absorption band; vibronic interaction takes the system into the relaxed excited state, which decays radiatively back to the ground state. The four rate equations which couple the ground and excited state spin populations contain terms due to optical pumping from the $M_s = \pm\frac{1}{2}$ ground-state levels $(n_\pm)$, spin–lattice relaxation (T_1 and T_1^e) in the ground and excited states, respectively, loss of spin memory (γ) and radiative decay (τ_R). In dynamic equilibrium we equate these rate equations to zero and deduce that the ground-state polarization $P_G = (n_+ - n_-)$, under conditions of perfect spin memory during optical pumping, is given by

$$P_G = \frac{P_S - (T_P/T_1)\tanh(g_g\mu_B B/2kT)}{1 + (T_P/T_1)}. \tag{12.44}$$

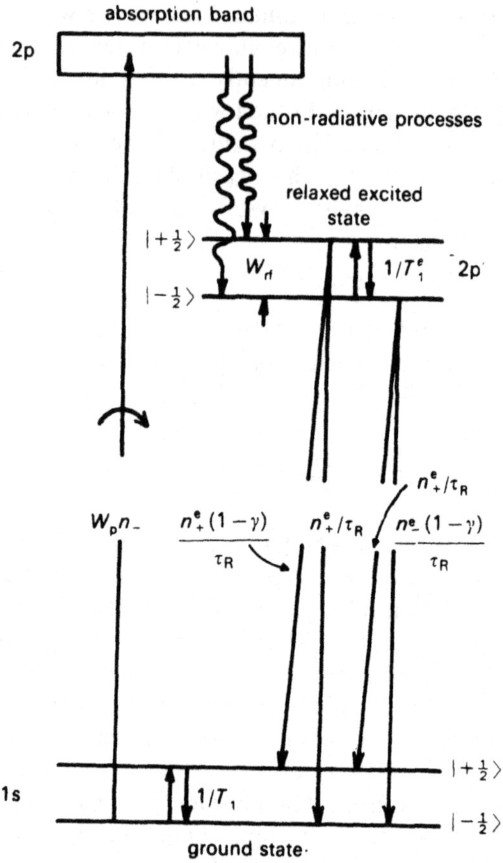

FIG. 12.35. The optical pumping cycle of F-centres: circularly polarized pump radiation transfers population of the $M_s = \pm\frac{1}{2}$ level of the 1s ground state to the unrelaxed $M_s = +\frac{1}{2}$ excited state, 2p. The combined effect of electron–phonon and spin–orbit couplings partially destroys memory of the ground-state spin polarization in proceeding to the relaxed excited state, 2p. The remaining spin memory is removed by microwave-induced spin–flip transitions between the $M_s = +\frac{1}{2}$ levels of the relaxed excited state (After Mollenauer *et al.* 1969).

The excited-state decay rate, T_p^{-1}, involves radiative decay, τ_R^{-1}, spin–lattice relaxation in the excited state, $(T_1^e)^{-1}$, and the loss of spin memory, γ. For the spin orientation to be maintained throughout the optical pumping cycle requires that $T_1^e \to \infty$, in which case $T_p \to \infty$ also and $P_G \to \tanh(g_g \mu_B / 2kT)$, the thermal equilibrium value. As spin memory is lost in the optical pumping cycle T_p becomes finite, and at high pump intensity $T_p \ll T_1$ so that $P_G \to P_s$, the saturated ground-state polarization. The extreme values of P_s are ± 1, depending upon the sense of circular polarization used to pump the ground-state spin system. Obviously we can use eqn (12.44) to determine both T_1 and

T_1^e via the dependence of P_G upon the magnetic field and temperature. Also, because the spin populations can be changed by microwave-induced transitions the measurement of MCD in the presence of spin memory can be used to detect ESR in both ground and excited states.

As discussed in Chapter 6 the MCD signal is strongly wavelength dependent. Mollenauer *et al.* (1969, 1971) used this to advantage in their studies of the optical pumping cycle of F-centres in potassium halides. The F-band was pumped using a mercury arc lamp, with an interference filter to select the positive peak in the MCD pattern (Fig. 6.14(c)). The ground-state polarization, P_G, shifted away from the thermal equilibrium value by the pump beam, is monitored with light set at the wavelength of the negative MCD peak. The probe beam is alternatively switched between σ_+- and σ_--polarizations by a piezo-optic modulator. If the pump light is pulsed on/off then the probe beam measures the time involved in re-establishing the thermal equilibrium value of the ground-state polarization (i.e. T_1). Such measurements of T_1 have advantages relative to pulsed microwave methods in that they probe the entire ground-state magnetization and that being non-resonant they may be used to measure T_1 over a very wide range of magnetic fields. Panepucci and Mollenauer (1968) made T_1 measurements in fields up

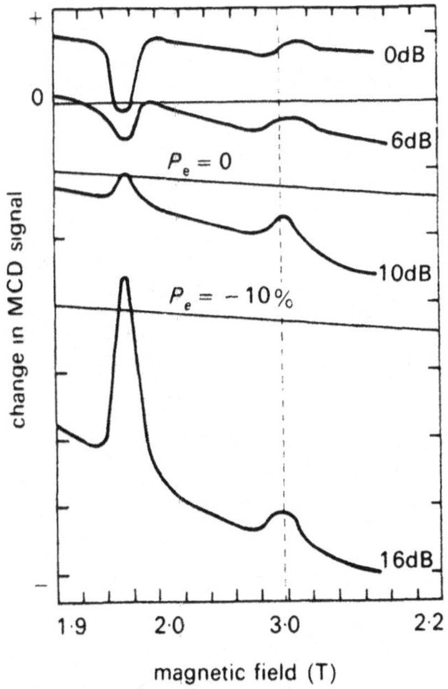

FIG. 12.36. The optically detected ESR spectra of F-centres in potassium chloride in the ground state (low-field line) and excited state (high-field line) measured at $T = 1.2$ K, $\nu = 52$ GHz with $B \parallel [100]$. (After Mollenauer *et al.* 1969.)

to 7 T. At low field ($B < 0.2$ T) the very long relaxation times were associated with F-centre clusters. Above 0.2 T the relaxation times obey

$$\frac{1}{T_1} = (\alpha B^3 + \beta B^5)\coth\left(\frac{g\mu_B B}{2kT}\right) \tag{12.45}$$

where α and β are constants. The first term is due to relaxation via phonon modulation of the hyperfine interaction, whereas the B^5 term is due to the direct process (Chapter 5). Since the latter term is weak for F-centres it can only be measured in fields greater than about 1.5 T. Similar studies of Tm^{2+} in alkaline earth halides also yielded a very weak direct process. This behaviour suggests that cross-relaxation via a fast-relaxing impurity is important at low temperature.

The same experimental arrangement, with the addition only of microwaves, may be used for detecting the excited state ESR via measurement of the MCD. As was the case with fluorescence detection of ODMR the sensitivity of this MCD experiment is quite remarkable. The weakest signals (F-centres in potassium bromide) measure an excited-state population of 3×10^5 spins or 10^3 spins per 0.1 mT linewidth. The method of Mollenauer *et al.* (1969, 1971) relied on the almost complete spin memory of F-centres during optical pumping. A resonant microwave field, B_1, partially removes this spin memory and changes the value of P_G measured by the probe beam. The ODMR lines shown in Fig. 12.37 were measured at a microwave frequency of 52 GHz. The low-field ESR line is due to F-centres in the ground state, whereas the high-field line corresponds to changes in P_G due to ESR transitions in the relaxed excited state. The observed ODMR lines are very broad and no hyperfine structure is observed. The measured ground state g-values were 1.976, 1.862, and 1.62 for potassium chloride, bromide, and iodide, respectively. Subsequently Mollenauer *et al.* (1969) extended these studies to the measurements of optically-detected electron nuclear double resonance (ODENDOR) of F-centres in potassium iodide. The ODENDOR lines were observed at RF frequencies very different from the known ENDOR frequencies of F-centres in the ground state. The spectra show that in the relaxed excited state, hyperfine interaction is due mainly to the second shell iodine neighbours with a Fermi contact term $A_0 \simeq 100$ MHz. This single hyperfine interaction is apparently much larger than that from all other shells and is almost twice that of F-centres in the ground state, confirming unequivocally the diffuse nature of the excited-state wavefunction of the F-centre.

The application of ODMR to luminescence processes in semiconductors has developed very rapidly (see for example Zakharchenya 1978; Lampel 1974; Davies 1976, 1985; Cavenett 1981). Pumping with circularly polarized bandgap light produces a measurable spin polarization of the conduction electrons, which polarization is reflected in the degree of circular polarization

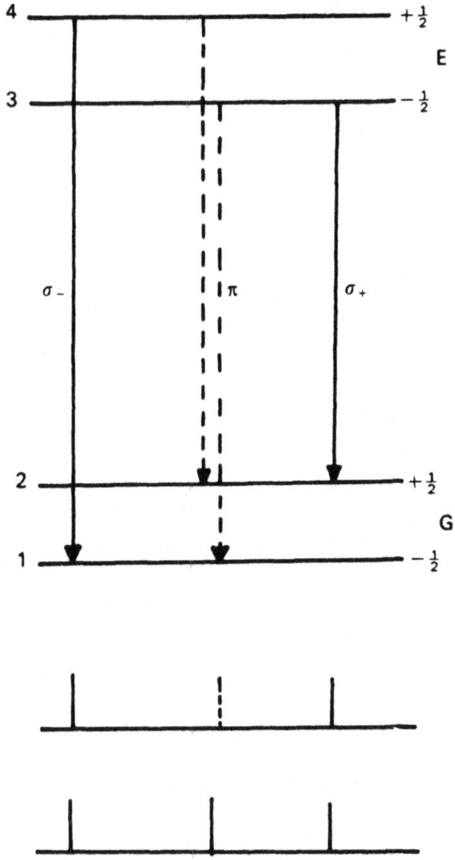

FIG. 12.37. Showing the circularly polarized ($\Delta M_s = \pm 1$) and linearly polarized emission transitions between spin doublet excited states (E) and ground states (G).

of the recombination luminescence. Conduction ESR may be detected as a change in the polarization of the emission. The technique has been successfully applied to gallium arsenide, $Ga_xAl_{1-x}As$ and $Ga_xIn_{1-x}As$ alloys, gallium antimonide, zinc telluride, and cadmium telluride. Since such measurements determine conduction electron *g*-values they provide an excellent test of current band theories (Herman and Weisbuch 1978).

12.7 Optical-microwave coherence spectroscopy

ODMR is detected via measurements of *incoherent* transient observables such as spin–lattice relaxation, electron transfer, and spectral diffusion. Generally such effects are referred to as population dynamics, since the spectra are revealed as changes in the excited state population induced by microwave-induced EPR transitions. However, the microwave field interacting resonantly with the atomic system produces in addition to population changes a coherent superposition state in which the phases of state vectors of all atoms are linked or well-defined. Thus we might also detect ODMR via *coherent* transient observables through the persistence of *phase coherence* in pulsed ODMR measurements.

12.7.1 Phase coherence in ODMR

One of the postulates of quantum mechanics asserts that at a fixed time, t, the state of a physical system may be defined by specifying a ket $|\psi(t)\rangle$ belonging to the state space Σ. Since Σ is a vector space this postulate implies a superposition principle, i.e. a linear combination of state vectors is also a state vector. An important consequence of this postulate is the appearance of superposition effects such as those which lead to wave–particle duality and we may anticipate the appearance of such effects in the coherence phenomena measured using ODMR. Suppose $|\psi_1(t)\rangle$ and $|\psi_2(t)\rangle$ are eigenstates of an observable A with eigenvalues of a_1 and a_2, respectively, at time t. The probability $P_1(a_n t)$ of finding a_n when A is measured at time t with the system in the state $|\psi_1(t)\rangle$ is just $P_1(a_n t) = |\langle u_n|\psi_1(t)\rangle|^2$, $|u_n\rangle$ being the normalized time-independent eigenvector of A corresponding to the eigenvalue a_n. P_1 is independent of t if $|\psi_1\rangle$ is an energy eigenstate. Similarly $P_2(a_n t)$ is given by $P_2(a_n t) = |\langle u_n|\psi_2(t)\rangle|^2$. Now consider the normalized linear superposition state

$$\psi(t) = \lambda_1\psi_1(t) + \lambda_2\psi_2(t) \tag{12.46}$$

where $\lambda_1^2 + \lambda_2^2 = 1$. The probability of finding a_n when A is measured with the system in this state is given by $P(a_n t) = |\langle u_n|\psi(t)\rangle|^2$, i.e.

$$\begin{aligned} P(a_n t) &= |\lambda_1\langle u_n|\psi_1(t)\rangle + \lambda_2\langle u_n|\psi_2(t)\rangle|^2 \\ &= \lambda_1^2 P_1(a_n(t)) + \lambda_2^2 P_2(a_n(t)) \\ &\quad + 2\mathrm{Re}\lambda_1\lambda_2^*\langle u_n|\psi_1(t)\rangle\langle u_n|\psi_2(t)\rangle^*. \end{aligned} \tag{12.47}$$

The first two terms are constants in time; the final term is the familiar oscillating superposition term. Consequently not only the moduli of λ_1 and λ_2 play a role, but also their relative phases enter explicitly into the physical predictions.

To illustrate how such superposition terms arise in ODMR, we consider the allowed electric dipole transitions between ground (G) and excited (E) state spin doublets ($S = \frac{1}{2}$) in a magnetic field parallel to the z-direction (Fig. 12.37). In the Zeeman orientation there are two oppositely polarized luminescence transitions: $|\mathrm{E}\rangle|+\frac{1}{2}\rangle \rightarrow |\mathrm{G}\rangle|-\frac{1}{2}\rangle$ is σ_- polarized and $|\mathrm{E}\rangle|-\frac{1}{2}\rangle \rightarrow |\mathrm{G}\rangle|+\frac{1}{2}\rangle$ is σ_+ polarized. In an orthogonal orientation (e.g. along y) these transitions are linearly polarized (σ_x) with E perpendicular to z. The $|E\rangle|\pm\frac{1}{2}\rangle|\rightarrow|G\rangle|\pm\frac{1}{2}\rangle$ transitions in the same orientation are linearly polarized with E parallel to z (i.e. π polarization). Application of resonant microwaves at frequency $\omega_0 = \omega_4 - \omega_3$, tends to equalize the populations of excited $|M_s = \pm\frac{1}{2}\rangle$ states and ODMR is detected as the change in intensity of circularly polarized light, $\sigma_{\pm}$, when viewing along the field or linearly polarized light σ_x when viewed along the y-direction. However, when in the excited state the spin system is coupled to the rotating microwave field and the system exists in the coherent superposition state

$$\psi = \lambda_3(t)\exp(-i\omega_3 t)|3\rangle + \lambda_4(t)\exp(-i\omega_4 t)|4\rangle \tag{12.48}$$

Thus the transition probability for σ_+ light seen by an observer along the y-axis, $P(x+iz)$ is just $|\langle E|x+iz|G\rangle|^2$ or

$$P(x+iz) = |\langle E|x+iz|G\rangle\langle E|x-iz|G\rangle| \tag{12.49}$$

$$= \lambda_4^*\lambda_3 \exp i\omega_0 t\langle 4|x+iz|1\rangle\langle 3|x-iz|2\rangle$$

$$+\text{constant terms.} \tag{12.50}$$

Experimentally we observe not the effects of a single atom but the ensemble average. Hence

$$\overline{P(x+iz)} = \mathrm{Re}\,[\lambda_4^*\lambda_3 \exp i\omega_0 t] + \lambda_4^*\lambda_4 + \lambda_3^*\lambda_3 \tag{12.51}$$

The significance of this equation is that there is a net macroscopic transverse polarization which oscillates in time at the Larmor frequency, ω_0. Thus the phase coherence in the excited spin levels is observable as a polarization modulation at frequency ω_0 with amplitude proportional to $\lambda_4^*\lambda_3$. This polarization modulation arises because of the relative amplitudes of the σ- and π-polarized light propagating in the y-direction.

Equation (12.51) suggests a number of experiments for detecting spin coherence in ODMR, the most obvious of which is to induce the polarization modulation of the emitted light using microwaves resonant with the Zeeman splitting in ground or excited state. The emitted light is then detected through a circular polarization analyser set normal to the magnetic field direction using a high-frequency phototube and phase-sensitive detection. Alternatively the precessing transverse magnetization may be induced by exciting with circularly polarized light modulated at the Larmor frequency. Both techniques were used by Chase (1968*b*) in studies of the Γ_8 excited state of substitutional Eu^{2+} in calcium fluoride in the X-band frequency range. As with other ODMR experiments, such observation of excited-state EPR is unaffected by vibronic interaction and the transverse magnetization modulation was detected over the broad optical emission band ($\Gamma_h \sim 1000\ \mathrm{cm}^{-1}$) of the Eu^{2+} ion. Similar measurements have been carried out in the ^2E state of Cr^{3+} ions in magnesium oxide (Henderson 1980, unpublished). Coherence spectroscopy may also be applicable to MCD of absorption spectra, since the partially absorbed light is modulated at the Larmor frequency by the precessing transverse magnetization.

Temporal evolution of a coherent state may be examined using microwave pulse sequences as in NMR. The system in the coherent state determined by λ_4 and λ_3 relaxes at switch-off, with some characteristic time, towards the original incoherent state. Population dynamics obviously plays a role in the loss of phase coherence, as also does any time-dependent internal interactions which may limit the persistence of phase coherence. The pure dephasing time T_2 arises out of fluctuations in the resonance frequency of the individual paramagnetic species, since for fluctuations which are random in space and time, the average value of the exponential factor in eqn (12.50) dies out so destroying phase coherence. If the microwave pulse is long relative to the combined radiative and non-radiative decay times of the emissive levels then one essentially measures the fluorescence decay time, τ, or the spin-lattice relaxation time, T_1, depending upon whether $T_1 \gtrless \tau$. Consequently it is advantageous to study coherence phenomena in electronic systems which are weakly coupled to the phonon spectrum of the crystal (T_1 long). Long-lived excited states, such as those discussed in Section 12.2, have been extensively used for coherence measurements.

Conventional ODMR is observed by changing the appropriate variable (B or λ) at a rate that is slow in comparison with the pulse frequency. However, in observing coherent spin transients B and λ are fixed and the ODMR signal is monitored at constant microwave frequency as a function of time after the end of the pulse sequence. Even in favourable cases the microwave-induced changes in light intensity are very small, typically less than a tenth of one per cent, and signal amplification using a lock-in or a boxcar detector is necessary. In a typical experiment the fluorescence level is measured at the completion of the microwave pulse sequence. The fluorescence decays back to its microwave-off equilibrium value in a time determined by the various relaxation processes, i.e. T_1. If we use a repetitive pulse sequence of frequency $\ll T_1^{-1}$ then the lock-in detector output signal is directly proportional to the initial fluorescence change. Various pulse sequences then allow one to study a variety of phenomena including transient nutation, stimulated spin echo, spin dephasing, and coherence memory.

12.7.2 Transient nutation in triplet state defects.

In transient nutation experiments the ODMR signal is monitored as a function of time after switching on the microwave field, the ODMR intensity at the trailing edge of the pulse being recorded as its length increases. The two-electron centres in calcium oxide and magnesium oxide provide extremely favourable conditions for the observation of spin coherence because of the large degree of spin alignment associated with the selective feeding and decay processes and the weak spin–lattice relaxation. The theory of triplet state coherence was developed by Breiland *et al.* (1975) using a density matrix formalism. Due to large crystal field splittings and slow relaxation processes, any pair of zero-field states may be used as the spin doublet coupled by the resonant microwave field: e.g. the (D–E) transition in Fig. 12.16 admixes states of y- and z-polarization. The temporal evolution of the population difference in a transient nutation experiment when the resonant microwave field is switched on at $t = 0$ then follows:

$$\frac{\Delta(t)}{\Delta(0)} = \cos \omega_1 t$$

in which $\omega_1 = \gamma B_{1z}$, γ being the gyromagnetic ratio and B_{1z} the z-component of the oscillating magnetic field. Hence the population difference nutates cosinusoidally between $\Delta(0)$ and $-\Delta(0)$, where $\Delta(0)$ represents the equilibrium value of the population difference before the application of the microwave field. At a time $t = \pi/\omega_1$ the populations of the y- and z-polarized states are interchanged giving a maximum change in the fluorescence. A pulse of this duration is referred to as a 'π-pulse'.

The density matrix approach also predicts coherence even when the microwave frequency, ω, is different from the Larmor frequency, ω_0, of the electron spins. Off resonance, the modulation frequency increases, and the modulation amplitude decreases. However, the first peak, still defines the period of a π-pulse even though the maximum intensity change does not correspond to a complete inversion in $|y\rangle$ and $|z\rangle$ populations. Obviously when lines are broadened *inhomogeneously*, e.g. by lattice strains, the off-resonance nutation is important because we must include the entire distribution of spin packets, $g(\Delta\omega)$. The spectrum is also *homogeneously* broadened by energy relaxation (T_1) and phase relaxation (T_2) processes. Since T_1 is in general long for triplet-state defects, only damping introduced by dephasing processes need to be considered. In this case $\Delta(\infty) \rightarrow 0$, the average value of the population difference in the undamped case. Furthermore, T_2 relaxation decreases the nutation frequency so that the π-pulse appears longer than in the absence of phase relaxation. Theoretical transient-nutation spectra are shown in Fig. 12.38; $(\Delta(0) - \Delta(t))/\Delta(0)$ is plotted here since this is the quantity measured by the ODMR method. For homogeneously broadened transitions the nutations are very heavily damped; over damping occurs when $T_2 < 1/2\omega$, resulting in a slow decrease in signal towards $\Delta(t) = \Delta(0)$ with no nutations.

If the effects of inhomogeneous line broadening are included the nutation is damped, but now the π-pulse is shortened relative to the undamped nutation. However, the time between successive maxima is almost unchanged from the undamped case. The comparison of nutation in the presence of homogeneous (Fig. 12.38(a)) and inhomogeneous (Fig. 12.38(b)) line-broadening shows that the shape of the nutation discriminates between the two broadening mech-

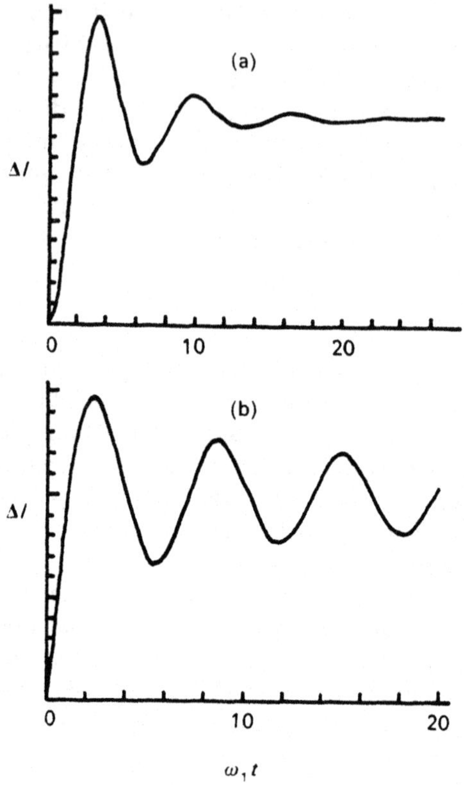

FIG. 12.38. Theoretical transient nutation spectra for triplet spin states in the presence of (a) homogeneous and (b) inhomogeneous broadening mechanisms.

anisms. Furthermore, by measuring the shape of the nutation, rather than just the duration of the π-pulse we can determine the frequency, ω_1, of the microwave field.

12.7.3 Optically detected spin echoes

The transient nutation experiment determines the pulse length of a $\frac{1}{2}\pi$-pulse (i.e. a pulse that produces the maximum coherence amplitude) and a π-pulse, since these are required to observe spin echoes. For the case of the F_2^{2+} centre in MgO (Fig. 12.39) we see from Fig. 12.15 that since the $B_{2u} \rightarrow A_g$ transition is an allowed transition whereas $A_u \rightarrow A_g$ is forbidden, there is a build-up of population in the A_u state. Thus the magnetic moments are aligned parallel to the z-direction giving a net moment $M_z = \Sigma_i m_{iz}$. The first microwave pulse applied so that B_1 is perpendicular to z rotates the net moment into alignment with the negative x-axis. At the end of the pulse, the different spin

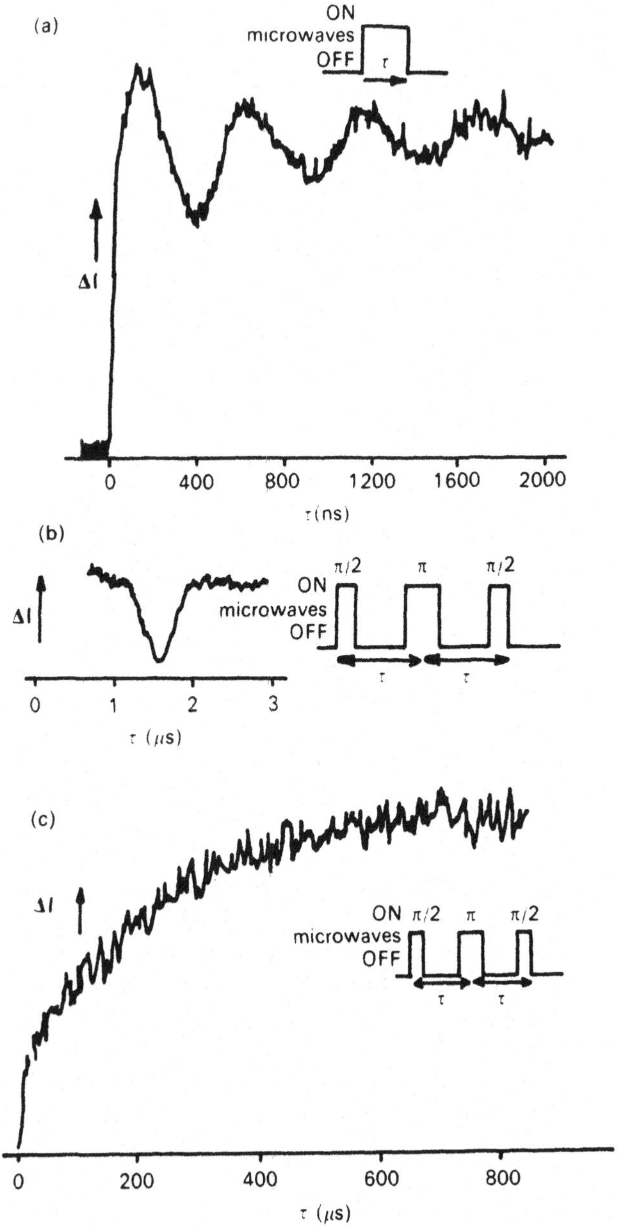

FIG. 12.39. Microwave coherent transients observed via the |D–E| transition in the $^3B_{1u}$ state of F_2^{2+}-centres in magnesium oxide. (a) transient nutation, (b) spin echo, and (c) decay of spin echo. (After Glasbeek *et al.* 1980b.)

packets making up the inhomogeneously broadened ODMR line precess at different rates in the xy-plane, so that the individual moments fan out in the time τ in this plane. A π-pulse, i.e. $\omega t = \pi$, causes the phase differences to be reversed in sign. In consequence in a time τ following this π-pulse the moments are refocused parallel to the positive x-axis so that a final $\frac{1}{2}\pi$ pulse brings the net moment back into coincidence with the z-direction, and re-establishes the population difference between the spin levels appropriate to the time $t = 0$. In other words the fluorescence intensity observed prior to the application of the microwaves is reproduced. Results representative of these types of experiment for the $^3B_{1u}$ state of the F_2^{2+}-centre in magnesium oxide are shown in Fig. 12.39 (Glasbeek *et al.* 1980b). Figure 12.39(a) shows transient nutation, from which the length of the $\frac{1}{2}\pi$-pulse is determined for observation of the spin echo signal (Fig. 12.39(b)). Finally the decay of the spin echo, measured by varying τ the length of the pulse separation is shown in Fig. 12.39(c).

In the absence of other dynamical processes which affect phase coherence, the spin echo decays with relaxation time, T_2, equal to the mean radiative decay time of the participating spin levels. However, more generally in solids triplet spins are modulated by fluctuating fields associated with nuclear or other electron spins. Such fluctuating fields can temporarily change the resonance frequency of a particular triplet spin, and in consequence disturb the phase relationships in the triplet spin system. Spin-echo decay by such a mechanism is generally non-exponential varying as $\exp -(2t/T_M)^x$, where T_M is the so-called phase-memory time and x is a parameter which depends on the nature of the relaxation mechanism. For both F-centres and F_2^{2+}-centres in calcium oxide the phase memory time T_M is about 120–150 μs (Glasbeek *et al.* 1980*a*; Gravesteijn and Glasbeek 1979). In both cases spin dephasing due to random local field fluctuations associated with spin–flip transitions on F^+-centres is dominant. By observing the effect of temperature or magnetic field on the coherence decay one may observe exchange narrowing within the F^+-centre spin ensemble and cross-relaxation between F^+-centre and F-centre spins (Glasbeek and Hond 1981). For F_2^{2+}-centres in magnesium oxide, spin-echo decay is determined by the fact that each F_2^{2+}-centre has on average one $^{25}Mg(I = \frac{5}{2})$ nuclide in its nearest-neighbour Mg^{2+} shell. Then one observes effects due to quantum beats arising from hyperfine interactions between the triplet spins and the fluctuating $I = \frac{5}{2}$ spins (Bovy and Glasbeek 1982).

References

Abragam, A. A. and Bleaney, B. (1970). *Electron Paramagnetic Resonance.* Oxford University Press.

Aegerter, M. A. and Lüty, F. (1970). *J. Luminescence* **1/2**, 624.

—— (1971). *Phys. Stat. Sol.* (b) **43**, 245.

Ahlers, F., Lohse, F. and Spaeth, J. M. (1982). *Solid State Comm.* **43**, 321.

Ahlers, F., Lohse, F., Spaeth, J. M. and Mollenauer, L. F. (1983). *Phys. Rev.* **B28**, 1249.

Anderson, C. H. (1974). In *Crystals with the Fluorite Structure* (ed. W. Hayes), p. 281. Oxford University Press.

Anderson, C. H. and Sabisky, E. (1966). *IEEE J. Quant. Elect.* QE**2**, 29.

Anderson, P. W. (1958). *Phys. Rev.* **109**, 1492.

Anderson, P. W., Halperin, B. W. and Varma, C. M. (1972). *Phil. Mag.* **25**, 1.

Andrews, L. J. and Hitelman, S. M. (1986). *J. Chem. Phys.* **84**, 5229; see also Andrews L. J. (1988). In *Spectroscopy of Solid State Laser-type Materials* (ed. B. Di Bartolo). Plenum Press, New York.

Andrews, L. J., Lempicki, A. and McCollum, B. C. (1981). *J. Chem. Phys.* **74**, 5526.

Andrews, L. J., McCollum, B. C., Stone, S., Guenther, D. E., Murphy, G. J. and Lempicki, A. (1984). *Development of Materials for a Luminescence Solar Collector* Final Report DOE/ER/04996-4, GTE Labs, Waltham, Mass.

Auzel, F. (1966). *C.R. Acad. Sci. Paris* **262**, 1016.

—— (1978). In *Luminescence of Inorganic Solids* (ed. B. Di Bartolo). Plenum Press, New York.

—— (1984a). In *Energy Transfer Processes in Condensed Matter* (ed. B. Di Bartolo), p. 497. Plenum Press, New York.

—— (1984b). *J. Luminescence* **31/32**, 759.

Bartram, R. H. and Stoneham, A. M. (1975). *Solid State Comm.* **17**, 1593; but see also (1985) *J. Phys.* C **18**, L549.

Bates, J. B. and Wood, R. F. (1975). *Solid State Comm.* **17**, 201.

Baumann, G. (1967). *Z. Phys.* **203**, 464.

Becquerel, E. (1867). *La Lumière, ses Causes et ses Effets.* Dunod, Paris.

Berg, A. A. and Dean, P. (1976). *Light Emitting Diodes.* Oxford University Press.

Bergin, F. J., Donegan, J. F., Glynn, T. J. and Imbusch, G. F. (1986). *J. Luminescence* **34**, 307; see also (1987). *J. Luminescence* **39**, 121.

Bethe, H. (1929). *Ann. Phys.* **3**, 133 (English translation, 'Splitting of terms in crystals', by Consultants Bureau Inc., New York).

Bethe, H. A. and Salpeter, E. E. (1957). *Quantum Mechanics of One and Two-Electron Atoms.* Springer-Verlag, Berlin.

Birgenau, R. J. (1969). *J. Chem. Phys.* **50**, 4282.

Blasse, G. (1978). In *Luminescence of Inorganic Solids* (ed. B. Di Bartolo). Plenum Press, New York.

—— (1984). In *Energy Transfer Processes in Condensed Matter* (ed. B. Di Bartolo), p. 251. Plenum Press, New York.

—— (1988). In *Spectroscopy of Solid State Laser-type Materials* (ed. B. Di Bartolo). Plenum Press, New York.

Blasse, G., Bleijenberg, K. C. and Krol, D. M. (1979). *J. Luminescence* **18/19**, 57.
Bloch, F. (1928). *Z. Phys.* **52**, 555.
Bloembergen, N. (1959). *Phys. Rev. Lett.* **2**, 84.
Blumberg, W. E., Eisinger, J. and Geschwind, S. (1963). *Phys. Rev.* **130**, 900.
Bogan, L. D. and Fitchen, D. B. (1970). *Phys. Rev.* **B1**, 4122.
Bohm, D. (1951). *Quantum Theory*, p. 467. Prentice-Hall Inc, Englewood Cliffs, NJ.
Born, M. and Oppenheimer, J. R. (1927). *Ann. Phys.* **84**, 457.
Boulon, G. (1984). In *Energy Transfer Processes in Condensed Matter* (ed. B. Di Bartolo), p. 603. Plenum Press, New York.
—— (1988). In *Spectroscopy of Solid State Laser-type Materials* (ed. B. Di Bartolo). Plenum Press, New York.
Bovy, M. W. L. and Glasbeek, M. (1982). *J. Chem. Phys.* **76**, 1676.
Brecher, C. and Riseberg, L. A. (1980). *Phys. Rev.* **B21**, 2607.
Breiland, W. G., Brenner, H. C. and Harris, C. B. (1975). *J. Chem. Phys.* **612**, 3458.
Brower, K. A. (1974). *Phys. Rev.* **B9**, 2607.
—— (1978). *Phys. Rev.* **B17**, 4130.
Brown, F. C. (1967). In *The Physics of Solids*. Benjamin, New York.
Brunwin, R. F., Cavenett, B. C., Davies, J. J. and Nicholls, J. E. (1976). *Solid State Comm.* **18**, 1283.
Brust, D., Bassani, G. F. and Philips, J. C. (1962). *Phys. Rev. Lett.* **9**, 94.
Brya, W. J., Geschwind, S. and Devlin, G. E. (1968). *Phys. Rev. Lett.* **21**, 1800.
Buisson, R. and Vial, J. C. (1981). *J. Physique-Lett.* **42**, L-115.
Buisson, J. P., Sadoc, A., Taurel, L. and Billardon, M. (1975). In *Light Scattering in Solids* (eds. M. Balkanski, R. Leite and S. P. Porto). Flammarion, Paris.
Burshtein, A. I. (1972). *Soviet Physics JETP* **31**, 882.
—— (1985). *J. Luminescence* **34**, 167.
Caird, J. A. (1986). In *Tunable Solid State Lasers II* (eds. A. B. Budgor, L. Esterowitz and L. G. DeShazer), p. 20. Springer-Verlag, Berlin.
Cardona, M. (1969). In *Solid State Phys.* Supp II, *Modulation Spectroscopy* (ed. F. Seitz, D. Turnbull, H. Ehrenreich). Academic Press, New York.
Carlson, B. C. and Rushbrooke, G. S. (1950). *Proc. Camb. Phil. Soc.* **46**, 626.
Carnall, W. T., Crosswhite, H. and Crosswhite, H. M. (1977). In *Energy Level Structure and Transition Probabilities of the Trivalent Lanthanides in* LaF_3. Argonne National Laboratory Report.
Carnall, W. T., Crosswhite, H., Crosswhite, H. M. and Conway, J. G. (1976). *J. Chem. Phys.* **64**, 3582.
Caspers, H. H., Rast, H. E. and Fry, J. L. (1970). *J. Chem. Phys.* **53**, 3208.
Castner, T. G. and Kanzig, W. (1957). *J. Phys. Chem. Sol.* **3**, 178.
Cavenett, B. C. (1981). *Adv. in Phys.* **30**, 475.
Cavenett, B. C., Brunwin, R. F. and Nicholls, J. E. (1977a). *Solid State Comm.* **22**, 71.
Cavenett, B. C., Dunstan, D. J., Brunwin, R. F. and Nicholls, J. E. (1977). *J. Phys.* C **10**, L361.
Cavenett, B. C., Dawson, P. and Morigaki, K. (1979). *J. Phys.* C **12**, L197.
Chase, L. L. (1968a). *Phys. Rev.* **168**A, 341.
—— (1968b). *Phys. Rev. Lett.* **21**, 888.
—— (1970). *Phys. Rev.* **B2**, 2308.
Chen, Y. and Abraham, M. M. (1975). *New Physics (Korean Physical Society)* **15**, 47.
Chen, Y., Williams, R. T. and Sibley, W. A. (1969). *Phys. Rev.* **182**, 960.

Chen, Y., Orera, V. M., Gonzalez, R. M. and Ballesteros, C. (1988). *J. Cryst. Def. Atmorph. Solids* **14**, 283.
Chiarotti, G., Grassano, U. M., Margaritondo, G. and Rosei, R. (1969). *Nuovo Cimento* **B64**, 159.
Cibert, J., Edel, P., Merle D'Aubigne, Y. and Romestain, R. (1979). *J. de Phys.* **40**, 1149.
Condon, E. U. and Shortley, G. H. (1935). In *Theory of Atomic Spectra*. Cambridge University Press.
Cone, R. L. and Melzer, R. S. (1975). *J. Chem. Phys.* **62**, 357.
Conway, J. G. (1976). *J. Chem. Phys.* **64**, 3582.
Corney, A. (1977). *Atomic and Laser Spectroscopy*. Clarendon Press, Oxford.
Craford, R. G., Shaw, R. W., Herzog, A. H. and Groves, W. O. (1972). *J. App. Phys.* **43**, 4075.
Crandall, R. S. (1965). *Phys. Rev.* **138A**, 1242.
Crosswhite, H. M. and Dieke, G. H. (1961). *J. Chem. Phys.* **35**, 1535.
Curie, D. (1963). *Luminescence in Crystals*. Methuen, London.
Davies, J. J. (1976). *Contemporary Phys.* **17**, 275.
—— (1985). *J. Crystal Growth*, **72**, 317 and references therein.
Dawson, R. and Pooley, D. (1969). *Phys. Stat. Sol.* **35**, 95.
Dawson, P., Morigaki, K. and Cavenett, B. (1979). In *Proc. Int. Conf. on Phys. of Semiconductors, Edinburgh* (ed. B. L. H. Wilson), p. 1023. Institute of Physics, London; see also Dawson, P., Killoran, N. and Cavenett, B. C. (1979). *Solid State Comm.* **32**, 1163.
Dawson, P., McDonagh, C. M., Henderson, B. and Welch, L. S. (1980). *J. Phys.* C **11**, L983.
De Rougement, F., Michau, V. and Frey, R. (1986). In *Tunable Solid State Lasers II* (eds. A. B. Budgor, L. Esterowitz and L. G. DeShazer), p. 175. Springer-Verlag, Berlin.
Dean, P. J. (1969). In *Applied Solid State Science* VI (ed. R. Wolfe and C. J. Kriessman), p. 1. Academic Press, New York.
—— (1973). *Progress in Solid State Chemistry* **8**, 1.
Dean, P. J., Schönherr, E. G. and Zetterstrom, R. B. (1970). *J. App. Phys.* **41**, 3434.
Delbecq, C. J., Hayes, W. and Yuster, P. H. (1961). *Phys. Rev.* **121**, 1043.
Delbecq, C. J., Hayes, W., O'Brien, M. C. M. and Yuster, P. (1963). *Proc. Roy. Soc.* **A271**, 243.
Delbecq, C. J., Gosh, A. K. and Yuster, P. H. (1966). *Phys. Rev.* **151**, 599.
—— (1967). *Phys. Rev.* **154**, 797.
Delsart, C. and Pelletier-Allard, N. (1973). *J. Phys.* C **6**, 1227.
Deutschbein, O. (1932). *Ann. Phys.* **20**, 828.
Dexter, D. L. (1953). *J. Chem. Phys.* **21**, 836.
Dexter, D. L., Klick, C. C. and Russell, G. A. (1955). *Phys. Rev.* **100**, 603.
Di Bartolo, B. (1968). In *Optical Interactions in Solids*. Wiley, New York.
—— (1984). In *Energy Transfer Processes in Condensed Matter* (ed. B. Di Bartolo) p. 103. Plenum Press, New York.
Dieke, G. H. (1968). In *Spectra and Energy Levels of Rare Earth Ions in Crystals*. Wiley Interscience, New York.
Dieke, G. H. and Leopold, L. (1957). *J. Opt. Soc. Am.* **47**, 944.
Dieke, G. H. and Pandey, B. (1964). *J. Chem. Phys.* **41**, 1952.
Dirac, P. A. M. (1927). *Quantum Mechanics*. Oxford University Press.

Donegan, J. F., Bergin, F. J., Imbusch, G. F. and Remeika, J. P. (1984). *J. Luminescence* **31/32**, 278.

Drude, P. (1902). *Ann. Phys. Leipzig* **7**, 712.

Dürr, U. and Brauch, U. (1986). In *Tunable Solid State Lasers II* (eds. A. B. Budgor, L. Esterowitz and L. G. DeShazer), p. 151. Springer-Verlag, Berlin.

Dürr, U., Brauch, U., Knierim, W. and Schiller, C. (1985). In *Tunable Solid State Lasers* (eds. P. Hammerling, A. B. Budgor and A. Pinto), p. 20. Springer-Verlag, Berlin.

Edel, P., Hennies, C., Merle d'Aubigne, Y., Romestain, R. and Twaroski, Y. (1972). *Phys. Rev. Lett.* **28**, 1268.

Edel, P., Ahlers, F. J., McDonagh, C. M., Henderson, B. and Spaeth, J. M. (1982). *J. Phys.* C **15**, 4913.

Eichler, H. J. (1986). *IEEE J. Quant. Electronics; Special Issue on Dynamic Gratings and Four Wave Mixing.* (August 1986)

Eichler, H. J., Pohl, D. W. and Günter, P. (1985). In *Laser Induced Dynamic Gratings.* Springer Series in Optical Sciences **50**. Springer-Verlag, Berlin.

Einstein, A. (1905). *Ann. Phys.* **17**, 132.

—— (1906). *ibid* **20**, 199.

Elliot, J. P. and Dawber, P. G. (1979). *Symmetry in Physics* Vols. 1 and 2. Macmillan, London.

Elliot, R. J., Thorpe, M. F., Imbusch, G. F., Loudon, R. and Parkinson, J. B. (1968). *Phys. Rev. Lett.* **21**, 147.

Elliot, S. R. (1984). In *Physics of Amorphous Materials.* Longman, London.

Englman, R. (1972). In *The Jahn–Teller Effect in Molecules and Crystals.* Wiley-Interscience, London. (The historic note by Teller in this book is also of interest.)

Engstrom, H. and Mollenauer, L. F. (1973). *Phys. Rev.* **B7**, 1616.

Eremenko, V. V. and Petrov, E. G. (1977). *Adv. in Phys.* **26**, 31.

Erickson, L. E. (1975). *Phys. Rev.* **B11**, 77.

—— (1977a). *Phys. Rev.* **16**, 4731.

—— (1977b). *Optics Comm.* **15**, 246.

Fairbank, W. M. Jr., Klauminzer, G. K. and Schawlow, A. L. (1975). *Phys. Rev.* **B11**, 60.

Feher, G., Hensel, J. C. and Gere, E. A. (1960). *Phys. Rev. Lett.* **5**, 309.

Feofilov, P. P. (1961). *The Physical Basis of Polarized Emission.* Consultants Bureau, New York.

Feofilov, P. P. and Trofimov, A. K. (1969). *Opt. Spectrosc. (USSR)* **27**, 291.

Ferguson, J., Wood, D. L. and Van Uitert, L. G. (1961). *J. Chem. Phys.* **51**, 2904.

Ferguson, J., Guggenheim, H. J. and Tanabe, Y. (1965). *Phys. Rev. Lett.* **14**, 737.

—— (1966). *J. Phys. Soc. Japan* **21**, 692.

Fetterman, H. R. (1968). Unpublished Ph.D. Thesis, Cornell University.

Fitchen, D. B. (1968). In *The Physics of Color Centers* (ed. W. B. Fowler), p. 293. Academic Press, New York.

Flaherty, J. M. and Di Bartolo, B. (1973). *Phys. Rev.* **B8**, 5232.

Fonger, W. H. and Struck, C. W. (1975). *Phys. Rev.* **B11**, 3251.

Förster, T. (1948). *Ann. Phys.* (Leipzig) **2**, 55.

Fowler, W. B. (1968). In *The Physics of Color Centers* (ed. W. B. Fowler), p. 53. Academic Press, New York.

Frenkel, J. I. (1936). *Phys. Z. Soviet Union* **9**, 158.

—— (1931). *Phys. Rev.* **37**, 17 and 1276.

Fritz, B. and Menke, E. (1965). *Sol. State Comm.* **3**, 61.
Fritz, B., Gerlach, J. and Gross, U. (1968). In *Localized Excitations in Solids* (ed. R. F. Wallis), p. 496. Plenum Press, New York.
Fröhlich, H. (1970). *Proc. 10th Conference on Physics of Semiconductors*, Cambridge, Mass.
Fuhrberg, P., Luhs, W., Struve, B. and Litfin, G. (1986). In *Tunable Solid State Lasers II* (eds. A. B. Budgor, L. Esterowitz and L. G. DeShazer), p. 113. Springer-Verlag, Berlin.
Fukuda, A. (1970). *Phys. Rev.* **B1**, 416.
Fukuda, A., Makishima, S., Mabuchi, T. and Onaka, R. (1967). *J. Phys. Chem. Sol.* **28**, 1763.
Fukuda, A., Matsushima, A. and Masunaga, S. (1976). *J. Luminescence* **12/13**, 139.
Fuller, R. G., Williams, R. T. and Kabler, M. N. (1970). *Phys. Rev. Lett.* **25**, 446.
Gächter, B. F. and Koningstein, J. A. (1974). *J. Chem. Phys.* **60**, 2003.
Gaunt, J. A. (1929). *Trans. Roy. Soc.* **A228**, 151 and 195.
Gebhardt, W. and Kuhnert, H. (1964). *Phys. Lett.* **11**, 15.
Gellerman, W., Lüty, F., Koch, K. P. and Litfin, G. (1980a). *Phys. Stat. Solids* **A57**, 411.
Gellerman, W., Lüty, F., Koch, K. P. and Welling, H. (1980b). *Opt. Comm.* **5**, 30.
Gellerman, W., Pollock, C. and Lüty, F. (1981). *Optics Comm.* **39**, 391.
Gellerman, W., Lüty, F., Wandt, D. and Welling, H. (1986). OSA Topical Meeting on Tunable Solid State Lasers Technical Digest p. 127; see also (1987) *Tunable Solid State Lasers II*, p. 251. Springer Series in Optical Science **52**, Springer-Verlag, Heidelberg.
Genack, A. Z., Weitz, D. A., Macfarlane, R. M., Shelby, R. M. and Schenzle, A. (1980). *Phys. Rev. Lett.* **45**, 438.
Genet, M., Delamoye, P., Edelstein, N. and Conway, J. (1977). *J. Chem. Phys.* **67**, 1670.
Geschwind, S. (1972). In *Electron Paramagnetic Resonance* (ed. S. Geschwind). Academic Press, New York.
Geschwind, S. and Remeika, J. P. (1961). *Phys. Rev.* **122**, 757.
Geschwind, S., Collins, R. J. and Schawlow, A. L. (1959). *Phys. Rev. Lett.* **3**, 545.
Geschwind, S., Kisliuk, P., Klein, M. P., Remeika, J. P. and Wood, D. L. (1962). *Phys. Rev.* **136**, 1684.
Geschwind, S., Devlin, G. E., Cohen, R. L. and Chinn, S. R. (1965). *Phys. Rev.* **137**A, 1087.
Gibson, A. F. and Elliott, R. (1974). *An Introduction to Solid State Physics and its Applications*. Macmillan, London.
Gires, P. and Meyer, J. (1961). *J. de Phys.* **22**, 832.
Glasbeek, M. and Hond, R. (1981). *Phys. Rev.* **23B**, 4220.
Glasbeek, M., Hond, R. and Zewail, A. H. (1980a). *Phys. Rev. Lett.* **45**, 744.
Glasbeek, M., Sitters, R. and Henderson, B. (1980b). *J. Phys.* C **13**, L1012.
Gondaira, K. and Tanabe, Y. (1966). *J. Phys. Soc. Japan* **21**, 1527.
Goovaerts, E., Andriessen, J., Nistor, S. V. and Schoemaker, D. (1981). *Phys. Rev.* **24B**, 29.
Goudsmit, S. and Uhlenbeck, G. (1926). *Z. Phys.* **35**, 618.
Gravesteijn, D. J. and Glasbeek, M. (1979). *Phys. Rev.* **19B**, 5549.
Gravesteijn, D. J., Scheigde, H. J. and Glasbeek, M. (1977). *Phys. Rev. Lett.* **39**, 105.
Greene, R. L., Sell, D. D., Feigelson, R. S., Imbusch, G. F. and Guggenheim, H. J. (1968). *Phys. Rev.* **B171**, 600.

Greenwood, N. N. (1968). *Ionic Crystals, Lattice Defects and Nonstoichiometry*, Butterworths, London.

Griffiths, J. H. E. and Orton, J. W. (1959). *Proc. Phys. Soc.* (Lond.) **78**, 948.

Griffith, J. S. (1961). *Theory of Transition Metal Ions*. Cambridge University Press.

Grosmann, M. (1963). In *Polarons and Excitons* (Proc. Scottish Universities Summer School, eds. C. G. Kuper and G. D. Whitfield). Oliver and Boyd, London.

Gurney, R. W. and Mott, N. F. (1937). *Proc. Phys. Soc.* **49** (extra part), 32.

Hagston, W. E. and Lowther, J. E. (1973). *Physics* **70**, 40.

Halliburton, L. E., Cowan, D. L. and Holroyd, L. V. (1975). *Phys. Rev.* **12B**, 3409.

Ham, F. S. (1965). *Phys. Rev.* **138**A, 1727.

—— (1972). In *Electron Paramagnetic Resonance* (ed. S. Geschwind). Plenum Press, New York.

Hamilton, D. R., Choyke, W. J. and Patrick, L. (1963). *Phys. Rev.* **131**, 127.

Hamilton, D. S., Seltzer, P. M. and Yen, W. M. (1977). *Phys. Rev.* **B16**, 1858.

Hänsch, T. W., Shahin, I. S. and Schawlow, A. L. (1971). *Phys. Rev. Lett.* **27**, 707.

—— (1972). *Nature* (*Phys. Sci.*) *London* **235**, 63.

Harris, E. A. and Yngvesson, K. S. (1968). *J. Phys.* C **1**, 990.

Hartree, D. R. (1928). *Proc. Camb. Phil. Soc.* **24**, 89, 111, and 426.

Hayes, W. (1978). *Proc. 3rd Specialised Colloque Ampère in Semi-conductors and Insulators* **3**, 121.

Hayes, W. and Loudon, R. (1978). *Scattering of Light by Crystals*. Wiley-Interscience, New York.

Hayes, W. and Stoneham, A. M. (1974). In *Crystals with the Fluorite Structure* (ed. W. Hayes). Clarendon Press, Oxford.

Hayes, W. and Stoneham, A. M. (1985). *Defects and Defect Processes in Non-metallic Solids*. J. Wiley, New York.

Hayes, W., Owen, I. B. and Walker, P. J. (1977). *J. Phys.* C **10**, 1751.

Hegarty, J. (1976). In Ph.D. Thesis, National University of Ireland, University College, Galway (unpublished).

Hegarty, J. and Yen, W. M. (1979). *Phys. Rev. Lett.* **43**, 1126.

Hegarty, J., Huber, D. L. and Yen, W. M. (1981). *Phys. Rev.* **B23**, 6271.

—— (1982). *Phys. Rev.* **B25**, 5638.

Heine, V. (1960). *Group Theory in Quantum Mechanics*. McGraw-Hill, New York.

Heitler, W. and London, F. (1927). *Z. Phys.* **44**, 455.

Henderson, B. (1972). *Defects in Crystalline Solids*. Arnold, London.

—— (1976). *J. Phys.* C **9**, 2185; *J. Phys.* C **9**, L579.

Henderson, B. and Garrison, A. H. (1973). *Adv. in Phys.* **22**, 423.

Henderson, B. and King, R. D. (1966). *Phil. Mag.* **13**, 1149.

Henderson, B. and McDonagh, C. M. (1980). *J. Phys.* C **13**, 5811.

Henderson, B., Stokowski, S. E. and Ensign, T. C. (1969). *Phys. Rev.* **183**, 826.

Henderson, B. and Tomlinson, A. C. (1969). *J. Phys. Chem. Sol.* **30**, 1801.

Henderson, B. and Wertz, J. E. (1977). *Defects in the Alkaline Earth Oxides*. Taylor and Francis, London.

Henderson, B., Chen, Y. and Sibley, W. A. (1972). *Phys. Rev.* **6B**, 4060.

Henderson, B., Yamaga, M., Glasbeek, M. and Sitters, R. (1981). *J. Phys.* C **14**, 2505.

Henry, C. E. and Slichter, C. P. (1968). In *Physics of Color Centers* (ed. W. B. Fowler), p. 351. Academic Press, New York.

Henry, C. E., Schnatterly, S. E. and Slichter, C. P. (1965). *Phys. Rev.* **137**A, 583.

Henry, M. O., Larkin, J. P. and Imbusch, G. F. (1976). *Phys. Rev.* **13B**, 1893.

Herman, R. C., Wallis, M. C. and Wallis, R. F. (1956). *Phys. Rev.* **103**, 87.
Hermann, C. and Weisbusch, C. (1978). *Semicond. and Insul.* **4**, 63.
Hesselring, W. H. and Wiersma, D. A. (1979). *Phys. Rev. Lett.* **43**, 991.
Hessler, J. P., Hegarty, J., Levey, C. G., Imbusch, G. F. and Yen, W. M. (1979). *J. Luminescence* **18/19**, 73.
Hirschfelder, J. O. (1938). *J. Chem. Phys.* **6**, 795.
Hirschfelder, J., Eyring, H. and Rosen, N. (1936). *J. Chem. Phys.* **4**, 121.
Hoffmann, D. M., Lohse, F., Paus, H. J., Smith, D. Y. and Spaeth, J. M. (1985). *J. Phys.* C **18**, 443.
Holmes, O. and McClure, D. S. (1957). *J. Chem. Phys.* **26**, 1686.
Holstein, T., Lyo, S. K. and Orbach, R. (1981). In *Laser Spectroscopy of Solids* (eds. W. M. Yen and P. M. Selzer), p. 39. Springer-Verlag, Berlin.
Hsu, D. and Skinner, J. L. (1984a). *J. Chem. Phys.* **81**, 1604.
—— (1984b). *J. Chem. Phys.* **81**, 5471.
Huang, K. and Rhys, A. (1950). *Proc. Roy. Soc.* **A204**, 406.
Huber, D. L. (1979). *Phys. Rev.* **B20**, 2307.
Huber, D. L., Hamilton, D. S. and Barnett, B. B. (1977). *Phys. Rev.* **B16**, 4642.
Huber, G. (1986). Conference on Tunable Solid State Lasers Technical Digest WA3, WB6, WB7.
Huber, G. and Peterman, K. (1985). In *Tunable Solid State Lasers* (eds. P. Hammerling, A. B. Budgor and A. Pinto) p. 11 and references therein. Springer-Verlag, Berlin.
Hüfner, S. (1978). *Optical Spectra of Transparent Rare-Earth Compounds.* Academic Press, New York.
Hughes, A. E. (1966). Unpublished Ph.D. Thesis, Oxford University.
—— (1970). *J. Phys.* C **3**, 627.
Hughes, A. E. and Henderson, B. (1972). In *Point Defects in Solids* (eds. J. H. Crawford and L. F. Slifkin). Plenum Press, New York.
Hughes, A. E. and Runciman, W. A. (1965). *Proc. Phys. Soc. London* **86**, 615.
Imbusch, G. F. and Geschwind, S. (1966). *Phys. Rev. Lett.* **17**, 238.
Imbusch, G. F., Yen, W. M., Schawlow, A. L., Devlin, G. E. and Remeika, J. P. (1964). *Phys. Rev.* **136A**, 481.
Imbusch, G. F., Chinn, S. R. and Geschwind, S. (1967). *Phys. Rev.* **161A**, 295.
Inokuti, M. and Hirayama, F. (1965). *J. Chem. Phys.* **43**, 1978.
Iverson, M. V. and Sibley, W. A. (1979). *J. Luminescence* **20**, 311.
Jacobsen, S. M., Smith, W. E., Reber, C. and Güdel, H. U. (1986). *J. Chem. Phys.* **84**, 5205.
Jahn, H. A. (1938). *Proc. Roy. Soc.* **A164**, 117.
Jahn, H. A. and Teller, E. (1937). *Proc. Roy. Soc.* **A161**, 220.
Jessop, P. E. and Szabo, A. (1980). *Optics Comm.* **33**, 301.
Jessop, P. E., Muramato, T. and Szabo, A. (1980). *Phys. Rev.* **B21**, 926.
Johannson, G., Lanzl, F., Modl, H., Von der Osten, W. and Waidelich (1968). *Z. Phys.* **210**, 1.
Johnson, L. F. and Guggenheim, H. J. (1967). *J. App. Phys.* **38**, 4837.
Johnson, L. F., Dietz, R. E. and Guggenheim, H. J. (1963). *Phys. Rev. Lett.* **11**, 318.
—— (1964). *App. Phys. Lett.* **5**, 2.
Johnson, L. F., Guggenheim, H. J. and Thomas, R. A. (1966). *Phys. Rev.* **149**, 179.
Johnson, L. F., Guggenheim, H. J., Bahnck, D. and Johnson, A. M. (1983). *Opt. Letters* **8**, 371.

Joosen, W., Leblans, M., Vanhimbeek, M., de Raedt, H., Goovaerts, E. and Schoemaker, D. (1988). *J. Cryst. Def. Amorph. Solids* **16**, 341.
Jørgensen, C. K. (1960). In *Absorption Spectra and Chemical Bonding in Complexes.* Pergamon Press, Oxford.
—— (1970). In *Progress in Inorganic Chemistry*, Vol. 12 (ed. S. J. Lidiard), Chapter 10, p. 39. Wiley Interscience, New York.
—— (1979). *J. Luminescence* **18/19**, 63.
Judd, B. R. (1962). *Phys. Rev.* **127**, 750.
—— (1963). *Operator Techniques in Atomic Spectroscopy.* McGraw-Hill, New York.
Kabler, M. N. (1964). *Phys. Rev.* **136**A, 1296.
—— (1972). In *Point Defects in Solids*, Vol. 1 (eds. J. H. Crawford and L. F. Slifkin). Plenum Press, New York.
Kallendonk, F. and Blasse, G. (1981). *J. Phys. Chem. Sol.* **43**, 481.
Kaminskii, A. A. (1981). *Laser Crystals.* Springer Series in Optical Sciences **14**. Springer-Verlag, Berlin.
—— (1985). *Phys. Stat. Sol.* **132**, 11.
Kaplyanskii, A. A. (1959). *Optics Spectrosc. USSR* **6**, 267.
—— (1964). *Optics Spectrosc. USSR* **16**, 329 and 557.
Kaplyanskii, A. A. and Przevuskii, A. K. (1967). *Soviet Phys.-Solid State* **9**, 190.
Keil, T. (1965). *Phys. Rev.* **140**, A601.
Keller, F. J., Murray, R. B., Weeks, R. A. and Abraham, M. M. (1967). *Phys. Rev.* **154**, 812.
Kemp, J. C. (1966). *J. Opt. Soc. Am.* **59**, 915.
Killoran, N., Cavenett, B. C. and Dean, P. J. (1981). *Solid State Comm.* **38**, 739.
Kisliuk, P., Chang, N. C., Scott, P. L. and Pryce, M. H. L. (1969). *Phys. Rev.* **184**, 367.
Kittel, C. (1966). *Introduction to Solid State Physics*, 3rd edn. Wiley, New York.
Klauminzer, G. K. (1970). Unpublished Ph.D. Thesis, Stanford University (M. L. Rep. No. 1878).
Klauminzer, G. K., Scott, P. L. and Moos, H. W. (1966). *Phys. Rev.* **142**, 248.
Klingshirn, C. (1984). In *Energy Transfer Processes in Condensed Matter* (ed. B. Di Bartolo), p. 235. Plenum Press, New York.
Knox, R. S. (1963). Theory of excitons. In *Solid State Physics, Suppl.* 5. Academic Press, New York.
Kohn, W. and Luttinger, J. M. (1955). *Phys. Rev.* **97**, 883 and **98**, 915.
Konitzer, J. D. and Markham, J. J. (1960). *J. Chem. Phys.* **32**, 843.
Krap, C. J., Glasbeek, M. and Van Woorst, J. D. W. (1978). *Phys. Rev.* **17**B, 61.
Krasinski, J., Papanestor, P., Pete, J. A. and Heller, D. F. (1986). In *Tunable Solid State Lasers II* (eds. A. B. Budgor, L. Esterowitz and L. G. DeShazer), p. 191. Springer-Verlag, Berlin.
Kronig, R. deL. and Penney, W. G. (1930). *Proc. Roy. Soc.* A**130**, 499.
Krupka, D. C. and Silsbee, R. H. (1964). *Phys. Rev. Lett.* **12**, 193.
Kushida, T. (1966). *J. Phys. Soc. Japan* **21**, 1331.
—— (1973). *J. Phys. Soc. Japan* **34**, 1318.
Kushida, T. and Takushi, E. (1975). *Phys. Rev.* **B12**, 824.
Lai, S. T., Chai, B. H. T., Long, M. and Shinn, M. D. (1987). In *Tunable Solid State Lasers II* (eds. A. B. Budgor, L. Esterowitz and L. G. DeShazer), p. 145; see also p. 76. Springer-Verlag, Berlin.
Lampel, G. (1974). *Proc. 11th Int. Conf. Phys. Semicond., Stuttgart.* Teubner, Stuttgart.
Larkin, J. P., Imbusch, G. F. and Dravnieks, F. (1973). *Phys. Rev.* **B7**, 495.

Lawson, C. M., Powell, R. C. and Zwicker, W. K. (1981). *Phys. Rev. Lett.* **46**, 1020.

Lax, M. (1954). *J. Chem. Phys.* **20**, 1752.

Lee, L-S, Rand, S. C. and Schawlow, A. L. (1984). *Phys. Rev.* **B29**, 6901.

Lee, K. M. (1983). Unpublished Ph.D. Thesis, Lehigh University.

Lee, K. M., Le Si Dang and Watkins, G. D. (1980). *Solid State Comm.* **35**, 527.

Lee, K. M., Le Si Dang, Watkins, G. D. and Choyke, W. J. (1981). *Solid State Comm.* **37**, 551.

Le Si Dang, Romestain, R., Merle d'Aubigné, Y. and Fukuda, A. (1977). *Phys. Rev. Lett.* **38**, 1539.

Le Si Dang, Merle d'Aubigné, Y. and Rasoalarison, Y. (1978). *J. de Phys.* **39**, 760.

Le Si Dang, Merle d'Aubigné, Y., Romestain, R. and Fukuda, A. (1978). *Solid State Comm.* **26**, 413.

Livanova, L. D., Saitkulov, I. G. and Stolov, A. L. (1969). *Soviet Phys.-Sol. State* **11**, 750.

Lorentz, H. A. (1909). *Theory of Electrons.* Teubner, Leipzig.

Loudon, R. (1966). *Adv. in Physics.* **17**, 243.

Lupei, V., Lupei, A., Georgescu, S. and Ursu, I. (1976). *J. Phys.* C **9**, 2619.

Lüty, F. (1968). In *The Physics of Color Centers* (ed. W. B. Fowler). Academic Press, New York.

Mabichi, T., Fukuda, A. and Onaka, R. (1966). *Science of Light* (Tokyo) **15**, 79.

McCall, S. L. and Hahn, E. L. (1969). *Phys. Rev.* **183**, 457.

McClure, D. S. (1959). *Electronic Spectra of Molecules and Ions in Crystals.* Academic Press, New York.

—— (1962). *J. Chem. Phys.* **36**, 2757.

—— (1975). In *Optical Properties of Ions in Solids* (ed. B. Di Bartolo), p. 259. Plenum Press, New York.

McCumber, D. E. (1964). *Phys. Rev.* **134**A, 299.

McCumber, D. E. and Sturge, M. D. (1963). *J. App. Phys.* **34**, 1682.

McDonagh, C. M., Dawson, P. and Henderson, B. (1980a). *J. Phys.* C **13**, 2191.

McDonagh, C. M., Henderson, B., Imbusch, G. F. and Dawson, P. (1980b). *J. Phys.* C **13**, 3309.

McDonagh, C. M., Henderson, B. and Imbusch, G. F. (1980c). *J. Phys.* C **13**, 6025.

Macfarlane, G. G., McLean, T. P., Quarrington, J. E. and Roberts, V. (1957). *Phys. Rev.* **108**, 1377.

Macfarlane, R. M. (1963). *J. Chem. Phys.* **39**, 3118.

—— (1967). *J. Chem. Phys.* **47**, 2066.

—— (1970). *Phys. Rev.* **B1**, 989.

Macfarlane, R. M. and Shelby, R. M. (1979). *Phys. Rev. Lett.* **42**, 788.

—— (1983). *Optics Comm.* **45**, 46.

Macfarlane, R. M., Wong, J. Y. and Sturge, M. D. (1968). *Phys. Rev.* **166**, 250.

Macfarlane, R. M., Shelby, R. M., Genack, A. Z. and Weitz, D. M. (1980). *Optics Lett.* **5**, 462.

Macfarlane, R. M., Harley, R. T. and Shelby, R. M. (1983). *Radn. Eff.* **72**, 1.

McLean, T. P. (1960). In *Progress in Semiconductors* **5** (ed. A. F. Gibson), p. 55. Heywood, London.

McShera, C., Colleran, P. J., Glynn, T. J. and Imbusch, G. F. (1982). *J. Luminescence* **28**, 41.

Maiman, T. (1960). *Nature* **187**, 493.

Manson, N. B. and Shah, G. A. (1977). *J. Phys.* C **10**, 1991.

Markham, J. J. (1966). *The F Center in Alkali Halides.* Academic Press, New York.

Maxwell, J. C. (1865). *Phil. Trans. Roy. Soc. (London)* **155**, 459.
—— (1873). *Treatise on Electricity and Magnetism.*
Melamed, N. T., De Sousa Barros, F., Vicarro, P. J. and Artman, J. O. (1972). *Phys. Rev.* **B5**, 3377.
Meltzer, R. S. and Cone, R. L. (1976). *Solid State Comm.* **20**, 553.
Merle d'Aubigné, Y. (1976). In *Defects and Their Structure in Non-Metallic Solids* (eds. B. Henderson and A. E. Hughes). Plenum Press, New York.
Merle d'Aubigné, Y., Le Si Dang, Romestain, R. and Wasiela, A. (1981). In *Recent Developments in Condensed Matter Physics*, Vol. 1 (ed. J. T. Devreese), p. 415. North-Holland, Amsterdam.
Mertz, J. L., Faulkner, R. A. and Dean, P. J. (1969). *Phys. Rev.* **188**, 1228.
Merzbacher, E. (1970). *Quantum Mechanics*, 2nd edn. Wiley, New York.
Miyakawa, T. and Dexter, D. L. (1970). *Phys. Rev.* **B1**, 2961.
Moerner, W. E., Schellenberg, F. M. and Bjorklund, G. C. (1982). *App. Phys.* **B28**, 263.
Moerner, W. E., Poxrowsky, P., Schellenberg, F. M. and Bjorklund, G. C. (1986). *Phys. Rev.* **B33**, 5702.
Mollenauer, L. F. (1965). Unpublished Ph.D. Thesis, Stanford University.
—— (1979). In *Quantum Electronics, B* (ed. C. L. Tang), Chapter 6. Academic Press, New York.
Mollenauer, L. F. and Bloom, D. M. (1979). *Opt. Lett.* **4**, 247.
Mollenauer, L. F. and Engstrom, R. (1973). *Phys. Rev.* **B7**, 1616.
Mollenauer, L. F. and Olson, D. H. (1974). *J. App. Phys.* **24**, 386.
Mollenauer, L. F. and Schawlow, A. L. (1968). *Phys. Rev.* **168**, 309.
Mollenauer, L. F. and Stolen, R. H. (1984). *Opt. Lett.* **9**, 13.
Mollenauer, L. F., Pan, S. and Yngveson, S. (1969). *Phys. Rev. Lett.* **23**, 683.
Mollenauer, L. F., Pan, S. and Winnacker, A. (1971). *Phys. Rev. Lett.* **26**, 1643.
Mollenauer, L. F., Stolen, R. H. and Gordon, J. P. (1980). *Phys. Rev. Lett.* **15**, 1095.
Mollenauer, L. F., Vieira, N. D. and Szeto, L. (1983a). *Phys. Rev.* **27B**, 5332.
Mollenauer, L. F., Wiesenfeld, J. M. and Ippen, E. P. (1983b). *Radn. Eff.* **72**, 73.
Moore, C. E. (1950). *Atomic Energy Levels.* US National Bureau of Standards Circ. 467 Vol. II, 16.
Moose, H. W. (1970). *J. Luminescence* **1**, **2**, 106.
Morigaki, K., Dawson, P. and Cavenett, B. C. (1978). *Solid State Comm.* **28**, 829.
Moulton, P. F. (1982). *Optical News* Nov/Dec, p. 9.
—— (1984). CLEO Technical Digest, WA2, p. 77.
—— (1985). In *Laser Handbook* (ed. M. Bass and M. L. Stitch), pp. 203–85. Elsevier.
—— (1986). *IEEE J. Quant. Elec.* QE-**21**, 1582.
Moulton, P. F. and Mooradian, A. (1979). *App. Phys. Lett.* **35**, 127.
Muramoto, T., Fukuda, Y. and Hashi, T. (1974). *Phys. Rev. Lett.* **48A**, 181.
Nakamura, A., Paget, D., Herman, C., Weisbusch, C., Lampel, G. and Cavenett, B. C. (1979). *Solid State Comm.* **30**, 411.
Nakazawa, E. and Shionoya, S. (1970). *Phys. Rev. Lett.* **25**, 1710.
Nelson, E. D., Wong, J. Y. and Schawlow, A. L. (1967). *Phys. Rev.* **156**, 298.
O'Connell, D. O., Henderson, B. and Bolton, J. D. (1981). *Solid State Comm.* **38**, 283 and 287.
O'Donnell, K. P., Lee, K. M. and Watkins, G. D. (1983). *Physica* **116B**, 258.
Ofelt, G. S. (1962). *J. Chem. Phys.* **37**, 511.
Orbach, R. (1975). In *Optical Properties of Ions in Solids* (ed. B. Di Bartolo), p. 355. Plenum Press, New York.

Ortiz, C., Macfarlane, R. M., Shelby, R. M., Lenth, W. and Bjorklund, G. C. (1981). *App. Phys.* **25**, 87.
Orton, J. W. (1968). *Electron Paramagnetic Resonance* Illiffe, London.
Panepucci, H. and Mollenauer, L. F. (1968). *Phys. Rev.* **178**, 589.
Pappalardo, R. G. (1978). In *Luminescence of Inorganic Solids* (ed. B. Di Bartolo) p. 175. Plenum Press, New York.
Pappalardo, R. G., Wood, D. L. and Linares, R. C. (1961a). *J. Chem. Phys.* **35**, 2041.
—— (1961b). *J. Chem. Phys.* **35**, 1460.
Parrot, R. (1978). In *Luminescence of Inorganic Solids* (ed. B. Di Bartolo), p. 393. Plenum Press, New York.
Paszek, A. (1978). Unpublished Ph.D. Thesis, Johns Hopkins University.
Patel, J. L., Davies, J. J. and Nicholls, J. E. (1981). *J. Phys.* C **14**, 1339.
Patrick, L. and Choyke, W. J. (1974). *Phys. Rev.* **B10**, 5091.
Pauling, L. (1939). *The Nature of The Chemical Bond.* Cornell University Press, Ithaca; see also (1967). *The Chemical Bond.* Cornell University Press, Ithaca.
Peckham, G. (1967). *Proc. Phys. Soc.* (London) **90**, 657.
Pelletier-Allard, N. and Pelletier, R. (1984). *J. Phys.* C **17**, 2129.
Phillip, H. R. and Ehrenreich, H. (1963). *Phys. Rev.* **129**, 1550.
Phillips, W. A. (1972). *J. Low Temp. Phys.* **7**, 351.
—— (1981). In *Amorphous Solids, Low Temperature Properties.* (ed. W. A. Phillips). Springer-Verlag, Berlin.
Pike, E. R. and Sarkar, S. (eds.) (1986). *Frontiers in Quantum Optics.* Adam Hilger, Bristol.
Pohl, R. W. (1937). *Proc. Phys. Soc.* **49** (extra part), 16.
Pollack, S. A. (1964). *J. Chem. Phys.* **40**, 2751.
Pollock, C. R., Pinto, J. F. and Efstratios, T. G. (1986). OSA Meeting on Tunable Solid State Lasers, Technical Digest, p. 130; see also (1987) *Tunable Solid State Lasers II*, p. 261. Springer Series in Optical Science **52**. Springer-Verlag, Heidelberg.
Porter, J. F. (1961). *Phys. Rev. Lett.* **7**, 414.
Pott, G. T. and McNicol, B. D. (1972). *J. Chem. Phys.* **56**, 5246.
Poulain, M., Lucas, J., Brun, P. and Drifford, M. (1977). *Colloques Internationaux du CNRS No. 255*, p. 257. Editions du CNRS, Paris.
Powell, R. C. (1986). *Tunable Solid State Lasers II* (eds. A. B. Budgor, L. Esterowitz and L. G. DeShazer), p. 5. Springer-Verlag, Berlin.
Powell, R. C., Di Bartolo, B., Birang, B. and Naiman, C. S. (1967). *Phys. Rev.* **155**, 296.
Preston, T. (1895). *The Theory of Light.* Macmillan, London.
Rabinowitch, E. and Belford, R. L. (1964). *Spectroscopy and Photochemistry of Uranyl Compounds.* Pergamon Press, Oxford.
Ranfagni, A. (1972). *Phys. Rev. Lett.* **28**, 743.
Ranfagni, A., Mugnai, D., Bacci, M., Villiani, G. and Fontana, M. P. (1983). *Adv. Phys.* **32**, 823.
Raman, C. V. (1928). *Indian J. Phys.* **2**, 1.
Ramdas, A. K. and Rodriguez, S. (1981). *Rep. on Prog. Phys.* **44**, 1207–1387.
Rashba, E. I. and Sturge, M. D. (1982). *Excitons.* North-Holland, Amsterdam.
Rast, H. E., Fry, J. L. and Caspers, H. H. (1967). *J. Chem. Phys.* **46**, 1460.
Rayleigh, Lord (Hon. J. W. Strutt) (1871). *Phil. Mag.* **x/i**, 107.
Rebane, K. K. (1970). *Impurity Spectra of Solids.* Plenum Press, New York.
Rebane, K. K. and Rebane, L. A. (1975). In *Optical Properties of Ions in Solids* (ed. B. Di Bartolo). Plenum Press, New York.

Reisfeld, R. (1984). In *Energy Transfer Processes in Condensed Matter* (ed. B. Di Bartolo), p. 521. Plenum Press, New York.
Reisfeld, R. and Jørgensen, C. K. (1975). In *Structure and Bonding* **22**, 123.
—— (1977). In *Lasers and Excited States of Rare Earths*, Springer-Verlag, Berlin, Heidelberg.
Riseberg, L. A. and Weber, M. J. (1975). In *Progress in Optics* **14** (ed. E. Wolf). North-Holland, Amsterdam.
Rius, G., Cox, R., Picard, P. and Santier, C. (1976). *Comptes Rend.* **271**, 724.
Romestain, R., Le Si Dang, Merle d'Aubigné, Y. and Fukuda, A. (1977). *Proc. 3rd Specialized Colloque Ampère in Semiconductors and Insulators* **3**, 175.
Rose, B. H. and Cowan, D. L. (1974). *Solid State Comm.* **15**, 775.
Runciman, W. A. (1965). *Proc. Phys. Soc.* **86**, 625.
Sangster, M. J. (1972). *Phys. Rev.* **B6**, 254.
Sangster, M. J. L., Peckham, G. and Saunderson, D. H. (1970). *J. Phys.* C **3**, 1026.
Schawlow, A. L. (1970). In *Lasers and Light* W. H. Freeman & Co.
Schawlow, A. L. and Devlin, G. E. (1961). *Phys. Rev. Lett.* **6**, 96.
Schawlow, A. L., Wood, D. L. and Clogston, A. M. (1959). *Phys. Rev. Lett.* **3**, 271.
Schawlow, A. L., Piksis, A. H. and Sugano, S. (1961). *Phys. Rev.* **122**, 1469.
Schnatterley, S. E. (1965). *Phys. Rev.* A **140**, 1364.
Schneider, I. and Marrone, M. J. (1979). *Opt. Lett.* **4**, 390.
—— (1981). *ibid* **6**, 627.
Schneider, I. and Pollock, C. R. (1983). *J. App. Phys.* **54**, 6193.
Schneider, I. and Moss, S. C. (1983). *Opt. Lett.* **8**, 7.
Schoemaker, D. (1976). In *Defects and Their Structure in Non-Metallic Solids* (eds. B. Henderson and A. E. Hughes). Plenum Press, New York.
Schulz du Bois, E. O. (1959). *Bell Syst. Tech. J.* **38**, 271.
Schuurmans, M. F. H. and Van Dijk, J. M. F. (1984). *Physica* **123B**, 131.
Scott, W. C. and Sturge, M. D. (1966). *Phys. Rev.* **146**, 262.
Seidel, H. (1963a). *Phys. Lett.* **6**, 150.
—— (1963b). *Phys. Lett.* **7**, 27.
Seidel, H. and Wolf, H. C. (1963). *Z. Phys.* **173**, 455.
—— (1968). In *Physics of Color Centers* (ed. W. B. Fowler). Academic Press, New York.
Seidel, H., Schwoerer, M. and Schmidt, D. (1965). *Z. Physik* **182**, 398.
Selzer, P. M. (1981). In *Laser Spectroscopy of Solids* (eds. W. M. Yen and P. M. Selzer), p. 113. Springer-Verlag, Berlin.
Selzer, P. M. and Yen, W. M. (1977). *Optics Lett.* **1**, 90.
Selzer, P. M., Huber, D. L., Hamilton, D. S., Yen, W. M. and Weber, M. J. (1976). *Phys. Rev. Lett.* **36**, 813.
Shelby, R. M. and Macfarlane, R. M. (1982). In *Picosecond Phenomena III*, p. 82. Springer-Verlag, Berlin.
Shinada, M., Sugano, S. and Kushida, T. (1966). *J. Phys. Soc. Japan* **21**, 1342.
Silsbee, R. H. (1956). *Phys. Rev.* **103**, 1675.
—— (1965). *Phys. Rev.* **138**, A180.
Slater, J. C. (1930). *Phys. Rev.* **35**, 210.
—— (1963). *Quantum Theory of Molecules and Solids*, Vol. I. McGraw-Hill, New York.
Smith, D. Y. and Spinolo, G. (1965). *Phys. Rev.* **140**, A2121.
Sonder, E. and Sibley, W. A. (1972). In *Point Defects in Solids* (eds. J. H. Crawford and L. F. Slifkin) Vol. 1, p. 201. Plenum Press, New York.

Spaeth, J. M. (1976). In *Defects and Their Structure in Non-Metallic Solids* (eds. B. Henderson and A. E. Hughes). Plenum Press, New York.
Stiles, L. F., Fontana, M. P. and Fitchen, D. B. (1969). *Solid State Comm.* **7**, 681.
—— (1970). *Phys. Rev.* **B2**, 2077.
Stokowski, S. E., Johnson, S. A. and Scott, P. L. (1966). *Phys. Rev.* **147**, 544.
Stoneham, A. M. (1985). *Theory of Defects in Solids*, 2nd edn. Oxford University Press.
Stolen, R. H., Mollenauer, L. F. and Tomlinson, W. T. (1983). *Opt. Lett.* **8**, 186.
Stout, J. W. (1959). *J. Chem. Phys.* **31**, 709.
Struck, C. W. and Fonger, W. H. (1970). *J. Chem. Phys.* **52**, 6364. *J. Luminescence* **1**, **2**, 456.
—— (1976). *J. Chem. Phys.* **64**, 1784.
Sturge, M. D. (1965). *Phys. Rev.* **140**, 880.
—— (1967). In *Solid State Physics* **20** (eds. F. Seitz, D. Turnbull and H. Ehrenreich), p. 92. Academic Press, New York.
—— (1970). *Phys. Rev.* **B1**, 1005.
—— (1973). *Phys. Rev.* **B8**, 6.
Sugano, S. (1962). *J. Chem. Phys.* **36**, 122.
Sugano, S. and Tanabe, Y. (1958). *J. Phys. Soc. Japan*, **13**, 880.
Sugano, S., Schawlow, A. L. and Varsanyi, F. (1960). *Phys. Rev.* **120**, 2045.
Sugano, S., Tanabe, Y. and Kamimura, H. (1970). *Multiplets of Transition-Metal Ions in Crystals.* Academic Press, New York.
Swank, R. L. and Brown, F. C. (1963). *Phys. Rev.* **130**, 34.
Szabo, A. (1970). *Phys. Rev. Lett.* **25**, 924.
—— (1975). *Phys. Rev.* **B11**, 4512.
Tallant, D. R. and Wright, J. C. (1975). *J. Chem. Phys.* **63**, 2074.
Tanabe, Y. and Kamimura, H. (1958). *J. Phys. Soc. Japan*, **13**, 394.
Tanabe, Y. and Sugano, S. (1954). *J. Phys. Soc. Japan* **9**, 753.
Tanabe, Y., Moriya, T. and Sugano, S. (1965). *Phys. Rev. Lett.* **15**, 1023.
Tanimoto, D. H., Zinicker, W. M. and Kemp, J. C. (1965). *Phys. Rev. Lett.* **14**, 645.
Tinkham, M. (1964). *Group Theory and Quantum Mechanics.* McGraw-Hill, New York.
Townes, C. H. and Schawlow, A. L. (1958). *Phys. Rev.* **112**, 1940.
Tyndall, J. W. (1869). *Notes of a Course of Nine Lectures on Light*, Royal Institution of Great Britain. Longmans, Green, London.
Van Doorn, C. Z. (1962). *Phillips Research Repts. Supp.* **4**.
Van der Ziel, J. P. and Van Uitert, L. G. (1969). *Phys. Rev.* **180**, 343.
—— (1973). *Phys. Rev.* **B8**, 1889.
Van Vechten, J. (1975). *J. Electron Mat.* **4**, 1159; see also *J. Electrochem. Soc.* **122**, 423.
Van Vleck, J. H. (1932). *The Theory of Electric and Magnetic Susceptibilities.* Oxford University Press.
Varsanyi, F. L. and Dieke, G. H. (1961). *Phys. Rev. Lett.* **7**, 442.
Varsanyi, F. L., Wood, D. L. and Schawlow, A. L. (1959). *Phys. Rev. Lett.* **3**, 544.
Von der Osten, W. (1976). In *Defects and Their Structure in Non-Metallic Solids* (eds. B. Henderson and A. E. Hughes). Plenum Press, New York.
Voron'ko, Yu. K., Osiko, V. V. and Shcherbakov, I. A. (1969). *Soviet Phys. JETP* **29**, 86.
Walling, J. C. (1982). *Laser Focus* **18**, 45.
Walling, J. C., Peterson, O. G., Jenssen, H. P., Morris, F. C. and O'Dell, E . W. (1980). *IEEE J. Quant. Elect.* **16**, 1302.
Wannier, G. H. (1937). *Phys. Rev.* **52**, 191.
Ware, W. R. (1983). In *Time Resolved Fluorescence Spectroscopy in Biochemistry and*

Biology (eds. R. B. Cundall and R. E. Dale), p. 23. NATO ASI Series A Life Sciences **69**, Plenum Press, New York.

Wasiela, A., Merle d'Aubigné, Y. and Romestain, R. (1980). *J. Phys.* C **13**, 1384.

Watts, R. K. (1975). In *Optical Properties of Ions in Solids* (ed. B. Di Bartolo), p. 307. Plenum Press, New York.

Weber, M. J. (1971). *Phys. Rev.* **B4**, 2932.

—— (1973). *Phys. Rev.* **B8**, 54.

—— (1976). *Phys. Rev. Lett.* **36**, 813.

—— (1981). In *Laser Spectroscopy of Solids* (eds. W. M. Yen and P. M. Selzer), p. 189. Springer-Verlag, Berlin.

—— (1986). (ed.) *CRC Handbook of Laser Science and Technology*. CRC Press, Boca Raton.

—— (1987). *J. Luminescence* **36**. (This special issue, edited by M. J. Weber, is devoted to the topic of optical linewidths in glass.)

Weissbluth, M. (1978). *Atoms and Molecules*. Academic Press, New York.

Welch, L. S. (1980). Unpublished Ph.D. Thesis, University of Reading.

Welch, L. S., Hughes, A. E., Pells, G. P., and Schoenberg, A. (1976). *J. Physique* **37**, C7, 198.

Welford, D. T. and Moulton, P. (1988). Private communication.

Wertheim, G. K., Butler, M. A., West, K. W. and Buchanan, D. N. E. (1974). *Rev. Sci. Ins.* **45**, 1369.

Wertz, J. E. and Auzins, P. (1957). *Phys. Rev.* **106**, 484.

Wertz, J. E. and Bolton, J. R. (1972). *Electron Spin Resonance*. McGraw-Hill, New York.

Wiesenfeld, J. M., Mollenauer, L. F. and Ippen, E. P. (1981). *Phys. Rev. Lett.* **47**, 1668.

Williams, F. E. (1951). *Phys. Rev.* **82**, 281.

Williams, G. P., Rosenblatt, G. H., Tuttle, A. E., Williams, R. T. and Chen, Y. (1988). *J. Cryst. Def. Amorph. Solids* **16**, 1209.

Wilson, B. A., Yen, W. M., Hegarty, J. and Imbusch, G. F. (1979). *Phys. Rev.* **B19**, 4238.

Wilson, B. A., Hegarty, J. and Yen, W. M. (1978). *Phys. Rev. Lett.* **41**, 268.

Wood, D. L. (1969). In *Optical Properties of Solids* (eds. S. Nudelman and S. S. Mitra). Plenum Press, New York.

Wood, D. L. and Remeika, J. P. (1967). *J. Chem. Phys.* **46**, 3595.

Wood, D. L., Ferguson, J., Knox, K. and Dillon, J. F. (1963). *J. Chem. Phys.* **39**, 890.

Wood, R. F. and Wilson, T. M. (1977). *Phys. Rev.* **15B**, 3700.

Woods, A. D. B., Brockhouse, B. N., Cowley, R. A. and Cochran, W. (1963). *Phys. Rev.* **131**, 1023.

Wooten, F. (1972). *Optical Properties of Solids*. Academic Press, New York.

Worlock, J. M. and Porto, S. P. S. (1965). *Phys. Rev. Lett.* **15**, 697.

Wright, J. C. (1977). *Anal. Chem.* **49**, 1690.

Wybourne, B. G. (1965). *Spectroscopic Properties of Rare Earths*. Wiley Interscience, New York.

Yen, W. M. (1986). In *Optical Spectroscopy of Glasses* (ed. I. Zschokke), p. 23. D. Reidel, Dordrecht.

Yen, W. M. and Selzer, P. M. (1981). In *Laser Spectroscopy of Solids* (eds. W. M. Yen and P. W. Selzer), p. 141. Springer-Verlag, Berlin.

Yokota, M. and Tanimoto, O. (1967). *J. Phys. Soc. Japan* **22**, 779.

Zakharchenya, B. (1978). *Semiconductors and Insulators* **4**, 35.

Zallen, R. (1983). *The Physics of Amorphous Solids*. Wiley, New York.

INDEX

9 780199 298624

Printed in the United States
By Bookmasters